STATISTICAL OPTIMIZATION FOR GEOMETRIC COMPUTATION

THEORY AND PRACTICE

KENICHI KANATANI

Department of Information Technology
Okayama University, Japan

DOVER PUBLICATIONS, INC.
Mineola, New York

Copyright

Bibliographical Note

This Dover edition, first published in 2005, is an unabridged and slightly corrected republication of the second (1999) impression of the work, originally published in the "Machine Intelligence and Pattern Recognition" series by Elsevier Science, Amsterdam, in 1996. An errata list has been added to the present edition on pp. xv–xviii.

Library of Congress Cataloging-in-Publication Data

Kanatani, Ken'ichi, 1947–
Statistical optimization for geometric computation : theory and practice / Kenichi Kanatani.
p. cm.
"Unabridged and slightly corrected republication of the 2nd (1999) impression of the work, originally published in the 'Machine intelligence and pattern recognition' series by Elsevier Science, Amsterdam, in 1996. An errata list has been added to the present edition on pp. xv–xviii."
Includes bibliographical references and index.
ISBN 0-486-44308-6 (pbk.)
1. Robotics. 2. Computer vision. 3. Mathematical statistics. I. Title.

TJ211.K354 2005
006.4'2'015195—dc22

2005042844

Manufactured in the United States of America
Dover Publications, Inc., 31 East 2nd Street, Mineola, N.Y. 11501

Preface

One of the central tasks of computer vision and robotics is building a 3-D model of the environment from sensor and image data. An intrinsic difficulty of this task is the fact that sensor and image data are not necessarily accurate. In the past, many techniques have been proposed to cope with this problem. One of the most effective solutions is to take advantage of our prior knowledge about the structure of the problem: we make use of such geometric relationships as collinearity, coplanarity, rigidity, and epipolar relation that should hold in the absence of noise; they are called *geometric constraints*. This book focuses on two such techniques that are the most fundamental. One is *geometric correction*: the data are optimally modified so as to satisfy the geometric constraints. The other is *parametric fitting*: the parameters of the geometric constraints are optimally estimated.

In order that a particular method be called "optimal," the theoretical bound on its attainable accuracy must be known; only methods which can attain it are qualified to be called optimal. In this book, we give many synthetic and real data examples to demonstrate that conventional methods are not optimal and how accuracy improves if truly optimal methods are employed.

However, computing optimal estimates alone is not sufficient; at the same time, we must evaluate in quantitative terms how reliable the resulting estimates are. The knowledge that a distance is optimally estimated to be 5m is of little use if we do not know whether the value is reliable within ±10cm or ±1m. This reliability issue has not received much attention in the past. In order to compute optimal estimates and evaluate their reliability, we need an efficient numerical algorithm. Since such estimation is based on our knowledge about the structure of the problem, we also need a criterion for testing if the assumed structure, or *model*, is correct. This book presents rigorous mathematical techniques for these purposes.

Thus, the subject of this book is closely related to traditional statistics. However, the main goal of traditional statistics is to infer the structure of a phenomenon from a number of sample data obtained by repeated measurements. In our applications, usually only *one* set of data is obtained by a measurement. Also, noise in electronic devices can be assumed to be fairly small, so we can apply first order analysis. Since geometric objects are represented by vectors and tensors in two and three dimensions, linear algebra, eigenvalue analysis in particular, plays a central role in our analysis. These features make our treatment very different from traditional statistics.

This book first summarizes the mathematics that is fundamental to our analysis and then presents techniques for optimal estimation and reliability evaluation by assuming that the noise is Gaussian. We also discuss numerical computation schemes called the *optimal filter* and *renormalization* and consider computational issues such as convergence of iterations and numerical stability. Then, we derive the *geometric information criterion* and apply it to such problems as stereo vision and 3-D motion analysis. Finally, the theoretical accuracy bound for a general non-Gaussian noise model is obtained in the form of the *Cramer-Rao lower bound* expressed in terms of the *Fisher information matrix*.

This book is an elaboration of the author's lecture note for a series of seminars he gave at the Department of Mathematical Engineering and Information Physics, University of Tokyo in 1994. The author thanks Shun-ichi Amari, Kokichi Sugihara, Koichiro Deguchi, and Ken Hayami of the Department of Mathematical Engineering and Information Physics, University of Tokyo and Keisuke Kinoshita of the ATR Human Information Processing Research Laboratories for detailed discussions. He also thanks Azriel Rosenfeld of the University of Maryland, who carefully read the entire manuscript, and Naoya Ohta of Gunma University, Yasushi Kanazawa of Gunma College of Technology, and all of the students in his laboratory for helping him with this laborious project. This work was in part supported by the Ministry of Education, Science, and Culture, Japan under a Grant in Aid for Scientific Research B (No. 07458067) and the Okawa Institute of Information and Telecommunication, Japan.

Gunma, Japan
January 1996

Kenichi Kanatani

Contents

Errata

p. 35 third line below eq. (2.49)

Error: Let $\{\boldsymbol{n}_1, ..., \boldsymbol{n}_m\}$ be

Correct: Let $\{\boldsymbol{n}_1, ..., \boldsymbol{n}_{n-m}\}$ be

p. 35 eq. (2.50)

Error: $\boldsymbol{P}_{\mathcal{N}} = \boldsymbol{I} - \sum_{i=1}^{m} \boldsymbol{n}_i \boldsymbol{n}_i^\top$

Correct: $\boldsymbol{P}_{\mathcal{N}} = \boldsymbol{I} - \sum_{i=1}^{n-m} \boldsymbol{n}_i \boldsymbol{n}_i^\top$

p. 42 first line below eq. (2.91)

Error: By multiplying the first equation by

Correct: By multiplying the second equation by

p. 44 first line

Error: matrix $\boldsymbol{P}_0$ implies

Correct: matrix $\boldsymbol{P}_1$ implies

p. 44 first line below eq. (2.105)

Error: where $\lambda_{\rm max}$ is the smallest

Correct: where $\lambda_{\rm max}$ is the largest

p. 46 eq. (2.117)

Error: $\leq \sum_{i=1}^{r} \lambda_{\rm max}^2 (\boldsymbol{u}_i, \boldsymbol{x})^2 =$

Correct: $\leq \sum_{i=1}^{n} \lambda_{\rm max}^2 (\boldsymbol{u}_i, \boldsymbol{x})^2 =$

p. 47 eq. (2.120)

Error: $\boldsymbol{x} \propto \boldsymbol{u}_{\rm max} + \mathcal{N}_{\boldsymbol{A}}$

Correct: $\boldsymbol{x} \propto \boldsymbol{u}_{\rm max}$

p. 47 three lines below eq. (2.120)

Error: The right-hand side means ... the singular value $\lambda_{\rm max}$.

Correct: *delete*

p. 60 eq. (2.205)

Error: $\boldsymbol{A}_s = A[\boldsymbol{A}]$.

Correct: $\boldsymbol{A}_a = A[\boldsymbol{A}]$.

p. 62 eq. (3.9)

Error: $X \sim E[X] + O(\frac{1}{\sqrt{N}})$,

Correct: $\bar{X} \sim E[X] + O(\frac{1}{\sqrt{N}})$,

p. 62 first line below eq. (3.9)

Error: and hence $X \sim E[X]$

Correct: and hence $\bar{X} \sim E[X]$

p. 66 third line below eq. (3.28)

Error: along a curve (Fig. 3.2b)

Correct: along a curve (Fig. 3.2c)

p. 66 fourth line below eq. (3.28)

Error: (Fig. 3.2c). If

Correct: (Fig. 3.2d). If

p. 72 second line below eq. (3.54)

Error: its characteristic function has

Correct: its moment generating function has

p. 77 fifth line below eq. (3.74)

Error: the null space of $\boldsymbol{A}$

Correct: the orthogonal complement of the null space of $\boldsymbol{A}$

p. 85 eq. (3.128)

Error: $+\boldsymbol{P}^{\mathcal{S}}_{\boldsymbol{\theta}} \int_{\mathcal{X}} \hat{\boldsymbol{\theta}}(\boldsymbol{x})(\nabla_{\boldsymbol{\theta}} \log p, \Delta\boldsymbol{\theta})p(\boldsymbol{x};\boldsymbol{\theta})d\boldsymbol{x}$

Correct: $+\boldsymbol{P}^{\mathcal{S}}_{\boldsymbol{\theta}} \int_{\mathcal{X}} (\hat{\boldsymbol{\theta}}(\boldsymbol{x}) - \boldsymbol{\theta})(\nabla_{\boldsymbol{\theta}} \log p, \Delta\boldsymbol{\theta})p(\boldsymbol{x};\boldsymbol{\theta})d\boldsymbol{x}$

p. 105 eq. (4.42)

Error: $(\boldsymbol{p}, \Delta\boldsymbol{p}) + (\boldsymbol{n}, \Delta\boldsymbol{n}) = 1$

Correct: $(\boldsymbol{p}, \Delta\boldsymbol{p}) + (\boldsymbol{n}, \Delta\boldsymbol{n}) = 0$

p. 109 first line below eq. (4.61)

Error: where $\boldsymbol{\nu} \propto \boldsymbol{n} \oplus (-1)$

Correct: where $\boldsymbol{\nu} \propto \boldsymbol{n} \oplus (-d)$

p. 117 first line above eq. (4.94)

Error: if $a < b$ $(a > b)$.

Correct: if $a > b$ $(a < b)$.

p. 117 second line from bottom

Error: called the *axis* and the *vertex*,

Correct: called the *vertex* and the *axis*,

p. 119 second line above eq. (4.99)

Error: and σ_1, σ_2, and σ_2
Correct: and σ_1, σ_2, and σ_3

p. 119 first line below eq. (4.99)

Error: principal axes in the the
Correct: principal axes in the

p. 119 second line below eq. (4.99)

Error: radii σ_1, σ_2, and σ_2
Correct: radii σ_1, σ_2, and σ_3

p. 126 eqs. (4.124)

Error: $\boldsymbol{\nu}' = N[\begin{pmatrix} \boldsymbol{R} & \boldsymbol{0} \\ -\boldsymbol{h}^\top \boldsymbol{R} & 1 \end{pmatrix} \boldsymbol{\nu}], \quad \boldsymbol{\nu} = N[\begin{pmatrix} \boldsymbol{R}^\top & \boldsymbol{0} \\ \boldsymbol{h}^\top & 1 \end{pmatrix} \boldsymbol{\nu}'].$

Correct: $\boldsymbol{\nu}' = N[\begin{pmatrix} \boldsymbol{R}^\top & \boldsymbol{0} \\ \boldsymbol{h}^\top & 1 \end{pmatrix} \boldsymbol{\nu}], \quad \boldsymbol{\nu} = N[\begin{pmatrix} \boldsymbol{R} & \boldsymbol{0} \\ -\boldsymbol{h}^\top \boldsymbol{R} & 1 \end{pmatrix} \boldsymbol{\nu}']$

p. 133 fourth line from top

Error: $F^{(k)}(\boldsymbol{u}_1, ..., \boldsymbol{u}_L)$
Correct: $F^{(k)}(\boldsymbol{u}_1, ..., \boldsymbol{u}_N)$

p. 133 first line below eq. (5.5)

Error: $F^{(k)}(\boldsymbol{u}_1, ..., \boldsymbol{u}_L)$
Correct: $F^{(k)}(\boldsymbol{u}_1, ..., \boldsymbol{u}_N)$

p. 152 eq. (5.117)

Error: $\hat{J} > \chi^2_{3,a}$
Correct: $\hat{J} > \chi^2_{4,a}$

p. 157 eq. (5.145)

Error: $\hat{\boldsymbol{\nu}} = N[\boldsymbol{\nu}_1 - \Delta\boldsymbol{\nu}_1] = N[\boldsymbol{\nu}_2 - \Delta\boldsymbol{\nu}_1]$
Correct: $\hat{\boldsymbol{\nu}} = N[\boldsymbol{\nu}_1 - \Delta\boldsymbol{\nu}_1] = N[\boldsymbol{\nu}_2 - \Delta\boldsymbol{\nu}_2]$

p. 180 Fig. 6.5

Error: Fig. 6.5(a) and Fig. 6.5(b) are in opposit order.
Correct: Interchange Fig. 6.5(a) with Fig. 6.5(b).

p. 230 last line

Error: decrease by two
Correct: decrease by three

p. 338 eq. (11.71)

Error: $\boldsymbol{G}\boldsymbol{G}^\top = (\boldsymbol{h} \times \boldsymbol{R})(\boldsymbol{h} \times \boldsymbol{R})^\top = (\boldsymbol{h} \times \boldsymbol{I})\boldsymbol{R}\boldsymbol{R}^\top(\boldsymbol{h} \times \boldsymbol{I})^\top(\boldsymbol{h} \times \boldsymbol{I})(\boldsymbol{h} \times \boldsymbol{I})^\top = \boldsymbol{P}_{\boldsymbol{h}}.$
Correct: $\boldsymbol{G}\boldsymbol{G}^\top = (\boldsymbol{h} \times \boldsymbol{R})(\boldsymbol{h} \times \boldsymbol{R})^\top = (\boldsymbol{h} \times \boldsymbol{I})\boldsymbol{R}\boldsymbol{R}^\top(\boldsymbol{h} \times \boldsymbol{I})^\top = (\boldsymbol{h} \times \boldsymbol{I})(\boldsymbol{h} \times \boldsymbol{I})^\top = \boldsymbol{P}_{\boldsymbol{h}}.$

p. 339 eq. (11.79)

Error: $\boldsymbol{K} = \boldsymbol{V}_{\boldsymbol{h}}\mathrm{diag}(1,1,0)\boldsymbol{V}_{\boldsymbol{h}}^\top\boldsymbol{R} = (\boldsymbol{v}_1. \boldsymbol{v}_2, \boldsymbol{h})\boldsymbol{\Lambda}(\boldsymbol{R}^\top\boldsymbol{v}_1, \boldsymbol{R}^\top\boldsymbol{v}_2, \boldsymbol{R}^\top\boldsymbol{v})^\top.$

Correct: $\boldsymbol{K} = \boldsymbol{V}_{\boldsymbol{h}}\mathrm{diag}(1,1,0)\boldsymbol{V}_{\boldsymbol{h}}^\top \boldsymbol{R} = (\boldsymbol{v}_1, \boldsymbol{v}_2, \boldsymbol{h})\boldsymbol{\Lambda}(\boldsymbol{R}^\top \boldsymbol{v}_1, \boldsymbol{R}^\top \boldsymbol{v}_2, \boldsymbol{R}^\top \boldsymbol{h})^\top.$

p. 425 eq. (13.24)

Error:

$$\begin{aligned}\sum_{\alpha=1}^{N} \|\hat{\boldsymbol{a}}_\alpha - \bar{\boldsymbol{a}}_\alpha\|^2_{V[\boldsymbol{a}_\alpha]} &= \sum_{\alpha=1}^{N} \|\hat{\boldsymbol{a}}_\alpha - \tilde{\boldsymbol{a}}_\alpha\|^2_{V[\boldsymbol{a}_\alpha]} + \sum_{\alpha=1}^{N} \|\tilde{\boldsymbol{a}}_\alpha - \bar{\boldsymbol{a}}_\alpha\|^2_{V[\boldsymbol{a}_\alpha]} \\ &= \sum_{\alpha=1}^{N} \|\hat{\boldsymbol{a}}_\alpha - \tilde{\boldsymbol{a}}_\alpha\|^2_{V[\boldsymbol{a}_\alpha]} + dN\end{aligned}$$

Correct:

$$\begin{aligned}E[\sum_{\alpha=1}^{N} \|\hat{\boldsymbol{a}}_\alpha - \bar{\boldsymbol{a}}_\alpha\|^2_{V[\boldsymbol{a}_\alpha]}] &= E[\sum_{\alpha=1}^{N} \|\hat{\boldsymbol{a}}_\alpha - \tilde{\boldsymbol{a}}_\alpha\|^2_{V[\boldsymbol{a}_\alpha]}] + E[\sum_{\alpha=1}^{N} \|\tilde{\boldsymbol{a}}_\alpha - \bar{\boldsymbol{a}}_\alpha\|^2_{V[\boldsymbol{a}_\alpha]}] \\ &= E[\sum_{\alpha=1}^{N} \|\hat{\boldsymbol{a}}_\alpha - \tilde{\boldsymbol{a}}_\alpha\|^2_{V[\boldsymbol{a}_\alpha]}] + dN\end{aligned}$$

Chapter 1

Introduction

This chapter introduces the aims, features, backgrounds, and organization of the subsequent chapters. In particular, the differences between our treatment and existing studies are described in detail. Also, related references are given for each subject. Since the discussion here refers to many technical issues in the subsequent chapters, readers are advised to go through this chapter for a general overview in the first reading and come back here later for detailed information.

1.1 The Aims of This Book

1.1.1 Statistical optimization for image and sensor data

Intelligent robots are expected to play a vital role in manufacturing industries and various hazardous environments such as nuclear power stations, outer space, and deep water. For autonomous robotic operations in an unknown environment, robots must first acquire a 3-D model of their workspace. The most fundamental source of 3-D information is *vision*. Today, the study of extracting 3-D information from video images and building a 3-D model of the scene, called *computer vision* or *image understanding*, is one of the research areas that attract the most attention all over the world [5, 6, 12, 47, 57, 65, 85, 90, 130, 135, 151, 185, 211, 227, 229]. Various other sensing techniques have also been developed—tactile sensing and ultrasonic range sensing, for example.

The crucial fact about image and sensor data is that *they are not necessarily accurate*. We discuss errors in range sensing in Section 10.3, but in the rest of the book we exclusively deal with points and lines detected from camera images. Errors in such image data originate from various sources. Camera images undergo optical distortion, called *aberration*, due to imperfections of the lens. They are also distorted by mechanical causes (e.g., misalignment of the lens and the array sensor) and electronic causes (e.g., difference between the vertical scanning rate and the horizontal sampling rate). Theoretically, such systematic distortions can be corrected by prior calibration. In reality, however, it is very difficult to remove them completely. Even if a perfect camera is used, images are usually inaccurate because of poor lighting, imperfect focusing, limited resolution, and various factors concerning the scene and objects in question (shades, shadows, specularity, reflection, etc.). Hence, points and lines detected by applying an image processing operation to the gray levels are not located where they should be. In this book, such deviations of points and lines are collectively termed "image noise".

Today, coping with the uncertainty of the data, whatever origins they have, is one of the greatest challenges to computer vision and robotics researchers, and many techniques have been proposed for this purpose. They are basically categorized into two types:

- *Traditional statistical approach*: the uncertainty of the data is conquered by *repeated measurements.*
- *Physics-oriented approach*: the uncertainty of the data is compensated for *by a priori knowledge about the structure of the problem.*

This book concentrates on the latter approach. Namely, we take advantage of the *geometric constraints* that measurement values should satisfy *if noise did not exist*. For point and line data, such constraints include collinearity and coplanarity of points, concurrency of lines, rigidity of motion, and what is known as the *epipolar constraint* for stereo vision and motion analysis. We focus on two of the most fundamental methods that make use of such constraints:

- *Geometric correction*: the data are optimally modified so as to satisfy the geometric constraints exactly.
- *Parametric fitting*: the parameters of the geometric constraints are optimally estimated.

The latter is not limited to fitting curves and surfaces to point data [20, 200, 160, 228] but covers a much wider range of problems such as 3-D motion analysis, as we will show later. In the following, we primarily consider points and lines in images, but in principle our statistical theory can also apply to other types of image data (e.g., gray levels of pixels) and sensor data obtained from devices other than cameras.

1.1.2 What are the issues?

For any type of statistical estimation, we encounter the following issues:

- *Accuracy bound.* In order to claim that a particular method is "optimal," one must know the theoretical bound on its attainable accuracy; only those methods which attain that bound are qualified to be called "optimal." If such methods are not known, we can still evaluate the performance of available methods by comparing their accuracy with the theoretical bound.
- *Reliability evaluation.* Computing an optimal estimate alone is not sufficient in real applications; we must at the same time evaluate the reliability of the resulting estimate in quantitative terms. If a robot does

not know about the reliability of the 3-D model according to which it is operating, the robot is unable to judge if its next action can achieve the given task within the required level of accuracy. For example, the knowledge that the distance to an object is optimally estimated to be 5m is of little use if the robot does not know whether the value is reliable within ±10cm or ±1m.

- *Efficiency of computation.* An efficient numerical algorithm must be available for computing optimal estimates and evaluating their reliability. In choosing algorithms, we must also take into consideration various computational issues other than efficiency, such as numerical instability due to rounding and approximation errors involved in the computation.

- *Plausibility of the model.* We need a criterion to judge if the assumed structure of the problem on which the estimation is based is correct. Given a sequence of points, for example, we can optimally fit a line to them by an efficient algorithm and evaluate the reliability of the resulting fit, *provided the points are assumed to be collinear in the absence of noise.* How can we confirm such a presumption?

In the past, the traditional statistical approach has been the main tool in dealing with noisy data, and various statistical techniques have been employed in many forms. However, they are mostly used to make *qualitative* decisions and judgments [35, 36, 77, 108, 111, 139, 204]. In such problems, probabilities are usually adjusted empirically so as to reflect subjective degrees of tolerance rather than physical noise characteristics. A typical example is a technique called *probabilistic* (or *stochastic*) *relaxation* [12, 151].

For numerical estimation problems, on the other hand, statistical aspects have not been considered very much. Rather, much attention has been paid to *algebraic* aspects—writing down equations that describe the geometric relationship between the assumed 3-D structure of the scene and the 2-D description of its projected image and then solving these equations. In the presence of noise, the equations obtained by substituting the data are often inconsistent with each other, so some kind of optimization is applied. However, the optimization criterion is chosen rather heuristically or merely for convenience of computation, and its performance is evaluated empirically by simulations using real and synthetic data. A typical example is a technique called *regularization* [158, 159], for which the regularizing parameter is adjusted on a trial-and-error basis. If statistical techniques are employed, they are often transplanted from textbooks on statistics. In this book, we give many synthetic and real data examples to demonstrate that conventional methods are not optimal and how accuracy improves if truly optimal methods are employed. We also show that the reliability of the resulting optimal solution can be evaluated in analytical terms.

Testing the validity of the assumed model is also very crucial in computer vision and robotics applications. However, it has been customary to make judgments based on a heuristic criterion. For example, a sequence of points are judged as collinear if the residual of line fitting is smaller than an arbitrarily set threshold. In this book, we present a systematic procedure for such a judgment in rigorous statistical terms.

1.1.3 Why is a new statistical theory necessary?

The aim of this book is to give a rigorous mathematical foundation to numerical optimization problems for computing 2-D and 3-D geometric quantities from inaccurate image and sensor data. To this end, we need a new theory of statistics, because the basic premises of traditional statistics do not apply to the problems we consider in this book: what is very important in traditional statistics is not so important in our problems, while what is very important in our problems has not been much recognized by statisticians.

One of the major reasons why traditional statistical theories are not suitable for our purpose is their rather narrowly defined framework of viewing statistical estimation as *inferring a structure by observing multiple data*. This is with a view to evaluating and comparing effects and procedures in domains that involve a large degree of uncertainty, such as medicine, biology, agriculture, manufacturing, sociology, economics, and politics [38, 42, 52, 53, 63, 115, 137, 230]. In such domains, the problem is usually translated into the mathematical language as *estimating the parameters involved in a probability distribution from multiple independent samples from it*. Although this framework is suitable in the above mentioned domains, statistical estimation problems in computer vision and robotics have many non-traditional elements.

In traditional statistics, errors are regarded as *uncontrollable*; the accuracy of estimation can be improved only by repeated measurements. However, repeating measurements is costly. Hence, if the accuracy is the same, those methods which require a smaller number of data are more desirable. In other words, methods whose accuracy improves rapidly as the number of data increases are more desirable than those with slow increase of accuracy. Thus, the study of *asymptotic* properties in the limit of a large number of data has been one of the central subjects in traditional statistics [18, 137].

In such engineering domains as computer vision and robotics, where electronic sensing devices are used, errors are usually very small and called *noise*. Moreover, they are *controllable*: accuracy can be improved, for example, by using high-resolution devices and controlling the environment (lighting, dust, temperature, humidity, vibration, etc.). However, such control is costly. Hence, if the accuracy is the same, those methods which allow higher levels of noise are more desirable. In other words, methods whose accuracy improves rapidly as the noise level decreases are more desirable than those with slow increase of accuracy. Thus, the study of the accuracy of estimation *in the limit of small noise* is very important. In this book, we assume that errors are small and apply first order approximations in various forms. In this sense, our analysis is essentially a *perturbation theory*.

In many engineering domains, repeating measurements under the same condition (which is easy) often produces the same result because the sources of inaccuracy in the device and the environment are fixed (but unknown). In such a domain, the basic premise of traditional statistics that *independent samples from the same distribution can be observed as many times as desired* does not hold. How can one do statistical estimation from only one set of data? One cannot take even the "sample average," which is the most fundamental statistic. The answer is, as we mentioned earlier, the use of *a priori knowledge* about the problem: we can make inferences from the degree to which the data deviate from the the geometric constraints that should hold if the data were accurate. We can also make use of partial information about the accuracy of the image and sensor devices. To a first approximation, the uncertainty of the data supplied by a particular measurement can be characterized by the *covariance matrix* inherent to the sensor, but we need not know it exactly if we incorporate a priori knowledge about the problem: it suffices to know the covariance matrix *up to scale*. In other words, all we need is *qualitative properties*, such as isotropy and homogeneity, of the error distribution.

Another difference from traditional statistics is the *geometric* nature of our problems. Traditional statistics is mainly concerned with individual *variables* involved in a phenomenon. In computer vision and robotics applications, however, we are concerned with geometric objects such as points, lines, and surfaces in two and three dimensions and their interrelations such as incidence, joins, and intersections. Such geometric objects and interrelations are described by vectors, tensors, and manifolds. Moreover, all procedures for statistical estimation must be written in a form that is *invariant to coordinate transformations*: the estimate obtained by applying a coordinate transformation to the data must coincide with the value obtained by applying the same coordinate transformation to the estimate. For example, the problem of fitting a line to approximately linearly correlated data is called *regression* in statistics, but the independent variables (called *abscissa variables*, *explanatory variables*, *controlled variables*, *predictor variables*, *covariate variables*, and by many other names) and the dependent variables (called *observed variables*, *data variables*, *response variables*, *outcome variables*, *ordinate variables*, and by many other names) must be distinguished. In contrast, the x and y coordinates are completely equivalent when we fit a line to approximately collinear point data; the coordinate axes are chosen merely for computational convenience, and the line to be fitted must be identical whatever coordinate system is used.

1.2 The Features of This Book

1.2.1 Theoretical accuracy bound

One of the most significant consequences of our analysis is that a *theoretical bound* on the attainable accuracy is given for geometric correction and parametric fitting problems in the form of the covariance matrix of the parameters to be estimated. The bound is first derived in terms of the covariance matrices of the data by assuming that the noise is Gaussian. Later, a rigorous mathematical proof is given for general non-Gaussian noise, for which the *Fisher information matrix* plays the role of the covariance matrices of the data.

The theoretical bound we derive corresponds to what is known as the *Cramer-Rao lower bound* in statistics, but our treatment is very different from traditional statistics. It is well known in statistics that as the number N of data increases, the variance/covariance of an estimate generally converges to zero at a rate of $O(1/N)$. Hence, the bound on accuracy is defined by the *asymptotic limit* of N times the variance/covariance as $N \to \infty$. However, sophisticated mathematical arguments are necessary for rigorously defining various types of convergence (week convergence, strong convergence, etc.). In this book, we fix the number N of data and analyze the first order behavior *in the limit of small noise*. Hence, the bound on accuracy is defined as the limit of $1/\epsilon^2$ times the variance/covariance as $\epsilon \to 0$ for an appropriately defined noise level ϵ. Such a *perturbation analysis* does not involve intricate mathematical subtleties about convergence.

With our perturbation approach, it can be shown that the ubiquitous least-squares optimization is not optimal; statistical bias exists in the least-squares solution in general. We can also show that the theoretical bound can be attained by *maximum likelihood estimation* in the first order if the distribution of the noise belongs to the *exponential family*, in which the Gaussian distribution is a typical example. Just as maximum likelihood estimation is given a special status in the traditional asymptotic theory of statistics, it also plays an important role in our perturbation analysis.

1.2.2 Evaluation of reliability and testing of hypotheses

The fact that the reliability of maximum likelihood estimation can be evaluated quantitatively implies, for example, that when we fit a line to a sequence of point data, we can compute not only an optimal fit but also the probability that the fitted line deviates from the true position by a specified amount. This fact provides a means to *visualize* the most likely deviations. In this book, we present two visualization techniques—the *standard confidence region* and the *primary deviation pair*. They are multi-dimensional analogues of the "confidence interval" for point estimation in statistics.

Being able to predict the probability of likely deviations means that we can derive a technique for *testing hypotheses*. Namely, if a prediction based on a hypothesis is not consistent with its observation to a significant degree, the hypothesis is rejected. For example, when we apply a minimization scheme for fitting a line to a sequence of points which are supposedly deviated from their original linear configuration by image noise, we can compute the probability distribution of the residual of the minimization. If the actual residual value is very large as compared with the predicted noise level of the sensor, the hypothesis that the original configuration of the points is linear is rejected. In this book, we give a rigorous statistical criterion for testing geometric hypotheses in the form of the χ^2 test.

However, such a χ^2 test is effective only when the accuracy of the sensor can be estimated a priori. If no information about the noise is available, we cannot tell whether the hypothesis is violated because of the noise or the hypothesis itself is wrong. Also, we need to set the *significance level* for doing a χ^2 test, but no theoretical basis exists for its choice (5% or 1%?). In this book, we show that the *goodness* of the assumed geometric constraint, or *model*, can be measured by the *geometric information criterion* obtained by modifying the AIC widely known in statistics [2]. By using the geometric information criterion, we can compare the plausibility of two models without introducing any arbitrarily set threshold.

1.2.3 Geometric models as manifolds

Parametric fitting may sound merely like fitting lines and surfaces to point data, but it has a much wider meaning. In this book, we primarily deals with such simple geometric primitives as points, lines, planes, conics, and quadrics in two and three dimensions, but our theory can be applied to *any* primitives as long as they can be specified by coordinates and parameters. In fact, any primitive can be viewed as a *point* in an appropriately defined parameter space; our theory holds in any dimensions. A relationship between primitives can be identified with a parameterized *manifold* in an abstract parameter space; we call such a manifold a *model*. The goal of parametric fitting is to find an optimal model that best fits the data points that represent individual primitives. The criterion of optimality is given in terms of the *Mahalanobis distance* in the parameter space.

Thus, inferring a true configuration that satisfies a constraint such as incidence, coincidence, collinearity, concurrency, coplanarity, parallelism, or orthogonality can be viewed as parametric fitting. From this viewpoint, the problem of 3-D reconstruction from stereo and motion images and optical flow is also parametric fitting: a 3-D structure is reconstructed by fitting a relationship called the *epipolar constraint* [47, 90, 135, 227].

Once a relationship is inferred by parametric fitting, the next stage is *geometric correction*. For example, after a line is fitted to non-collinear points, the individual points are optimally moved onto the fitted line. If the epipolar geometry is fitted to stereo or motion data, they are optimally corrected so as to satisfy the fitted geometry. Mathematically, this problem can be viewed as an optimal *projection* of points that represent primitives onto a manifold in an abstract space, and the criterion of optimality is given in terms of the Mahalanobis distance. This type of geometric correction not only increases the accuracy of 3-D reconstruction but also allows us to *evaluate its reliability*, since the amount of such correction indicates the degree of inaccuracy of the data.

Geometric correction is also important if we adopt a computational technique called *linearization* for 3-D motion analysis. It has been widely believed that this technique sacrifices accuracy at the cost of computational convenience [189, 191, 207]. In this book, we point out that linearization does not reduce the accuracy *if the geometric correction is optimally applied.*

In the past, attempts have been made to view motion and optical flow analysis as traditional statistical estimation and apply standard statistical techniques [222, 233]. However, explicit and analytically closed expressions are difficult to obtain because what are known in statistics as *nuisance parameters* are involved. Viewing the problem as parametric fitting and geometric correction in an abstract space, we can systematically derive an optimal estimation criterion, an efficient numerical scheme, a theoretical bound on accuracy, and statistical tests for various types of hypotheses. In this respect, our approach is very different from all existing treatments of motion and optical flow analysis.

1.2.4 Numerical schemes for optimization

Presenting a mathematical theory is not the only purpose of this book. Giving consideration to making the theory applicable in real situations is also a major theme. The use of the *rank-constrained generalized inverse* to prevent numerical instability and increase the robustness of the solution is one example. Numerical schemes for computing an optimal solution are also studied in detail. Since an optimal solution is usually given by nonlinear optimization, which is time-consuming if solved by numerical search, we devise two simplified schemes for computing an approximately optimal solution: the *optimal filter* and *renormalization.* The latter plays a major role in this book. The renormalization procedure requires no *a priori* knowledge of the noise level; it is estimated *a posteriori* as a result of renormalization. It consists of iterated computations of eigenvalues and eigenvectors and bias-correction procedures. The accuracy of renormalization is shown to attain the theoretical bound in the first order. Many simulations and real image examples are given to show that renormalization is superior to conventional methods.

1.3 Organization and Background

1.3.1 Fundamentals of linear algebra

One of the main characteristics of our analysis is the *geometric* nature of the problem: the data are vectors and tensors in two and three dimensions that represent 2-D and 3-D objects. Hence, *linear algebra* plays a fundamental role. Chapter 2 summarizes vector and matrix calculus, the eigenvalue problem, the singular value problem, the generalized inverse, and matrix and tensor algebra. Since the materials presented here are well established facts or their easy derivatives, theorems and propositions are stated without proofs. Many of them are proved in Kanatani [90] in the form of the answers to the exercises. The generalized inverse that will be used in this book is only of the *Moore-Penrose type* (see [171] for other types). Readers who want a more advanced treatment of 3-D rotations and related subjects, such as Lie groups and Lie algebras, are advised to read Kanatani [85].

Along with fundamentals of linear algebra, topics concerning linear equations and optimization are also presented in Chapter 2: techniques for robustly solving indeterminate or ill-posed equations are summarized, and analytical solutions are derived for least-squares optimization, constrained quadratic optimization, and optimal rotation fitting. The optimal rotation fitting problem has been studied by many researchers in many different forms [11, 67, 69, 90, 93, 212]. The solution given in Chapter 2 is based on Kanatani [90, 93].

1.3.2 Probabilities and statistical estimation

Chapter 3 summarizes basic concepts of probability and statistics such as mean, variance, covariance, Gaussian distribution, χ^2 distribution, and the χ^2 test. Also, various principles of statistical estimation including *maximum likelihood estimation* and *maximum a posteriori probability estimation* (*Bayesian estimation*) are described by assuming that the noise is Gaussian.

Then, the (discrete-time) *Kalman filter* is derived by applying maximum a posteriori probability estimation to a linear dynamical system with Gaussian noise. The Kalman filter is often defined as *minimum mean square estimation* by means of *orthogonal projection*, since this formalism can be applied to non-Gaussian processes as well. However, minimum mean square estimation is identical with maximum a posteriori probability estimation if all variables are Gaussian, and the derivation of the Kalman filter is much easier for Gaussian processes. The details of the Kalman filter as well as the (continuous-time) *Kalman-Bucy filter* are left to the literature [10, 29, 33, 54, 74, 78, 79, 138, 140].

Chapter 3 also gives a general formulation of statistical estimation that does not depend on Gaussian properties. First, the *Cramer-Rao lower bound* is derived for the covariance matrix of an arbitrary unbiased estimator in terms of the *Fisher information matrix*. Then, asymptotic properties of the maximum likelihood estimator and the role of the *exponential family* of distributions are discussed without going into details. Finally, we derive the *AIC* (*Akaike information criterion*) that measures the goodness of a statistical model. The AIC was proposed by Akaike [2] and has been widely used for selecting a statistical model for a given random phenomenon without introducing an arbitrarily set threshold such as the *significance level* of the χ^2 test.

Our analysis has two non-traditional elements. Firstly, we consider probability distributions over a *manifold* defined by geometric constraints, so fundamentals of manifolds and tangent spaces are briefly summarized. Detailed discussions, *transversality* in particular, are found in books on *catastrophe theory* [166, 201]. Statistical treatment of random variables that are constrained to be in a manifold is very difficult even for a very simple one such as a sphere in three dimensions [82, 129]. In order to simplify the argument, we consider only *local distributions*, assuming that the noise is very small. This is related to the second feature of our analysis: we are mainly interested in the statistical behavior of estimates *in the limit of small noise* as opposed to their behavior *in the limit of a large number of data*, which is the main concern of traditional statistical analysis.

There exists a vast of amount of introductory literature on probability and statistics. Classical textbooks include [37, 38, 49, 52, 53, 62, 63, 104, 107, 115, 170, 230]. Statistical analysis involving generalized linear models, the Fisher information matrix, and the exponential family of distributions can be found in [42, 137]. Recently, much attention of statisticians has been drawn to geometric treatments of statistical distributions: a parameterized probability density is identified with a point in the parameter space, and statistical problems are interpreted in such geometric terms as *Riemannian metrics* and *affine connections*. This approach is called *statistical geometry* [7, 14, 15, 143]. Although this book also takes a geometric approach, we do not need such sophisticated concepts: all we need is *tangent spaces* to manifolds and *projection operations* onto them.

1.3.3 Representation of geometric primitives

Chapter 4 discusses mathematical representations of basic geometric primitives in two and three dimensions such as points, lines, planes, conics, and quadrics. These primitives are the main ingredients of *projective geometry* [90, 183, 184]. However, the representations established in projective geometry are defined with a view to making the mathematical treatment *consistent* without allowing any anomalies. For example, ordinary points and ideal points (points at infinity) are treated in the same way in terms of *homogeneous coordinates*. Hence, two lines always meet at a single (ordinary or ideal) point in two dimensions. Another characteristic of projective geometry is its *duality*: points and lines are treated in an identical manner in two dimensions, and so are points and planes in three dimensions. This implies that theorems and statements concerning points, lines, and planes are automatically extended to their duals.

In engineering applications such as computer vision and robotics, ordinary objects and ideal objects *need* to be distinguished, because only ordinary objects can be measured by sensors in real environments. Also, the duality concerning points, lines, and planes does *not* hold in real environments, because the error behavior is different for an object and its dual. In image analysis, for example, feature lines are defined by edge detection followed by line fitting, while feature points are usually defined as the intersections of detected feature lines. Since the error behavior of lines thus defined and the error behavior of points computed from them are very different, the mathematical elegance of projective geometry is destroyed, and representations of geometric primitives should reflect the way they are computed from sensor data.

In order to describe geometric primitives and their relationships in real environments, one must therefore reformulate projective geometry from a computational point of view. This was done by Kanatani [86, 90, 91, 94, 95], who called the resulting formalism *computational projective geometry*. The treatment in Chapter 4 is a hybrid of projective geometry and real computation, but the emphasis is shifted more toward *computational aspects* than toward the role of projective geometry. Moreover, the same primitive is often given multiple representations; one is useful for error analysis, while another is useful for describing geometric relationships to other primitives.

Almost all the representations used in this book have *inherent constraints* in one way or another, and the number of parameters that specify a primitive is generally larger than its true degrees of freedom. As a result, error behavior is described by a *singular* covariance matrix having a *null space*, and dealing with null spaces is one of the main characteristics of our analysis. In Chapter 4, we list formulae to convert expressions for geometric properties and error behavior from one representation to another and from one coordinate system to another.

Then, *perspective projection* is introduced as a physical interpretation of the homogeneous coordinates. This is the key relationship between 3-D objects and their 2-D descriptions. In order to apply 3-D analysis based on perspective projection to real camera images, the camera is assumed to be *calibrated*, i.e., its imaging geometry is known and modeled as perspective projection. Techniques for camera calibration are found in [13, 32, 44, 56, 59, 57, 88, 98, 110, 116, 117, 136, 156, 199, 210, 213, 214, 215, 218, 223].

Finally, we give a brief account of *conics* and *quadrics*. A conic (also referred to as a *conic locus* or *conic section*) is a plane curve defined by a quadratic equation in the coordinates. Conics are classified into *ellipses* (including circles), *parabolas*, *hyperbolas*, and their various degeneracies. If a robot is to operate in an industrial environment (say, in a nuclear power station), it must recognize gauges, meters, dials, and handles, most of which are circular, and circles are perspectively projected into ellipses. Hence, ellipses, or conics in general, are widely recognized as one of the most fundamental features in the study of computer vision and robotics [50, 169, 121, 178]. Detected conics can provide not only clues to object recognition; if the observed conic in an image is known to be a perspective projection of a conic in the scene of a known shape, its 3-D geometry can be computed analytically [51, 90, 97, 128, 180, 181].

A quadric is a space surface defined by a quadratic equation in the coordinates. Quadrics are classified into *ellipsoids* (including spheres), *paraboloids*, *hyperboloids*, and their various degeneracies. As compared with conics, the role of quadrics appears to be minor in computer vision and robotics applications, since ellipsoidal objects are rare in real environments. However, quadrics play an important role for *visualizing* the reliability of computation in three dimensions. We also present a technique for visualizing the reliability of objects that have more than three degrees of freedom by means of the *primary deviation pair*.

In Chapter 4, we only deal with points, lines, planes, conics, and quadrics in two and three dimensions, but this is simply because they are the most frequently encountered primitives in robotics applications. Indeed, the statistical estimation theory described in this book can be applied to any primitives if they are identified with points in an abstract parameter space and their interrelationships are represented by manifolds in that space.

1.3.4 Geometric correction

Chapter 5 discusses the *geometric correction problem* in general. A typical problem is to move data points so that they are on a particular line, curve, plane, or surface. Such a correction is necessary if the points we are observing are known to be on a specific line, curve, plane, or surface in the absence of noise. A naive idea is to project each point to the "nearest point" on the line, etc. However, this can be justified only when the noise distribution is isotropic. Otherwise, the correction should be such that it compensates for the most likely deviation. This problem is not limited to points; it can also apply to lines, planes, and more complex objects for imposing geometric constraints on their configurations.

If the noise is Gaussian, an optimal solution is obtained by minimizing the *Mahalanobis distance*: the solution is the "nearest point" on the constraint surface in the parameter space measured in the Mahalanobis distance. We derive an explicit expression for the optimal solution, for which the noise level need not be known: it can be estimated *a posteriori* by analyzing the statistical behavior of the residual of the minimization. We also show that in the course of computing an optimal solution, its *a posteriori covariance matrix* can be is automatically evaluated.

If the noise level can be predicted in advance, we can *test* the existence of the constraint: the hypothesis that the observed object is in a certain configuration is rejected if the residual of the minimization is very large as compared with the predicted noise level. We formalize this process as the χ^2 *test*.

After giving a general theory, we apply it to the problem of imposing coincidence and incidence on points, lines, conics, and planes. We also study the problem of imposing orthogonality on orientations. Applications of this type of geometric correction in reconstructing 3-D structures from stereo images for robot navigation are found in [161, 162, 163, 165].

1.3.5 3-D computation by stereo vision

In Chapter 6, we study *stereo vision*—3-D reconstruction by triangulation from two (or more) images. Since the principle is very simple, stereo vision has become one of the most widely used means of 3-D sensing for autonomous manipulation and navigation by robots. In the past, studies of stereo vision have been almost entirely concentrated on *matching* of the two stereo images. This is because establishing correspondences between the two images is a very difficult task to automate efficiently, and many matching techniques have been proposed and tested [47, 57, 227]. In contrast to matching, 3-D reconstruction appears rather trivial. However, since image data contain noise, the reconstructed 3-D structure is not necessarily accurate. Hence, it is very important to evaluate the reliability of the reconstructed 3-D structure. Otherwise, robots are unable to take appropriate actions to achieve given tasks with required precision. This issue has been discussed by only a few researchers (e.g., see [19]). We concentrate on this issue and evaluate the covariance matrices of the reconstructed points and lines by applying the theory of geometric correction given in Chapter 5.

In Chapter 6, we first discuss the *epipolar constraint* of stereo vision by introducing the concepts of *epipoles* and *epipolars* and show that the degree to which the required epipolar equation is not satisfied provides an estimate of the noise level, from which the reliability of the reconstructed 3-D structure can be evaluated. This result is based on Kanazawa and Kanatani [101]. It turns out that this analysis is equivalent to introducing a *Riemannian metric* into the 3-D scene by projecting the direct product of two images onto the 3-D manifold defined by the epipolar equation. Applying the theory of geometric correction given in Chapter 5, we also present an optimal scheme for imposing the constraint that the reconstructed points and lines must be on a specified planar surface. Finally, we analyze the errors in 3-D reconstruction due to camera calibration errors (e.g., see [39, 80, 186] for stereo camera calibration).

1.3.6 Parametric fitting

In Chapter 7, we study methods for fitting a geometric object to multiple instances of another geometric object in an optimal manner in the presence of noise. A typical example is fitting a line to a sequence of points. Line fitting is one of the most important steps in computer vision, and various techniques for it have been studied in the past [81, 106, 219, 235]. This is because the first step of image analysis is detecting *edges*, i.e., sequences of pixels constituting boundaries that separate objects from the background. They are detected by applying a filter called an *edge operator*. Since man-made objects in robotic workspaces usually have linear boundaries, many objects in the image can be located by fitting straight lines to detected edge pixels. Then, their shapes and locations in the scene are inferred by various technique, e.g., by computing *vanishing points* and *focuses of expansion* [24, 90]. Since errors in line fitting propagate to the final 3-D reconstruction, the reliability of the reconstructed 3-D shape can be evaluated if the reliability of line fitting is quantitatively evaluated. In Chapter 7, we analyze the statistical behavior of a line fitted to a sequence of edge pixels. See [23, 87, 88, 91, 94, 187, 220] for various types of statistical inference based on line fitting errors.

The theory in Chapter 7 is not limited to line fitting. We generalize the theory of parametric fitting to deal with arbitrary geometric objects in abstract terms. The goal is not only obtaining an optimal fit but also evaluating its reliability in statistical terms. First, the criterion for parametric fitting is derived as maximum likelihood estimation by assuming that the noise is Gaussian. The covariance matrix of the resulting estimate is evaluated, and the statistical behavior of the residual of optimization is analyzed. This analysis leads to the χ^2 test for the hypothesis that the observed objects are in a special configuration. Then, we study various types of fitting problem in two and three dimensions such as finding an optimal average, estimating a common intersection, and fitting a common line or plane.

1.3.7 Optimal filter and renormalization

In Chapters 8 and 9, we study numerical methods for efficiently computing the optimal solution of the parametric fitting problem. In Chapter 8, we construct a filter that optimally updates the estimate each time a new datum is read; we call it simply the *optimal filter*. The update rule is derived for Gaussian noise by adopting the *Bayesian* standpoint and applying *maximum a posteriori probability estimation*. Various assumptions and approximations involved in the derivation are elucidated, and the philosophical implications of the Bayesian approach are also discussed. The update rule is simplified by introducing the *effective gradient approximation*.

The Kalman filter was originally derived for dynamical linear systems, but it can also be applied to nonlinear dynamical systems by introducing linear approximation into the system equations. The resulting filter is known as the *extended Kalman filter* [10, 29, 33, 54, 140]. In Chapter 8, we show that the optimal filter is also obtained from the extended Kalman filter if the parametric fitting problem is identified with a nonlinear dynamical system *with no state transition* (i.e., a nonlinear "static" system). For this reason, the optimal filter described in Chapter 8 is often called the "extended Kalman filter" [61, 161, 162, 163, 165]. Porrill [164] pointed out the fact that the extended Kalman filter yields a statistically biased solution if the standard procedure for linear approximation is applied. He also proposed a method for correcting the bias, which is a special version of the effective gradient approximation described in Chapter 8.

Chapter 9 focuses on the parametric fitting problem for linear equations. To a first approximation, the problem reduces to least-squares fitting, for which the solution can be obtained by solving an eigenvalue problem; we call the resulting solution the *eigenvector fit*. We first show that this least-squares approximation introduces statistical bias into the solution. After analyzing the statistical bias in quantitative terms, we present a simple correction scheme for cancelling this bias; we call the resulting solution the *unbiased eigenvector fit*. In order to compute the unbiased eigenvector fit, however, we need to estimate the noise level precisely, which is usually difficult. In order to avoid this difficulty, we present a scheme for computing an unbiased solution, called the *generalized eigenvector fit*, without assuming a priori knowledge of the noise level.

The computation is further simplified into an iterative form called *renormalization*. In contrast to the optimal filter, no initial estimate is necessary for renormalization. The noise level is estimated *a posteriori* as a result of renormalization, and the covariance matrix of the computed estimate is automatically obtained. We then discuss a procedure called *linearization*, which allows us to apply renormalization to nonlinear equations. Finally, we define *second order renormalization* that removes statistical bias up to higher order terms.

1.3.8 Applications of geometric estimation

Chapter 10 illustrates the renormalization technique by solving the problems of line fitting and conic fitting in two dimensions and plane fitting in three dimensions. The computation for line fitting by renormalization is straightforward. If the image noise is uniform and isotropic, the least-squares scheme gives an optimal fit. This means that accuracy is not increased by the use of renormalization any further. However, the advantage of renormalization is that in the course of the fitting computation the reliability of the fit is automatically evaluated in the form of the covariance matrix. We will demonstrate this by showing the standard confidence region and the primary deviation pair that visualize the reliability of the solution computed from simulations and real image data.

Next, we study the problem of conic fitting. As pointed out earlier, conics are important image features for computer vision, and if the observed conic in an image is known to be a perspective projection of a conic in the scene of a known shape, its 3-D geometry can be computed analytically [51, 90, 97, 121, 128, 178, 180, 181]. In order to do such an analysis, the conic must be given a mathematical representation in advance. This is done by applying an edge operator to a gray-level image, detecting object boundaries as edge segments, and fitting a quadratic curve to those edge segments which supposedly constitute a boundary of a conic region. Numerous conic fitting techniques have been proposed in the past [4, 20, 21, 34, 41, 45, 55, 145, 155, 167, 179, 182], but most of them are least-squares schemes with different parameterizations and criteria, and little consideration has been given to the statistical behavior of image noise. An exception is Porrill [164], who devised an iterative filtering scheme, which he called the "extended Kalman filter." Pointing out the existence of statistical bias in the solution if the filter was formulated naively, he proposed a bias correction procedure, which is equivalent to the "effective gradient approximation" introduced in Chapter 8. Kanatani [92] presented a prototype of renormalization, from which the formulation in this book has evolved. As in the case of line fitting, the reliability of the fit is evaluated in the course of the the fitting computation.

Finally, we study the problem of fitting a planar surface to two types of data. Planar surface fitting is a very important process in an indoor robotic workspace, where many objects, including walls, ceilings, and floors, have planar surfaces. First, we assume that the 3-D positions of feature points that are known to be coplanar in the scene are measured by a *range finder*. In this case, the uncertainty of the data has a special form, according to which the covariance matrix of the data is modeled. The computation is straightforward, and the reliability of the fit is automatically evaluated in the form of the covariance matrix in the course of the fitting computation. Line and plane fitting to range data has been studied by many researchers in the past [22], but the reliability of the fit has been evaluated only in an ad hoc manner [197]. The method described here is based on Kanazawa and Kanatani [102].

Next, we consider the problem of reconstructing a planar surface from stereo images of feature points that are known to be coplanar in the scene. An indirect but straightforward method is first computing the 3-D positions of individual feature points separately and then fitting a planar surface to them. However, it is expected that the reliability of the 3-D reconstruction can be enhanced if the knowledge that the feature points are coplanar is incorporated from the beginning. With this motivation, we optimally reconstruct a planar surface directly from stereo correspondence pairs. Applying the theory of geometric correction given in Chapter 5, we optimally correct corresponding pairs of feature points in the stereo images in such a way that they define exactly coplanar points in the scene. Then, the parameters of the plane on which they lie are computed by renormalization. Here again, the reliability of the fit is evaluated in the course of the fitting computation. This result is based on Kanazawa and Kanatani [100].

1.3.9 3-D motion analysis

Mathematical analysis of 3-D rigid motion estimation, known as *shape* (or *structure*) *from motion*, was initiated by Ullman [211], who presented a basic mathematical framework that has had a lasting influence over the subsequent computer vision research. Roach and Aggarwal [177] applied this framework to real images and obtained the solution by numerical search. Nagel [144] presented a semi-analytical formulation, reducing the problem to solving a single nonlinear equation. A complete analytical solution for eight feature points was independently given by Longuet-Higgins [122] and Tsai and Huang [207]. The solution of Longuet-Higgins was based on elementary vector calculus, while the solution of Tsai and Huang involved singular value decomposition. Zhuang et al. [238] combined them into a simplified eight-point algorithm. Zhuang [236] also discussed the uniqueness issue. All these algorithms first compute the *essential matrix* from the *epipolar equation* and then compute the *motion parameters* from it. This technique is called *linearization*, and algorithms that use it are called *linearized algorithms*. Huang and Faugeras [73] pointed out that a matrix can be an essential matrix for some motion if and only if it has singular values 1, 1, and 0. Linear algorithms compute the essential matrix without considering this *decomposability condition*.

Since the essential matrix has five degrees of freedom, a 3-D interpretation can be determined, in principle, from five feature points. Using a numerical technique called the *homotopy method*, Netravali et al. [150] showed the existence of at most ten solutions. Arguing from the standpoint of projective geometry, Faugeras and Maybank [48] also showed that at most ten solutions can be obtained from five feature points. They reduced the problem to solving an algebraic equation of degree ten and solved it by symbolic algebra software. Using the quaternion representation of 3-D rotation, Jerian and Jain [75] reduced the problem to solving the resultant of degree 16 of a pair of polynomials of degree 4 in two variables and computed the solution by symbolic algebra software. Other proposed techniques include [113, 103]. Jerian and Jain [76] exhaustively reviewed algorithms known by that time and compared their performances for noisy data.

However, all these algorithms are constructed on the assumption that all data are exact. Hence, they are all *fragile* in the sense that inconsistencies arise in the presence of noise (e.g., the solution becomes different, depending on which of the theoretically equivalent relationships are used). A noise robust algorithm was presented by Weng et al. [226], who estimated the essential matrix from the epipolar equation by least squares and then computed the motion parameters by least squares. Spetsakis and Aloimonos [188, 189, 190] applied direct optimization to the epipolar equation without computing the essential matrix.

Although error analyses have been given for 3-D motion analysis by several researchers, most of the studies were empirical and qualitative, e.g., estimating approximate orders of errors and conducting simulations with noisy data [112, 157]. A notable exception is Weng et al. [226], who analyzed the perturbation of the essential matrix and the resulting motion parameters in detail. The fact that the least-squares solution based on the epipolar equation is statistically biased has also been recognized [3, 189, 190] and analyzed in detail by Kanatani [89, 90]. Spetsakis [191] conducted a statistical error analysis on a heuristic basis. Weng et al. [222] presented an optimal algorithm by identifying the problem with a traditional statistical estimation problem.

Planar surface motion has also been studied by many psychologists from the viewpoint of human visual perception, and the fact that multiple 3-D interpretations are possible was pointed out by Hay [58] as early as in 1966. A rigorous mathematical analysis was given by Tsai and Huang [206, 208] and Tsai et al. [209] by using singular value decomposition. A complete solution was presented by Longuet-Higgins [125], and errors analysis was done by Weng et al. [221]. These results are summarized in Kanatani [47, 90, 135, 227]. Hu and Ahuja [72] extended the analysis to multiple planar surface images.

Chapter 11 presents a statistically optimal algorithm for computing the 3-D camera motion and the object shape from corresponding points over two views. We first study the case in which the feature points in the scene are in general position and then the case in which they are known to be coplanar. Our analysis is distinct from all existing studies in many respects. First of all, the *reliability* of the computed motion parameters and the reconstructed depths is evaluated in the form of their covariance matrices, and the *theoretical bound* on their attainable accuracy is obtained in an *explicit* form. This is made possible by viewing 3-D motion analysis as a correction and fitting problem and applying the theories established in Chapters 5 and 7. This viewpoint is in a sharp contrast to that of Weng et al. [222], who treated the problem as a traditional statistical estimation problem. A germ of our approach was found in the work of Trivedi [205], but his idea has not been noticed very much: the present theory is in a sense a full-fledged version of his idea. The analysis of general motion given here is based on Kanatani [96], and the analysis of planar surface motion is based on Kanatani and Takeda [99].

Since we first compute the essential matrix and then decompose it into the motion parameters, our algorithm can be classified as a linear algorithm. In the past, the linearization technique has often been rejected as having poor accuracy as compared with direct nonlinear optimization. We point out that this is because statistical aspects have not been considered fully: we demonstrate that linearization does not reduce the accuracy *if the essential matrix is optimally computed and optimally corrected.* The renormalization procedure produces not only a statistically optimal estimate of the essential matrix but also its *covariance tensor*, according to which the computed essential matrix is optimally corrected so as to satisfy the decomposability condition.

Another new viewpoint given in Chapter 11 is the introduction of two types of *statistical test.* One is the *rotation test.* Prior to 3-D analysis, we need to test if the camera motion is a pure rotation, in which case no 3-D information is obtained. This is done by first hypothesizing that the camera motion is a pure rotation and then testing whether the observed images support that hypothesis to a statistically admissible degree. Since no 3-D information is obtained if the camera motion is a pure rotation, the degree to which the observed images support the rotation hypothesis can be viewed as defining the *information* in the motion images. By computing the information in observed images, we can predict the degree of robustness of the resulting 3-D reconstruction.

The other test is the *planarity test*: we need to test if the object is a planar surface, in which case a separate algorithm must be used. In the past, various ad hoc criteria have been used for this test. For example, since the algorithm for general motion breaks down if the object is exactly planar, it has been customary to switch to the planar surface algorithm *only when computational difficulties are encountered* (e.g., if theoretically nonzero quantities approach zero), and the judgment about this has been made arbitrarily. Our approach is the opposite: we first use the planar surface algorithm and switch to the general motion algorithm *only when the assumption of planarity is questioned.* We make this judgment by the χ^2 test, hypothesizing that the object is a planar surface and testing if the observed data support that hypothesis.

3-D interpretation of the scene becomes more reliable and more realistic if a sequence of image frames is incorporated [71, 198]. However, the discussion in this book is limited to two-view analysis, because multi-view analysis raises many mathematical and technical problems that are outside the scope of this book. One approach is the use of the *extended Kalman filter* with non-linear internal dynamics (the optimal filter we discuss in Chapter 8 has no internal dynamics); see [25, 26, 27, 28, 47, 114, 131, 164, 227, 231]. Thomas et al. [202] presented a related statistical analysis. Another approach includes considering the physics of the 3-D motion in question [225] and using stereo image sequences [224, 234]. Tomasi and Kanade [203] proposed a 3-D analysis method called *factorization* for a sequence of orthographically projected images.

Chapter 11 includes a brief discussion about the *critical surface* that gives rise to ambiguity of 3-D interpretation. Longuet-Higgins [123] pointed out that all linear algorithms for determining the essential matrix fail to yield a unique solution if and only if all the feature points are on a special quadric surface, which is called the *weak critical surface* in this book. Horn [68] studied the problem of two-camera registration, which is different from 3-D motion analysis in appearance but has essentially the same mathematical structure. He also discussed ambiguity of interpretation and showed that the critical surface must be a hyperboloid of one sheet or its degeneracy. A more direct analysis was given by Longuet-Higgins [126], who analyzed the condition that the critical surface degenerates into two planar surfaces or takes on other special forms. Further analyses were done by Negahdaripour [147, 148] and Maybank [133, 134, 135], giving detailed classifications of possible types of the critical surface. Their results are concisely summarized in Kanatani [90].

In this book, we concentrate on 3-D analysis of motion images and do not deal with image processing techniques for tracking feature points, on which research is still in progress [47, 57, 227].

1.3.10 3-D interpretation of optical flow

If a sequence of images is taken by a smoothly moving camera, the difference between two consecutive frames is very small. Small displacements of points define a dense "flow," called *optical flow*, on the image plane. In general, the computation of 3-D interpretation based on optical flow is likely to be sensitively affected by image noise. Also, detecting accurate optical flow is very difficult. However, optical flow has the advantage that the flow can be detected densely (usually at each pixel) over the entire image by an image processing technique, while it is in general very difficult to detect point-to-point correspondences between two images—particularly so if the interframe motion is large. Hence, the use of optical flow is expected to have practical significance only if an appropriate optimization technique is available.

In Chapter 12, we first summarize the principle for optimally detecting optical flow from two gray-level images. A prototype of optical flow detection technique is due to Horn and Schunck [70]. However, it has a flaw in that it uses the *smoothness constraint*, which is a special case of the heuristics called *regularization* [158, 159]. If the detected flow is to be used for image segmentation (e.g., for separating moving objects from the stationary scene), the smoothed (or *regularized*) solution will often blur the motion boundaries. If the purpose is 3-D reconstruction, smoothing does not increase the accuracy of the reconstructed 3-D shape, because smoothing is essentially *interpolation* in the image. Rather, we should reconstruct only those 3-D points which produce reliable optical flow. We should then interpolate them *in the scene* when and only when some knowledge about the true shape of the object (e.g., planarity) is available. In Chapter 12, we apply the theory of parametric fitting given in Chapter 7 to optical flow detection. However, since research on optical flow detection is still in progress [47, 57, 227], we do not go into details. A comprehensive review of existing techniques is given in [16]. The discussion in this book is limited to a single optical flow image observed by one camera in motion. 3-D analysis from stereo optical flow images observed by two or more cameras in motion is found in [105, 118, 142, 216].

We then turn to 3-D reconstruction from optical flow, which has also been studied in various forms in the past. If the object is a planar surface, the solution can be computed in an analytical form although multiple solutions exist [83, 84, 124, 192]. If the object surface is expressed as a polynomial or a collection of planar patches, the problem reduces to estimating the coefficients [1, 193, 217]. An analytical solution can be obtained if spatial derivatives of the flow velocity are used [127]. Since the flow due to camera rotation is depth independent [85], thereby globally continuous and smooth, a sudden change of the flow over a small number of pixels implies the existence of a translational motion and a depth discontinuity; this is the phenomenon called *motion parallax*, from which the translational velocity can be estimated [127, 172]. More systematically, the translation velocity can be determined by subtracting the effect of camera rotation in such a way that the resulting flow has a common *focus of expansion* [168]. A more direct approach is to do numerical search for minimizing the sum of the squares of the differences between the observed flow and the expected theoretical expression [30, 43, 60, 232]. Zhuang et al. [237] derived a linear algorithm similar to the corresponding finite motion algorithm.

As in the case of finite motion, it has been pointed out that the solution based on least-squares optimization is likely to be systematically biased [40]. Tagawa et al. [195, 196] proposed an iterative method to remove statistical bias, which was a prototype of the renormalization procedure described in this book. Endoh et al. [46] discussed the asymptotic accuracy behavior. A theoretical bound on accuracy based on the Cramer-Rao inequality was studied by Young and Chellappa [233], who regarded optical flow analysis as a statistical estimation problem in the traditional form. Various aspects about the accuracy and robustness of the solution have been studied in many different forms [31, 141, 149, 194].

Mathematically, optical flow is simply an infinitesimal limit of a finite image motion, so all procedures for 3-D interpretation of optical flow should be obtained from those for finite motion by taking the limit as the interframe time approaches zero. In fact, the analysis of 3-D interpretation of optical flow given in Chapter 12 is exactly parallel to the finite motion analysis given in Chapter 11: we first study the case in which the object has a general shape and then the case in which the object is a planar surface. We also describe a theoretical bound on the attainable accuracy and present a computational scheme by using renormalization and linearization. The analysis given here is based on Ohta and Kanatani [154].

The rotation test and the planarity test are also discussed in the same way as in the case of finite motion. Finally, we study the error introduced by identifying the interframe displacement by a continuous flow and point out that the ratio of image noise to optical flow is a very deceptive measure for predicting the accuracy of the resulting 3-D interpretation.

A brief description of the critical surface of optical flow is also given here. Its definition and geometric properties are the same as those of finite motion, but historically the existence of the critical surface was first pointed out for optical flow. Maybank [132] showed that such a surface should necessarily be a quadric surface passing though the viewpoint. Horn [66] introduced the term "critical surface" and showed that the critical surface must generally be a ruled quadric and hence a hyperboloid of one sheet or its degeneracy. Negahdaripour [146] exhaustively classified the possible types of critical surface. Their results are concisely summarized in Kanatani [90].

1.3.11 Information criterion for model selection

In order to apply the techniques for geometric correction and parametric fitting, one needs to know the *geometric model*, i.e., the constraints and hypotheses that should hold in the absence of noise. But how can one prefer one geometric model to another? The problem of model selection is very important in many engineering domains where probabilities are involved, and two approaches have attracted attention: one is the *AIC* (*Akaike information criterion*) [2]; the other is the *MDL* (*minimum description length*) *principle* [173, 174, 175, 176]. The AIC favors a model whose maximum likelihood estimator has a *minimum residual for future data*, while the MDL principle favors a model that can be described in a *code of minimum length*.

Since a geometric model is defined as a *manifold*, the complexity of the model is evaluated not only by its *degree of freedom* but also by such invariant quantities as the *dimension* and *codimension* of the manifold. In Chapter 13, we first summarize what we call the *Mahalanobis geometry* of maximum likelihood estimation and derive the *geometric information criterion* by modifying the AIC. This criterion in its original form requires exact knowledge of the noise level. In order to avoid this, we present a *comparison criterion* for evaluating the goodness of one model relative to another without using any arbitrarily set threshold such as the significance level of the χ^2 test. We derive explicit expressions for the comparison criterion for point data in two and three dimensions, 3-D reconstruction by stereo vision, 3-D motion analysis, and 3-D interpretation of optical flow. At the same time, we elucidate the geometric structures of these problems and their interrelationships—in particular, the ambiguity and singularity of 3-D reconstruction when the object is a planar surface or the camera motion is a pure rotation.

1.3.12 General theory of geometric estimation

Chapter 14 presents a rigorous mathematical foundation for the theory of geometric correction given in Chapter 5 and the theory of parametric fitting given in Chapter 7. For both, the problem is stated in general terms without assuming Gaussian noise: the role of the covariance matrix for a

Gaussian distribution is played by the *Fisher information matrix*. We derive a lower bound, which corresponds to the *Cramer-Rao lower bound* in traditional statistics, on the covariance matrix of the unbiased estimator of the parameter. Then, the maximum likelihood estimator is proved to attain it in the first order if the problem belongs to the *exponential family*. Finally, the maximum likelihood estimation process is expressed in a computationally convenient form, where the rank-constrained generalized inverse is used to deal with the ill-posedness of the problem and the numerical instability of the solution. The theme of Chapter 14 is essentially statistical estimation, but our treatment has non-traditional elements for the reasons pointed out in Section 1.1.3.

The statistical problem closely related to the parametric fitting problem is what is known as the *Neyman-Scott problem* [152]: observing multiple data, each having a distribution characterized by a common parameter (called the *structure parameter* or the *parameter of interest*) and a distinct parameter (called the *nuisance parameter*) that depends on each observation, one has to estimate the structural parameter without knowing the nuisance parameters [18]. In the parametric fitting problem, the true values of image/sensor data can be viewed as nuisance parameters, since the goal is to estimate the fitting parameter that characterizes the 2-D/3-D structure without knowing the exact values of the sensor data. However, the probability densities of the sensor data do not involve the fitting parameter. Hence, various mathematical concepts and tools developed for the Neyman-Scott problem [8, 9, 17, 109, 119, 120] cannot be applied directly in the form given in the literature of statistics.

1.4 The Analytical Mind: Strength and Weakness

1.4.1 Criticisms of statistical approaches

In concluding this chapter, let us note the fact that there exist strong criticisms of statistical (or in general analytical) theories of robotics and computer vision. Since any theory is essentially an *idealization* of reality, there is always something in the real world that the theory does not cover, and statistical theories of robotics and computer vision have often been criticized for this very reason. Strong arguments against statistical theories include the following:

- In order to apply a statistical theory, one must model the noise in a mathematically tractable form, say, assuming that the noise is independent, uniform, homogeneous, isotropic, Gaussian, etc. However, these are mathematical artifacts: noise in reality is more or less correlated, non-uniform, non-homogeneous, non-isotropic, and non-Gaussian. Hence, one does not know how reliable the analytical results based on these assumptions are.

- A statistically optimal solution is usually obtained by nonlinear optimization in a very complicated form, requiring a lot of computations. However, a human can easily perceive 3-D structures using vision and can take appropriate actions with sufficient precision. It is difficult to imagine that the brain conducts complicated optimization computations, such as renormalization, each time a human sees or does something. Rather, human perception seems to be based on ingenious combinations of experiences (data) and heuristics (rules). Hence, it should be possible to build an intelligent robot that has a human's versatility without doing rigorous mathematical analysis. We should learn from humans rather than relying on mathematics.

- Uncertainty in the real world is not merely quantitative. There is always a possibility that an unexpected circumstance happens—sudden appearance or disappearance of an object, for example. Since it is difficult to model the uncertainty in the real world completely, it should be more realistic to adopt heuristics and avoid rigorous mathematical assumptions. Experience has shown that empirically adjusted systems often outperform those designed by precise mathematical analysis; the latter do well only when the assumptions are exactly met.

Those who criticize the statistical approach advocate such non-analytical paradigms as *artificial intelligence*, *fuzzy inference*, and *neuro-computing*. In the domain of computer vision, such compromises as *qualitative vision* and *purposive vision* have been proposed as alternatives to rigorous mathematical analysis. However, just as the importance of such paradigms cannot be denied, the importance of analytical methods cannot be denied either, because advantages and disadvantages are both sides of the same coin. Rigid and inflexible as it is, an analytically designed system has the advantage that *it can faithfully reveal the insufficiency of the model*. Namely, its poor behavior implies the existence of new factors in the environment that are not properly modeled in the analysis, and the study of these becomes the next goal, an analysis guiding us to a new analysis. In contrast, such insufficiency is often concealed by adroit behavior of a heuristically designed system, and its long-term performance is unpredictable—it may suddenly break down when its versatility capacity reaches a limit. This is an inescapable destiny of a system that *learns* from humans. By a careful mathematical analysis, *humans* can learn much about the mechanisms that underlie seemingly simple phenomena.

1.4.2 Gaussian noise assumption and outlier detection

In our statistical analysis, the covariance matrix plays a central role as a measure of uncertainty. This is practically equivalent to assuming that the distribution of the noise is either Gaussian or approximately Gaussian. We develop a general non-Gaussian theory in Chapter 14, but the noise distribution is required to share many properties with the Gaussian distribution—smoothness and unimodality, for instance. Although assuming Gaussian or Gaussian-like noise makes mathematical analysis very easy, this certainly sets a limitation on potential applications of our theory.

A typical situation in which this Gaussian noise assumption is violated is when the noise is due to systematic bias in the data acquisition procedure. For example, lens aberration causes a systematic distortion of the image, e.g., symmetric elongation or contraction around its center. The use of wrong camera parameters estimated by poor calibration is another source of systematic errors. In processing images, error characteristics are different from operation to operation. For example, if the epipolar constraint is used for finding point correspondences between stereo images, the error characteristics are very different along the epipolar and in the direction orthogonal to it.

In this book, deviations of points and lines from their supposed positions are collectively regarded as "noise" irrespective of their sources. The Gaussian noise assumption is a good approximation if a lot of independent noise sources are involved, each having a small effect. This is a consequence of the *central limit theorem*. In contrast, the Gaussian noise assumption poses a difficulty if a small number of very different noise sources coexist. If only one noise source exists, the noise characteristics can be estimated by repeating measurements. If multiple sources exist and one source is dominant, data from a dominant source are called *inliers*; those from other sources are called *outliers*. In order to do statistical estimation, one must judge which data are inliers and which are outliers. This can be done, for example, by hypothesizing that particular data are inliers and testing that hypothesis by various statistical tests. Such a procedure is called *outlier detection*.

This book does not deal with outlier detection at all. This is because it requires treatments that are very difficult to generalize. However, it should be emphasized that the theory of statistical estimation and model selection presented in this book provides a theoretical foundation for developing such techniques. See [139, 204, 219] for outlier detection in computer vision problems.

1.4.3 Remaining problems

In this book, we concentrate only on *numerical computation* by assuming that necessary data are provided by image processing and other sensing operations. We do not go into the details of image processing techniques involving gray levels and colors such as *edge detection*, *stereo matching*, and *feature point tracking* (see [47, 57, 185, 227] for these topics). We also do not consider 3-D analysis based on the gray levels of images such as *shape from shading* and *photometric stereo* (see [65]). Since the focus is on numerical computation, we do not discuss paradigms of image understanding and computer vision and relationships with human perception. See [5, 6, 12, 130, 151, 211, 229] for such arguments.

Chapter 2

Fundamentals of Linear Algebra

This chapter presents fundamentals of linear algebra that will be necessary in subsequent chapters. Also, the symbols and terminologies that will be used throughout this book are defined here. Since the materials presented here are well established facts or their easy derivatives, theorems and propositions are listed without proofs; readers should refer to standard textbooks on mathematics for the details.

2.1 Vector and Matrix Calculus

2.1.1 Vectors and matrices

Throughout this book, geometric quantities such as vectors and tensors are described with respect to a *Cartesian coordinate system*, the coordinate axes being mutually orthogonal and having the same unit of length[1]. We also assume that the coordinate system is *right-handed*[2].

By a *vector*, we mean a column of real numbers[3]. Vectors are denoted by lowercase boldface letters such as $\boldsymbol{a}$, $\boldsymbol{b}$, $\boldsymbol{u}$, and $\boldsymbol{v}$; their components are written in the corresponding lowercase italic letters. A vector whose components are a_1, a_2, ..., a_n is also denoted by (a_i), $i = 1, ..., n$; the number n of the components is called the *dimension* of this vector. If the dimension is understood, notations such as (a_i) are used. In the following, an n-dimensional vector is referred to as an *n-vector*. The vector whose components are all 0 is called the *zero vector* and denoted by $\mathbf{0}$ (the dimension is usually implied by the context).

A *matrix* is an array of real numbers. Matrices are denoted by uppercase boldface letters such as $\boldsymbol{A}$, $\boldsymbol{B}$, $\boldsymbol{S}$, and $\boldsymbol{T}$; their elements are written in the corresponding uppercase italic letters. A matrix is also defined by its elements as (A_{ij}), $i = 1, ..., m$, $j = 1, ..., n$; such a matrix is said to be of *type mn*. In the following, a matrix of type mn is referred to as an *mn-matrix*; if $m = n$, it is also called a *square matrix* or simply *n-dimensional matrix*. If the type is

[1] This is only an intuitive definition, since "orthogonality" and "length" are later defined in terms of coordinates. To be strict, we need to start with axioms of one kind or another (we do not go into the details).

[2] In three dimensions, a Cartesian coordinate system is *right-handed* if the x-, y-, and z-axes have the same orientations as the thumb, the forefinger, and the middle finger, respectively, of a right hand. Otherwise, the coordinate system is *left-handed*. In other dimensions, the handedness, or the *parity*, can be defined arbitrarily: if a coordinate system is right-handed, its mirror image is left-handed (we do not go into the details).

[3] We do not deal with complex numbers in this book.

understood, notations such as (A_{ij}) are used. The matrix whose elements are all 0 is called the *zero matrix* and denoted by $\boldsymbol{O}$ (the type is usually implied by the context). If not explicitly stated, the type is understood to be nn in *this* chapter but 33 *in the rest of this book*.

The *unit matrix* is denoted by $\boldsymbol{I}$; its elements are written as δ_{ij} (not I_{ij}); the dimension is usually implied by the context. The symbol δ_{ij}, which takes value 1 for $i = j$ and 0 otherwise, is called the *Kronecker delta*. Addition and subtraction of matrices and multiplication of a matrix by a scalar, vector, or matrix are defined in the standard way.

The *trace* of nn-matrix $\boldsymbol{A} = (A_{ij})$ is the sum $\sum_{i=1}^{n} A_{ii}$ of its diagonal elements and is denoted by $\mathrm{tr}\boldsymbol{A}$. Evidently, $\mathrm{tr}\boldsymbol{I} = n$. The transpose of a vector or matrix is denoted by superscript $\top$. A matrix $\boldsymbol{A}$ is *symmetric* if $\boldsymbol{A} = \boldsymbol{A}^\top$. We say that a matrix is of *type* (nn) or an (nn)-*matrix* if it is an n-dimensional symmetric matrix. A matrix $\boldsymbol{A}$ is *antisymmetric* (or *skew-symmetric*) if $\boldsymbol{A} = -\boldsymbol{A}^\top$. We say that a matrix is of *type* $[nn]$ or $[nn]$-*matrix* if it is an n-dimensional antisymmetric matrix. Note the following expression, which is sometimes called the *outer product* of vectors $\boldsymbol{a}$ and $\boldsymbol{b}$:

$$\boldsymbol{a}\boldsymbol{b}^\top = (a_i b_j) = \begin{pmatrix} a_1 b_1 & a_1 b_2 & \cdots & a_1 b_n \\ a_2 b_1 & a_2 b_2 & \cdots & a_2 b_n \\ \vdots & \vdots & \cdots & \vdots \\ a_n b_1 & a_n b_2 & \cdots & a_n b_n \end{pmatrix}. \tag{2.1}$$

The following identities are very familiar:

$$(\boldsymbol{A}^\top)^\top = \boldsymbol{A}, \qquad (\boldsymbol{A}\boldsymbol{B})^\top = \boldsymbol{B}^\top \boldsymbol{A}^\top,$$
$$\mathrm{tr}(\boldsymbol{A}^\top) = \mathrm{tr}\boldsymbol{A}, \qquad \mathrm{tr}(\boldsymbol{A}\boldsymbol{B}) = \mathrm{tr}(\boldsymbol{B}\boldsymbol{A}). \tag{2.2}$$

The *inner product* of vectors $\boldsymbol{a} = (a_i)$ and $\boldsymbol{b} = (b_i)$ is defined by

$$(\boldsymbol{a}, \boldsymbol{b}) = \boldsymbol{a}^\top \boldsymbol{b} = \sum_{i=1}^{n} a_i b_i. \tag{2.3}$$

Evidently, $(\boldsymbol{a}, \boldsymbol{b}) = (\boldsymbol{b}, \boldsymbol{a})$. Vectors $\boldsymbol{a}$ and $\boldsymbol{b}$ are said to be *orthogonal* if $(\boldsymbol{a}, \boldsymbol{b}) = 0$. The following identities are easily confirmed:

$$(\boldsymbol{a}, \boldsymbol{T}\boldsymbol{b}) = (\boldsymbol{T}^\top \boldsymbol{a}, \boldsymbol{b}), \qquad \mathrm{tr}(\boldsymbol{a}\boldsymbol{b}^\top) = (\boldsymbol{a}, \boldsymbol{b}). \tag{2.4}$$

The matrix consisting of vectors $\boldsymbol{a}_1$, $\boldsymbol{a}_2$, ..., $\boldsymbol{a}_n$ as its columns in that order is denoted by $(\boldsymbol{a}_1, \boldsymbol{a}_2, ..., \boldsymbol{a}_n)$. If

$$\boldsymbol{A} = (\boldsymbol{a}_1, \boldsymbol{a}_2, ..., \boldsymbol{a}_n), \qquad \boldsymbol{B} = (\boldsymbol{b}_1, \boldsymbol{b}_2, ..., \boldsymbol{b}_n), \tag{2.5}$$

the following identities hold:

$$\boldsymbol{A}\boldsymbol{B}^\top = \sum_{i=1}^{n} \boldsymbol{a}_i \boldsymbol{b}_i^\top,$$

$$A^\top B = \begin{pmatrix} (a_1, b_1) & (a_1, b_2) & \cdots & (a_1, b_n) \\ (a_2, b_1) & (a_2, b_2) & \cdots & (a_2, b_n) \\ \vdots & \vdots & \cdots & \vdots \\ (a_n, b_1) & (a_n, b_2) & \cdots & (a_n, b_n) \end{pmatrix}. \tag{2.6}$$

The *norm*[4] and the *normalization operator* $N[\,\cdot\,]$ are defined as follows:

$$\|a\| = \sqrt{(a, a)} = \sqrt{\sum_{i=1}^{n} a_i{}^2}, \qquad N[a] = \frac{a}{\|a\|}. \tag{2.7}$$

A *unit vector* is a vector of unit norm. A set of vectors $\{u_1, \ldots, u_r\}$ is said to be *orthonormal* if its members are all unit vectors and orthogonal to each other: $(u_i, u_j) = \delta_{ij}$.

The following *Schwarz inequality* holds:

$$-\|a\| \cdot \|b\| \le (a, b) \le \|a\| \cdot \|b\|. \tag{2.8}$$

Equality holds if vectors a and b are *parallel*, meaning that there exists a real number t such that $a = tb$ or $b = 0$. The Schwarz inequality implies the following *triangle inequality* with the same equality condition:

$$\|a + b\| \le \|a\| + \|b\|. \tag{2.9}$$

2.1.2 Determinant and inverse

The *determinant* of a square matrix $A = (A_{ij})$, denoted by $\det A$ or $|A|$, is defined by

$$\det A = \sum_{i_1, \ldots, i_n = 1}^{n} \epsilon_{i_1 \cdots i_n} A_{1 i_1} \cdots A_{n i_n}, \tag{2.10}$$

where $\epsilon_{i_1 \cdots i_n}$ is the *signature symbol* defined by

$$\epsilon_{i_1 i_2 \cdots i_n} = \begin{cases} 1 & \text{if } (i_1 i_2 \cdots i_n) \text{ is an even permutation of } (12 \cdots n), \\ -1 & \text{if } (i_1 i_2 \cdots i_n) \text{ is an odd permutation of } (12 \cdots n), \\ 0 & \text{otherwise.} \end{cases} \tag{2.11}$$

Evidently, $\det I = 1$. The following identity holds:

$$\det(AB) = \det A \det B. \tag{2.12}$$

[4] This norm is called the *Euclidean norm* (or the *2-norm*). In general, the norm $\|\mathbf{a}\|$ can be defined arbitrarily as long as (i) $\|\mathbf{a}\| \ge 0$, equality holding if and only if $\mathbf{a} = \mathbf{0}$, (ii) $\|c\mathbf{a}\| = |c| \cdot \|\mathbf{a}\|$ for any scalar c, and (iii) the triangle inequality (2.9) holds. There exist other definitions that satisfy these—the *1-norm* $\|\mathbf{a}\|_1 = \Sigma_{i=1}^n |a_i|$ and the *∞-norm* $\|\mathbf{a}\|_\infty = \max_i |a_i|$, for instance. They can be generalized into the *Minkowski norm* (or the *p-norm*) $\|\mathbf{a}\|_p = \sqrt[p]{\Sigma_{i=1}^n |a_i|^p}$ for $1 \le p \le \infty$; the 1-norm, the 2-norm, and the ∞-norm are special cases of the Minkowski norm for $p = 1, 2, \infty$, respectively.

Replacing A_{ij} by $\delta_{ij} + \varepsilon A_{ij}$ in eq. (2.10) and expanding it in ε, we obtain

$$\det(\boldsymbol{I} + \varepsilon \boldsymbol{A}) = 1 + \varepsilon \mathrm{tr} \boldsymbol{A} + O(\varepsilon^2), \tag{2.13}$$

where the *order symbol* $O(\cdots)$ denotes terms having order the same as or higher than $\cdots$.

Let $\boldsymbol{A}^{(ij)}$ be the matrix obtained from a square matrix $\boldsymbol{A} = (A_{ij})$ by removing the ith row and the jth column. The determinant $\det \boldsymbol{A}$ is expanded in the form

$$\det \boldsymbol{A} = \sum_{i=1}^{n} (-1)^{i+j} A_{ij} \det \boldsymbol{A}^{(ij)} = \sum_{j=1}^{n} (-1)^{i+j} A_{ij} \det \boldsymbol{A}^{(ij)}. \tag{2.14}$$

This is called the *cofactor expansion formula*. The *cofactor* (or *adjugate*) *matrix* $\boldsymbol{A}^\dagger = (A^\dagger_{ij})$ of $\boldsymbol{A}$ is defined by

$$A^\dagger_{ij} = (-1)^{i+j} \det \boldsymbol{A}^{(ji)}. \tag{2.15}$$

Eq. (2.14) can be rewritten as

$$\boldsymbol{A}\boldsymbol{A}^\dagger = \boldsymbol{A}^\dagger \boldsymbol{A} = (\det \boldsymbol{A})\boldsymbol{I}. \tag{2.16}$$

The following identity holds:

$$\det(\boldsymbol{A} + \varepsilon \boldsymbol{B}) = \det \boldsymbol{A} + \varepsilon \mathrm{tr}(\boldsymbol{A}^\dagger \boldsymbol{B}) + O(\varepsilon^2). \tag{2.17}$$

The elements of the cofactor matrix $\boldsymbol{A}^\dagger$ of nn-matrix $\boldsymbol{A}$ are all polynomials of degree $n-1$ in the elements of $\boldsymbol{A}$. In three dimensions, the cofactor matrix of $\boldsymbol{A} = (A_{ij})$ has the following form:

$$\boldsymbol{A}^\dagger = \begin{pmatrix} A_{22}A_{33} - A_{32}A_{23} & A_{32}A_{13} - A_{12}A_{33} & A_{12}A_{23} - A_{22}A_{13} \\ A_{23}A_{31} - A_{33}A_{21} & A_{33}A_{11} - A_{13}A_{31} & A_{13}A_{21} - A_{23}A_{11} \\ A_{21}A_{32} - A_{31}A_{22} & A_{31}A_{12} - A_{11}A_{32} & A_{11}A_{22} - A_{21}A_{12} \end{pmatrix}. \tag{2.18}$$

The *inverse* $\boldsymbol{A}^{-1}$ of a square matrix $\boldsymbol{A}$ is defined by

$$\boldsymbol{A}\boldsymbol{A}^{-1} = \boldsymbol{A}^{-1}\boldsymbol{A} = \boldsymbol{I}, \tag{2.19}$$

if such an $\boldsymbol{A}^{-1}$ exists. A square matrix is *singular* if its inverse does not exist, and *nonsingular* (or *of full rank*) otherwise. Eq. (2.16) implies that if $\boldsymbol{A}$ is nonsingular, its inverse $\boldsymbol{A}^{-1}$ is given by

$$\boldsymbol{A}^{-1} = \frac{\boldsymbol{A}^\dagger}{\det \boldsymbol{A}}. \tag{2.20}$$

If we define $\boldsymbol{A}^0 = \boldsymbol{I}$, the following identities hold for nonsingular matrices (k is a nonnegative integer):

$$(\boldsymbol{A}^{-1})^{-1} = \boldsymbol{A}, \quad (\boldsymbol{A}\boldsymbol{B})^{-1} = \boldsymbol{B}^{-1}\boldsymbol{A}^{-1}, \quad (\boldsymbol{A}^{-1})^k = (\boldsymbol{A}^k)^{-1},$$

$$(\boldsymbol{A}^\top)^{-1} = (\boldsymbol{A}^{-1})^\top, \qquad \det \boldsymbol{A}^{-1} = \frac{1}{\det \boldsymbol{A}}. \tag{2.21}$$

The third identity implies that matrix $(\boldsymbol{A}^{-1})^k$ can be unambiguously denoted by $\boldsymbol{A}^{-k}$. Note that the determinant and the inverse are defined only for square matrices.

Let $\boldsymbol{A}$ be a nonsingular nn-matrix, and $\boldsymbol{B}$ a nonsingular mm-matrix. Let $\boldsymbol{S}$ and $\boldsymbol{T}$ be nm-matrices. The following *matrix inversion formula* holds, provided that the inverses involved all exist:

$$(\boldsymbol{A} + \boldsymbol{S}\boldsymbol{B}\boldsymbol{T}^\top)^{-1} = \boldsymbol{A}^{-1} - \boldsymbol{A}^{-1}\boldsymbol{S}(\boldsymbol{B}^{-1} + \boldsymbol{T}^\top\boldsymbol{A}^{-1}\boldsymbol{S})^{-1}\boldsymbol{T}^\top\boldsymbol{A}^{-1}. \tag{2.22}$$

If $m = 1$, the nm-matrices $\boldsymbol{S}$ and $\boldsymbol{T}$ are n-vectors, and the mm-matrix $\boldsymbol{B}$ is a scalar. If we let $\boldsymbol{B} = 1$ and write $\boldsymbol{S}$ and $\boldsymbol{T}$ as $\boldsymbol{s}$ and $\boldsymbol{t}$, respectively, the above formula reduces to

$$(\boldsymbol{A} + \boldsymbol{s}\boldsymbol{t}^\top)^{-1} = \boldsymbol{A}^{-1} - \frac{\boldsymbol{A}^{-1}\boldsymbol{s}\boldsymbol{t}^\top\boldsymbol{A}^{-1}}{1 + (\boldsymbol{t}, \boldsymbol{A}^{-1}\boldsymbol{s})}. \tag{2.23}$$

For $\boldsymbol{A} = \boldsymbol{I}$, we obtain

$$(\boldsymbol{I} + \boldsymbol{s}\boldsymbol{t}^\top)^{-1} = \boldsymbol{I} - \frac{\boldsymbol{s}\boldsymbol{t}^\top}{1 + (\boldsymbol{s}, \boldsymbol{t})}. \tag{2.24}$$

2.1.3 Vector product in three dimensions

In three dimensions, the signature symbol defined by eq. (2.11) is often referred to as the *Eddington epsilon*[5]. It satisfies the following identity:

$$\sum_{m=1}^{3} \epsilon_{ijm}\epsilon_{klm} = \delta_{ik}\delta_{jl} - \delta_{il}\delta_{jk}. \tag{2.25}$$

The *vector* (or *exterior*) *product* of 3-vectors $\boldsymbol{a} = (a_i)$ and $\boldsymbol{b} = (b_i)$ is defined by

$$\boldsymbol{a} \times \boldsymbol{b} = \Big(\sum_{j,k=1}^{3} \epsilon_{ijk} a_j b_k\Big) = \begin{pmatrix} a_2 b_3 - a_3 b_2 \\ a_3 b_1 - a_1 b_3 \\ a_1 b_2 - a_2 b_1 \end{pmatrix}. \tag{2.26}$$

Evidently,

$$\boldsymbol{a} \times \boldsymbol{b} = -\boldsymbol{b} \times \boldsymbol{a}, \qquad \boldsymbol{a} \times \boldsymbol{a} = \boldsymbol{0},$$
$$(\boldsymbol{b}, \boldsymbol{a} \times \boldsymbol{b}) = (\boldsymbol{a}, \boldsymbol{a} \times \boldsymbol{b}) = 0. \tag{2.27}$$

The following identities, known as the *Lagrange formulae*, are direct consequences of eq. (2.25):

$$\boldsymbol{a} \times (\boldsymbol{b} \times \boldsymbol{c}) = (\boldsymbol{a}, \boldsymbol{c})\boldsymbol{b} - (\boldsymbol{a}, \boldsymbol{b})\boldsymbol{c},$$

[5]Some authors use different terminologies such as the *Levi-Civita symbol*.

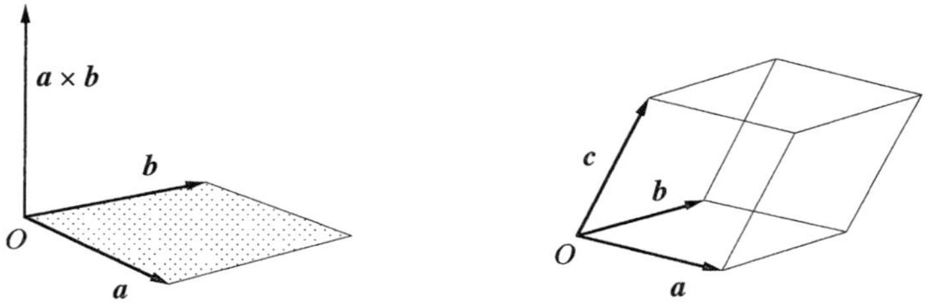

(a) (b)

Fig. 2.1. (a) Vector product. (b) Scalar triple product.

$$(\boldsymbol{a} \times \boldsymbol{b}) \times \boldsymbol{c} = (\boldsymbol{a}, \boldsymbol{c})\boldsymbol{b} - (\boldsymbol{b}, \boldsymbol{c})\boldsymbol{a}. \tag{2.28}$$

The expressions $\boldsymbol{a} \times (\boldsymbol{b} \times \boldsymbol{c})$ and $(\boldsymbol{a} \times \boldsymbol{b}) \times \boldsymbol{c}$ are called *vector triple products*. The following identities also hold:

$$(\boldsymbol{a} \times \boldsymbol{b}, \boldsymbol{c} \times \boldsymbol{d}) = (\boldsymbol{a}, \boldsymbol{c})(\boldsymbol{b}, \boldsymbol{d}) - (\boldsymbol{a}, \boldsymbol{d})(\boldsymbol{b}, \boldsymbol{c}), \tag{2.29}$$

$$\|\boldsymbol{a} \times \boldsymbol{b}\|^2 = \|\boldsymbol{a}\|^2 \|\boldsymbol{b}\|^2 - (\boldsymbol{a}, \boldsymbol{b})^2. \tag{2.30}$$

If 3-vectors $\boldsymbol{a}$ and $\boldsymbol{b}$ make angle θ, we have

$$(\boldsymbol{a}, \boldsymbol{b}) = \|\boldsymbol{a}\| \cdot \|\boldsymbol{b}\| \cos\theta, \qquad \|\boldsymbol{a} \times \boldsymbol{b}\| = \|\boldsymbol{a}\| \cdot \|\boldsymbol{b}\| \sin\theta. \tag{2.31}$$

Eq. (2.30) states the well-known trigonometric identity $\cos^2\theta + \sin^2\theta = 1$. From eq. (2.26), the third of eqs. (2.27), and the second of eqs. (2.31), we can visualize $\boldsymbol{a} \times \boldsymbol{b}$ as a vector normal to the plane defined by $\boldsymbol{a}$ and $\boldsymbol{b}$; the length of $\boldsymbol{a} \times \boldsymbol{b}$ equals the area of the parallelogram made by $\boldsymbol{a}$ and $\boldsymbol{b}$ (Fig. 2.1a).

The *scalar triple product* $|\boldsymbol{a}, \boldsymbol{b}, \boldsymbol{c}|$ of 3-vectors $\boldsymbol{a}$, $\boldsymbol{b}$, and $\boldsymbol{c}$ is the determinant of the matrix $(\boldsymbol{a}, \boldsymbol{b}, \boldsymbol{c})$ having $\boldsymbol{a}$, $\boldsymbol{b}$, $\boldsymbol{c}$ as its columns in that order. We say that three 3-vectors $\{\boldsymbol{a}, \boldsymbol{b}, \boldsymbol{c}\}$ are a *right-handed* system if $|\boldsymbol{a}, \boldsymbol{b}, \boldsymbol{c}| > 0$ and a *left-handed* system if $|\boldsymbol{a}, \boldsymbol{b}, \boldsymbol{c}| < 0$. The scalar triple product $|\boldsymbol{a}, \boldsymbol{b}, \boldsymbol{c}|$ equals the signed volume of the parallelepiped defined by $\boldsymbol{a}$, $\boldsymbol{b}$, and $\boldsymbol{c}$ (Fig. 2.1b); the volume is positive if the three vectors are a right-handed system in that order and negative if they are a left-handed system. The equality $|\boldsymbol{a}, \boldsymbol{b}, \boldsymbol{c}| = 0$ holds if and only if $\boldsymbol{a}$, $\boldsymbol{b}$, and $\boldsymbol{c}$ are *coplanar*, i.e., if they all lie on a common plane.

We can also write

$$|\boldsymbol{a}, \boldsymbol{b}, \boldsymbol{c}| = (\boldsymbol{a} \times \boldsymbol{b}, \boldsymbol{c}) = (\boldsymbol{b} \times \boldsymbol{c}, \boldsymbol{a}) = (\boldsymbol{c} \times \boldsymbol{a}, \boldsymbol{b}). \tag{2.32}$$

Since $|\boldsymbol{a}, \boldsymbol{b}, \boldsymbol{a} \times \boldsymbol{b}| = \|\boldsymbol{a} \times \boldsymbol{b}\|^2$, the vector product $\boldsymbol{a} \times \boldsymbol{b}$ is oriented, if it is not $\boldsymbol{0}$, in such a way that $\{\boldsymbol{a}, \boldsymbol{b}, \boldsymbol{a} \times \boldsymbol{b}\}$ form a right-handed system (Fig. 2.1a).

The following identity also holds:

$$(\boldsymbol{a} \times \boldsymbol{b}) \times (\boldsymbol{c} \times \boldsymbol{d}) = |\boldsymbol{a}, \boldsymbol{b}, \boldsymbol{d}|\boldsymbol{c} - |\boldsymbol{a}, \boldsymbol{b}, \boldsymbol{c}|\boldsymbol{d} = |\boldsymbol{a}, \boldsymbol{c}, \boldsymbol{d}|\boldsymbol{b} - |\boldsymbol{b}, \boldsymbol{c}, \boldsymbol{d}|\boldsymbol{a}. \tag{2.33}$$

Taking the determinant of $(\boldsymbol{a}, \boldsymbol{b}, \boldsymbol{c})(\boldsymbol{a}, \boldsymbol{b}, \boldsymbol{c})^\top$ (see eq. (2.12)), we obtain

$$|\boldsymbol{a}, \boldsymbol{b}, \boldsymbol{c}|^2 = \begin{vmatrix} \|\boldsymbol{a}\|^2 & (\boldsymbol{a}, \boldsymbol{b}) & (\boldsymbol{a}, \boldsymbol{c}) \\ (\boldsymbol{b}, \boldsymbol{a}) & \|\boldsymbol{b}\|^2 & (\boldsymbol{b}, \boldsymbol{c}) \\ (\boldsymbol{c}, \boldsymbol{a}) & (\boldsymbol{c}, \boldsymbol{b}) & \|\boldsymbol{c}\|^2 \end{vmatrix}. \tag{2.34}$$

The *vector* (or *exterior*) *product* of 3-vector $\boldsymbol{a}$ and 33-matrix $\boldsymbol{T} = (\boldsymbol{t}_1, \boldsymbol{t}_2, \boldsymbol{t}_3)$ is defined by

$$\boldsymbol{a} \times \boldsymbol{T} = (\boldsymbol{a} \times \boldsymbol{t}_1, \boldsymbol{a} \times \boldsymbol{t}_2, \boldsymbol{a} \times \boldsymbol{t}_3). \tag{2.35}$$

From this definition, the following identities are obtained:

$$\boldsymbol{a} \times (\boldsymbol{T}\boldsymbol{b}) = (\boldsymbol{a} \times \boldsymbol{T})\boldsymbol{b},$$

$$\boldsymbol{a} \times \boldsymbol{I} = \begin{pmatrix} 0 & -a_3 & a_2 \\ a_3 & 0 & -a_1 \\ -a_2 & a_1 & 0 \end{pmatrix}, \qquad (\boldsymbol{a} \times \boldsymbol{I})^\top = -\boldsymbol{a} \times \boldsymbol{I}. \tag{2.36}$$

The matrix $\boldsymbol{a} \times \boldsymbol{I}$ is called the *antisymmetric matrix associated with* the 3-vector $\boldsymbol{a}$. The following identity is an alternative expression to the Lagrange formulae (2.28):

$$(\boldsymbol{a} \times \boldsymbol{I})(\boldsymbol{b} \times \boldsymbol{I})^\top = (\boldsymbol{a}, \boldsymbol{b})\boldsymbol{I} - \boldsymbol{b}\boldsymbol{a}^\top. \tag{2.37}$$

The vector (or exterior) product of 33-matrix $\boldsymbol{T}$ and 3-vector $\boldsymbol{b}$ is defined by

$$\boldsymbol{T} \times \boldsymbol{b} = \boldsymbol{T}(\boldsymbol{b} \times \boldsymbol{I})^\top. \tag{2.38}$$

This definition implies the following identities:

$$(\boldsymbol{a} \times \boldsymbol{T})^\top = \boldsymbol{T}^\top \times \boldsymbol{a}, \qquad (\boldsymbol{T} \times \boldsymbol{b})^\top = \boldsymbol{b} \times \boldsymbol{T}^\top,$$

$$(\boldsymbol{T} \times \boldsymbol{b})\boldsymbol{c} = \boldsymbol{T}(\boldsymbol{c} \times \boldsymbol{b}). \tag{2.39}$$

It is easy to confirm that

$$(\boldsymbol{a} \times \boldsymbol{T}) \times \boldsymbol{b} = \boldsymbol{a} \times (\boldsymbol{T} \times \boldsymbol{b}), \tag{2.40}$$

which can be written unambiguously as $\boldsymbol{a} \times \boldsymbol{T} \times \boldsymbol{b}$. We also have

$$(\boldsymbol{a} \times \boldsymbol{T} \times \boldsymbol{b})^\top = \boldsymbol{b} \times \boldsymbol{T}^\top \times \boldsymbol{a}. \tag{2.41}$$

Eq. (2.37) now reads

$$\boldsymbol{a} \times \boldsymbol{I} \times \boldsymbol{b} = (\boldsymbol{a}, \boldsymbol{b})\boldsymbol{I} - \boldsymbol{b}\boldsymbol{a}^\top. \tag{2.42}$$

The following identities are also important:

$$(\boldsymbol{a} \times \boldsymbol{b})(\boldsymbol{c} \times \boldsymbol{d})^\top = \boldsymbol{a} \times (\boldsymbol{b}\boldsymbol{d}^\top) \times \boldsymbol{c} = \boldsymbol{b} \times (\boldsymbol{a}\boldsymbol{c}^\top) \times \boldsymbol{d}, \tag{2.43}$$

$$(\boldsymbol{a} \times \boldsymbol{b}, \boldsymbol{T}(\boldsymbol{c} \times \boldsymbol{d})) = (\boldsymbol{a}, (\boldsymbol{b} \times \boldsymbol{T} \times \boldsymbol{d})\boldsymbol{c}) = (\boldsymbol{b}, (\boldsymbol{a} \times \boldsymbol{T} \times \boldsymbol{c})\boldsymbol{d}). \tag{2.44}$$

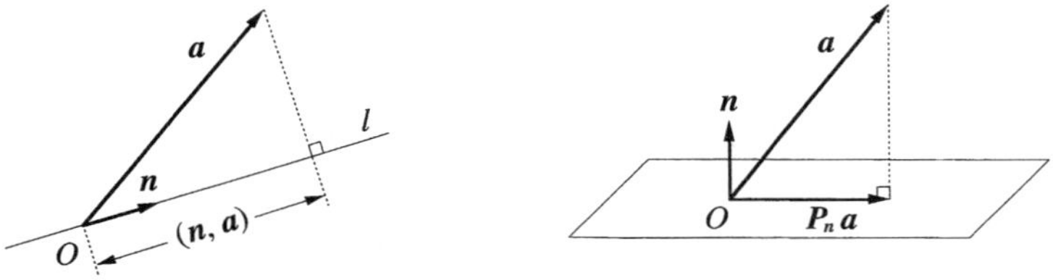

(a) (b)

Fig. 2.2. (a) Projection onto a line. (b) Projection onto a plane.

The *exterior product* $[\boldsymbol{A} \times \boldsymbol{B}]$ of 33-matrices $\boldsymbol{A} = (A_{ij})$ and $\boldsymbol{B} = (B_{ij})$ is a 33-matrix defined as follows[6]:

$$[\boldsymbol{A} \times \boldsymbol{B}]_{ij} = \sum_{k,l,m,n=1}^{3} \epsilon_{ikl}\epsilon_{jmn} A_{km} B_{ln}. \tag{2.45}$$

If $\boldsymbol{A}$ and $\boldsymbol{B}$ are both symmetric, their exterior product $[\boldsymbol{A} \times \boldsymbol{B}]$ is also symmetric.

2.1.4 Projection matrices

If a vector $\boldsymbol{a}$ is projected orthogonally onto a line l that extends along a unit vector $\boldsymbol{n}$, it defines on l a segment of signed length $(\boldsymbol{n}, \boldsymbol{a})$ (Fig. 2.2a); it is positive in the direction $\boldsymbol{n}$ and negative in the direction $-\boldsymbol{n}$. The vector $\boldsymbol{a}$ is decomposed into the component $(\boldsymbol{n}, \boldsymbol{a})\boldsymbol{n}$ parallel to l and the component $\boldsymbol{a} - (\boldsymbol{n}, \boldsymbol{a})\boldsymbol{n}$ $(= (\boldsymbol{I} - \boldsymbol{n}\boldsymbol{n}^\top)\boldsymbol{a})$ orthogonal it. Let $\{\boldsymbol{n}\}_L$ be the one-dimensional subspace defined by unit vector $\boldsymbol{n}$, and $\{\boldsymbol{n}\}_L^\perp$ its *orthogonal complement*—the set of all vectors orthogonal to $\boldsymbol{n}$. The projection of a vector $\boldsymbol{a}$ onto $\{\boldsymbol{n}\}_L^\perp$ is written as $\boldsymbol{P}_{\boldsymbol{n}}\boldsymbol{a}$ (Fig. 2.2b). The matrix $\boldsymbol{P}_{\boldsymbol{n}}$ is defined by

$$\boldsymbol{P}_{\boldsymbol{n}} = \boldsymbol{I} - \boldsymbol{n}\boldsymbol{n}^\top, \tag{2.46}$$

and called the *projection matrix* onto the plane orthogonal to $\boldsymbol{n}$, or the projection matrix *along* $\boldsymbol{n}$. The following identities are easily confirmed:

$$\boldsymbol{P}_{\boldsymbol{n}} = \boldsymbol{P}_{\boldsymbol{n}}^\top, \qquad \boldsymbol{P}_{\boldsymbol{n}}^2 = \boldsymbol{P}_{\boldsymbol{n}},$$

$$\det \boldsymbol{P}_{\boldsymbol{n}} = 0, \qquad \mathrm{tr}\boldsymbol{P}_{\boldsymbol{n}} = n - 1, \qquad \|\boldsymbol{P}_{\boldsymbol{n}}\| = \sqrt{n-1}. \tag{2.47}$$

Here, the matrix norm $\|\cdot\|$ is defined by $\|\boldsymbol{A}\| = \sqrt{\sum_{i=1}^{m}\sum_{j=1}^{n} A_{ij}{}^2}$ for mn-matrix $\boldsymbol{A} = (A_{ij})$. In three dimensions, eq. (2.42) implies the following identity for unit vector $\boldsymbol{n}$:

$$\boldsymbol{n} \times \boldsymbol{I} \times \boldsymbol{n} = (\boldsymbol{n} \times \boldsymbol{I})(\boldsymbol{n} \times \boldsymbol{I})^\top = \boldsymbol{P}_{\boldsymbol{n}}. \tag{2.48}$$

[6]For example, $[\mathbf{A} \times \mathbf{B}]_{11} = A_{22}B_{33} - A_{32}B_{23} - A_{23}B_{32} + A_{33}B_{22}$.

The projection matrix can be generalized as follows. Let the symbol $\mathcal{R}^n$ denote the n-dimensional space of all n-vectors. Let $\mathcal{S}$ be an m-dimensional subspace of $\mathcal{R}^n$, and $\mathcal{N}$ $(= \mathcal{S}^\perp)$ its orthogonal complement—the set of all vectors that are orthogonal to every vector in $\mathcal{S}$. The *orthogonal projection*[7] $\boldsymbol{P}_\mathcal{N}$ onto $\mathcal{S}$ is a linear mapping such that for an arbitrary vector $\boldsymbol{v} \in \mathcal{R}^n$

$$\boldsymbol{P}_\mathcal{N}\boldsymbol{v} \in \mathcal{S}, \qquad \boldsymbol{v} - \boldsymbol{P}_\mathcal{N}\boldsymbol{v} \in \mathcal{N}. \tag{2.49}$$

In other words, $\boldsymbol{P}_\mathcal{N}$ is the operator that *removes* the component in $\mathcal{N}$. We also use an alternative notation $\boldsymbol{P}^\mathcal{S}$ when we want to indicate the space to be projected explicitly. Let $\{\boldsymbol{n}_1, ..., \boldsymbol{n}_m\}$ be an orthonormal basis of $\mathcal{N}$. The orthogonal projection $\boldsymbol{P}_\mathcal{N}$ has the following matrix expression:

$$\boldsymbol{P}_\mathcal{N} = \boldsymbol{I} - \sum_{i=1}^{m} \boldsymbol{n}_i\boldsymbol{n}_i^\top. \tag{2.50}$$

Eqs. (2.47) can be generalized as follows:

$$\boldsymbol{P}_\mathcal{N} = \boldsymbol{P}_\mathcal{N}^\top, \qquad \boldsymbol{P}_\mathcal{N}^2 = \boldsymbol{P}_\mathcal{N},$$

$$\det \boldsymbol{P}_\mathcal{N} = 0, \quad \mathrm{tr}\boldsymbol{P}_\mathcal{N} = n - m, \quad \|\boldsymbol{P}_\mathcal{N}\| = \sqrt{n-m}. \tag{2.51}$$

2.1.5 Orthogonal matrices and rotations

Matrix $\boldsymbol{R}$ is *orthogonal* if one of the following conditions holds (all are equivalent to each other):

$$\boldsymbol{R}\boldsymbol{R}^\top = \boldsymbol{I}, \quad \boldsymbol{R}^\top\boldsymbol{R} = \boldsymbol{I}, \quad \boldsymbol{R}^{-1} = \boldsymbol{R}^\top. \tag{2.52}$$

Equivalently, matrix $\boldsymbol{R} = (\boldsymbol{r}_1, ..., \boldsymbol{r}_n)$ is orthogonal if and only if its columns form an orthonormal set of vectors: $(\boldsymbol{r}_i, \boldsymbol{r}_j) = \delta_{ij}$.

For an orthogonal matrix $\boldsymbol{R}$ and vectors $\boldsymbol{a}$ and $\boldsymbol{b}$, we have

$$(\boldsymbol{Ra}, \boldsymbol{Rb}) = (\boldsymbol{a}, \boldsymbol{b}), \qquad \|\boldsymbol{Ra}\| = \|\boldsymbol{a}\|. \tag{2.53}$$

The second equation implies that the length of a vector is unchanged after multiplication by an orthogonal matrix. The first one together with eqs. (2.31) implies that in three dimensions the angle that two vectors make is also unchanged.

Applying eq. (2.12) to eqs. (2.52), we see that $\det \boldsymbol{R} = \pm 1$ for an orthogonal matrix $\boldsymbol{R}$. If $\det \boldsymbol{R} = 1$, the orthogonal matrix $\boldsymbol{R}$ is said to be a *rotation*

[7] The notation given here is non-traditional: the projection onto subspace $\mathcal{S}$ is usually denoted by $\mathbf{P}_\mathcal{S}$. Our definition is in conformity to the notation $\mathbf{P_n}$ given by eq. (2.46).

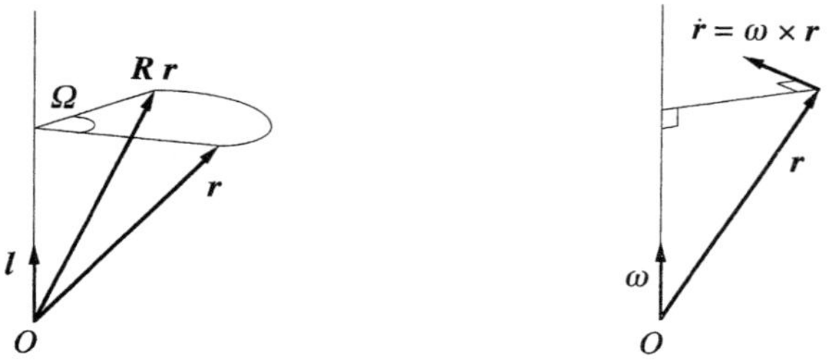

(a) (b)

Fig. 2.3. (a) Axis and angle of rotation. (b) Instantaneous rotation.

matrix[8]. In three dimensions, the orthonormal Cartesian coordinate basis vectors are

$$\boldsymbol{i} = \begin{pmatrix} 1 \\ 0 \\ 0 \end{pmatrix}, \quad \boldsymbol{j} = \begin{pmatrix} 0 \\ 1 \\ 0 \end{pmatrix}, \quad \boldsymbol{k} = \begin{pmatrix} 0 \\ 0 \\ 1 \end{pmatrix}. \tag{2.54}$$

The columns of a three-dimensional rotation matrix $\boldsymbol{R} = (\boldsymbol{r}_1, \boldsymbol{r}_2, \boldsymbol{r}_3)$ define a right-handed orthonormal system $\{\boldsymbol{r}_1, \boldsymbol{r}_2, \boldsymbol{r}_3\}$. The matrix $\boldsymbol{R}$ maps the coordinate basis $\{\boldsymbol{i}, \boldsymbol{j}, \boldsymbol{k}\}$ to $\{\boldsymbol{r}_1, \boldsymbol{r}_2, \boldsymbol{r}_3\}$. Such a map is realized as a rotation along an *axis* $\boldsymbol{l}$ by an *angle* Ω of rotation (*Euler's theorem*; Fig. 2.3a). The axis $\boldsymbol{l}$ (unit vector) and the angle Ω (measured in the screw sense) of rotation $\boldsymbol{R}$ are computed as follows:

$$\boldsymbol{l} = N[\begin{pmatrix} R_{32} - R_{23} \\ R_{13} - R_{31} \\ R_{21} - R_{12} \end{pmatrix}], \qquad \Omega = \cos^{-1} \frac{\mathrm{tr}\boldsymbol{R} - 1}{2}. \tag{2.55}$$

Conversely, an axis $\boldsymbol{l}$ and an angle Ω define a rotation $\boldsymbol{R}$ in the following form:

$$\boldsymbol{R} = \begin{pmatrix} \cos\Omega + {l_1}^2(1-\cos\Omega) & l_1 l_2(1-\cos\Omega) - l_3\sin\Omega & l_1 l_3(1-\cos\Omega) + l_2\sin\Omega \\ l_2 l_1(1-\cos\Omega) + l_3\sin\Omega & \cos\Omega + {l_2}^2(1-\cos\Omega) & l_2 l_3(1-\cos\Omega) - l_1\sin\Omega \\ l_3 l_1(1-\cos\Omega) - l_2\sin\Omega & l_3 l_2(1-\cos\Omega) + l_1\sin\Omega & \cos\Omega + {l_3}^2(1-\cos\Omega) \end{pmatrix}. \tag{2.56}$$

From this equation, we see that a rotation around unit vector $\boldsymbol{l}$ by a small angle $\Delta\Omega$ is expressed in the form

$$\boldsymbol{R} = \boldsymbol{I} + \Delta\Omega\boldsymbol{l} \times \boldsymbol{I} + O(\Delta\Omega^2), \tag{2.57}$$

[8]The set of all n-dimensional rotation matrices forms a group, denoted by $SO(n)$, under matrix multiplication. It is a subgroup of $O(n)$, the group consisting of all n-dimensional orthogonal matrices. The group consisting of all nonsingular nn-matrices is denoted by $GL(n)$, and the group consisting of all nn-matrices of determinant 1 is denoted by $SL(n)$.

which implies that for a continuous rotation there exists a vector $\boldsymbol{\omega}$ such that $\Delta\Omega\boldsymbol{l} = \boldsymbol{\omega}\Delta t + O(\Delta t^2)$ for a short lapse of time Δt. The vector $\boldsymbol{\omega}$ is called the *rotation velocity*: Its orientation $N[\boldsymbol{\omega}]$ defines the *instantaneous axis* of rotation; its norm $\|\boldsymbol{\omega}\|$ defines the *angular velocity*. Eq. (2.57) reads

$$\boldsymbol{R} = \boldsymbol{I} + \boldsymbol{\omega} \times \boldsymbol{I}\Delta t + O(\Delta t^2). \tag{2.58}$$

Hence, the velocity $\dot{\boldsymbol{r}} = \lim_{\Delta t \to 0}(\boldsymbol{R}\boldsymbol{r} - \boldsymbol{r})/\Delta t$ of vector $\boldsymbol{R}\boldsymbol{r}$ at $\Delta t = 0$ has the form

$$\dot{\boldsymbol{r}} = \boldsymbol{\omega} \times \boldsymbol{r}, \tag{2.59}$$

and is orthogonal to both $\boldsymbol{r}$ and $\boldsymbol{\omega}$ (Fig. 2.3b).

2.2 Eigenvalue Problem

2.2.1 Spectral decomposition

An *eigenvector* of an (nn)-matrix[9] $\boldsymbol{A}$ for *eigenvalue* λ is a nonzero vector $\boldsymbol{u}$ such that

$$\boldsymbol{A}\boldsymbol{u} = \lambda\boldsymbol{u}. \tag{2.60}$$

This equation can be rewritten as $(\lambda\boldsymbol{I} - \boldsymbol{A})\boldsymbol{u} = \boldsymbol{0}$, which has a non-trivial solution if and only if the function

$$\phi_{\boldsymbol{A}}(\lambda) = |\lambda\boldsymbol{I} - \boldsymbol{A}| \tag{2.61}$$

has a zero: $\phi_{\boldsymbol{A}}(\lambda) = 0$. The function $\phi_{\boldsymbol{A}}(\lambda)$ is an nth degree polynomial in λ and called the *characteristic polynomial* of $\boldsymbol{A}$. The equation $\phi_{\boldsymbol{A}}(\lambda) = 0$ is called the *characteristic equation* and has n roots $\{\lambda_i\}$ (with multiplicities counted). The number of nonzero eigenvalues (with multiplicities counted) is called the *rank* of the (nn)-matrix $\boldsymbol{A}$.

The eigenvalues $\{\lambda_i\}$ of an (nn)-matrix $\boldsymbol{A}$ are all real. The corresponding set $\{\boldsymbol{u}_i\}$ of unit eigenvectors can be chosen to be an orthonormal set. Let us call the set $\{\boldsymbol{u}_i\}$ so defined the *eigensystem* of the (nn)-matrix $\boldsymbol{A}$. An (nn)-matrix $\boldsymbol{A}$ is expressed in terms of its eigenvalues $\{\lambda_i\}$ and eigensystem $\{\boldsymbol{u}_i\}$ in the form

$$\boldsymbol{A} = \sum_{i=1}^{n} \lambda_i \boldsymbol{u}_i \boldsymbol{u}_i^\top, \tag{2.62}$$

which is called the *spectral* (or *eigenvalue*) *decomposition* of $\boldsymbol{A}$. In particular, the identity

$$\sum_{i=1}^{n} \boldsymbol{u}_i \boldsymbol{u}_i^\top = \boldsymbol{I} \tag{2.63}$$

[9] Recall that by an (nn)-matrix we mean an n-dimensional symmetric matrix (see Section 2.1.1). Eigenvalues and eigenvectors are defined for a general (non-symmetric) matrix in exactly the same way, but in this book we deal with eigenvalues and eigenvectors of symmetric matrices only.

holds for an arbitrary orthonormal system $\{\boldsymbol{u}_i\}$. From $(\boldsymbol{x}, \sum_{i=1}^n \boldsymbol{u}_i\boldsymbol{u}_i^\top \boldsymbol{x}) = (\boldsymbol{x}, \boldsymbol{I}\boldsymbol{x})$, we obtain the following identity for an arbitrary vector and an arbitrary orthonormal system $\{\boldsymbol{u}_i\}$:

$$\sum_{i=1}^n (\boldsymbol{u}_i, \boldsymbol{x})^2 = \|\boldsymbol{x}\|^2. \tag{2.64}$$

Let $\{\lambda_i\}$ be the eigenvalues of (nn)-matrix $\boldsymbol{A}$, and $\{\boldsymbol{u}_i\}$ the corresponding eigensystem. Since $\{\boldsymbol{u}_i\}$ is an orthonormal system, the matrix $\boldsymbol{U} = (\boldsymbol{u}_1, \boldsymbol{u}_2, \cdots, \boldsymbol{u}_n)$ is orthogonal. Eq. (2.62) is equivalent to

$$\boldsymbol{A} = \boldsymbol{U}\boldsymbol{\Lambda}\boldsymbol{U}^\top, \tag{2.65}$$

where $\boldsymbol{\Lambda}$ is the diagonal matrix with diagonal elements λ_1, λ_2, ..., λ_n in that order; we write

$$\boldsymbol{\Lambda} = \mathrm{diag}(\lambda_1, \lambda_2, ..., \lambda_n). \tag{2.66}$$

From eq. (2.65), we obtain

$$\boldsymbol{U}^\top \boldsymbol{A}\boldsymbol{U} = \boldsymbol{\Lambda}, \tag{2.67}$$

which is called the *diagonalization* of $\boldsymbol{A}$. Applying the fourth of eqs. (2.2) and eq. (2.12) to eq. (2.65), we obtain the following identities:

$$\mathrm{tr}\boldsymbol{A} = \sum_{i=1}^n \lambda_i, \qquad \det \boldsymbol{A} = \prod_{i=1}^n \lambda_i. \tag{2.68}$$

From the spectral decomposition (2.62), the kth power $\boldsymbol{A}^k$ for an arbitrary integer $k > 0$ is given by

$$\boldsymbol{A}^k = \sum_{i=1}^n \lambda_i^k \boldsymbol{u}_i\boldsymbol{u}_i^\top. \tag{2.69}$$

This can be extended to an arbitrary polynomial $p(x)$:

$$p(\boldsymbol{A}) = \sum_{i=1}^n p(\lambda_i)\boldsymbol{u}_i\boldsymbol{u}_i^\top. \tag{2.70}$$

If $\boldsymbol{A}$ is of full rank, its inverse $\boldsymbol{A}^{-1}$ is given by

$$\boldsymbol{A}^{-1} = \sum_{i=1}^n \frac{1}{\lambda_i}\boldsymbol{u}_i\boldsymbol{u}_i^\top. \tag{2.71}$$

This can be extended to an arbitrary negative power of $\boldsymbol{A}$ (see the third of eqs. (2.21)):

$$\boldsymbol{A}^{-k} = \sum_{i=1}^n \frac{1}{\lambda_i^k}\boldsymbol{u}_i\boldsymbol{u}_i^\top. \tag{2.72}$$

2.2.2 Generalized inverse

An (nn)-matrix $\boldsymbol{A}$ is *positive definite* if its eigenvalues are all positive, and is *positive semi-definite* if its eigenvalues are all nonnegative; it is *negative definite* if its eigenvalues are all negative, and is *negative semi-definite* if its eigenvalues are all nonpositive.

For a positive semi-definite (nn)-matrix $\boldsymbol{A}$, eq. (2.69) can be extended to arbitrary non-integer powers $\boldsymbol{A}^q$, $q > 0$. In particular, the "square root" $\sqrt{\boldsymbol{A}}$ of $\boldsymbol{A}$ is defined by

$$\sqrt{\boldsymbol{A}} = \sum_{i=1}^{n} \sqrt{\lambda_i} \boldsymbol{u}_i \boldsymbol{u}_i^\top. \tag{2.73}$$

It is easy to see that $(\sqrt{\boldsymbol{A}})^2 = \boldsymbol{A}$. If $\boldsymbol{A}$ is positive definite, eq. (2.69) can be extended to arbitrary negative non-integer powers such as $\boldsymbol{A}^{-2/3}$.

Let $\{\boldsymbol{r}_1, ..., \boldsymbol{r}_l\}_L$ denote the linear subspace *spanned* (or *generated*) by $\boldsymbol{r}_1$, ..., $\boldsymbol{r}_l$, i.e., the set of all vectors that can be expressed as a linear combination $\sum_{i=1}^{l} c_i \boldsymbol{r}_i$ for some real numbers c_1, ..., c_l. A positive semi-definite (nn)-matrix of rank r ($\leq n$) has the following spectral decomposition:

$$\boldsymbol{A} = \sum_{i=1}^{r} \lambda_i \boldsymbol{u}_i \boldsymbol{u}_i^\top, \qquad \lambda_i > 0, \quad i = 1, ..., r. \tag{2.74}$$

Let the symbol $\mathcal{R}^n$ denote the n-dimensional space of all n-vectors. The r-dimensional subspace

$$\mathcal{R}_{\boldsymbol{A}} = \{\boldsymbol{u}_1, ..., \boldsymbol{u}_r\}_L \subset \mathcal{R}^n \tag{2.75}$$

is called the *range* (or *image space*) of $\boldsymbol{A}$, for which the set $\{\boldsymbol{u}_1, ..., \boldsymbol{u}_r\}$ is an orthonormal basis. The $(n-r)$-dimensional subspace

$$\mathcal{N}_{\boldsymbol{A}} = \{\boldsymbol{u}_{r+1}, ..., \boldsymbol{u}_n\}_L \subset \mathcal{R}^n \tag{2.76}$$

is called the *null space* of $\boldsymbol{A}$, for which the set $\{\boldsymbol{u}_{r+1}, ..., \boldsymbol{u}_n\}$ is an orthonormal basis. The n-dimensional space is the direct sum of $\mathcal{R}_{\boldsymbol{A}}$ and $\mathcal{N}_{\boldsymbol{A}}$, each being the orthogonal complement of the other:

$$\mathcal{R}^n = \mathcal{R}_{\boldsymbol{A}} \oplus \mathcal{N}_{\boldsymbol{A}}, \qquad \mathcal{R}_{\boldsymbol{A}} \perp \mathcal{N}_{\boldsymbol{A}}. \tag{2.77}$$

This definition implies

$$\boldsymbol{P}_{\mathcal{N}_A} \boldsymbol{A} = \boldsymbol{A} \boldsymbol{P}_{\mathcal{N}_A} = \boldsymbol{A}. \tag{2.78}$$

The (*Moore-Penrose*) *generalized* (or *pseudo*) *inverse*[10] $\boldsymbol{A}^-$ of $\boldsymbol{A}$ is defined

[10]The Moore-Penrose generalized inverse is often denoted by $\mathbf{A}^+$ in order to distinguish it from the generalized inverse in general, which is defined as the matrix $\mathbf{X}$ that satisfies $\mathbf{AXA} = \mathbf{A}$ and denoted by $\mathbf{A}^-$. The generalized inverse we use throughout this book is always the Moore-Penrose type, so we adopt the generic symbol $\mathbf{A}^-$. The symbol $\mathbf{A}^+$ will be given another meaning (see Section 2.2.6).

by

$$A^- = \sum_{i=1}^{r} \frac{1}{\lambda_i} u_i u_i^\top. \tag{2.79}$$

Evidently, the generalized inverse A^- coincides with the inverse A^{-1} if A is of full rank. From this definition, the following relationships are obtained (see eqs. (2.50) and (2.63)):

$$(A^-)^- = A, \qquad P_{\mathcal{N}_A} A^- = A^- P_{\mathcal{N}_A} = A^-,$$
$$A^- A = AA^- = P_{\mathcal{N}_A}. \tag{2.80}$$

From eqs. (2.78) and (2.80), we obtain

$$AA^- A = A, \qquad A^- AA^- = A^-. \tag{2.81}$$

The rank and the generalized inverse of a matrix are well defined concepts in a mathematical sense only; it rarely occurs in finite precision numerical computation that some eigenvalues are precisely zero. In computing the generalized inverse numerically, the rank of the matrix should be predicted by a theoretical analysis first. Then, the matrix should be modified so that it has the desired rank. Let A be a positive semi-definite (nn)-matrix of rank r; let $A = \sum_{i=1}^{r} \lambda_i u_i u_i^\top$, $\lambda_1 \geq \cdots \geq \lambda_r > 0$, be its spectral decomposition. Its *rank-constrained generalized inverse* $(A)_{r'}^-$ of rank r' $(\leq r)$ is defined by

$$(A)_{r'}^- = \sum_{i=1}^{r'} \frac{1}{\lambda_i} u_i u_i^\top. \tag{2.82}$$

From this definition, the following identities are obtained:

$$(A)_{r'}^- A = A(A)_{r'}^- = P_{\mathcal{N}_{(A)_{r'}^-}}, \qquad (A)_{r'}^- A(A)_{r'}^- = (A)_{r'}^-. \tag{2.83}$$

Let A be an (nn)-matrix, and B an (mm)-matrix. Let S and T be nm-matrices. Even if A and B are not of full rank, the matrix inversion formula (2.22) holds in the form

$$(A + P_{\mathcal{N}_A} SBT^\top P_{\mathcal{N}_A})^- = A^- - A^- S(B^- + P_{\mathcal{N}_B} T^\top A^- SP_{\mathcal{N}_B})^- T^\top A^-, \tag{2.84}$$

provided that matrix $A + P_{\mathcal{N}_A} SBT^\top P_{\mathcal{N}_A}$ has the same rank as A and matrix $B^- + P_{\mathcal{N}_B} T^\top A^- SP_{\mathcal{N}_B}$ has the same rank as B^-. We call eq. (2.84) the *generalized matrix inversion formula*.

2.2.3 Rayleigh quotient and quadratic form

For an (nn)-matrix A, the expression $(u, Au)/\|u\|^2$ is called the *Rayleigh quotient* of vector u for A. Let $\lambda_{\min}$ and $\lambda_{\max}$ be, respectively, the largest and the smallest eigenvalues of $\boldsymbol{A}$. The following inequality holds for an arbitrary nonzero vector $\boldsymbol{u}$:

$$\lambda_{\min} \leq \frac{(\boldsymbol{u}, \boldsymbol{A}\boldsymbol{u})}{\|\boldsymbol{u}\|^2} \leq \lambda_{\max}. \tag{2.85}$$

The left equality holds if $\boldsymbol{u}$ is an eigenvector of $\boldsymbol{A}$ for eigenvalue $\lambda_{\min}$; the right equality holds if $\boldsymbol{u}$ is an eigenvector for eigenvalue $\lambda_{\max}$.

The Rayleigh quotient $(\boldsymbol{u}, \boldsymbol{A}\boldsymbol{u})/\|\boldsymbol{u}\|^2$ is invariant to multiplication of $\boldsymbol{u}$ by a constant and hence is a function of the orientation of $\boldsymbol{u}$: if we put $\boldsymbol{n} = N[\boldsymbol{u}]$, then $(\boldsymbol{u}, \boldsymbol{A}\boldsymbol{u})/\|\boldsymbol{u}\|^2 = (\boldsymbol{n}, \boldsymbol{A}\boldsymbol{n})$, which is called the *quadratic form* in $\boldsymbol{n}$ for $\boldsymbol{A}$. Eq. (2.85) implies

$$\min_{\|\mathbf{n}\|=1} (\boldsymbol{n}, \boldsymbol{A}\boldsymbol{n}) = \lambda_{\min}, \qquad \max_{\|\mathbf{n}\|=1} (\boldsymbol{n}, \boldsymbol{A}\boldsymbol{n}) = \lambda_{\max}. \tag{2.86}$$

The minimum is attained by any unit eigenvector $\boldsymbol{n}$ of $\boldsymbol{A}$ for eigenvalue $\lambda_{\min}$; the maximum is attained by any unit eigenvector $\boldsymbol{n}$ for eigenvalue $\lambda_{\max}$. It follows that an (nn)-matrix $\boldsymbol{A}$ is positive definite if and only if $(\boldsymbol{r}, \boldsymbol{A}\boldsymbol{r}) > 0$ for an arbitrary nonzero vector $\boldsymbol{r}$; it is positive semi-definite if and only if $(\boldsymbol{r}, \boldsymbol{A}\boldsymbol{r}) \geq 0$ for an arbitrary n-vector $\boldsymbol{r}$.

For an arbitrary mn-matrix $\boldsymbol{B}$, the matrix $\boldsymbol{B}^\top\boldsymbol{B}$ is symmetric (see the second of eq. (2.2)). It is also positive semi-definite since $(\boldsymbol{r}, \boldsymbol{B}^\top\boldsymbol{B}\boldsymbol{r}) = \|\boldsymbol{B}\boldsymbol{r}\|^2 \geq 0$ for an arbitrary n-vector $\boldsymbol{r}$. If $\boldsymbol{B}$ is an nn-matrix of full rank, equality holds if and only if $\boldsymbol{r} = \boldsymbol{0}$. For an (nn)-matrix $\boldsymbol{A}$, its square root $\sqrt{\boldsymbol{A}}$ is also symmetric (see eq. (2.73)). We can also write $\boldsymbol{A} = \sqrt{\boldsymbol{A}}^\top\sqrt{\boldsymbol{A}}$. From these observations, we conclude the following:

- Matrix $\boldsymbol{A}$ is positive semi-definite if and only if there exists a matrix $\boldsymbol{B}$ such that $\boldsymbol{A} = \boldsymbol{B}^\top\boldsymbol{B}$.

- Matrix $\boldsymbol{A}$ is positive definite if and only if there exists a nonsingular matrix $\boldsymbol{B}$ such that $\boldsymbol{A} = \boldsymbol{B}^\top\boldsymbol{B}$.

- If $\boldsymbol{A}$ is a positive semi-definite (nn)-matrix, matrix $\boldsymbol{B}^\top\boldsymbol{A}\boldsymbol{B}$ is a positive semi-definite (mm)-matrix for any nm-matrix $\boldsymbol{B}$.

2.2.4 Nonsingular generalized eigenvalue problem

Let $\boldsymbol{A}$ be an (nn)-matrix, and $\boldsymbol{G}$ a positive semi-definite (nn)-matrix. If there exists a nonzero vector $\boldsymbol{u}$ and a scalar λ such that

$$\boldsymbol{A}\boldsymbol{u} = \lambda\boldsymbol{G}\boldsymbol{u}, \tag{2.87}$$

the scalar λ is called the *generalized eigenvalue* of $\boldsymbol{A}$ with respect to $\boldsymbol{G}$; the vector $\boldsymbol{u}$ is called the corresponding *generalized eigenvector*. The problem of computing such $\boldsymbol{u}$ and λ is said to be *nonsingular* if $\boldsymbol{G}$ is of full rank, and *singular* otherwise.

Consider the nonsingular generalized eigenvalue problem. Eq. (2.87) can be rewritten as $(\lambda\boldsymbol{G} - \boldsymbol{A})\boldsymbol{u} = \boldsymbol{0}$, which has a nonzero solution $\boldsymbol{u}$ if and only if function

$$\phi_{\boldsymbol{A},\boldsymbol{G}}(\lambda) = |\lambda\boldsymbol{G} - \boldsymbol{A}| \tag{2.88}$$

has a zero: $\phi_{\boldsymbol{A},\boldsymbol{G}}(\lambda) = 0$. The function $\phi_{\boldsymbol{A},\boldsymbol{G}}(\lambda)$ is an nth degree polynomial in λ and is called the *generalized characteristic polynomial* of $\boldsymbol{A}$ with respect to $\boldsymbol{G}$. The equation $\phi_{\boldsymbol{A},\boldsymbol{G}}(\lambda) = 0$ is called the *generalized characteristic equation* of $\boldsymbol{A}$ with respect to $\boldsymbol{G}$ and has n roots $\{\lambda_i\}$ (with multiplicities counted). The generalized eigenvalue problem with respect to $\boldsymbol{I}$ reduces to the usual eigenvalue problem.

The generalized eigenvalues $\{\lambda_i\}$ of $\boldsymbol{A}$ with respect to $\boldsymbol{G}$ are all real. The corresponding generalized eigenvectors $\{\boldsymbol{u}_i\}$ can be chosen so that

$$(\boldsymbol{u}_i, \boldsymbol{G}\boldsymbol{u}_j) = \delta_{ij}, \tag{2.89}$$

which implies

$$(\boldsymbol{u}_i, \boldsymbol{A}\boldsymbol{u}_j) = \lambda_j\delta_{ij}. \tag{2.90}$$

Let us call the set $\{\boldsymbol{u}_i\}$ so defined the *generalized eigensystem* of the (nn)-matrix with respect to the positive definite (nn)-matrix $\boldsymbol{G}$. Let $\boldsymbol{U} = (\boldsymbol{u}_1, ..., \boldsymbol{u}_n)$ and $\boldsymbol{\Lambda} = \text{diag}(\lambda_1, ..., \lambda_n)$, respectively. Eqs. (2.89) and (2.90) can be rewritten as

$$\boldsymbol{U}^\top\boldsymbol{G}\boldsymbol{U} = \boldsymbol{I}, \qquad \boldsymbol{U}^\top\boldsymbol{A}\boldsymbol{U} = \boldsymbol{\Lambda}. \tag{2.91}$$

By multiplying the first equation by $\boldsymbol{G}\boldsymbol{U}$ from the left and $\boldsymbol{U}^\top\boldsymbol{G}$ from the right, the following *generalized spectral decomposition* is obtained:

$$\boldsymbol{A} = \boldsymbol{G}\boldsymbol{U}\boldsymbol{\Lambda}\boldsymbol{U}^\top\boldsymbol{G} = \sum_{i=1}^{n}\lambda_i(\boldsymbol{G}\boldsymbol{u}_i)(\boldsymbol{G}\boldsymbol{u}_i)^\top. \tag{2.92}$$

The number of nonzero generalized eigenvalues is equal to the rank of $\boldsymbol{A}$. If $\boldsymbol{A}$ is positive definite, $\{\lambda_i\}$ are all positive; if $\boldsymbol{A}$ is positive semi-definite, $\{\lambda_i\}$ are all nonnegative.

The generalized eigenvalue problem $\boldsymbol{A}\boldsymbol{u} = \lambda\boldsymbol{G}\boldsymbol{u}$ reduces to an ordinary eigenvalue problem as follows. Let $\boldsymbol{C} = \boldsymbol{G}^{-1/2}$ and $\tilde{\boldsymbol{u}} = \boldsymbol{C}^{-1}\boldsymbol{u}$ (see eqs. (2.71) and (2.73)). It is easy to see that eq. (2.87) can be written as

$$\tilde{\boldsymbol{A}}\tilde{\boldsymbol{u}} = \lambda\tilde{\boldsymbol{u}}, \qquad \tilde{\boldsymbol{A}} = \boldsymbol{C}\boldsymbol{A}\boldsymbol{C}. \tag{2.93}$$

If an eigenvector $\tilde{\boldsymbol{u}}$ of $\tilde{\boldsymbol{A}}$ is computed, the corresponding generalized eigenvector is given by

$$\boldsymbol{u} = \boldsymbol{C}\tilde{\boldsymbol{u}}. \tag{2.94}$$

The expression $(\boldsymbol{u}, \boldsymbol{A}\boldsymbol{u})/(\boldsymbol{u}, \boldsymbol{G}\boldsymbol{u})$ for an (nn)-matrix $\boldsymbol{A}$ and a positive definite (nn)-matrix $\boldsymbol{G}$ is called the *generalized Rayleigh quotient* of $\boldsymbol{u}$. It satisfies

$$\lambda_{\min} \leq \frac{(\boldsymbol{u}, \boldsymbol{A}\boldsymbol{u})}{(\boldsymbol{u}, \boldsymbol{G}\boldsymbol{u})} \leq \lambda_{\max}, \tag{2.95}$$

where $\lambda_{\min}$ and $\lambda_{\max}$ are, respectively, the largest and the smallest generalized eigenvalues of $\boldsymbol{A}$ with respect to $\boldsymbol{G}$. The left equality holds if $\boldsymbol{u}$ is a generalized eigenvector of $\boldsymbol{A}$ for the generalized eigenvalue $\lambda_{\min}$; the right equality holds if $\boldsymbol{u}$ is a generalized eigenvector for the generalized eigenvalue $\lambda_{\max}$.

2.2.5 Singular generalized eigenvalue problem

Consider the singular generalized eigenvalue problem of an (nn)-matrix $\boldsymbol{A}$ with respect to a positive semi-definite (nn)-matrix $\boldsymbol{G}$ of rank m $(< n)$. Let $\{\boldsymbol{v}_1, ..., \boldsymbol{v}_m\}$ be an orthonormal basis of the range $\mathcal{R}_{\boldsymbol{G}}$ of $\boldsymbol{G}$, and $\{\boldsymbol{v}_{m+1}, ..., \boldsymbol{v}_n\}$ an orthonormal basis of its null space $\mathcal{N}_{\boldsymbol{G}}$. Define an nm-matrix $\boldsymbol{P}_1$ and an $n(n-m)$-matrix $\boldsymbol{P}_0$ by

$$\boldsymbol{P}_1 = (\boldsymbol{v}_1, ..., \boldsymbol{v}_m), \qquad \boldsymbol{P}_0 = (\boldsymbol{v}_{m+1}, ..., \boldsymbol{v}_n). \tag{2.96}$$

Then,

$$\boldsymbol{P}_1^\top \boldsymbol{P}_1 = \boldsymbol{I}, \qquad \boldsymbol{P}_1^\top \boldsymbol{P}_0 = \boldsymbol{O}, \qquad \boldsymbol{P}_0^\top \boldsymbol{P}_0 = \boldsymbol{I}. \tag{2.97}$$

Here, we only consider the case where $\boldsymbol{P}_0^\top \boldsymbol{A} \boldsymbol{P}_0$ is nonsingular[11]. Since $\mathcal{R}^n = \mathcal{R}_{\boldsymbol{G}} \oplus \mathcal{N}_{\boldsymbol{G}}$, an arbitrary n-vector can be uniquely written in the form

$$\boldsymbol{u} = \boldsymbol{P}_1 \boldsymbol{x} + \boldsymbol{P}_0 \boldsymbol{y}, \tag{2.98}$$

where $\boldsymbol{x}$ is an m-vector and $\boldsymbol{y}$ is an $(n-m)$-vector. Eqs. (2.97) imply that $\boldsymbol{x}$ and $\boldsymbol{y}$ are respectively given by

$$\boldsymbol{x} = \boldsymbol{P}_1^\top \boldsymbol{u}, \qquad \boldsymbol{y} = \boldsymbol{P}_0^\top \boldsymbol{u}. \tag{2.99}$$

Substituting eq. (2.98) into eq. (2.87) and noting the identities $\boldsymbol{G}\boldsymbol{P}_0 = \boldsymbol{O}$ and $\boldsymbol{P}_0^\top \boldsymbol{G} = \boldsymbol{O}$, we can split eq. (2.87) into the following two equations:

$$\boldsymbol{A}^* \boldsymbol{x} = \lambda \boldsymbol{G}^* \boldsymbol{x}, \qquad \boldsymbol{y} = \boldsymbol{B}^* \boldsymbol{x}. \tag{2.100}$$

Here, $\boldsymbol{A}^*$ and $\boldsymbol{G}^*$ are (mm)-matrices; $\boldsymbol{B}^*$ is an $(n-m)m$-matrix. They are defined by

$$\begin{aligned} \boldsymbol{A}^* &= \boldsymbol{P}_1^\top \boldsymbol{A} \boldsymbol{P}_1 - \boldsymbol{P}_1^\top \boldsymbol{A} \boldsymbol{P}_0 \boldsymbol{C}^{*-1} \boldsymbol{P}_0^\top \boldsymbol{A} \boldsymbol{P}_1, \\ \boldsymbol{G}^* &= \boldsymbol{P}_1^\top \boldsymbol{G} \boldsymbol{P}_1, \qquad \boldsymbol{B}^* = -\boldsymbol{C}^{*-1} \boldsymbol{P}_0^\top \boldsymbol{A} \boldsymbol{P}_1, \end{aligned} \tag{2.101}$$

where $\boldsymbol{C}^*$ is an $(n-m)(n-m)$-matrix defined by

$$\boldsymbol{C}^* = \boldsymbol{P}_0^\top \boldsymbol{A} \boldsymbol{P}_0. \tag{2.102}$$

[11] This is always true if $\mathbf{A}$ is positive definite or negative definite.

The definition of the matrix $\boldsymbol{P}_0$ implies that the matrix $\boldsymbol{G}^*$ is positive definite. Hence, the first of eqs. (2.100) is a nonsingular generalized eigenvalue problem.

The generalized Rayleigh quotient of $\boldsymbol{A}$ with respect to $\boldsymbol{G}$ for $\boldsymbol{u} \notin \mathcal{N}_{\boldsymbol{G}}$ (i.e., $\boldsymbol{x} \neq \boldsymbol{0}$) can be written as follows:

$$\frac{(\boldsymbol{u}, \boldsymbol{A}\boldsymbol{u})}{(\boldsymbol{u}, \boldsymbol{G}\boldsymbol{u})} = \frac{(\boldsymbol{x}, \boldsymbol{A}^*\boldsymbol{x}) + (\boldsymbol{y} - \boldsymbol{B}^*\boldsymbol{x}, \boldsymbol{C}^*(\boldsymbol{y} - \boldsymbol{B}^*\boldsymbol{x}))}{(\boldsymbol{x}, \boldsymbol{G}^*\boldsymbol{x})}. \tag{2.103}$$

If $\boldsymbol{C}^*$ is positive definite[12], we observe that

$$\frac{(\boldsymbol{u}, \boldsymbol{A}\boldsymbol{u})}{(\boldsymbol{u}, \boldsymbol{G}\boldsymbol{u})} \geq \frac{(\boldsymbol{x}, \boldsymbol{A}^*\boldsymbol{x})}{(\boldsymbol{x}, \boldsymbol{G}^*\boldsymbol{x})} \geq \lambda_{\min}, \tag{2.104}$$

where $\lambda_{\min}$ is the smallest generalized eigenvalue of $\boldsymbol{A}$ with respect to $\boldsymbol{G}$ (see eqs. (2.100)). Equality holds if $\boldsymbol{u}$ is the corresponding generalized eigenvector. If $\boldsymbol{C}^*$ is negative definite[13], we observe that

$$\frac{(\boldsymbol{u}, \boldsymbol{A}\boldsymbol{u})}{(\boldsymbol{u}, \boldsymbol{G}\boldsymbol{u})} \leq \frac{(\boldsymbol{x}, \boldsymbol{A}^*\boldsymbol{x})}{(\boldsymbol{x}, \boldsymbol{G}^*\boldsymbol{x})} \leq \lambda_{\max}, \tag{2.105}$$

where $\lambda_{\max}$ is the smallest generalized eigenvalue of $\boldsymbol{A}$ with respect to $\boldsymbol{G}$. Equality holds if $\boldsymbol{u}$ is the corresponding generalized eigenvector.

2.2.6 Perturbation theorem

Let $\boldsymbol{A}$ and $\boldsymbol{D}$ be (nn)-matrices. Let $\{\lambda_i\}$ be the eigenvalues of $\boldsymbol{A}$, and $\{\boldsymbol{u}_i\}$ the corresponding eigensystem:

$$\boldsymbol{A}\boldsymbol{u}_i = \lambda_i \boldsymbol{u}_i, \qquad (\boldsymbol{u}_i, \boldsymbol{u}_j) = \delta_{ij}. \tag{2.106}$$

Consider a perturbed matrix

$$\boldsymbol{A}' = \boldsymbol{A} + \epsilon \boldsymbol{D} \tag{2.107}$$

for a small ϵ. Let $\{\lambda_i{}'\}$ and $\{\boldsymbol{u}_i'\}$ be, respectively, the eigenvalues and the eigensystem of $\boldsymbol{A}'$ corresponding to $\{\lambda_i\}$ and $\{\boldsymbol{u}_i\}$. The following relations hold (the *perturbation theorem*):

$$\lambda_i{}' = \lambda_i + \epsilon(\boldsymbol{u}_i, \boldsymbol{D}\boldsymbol{u}_i) + O(\epsilon^2), \tag{2.108}$$

$$\boldsymbol{u}_i' = \boldsymbol{u}_i + \epsilon \sum_{j \neq i} \frac{(\boldsymbol{u}_j, \boldsymbol{D}\boldsymbol{u}_i)\boldsymbol{u}_j}{\lambda_i - \lambda_j} + O(\epsilon^2). \tag{2.109}$$

Let $\boldsymbol{u}_n$ be the unit eigenvector of $\boldsymbol{A}$ for the smallest eigenvalue λ_n, which is assumed to be a simple root. Let $\{\boldsymbol{u}_i\}$ be the eigensystem of $\boldsymbol{A}$ defined so

[12] This is always true if $\mathbf{A}$ is positive definite.
[13] This is always true if $\mathbf{A}$ is negative definite.

that the corresponding eigenvalues are $\lambda_1 \geq \cdots \geq \lambda_{n-1} > \lambda_n$. Define matrix $\boldsymbol{A}^+$ by

$$\boldsymbol{A}^+ = \sum_{i=1}^{n-1} \frac{\boldsymbol{u}_i \boldsymbol{u}_i^\top}{\lambda_i - \lambda_n}. \tag{2.110}$$

This is a positive semi-definite matrix having eigenvalues $\{1/(\lambda_i - \lambda_n)\}$ for the same eigensystem $\{\boldsymbol{u}_i\}$. If $\lambda_n = 0$, the matrix $\boldsymbol{A}^+$ coincides with the generalized inverse $\boldsymbol{A}^-$. Eq. (2.109) can be rewritten as

$$\boldsymbol{u}_n' = \boldsymbol{u}_n - \epsilon \boldsymbol{A}^+ \boldsymbol{D} \boldsymbol{u}_n + O(\epsilon^2). \tag{2.111}$$

Let $\boldsymbol{A}$ and $\boldsymbol{D}$ be (nn)-matrices, and $\boldsymbol{G}$ a positive definite (nn)-matrix. Let $\{\lambda_i\}$ be the generalized eigenvalues of $\boldsymbol{A}$ with respect to $\boldsymbol{G}$, and $\{\boldsymbol{u}_i\}$ the corresponding generalized eigensystem:

$$\boldsymbol{A}\boldsymbol{u}_i = \lambda_i \boldsymbol{G}\boldsymbol{u}_i, \qquad (\boldsymbol{u}_i, \boldsymbol{G}\boldsymbol{u}_j) = \delta_{ij}. \tag{2.112}$$

If $\boldsymbol{A}$ is perturbed in the form of eq. (2.107), the perturbation theorem holds in the same form. Eq. (2.111) also holds if $\{\lambda_i\}$ in eq. (2.110) are interpreted as generalized eigenvalues of $\boldsymbol{A}$ with respect to $\boldsymbol{G}$.

2.3 Linear Systems and Optimization

2.3.1 Singular value decomposition and generalized inverse

If $\boldsymbol{A}$ is an mn-matrix, $\boldsymbol{A}^\top \boldsymbol{A}$ is a positive semi-definite (nn)-matrix, and $\boldsymbol{A}\boldsymbol{A}^\top$ is a positive semi-definite (mm)-matrix. They share the same nonzero eigenvalues $\sigma_1 \geq \sigma_2 \geq \cdots \geq \sigma_r$ (> 0), $r \leq \min(m, n)$. The number r is called the *rank* of $\boldsymbol{A}$. Let $\lambda_i = \sqrt{\sigma_i}$, $i = 1, \ldots, r$, and $\lambda_i = 0$, $i = r+1, \ldots, \max(m, n)$. It can be shown that orthonormal systems $\{\boldsymbol{u}_i\}$, $i = 1, \ldots, n$, and $\{\boldsymbol{v}_i\}$, $i = 1, \ldots, m$, exist such that

- $\boldsymbol{A}\boldsymbol{u}_i = \lambda_i \boldsymbol{v}_i$, $i = 1, \ldots, \min(m, n)$.
- $\{\boldsymbol{u}_i\}$, $i = 1, \ldots, n$, is the eigensystem of $\boldsymbol{A}^\top \boldsymbol{A}$ for eigenvalues $\{\lambda_i^2\}$, $i = 1, \ldots, n$.
- $\{\boldsymbol{v}_i\}$, $i = 1, \ldots, m$, is the eigensystem of $\boldsymbol{A}\boldsymbol{A}^\top$ for eigenvalues $\{\lambda_i^2\}$, $i = 1, \ldots, m$.

Matrix $\boldsymbol{A}$ is expressed in terms of $\{\boldsymbol{u}_i\}$, $\{\boldsymbol{v}_i\}$, and $\{\lambda_i\}$ in the form

$$\boldsymbol{A} = \sum_{i=1}^{r} \lambda_i \boldsymbol{v}_i \boldsymbol{u}_i^\top. \tag{2.113}$$

This is called the *singular value decomposition* of $\boldsymbol{A}$; the values $\{\lambda_i\}$, $i = 1, \ldots, \min(m, n)$, are called the *singular values* of $\boldsymbol{A}$. Let us call $\{\boldsymbol{u}_i\}$, $i = 1, \ldots, n$,

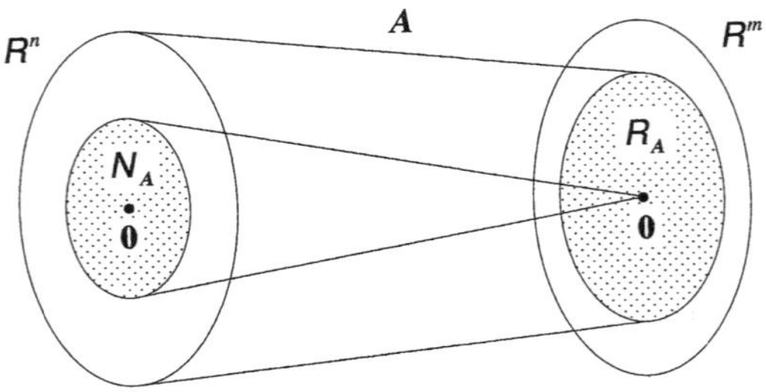

Fig. 2.4. The range $\mathcal{R}_{\boldsymbol{A}}$ and the null space $\mathcal{N}_{\boldsymbol{A}}$ of linear mapping $\mathbf{A}$.

and $\{\boldsymbol{v}_i\}$, $i = 1, ..., m$, the *right orthonormal system* and the *left orthonormal system* of $\boldsymbol{A}$, respectively.

If we define orthogonal matrices $\boldsymbol{U} = (\boldsymbol{u}_1, \boldsymbol{u}_2, ..., \boldsymbol{u}_n)$ and $\boldsymbol{V} = (\boldsymbol{v}_1, \boldsymbol{v}_2, ..., \boldsymbol{v}_m)$, eq. (2.113) can be rewritten in the form

$$\boldsymbol{A} = \boldsymbol{V}\boldsymbol{\Lambda}\boldsymbol{U}^\top, \tag{2.114}$$

where $\boldsymbol{\Lambda}$ is an mn matrix whose first r diagonal elements are λ_1, ..., λ_r in that order and whose other elements are all zero. If $m = n$, matrix $\boldsymbol{\Lambda}$ is diagonal.

The r-dimensional linear subspace

$$\mathcal{R}_{\boldsymbol{A}} = \{\boldsymbol{v}_1, ..., \boldsymbol{v}_r\}_L \subset \mathcal{R}^m \tag{2.115}$$

is called the *range* (or *image space*) of $\boldsymbol{A}$: for any m-vector $\boldsymbol{y} \in \mathcal{R}_{\boldsymbol{A}}$, there exists an n-vector $\boldsymbol{x}$ such that $\boldsymbol{y} = \boldsymbol{A}\boldsymbol{x}$ (Fig. 2.4). The $(n-r)$-dimensional linear subspace

$$\mathcal{N}_{\boldsymbol{A}} = \{\boldsymbol{u}_{r+1}, ..., \boldsymbol{u}_n\}_L \subset \mathcal{R}^n \tag{2.116}$$

is called the *null space* (or *kernel*) of $\boldsymbol{A}$: $\boldsymbol{A}\boldsymbol{x} = \boldsymbol{0}$ for any n-vector $\boldsymbol{x} \in \mathcal{N}_{\boldsymbol{A}}$ (Fig. 2.4). If $\boldsymbol{A}$ is symmetric, its right and left orthonormal systems coincide with its eigensystem, and its singular value decomposition coincides with its spectral decomposition (see eq. (2.62)).

Since $\{\boldsymbol{u}_i\}$ is an orthonormal system, eq. (2.64) holds for an arbitrary n-vector $\boldsymbol{x}$. Let $\lambda_{\max}$ be the maximum singular value. Since $\{\boldsymbol{v}_i\}$ is also an orthonormal system, we see from eq. (2.113) that

$$\|\boldsymbol{A}\boldsymbol{x}\|^2 = \|\sum_{i=1}^{r} \lambda_i(\boldsymbol{u}_i, \boldsymbol{x})\boldsymbol{v}_i\|^2 = \sum_{i=1}^{r} \lambda_i^2(\boldsymbol{u}_i, \boldsymbol{x})^2 \le \sum_{i=1}^{r} \lambda_{\max}^2(\boldsymbol{u}_i, \boldsymbol{x})^2 = \lambda_{\max}^2\|\boldsymbol{x}\|^2. \tag{2.117}$$

Hence, if we define the *spectral norm* (or the *natural norm*) of $\boldsymbol{A}$ by

$$\|\boldsymbol{A}\|_s = \lambda_{\max}, \tag{2.118}$$

eq. (2.117) implies the following inequality:

$$\|\boldsymbol{A}\boldsymbol{x}\| \leq \|\boldsymbol{A}\|_s \|\boldsymbol{x}\|. \tag{2.119}$$

Equality holds for

$$\boldsymbol{x} \propto \boldsymbol{u}_{\max} + \mathcal{N}_{\boldsymbol{A}}. \tag{2.120}$$

The right-hand side means the first term plus any element of $\mathcal{N}_{\boldsymbol{A}}$ (and such a form only), and $\boldsymbol{u}_{\max}$ is the vector $\boldsymbol{u}_i$ corresponding to the singular value $\lambda_{\max}$.

Let eq. (2.113) be the singular value decomposition of matrix $\boldsymbol{A}$. Its (*Moore-Penrose*) *generalized inverse* is defined by

$$\boldsymbol{A}^- = \sum_{i=1}^{r} \frac{\boldsymbol{u}_i \boldsymbol{v}_i^\top}{\lambda_i}. \tag{2.121}$$

Evidently, the generalized inverse $\boldsymbol{A}^-$ coincides with the inverse $\boldsymbol{A}^{-1}$ if $\boldsymbol{A}$ is nonsingular. In correspondence with eq. (2.78) and eqs. (2.80), the following relationships hold:

$$(\boldsymbol{A}^-)^- = \boldsymbol{A}, \qquad \boldsymbol{A}^-\boldsymbol{A} = \boldsymbol{P}_{\mathcal{N}_A}, \qquad \boldsymbol{A}\boldsymbol{A}^- = \boldsymbol{P}^{\mathcal{R}_A},$$

$$\boldsymbol{P}^{\mathcal{R}_A}\boldsymbol{A} = \boldsymbol{A}\boldsymbol{P}_{\mathcal{N}_A} = \boldsymbol{A}, \qquad \boldsymbol{P}_{\mathcal{N}_A}\boldsymbol{A}^- = \boldsymbol{A}^-\boldsymbol{P}^{\mathcal{R}_A} = \boldsymbol{A}^-. \tag{2.122}$$

Here, $\boldsymbol{P}^{\mathcal{R}_A}$ $(= \boldsymbol{P}_{\mathcal{R}_A^\perp})$ and $\boldsymbol{P}_{\mathcal{N}_A}$ $(= \boldsymbol{P}^{\mathcal{N}_A^\perp})$ are the projection matrices onto $\mathcal{R}_{\boldsymbol{A}}$ and $\mathcal{N}_{\boldsymbol{A}}^\perp$, respectively. From the above equations, we obtain

$$\boldsymbol{A}\boldsymbol{A}^-\boldsymbol{A} = \boldsymbol{A}, \qquad \boldsymbol{A}^-\boldsymbol{A}\boldsymbol{A}^- = \boldsymbol{A}^-. \tag{2.123}$$

The *rank-constrained generalized inverse* $(\boldsymbol{A})_{r'}^-$ of rank r' $(\leq r)$ is defined by

$$(\boldsymbol{A})_{r'}^- = \sum_{i=1}^{r'} \frac{\boldsymbol{u}_i \boldsymbol{v}_i^\top}{\lambda_i}, \tag{2.124}$$

and the following relations hold:

$$(\boldsymbol{A})_{r'}^-\boldsymbol{A} = \boldsymbol{P}^{\mathcal{R}_{(A)_{r'}^-}}, \qquad \boldsymbol{A}(\boldsymbol{A})_{r'}^- = \boldsymbol{P}_{\mathcal{N}_{(A)_{r'}^-}},$$

$$(\boldsymbol{A})_{r'}^-\boldsymbol{A}(\boldsymbol{A})_{r'}^- = (\boldsymbol{A})_{r'}^-. \tag{2.125}$$

2.3.2 Linear equations

Let $\boldsymbol{A}$ be an mn-matrix, and $\boldsymbol{b}$ an m-vector. Consider the following linear equation for n-vector $\boldsymbol{x}$:

$$\boldsymbol{A}\boldsymbol{x} = \boldsymbol{b}. \tag{2.126}$$

The following is the fundamental theorem for linear equations:

- The solution exists if and only if $\boldsymbol{b} \in \mathcal{R}_{\boldsymbol{A}}$ (or $\boldsymbol{P}_{\mathcal{R}_{\boldsymbol{A}}}\boldsymbol{b} = \boldsymbol{0}$).
- If the solution exists, it is unique if and only if $\mathcal{N}_{\boldsymbol{A}} = \{\boldsymbol{0}\}$.

The problem (2.126) is said to be *consistent* (or *solvable*) when $\boldsymbol{b} \in \mathcal{R}_{\boldsymbol{A}}$, and *inconsistent* (or *unsolvable*) otherwise; if it is consistent, it is said to be *determinate* when $\mathcal{N}_{\boldsymbol{A}} = \{\boldsymbol{0}\}$, and *indeterminate* otherwise.

If eq. (2.126) is solvable, the solution can be explicitly written in the following form:

$$\boldsymbol{x} = \boldsymbol{A}^{-}\boldsymbol{b} + \mathcal{N}_{\boldsymbol{A}}. \tag{2.127}$$

If $\boldsymbol{A}$ is nonsingular, the solution is given by

$$\boldsymbol{x} = \boldsymbol{A}^{-1}\boldsymbol{b} = \frac{\boldsymbol{A}^{\dagger}\boldsymbol{b}}{\det \boldsymbol{A}}, \tag{2.128}$$

where $\boldsymbol{A}^{\dagger}$ is the cofactor matrix of $\boldsymbol{A}$ (see eq. (2.20)). Let $\boldsymbol{A} = (\boldsymbol{a}_1, ..., \boldsymbol{a}_n)$. From the cofactor expansion formula (2.14), the following *Cramer formula* is obtained:

$$x_i = \frac{|\boldsymbol{a}_1, ..., \overset{(i)}{\boldsymbol{b}}, ..., \boldsymbol{a}_n|}{\det \boldsymbol{A}}. \tag{2.129}$$

The numerator on the right-hand side is the determinant of the matrix obtained by replacing the ith column of $\boldsymbol{A}$ by $\boldsymbol{b}$.

If $\det \boldsymbol{A}$ is very close to 0, a small perturbation of $\boldsymbol{b}$ can causes a large perturbation to the solution $\boldsymbol{x}$. If this occurs, the linear equation (2.126) is said to be *ill-conditioned*; otherwise, it is *well-conditioned*. If $\boldsymbol{b}$ is perturbed into $\boldsymbol{b} + \Delta\boldsymbol{b}$, the solution $\boldsymbol{x} = \boldsymbol{A}^{-1}\boldsymbol{b}$ is perturbed by $\Delta\boldsymbol{x} = \boldsymbol{A}^{-1}\Delta\boldsymbol{b}$. Applying eq. (2.119), we obtain $\|\Delta\boldsymbol{x}\| \leq \|\boldsymbol{A}^{-1}\|_s\|\Delta\boldsymbol{b}\|$. From eq. (2.126), we have $\|\boldsymbol{b}\| \leq \|\boldsymbol{A}\|_s\|\boldsymbol{x}\|$. Combining these, we obtain

$$\frac{\|\Delta\boldsymbol{x}\|}{\|\boldsymbol{x}\|} \leq \mathrm{cond}(\boldsymbol{A})\frac{\|\Delta\boldsymbol{b}\|}{\|\boldsymbol{b}\|}, \tag{2.130}$$

where

$$\mathrm{cond}(\boldsymbol{A}) = \|\boldsymbol{A}\|_s\|\boldsymbol{A}^{-1}\|_s = \frac{\lambda_{\max}}{\lambda_{\min}}. \tag{2.131}$$

Here, $\lambda_{\max}$ and $\lambda_{\min}$ are the largest and the smallest singular values of $\boldsymbol{A}$, respectively (see eq. (2.118)). The number $\mathrm{cond}(\boldsymbol{A})$ is called the *condition number*[14] and measures the ill-posedness of the linear equation (2.126)—the equation becomes more ill-conditioned as $\mathrm{cond}(\boldsymbol{A})$ becomes larger.

Suppose eq. (2.126) is consistent but only r ($\leq m$) of the m component equations are independent, i.e., the matrix $\boldsymbol{A}$ has rank r. Theoretically, the

[14]The condition number can also be defined for a singular matrix $\mathbf{A}$ in the form $\mathrm{cond}(\mathbf{A}) = \|\mathbf{A}\|_s\|\mathbf{A}^{-}\|_s = \lambda_{\max}/\lambda_{\min}$, where $\lambda_{\max}$ and $\lambda_{\min}$ are, respectively, the largest and the smallest of the nonnegative singular values of $\mathbf{A}$.

solution is given in the form of eq. (2.127). However, if the elements of the matrix $\boldsymbol{A}$ and the components of the vector $\boldsymbol{b}$ are supplied by a physical measurement, all the m equations may be independent because of noise. As a result, eq. (2.126) may become ill-conditioned or inconsistent. In such a case, a well-conditioned equation that gives a good approximation to $\boldsymbol{x}$ is obtained by "projecting" both sides of eq. (2.126) onto the eigenspace of $\boldsymbol{A}$ defined by the largest r singular values. The solution of the projected equation is given in terms of the rank-constrained generalized inverse in the form

$$\hat{\boldsymbol{x}} = (\boldsymbol{A})_r^- \boldsymbol{b} + \mathcal{N}_{(\boldsymbol{A})_r^-}. \tag{2.132}$$

The rank r is estimated either by an a priori theoretical analysis or by appropriately thresholding the singular values of $\boldsymbol{A}$ a posteriori.

2.3.3 *Quadratic optimization*

A. Least-squares optimization

Let $\boldsymbol{A}$ be an mn-matrix, and $\boldsymbol{b}$ an m-vector. Consider the *least-squares optimization* for n-vector $\boldsymbol{x}$ in the form

$$J[\boldsymbol{x}] = \|\boldsymbol{A}\boldsymbol{x} - \boldsymbol{b}\|^2 \rightarrow \min. \tag{2.133}$$

Application of the singular value decomposition to $\boldsymbol{A}$ yields the general solution in the following form:

$$\hat{\boldsymbol{x}} = \boldsymbol{A}^- \boldsymbol{b} + \mathcal{N}_{\boldsymbol{A}}. \tag{2.134}$$

If $\boldsymbol{x}$ is constrained to be in $\mathcal{N}_{\boldsymbol{A}}^{\perp}$, the solution is uniquely given by $\hat{\boldsymbol{x}} = \boldsymbol{A}^- \boldsymbol{b}$. The *residual* $J[\hat{\boldsymbol{x}}]$ is given by

$$J[\hat{\boldsymbol{x}}] = \|\boldsymbol{P}_{\mathcal{R}_A} \boldsymbol{b}\|^2. \tag{2.135}$$

Evidently, the residual is 0 if and only if $\boldsymbol{A}\boldsymbol{x} = \boldsymbol{b}$ is solvable.

B. Unconstrained quadratic optimization

Let $\boldsymbol{C}$ be a positive semi-definite (nn)-matrix, and $\boldsymbol{d}$ an n-vector. Consider the quadratic optimization for n-vector $\boldsymbol{x}$ in the form

$$J[\boldsymbol{x}] = \frac{1}{2}(\boldsymbol{x}, \boldsymbol{C}\boldsymbol{x}) + (\boldsymbol{d}, \boldsymbol{x}) \rightarrow \min. \tag{2.136}$$

If $\boldsymbol{x}$ is constrained to be in $\mathcal{N}_{\boldsymbol{C}}^{\perp}$, the solution is uniquely given in the following form:

$$\hat{\boldsymbol{x}} = -\boldsymbol{C}^- \boldsymbol{d}. \tag{2.137}$$

The residual is

$$J[\hat{\boldsymbol{x}}] = -\frac{1}{2}(\boldsymbol{d}, \boldsymbol{C}^- \boldsymbol{d}). \tag{2.138}$$

C. Constrained quadratic optimization

Let $\boldsymbol{S}$ be a positive semi-definite (nn)-matrix. Consider the quadratic optimization for n-vector $\boldsymbol{x}$ in the form

$$J[\boldsymbol{x}] = \frac{1}{2}(\boldsymbol{x}, \boldsymbol{S}\boldsymbol{x}) \to \min. \tag{2.139}$$

Evidently, $\boldsymbol{x} = \boldsymbol{0}$ is a solution (but not necessarily unique) if no constraint is imposed on $\boldsymbol{x}$. The following three types of constraint are important:

- If $\boldsymbol{x}$ is constrained to be a unit vector ($\|\boldsymbol{x}\| = 1$), the solution is given by any unit eigenvector $\hat{\boldsymbol{x}}$ of $\boldsymbol{S}$ for the smallest eigenvalue $\lambda_{\min}$ (see eqs. (2.86)); the residual is $J[\hat{\boldsymbol{x}}] = \lambda_{\min}$ (see eq. (2.95)).
- If $\boldsymbol{x}$ is constrained by $(\boldsymbol{x}, \boldsymbol{G}\boldsymbol{x}) = 1$ for a positive definite (nn)-matrix $\boldsymbol{G}$, the solution is given by any unit generalized eigenvector $\hat{\boldsymbol{x}} \in \mathcal{N}_{\boldsymbol{S}}^{\perp}$ of $\boldsymbol{S}$ with respect to $\boldsymbol{G}$ for the smallest generalized eigenvalue $\lambda_{\min}$; the residual is $J[\hat{\boldsymbol{x}}] = \lambda_{\min}$. If $\boldsymbol{S}$ is of full rank, the same conclusion is obtained even though $\boldsymbol{G}$ is not of full rank (see eq. (2.104)).
- Suppose $\boldsymbol{x}$ is constrained by a linear equation $\boldsymbol{A}\boldsymbol{x} = \boldsymbol{b}$, where $\boldsymbol{A}$ is an mn-matrix and $\boldsymbol{b}$ is an m-vector. If

  1. $\boldsymbol{x}$ is constrained to be in $\mathcal{N}_{\boldsymbol{S}}^{\perp}$, and
  2. the constraint $\boldsymbol{A}\boldsymbol{x} = \boldsymbol{b}$ is *satisfiable* for $\boldsymbol{x} \in \mathcal{N}_{\boldsymbol{S}}^{\perp}$, i.e., at least one $\boldsymbol{x}_0 \in \mathcal{N}_{\boldsymbol{S}}^{\perp}$ exists such that $\boldsymbol{A}\boldsymbol{x}_0 = \boldsymbol{b}$,

  then the solution is uniquely given in the following form:

  $$\hat{\boldsymbol{x}} = \boldsymbol{S}^{-}\boldsymbol{A}^{\top}(\boldsymbol{A}\boldsymbol{S}^{-}\boldsymbol{A}^{\top})^{-}\boldsymbol{b}. \tag{2.140}$$

  The residual is

  $$J[\hat{\boldsymbol{x}}] = \frac{1}{2}(\boldsymbol{b}, (\boldsymbol{A}\boldsymbol{S}^{-}\boldsymbol{A}^{\top})^{-}\boldsymbol{b}). \tag{2.141}$$

2.3.4 Matrix inner product and matrix norm

The *matrix inner product* of mn-matrices $\boldsymbol{A} = (A_{ij})$ and $\boldsymbol{B} = (B_{ij})$ is defined by

$$(\boldsymbol{A}; \boldsymbol{B}) = \mathrm{tr}(\boldsymbol{A}^{\top}\boldsymbol{B}) = \mathrm{tr}(\boldsymbol{A}\boldsymbol{B}^{\top}) = \sum_{i=1}^{m}\sum_{j=1}^{n} A_{ij}B_{ij}. \tag{2.142}$$

Evidently, $(\boldsymbol{A}; \boldsymbol{B}) = (\boldsymbol{B}; \boldsymbol{A})$. If $(\boldsymbol{A}; \boldsymbol{B}) = 0$, matrices $\boldsymbol{A}$ and $\boldsymbol{B}$ are said to be *orthogonal*. An (nn)-matrix is orthogonal to any $[nn]$-matrix; an $[nn]$-matrix is orthogonal to any (nn)-matrix. The following identities are easy to prove:

$$(\boldsymbol{A}; \boldsymbol{B}\boldsymbol{C}) = (\boldsymbol{B}^{\top}\boldsymbol{A}; \boldsymbol{C}) = (\boldsymbol{A}\boldsymbol{C}^{\top}; \boldsymbol{B}),$$

$$(a, Ab) = (ab^\top; A), \qquad (ab^\top; cd^\top) = (a, c)(b, d). \tag{2.143}$$

The (*Euclidean*) *matrix norm*[15] of an mn-matrix is defined by

$$\|A\| = \sqrt{(A; A)} = \sqrt{\sum_{i=1}^{m} \sum_{j=1}^{n} A_{ij}^2}. \tag{2.144}$$

We define the *normalization* $N[\,\cdot\,]$ of an nn-matrix A as follows (see the second of eqs. (2.7)):

$$N[A] = \frac{A}{\|A\|}. \tag{2.145}$$

The *Schwarz inequality* and the *triangle inequality* hold in the same way as in the case of vectors:

$$-\|A\| \cdot \|B\| \le (A; B) \le \|A\| \cdot \|B\|, \tag{2.146}$$

$$\|A + B\| \le \|A\| + \|B\|. \tag{2.147}$$

In both inequalities, equality holds if and only if there exists a real number t such that $A = tB$ or $B = O$.

Let U be an n-dimensional orthogonal matrix. From eqs. (2.52) and the first of eqs. (2.143), it is immediately seen that for arbitrary nn-matrices A and B

$$(UA; UB) = (AU; BU) = (A; B). \tag{2.148}$$

Letting $A = B$, we obtain

$$\|UA\| = \|AU\| = \|A\|. \tag{2.149}$$

Further letting $A = I$, we see that

$$\|U\| = \sqrt{n}. \tag{2.150}$$

A nonsingular nn-matrix T defines a mapping from an nn-matrix A to an nn-matrix in the form

$$A' = T^{-1}AT. \tag{2.151}$$

[15] Some authors use different terminologies such as the *Frobenius norm*, the *Schur norm*, and the *Schmidt norm*. In general, the norm $\|\mathbf{A}\|$ can be defined arbitrarily as long as (i) $\|\mathbf{A}\| \ge 0$, equality holding if and only if $\mathbf{A} = \mathbf{O}$, (ii) $\|c\mathbf{A}\| = |c| \cdot \|\mathbf{A}\|$ for any scalar c, and (iii) the triangle inequality (2.147) holds. There exist other definitions that satisfy these—the *1-norm* $\|\mathbf{A}\|_1 = \Sigma_{i=1}^n \max_j |A_{ij}|$, the ∞*-norm* $\|\mathbf{A}\|_\infty = \Sigma_{j=1}^n \max_i |A_{ij}|$, and the *spectral norm* $\|\mathbf{A}\|_s$ defined by eq. (2.118), for instance. If $\|\mathbf{Ax}\| \le \|\mathbf{A}\| \cdot \|\mathbf{x}\|$ holds, the matrix norm $\|\mathbf{A}\|$ is said to be *consistent* with the vector norm $\|\mathbf{x}\|$. The spectral norm $\|\mathbf{A}\|_s$ is consistent with the Euclidean norm $\|\mathbf{x}\|$, and the 1-norm $\|\mathbf{A}\|_1$ and the ∞-norm $\|\mathbf{A}\|_\infty$ are consistent with the 1-norm $\|\mathbf{x}\|_1$ and the ∞-norm $\|\mathbf{x}\|_\infty$, respectively (see Footnote 4 in Section 2.1).

This is a one-to-one and onto mapping and is called the *similarity transformation*[16].

A function $f(\cdot)$ of a matrix is called an *invariant* with respect to similarity transformations if $f(\boldsymbol{A}') = f(\boldsymbol{A})$ for an arbitrary nonsingular matrix $\boldsymbol{T}$. The trace and the determinant are typical invariants:

$$\mathrm{tr}(\boldsymbol{T}^{-1}\boldsymbol{A}\boldsymbol{T}) = \mathrm{tr}\boldsymbol{A}, \qquad \det(\boldsymbol{T}^{-1}\boldsymbol{A}\boldsymbol{T}) = \det \boldsymbol{A}. \tag{2.152}$$

Eq. (2.67) implies that any symmetric matrix is mapped to a diagonal matrix by an appropriate similarity transformation; the transformation is defined by an orthogonal matrix. Hence, if $\boldsymbol{A}$ is a symmetric matrix with eigenvalues $\{\lambda_i\}$, any invariant with respect to similarity transformations is a function of $\{\lambda_i\}$. Eqs. (2.67) and (2.149) imply that

$$\|\boldsymbol{A}\| = \sqrt{\sum_{i=1}^{n} \lambda_i{}^2}. \tag{2.153}$$

Hence, $\|\boldsymbol{A}\|$ is also an invariant with respect to similarity transformation.

In three dimensions, $\mathrm{tr}\boldsymbol{A}$, $\det \boldsymbol{A}$, and $\|\boldsymbol{A}\|$ can uniquely determine the three eigenvalues $\{\lambda_1, \lambda_2, \lambda_3\}$ of a (33)-matrix $\boldsymbol{A}$ (see eqs. (2.68)). Hence, the three invariants $\{\mathrm{tr}\boldsymbol{A}, \det \boldsymbol{A}, \|\boldsymbol{A}\|\}$ are an *invariant basis* in the sense that any invariant can be expressed in terms of them.

A nonsingular nn-matrix $\boldsymbol{T}$ defines a mapping from an (nn)-matrix $\boldsymbol{A}$ to an (nn)-matrix in the form

$$\boldsymbol{A}' = \boldsymbol{T}^\top \boldsymbol{A}\boldsymbol{T}. \tag{2.154}$$

This is a one-to-one and onto mapping and called the *congruence transformation*[17]. The pair (p, q) consisting of the number p of positive eigenvalues and the number q of negative eigenvalues of an (nn)-matrix $\boldsymbol{A}$ is called the *signature* of $\boldsymbol{A}$. Under a congruence transformation, the signature does not change (*Sylvester's law of inertia*). Hence, the rank is also preserved. It follows that a positive definite symmetric matrix is always transformed to a positive definite symmetric matrix; a positive semi-definite symmetric matrix is always transformed to a positive semi-definite matrix of the same rank.

The congruence transformation defined by an orthogonal matrix $\boldsymbol{U}$ coincides with the similarity transformation defined by $\boldsymbol{U}$, and the matrix inner product and the matrix norm are also preserved:

$$(\boldsymbol{U}^\top \boldsymbol{A}\boldsymbol{U}; \boldsymbol{U}^\top \boldsymbol{B}\boldsymbol{U}) = (\boldsymbol{A}; \boldsymbol{B}), \qquad \|\boldsymbol{U}^\top \boldsymbol{A}\boldsymbol{U}\| = \|\boldsymbol{A}\|. \tag{2.155}$$

[16] Similarity transformations define a group of transformations isomorphic to $GL(n)$, the group of nonsingular matrices under multiplication.

[17] Congruence transformations define a group of transformations isomorphic to $GL(n)$, the group of nonsingular matrices under multiplication.

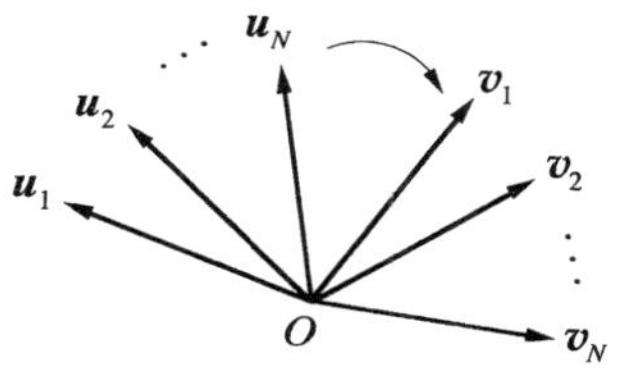

Fig. 2.5. Finding a rotation that maps one set of vectors to another.

2.3.5 *Optimal rotation fitting*

Let $\{\boldsymbol{u}_\alpha\}$ and $\{\boldsymbol{v}_\alpha\}$, $\alpha = 1, ..., N$, be two sets of n-vectors. Consider the problem of finding a rotation $\boldsymbol{R}$ such that

$$\sum_{\alpha=1}^{N} W_\alpha \|\boldsymbol{u}_\alpha - \boldsymbol{R}\boldsymbol{v}_\alpha\|^2 \to \min, \tag{2.156}$$

where W_α are nonnegative weights (Fig. 2.5). Since $\|\boldsymbol{R}\boldsymbol{v}_\alpha\| = \|\boldsymbol{v}_\alpha\|$, the right-hand side can be rewritten as $\sum_{\alpha=1}^{N} W_\alpha \|\boldsymbol{u}_\alpha\|^2 - 2\sum_{\alpha=1}^{N} W_\alpha (\boldsymbol{u}_\alpha, \boldsymbol{R}\boldsymbol{v}_\alpha)$ $+ \sum_{\alpha=1}^{N} W_\alpha \|\boldsymbol{v}_\alpha\|^2$. Hence, if we define the *correlation matrix*

$$\boldsymbol{A} = \sum_{\alpha=1}^{N} W_\alpha \boldsymbol{u}_\alpha \boldsymbol{v}_\alpha^\top, \tag{2.157}$$

the problem can be rewritten as follows (see the second of eqs. (2.143)):

$$(\boldsymbol{A}; \boldsymbol{R}) \to \max. \tag{2.158}$$

This problem can also be viewed as finding a rotation matrix $\boldsymbol{R}$ that is the closest to a given matrix $\boldsymbol{A}$ in the matrix norm:

$$\|\boldsymbol{R} - \boldsymbol{A}\| \to \min. \tag{2.159}$$

In fact, eqs. (2.144) and (2.150) imply that $\|\boldsymbol{R}-\boldsymbol{A}\|^2 = \|\boldsymbol{R}\|^2 - 2(\boldsymbol{R}; \boldsymbol{A}) + \|\boldsymbol{A}\|^2$ $= n - 2(\boldsymbol{A}; \boldsymbol{R}) + \|\boldsymbol{A}\|^2$, so minimizing $\|\boldsymbol{R} - \boldsymbol{A}\|$ is equivalent to maximizing $(\boldsymbol{A}; \boldsymbol{R})$

Let $\boldsymbol{A} = \boldsymbol{V}\boldsymbol{\Lambda}\boldsymbol{U}^\top$ be the singular value decomposition of $\boldsymbol{A}$. The solution of the optimization (2.159) is given by

$$\boldsymbol{R} = \boldsymbol{V}\mathrm{diag}(1, ..., 1, \det(\boldsymbol{V}\boldsymbol{U}^\top))\boldsymbol{U}^\top. \tag{2.160}$$

If the optimization is conducted over orthogonal matrices (i.e., if $\det \boldsymbol{R} = 1$ is not required), the solution is given by

$$\boldsymbol{R} = \boldsymbol{V}\boldsymbol{U}^\top. \tag{2.161}$$

2.4 Matrix and Tensor Algebra

2.4.1 *Direct sum and tensor product*

For an m-vector $\boldsymbol{a} = (a_i)$ and an n-vector $\boldsymbol{b} = (b_i)$, the $(m+n)$-vector $(a_1, ..., a_m, b_1, ..., b_n)^\top$ is called the *direct sum* of $\boldsymbol{a}$ and $\boldsymbol{b}$ and denoted by $\boldsymbol{a} \oplus \boldsymbol{b}$. For an mm-matrix $\boldsymbol{A}$ and an nn-matrix $\boldsymbol{B}$, the $(m+n)(m+n)$-matrix that has $\boldsymbol{A}$ and $\boldsymbol{B}$ as diagonal blocks in that order and zero elements elsewhere is called the *direct sum* of $\boldsymbol{A}$ and $\boldsymbol{B}$ and denoted by $\boldsymbol{A} \oplus \boldsymbol{B}$. Direct sums of more than two vectors or more than two matrices are defined similarly:

$$\boldsymbol{a} \oplus \cdots \oplus \boldsymbol{b} = \begin{pmatrix} \boldsymbol{a} \\ \vdots \\ \boldsymbol{b} \end{pmatrix}, \qquad \boldsymbol{A} \oplus \cdots \oplus \boldsymbol{B} = \begin{pmatrix} \boldsymbol{A} & & \\ & \ddots & \\ & & \boldsymbol{B} \end{pmatrix}. \tag{2.162}$$

Let $\boldsymbol{A}$ be an mm-matrix, and $\boldsymbol{B}$ an nn-matrix. Let $\boldsymbol{u}$ and $\boldsymbol{a}$ be m-vectors, and $\boldsymbol{v}$ and $\boldsymbol{b}$ n-vectors. The following relations are obvious:

$$(\boldsymbol{A} \oplus \boldsymbol{B})(\boldsymbol{u} \oplus \boldsymbol{v}) = (\boldsymbol{A}\boldsymbol{u}) \oplus (\boldsymbol{B}\boldsymbol{v}),$$

$$(\boldsymbol{a} \oplus \boldsymbol{b}, \boldsymbol{u} \oplus \boldsymbol{v}) = (\boldsymbol{a}, \boldsymbol{u}) + (\boldsymbol{b}, \boldsymbol{v}). \tag{2.163}$$

A set of real numbers $\mathcal{T} = (T_{i_1 i_2 \cdots i_r})$, $i_1, i_2, ..., i_r = 1, ..., n$, with r indices running over n-dimensional coordinates is called a *tensor* of *dimension* n and *degree* r. If each index corresponds to coordinates of a different dimensionality, $\mathcal{T}$ is called a tensor of *mixed dimensions* or a *mixed tensor*. If index i_k runs over 1, ..., n_k for $k = 1, ..., r$, the tensor is said to be of *type* $n_1 n_2 \cdots n_r$. A tensor of type $n_1 n_2 \cdots n_r$ is also referred to as an $n_1 n_2 \cdots n_r$-tensor. If $T_{i_1 i_2 \cdots i_r}$ is symmetric with respect to indices i_k and i_{k+1}, the type is written as $i_1 \cdots (i_k i_{k+1}) \cdots i_r$; If $T_{i_1 i_2 \cdots i_r}$ is antisymmetric with respect to indices i_k and i_{k+1}, the type is written as $i_1 \cdots [i_k i_{k+1}] \cdots i_r$; Scalars, vectors, and matrices are tensors of degrees 0, 1, and 2, respectively.

The *tensor product* of tensor $\mathcal{A} = (A_{i_1 \cdots i_r})$ of degree r and tensor $\mathcal{B} = (B_{i_1 \cdots i_s})$ of degree s is a tensor $\mathcal{C} = (C_{i_1 \cdots i_{r+s}})$ of degree $r+s$ defined by

$$C_{i_1 \cdots i_{r+s}} = A_{i_1 \cdots i_r} B_{i_1 \cdots i_s}. \tag{2.164}$$

This is symbolically written as

$$\mathcal{C} = \mathcal{A} \otimes \mathcal{B}. \tag{2.165}$$

The following identities hold for scalar c and vectors $\boldsymbol{a}$ and $\boldsymbol{b}$:

$$c \otimes \boldsymbol{u} = c\boldsymbol{u}, \qquad \boldsymbol{a} \otimes \boldsymbol{b} = \boldsymbol{a}\boldsymbol{b}^\top. \tag{2.166}$$

2.4.2 Cast in three dimensions

A. 33-matrices

The elements of a 33-matrix $\boldsymbol{A} = (A_{ij})$ are rearranged into a 9-vector

$$\boldsymbol{a} = \begin{pmatrix} A_{11} \\ A_{12} \\ \vdots \\ A_{33} \end{pmatrix}, \tag{2.167}$$

which can be written as $\boldsymbol{a} = (a_\kappa)$ with

$$a_\kappa = A_{(\kappa-1)\mathrm{div}3+1,(\kappa-1)\mathrm{mod}3+1}. \tag{2.168}$$

The symbols 'div' and 'mod' denote integer division and integer remainder, respectively. Conversely, a 9-vector $\boldsymbol{a} = (a_\kappa)$ is rearranged into a 33-matrix

$$\boldsymbol{A} = \begin{pmatrix} a_1 & a_2 & a_3 \\ a_4 & a_5 & a_6 \\ a_7 & a_8 & a_9 \end{pmatrix}, \tag{2.169}$$

which can be written as $\boldsymbol{A} = (A_{ij})$ with

$$A_{ij} = a_{3(i-1)+j}. \tag{2.170}$$

The above *type transformation* or *cast* is denoted by

$$\boldsymbol{a} = \mathrm{type}_9[\boldsymbol{A}], \qquad \boldsymbol{A} = \mathrm{type}_{33}[\boldsymbol{a}]. \tag{2.171}$$

The norm is preserved by cast:

$$\|\boldsymbol{a}\| = \|\boldsymbol{A}\|. \tag{2.172}$$

The left-hand side designates the vector norm, whereas the right-hand side designates the matrix norm. The cast can be extended to tensors:

- A 3333-tensor $\mathcal{T} = (T_{ijkl})$ is cast, by rearranging the elements with respect to the indices i and j, into a mixed tensor ${}^*\mathcal{T} = ({}^*T_{\kappa kl})$ of type 933, which is denoted by $\mathrm{type}_{933}[\mathcal{T}]$; the inverse cast is $\mathcal{T} = \mathrm{type}_{3333}[{}^*\mathcal{T}]$.
- A 3333-tensor $\mathcal{T} = (T_{ijkl})$ is cast into a tensor $\mathcal{T}^* = (T^*_{ij\kappa})$ of type 339, which is denoted by $\mathrm{type}_{339}[\mathcal{T}]$; the inverse cast is $\mathcal{T} = \mathrm{type}_{3333}[\mathcal{T}^*]$.
- If both operations are applied, $\mathcal{T} = (T_{ijkl})$ is cast into a 99-matrix $\boldsymbol{T} = (T_{\kappa\lambda})$, which is denoted by $\mathrm{type}_{99}[\mathcal{T}]$; the inverse cast is $\mathcal{T} = \mathrm{type}_{3333}[\boldsymbol{T}]$.

B. (33)-matrices

The elements of a (33)-matrix $\boldsymbol{S} = (S_{ij})$ are rearranged into a 6-vector

$$\boldsymbol{s} = \begin{pmatrix} S_{11} \\ S_{22} \\ S_{33} \\ \sqrt{2}S_{23} \\ \sqrt{2}S_{31} \\ \sqrt{2}S_{12} \end{pmatrix}. \tag{2.173}$$

Conversely, a 6-vector $\boldsymbol{s} = (s_\kappa)$ is rearranged into a (33)-matrix

$$\boldsymbol{S} = \frac{1}{\sqrt{2}} \begin{pmatrix} \sqrt{2}s_1 & s_6 & s_5 \\ s_6 & \sqrt{2}s_2 & s_4 \\ s_5 & s_4 & \sqrt{2}s_3 \end{pmatrix}. \tag{2.174}$$

This cast is denoted by

$$\boldsymbol{s} = \mathrm{type}_6[\boldsymbol{S}], \qquad \boldsymbol{S} = \mathrm{type}_{(33)}[\boldsymbol{s}]. \tag{2.175}$$

The norm is preserved by cast:

$$\|\boldsymbol{s}\| = \|\boldsymbol{S}\|. \tag{2.176}$$

The cast can be extended to tensors:

- A (33)33-tensor $\mathcal{L} = (L_{ijkl})$ is cast, by rearranging the elements with respect to the indices i and j, into a mixed tensor ${}^*\mathcal{L} = ({}^*L_{\kappa kl})$ of type 633, which is denoted by $\mathrm{type}_{633}[\mathcal{L}]$; the inverse cast is $\mathcal{L} = \mathrm{type}_{(33)33}[{}^*\mathcal{L}]$.
- A 33(33)-tensor $\mathcal{N} = (S_{ijkl})$ is cast to a mixed tensor $\mathcal{N}^* = (S^*_{ij\kappa})$ of type 336, which is denoted by $\mathrm{type}_{336}[\mathcal{N}]$; the inverse cast is $\mathcal{N} = \mathrm{type}_{33(33)}[\mathcal{N}^*]$.
- If both operations are applied, a (33)(33)-tensor $\mathcal{M} = (M_{ijkl})$ is cast to a 66-matrix $\boldsymbol{M} = (M_{\kappa\lambda})$, which is denoted by $\mathrm{type}_{66}[\mathcal{M}]$. In elements,

$$\boldsymbol{M} = \begin{pmatrix} M_{1111} & M_{1122} & M_{1133} & \sqrt{2}M_{1123} & \sqrt{2}M_{1131} & \sqrt{2}M_{1112} \\ M_{2211} & M_{2222} & M_{2233} & \sqrt{2}M_{2223} & \sqrt{2}M_{2231} & \sqrt{2}M_{2212} \\ M_{3311} & M_{3322} & M_{3333} & \sqrt{2}M_{3323} & \sqrt{2}M_{3331} & \sqrt{2}M_{3312} \\ \sqrt{2}M_{2311} & \sqrt{2}M_{2322} & \sqrt{2}M_{2333} & 2M_{2323} & 2M_{2331} & 2M_{2312} \\ \sqrt{2}M_{3111} & \sqrt{2}M_{3122} & \sqrt{2}M_{3133} & 2M_{3123} & 2M_{3131} & 2M_{3112} \\ \sqrt{2}M_{1211} & \sqrt{2}M_{1222} & \sqrt{2}M_{1233} & 2M_{1223} & 2M_{1231} & 2M_{1212} \end{pmatrix}. \tag{2.177}$$

The inverse cast is $\mathcal{M} = \mathrm{type}_{(33)(33)}[\boldsymbol{M}]$.

C. [33]-matrices

The elements of a [33]-matrix $\boldsymbol{W} = (W_{ij})$ are rearranged into a 3-vector

$$\boldsymbol{w} = \begin{pmatrix} W_{32} \\ W_{13} \\ W_{21} \end{pmatrix}, \tag{2.178}$$

which can be written as $\boldsymbol{w} = (w_\kappa)$ with

$$w_\kappa = -\frac{1}{2}\sum_{i,j=1}^{3} \epsilon_{\kappa ij} W_{ij}. \tag{2.179}$$

Conversely, a 3-vector $\boldsymbol{w} = (w_\kappa)$ is rearranged into a [33]-matrix

$$\boldsymbol{W} = \begin{pmatrix} 0 & -w_3 & w_2 \\ w_3 & 0 & -w_1 \\ -w_2 & w_1 & 0 \end{pmatrix} = \boldsymbol{w} \times \boldsymbol{I}, \tag{2.180}$$

which can be written as $\boldsymbol{W} = (W_{ij})$ with

$$W_{ij} = -\sum_{k=1}^{3} \epsilon_{ij\kappa} w_\kappa. \tag{2.181}$$

This cast is denoted by

$$\boldsymbol{w} = \mathrm{type}_3[\boldsymbol{W}], \qquad \boldsymbol{W} = \mathrm{type}_{[33]}[\boldsymbol{w}]. \tag{2.182}$$

The following identities hold, where $\boldsymbol{r}$ is an arbitrary 3-vector:

$$\|\boldsymbol{W}\| = \sqrt{2}\|\boldsymbol{w}\|, \qquad \boldsymbol{W}\boldsymbol{r} = \boldsymbol{w} \times \boldsymbol{r}. \tag{2.183}$$

The cast can be extended to tensors:

- A [33]33-tensor $\mathcal{P} = (P_{ijkl})$ is cast, by rearranging the elements with respect to the indices i and j, into a mixed tensor ${}^*\mathcal{P} = ({}^*P_{\kappa kl})$ of type 333, which is denoted by $\mathrm{type}_{333}[\mathcal{P}]$; the inverse cast is $\mathcal{P} = \mathrm{type}_{[33]33}[{}^*\mathcal{P}]$.
- A 33[33]-tensor $\mathcal{Q} = (Q_{ijkl})$ is cast to a mixed tensor $\mathcal{Q}^* = (Q^*_{ij\kappa})$ of type 333, which is denoted by $\mathrm{type}_{333}[\mathcal{Q}]$; the inverse cast is $\mathcal{Q} = \mathrm{type}_{33[33]}[\mathcal{Q}^*]$.
- If both operations are applied, a [33][33]-tensor $\mathcal{R} = (R_{ijkl})$ is cast to a 33-matrix $\boldsymbol{R} = (R_{\kappa\lambda})$, which is denoted by $\mathrm{type}_{33}[\mathcal{R}]$. In elements,

$$\boldsymbol{R} = \begin{pmatrix} R_{3232} & R_{3213} & R_{3221} \\ R_{1332} & R_{1313} & R_{1321} \\ R_{2132} & R_{2113} & R_{2121} \end{pmatrix}. \tag{2.184}$$

The inverse cast is $\mathcal{R} = \mathrm{type}_{[33][33]}[\boldsymbol{R}]$.

2.4.3 Linear mapping of matrices in three dimensions

A. 33-matrices

A 3333-tensor $\mathcal{T} = (T_{ijkl})$ defines a linear mapping from a 33-matrix to a 33-matrix: matrix $\boldsymbol{A} = (A_{ij})$ is mapped to matrix $\boldsymbol{A}' = (A'_{ij})$ in the form

$$A'_{ij} = \sum_{k,l=1}^{3} T_{ijkl}A_{kl}. \tag{2.185}$$

This mapping is denoted by

$$\boldsymbol{A}' = \mathcal{T}\boldsymbol{A}. \tag{2.186}$$

The identity mapping $\mathcal{I} = (I_{ijkl})$ is given by

$$I_{ijkl} = \delta_{ik}\delta_{jl}. \tag{2.187}$$

The similarity transformation $\boldsymbol{A}' = \boldsymbol{T}^{-1}\boldsymbol{A}\boldsymbol{T}$ defined by a nonsingular matrix $\boldsymbol{T} = (T_{ij})$ maps a 33-matrix $\boldsymbol{A}$ to a 33-matrix (see eq. (2.151)). This mapping can be written as $\boldsymbol{A}' = \mathcal{T}\boldsymbol{A}$, where the tensor $\mathcal{T} = (T_{ijkl})$ is defined by

$$T_{ijkl} = T^{-1}_{ik}T_{lj}. \tag{2.188}$$

Here, T^{-1}_{ik} denotes the (ik) element of $\boldsymbol{T}^{-1}$.

If a 3333-tensor $\mathcal{T}$ is cast into a 99-matrix $\boldsymbol{T}$ and if 33-matrices $\boldsymbol{A}$ and $\boldsymbol{A}'$ are cast into 9-vectors $\boldsymbol{a}$ and $\boldsymbol{a}'$, respectively, the mapping $\boldsymbol{A}' = \mathcal{T}\boldsymbol{A}$ is identified with

$$\boldsymbol{a}' = \boldsymbol{T}\boldsymbol{a}, \tag{2.189}$$

which is a linear mapping from a 9-vector $\boldsymbol{a}$ to a 9-vector $\boldsymbol{a}'$. Hence, the mapping $\mathcal{T}$ is nonsingular if and only if the 99-matrix $\boldsymbol{T}$ obtained by cast is nonsingular. The inverse $\mathcal{T}^{-1}$ of a nonsingular mapping $\mathcal{T}$ is given through the cast:

$$\mathcal{T}^{-1} = \text{type}_{3333}[\text{type}_{99}[\mathcal{T}]^{-1}]. \tag{2.190}$$

If mapping $\mathcal{T}$ is singular, its generalized inverse is also defined through the same cast:

$$\mathcal{T}^{-} = \text{type}_{3333}[\text{type}_{99}[\mathcal{T}]^{-}]. \tag{2.191}$$

A 33-matrix $\boldsymbol{A}$ is an *eigenmatrix* of a 3333-tensor $\mathcal{T}$ for eigenvalue λ if

$$\mathcal{T}\boldsymbol{A} = \lambda\boldsymbol{A}. \tag{2.192}$$

Eigenvalues and eigenmatrices are computed by solving the eigenvalue problem of the (99)-matrix obtained by cast: if $\boldsymbol{T} = \text{type}_{99}[\mathcal{T}]$ and $\boldsymbol{a} = \text{type}_9[\boldsymbol{A}]$, eq. (2.192) reads

$$\boldsymbol{T}\boldsymbol{a} = \lambda\boldsymbol{a}. \tag{2.193}$$

B. (33)-matrices

A (33)(33)-tensor $\mathcal{M} = (M_{ijkl})$ defines a linear mapping from a (33)-matrix to a (33)-matrix: matrix $\boldsymbol{S}$ is mapped to matrix $\boldsymbol{S}' = \mathcal{M}\boldsymbol{S}$ in the form eq. (2.185). The identity mapping $\mathcal{I} = (I_{ijkl})$ is given by

$$I_{ijkl} = \frac{1}{2}(\delta_{ik}\delta_{jl} + \delta_{jk}\delta_{il}). \tag{2.194}$$

The congruence transformation $\boldsymbol{S}' = \boldsymbol{T}^{-1}\boldsymbol{S}\boldsymbol{T}$ defined by a nonsingular 33-matrix $\boldsymbol{T} = (T_{ij})$ maps a (33)-matrix $\boldsymbol{S}$ to a (33)-matrix (see eq. (2.154)). This mapping can be written as $\boldsymbol{S}' = \mathcal{M}\boldsymbol{S}$, where the tensor $\mathcal{M} = (M_{ijkl})$ is defined by

$$M_{ijkl} = \frac{1}{2}(T_{ki}T_{lj} + T_{kj}T_{li}). \tag{2.195}$$

If a (33)(33)-tensor $\mathcal{M}$ is cast into a 66-matrix $\boldsymbol{M}$ and if (33)-matrices $\boldsymbol{S}$ and $\boldsymbol{S}'$ are cast into 6-vectors $\boldsymbol{s}$ and $\boldsymbol{s}'$, respectively, the mapping $\boldsymbol{S}' = \mathcal{M}\boldsymbol{S}$ is identified with

$$\boldsymbol{s}' = \boldsymbol{M}\boldsymbol{s}, \tag{2.196}$$

which is a linear mapping from 6-vector $\boldsymbol{s}$ to 6-vector $\boldsymbol{s}'$. Hence, the mapping $\mathcal{M}$ is nonsingular if and only if the 66-matrix $\boldsymbol{M}$ obtained by cast is nonsingular. The inverse $\mathcal{M}^{-1}$ and the generalized inverse $\mathcal{M}^{-}$ are defined through the cast:

$$\mathcal{M}^{-1} = \text{type}_{(33)(33)}[\text{type}_{66}[\mathcal{M}]^{-1}], \tag{2.197}$$

$$\mathcal{M}^{-} = \text{type}_{(33)(33)}[\text{type}_{66}[\mathcal{M}]^{-}]. \tag{2.198}$$

Eigenvalues and eigenmatrices are also defined and computed through the cast.

C. [33]-matrices

If a [33][33]-tensor $\mathcal{R}$ is cast into a 33-matrix $\boldsymbol{R}$ and if [33]-matrices $\boldsymbol{W}$ and $\boldsymbol{W}'$ are cast into 3-vectors $\boldsymbol{w}$ and $\boldsymbol{w}'$, respectively, the mapping $\boldsymbol{W}' = \mathcal{R}\boldsymbol{W}$ is identified with

$$\boldsymbol{w}' = 2\boldsymbol{R}\boldsymbol{w}, \tag{2.199}$$

which is a linear mapping from 3-vector $\boldsymbol{w}$ to 3-vector $\boldsymbol{w}'$. Hence, the mapping $\mathcal{R}$ is nonsingular if and only if the 33-matrix $\boldsymbol{R}$ obtained by cast is nonsingular. The inverse $\mathcal{R}^{-1}$ and the generalized inverse $\mathcal{R}^{-}$ are defined through the cast:

$$\mathcal{R}^{-1} = \frac{1}{4}\text{type}_{[33][33]}[\text{type}_{33}[\mathcal{R}]^{-1}], \tag{2.200}$$

$$\mathcal{R}^{-} = \frac{1}{4}\text{type}_{[33][33]}[\text{type}_{33}[\mathcal{R}]^{-}]. \tag{2.201}$$

Eigenvalues and eigenmatrices are also defined and computed through the cast.

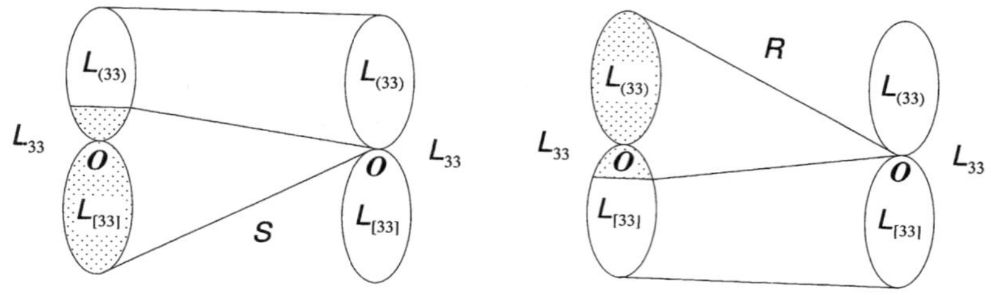

(a) (b)

Fig. 2.6. (a) Linear mapping defined by (33)(33)-tensor $\mathcal{S}$. (a) Linear mapping defined by [33][33]-tensor $\mathcal{R}$.

D. Matrix spaces

The nine-dimensional linear space $\mathcal{L}_{33}$ of all 33-matrices is the direct sum of the six-dimensional subspace $\mathcal{L}_{(33)}$ of all (33)-matrices and the three-dimensional subspace $\mathcal{L}_{[33]}$ of all [33]-matrices (Fig. 2.6). The two subspaces are orthogonal complements of each other (see Section 2.3.4):

$$\mathcal{L}_{33} = \mathcal{L}_{(33)} \oplus \mathcal{L}_{[33]}, \qquad \mathcal{L}_{(33)} \perp \mathcal{L}_{[33]}. \tag{2.202}$$

This is because any 33-matrix $\boldsymbol{A}$ is uniquely decomposed into a (33)-matrix $\boldsymbol{A}_s$ and a [33]-matrix $\boldsymbol{A}_a$:

$$\boldsymbol{A} = \boldsymbol{A}_s + \boldsymbol{A}_a, \qquad (\boldsymbol{A}_s; \boldsymbol{A}_a) = 0, \tag{2.203}$$

$$\boldsymbol{A}_s = \mathcal{S}[\boldsymbol{A}], \qquad \boldsymbol{A}_s = \mathcal{A}[\boldsymbol{A}]. \tag{2.204}$$

Here, the *symmetrization operator* $\mathcal{S}[\,\cdot\,]$ and the *antisymmetrization operator* $\mathcal{A}[\,\cdot\,]$ are defined as follows:

$$\mathcal{S}[\boldsymbol{A}] = \frac{1}{2}(\boldsymbol{A} + \boldsymbol{A}^\top), \qquad \mathcal{A}[\boldsymbol{A}] = \frac{1}{2}(\boldsymbol{A} - \boldsymbol{A}^\top). \tag{2.205}$$

We observe the following:

- If a (33)(33)-tensor $\mathcal{S}$ is viewed as a 3333-tensor, the linear mapping it defines is singular: its null space includes $\mathcal{L}_{[33]}$, and its range is a subspace of $\mathcal{L}_{(33)}$ (Fig. 2.6a). Hence, it always has eigenvalue 0, whose multiplicity is at least 3.

- If a [33][33]-tensor $\mathcal{R}$ is viewed as a 3333-tensor, the linear mapping it defines is also singular: its null space includes $\mathcal{L}_{(33)}$, and its range is a subspace of $\mathcal{L}_{[33]}$ (Fig. 2.6b). Hence, it always has eigenvalue 0, whose multiplicity is at least 6.

Chapter 3

Probabilities and Statistical Estimation

This chapter summarizes mathematical fundamentals of probabilities and statistical estimation. Since the facts established here are directly connected with the analysis in the subsequent chapters, brief derivations are given to most propositions, but those which require lengthy and subtle mathematical arguments are stated without proofs.

3.1 Probability Distributions

3.1.1 Mean, variance, and covariance

Let x be a scalar random variable, and $p(x)$ its probability density defined for real x. The *expectation* $E[x]$ and the *variance* $V[x]$ of x are defined by

$$E[x] = \int_{-\infty}^{\infty} xp(x)dx, \qquad V[x] = \int_{-\infty}^{\infty} (x - E[x])^2 p(x)dx. \tag{3.1}$$

By definition, the variance $V[x]$ is nonnegative. Two random variables x and y are *independent* of each other if their joint probability density $p(x,y)$ has the form $p(x,y) = p_x(x)p_y(y)$. The *covariance* (or *correlation*) of x and y is defined by

$$V[x,y] = E[(x - E[x])(y - E[y])]. \tag{3.2}$$

Random variables x and y are said to be *uncorrelated* if $V[x,y] = 0$. Independent random variables are always uncorrelated, but the converse does not necessarily hold.

Let $\boldsymbol{x}$ be an n-vector random variable, and $p(\boldsymbol{x})$ its probability density defined in the entire n-dimensional space $\mathcal{R}^n$. The expectation $E[\boldsymbol{x}]$ and the *variance-covariance matrix* (or simply *covariance matrix*) $V[\boldsymbol{x}]$ of $\boldsymbol{x}$ are defined by

$$E[\boldsymbol{x}] = \int_{\mathcal{R}^n} \boldsymbol{x}p(\boldsymbol{x})d\boldsymbol{x},$$

$$V[\boldsymbol{x}] = \int_{\mathcal{R}^n} (\boldsymbol{x} - E[\boldsymbol{x}])(\boldsymbol{x} - E[\boldsymbol{x}])^\top p(\boldsymbol{x})d\boldsymbol{x}. \tag{3.3}$$

The covariance matrix $V[\boldsymbol{x}]$ is always positive semi-definite, since

$$(\boldsymbol{a}, V[\boldsymbol{x}]\boldsymbol{a}) = E[(\boldsymbol{x} - E[\boldsymbol{x}], \boldsymbol{a})^2] \geq 0 \tag{3.4}$$

for an arbitrary n-vector $\boldsymbol{a}$. The covariance matrix $V[\boldsymbol{x}]$ is diagonal if and only if the components of $\boldsymbol{x}$ are uncorrelated to each other. The *variance-covariance tensor* (or simply *covariance tensor*) $\mathcal{V}[\boldsymbol{X}]$ of an mn-matrix random variable $\boldsymbol{X}$ is an $mnmn$-tensor defined by

$$\mathcal{V}[\boldsymbol{X}] = E[(\boldsymbol{X} - E[\boldsymbol{X}]) \otimes (\boldsymbol{X} - E[\boldsymbol{X}])]. \tag{3.5}$$

This is also positive semi-definite, since

$$(\boldsymbol{A}, \mathcal{V}[\boldsymbol{X}]\boldsymbol{A}) = E[(\boldsymbol{X} - E[\boldsymbol{X}]; \boldsymbol{A})^2] \geq 0 \tag{3.6}$$

for an arbitrary mn-matrix $\boldsymbol{A}$. Quantities $E[x^2]$, $E[\|\boldsymbol{x}\|^2]$, and $E[\|\boldsymbol{X}\|^2]$ are called the *mean squares* of x, $\boldsymbol{x}$, and $\boldsymbol{X}$, respectively. Their square roots are called the *root mean squares* of the respective random variables.

Let X_1, ..., X_N be independent samples (they may be scalars, vectors, or matrices) from a distribution that has mean $E[X]$ and variance or covariance matrix/tensor $V[X]$. Their *average*

$$\bar{X} = \frac{1}{N}\sum_{\alpha=1}^{N} X_\alpha \tag{3.7}$$

has the following expectation and variance or covariance matrix/tensor:

$$E[\bar{X}] = E[X], \qquad V[\bar{X}] = \frac{1}{N}V[X]. \tag{3.8}$$

This implies that for a large number N

$$X \sim E[X] + O(\frac{1}{\sqrt{N}}), \tag{3.9}$$

and hence $X \sim E[X]$ in the asymptotic limit $N \to \infty$, where the symbol "$\sim$" indicates that the statistical behavior is similar. This fact is known as the *law of large numbers*[1].

If $\boldsymbol{x}$ and $\boldsymbol{y}$ are random variables, their direct sum $\boldsymbol{x} \oplus \boldsymbol{y}$ is also a random variable. Its expectation is

$$E[\boldsymbol{x} \oplus \boldsymbol{y}] = E[\boldsymbol{x}] \oplus E[\boldsymbol{y}]. \tag{3.10}$$

The covariance matrix of $\boldsymbol{x} \oplus \boldsymbol{y}$ has the form

$$V[\boldsymbol{x} \oplus \boldsymbol{y}] = \begin{pmatrix} V[\boldsymbol{x}] & V[\boldsymbol{x}, \boldsymbol{y}] \\ V[\boldsymbol{y}, \boldsymbol{x}] & V[\boldsymbol{y}] \end{pmatrix}, \tag{3.11}$$

where

$$V[\boldsymbol{x}, \boldsymbol{y}] = E[(\boldsymbol{x} - E[\boldsymbol{x}])(\boldsymbol{y} - E[\boldsymbol{y}])^\top],$$

[1]The law of large numbers can be stated in many different ways; the precise meaning of the symbol "$\sim$" differs in each case (we omit the details).

$$V[\boldsymbol{y}, \boldsymbol{x}] = E[(\boldsymbol{y} - E[\boldsymbol{y}])(\boldsymbol{x} - E[\boldsymbol{x}])^\top] = V[\boldsymbol{x}, \boldsymbol{y}]^\top. \tag{3.12}$$

If $\boldsymbol{x}$ and $\boldsymbol{y}$ are independent, then $V[\boldsymbol{x}, \boldsymbol{y}] = \boldsymbol{O}$ and $V[\boldsymbol{y}, \boldsymbol{x}] = \boldsymbol{O}$, and hence the covariance matrix of $\boldsymbol{x} \oplus \boldsymbol{y}$ has the form

$$V[\boldsymbol{x} \oplus \boldsymbol{y}] = V[\boldsymbol{x}] \oplus V[\boldsymbol{y}]. \tag{3.13}$$

If $\boldsymbol{x}$ is an n-vector random variable and $\boldsymbol{A}$ is an mn-matrix, then $\boldsymbol{y} = \boldsymbol{A}\boldsymbol{x}$ is an m-vector random variable; its expectation and covariance matrix are

$$E[\boldsymbol{y}] = \boldsymbol{A}E[\boldsymbol{x}], \qquad V[\boldsymbol{y}] = \boldsymbol{A}V[\boldsymbol{x}]\boldsymbol{A}^\top. \tag{3.14}$$

Let $\boldsymbol{y} = \boldsymbol{y}(\boldsymbol{x})$ be an m-vector function of n-vector $\boldsymbol{x}$. If $\boldsymbol{x}$ is a random variable, then $\boldsymbol{y}$ is also a random variable. If we write $\bar{\boldsymbol{x}} = E[\boldsymbol{x}]$ and $\boldsymbol{x} = \bar{\boldsymbol{x}} + \Delta\boldsymbol{x}$, the deviation $\Delta\boldsymbol{x}$ is a random variable of mean $\boldsymbol{0}$. If we write $\bar{\boldsymbol{y}} = E[\boldsymbol{y}]$ and $\boldsymbol{y} = \bar{\boldsymbol{y}} + \Delta\boldsymbol{y}$, we obtain to a first approximation

$$\bar{\boldsymbol{y}} = \boldsymbol{y}(\bar{\boldsymbol{x}}), \qquad \Delta\boldsymbol{y} = \left.\frac{\partial \boldsymbol{y}}{\partial \boldsymbol{x}}\right|_{\bar{\boldsymbol{x}}} \Delta\boldsymbol{x}, \tag{3.15}$$

where $\partial\boldsymbol{y}/\partial\boldsymbol{x}|_{\bar{\boldsymbol{x}}}$ is an mn-matrix whose (ij) element is $\partial y_i/\partial x_j$ evaluated at $\bar{\boldsymbol{x}}$. To a first approximation, the covariance matrix of $\boldsymbol{y}$ can be written as

$$V[\boldsymbol{y}] = \left.\frac{\partial \boldsymbol{y}}{\partial \boldsymbol{x}}\right|_{\bar{\boldsymbol{x}}} V[\boldsymbol{x}] \left.\frac{\partial \boldsymbol{y}}{\partial \boldsymbol{x}}\right|_{\bar{\boldsymbol{x}}}^\top. \tag{3.16}$$

Let $\boldsymbol{x}$ be an n-vector random variable, and let $\boldsymbol{n} = N[\boldsymbol{x}]$. If we write $\bar{\boldsymbol{x}} = E[\boldsymbol{x}]$, $\boldsymbol{x} = \bar{\boldsymbol{x}} + \Delta\boldsymbol{x}$, $\bar{\boldsymbol{n}} = E[\boldsymbol{n}]$, and $\boldsymbol{n} = \bar{\boldsymbol{n}} + \Delta\boldsymbol{n}$, we obtain to a first approximation

$$\bar{\boldsymbol{n}} = N[\bar{\boldsymbol{x}}], \qquad \Delta\boldsymbol{n} = \frac{1}{\|\bar{\boldsymbol{x}}\|}\boldsymbol{P}_{\bar{\boldsymbol{n}}}\Delta\boldsymbol{x}, \tag{3.17}$$

where $\boldsymbol{P}_{\bar{\boldsymbol{n}}}$ is the projection matrix along $\bar{\boldsymbol{n}}$. To a first approximation, the covariance matrix of $\boldsymbol{n}$ can be written as

$$V[\boldsymbol{n}] = \frac{1}{\|\bar{\boldsymbol{x}}\|^2}\boldsymbol{P}_{\bar{\boldsymbol{n}}}V[\boldsymbol{x}]\boldsymbol{P}_{\bar{\boldsymbol{n}}}. \tag{3.18}$$

3.1.2 Geometry of probability distributions

Let $\boldsymbol{x}$ be an n-vector random variable, and let $\bar{\boldsymbol{x}} = E[\boldsymbol{x}]$. The deviation $\Delta\boldsymbol{x} = \boldsymbol{x} - \bar{\boldsymbol{x}}$, often called "error", is an n-vector random variable of mean $\boldsymbol{0}$. We can write

$$\boldsymbol{x} = \bar{\boldsymbol{x}} + \Delta\boldsymbol{x}, \qquad V[\boldsymbol{x}] = E[\Delta\boldsymbol{x}\Delta\boldsymbol{x}^\top]. \tag{3.19}$$

The mean square $E[\|\Delta\boldsymbol{x}\|^2]$ of the error $\Delta\boldsymbol{x}$ is given by the trace of the covariance matrix $V[\boldsymbol{x}]$:

$$E[\|\Delta\boldsymbol{x}\|^2] = \mathrm{tr}V[\boldsymbol{x}]. \tag{3.20}$$

The spectral decomposition of the covariance matrix $V[\boldsymbol{x}]$ has the following form (see eq. (2.62)):

$$V[\boldsymbol{x}] = \sum_{i=1}^{n} \sigma_i^2 \boldsymbol{u}_i \boldsymbol{u}_i^\top, \qquad \sigma_1 \geq \cdots \geq \sigma_n \geq 0. \tag{3.21}$$

The vector $\boldsymbol{u}_1$ indicates the orientation in which the error in $\boldsymbol{x}$ is most likely to occur, and σ_1^2 is the variance in that orientation. In fact, for an arbitrary unit vector $\boldsymbol{u}$

$$V[(\boldsymbol{u}, \boldsymbol{x})] = E[(\boldsymbol{u}, \Delta\boldsymbol{x})^2] = (\boldsymbol{u}, E[\Delta\boldsymbol{x}\Delta\boldsymbol{x}^\top]\boldsymbol{u}) = (\boldsymbol{u}, V[\boldsymbol{x}]\boldsymbol{u}), \tag{3.22}$$

which is maximized by $\boldsymbol{u}_1$; the maximum value is σ_1^2 (see eqs. (2.86)). We can also see that for each i *the eigenvalue* σ_i^2 *indicates the variance of the error* $\Delta\boldsymbol{x}$ *in orientation* $\boldsymbol{u}_i$. Since $\{\boldsymbol{u}_i\}$ is an orthonormal system, the error $\Delta\boldsymbol{x}$ can be expressed in the form

$$\Delta\boldsymbol{x} = \sum_{i=1}^{n} \Delta x_i \boldsymbol{u}_i, \qquad \Delta x_i = (\Delta\boldsymbol{x}, \boldsymbol{u}_i). \tag{3.23}$$

It follows from eq. (3.21) that *the distribution in each orientation is uncorrelated*:

$$E[\Delta x_i \Delta x_j] = (\boldsymbol{u}_i, V[\boldsymbol{x}]\boldsymbol{u}_j) = \sigma_i^2 \delta_{ij}. \tag{3.24}$$

If the distribution of the error $\Delta\boldsymbol{x}$ is *isotropic*, i.e., its occurrence is equally likely in every orientation, we have $\sigma_1^2 = \cdots = \sigma_n^2$. Eq. (2.63) implies that the covariance matrix $V[\boldsymbol{x}]$ has the form

$$V[\boldsymbol{x}] = \frac{\sigma^2}{n}\boldsymbol{I}, \qquad \sigma^2 = E[\|\Delta\boldsymbol{x}\|^2]. \tag{3.25}$$

Let $\{\boldsymbol{u}_1, ..., \boldsymbol{u}_n\}$ be an orthonormal basis of $\mathcal{R}^n$. If the distribution of $\Delta\boldsymbol{x}$ is restricted to the r-dimensional subspace $\mathcal{S} = \{\boldsymbol{u}_1, ..., \boldsymbol{u}_r\}_L \subset \mathcal{R}^n$, the covariance matrix $V[\boldsymbol{x}]$ is singular:

$$V[\boldsymbol{x}] = \sum_{i=1}^{r} \sigma_i^2 \boldsymbol{u}_i \boldsymbol{u}_i^\top. \tag{3.26}$$

The null space of $V[\boldsymbol{x}]$ is the orthogonal complement $\mathcal{N} = \{\boldsymbol{u}_{r+1}, ..., \boldsymbol{u}_n\}_L \subset \mathcal{R}^n$ of $\mathcal{S}$. If the distribution is isotropic in $\mathcal{S}$, the covariance matrix $V[\boldsymbol{x}]$ has the following form (see eqs. (2.50) and (2.51)):

$$V[\boldsymbol{x}] = \frac{\sigma^2}{r}\boldsymbol{P}_{\mathcal{N}}, \qquad \sigma^2 = E[\|\Delta\boldsymbol{x}\|^2]. \tag{3.27}$$

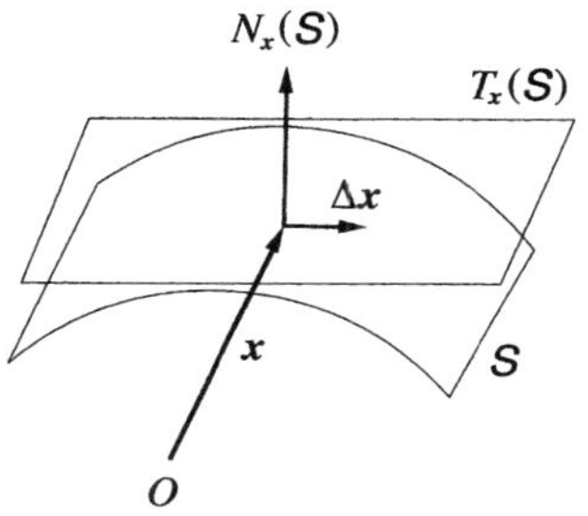

Fig. 3.1. The tangent space $T_{\mathbf{x}}(\mathcal{S})$ and the normal space $T_{\mathbf{x}}(\mathcal{S})$ to $\mathcal{S}$ at $\mathbf{x}$ in three-dimensions.

3.2 Manifolds and Local Distributions

3.2.1 Manifolds and tangent spaces

A (*differential*) *manifold* $\mathcal{S}$ in an n-dimensional space $\mathcal{R}^n$ is a subset of $\mathcal{R}^n$ such that for each point $\boldsymbol{x} \in \mathcal{S}$ there exists a *diffeomorphism* (a smooth differential mapping) between a neighborhood of $\boldsymbol{x}$ in $\mathcal{S}$ and an open subset of $\mathcal{R}^m$ for some m, which is called the *dimension* of the manifold $\mathcal{S}$. A manifold is a generalization of a smooth curve in two dimensions and a smooth surface in three dimensions (the exact definition of a manifold is omitted). An n-dimensional space $\mathcal{R}^n$ is itself an n-dimensional manifold. If an m'-dimensional manifold $\mathcal{S}'$ is a subset of an m-dimensional manifold $\mathcal{S}$, manifold $\mathcal{S}'$ is said to be a *submanifold* of $\mathcal{S}$ of *codimension* $m - m'$. In particular, an m-dimensional manifold in $\mathcal{R}^m$ is a submanifold of $\mathcal{R}^n$ of codimension $n - m$.

Manifolds are often defined by equations[2]. For example, an $(n-1)$-dimensional unit sphere in n dimensions centered at the coordinate origin, denoted by S^{n-1}, is defined by $\|\boldsymbol{x}\|^2 = 1$. An equation $f(\boldsymbol{x}) = 0$ is *nonsingular* if it defines a manifold of codimension 1, and is *singular* otherwise. For example, equation $\|\boldsymbol{x}\|^2 = 0$ is singular[3], because the manifold it defines is the coordinate origin O, which is zero-dimensional (i.e., of codimension n).

Let $\boldsymbol{x}$ be a point in an m-dimensional manifold $\mathcal{S} \subset \mathcal{R}^n$. The set of all infinitesimally small vectors $\Delta\boldsymbol{x} \in \mathcal{R}^n$ such that $\boldsymbol{x} + \Delta\boldsymbol{x} \in \mathcal{S}$ forms an m-dimensional linear space to a first approximation. This linear space is denoted by $T_{\boldsymbol{x}}(\mathcal{S})$ and called the *tangent space*[4] to $\mathcal{S}$ at $\boldsymbol{x}$ (Fig. 3.1). A tangent space is a generalization of a tangent line to a curve in two and three dimensions and a tangent plane to a surface in three dimensions. The orthogonal complement

[2] A manifold is also called an *algebraic variety* if it can be defined by polynomial equations $F^{(k)}(\mathbf{x}) = 0$, $k = 1, ..., L$.

[3] In this book, we always consider *real* spaces; in a complex space, equation $\|\mathbf{x}\|^2 = 0$ defines an $(n-1)$-dimensional *imaginary* surface.

[4] The collection of all $T_{\mathbf{x}}(\mathcal{S})$, $\mathbf{x} \in \mathcal{S}$, is called the *tangent bundle* of $\mathcal{S}$.

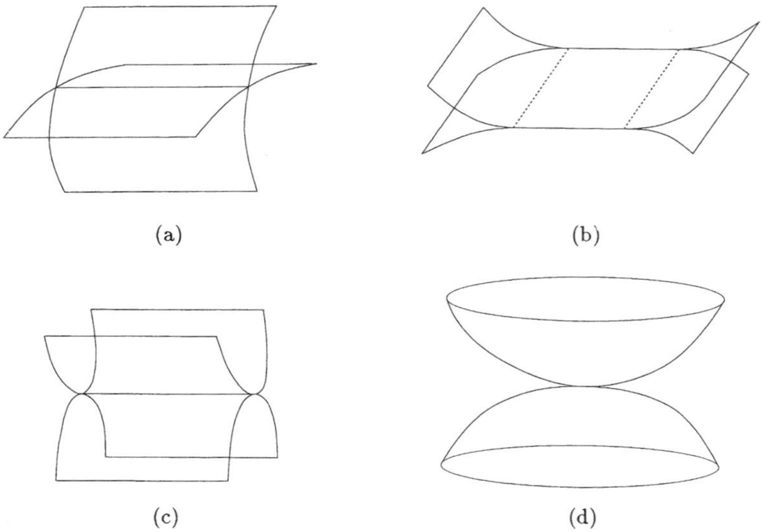

Fig. 3.2. (a) Two surfaces intersecting transversally. (b)–(d) Two surfaces meeting non-transversally.

of $T_{\boldsymbol{x}}(\mathcal{S}) \subset \mathcal{R}^n$, denoted by $N_{\boldsymbol{x}}(\mathcal{S})$, is an $(n-m)$-dimensional linear space and is called the *normal space* to $\mathcal{S}$. If $f(\boldsymbol{x}) = 0$ is a nonsingular equation that defines a manifold of codimension 1, it has an $(n-1)$-dimensional tangent space $T_{\boldsymbol{x}}(\mathcal{S}) = \{\nabla f\}_L^\perp$ and a one-dimensional normal space $N_{\boldsymbol{x}}(\mathcal{S}) = \{\nabla f\}_L$, where $\nabla f = (\partial f/\partial x_1, ..., \partial f/\partial x_n)^\top$.

If an m-dimensional manifold $\mathcal{S}$ and an m'-dimensional manifold $\mathcal{S}'$ meet in $\mathcal{R}^n$ and if $m+m' \geq n$, their intersection $\mathcal{S}\cap\mathcal{S}'$ is in general an $(m+m'-n)$-dimensional manifold. Equivalently put, if a manifold $\mathcal{S}$ of codimension l and a manifold $\mathcal{S}'$ of codimension l' meet in $\mathcal{R}^n$ and if $l+l' \leq n$, their intersection $\mathcal{S}\cap\mathcal{S}'$ is in general a manifold of codimension $l+l'$. If the following condition is satisfied in addition, manifolds $\mathcal{S}$ and $\mathcal{S}'$ are said to intersect *transversally*:

$$T_{\boldsymbol{x}}(\mathcal{S}\cap\mathcal{S}') = T_{\boldsymbol{x}}(\mathcal{S}) \cap T_{\boldsymbol{x}}(\mathcal{S}'). \tag{3.28}$$

For example, two surfaces in three dimensions intersect transversally if they cut each other along a curve (Fig. 3.2a), but they do not if they overlap (Fig. 3.2b) or touch each other along a curve (Fig. 3.2b) or at a single point (Fig. 3.2c). If two manifolds intersect transversally, their intersection is *structurally stable* in the sense that its dimension (or codimension) is preserved if the two manifolds are infinitesimally perturbed in an arbitrary manner.

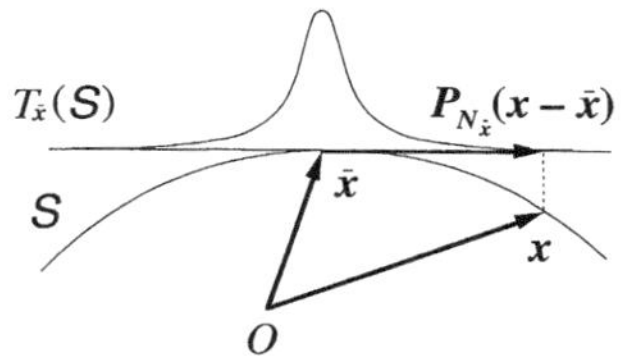

Fig. 3.3. A local distribution.

3.2.2 Local distributions

Let $\boldsymbol{x}$ be an n-vector random variable whose distribution is restricted to an m-dimensional manifold $\mathcal{S} \subset \mathcal{R}^n$. In general, mathematical treatment is very difficult if a random variable is constrained to be in a manifold. For example, if $\boldsymbol{x}$ is a unit 3-vector, its probability density $p(\boldsymbol{x})$ is defined over a unit sphere S^2, but its expectation $E[\boldsymbol{x}] = \int_{S^2} \boldsymbol{x}p(\boldsymbol{x})d\boldsymbol{x}$ is generally inside the sphere. In this book, whenever we consider a probability distribution of a random variable $\boldsymbol{x}$ constrained to be in a manifold $\mathcal{S}$, we assume that $\boldsymbol{x}$ has a *local distribution* in the sense that the distribution is sufficiently localized around some point $\bar{\boldsymbol{x}} \in \mathcal{S}$ and hence to a first approximation the domain of the distribution can be identified with the tangent space $T_{\bar{\boldsymbol{x}}}(\mathcal{S})$ at $\bar{\boldsymbol{x}}$ (Fig. 3.3). We choose the point $\bar{\boldsymbol{x}}$ in such a way that

$$\boldsymbol{P}^{\mathcal{S}}_{\bar{\boldsymbol{x}}} E[\boldsymbol{x} - \bar{\boldsymbol{x}}] = \boldsymbol{0}, \tag{3.29}$$

and identify the covariance matrix of $\boldsymbol{x}$ with

$$V[\boldsymbol{x}] = \boldsymbol{P}^{\mathcal{S}}_{\bar{\boldsymbol{x}}} E[(\boldsymbol{x} - \bar{\boldsymbol{x}})(\boldsymbol{x} - \bar{\boldsymbol{x}})^\top] \boldsymbol{P}^{\mathcal{S}}_{\bar{\boldsymbol{x}}}, \tag{3.30}$$

where $\boldsymbol{P}^{\mathcal{S}}_{\bar{\boldsymbol{x}}}$ is the projection matrix onto the tangent space $T_{\bar{\boldsymbol{x}}}(\mathcal{S})$ at $\bar{\boldsymbol{x}}$. We often call $\bar{\boldsymbol{x}}$ simply "the true value of $\boldsymbol{x}$". Eq. (3.30) implies that the range and the null space of $V[\boldsymbol{x}]$ coincide with the tangent space $T_{\bar{\boldsymbol{x}}}(\mathcal{S})$ and the normal space $N_{\bar{\boldsymbol{x}}}(\mathcal{S})$, respectively. For brevity, we call the null space of the covariance matrix $V[\boldsymbol{x}]$ of $\boldsymbol{x}$ simply "the null space of $\boldsymbol{x}$".

3.2.3 Covariance matrix of a 3-D rotation

Consider a local distribution of three-dimensional rotations[5] around $\bar{\boldsymbol{R}}$. Namely, we regard a three-dimensional rotation matrix $\boldsymbol{R}$ as a random variable perturbed from $\bar{\boldsymbol{R}}$ by a small amount $\Delta\boldsymbol{R}$ in the form $\boldsymbol{R} = \bar{\boldsymbol{R}} + \Delta\boldsymbol{R}$. Since $\boldsymbol{R}$ and $\bar{\boldsymbol{R}}$ are both rotations, the transformation from $\bar{\boldsymbol{R}}$ to $\boldsymbol{R}$ is also

[5] A three-dimensional rotation can also be represented by a 4-vector $\mathbf{q}$, called *quaternion*, constrained to be on a 3-dimensional unit sphere S^3 in four dimensions. Hence, the distribution of three-dimensional rotations can also be thought of as defined over S^3.

a rotation around some axis by a small angle. Since a small rotation has the form given by eq. (2.57), we can write $\boldsymbol{R} = (\boldsymbol{I} + \Delta\Omega \boldsymbol{l} \times \boldsymbol{I} + O(\Delta\Omega^2))\bar{\boldsymbol{R}}$, or

$$\boldsymbol{R} = \bar{\boldsymbol{R}} + \Delta\Omega \boldsymbol{l} \times \bar{\boldsymbol{R}} + O(\Delta\Omega^2). \tag{3.31}$$

Hence, to a first approximation

$$\Delta\boldsymbol{R} = \Delta\Omega \boldsymbol{l} \times \bar{\boldsymbol{R}}. \tag{3.32}$$

We define the *covariance matrix* of rotation $\boldsymbol{R}$ by

$$V[\boldsymbol{R}] = E[\Delta\Omega^2 \boldsymbol{l}\boldsymbol{l}^\top]. \tag{3.33}$$

The unit eigenvector of $V[\boldsymbol{R}]$ for the largest eigenvalue indicates the axis around which perturbation is most likely to occur. The corresponding eigenvalue indicates the mean square of the angle of rotation around that axis. The mean square of the angle of perturbed rotation in total is

$$E[\Delta\Omega^2] = \mathrm{tr}V[\boldsymbol{R}]. \tag{3.34}$$

In particular, if the perturbation is equally likely to occur for every axis orientation, the covariance matrix $V[\boldsymbol{R}]$ has the form

$$V[\boldsymbol{R}] = \frac{\sigma_\Omega^2}{3}\boldsymbol{I}, \qquad \sigma_\Omega^2 = E[\Delta\Omega^2]. \tag{3.35}$$

3.3 Gaussian Distributions and χ^2 Distributions

3.3.1 Gaussian distributions

The most fundamental probability distribution of an n-vector random variable is the multidimensional *Gaussian distribution* (or *normal distribution*). We say that n-vector $\boldsymbol{x}$ is a *Gaussian* random variable if it has a multidimensional Gaussian distribution. If it has mean $\boldsymbol{m}$ and covariance matrix $\boldsymbol{\Sigma}$ of full rank, the probability density has the form

$$p(\boldsymbol{x}) = \frac{1}{\sqrt{(2\pi)^n|\boldsymbol{\Sigma}|}} e^{-(\boldsymbol{x}-\boldsymbol{m}, \boldsymbol{\Sigma}^{-1}(\boldsymbol{x}-\boldsymbol{m}))/2}, \tag{3.36}$$

which defines a distribution over the entire n-dimensional space $\mathcal{R}^n$. It is easy to confirm that

$$\int_{\mathcal{R}^n} p(\boldsymbol{x})d\boldsymbol{x} = 1, \qquad E[\boldsymbol{x}] = \int_{\mathcal{R}^n} \boldsymbol{x}p(\boldsymbol{x})d\boldsymbol{x} = \boldsymbol{m},$$

$$V[\boldsymbol{x}] = \int_{\mathcal{R}^n} (\boldsymbol{x}-\boldsymbol{m})(\boldsymbol{x}-\boldsymbol{m})^\top p(\boldsymbol{x})d\boldsymbol{x} = \boldsymbol{\Sigma}. \tag{3.37}$$

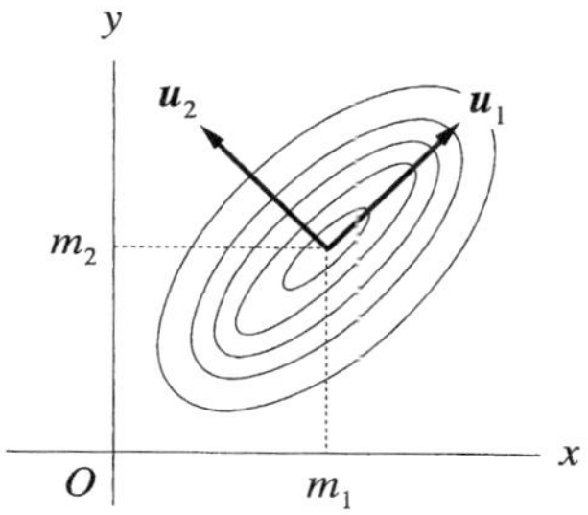

Fig. 3.4. Contours along which probability density is constant.

The probability density is constant on the surface in $\mathcal{R}^n$ defined by

$$(\boldsymbol{x} - \boldsymbol{m}, \boldsymbol{\Sigma}^{-1}(\boldsymbol{x} - \boldsymbol{m})) = c \tag{3.38}$$

for a positive constant c (Fig. 3.4).

Suppose $\boldsymbol{x}$ is decomposed into the form $\boldsymbol{x} = \boldsymbol{x}_1 \oplus \boldsymbol{x}_2$, and let $\boldsymbol{m} = \boldsymbol{m}_1 \oplus \boldsymbol{m}_2$ be the corresponding decomposition of the mean $\boldsymbol{m}$. If $\boldsymbol{x}_1$ and $\boldsymbol{x}_2$ are uncorrelated to each other, the covariance matrix $\boldsymbol{\Sigma}$ is decomposed into the form $\boldsymbol{\Sigma} = \boldsymbol{\Sigma}_1 \oplus \boldsymbol{\Sigma}_2$, where $\boldsymbol{\Sigma}_1$ and $\boldsymbol{\Sigma}_2$ are the covariance matrices of $\boldsymbol{x}_1$ and $\boldsymbol{x}_2$, respectively. Then, eq. (3.36) has the form

$$\begin{aligned} p(\boldsymbol{x}) = & \frac{1}{\sqrt{(2\pi)^n|\boldsymbol{\Sigma}_1|}} e^{-(\boldsymbol{x}_1 - \boldsymbol{m}_1, \boldsymbol{\Sigma}_1{}^{-1}(\boldsymbol{x}_1 - \boldsymbol{m}_1))/2} \\ & \times \frac{1}{\sqrt{(2\pi)^n|\boldsymbol{\Sigma}_2|}} e^{-(\boldsymbol{x}_2 - \boldsymbol{m}_2, \boldsymbol{\Sigma}_2{}^{-1}(\boldsymbol{x}_2 - \boldsymbol{m}_2))/2}. \end{aligned} \tag{3.39}$$

This means that *uncorrelated Gaussian random variables are always independent of each other*.

In one dimension, the probability density reduces to

$$p(x) = \frac{1}{\sqrt{2\pi}\sigma} e^{-(x-m)^2/2\sigma^2}, \tag{3.40}$$

where σ^2 (> 0) is the variance of x (Fig. 3.5). The value σ is called the *standard deviation*. Let us call the pair

$$\{m + \sigma, m - \sigma\} \tag{3.41}$$

the *standard deviation pair*, and the interval

$$[m - \sigma, m + \sigma] \tag{3.42}$$

the *standard confidence interval*. The probability that x falls into the standard confidence interval is about 68.27%. If $m = 0$ and $\sigma = 1$, the distribution is called the *standard Gaussian* (or *normal*) *distribution*.

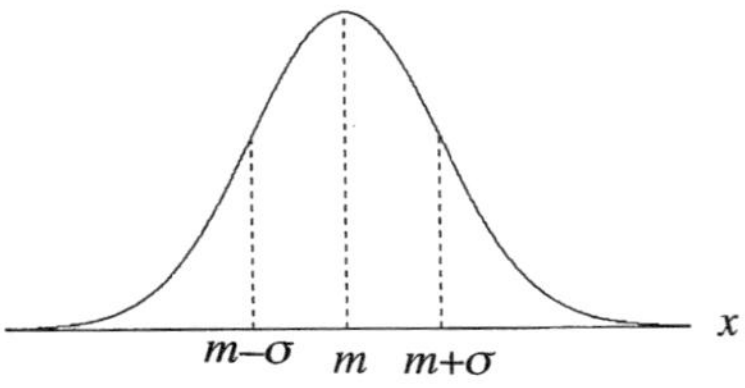

Fig. 3.5. One-dimensional Gaussian distribution.

Gaussian distribution of mean $\boldsymbol{m}$ and covariance matrix $\boldsymbol{\Sigma}$ of rank r ($< n$) is defined as follows. Consider the case $\boldsymbol{m} = \boldsymbol{0}$ first. Let

$$\boldsymbol{\Sigma} = \sum_{i=1}^{n} \sigma_i{}^2 \boldsymbol{u}_i \boldsymbol{u}_i, \qquad \sigma_1 \geq \cdots \geq \sigma_r > \sigma_{r+1} = \cdots = \sigma_n (= 0), \tag{3.43}$$

be the spectral decomposition of $\boldsymbol{\Sigma}$. From the discussion in Section 3.1.2, we see that $\boldsymbol{x}$ cannot deviate in the orientations $\boldsymbol{u}_{r+1}$, ..., $\boldsymbol{u}_n$. Since $\{\boldsymbol{u}_i\}$ is an orthonormal basis of $\mathcal{R}^n$, vector $\boldsymbol{x}$ is expressed in the form

$$\boldsymbol{x} = \sum_{i=1}^{r} x_i \boldsymbol{u}_i, \qquad x_i = (\boldsymbol{x}, \boldsymbol{u}_i). \tag{3.44}$$

It follows that the distribution is limited to the range $\mathcal{R}_{\boldsymbol{\Sigma}} = \{\boldsymbol{u}_1, ..., \boldsymbol{u}_r\}_L$ of the covariance matrix $\boldsymbol{\Sigma}$. The components x_1, ..., x_r have an r-dimensional Gaussian distribution with density

$$p(x_1, ..., x_r) = \frac{1}{\sqrt{(2\pi)^r \prod_{i=1}^{r} \sigma_i}} e^{-\sum_{i=1}^{r} x_i{}^2/2\sigma_i{}^2}, \tag{3.45}$$

which can be rewritten in the form

$$p(\boldsymbol{x}) = \frac{1}{\sqrt{(2\pi)^r |\boldsymbol{\Sigma}|_+}} e^{-(\boldsymbol{x}, \boldsymbol{\Sigma}^- \boldsymbol{x})/2}, \tag{3.46}$$

where $|\boldsymbol{\Sigma}|_+ = \prod_{i=1}^{r} \sigma_i^2$ is the product of all positive eigenvalues of $\boldsymbol{\Sigma}$. Eq. (3.46) defines a probability distribution only in the r-dimensional subspace $\mathcal{R}_{\boldsymbol{\Sigma}}$. Hence,

$$\int_{\mathcal{R}_\Sigma} p(\boldsymbol{x}) d\boldsymbol{x} = 1, \qquad E[\boldsymbol{x}] = \int_{\mathcal{R}_\Sigma} \boldsymbol{x} p(\boldsymbol{x}) d\boldsymbol{x} = \boldsymbol{0},$$

$$V[\boldsymbol{x}] = \int_{\mathcal{R}_\Sigma} \boldsymbol{x}\boldsymbol{x}^\top p(\boldsymbol{x}) d\boldsymbol{x} = \boldsymbol{\Sigma}. \tag{3.47}$$

The Gaussian distribution for $\boldsymbol{m} \neq \boldsymbol{0}$ is defined by replacing $\boldsymbol{x}$ by $\boldsymbol{x} - \boldsymbol{m}$.

If the covariance matrix $\boldsymbol{\Sigma}$ has the spectral decomposition in the form of eq. (3.43), $\boldsymbol{u}_1$ indicates the orientation of the most likely deviation (see Fig. 3.4); σ_1 is the standard deviation in that orientation. Hence, the probability that $-\sigma_1 \leq (\boldsymbol{u}_1, \boldsymbol{x} - \boldsymbol{m}) \leq \sigma_1$ is about 68.27%. Let us call

$$\{\boldsymbol{m} + \sigma_1 \boldsymbol{u}_1, \boldsymbol{m} - \sigma_1 \boldsymbol{u}_1\} \tag{3.48}$$

the *primary deviation pair*, which indicates in which orientation the deviation is most likely to occur.

The Gaussian distribution plays a special role in statistics for many reasons, of which the most important is the fact that if X_1, ..., X_N are independent samples from a distribution of mean zero and variance/covariance matrix $V[X]$, the average $\bar{X} = \sum_{\alpha=1}^{N} X_\alpha / N$ is asymptotically a Gaussian random variable of mean zero and variance/covariance matrix $V[X]/N$ under a mild regularity condition. This fact is known as the *central limit theorem*[6]. Other important properties of the Gaussian distribution include the following:

- If $\boldsymbol{x}$ is an n-vector Gaussian random variable of mean $\boldsymbol{m}$ and covariance matrix $\boldsymbol{\Sigma}$, m-vector $\boldsymbol{y} = \boldsymbol{A}\boldsymbol{x}$ for an arbitrary mn-matrix $\boldsymbol{A}$ is also a Gaussian random variable of mean $\boldsymbol{A}\boldsymbol{m}$ and covariance matrix $\boldsymbol{A}\boldsymbol{\Sigma}\boldsymbol{A}^\top$ (see eqs. (3.14)).

- Each component of a vector Gaussian random variable $\boldsymbol{x}$ is independent and has the standard Gaussian distribution if and only if $\boldsymbol{m} = \boldsymbol{0}$ and $\boldsymbol{\Sigma} = \boldsymbol{I}$.

- If each component of $\boldsymbol{x}$ is independent and has the standard Gaussian distribution, each component of vector $\boldsymbol{y} = \boldsymbol{A}\boldsymbol{x}$ is independent and has the standard Gaussian distribution if and only if $\boldsymbol{A}\boldsymbol{A}^\top = \boldsymbol{I}$.

Since the Gaussian distribution is defined over the entire n-dimensional space $\mathcal{R}^n$ or its linear subspace, the probability tails away infinitely. However, we can define a Gaussian distribution over an arbitrary manifold $\mathcal{S} \subset \mathcal{R}^n$ if the distribution is sufficiently localized around one point $\bar{\boldsymbol{x}} \in \mathcal{S}$ and hence the domain of the distribution can be identified with the tangent space $T_{\bar{\boldsymbol{x}}}(\mathcal{S})$ (see Section 3.2.2). Namely, the distribution can be regarded as *locally Gaussian* if it has a probability density in the form

$$p(\boldsymbol{x}) = Ce^{-(\boldsymbol{x}-\bar{\boldsymbol{x}}, \boldsymbol{\Sigma}^{-}(\boldsymbol{x}-\bar{\boldsymbol{x}}))/2}, \tag{3.49}$$

where C is the normalization constant. The mean $\bar{\boldsymbol{x}}$ and the covariance matrix $\boldsymbol{\Sigma}$ are assumed to satisfy the following relations:

$$\int_{\mathcal{S}} p(\boldsymbol{x}) d\boldsymbol{x} = 1, \qquad \boldsymbol{P}^{\mathcal{S}}_{\bar{\boldsymbol{x}}} \int_{\mathcal{S}} (\boldsymbol{x} - \bar{\boldsymbol{x}}) p(\boldsymbol{x}) d\boldsymbol{x} = \boldsymbol{0},$$

[6] We omit the exact statement of the theorem and the proof.

$$\boldsymbol{P}^{S}_{\bar{\boldsymbol{x}}} \int_{S} (\boldsymbol{x} - \bar{\boldsymbol{x}})(\boldsymbol{x} - \bar{\boldsymbol{x}})^{\top} p(\boldsymbol{x}) d\boldsymbol{x} \boldsymbol{P}^{S}_{\bar{\boldsymbol{x}}} = \boldsymbol{\Sigma}. \tag{3.50}$$

Here, $\boldsymbol{P}^{S}_{\bar{\boldsymbol{x}}}$ is the projection matrix onto the tangent space $T_{\bar{\boldsymbol{x}}}(\mathcal{S})$ at $\bar{\boldsymbol{x}}$.

3.3.2 Moment generating functions and moments

The *moment generating function* of a scalar random variable x is defined by[7]

$$\Phi(\theta) = E[e^{x\theta}] = \sum_{k=0}^{\infty} \frac{E[x^k]}{k!} \theta^k. \tag{3.51}$$

If x is a Gaussian random variable of mean 0 and variance σ^2, its moment generating function has the following form:

$$\Phi(\theta) = e^{\boldsymbol{\Sigma}^2 \theta^2/2} = \sum_{k=0}^{\infty} \frac{\boldsymbol{\Sigma}^{2k}}{2^k k!} \theta^{2k}. \tag{3.52}$$

Comparing this with eq. (3.51) term by term, we obtain the kth *moment* $E[x^k]$ in the following form:

$$E[x^k] = \begin{cases} \dfrac{k!\sigma^k}{2^{k/2}(k/2)!}, & k = 0, 2, 4, 6, ..., \\ 0, & k = 1, 3, 5, 7, ... \end{cases} \tag{3.53}$$

The moment generating function of a vector random variable $\boldsymbol{x}$ is defined by[8]

$$\Phi(\boldsymbol{\theta}) = E[e^{(\boldsymbol{x}, \boldsymbol{\theta})}] = \sum_{k=0}^{\infty} \frac{1}{k!} E[(\boldsymbol{x}, \boldsymbol{\theta})^k], \tag{3.54}$$

where the argument $\boldsymbol{\theta}$ is also an n-vector. If $\boldsymbol{x}$ is a Gaussian random variable of mean $\boldsymbol{0}$ and covariance matrix $\boldsymbol{\Sigma}$, its characteristic function has the following form:

$$\Phi(\boldsymbol{\theta}) = e^{(\boldsymbol{\theta}, \boldsymbol{\Sigma}\boldsymbol{\theta})/2} = \sum_{k=0}^{\infty} \frac{1}{2^k k!} (\boldsymbol{\theta}, \boldsymbol{\Sigma}\boldsymbol{\theta})^k. \tag{3.55}$$

Comparing eqs. (3.54) and (3.55), we obtain the expressions for the (*multidimensional*) *moments* $E[x_{i_1} x_{i_2} \cdots x_{i_r}]$. For example,

$$E[x_i] = 0, \qquad E[x_i x_j] = \Sigma_{ij}, \qquad E[x_i x_j x_k] = 0,$$

$$E[x_i x_j x_k x_l] = \Sigma_{ij}\Sigma_{kl} + \Sigma_{ik}\Sigma_{jl} + \Sigma_{il}\Sigma_{jk}, \qquad E[x_i x_j x_k x_l x_m] = 0. \tag{3.56}$$

---

[7] The function $\Phi(i\theta)$ is called the *characteristic function* (i is the imaginary unit). It is simply the Fourier transform of the probability density $p(x)$ of x.

[8] Function $\Phi(i\theta)$ is also called the *characteristic function*. It is the multidimensional Fourier transform of the probability density $p(\mathrm{x})$ of x.

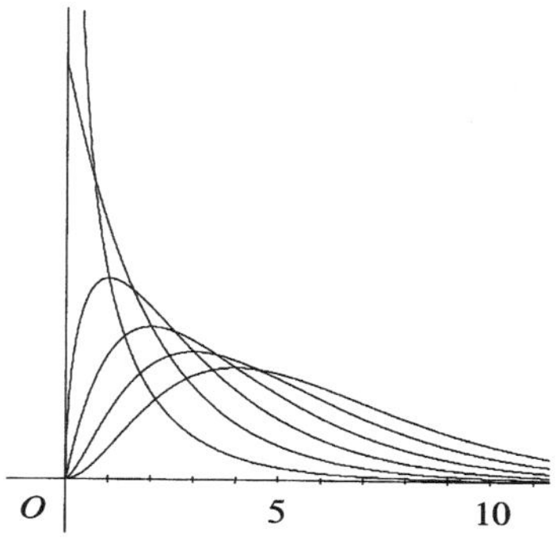

Fig. 3.6. χ^2 distribution with r degrees of freedom.

3.3.3 χ^2 distributions

If x_1, ..., x_r are independent samples from the standard Gaussian distribution, the distribution of

$$R = x_1{}^2 + \cdots + x_r{}^2 \tag{3.57}$$

is called the χ^2 *distribution* with r *degrees of freedom*. We call a random variable which has the χ^2 distribution simply a χ^2 *variable*. Its probability density is defined over $[0, \infty)$ in the form

$$p_r(R) = \frac{1}{2^{r/2}\Gamma(r/2)} R^{r/2-1} e^{-R/2}, \tag{3.58}$$

where $\Gamma(n) = \int_0^\infty t^{n-1}e^{-t}dt$ is the *Gamma function*[9] (Fig. 3.6). The mean and the variance of this distribution are

$$E[R] = r, \qquad V[R] = 2r. \tag{3.59}$$

The density $p_r(R)$ takes its maximum at $R = r - 2$. The important facts concerning the χ^2 distribution include the following:

- If R_1, ..., R_N are independent χ^2 variables with r_1, ..., r_N degrees of freedom, respectively, the sum

$$R = R_1 + \cdots + R_N \tag{3.60}$$

  is a χ^2 variable with $r_1 + \cdots + r_N$ degrees of freedom.

- If $\boldsymbol{x}$ is a Gaussian random variable of mean $\mathbf{0}$ and covariance matrix $\boldsymbol{\Sigma}$ of rank r, the quadratic form

$$R = (\boldsymbol{x}, \boldsymbol{\Sigma}^{-}\boldsymbol{x}) \tag{3.61}$$

[9] $\Gamma(n+1) = n!$ and $\Gamma(n+1/2) = (2n)!\sqrt{\pi}/2^{2n}n!$ for nonnegative integers n.

is a χ^2 variable with r degrees of freedom.

- The probability that a Gaussian random variable of mean $\mathbf{0}$ and covariance matrix $\boldsymbol{\Sigma}$ of rank r satisfies

$$(\boldsymbol{x}, \boldsymbol{\Sigma}^{-}\boldsymbol{x}) \leq 1 \tag{3.62}$$

is equal to[10] $\int_0^1 p_r(R)dR$.

- If $\boldsymbol{x}_\alpha$, $\alpha = 1, ..., N$, are independent Gaussian random variables, each having mean $\mathbf{0}$ and covariance matrix $\boldsymbol{\Sigma}_\alpha$ of rank r_α, the sum

$$R = \sum_{\alpha=1}^{N} (\boldsymbol{x}_\alpha, \boldsymbol{\Sigma}_\alpha^{-}\boldsymbol{x}_\alpha) \tag{3.63}$$

is a χ^2 variable with $\sum_{\alpha=1}^{N} r_\alpha$ degrees of freedom.

- Let n-vector $\boldsymbol{x}$ and m-vector $\boldsymbol{y}$ be Gaussian random variables of mean $\mathbf{0}$, and let $\boldsymbol{\Sigma}_{\boldsymbol{x}}$ and $\boldsymbol{\Sigma}_{\boldsymbol{y}}$ be their respective covariance matrices. Let n and r ($\leq n$) be the ranks of $\boldsymbol{\Sigma}_{\boldsymbol{x}}$ and $\boldsymbol{\Sigma}_{\boldsymbol{y}}$, respectively. If there exists an mn-matrix $\boldsymbol{A}$ such that $\boldsymbol{y} = \boldsymbol{A}\boldsymbol{x}$, the difference

$$R = (\boldsymbol{x}, \boldsymbol{\Sigma}_{\boldsymbol{x}}^{-1}\boldsymbol{x}) - (\boldsymbol{y}, \boldsymbol{\Sigma}_{\boldsymbol{y}}^{-}\boldsymbol{y}) \tag{3.64}$$

is a χ^2 variable with $n - r$ degrees of freedom (*Cochran's theorem*[11]).

3.3.4 Mahalanobis distance and χ^2 test

Let n-vector $\boldsymbol{x}$ be a Gaussian random variable of mean $\mathbf{0}$ and covariance matrix $\boldsymbol{\Sigma}$. If $\boldsymbol{\Sigma}$ is of full rank, we can define a norm[12] of $\boldsymbol{x}$ by

$$\|\boldsymbol{x}\|_{\boldsymbol{\Sigma}} = \sqrt{(\boldsymbol{x}, \boldsymbol{\Sigma}^{-1}\boldsymbol{x})}. \tag{3.65}$$

Equidistant points from the origin in this norm have equal probability densities, and the probability density at $\boldsymbol{x}$ becomes smaller as $\|\boldsymbol{x}\|_{\boldsymbol{\Sigma}}$ becomes larger. The value $\|\boldsymbol{x}\|_{\boldsymbol{\Sigma}}$ is called the *Mahalanobis distance* of $\boldsymbol{x}$ from the origin. If $\boldsymbol{x}$ is randomly chosen, $\|\boldsymbol{x}\|_{\boldsymbol{\Sigma}}^2$ is a χ^2 variable with n degrees of freedom.

If $\boldsymbol{\Sigma}$ has rank r ($< n$), we can define a pseudo-norm[13]

$$\|\boldsymbol{x}\|_{\boldsymbol{\Sigma}} = \sqrt{(\boldsymbol{x}, \boldsymbol{\Sigma}^{-}\boldsymbol{x})}, \tag{3.66}$$

---

[10] In four decimal digits, this equals 0.6827, 0.3935, 0.1987, 0.0902, 0.0374 for $r = 1, 2, 3, 4, 5$, respectively.

[11] To be exact, this is a special case of Cochran's theorem.

[12] For any positive definite symmetric matrix Σ, eq. (3.65) defines a norm in the strict mathematical sense described in Footnote 4 in Section 2.1.1.

[13] This is not a norm in the strict mathematical sense because the triangle inequality (2.9) does not hold; see eq. (3.67).

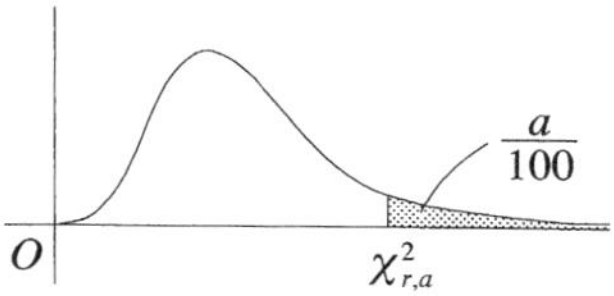

Fig. 3.7. χ^2 test with significance level $a\%$.

which is also called the Mahalanobis distance. Since $\|\boldsymbol{x}\|_{\boldsymbol{\Sigma}} = 0$ for $\boldsymbol{x} \in \mathcal{N}_{\boldsymbol{\Sigma}}$, eq. (3.66) defines a distance in the usual sense only in the range $\mathcal{R}_{\boldsymbol{\Sigma}}$ of $\boldsymbol{\Sigma}$; for $\boldsymbol{x}_1 \in \mathcal{R}_{\boldsymbol{\Sigma}}$ and $\boldsymbol{x}_2 \in \mathcal{N}_{\boldsymbol{\Sigma}}$, we have

$$\|\boldsymbol{x}_1 + \boldsymbol{x}_2\|_{\boldsymbol{\Sigma}} = \|\boldsymbol{x}_1\|_{\boldsymbol{\Sigma}}. \tag{3.67}$$

If $\boldsymbol{x}$ is randomly chosen, $\|\boldsymbol{x}\|^2_{\boldsymbol{\Sigma}}$ is a χ^2 variable with r degrees of freedom.

The χ^2 distribution provides a simple means to *test* hypotheses. In many problems, we can define a random variable R in the form of eq. (3.57), where each x_i may not have zero mean. The expectation of R become the smallest when all x_i have zero means. Suppose all x_i have zero means if and only if some condition is satisfied. This condition is regarded as a *hypothesis* and can be tested by observing a sampled value of R: the hypothesis is rejected if it is very large to an inadmissible degree. An exact procedure is as follows. Let R be a sample from a χ^2 distribution with r degrees of freedom under the hypothesis. The hypothesis is rejected with *significance level* $a\%$ (or with *confidence level* $(100 - a)\%$) if it falls into the *rejection region* $(\chi^2_{r,a}, \infty)$ and is regarded as *acceptable*[14] otherwise (Fig. 3.7). The threshold value $\chi^2_{r,a}$ is called the *$a\%$ significance value* of χ^2 with r degrees of freedom and defined in such a way that[15]

$$\int_{\chi^2_{r,a}}^{\infty} p_r(R)dR = \frac{a}{100}. \tag{3.68}$$

Thus, the hypothesis is rejected with significance level $a\%$ if

$$R > \chi^2_{r,a}. \tag{3.69}$$

This procedure is called the χ^2 *test* and frequently appears in many practical problems—in particular when least-squares optimization based on the Mahalanobis distance is used, since the residual of optimization is usually a χ^2 variable if the noise is Gaussian (see the next section).

[14] Note that we do not say that the hypothesis is *accepted*. Being *acceptable* means that *there exists no evidence strong enough to reject it*.

[15] If r is large, say $r > 30$, the approximation $\chi^2_{r,a} \approx (N_a + \sqrt{2r-1})^2/2$ holds, where the number N_a is defined in such a way that a standard Gaussian random variable falls in the interval (N_a, ∞) with probability $a/100$.

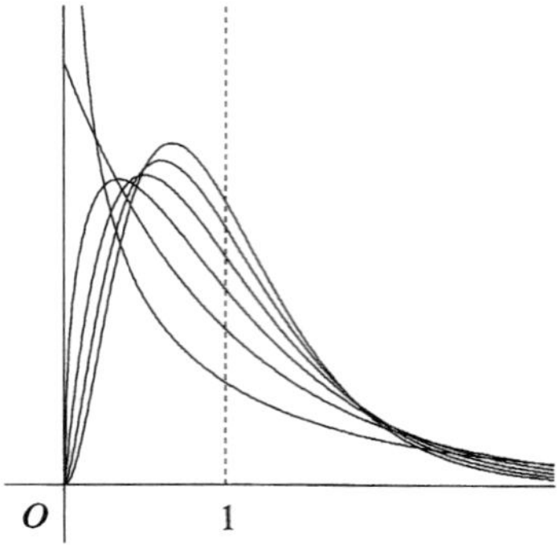

Fig. 3.8. Modified χ^2 distribution with r degrees of freedom.

If R is a χ^2 variable with r degrees of freedom, the distribution of

$$s = \frac{R}{r} \tag{3.70}$$

is sometimes called the *modified* χ^2 *distribution*[16] with r degrees of freedom. The probability density of s is given by $rp_r(rs)$, where $p_r(R)$ is the χ^2 probability density given by eq. (3.58) (Fig. 3.8). Eq. (3.59) implies that its expectation and variance are

$$E[s] = 1, \qquad V[s] = \frac{2}{r}. \tag{3.71}$$

In terms of the modified χ^2 variable s, the χ^2 test given by eq. (3.69) can be rewritten as

$$s > \frac{\chi^2_{r,a}}{r}. \tag{3.72}$$

The χ^2 test usually takes this form when the magnitude of the noise is estimated and compared with its presumed value, as will be shown in later chapters.

3.4 Statistical Estimation for Gaussian Models

3.4.1 Maximum likelihood estimation

Let $\boldsymbol{x}$ be an n-vector, and $\boldsymbol{A}$ an mn-matrix. Let $\boldsymbol{\epsilon}$ be an m-vector Gaussian random variable of mean $\boldsymbol{0}$ and covariance matrix $\boldsymbol{\Sigma}$ of full rank. Then

$$\boldsymbol{y} = \boldsymbol{A}\boldsymbol{x} + \boldsymbol{\epsilon} \tag{3.73}$$

[16] This terminology is not widely used because the only difference from the χ^2 distribution is scaling. However, this distribution plays an essential role in the problems we study in this book, as we will see later.

is an m-vector Gaussian random variable with mean $\boldsymbol{Ax}$ and covariance matrix $\boldsymbol{\Sigma}$. Hence, the probability density of $\boldsymbol{y}$ is given by

$$p(\boldsymbol{y}) = \frac{1}{\sqrt{(2\pi)^n|\boldsymbol{\Sigma}|}} e^{-(\boldsymbol{y}-\boldsymbol{Ax},\boldsymbol{\Sigma}^{-1}(\boldsymbol{y}-\boldsymbol{Ax}))/2}. \tag{3.74}$$

Consider the problem of estimating the parameter $\boldsymbol{x}$ from a sampled value $\boldsymbol{y}$. Namely, we want to find a function $\hat{\boldsymbol{x}}(\boldsymbol{y})$ that gives an estimate of $\boldsymbol{x}$ for a given $\boldsymbol{y}$. Such a function is called an *estimator*. Evidently, any value $\boldsymbol{x}'$ such that $\boldsymbol{Ax}' = \boldsymbol{0}$ can be added to $\boldsymbol{x}$. In order to remove this indeterminacy, we assume that $\boldsymbol{x}$ is constrained to be in $\mathcal{N}_{\boldsymbol{A}}^{\perp}$, the null space of matrix $\boldsymbol{A}$. *Maximum likelihood estimation* seeks the value $\boldsymbol{x}$ that maximizes the probability density $p(\boldsymbol{y})$, which is called the *likelihood* when viewed as a function of the observed value $\boldsymbol{y}$. The problem reduces to minimizing the Mahalanobis distance $\|\boldsymbol{y} - \boldsymbol{Ax}\|_{\boldsymbol{\Sigma}}$, i.e.,

$$J[\boldsymbol{x}] = (\boldsymbol{y} - \boldsymbol{Ax}, \boldsymbol{\Sigma}^{-1}(\boldsymbol{y} - \boldsymbol{Ax})) \to \min \tag{3.75}$$

under the constraint $\boldsymbol{x} \in \mathcal{N}_{\boldsymbol{A}}^{\perp}$. The solution, which is called the *maximum likelihood estimator*, is obtained in the following form (see eqs. (2.136) and (2.137)):

$$\hat{\boldsymbol{x}} = (\boldsymbol{A}^\top \boldsymbol{\Sigma}^{-1} \boldsymbol{A})^- \boldsymbol{A}^\top \boldsymbol{\Sigma}^{-1} \boldsymbol{y}. \tag{3.76}$$

Its expectation and covariance matrix are

$$E[\hat{\boldsymbol{x}}] = \boldsymbol{x}, \qquad V[\hat{\boldsymbol{x}}] = (\boldsymbol{A}^\top \boldsymbol{\Sigma}^{-1} \boldsymbol{A})^-. \tag{3.77}$$

An estimator is *unbiased* if its expectation coincides with the true value. The first of eqs. (3.77) implies that the maximum likelihood estimator $\hat{\boldsymbol{x}}$ is unbiased. The residual $J[\hat{\boldsymbol{x}}]$ of the function $J[\boldsymbol{x}]$ given by eq. (3.75) can be written as follows (see eq. 2.138)):

$$J[\hat{\boldsymbol{x}}] = (\boldsymbol{y}, \boldsymbol{\Sigma}^{-1}\boldsymbol{y}) - (\hat{\boldsymbol{x}}, \boldsymbol{A}^\top \boldsymbol{\Sigma}^{-1} \boldsymbol{A}\hat{\boldsymbol{x}}). \tag{3.78}$$

This is a χ^2 variable with $n - m'$ degrees of freedom, where $m' = \mathrm{rank}\boldsymbol{A}$ (see eq. (3.64)).

If each component of $\boldsymbol{\epsilon}$ distributes independently and isotropically with the same root mean square ϵ, the covariance matrix of $\boldsymbol{\epsilon}$ has the form $V[\boldsymbol{\epsilon}] = \epsilon^2 \boldsymbol{I}$ (see eqs. (3.25)). Hence, eq. (3.75) reduces to the least-squares optimization

$$\|\boldsymbol{y} - \boldsymbol{Ax}\|^2 \to \min, \tag{3.79}$$

and the maximum likelihood estimator $\hat{\boldsymbol{x}}$ is given as follows (see eq. (2.134)):

$$\hat{\boldsymbol{x}} = \boldsymbol{A}^- \boldsymbol{y}. \tag{3.80}$$

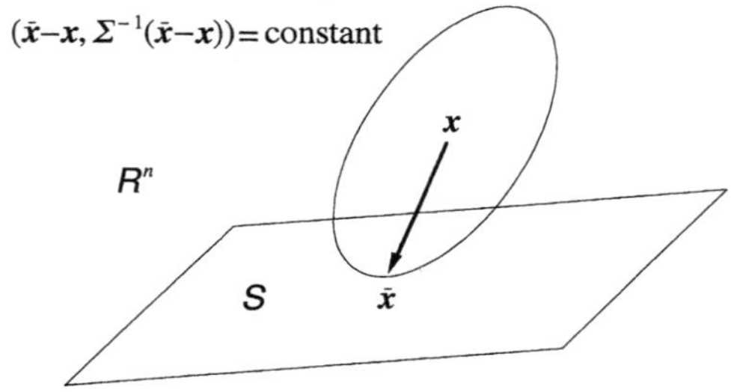

Fig. 3.9. The point $\bar{\mathbf{x}} \in \mathcal{S}$ that minimizes $J[\bar{\mathbf{x}}]$ is the tangent point of the equilikelihood surface to $\mathcal{S}$.

3.4.2 Optimization with linear constraints

Let $\boldsymbol{x}$ be an n-vector Gaussian random variable with an unknown mean $\bar{\boldsymbol{x}}$ and a known covariance matrix $\boldsymbol{\Sigma}$ of full rank. Suppose the mean $\bar{\boldsymbol{x}}$ satisfies a linear constraint $\boldsymbol{A}\boldsymbol{x} = \boldsymbol{b}$. Consider the problem of estimating $\bar{\boldsymbol{x}}$ from a sampled value $\boldsymbol{x}$. Since the probability density of $\boldsymbol{x}$ has the form

$$p(\boldsymbol{x}) = \frac{1}{\sqrt{(2\pi)^n|\boldsymbol{\Sigma}|}} e^{-(\boldsymbol{x}-\bar{\boldsymbol{x}}, \boldsymbol{\Sigma}^{-1}(\boldsymbol{x}-\bar{\boldsymbol{x}}))/2}, \tag{3.81}$$

the maximum likelihood estimator for $\bar{\boldsymbol{x}}$ is obtained by the minimization

$$J[\bar{\boldsymbol{x}}] = (\boldsymbol{x} - \bar{\boldsymbol{x}}, \boldsymbol{\Sigma}^{-1}(\boldsymbol{x} - \bar{\boldsymbol{x}})) \to \min \tag{3.82}$$

under the constraint $\boldsymbol{A}\bar{\boldsymbol{x}} = \boldsymbol{b}$.

In geometric terms, this problem can be interpreted as follows. The constraint $\boldsymbol{A}\boldsymbol{x} = \boldsymbol{b}$ defines an *affine subspace*[17] $\mathcal{S}$ in $\mathcal{R}^n$. The minimization (3.82) means that $\bar{\boldsymbol{x}}$ is a point in $\mathcal{S}$ that has the *shortest Mahalanobis distance* $\|\bar{\boldsymbol{x}} - \bar{\boldsymbol{x}}\|_{\boldsymbol{\Sigma}}$ from the sampled position $\boldsymbol{x}$. Since the equilikelihood surface (the set of all $\bar{\boldsymbol{x}}$ for which $J[\bar{\boldsymbol{x}}]$ = constant) is an "ellipsoid" in $\mathcal{R}^n$, the point $\bar{\boldsymbol{x}}$ that minimizes $J[\bar{\boldsymbol{x}}]$ in $\mathcal{S}$ is the "tangent point" of this ellipsoid to $\mathcal{S}$; all other points in $\mathcal{S}$ should be outside that ellipsoid (Fig. 3.9).

If we let $\Delta\boldsymbol{x} = \boldsymbol{x} - \bar{\boldsymbol{x}}$, eq. (3.82) reduces to the minimization

$$(\Delta\boldsymbol{x}, \boldsymbol{\Sigma}^{-1}\Delta\boldsymbol{x}) \to \min \tag{3.83}$$

under the constraint

$$\boldsymbol{A}\Delta\boldsymbol{x} = \boldsymbol{A}\boldsymbol{x} - \boldsymbol{b}. \tag{3.84}$$

If this constraint is satisfiable, the maximum likelihood estimator $\hat{\boldsymbol{x}}$ $(= \boldsymbol{x} - \Delta\boldsymbol{x})$ is obtained as follows (see eq. (2.140)):

$$\hat{\boldsymbol{x}} = \boldsymbol{x} - \boldsymbol{\Sigma}\boldsymbol{A}^\top(\boldsymbol{A}\boldsymbol{\Sigma}\boldsymbol{A}^\top)^-(\boldsymbol{A}\boldsymbol{x} - \boldsymbol{b}). \tag{3.85}$$

[17] An *affine subspace* of $\mathcal{R}^n$ is a subset of $\mathcal{R}^n$ obtained by translating a linear subspace in $\mathcal{R}^n$.

Its expectation and covariance matrix are

$$E[\hat{\boldsymbol{x}}] = \bar{\boldsymbol{x}}, \qquad V[\hat{\boldsymbol{x}}] = \boldsymbol{\Sigma} - \boldsymbol{\Sigma}\boldsymbol{A}^\top(\boldsymbol{A}\boldsymbol{\Sigma}\boldsymbol{A}^\top)^-\boldsymbol{A}\boldsymbol{\Sigma}. \tag{3.86}$$

Hence, the maximum likelihood estimator $\hat{\boldsymbol{x}}$ is unbiased. The residual $J[\hat{\boldsymbol{x}}]$ of the function $J[\boldsymbol{x}]$ given by eq. (3.82) can be written as

$$J[\hat{\boldsymbol{x}}] = (\boldsymbol{A}\boldsymbol{x} - \boldsymbol{b}, (\boldsymbol{A}\boldsymbol{\Sigma}\boldsymbol{A}^\top)^-(\boldsymbol{A}\boldsymbol{x} - \boldsymbol{b})). \tag{3.87}$$

This is a χ^2 variable with m' degrees of freedom, where $m' = \mathrm{rank}(\boldsymbol{A}\boldsymbol{\Sigma}\boldsymbol{A}^\top)$ $(= \mathrm{rank}\boldsymbol{A})$ (see eq. (3.61)).

3.4.3 Maximum a posteriori probability estimation

Let $\boldsymbol{x}$ be an n-vector, and $\boldsymbol{A}$ an mn-matrix. Let $\boldsymbol{y}$ be an m-vector Gaussian random variable of mean $\bar{\boldsymbol{y}}$ and covariance matrix $\boldsymbol{\Sigma}_{\boldsymbol{y}}$ of full rank. Consider the following linear model:

$$\boldsymbol{z} = \boldsymbol{A}\boldsymbol{x} + \boldsymbol{y}. \tag{3.88}$$

We want to estimate the parameter $\boldsymbol{x}$ from a sampled value of m-vector $\boldsymbol{z}$. Suppose the parameter $\boldsymbol{x}$ has an *a priori distribution* with mean $\bar{\boldsymbol{x}}$ and covariance matrix $\boldsymbol{\Sigma}_{\boldsymbol{x}}$ of full rank. The a priori probability density (or *prior*) of $\boldsymbol{x}$ is

$$p(\boldsymbol{x}) = \frac{1}{\sqrt{(2\pi)^n|\boldsymbol{\Sigma}_{\boldsymbol{x}}|}} e^{-(\boldsymbol{x}-\bar{\boldsymbol{x}}, \boldsymbol{\Sigma}_{\boldsymbol{x}}^{-1}(\boldsymbol{x}-\bar{\boldsymbol{x}}))/2}. \tag{3.89}$$

The probability density of $\boldsymbol{y}$ is

$$p(\boldsymbol{y}) = \frac{1}{\sqrt{(2\pi)^m|\boldsymbol{\Sigma}_{\boldsymbol{y}}|}} e^{-(\boldsymbol{y}-\bar{\boldsymbol{y}}, \boldsymbol{\Sigma}_{\boldsymbol{y}}^{-1}(\boldsymbol{y}-\bar{\boldsymbol{y}}))/2}. \tag{3.90}$$

For a particular value of $\boldsymbol{x}$, the m-vector $\boldsymbol{z}$ defined by eq. (3.88) is a Gaussian random variable of mean $\boldsymbol{A}\boldsymbol{x}+\bar{\boldsymbol{y}}$ and covariance matrix $\boldsymbol{\Sigma}_{\boldsymbol{y}}$. Hence, the *conditional probability density* $p(\boldsymbol{z}|\boldsymbol{x})$ of $\boldsymbol{z}$ conditioned on $\boldsymbol{x}$ is

$$p(\boldsymbol{z}|\boldsymbol{x}) = \frac{1}{\sqrt{(2\pi)^m|\boldsymbol{\Sigma}_{\boldsymbol{y}}|}} e^{-(\boldsymbol{z}-\boldsymbol{A}\boldsymbol{x}-\bar{\boldsymbol{y}}, \boldsymbol{\Sigma}_{\boldsymbol{y}}^{-1}(\boldsymbol{z}-\boldsymbol{A}\boldsymbol{x}-\bar{\boldsymbol{y}}))/2}. \tag{3.91}$$

The *marginal probability density* $p(\boldsymbol{z})$ of $\boldsymbol{z}$ is defined by

$$p(\boldsymbol{z}) = \int p(\boldsymbol{z}|\boldsymbol{x})p(\boldsymbol{x})d\boldsymbol{x}, \tag{3.92}$$

which is computed indirectly as follows. From eq. (3.88), the expectation and the covariance matrix of $\boldsymbol{z}$ are given by

$$E[\boldsymbol{z}] = \boldsymbol{A}\bar{\boldsymbol{x}} + \bar{\boldsymbol{y}}, \qquad V[\boldsymbol{z}] = \boldsymbol{A}\boldsymbol{\Sigma}_{\boldsymbol{x}}\boldsymbol{A}^\top + \boldsymbol{\Sigma}_{\boldsymbol{y}}. \tag{3.93}$$

Hence, the marginal probability density $p(\boldsymbol{z})$ should have the form

$$p(\boldsymbol{z}) = \frac{1}{\sqrt{(2\pi)^m|\boldsymbol{A}\boldsymbol{\Sigma}_{\boldsymbol{x}}\boldsymbol{A}^\top + \boldsymbol{\Sigma}_{\boldsymbol{y}}|}} e^{-(\boldsymbol{z}-\boldsymbol{A}\bar{\boldsymbol{x}}-\bar{\boldsymbol{y}},(\boldsymbol{A}\boldsymbol{\Sigma}_{\boldsymbol{x}}\boldsymbol{A}^\top+\boldsymbol{\Sigma}_{\boldsymbol{y}})^{-1}(\boldsymbol{z}-\boldsymbol{A}\bar{\boldsymbol{x}}-\bar{\boldsymbol{y}}))/2}. \tag{3.94}$$

The *a posteriori probability density* (or *posterior*) $p(\boldsymbol{x}|\boldsymbol{z})$ of $\boldsymbol{x}$ *conditioned on* $\boldsymbol{z}$ is defined by

$$p(\boldsymbol{x}|\boldsymbol{z}) = \frac{p(\boldsymbol{z}|\boldsymbol{x})p(\boldsymbol{x})}{p(\boldsymbol{z})}, \tag{3.95}$$

which is known as the *Bayes formula*. *Maximum a posteriori probability estimation* (often called *Bayes estimation*) seeks the value of $\boldsymbol{x}$ that maximizes the a posteriori probability density $p(\boldsymbol{x}|\boldsymbol{z})$. Maximum likelihood estimation is a special case obtained by setting $p(\boldsymbol{x})$ = constant.

If eqs. (3.89), (3.91), and (3.94) are substituted into eq. (3.95), the a posteriori probability density is obtained in the form

$$p(\boldsymbol{x}|\boldsymbol{z}) = \sqrt{\frac{|\boldsymbol{\Sigma}_{\boldsymbol{x}}^{-1} + \boldsymbol{A}^\top\boldsymbol{\Sigma}_{\boldsymbol{y}}^{-1}\boldsymbol{A}|}{(2\pi)^n}} e^{-(\boldsymbol{x}-\hat{\boldsymbol{x}},(\boldsymbol{\Sigma}_{\boldsymbol{x}}^{-1}+\boldsymbol{A}^\top\boldsymbol{\Sigma}_{\boldsymbol{y}}^{-1}\boldsymbol{A})(\boldsymbol{x}-\hat{\boldsymbol{x}}))/2}, \tag{3.96}$$

where the vector $\hat{\boldsymbol{x}}$ is defined as follows (see the matrix inversion formula (2.22)):

$$\begin{aligned}\hat{\boldsymbol{x}} &= \bar{\boldsymbol{x}} + (\boldsymbol{\Sigma}_{\boldsymbol{x}}^{-1} + \boldsymbol{A}^\top\boldsymbol{\Sigma}_{\boldsymbol{y}}^{-1}\boldsymbol{A})^{-1}\boldsymbol{A}^\top\boldsymbol{\Sigma}_{\boldsymbol{y}}^{-1}(\boldsymbol{z} - (\boldsymbol{A}\bar{\boldsymbol{x}} + \bar{\boldsymbol{y}})) \\ &= \bar{\boldsymbol{x}} + \boldsymbol{\Sigma}_{\boldsymbol{x}}\boldsymbol{A}^\top(\boldsymbol{A}\boldsymbol{\Sigma}_{\boldsymbol{x}}\boldsymbol{A}^\top + \boldsymbol{\Sigma}_{\boldsymbol{y}})^{-1}(\boldsymbol{z} - (\boldsymbol{A}\bar{\boldsymbol{x}} + \bar{\boldsymbol{y}})).\end{aligned} \tag{3.97}$$

Thus, the a posteriori probability density $p(\boldsymbol{x}|\boldsymbol{z})$ defines a Gaussian distribution of $\boldsymbol{x}$. If we write its mean and covariance matrix as $E[\boldsymbol{x}|\boldsymbol{z}]$ and $V[\boldsymbol{x}|\boldsymbol{z}]$, respectively, we have

$$V[\boldsymbol{x}|\boldsymbol{z}] = (\boldsymbol{\Sigma}_{\boldsymbol{x}}^{-1} + \boldsymbol{A}^\top\boldsymbol{\Sigma}_{\boldsymbol{y}}^{-1}\boldsymbol{A})^{-1}, \tag{3.98}$$

$$E[\boldsymbol{x}|\boldsymbol{z}] = \bar{\boldsymbol{x}} + V[\boldsymbol{x}|\boldsymbol{z}]\boldsymbol{A}^\top\boldsymbol{\Sigma}_{\boldsymbol{y}}^{-1}(\boldsymbol{z} - (\boldsymbol{A}\bar{\boldsymbol{x}} + \bar{\boldsymbol{y}})). \tag{3.99}$$

Evidently, $p(\boldsymbol{x}|\boldsymbol{z})$ is maximized by $\boldsymbol{x} = \hat{\boldsymbol{x}}$ $(= E[\boldsymbol{x}|\boldsymbol{z}])$ given by eq. (3.97). We also see that $E[\hat{\boldsymbol{x}}] = \bar{\boldsymbol{x}}$, i.e., the maximum a posteriori probability estimator $\hat{\boldsymbol{x}}$ is unbiased. The uncertainty of $\boldsymbol{x}$ that still remains after $\boldsymbol{z}$ is observed is described by the *a posteriori covariance matrix* $V[\boldsymbol{x}|\boldsymbol{z}]$ given by eq. (3.98).

The marginal probability density $p(\boldsymbol{z})$ in the denominator in the Bayes formula (3.95) does not involve $\boldsymbol{x}$. Hence, maximizing $p(\boldsymbol{x}|\boldsymbol{z})$ is equivalent to maximizing $p(\boldsymbol{z}|\boldsymbol{x})p(\boldsymbol{x})$, which in turn is equivalent to maximizing $\log p(\boldsymbol{x}) + \log p(\boldsymbol{z}|\boldsymbol{x})$. If eqs. (3.89) and (3.91) are substituted into this, the problem can be written in the following form:

$$J[\boldsymbol{x}] = (\boldsymbol{x}-\bar{\boldsymbol{x}}, \boldsymbol{\Sigma}_{\boldsymbol{x}}^{-1}(\boldsymbol{x}-\bar{\boldsymbol{x}})) + (\boldsymbol{z}-\boldsymbol{A}\boldsymbol{x}-\bar{\boldsymbol{y}}, \boldsymbol{\Sigma}_{\boldsymbol{y}}^{-1}(\boldsymbol{z}-\boldsymbol{A}\boldsymbol{x}-\bar{\boldsymbol{y}})) \to \min. \tag{3.100}$$

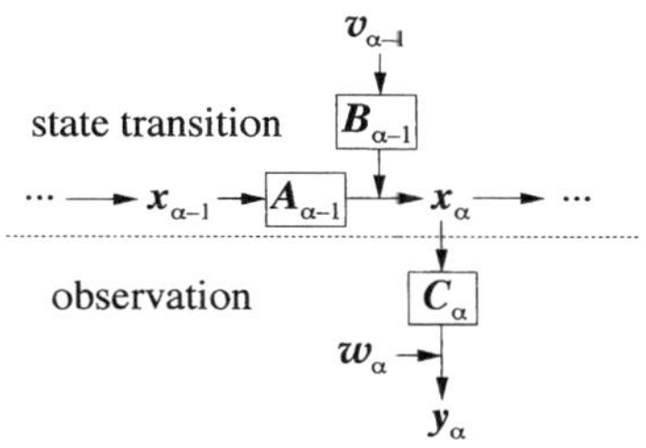

Fig. 3.10. State transition of a linear dynamical system.

Hence, the problem is viewed as minimizing the square sum of the Mahalanobis distances:

$$J[\boldsymbol{x}] = \|\boldsymbol{x} - \bar{\boldsymbol{x}}\|^2_{\boldsymbol{\Sigma}_{\boldsymbol{x}}} + \|\boldsymbol{z} - \boldsymbol{A}\boldsymbol{x} - \bar{\boldsymbol{y}}\|^2_{\boldsymbol{\Sigma}_{\boldsymbol{y}}} \to \min. \tag{3.101}$$

The residual $J[\hat{\boldsymbol{x}}]$ can be written as

$$J[\hat{\boldsymbol{x}}] = (\boldsymbol{z} - \boldsymbol{A}\bar{\boldsymbol{x}} - \bar{\boldsymbol{y}}, (\boldsymbol{A}\boldsymbol{\Sigma}_{\boldsymbol{x}}\boldsymbol{A}^\top + \boldsymbol{\Sigma}_{\boldsymbol{y}})^{-1}(\boldsymbol{z} - \boldsymbol{A}\bar{\boldsymbol{x}} - \bar{\boldsymbol{y}})). \tag{3.102}$$

This is a χ^2 variable with m degrees of freedom. The marginal probability density $p(\boldsymbol{z})$ given by eq. (3.94) has the form

$$p(\boldsymbol{z}) = \text{constant} \times e^{-J[\hat{\boldsymbol{x}}]/2}. \tag{3.103}$$

3.4.4 Kalman filter

The *Kalman filter* is an iterative linear update procedure for maximum a posteriori probability estimation when the parameters to be estimated change as time progresses in the form of a *linear dynamical system*. Let n-vector $\boldsymbol{x}_\alpha$ be the *state vector* at time α, and L-vector $\boldsymbol{y}_\alpha$ the *observation vector* at time α. The process of state transition and observation is described by

$$\boldsymbol{x}_\alpha = \boldsymbol{A}_{\alpha-1}\boldsymbol{x}_{\alpha-1} + \boldsymbol{B}_{\alpha-1}\boldsymbol{v}_{\alpha-1}, \tag{3.104}$$

$$\boldsymbol{y}_\alpha = \boldsymbol{C}_\alpha\boldsymbol{x}_\alpha + \boldsymbol{w}_\alpha, \tag{3.105}$$

where $\boldsymbol{A}_{\alpha-1}$, $\boldsymbol{B}_{\alpha-1}$, and $\boldsymbol{C}_\alpha$ are constant matrices (Fig. 3.10). Vectors $\boldsymbol{v}_\alpha$ and $\boldsymbol{w}_\alpha$ are assumed to be independent Gaussian random variables, and their expectations $E[\boldsymbol{v}_\alpha]$ and $E[\boldsymbol{w}_\alpha]$ and covariance matrices $V[\boldsymbol{v}_\alpha]$ and $V[\boldsymbol{w}_\alpha]$ are assumed to be known. Furthermore, the covariance matrix $V[\boldsymbol{w}_\alpha]$ is assumed to be of full rank.

With this setting, the Kalman filter computes the estimator $\hat{\boldsymbol{x}}_\alpha$ of the state vector $\boldsymbol{x}_\alpha$ at time α and its covariance matrix $V[\hat{\boldsymbol{x}}_\alpha]$ by iterating maximum a posteriori probability estimation. The update rule is derived as follows.

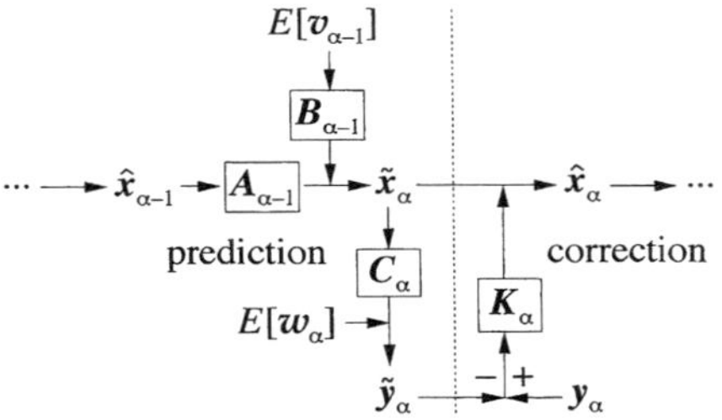

Fig. 3.11. Kalman filter.

Assume that $\boldsymbol{x}_{\alpha-1}$ is a Gaussian random variable of mean $\hat{\boldsymbol{x}}_{\alpha-1}$ and covariance matrix $V[\hat{\boldsymbol{x}}_{\alpha-1}]$. Since eq. (3.104) is linear in $\boldsymbol{x}_{\alpha-1}$ and $\boldsymbol{v}_{\alpha-1}$, the state vector $\boldsymbol{x}_\alpha$ at time α is also a Gaussian random variable. Let $\tilde{\boldsymbol{x}}_\alpha$ and $\tilde{V}[\boldsymbol{x}_\alpha]$ be, respectively, the mean and the covariance matrix of that distribution. They are computed from eqs. (3.88) and (3.93) as follows (read $\boldsymbol{z}$, $\boldsymbol{A}$, $\boldsymbol{x}$, and $\boldsymbol{y}$ in eqs. (3.93) as $\boldsymbol{x}_\alpha$, $\boldsymbol{A}_{\alpha-1}$, $\boldsymbol{x}_{\alpha-1}$, and $\boldsymbol{B}_{\alpha-1}\boldsymbol{v}_{\alpha-1}$, respectively):

$$\tilde{\boldsymbol{x}}_\alpha = \boldsymbol{A}_{\alpha-1}\hat{\boldsymbol{x}}_{\alpha-1} + \boldsymbol{B}_{\alpha-1}E[\boldsymbol{v}_{\alpha-1}], \tag{3.106}$$

$$\tilde{V}[\hat{\boldsymbol{x}}_\alpha] = \boldsymbol{A}_{\alpha-1}V[\hat{\boldsymbol{x}}_{\alpha-1}]\boldsymbol{A}_{\alpha-1}^\top + \boldsymbol{B}_{\alpha-1}V[\boldsymbol{v}_{\alpha-1}]\boldsymbol{B}_{\alpha-1}^\top. \tag{3.107}$$

If the value $\boldsymbol{y}_\alpha$ determined by eq. (3.105) is observed, the a posteriori probability distribution of $\boldsymbol{x}_\alpha$ is also Gaussian. Let $\hat{\boldsymbol{x}}_\alpha$ and $V[\hat{\boldsymbol{x}}_\alpha]$ be, respectively, the mean and the covariance matrix of that distribution. They are computed as follows (read $\boldsymbol{z}$, $\boldsymbol{A}$, $\boldsymbol{x}$, and $\boldsymbol{y}$ in eqs. (3.98), and (3.99) as $\boldsymbol{y}_\alpha$, $\boldsymbol{C}_\alpha$, $\boldsymbol{x}_\alpha$, and $\boldsymbol{w}_\alpha$, respectively):

$$\hat{\boldsymbol{x}}_\alpha = \tilde{\boldsymbol{x}}_\alpha + V[\hat{\boldsymbol{x}}_\alpha]\boldsymbol{C}_\alpha^\top V[\boldsymbol{w}_\alpha]^{-1}(\boldsymbol{y}_\alpha - (\boldsymbol{C}_\alpha\tilde{\boldsymbol{x}}_\alpha + E[\boldsymbol{w}_\alpha])), \tag{3.108}$$

$$V[\hat{\boldsymbol{x}}_\alpha] = \left(\tilde{V}[\hat{\boldsymbol{x}}_\alpha]^{-1} + \boldsymbol{C}_\alpha^\top V[\boldsymbol{w}_\alpha]^{-1}\boldsymbol{C}_\alpha\right)^{-1}. \tag{3.109}$$

Eqs. (3.106)–(3.109) define the Kalman filter for computing $\hat{\boldsymbol{x}}_\alpha$ and $V[\hat{\boldsymbol{x}}_\alpha]$ from $\hat{\boldsymbol{x}}_{\alpha-1}$ and $V[\hat{\boldsymbol{x}}_{\alpha-1}]$.

Eq. (3.108) can also be written in the following form:

$$\hat{\boldsymbol{x}}_\alpha = \tilde{\boldsymbol{x}}_\alpha + \boldsymbol{K}_\alpha(\boldsymbol{y}_\alpha - \tilde{\boldsymbol{y}}_\alpha), \tag{3.110}$$

$$\boldsymbol{K}_\alpha = V[\hat{\boldsymbol{x}}_\alpha]\boldsymbol{C}_\alpha^\top V[\boldsymbol{w}_\alpha]^{-1}, \qquad \tilde{\boldsymbol{y}}_\alpha = \boldsymbol{C}_\alpha\tilde{\boldsymbol{x}}_\alpha + E[\boldsymbol{w}_\alpha]. \tag{3.111}$$

Since $\tilde{\boldsymbol{x}}_\alpha$ and $\tilde{\boldsymbol{y}}_\alpha$ are maximum likelihood estimators of $\boldsymbol{x}_\alpha$ and $\boldsymbol{y}_\alpha$ before the actual value $\boldsymbol{y}_\alpha$ is observed, eq. (3.110) can be viewed as correcting the predicted value $\tilde{\boldsymbol{x}}_\alpha$ by feeding back the difference between the actual observation $\boldsymbol{y}_\alpha$ and its estimator $\tilde{\boldsymbol{y}}_\alpha$ (Fig. 3.11). In this sense, the matrix $\boldsymbol{K}_\alpha$ is often referred to as the *Kalman gain*. The difference $\boldsymbol{y}_\alpha - \tilde{\boldsymbol{y}}_\alpha$ is independent of $\boldsymbol{y}_{\alpha-1}$, $\boldsymbol{y}_{\alpha-2}$, ..., and has mean $\boldsymbol{0}$; it is called the *innovation* of $\boldsymbol{y}_\alpha$.

In the above formulation, the Kalman filter is derived as maximum a posteriori probability estimation on the assumption that all the variables are Gaussian. However, the same Kalman filter can be obtained without assuming Gaussian distributions: if we adopt the criterion of *minimum mean square estimation*, we can obtain eqs. (3.106)–(3.109) by *orthogonal projection*[18] of the state vector onto the affine subspace defined by the observation vector (we omit the details).

3.5 General Statistical Estimation

3.5.1 Score and Fisher information matrix

We now study the problem of statistical estimation in the general case where the distribution of the data or noise is not necessarily Gaussian and the statistical model is not necessarily linear. In abstract terms, the problem is stated as follows. Let $\boldsymbol{x}$ be a random variable that has a probability density $p(\boldsymbol{x};\boldsymbol{\theta})$ parameterized by $\boldsymbol{\theta}$. The problem is estimating the parameter $\boldsymbol{\theta}$ by observing random samples from that distribution.

We assume that $\boldsymbol{x}$ is an n-vector constrained to be in an n'-dimensional manifold $\mathcal{X} \subset \mathcal{R}^n$ and the parameter $\boldsymbol{\theta}$ is an m-vector constrained to be in an m'-dimensional manifold $\mathcal{S} \subset \mathcal{R}^m$. The probability density $p(\boldsymbol{x};\boldsymbol{\theta})$ is assumed to be continuous and continuously differentiable with respect to both $\boldsymbol{x}$ and $\boldsymbol{\theta}$ an arbitrary number of times. We also assume that $p(\boldsymbol{x};\boldsymbol{\theta}) > 0$ for all $\boldsymbol{x} \in \mathcal{X}$. Furthermore, differentiation $\nabla_{\boldsymbol{\theta}}(\,\cdot\,)$ $(= (\partial(\,\cdot\,)/\partial\theta_1, ..., \partial(\,\cdot\,)/\partial\theta_n)^\top)$ with respect to $\boldsymbol{\theta}$ and integration $\int d\boldsymbol{x}$ with respect to $\boldsymbol{x}$ are assumed to be interchangeable for any expression of $p(\boldsymbol{x};\boldsymbol{\theta})$ (as long as the integration exists).

The probability density $p(\boldsymbol{x};\boldsymbol{\theta})$ is not defined for $\boldsymbol{\theta} \notin \mathcal{S}$. For the convenience of analysis, however, we extend it to $\boldsymbol{\theta} \notin \mathcal{S}$ in such a way that $p(\boldsymbol{x};\boldsymbol{\theta}+\Delta\boldsymbol{\theta}) = p(\boldsymbol{x};\boldsymbol{\theta}) + O(\Delta\boldsymbol{\theta})^2$ for all $\boldsymbol{\theta} \in \mathcal{S}$ and $\Delta\boldsymbol{\theta} \in T_{\boldsymbol{\theta}}(\mathcal{S})^\perp$. Here, $T_{\boldsymbol{\theta}}(\mathcal{S})$ is the tangent space to manifold $\mathcal{S}$ at $\boldsymbol{\theta}$ (see Section 3.2.1); $O(\Delta\boldsymbol{\theta})^2$ denotes terms of degree 2 or higher in the components of $\Delta\boldsymbol{\theta}$. In intuitive terms, the probability density is "constant in the normal direction" to $\mathcal{S}$ in $\mathcal{R}^m$. This assumption implies the identity

$$\nabla_{\boldsymbol{\theta}} p(\boldsymbol{x};\boldsymbol{\theta}) \in T_{\boldsymbol{\theta}}(\mathcal{S}). \tag{3.112}$$

Define an m-vector random variable $\boldsymbol{l}$ by

$$\boldsymbol{l} = \nabla_{\boldsymbol{\theta}} \log p(\boldsymbol{x};\boldsymbol{\theta}). \tag{3.113}$$

This vector is called the *score* of $\boldsymbol{x}$ with respect to the parameter $\boldsymbol{\theta}$. Since

[18]Orthogonality is defined in the statistical sense as having no correlation. We omit the details.

$p(\boldsymbol{x};\boldsymbol{\theta})$ is a probability density, the normalization condition

$$\int_{\mathcal{X}} p(\boldsymbol{x};\boldsymbol{\theta})d\boldsymbol{x} = 1 \tag{3.114}$$

holds for any $\boldsymbol{\theta} \in \mathcal{S}$. It follows that if $\boldsymbol{\theta}$ is perturbed into $\boldsymbol{\theta}+\Delta\boldsymbol{\theta}$ in such a way that $\boldsymbol{\theta}+\Delta\boldsymbol{\theta} \in \mathcal{S}$, the first variation of the left-hand side of eq. (3.114) must be 0. The constraint $\boldsymbol{\theta}+\Delta\boldsymbol{\theta} \in \mathcal{S}$ requires $\Delta\boldsymbol{\theta} \in T_{\boldsymbol{\theta}}(\mathcal{S})$ to a first approximation. If we use the *logarithmic differentiation formula*

$$\nabla_{\boldsymbol{\theta}} p(\boldsymbol{x};\boldsymbol{\theta}) = p(\boldsymbol{x};\boldsymbol{\theta})\nabla_{\boldsymbol{\theta}} \log p(\boldsymbol{x};\boldsymbol{\theta}), \tag{3.115}$$

the first variation of the left-hand side of eq. (3.114) is

$$\int_{\mathcal{X}} p(\boldsymbol{x};\boldsymbol{\theta})(\nabla_{\boldsymbol{\theta}} \log p(\boldsymbol{x};\boldsymbol{\theta}), \Delta\boldsymbol{\theta})d\boldsymbol{x} = (E[\boldsymbol{l}], \Delta\boldsymbol{\theta}), \tag{3.116}$$

where we have invoked the assumption that differentiation and integration are interchangeable. Since eq. (3.116) must vanish for an arbitrary $\Delta\boldsymbol{\theta} \in T_{\boldsymbol{\theta}}(\mathcal{S})$, we conclude that $E[\boldsymbol{l}] \in T_{\boldsymbol{\theta}}(\mathcal{S})^{\perp}$. However, eq. (3.112) implies that $E[\boldsymbol{l}] \in T_{\boldsymbol{\theta}}(\mathcal{S})$. It follows that the score $\boldsymbol{l}$ is a random variable of mean $\boldsymbol{0}$:

$$E[\boldsymbol{l}] = \boldsymbol{0}. \tag{3.117}$$

The *Fisher information matrix* is defined by

$$\boldsymbol{J} = E[\boldsymbol{l}\boldsymbol{l}^{\top}]. \tag{3.118}$$

Taking the expectation of the identity

$$\frac{\partial^2 \log p}{\partial\theta_i \partial\theta_j} = -\frac{\partial \log p}{\partial\theta_i}\frac{\partial \log p}{\partial\theta_j} + \frac{1}{p}\frac{\partial^2 p}{\partial\theta_i \partial\theta_j}, \tag{3.119}$$

and noting that differentiation and integration are interchangeable, we obtain

$$\begin{aligned}\int_{\mathcal{X}} \frac{\partial^2 \log p}{\partial\theta_i \partial\theta_j} p d\boldsymbol{x} &= -\int_{\mathcal{X}} \frac{\partial \log p}{\partial\theta_i}\frac{\partial \log p}{\partial\theta_j} p d\boldsymbol{x} + \int_{\mathcal{X}} \frac{\partial^2}{\partial\theta_i \partial\theta_j} p d\boldsymbol{x} \\ &= -\int_{\mathcal{X}} l_i l_j p d\boldsymbol{x} + \frac{\partial^2}{\partial\theta_i \partial\theta_j}\int_{\mathcal{X}} p d\boldsymbol{x} = -E[l_i l_j]. \end{aligned} \tag{3.120}$$

Hence, if an (mm)-matrix $\boldsymbol{L}$ is defined by

$$\boldsymbol{L} = -\nabla^2_{\boldsymbol{\theta}} \log p(\boldsymbol{x};\boldsymbol{\theta}), \tag{3.121}$$

where $\nabla^2_{\boldsymbol{\theta}}(\,\cdot\,)$ denotes a matrix whose (ij) element is $\partial^2(\,\cdot\,)/\partial\theta_i\partial\theta_j$, the Fisher information matrix is expressed in the following form:

$$\boldsymbol{J} = E[\boldsymbol{L}]. \tag{3.122}$$

Since $\boldsymbol{l} \in T_{\boldsymbol{\theta}}(\mathcal{S})$, we have

$$\boldsymbol{P}_{\boldsymbol{\theta}}^{\mathcal{S}} \boldsymbol{J} = \boldsymbol{J} \boldsymbol{P}_{\boldsymbol{\theta}}^{\mathcal{S}} = \boldsymbol{J}, \tag{3.123}$$

where $\boldsymbol{P}_{\boldsymbol{\theta}}^{\mathcal{S}}$ is the projection matrix onto $T_{\boldsymbol{\theta}}(\mathcal{S})$. Hence, the rank of $\boldsymbol{J}$ is at most m'. We say that the distribution $p(\boldsymbol{x}; \boldsymbol{\theta})$ is *regular* if $\boldsymbol{J}$ has rank m, which means that $\boldsymbol{l}$ can take all orientations in $T_{\boldsymbol{\theta}}(\mathcal{S})$ if $\boldsymbol{x}$ is appropriately chosen. In this book, we consider only regular distributions. Since the range of $\boldsymbol{J}$ coincides with $T_{\boldsymbol{\theta}}(\mathcal{S})$ for a regular distribution, the following identity holds:

$$\boldsymbol{J}\boldsymbol{J}^{-} = \boldsymbol{J}^{-}\boldsymbol{J} = \boldsymbol{P}_{\boldsymbol{\theta}}^{\mathcal{S}}. \tag{3.124}$$

If the distribution is locally Gaussian and has the probability density given by eq. (3.49), the score $\boldsymbol{l}$ and the Fisher information matrix $\boldsymbol{J}$ have the following form:

$$\boldsymbol{l} = \boldsymbol{\Sigma}^{-}(\boldsymbol{x} - \bar{\boldsymbol{x}}), \qquad \boldsymbol{L} = \boldsymbol{J} = \boldsymbol{\Sigma}^{-}. \tag{3.125}$$

3.5.2 Unbiased estimator and Cramer-Rao lower bound

Let $\boldsymbol{x} \in \mathcal{X}$ be a sample from a distribution which has a probability density $p(\boldsymbol{x}; \boldsymbol{\theta})$ parameterized by $\boldsymbol{\theta} \in \mathcal{S}$. Let $\hat{\boldsymbol{\theta}}(\boldsymbol{x})$ be an *estimator* of $\boldsymbol{\theta}$, i.e., a function of $\boldsymbol{x}$ that returns an estimate of $\boldsymbol{\theta}$. The estimator $\hat{\boldsymbol{\theta}}(\boldsymbol{x})$ is assumed to satisfy the constraint on $\boldsymbol{\theta}$: $\hat{\boldsymbol{\theta}}(\boldsymbol{x}) \in \mathcal{S}$ for any $\boldsymbol{x} \in \mathcal{X}$. The estimator $\hat{\boldsymbol{\theta}}(\boldsymbol{x})$ is *unbiased* if[19]

$$\boldsymbol{P}_{\boldsymbol{\theta}}^{\mathcal{S}} \int_{\mathcal{X}} (\hat{\boldsymbol{\theta}}(\boldsymbol{x}) - \boldsymbol{\theta}) p(\boldsymbol{x}; \boldsymbol{\theta}) d\boldsymbol{x} = \boldsymbol{0}, \tag{3.126}$$

where $\boldsymbol{P}_{\boldsymbol{\theta}}^{\mathcal{S}}$ is the projection matrix onto $T_{\boldsymbol{\theta}}(\mathcal{S})$ (see eq. (3.29)). The covariance matrix of the estimator $\hat{\boldsymbol{\theta}}(\boldsymbol{x})$ is defined as follows (see eq. (3.30)):

$$V[\hat{\boldsymbol{\theta}}] = \boldsymbol{P}_{\boldsymbol{\theta}}^{\mathcal{S}} \int_{\mathcal{X}} (\hat{\boldsymbol{\theta}}(\boldsymbol{x}) - \boldsymbol{\theta})(\hat{\boldsymbol{\theta}}(\boldsymbol{x}) - \boldsymbol{\theta})^{\top} p(\boldsymbol{x}; \boldsymbol{\theta}) d\boldsymbol{x} \boldsymbol{P}_{\boldsymbol{\theta}}^{\mathcal{S}}. \tag{3.127}$$

Since eq. (3.126) holds identically for any $\boldsymbol{\theta} \in \mathcal{S}$, the first variation of the left-hand side of eq. (3.126) must be $\boldsymbol{0}$ if $\boldsymbol{\theta}$ is perturbed into $\boldsymbol{\theta} + \Delta\boldsymbol{\theta}$ for $\Delta\boldsymbol{\theta} \in T_{\boldsymbol{\theta}}(\mathcal{S})$. If we use the logarithmic differentiation formula (3.115) and interchange differentiation and integration, the first variation of the left-hand side of eq. (3.126) can be written as

$$-\boldsymbol{P}_{\boldsymbol{\theta}}^{\mathcal{S}} \int_{\mathcal{X}} \Delta\boldsymbol{\theta} p(\boldsymbol{x}; \boldsymbol{\theta}) d\boldsymbol{x} + \boldsymbol{P}_{\boldsymbol{\theta}}^{\mathcal{S}} \int_{\mathcal{X}} \hat{\boldsymbol{\theta}}(\boldsymbol{x}) (\nabla_{\boldsymbol{\theta}} \log p, \Delta\boldsymbol{\theta}) p(\boldsymbol{x}; \boldsymbol{\theta}) d\boldsymbol{x}$$
$$= -\Delta\boldsymbol{\theta} + E[\boldsymbol{P}_{\boldsymbol{\theta}}^{\mathcal{S}} \hat{\boldsymbol{\theta}}(\boldsymbol{x}) \boldsymbol{l}^{\top}] \Delta\boldsymbol{\theta}, \tag{3.128}$$

[19] It seems that unbiasedness can be defined by $E[\hat{\theta}] = \theta$, but $E[\hat{\theta}]$ may be outside $\mathcal{S}$ if $\mathcal{S}$ is "curved" (see Section 3.2.2).

where we have employed the approximation $\boldsymbol{P}^{\mathcal{S}}_{\boldsymbol{\theta}+\Delta\boldsymbol{\theta}}E[\hat{\boldsymbol{\theta}}-\boldsymbol{\theta}] \approx \boldsymbol{0}$. The exact expression for $\boldsymbol{P}^{\mathcal{S}}_{\boldsymbol{\theta}+\Delta\boldsymbol{\theta}}E[\hat{\boldsymbol{\theta}}-\boldsymbol{\theta}]$ involves the *second fundamental form* of the manifold $\mathcal{S}$ at $\boldsymbol{\theta}$. Roughly speaking, it has the order of $\sigma^2\Delta\boldsymbol{\theta}/R^2$, where σ is the "standard deviation" of $\hat{\boldsymbol{\theta}}$ in $\mathcal{S}$ and R is the "radius of curvature" of $\mathcal{S}$ (we omit the exact analysis). Here, we simply assume that the manifold $\mathcal{S}$ is sufficiently "flat" in the domain within which the distribution of $\hat{\boldsymbol{\theta}}$ is localized. Throughout this book, this is always assumed whenever we talk about local distributions.

Eq. (3.128) must vanish for an arbitrary $\Delta\boldsymbol{\theta} \in T_{\boldsymbol{\theta}}(\mathcal{S})$. Since $E[\boldsymbol{P}^{\mathcal{S}}_{\boldsymbol{\theta}}\boldsymbol{\theta}\boldsymbol{l}^\top] = \boldsymbol{P}^{\mathcal{S}}_{\boldsymbol{\theta}}\boldsymbol{\theta}E[\boldsymbol{l}]^\top = \boldsymbol{0}$, this condition is written as

$$E[\boldsymbol{P}^{\mathcal{S}}_{\boldsymbol{\theta}}(\hat{\boldsymbol{\theta}}-\boldsymbol{\theta})\boldsymbol{l}^\top]\Delta\boldsymbol{\theta} = \Delta\boldsymbol{\theta}, \qquad \Delta\boldsymbol{\theta} \in \mathcal{S}. \tag{3.129}$$

If we let $\Delta\boldsymbol{\theta} = \boldsymbol{P}^{\mathcal{S}}_{\boldsymbol{\theta}}\Delta\boldsymbol{\epsilon}$, then $\Delta\boldsymbol{\theta} \in \mathcal{S}$ for an arbitrary $\Delta\boldsymbol{\epsilon} \in \mathcal{R}^m$. Hence, the identity

$$E[\boldsymbol{P}^{\mathcal{S}}_{\boldsymbol{\theta}}(\hat{\boldsymbol{\theta}}-\boldsymbol{\theta})(\boldsymbol{P}^{\mathcal{S}}_{\boldsymbol{\theta}}\boldsymbol{l})^\top]\Delta\boldsymbol{\epsilon} = \boldsymbol{P}^{\mathcal{S}}_{\boldsymbol{\theta}}\Delta\boldsymbol{\epsilon} \tag{3.130}$$

must hold identically for $\boldsymbol{\epsilon} \in \mathcal{R}^m$. Since $\boldsymbol{l} \in T_{\boldsymbol{\theta}}(\mathcal{S})$ and hence $\boldsymbol{P}^{\mathcal{S}}_{\boldsymbol{\theta}}\boldsymbol{l} = \boldsymbol{l}$, eq. (3.130) implies

$$E[\boldsymbol{P}^{\mathcal{S}}_{\boldsymbol{\theta}}(\hat{\boldsymbol{\theta}}-\boldsymbol{\theta})\boldsymbol{l}^\top] = \boldsymbol{P}^{\mathcal{S}}_{\boldsymbol{\theta}}. \tag{3.131}$$

Combining this with eqs. (3.127) and (3.118), we obtain the following relationship:

$$E[\begin{pmatrix} \boldsymbol{P}^{\mathcal{S}}_{\boldsymbol{\theta}}(\hat{\boldsymbol{\theta}}-\boldsymbol{\theta}) \\ \boldsymbol{l} \end{pmatrix}\begin{pmatrix} \boldsymbol{P}^{\mathcal{S}}_{\boldsymbol{\theta}}(\hat{\boldsymbol{\theta}}-\boldsymbol{\theta}) \\ \boldsymbol{l} \end{pmatrix}^\top] = \begin{pmatrix} V[\hat{\boldsymbol{\theta}}] & \boldsymbol{P}^{\mathcal{S}}_{\boldsymbol{\theta}} \\ \boldsymbol{P}^{\mathcal{S}}_{\boldsymbol{\theta}} & \boldsymbol{J} \end{pmatrix}. \tag{3.132}$$

Since the left-hand side is positive semi-definite, the following matrix is also positive semi-definite (see Section 2.2.3):

$$\begin{pmatrix} \boldsymbol{P}^{\mathcal{S}}_{\boldsymbol{\theta}} & -\boldsymbol{J}^- \\ & \boldsymbol{J}^- \end{pmatrix}\begin{pmatrix} V[\hat{\boldsymbol{\theta}}] & \boldsymbol{P}^{\mathcal{S}}_{\boldsymbol{\theta}} \\ \boldsymbol{P}^{\mathcal{S}}_{\boldsymbol{\theta}} & \boldsymbol{J} \end{pmatrix}\begin{pmatrix} \boldsymbol{P}^{\mathcal{S}}_{\boldsymbol{\theta}} & \\ -\boldsymbol{J}^- & \boldsymbol{J}^- \end{pmatrix} = \begin{pmatrix} V[\hat{\boldsymbol{\theta}}]-\boldsymbol{J}^- & \\ & \boldsymbol{J}^- \end{pmatrix}. \tag{3.133}$$

Here, we have assumed that the distribution is regular and hence eqs. (3.123) and (3.124) hold. Since $\boldsymbol{J}^-$ is positive semi-definite, the positive semi-definiteness of eq. (3.133) implies

$$V[\hat{\boldsymbol{\theta}}] \succeq \boldsymbol{J}^-, \tag{3.134}$$

where $\boldsymbol{A} \succeq \boldsymbol{B}$ means $\boldsymbol{A}-\boldsymbol{B}$ is positive semi-definite. Eq. (3.134) is called the *Cramer-Rao inequality* and gives a lower bound, called the *Cramer-Rao lower bound*, on the covariance matrix $V[\hat{\boldsymbol{\theta}}]$ of an arbitrary unbiased estimator $\hat{\boldsymbol{\theta}}$. If this bound is attained, the estimator $\hat{\boldsymbol{\theta}}$ is said to be *efficient*.

Suppose N independent samples $\boldsymbol{x}_1, \ldots, \boldsymbol{x}_N$ are observed from a distribution whose probability density is $p(\boldsymbol{x};\boldsymbol{\theta})$. Let $\hat{\boldsymbol{\theta}}(\boldsymbol{x}_1,\ldots,\boldsymbol{x}_N)$ be an estimator of $\boldsymbol{\theta}$. If we consider the direct sum

$$\tilde{\boldsymbol{x}} = \boldsymbol{x}_1 \oplus \cdots \oplus \boldsymbol{x}_N, \tag{3.135}$$

the independence of $\boldsymbol{x}_1, \ldots, \boldsymbol{x}_N$ implies that $\tilde{\boldsymbol{x}}$ has the probability density

$$\tilde{p}(\tilde{\boldsymbol{x}};\boldsymbol{\theta}) = p(\boldsymbol{x}_1;\boldsymbol{\theta})\cdots p(\boldsymbol{x}_N;\boldsymbol{\theta}). \tag{3.136}$$

Since the estimator $\hat{\boldsymbol{\theta}}(\boldsymbol{x}_1,\ldots,\boldsymbol{x}_N)$ can be viewed as an estimator $\hat{\boldsymbol{\theta}}(\tilde{\boldsymbol{x}})$ of $\tilde{\boldsymbol{x}}$, the argument described earlier can be applied. The score of $\tilde{\boldsymbol{x}}$ is

$$\tilde{\boldsymbol{l}} = \nabla_{\boldsymbol{\theta}} \log \tilde{p}(\tilde{\boldsymbol{x}};\boldsymbol{\theta}) = \nabla_{\boldsymbol{\theta}} \sum_{\alpha=1}^{N} \log p(\boldsymbol{x}_\alpha;\boldsymbol{\theta}) = \sum_{\alpha=1}^{N} \boldsymbol{l}_\alpha, \tag{3.137}$$

where $\boldsymbol{l}_\alpha$ is the score of $\boldsymbol{x}_\alpha$. Since $\{\boldsymbol{x}_\alpha\}$ are independent, the Fisher information matrix of $\tilde{\boldsymbol{x}}$ is

$$\tilde{\boldsymbol{J}} = E[\left(\sum_{\alpha=1}^{N} \boldsymbol{l}_\alpha\right)\left(\sum_{\beta=1}^{N} \boldsymbol{l}_\beta\right)^\top] = \sum_{\alpha,\beta=1}^{N} E[\boldsymbol{l}_\alpha \boldsymbol{l}_\beta^\top] = N\boldsymbol{J}, \tag{3.138}$$

where $\boldsymbol{J}$ is the Fisher information matrix for $p(\boldsymbol{x};\boldsymbol{\theta})$. Consequently, the Cramer-Rao lower bound is given in the following form:

$$V[\hat{\boldsymbol{\theta}}] \succeq \frac{1}{N}\boldsymbol{J}^-. \tag{3.139}$$

In particular, if $\boldsymbol{\theta}$ is a scalar, we have

$$V[\hat{\boldsymbol{\theta}}] \geq \frac{1}{NE[(\partial \log p/\partial\theta)^2]} = -\frac{1}{NE[\partial^2 \log p/\partial\theta^2]}. \tag{3.140}$$

3.6 Maximum Likelihood Estimation

3.6.1 *Maximum likelihood estimator and the exponential family*

Given a sample $\boldsymbol{x} \in \mathcal{X}$ from a distribution which has a probability density $p(\boldsymbol{x};\boldsymbol{\theta})$ parameterized by $\boldsymbol{\theta} \in \mathcal{S}$, the *maximum likelihood estimator* $\hat{\boldsymbol{\theta}}$ is the value of $\boldsymbol{\theta}$ that maximizes the *likelihood* $p(\boldsymbol{x};\boldsymbol{\theta})$, i.e., the probability density viewed as a function of $\boldsymbol{\theta}$ by substituting the sampled value. Hence, the maximum likelihood estimator $\hat{\boldsymbol{\theta}}$ is the solution of the minimization

$$J = -2\log p(\boldsymbol{x};\boldsymbol{\theta}) \rightarrow \min. \tag{3.141}$$

The probability density $p(\boldsymbol{x};\boldsymbol{\theta})$ is assumed to have the properties described in preceding section.

In order to distinguish variables from their true values, we regard $\boldsymbol{\theta}$ as a variable and write its true value as $\bar{\boldsymbol{\theta}}$. With the expectation that the maximum likelihood estimator $\hat{\boldsymbol{\theta}}$ is close to $\bar{\boldsymbol{\theta}}$, we write

$$\boldsymbol{\theta} = \bar{\boldsymbol{\theta}} + \Delta\boldsymbol{\theta}. \tag{3.142}$$

The constraint $\boldsymbol{\theta} \in \mathcal{S}$ requires $\Delta\boldsymbol{\theta} \in T_{\bar{\boldsymbol{\theta}}}(\mathcal{S})$ to a first approximation. Substituting eq. (3.142) into the function J in eq. (3.141) and expanding it in the neighborhood of $\bar{\boldsymbol{\theta}}$, we obtain

$$J = -2\log p(\boldsymbol{x}; \bar{\boldsymbol{\theta}}) - 2(\bar{\boldsymbol{l}}, \Delta\boldsymbol{\theta}) + (\Delta\boldsymbol{\theta}, \bar{\boldsymbol{L}}\Delta\boldsymbol{\theta}) + O(\Delta\boldsymbol{\theta})^3, \tag{3.143}$$

where $\bar{\boldsymbol{l}}$ is the score defined by eq. (3.113) and $\bar{\boldsymbol{L}}$ is the matrix defined by eq. (3.121). The bar indicates that the value is evaluated at $\bar{\boldsymbol{\theta}}$.

The assumption (3.112) implies that

$$\boldsymbol{P}^{\mathcal{S}}_{\bar{\boldsymbol{\theta}}}\bar{\boldsymbol{L}} = \bar{\boldsymbol{L}}\boldsymbol{P}^{\mathcal{S}}_{\bar{\boldsymbol{\theta}}} = \bar{\boldsymbol{L}}, \tag{3.144}$$

where $\boldsymbol{P}^{\mathcal{S}}_{\bar{\boldsymbol{\theta}}}$ is the projection matrix onto $T_{\bar{\boldsymbol{\theta}}}(\mathcal{S})$. It follows that the rank of $\bar{\boldsymbol{L}}$ is at most the dimension m' of the tangent space $T_{\bar{\boldsymbol{\theta}}}(\mathcal{S})$. Here, we assume that the rank of $\bar{\boldsymbol{L}}$ is exactly m' so as to guarantee the unique existence of the value of $\boldsymbol{\theta} \in \mathcal{S}$ that minimizes J. Then, the range of $\bar{\boldsymbol{L}}$ coincides with $T_{\bar{\boldsymbol{\theta}}}(\mathcal{S})$.

If the term $O(\Delta\boldsymbol{\theta})^3$ in eq. (3.143) is ignored, the function J is minimized under the constraint $\Delta\boldsymbol{\theta} \in T_{\bar{\boldsymbol{\theta}}}(\mathcal{S})$ by the following value (see eq. (2.137)):

$$\Delta\hat{\boldsymbol{\theta}} = -\bar{\boldsymbol{L}}^{-}\bar{\boldsymbol{l}}. \tag{3.145}$$

It follows that the maximum likelihood estimator is given by $\hat{\boldsymbol{\theta}} = \bar{\boldsymbol{\theta}} + \Delta\boldsymbol{\theta}$. From eqs. (3.117) and (3.144), we see that $\boldsymbol{P}_{\mathcal{N}_{\bar{\boldsymbol{\theta}}}} E[\hat{\boldsymbol{\theta}} - \bar{\boldsymbol{\theta}}] = \boldsymbol{0}$, which means that *the maximum likelihood estimator is unbiased in the first order*[20].

A probability distribution which has a probability density $p(\boldsymbol{x}; \boldsymbol{\theta})$ parameterized by $\boldsymbol{\theta}$ is said to belong to the *exponential family* if $p(\boldsymbol{x}; \boldsymbol{\theta})$ can be expressed in terms of a vector function $\boldsymbol{f}(\boldsymbol{x})$ and scalar functions $C(\boldsymbol{\theta})$ and $g(\boldsymbol{x})$ in the form[21]

$$p(\boldsymbol{x}; \boldsymbol{\theta}) = C(\boldsymbol{\theta}) \exp[(\boldsymbol{f}(\boldsymbol{x}), \boldsymbol{\theta}) + g(\boldsymbol{x})]. \tag{3.146}$$

Many probability distributions that appear in practical applications have probability densities in this form, and the Gaussian distribution is a typical example[22]. For a distribution of the exponential family, the matrix $\bar{\boldsymbol{L}}$

[20] The proviso "in the first order" means that the result is obtained by ignoring high order terms in $\Delta\theta$.

[21] If the exponent on the right-hand side of eq. (3.146) has the form $(\mathbf{f}(\mathbf{x}), \mathbf{h}(\theta)) + g(\mathbf{x})$ for some vector function $\mathbf{h}(\cdot)$, the distribution is said to belong to the *curved exponential family*. If $\mathbf{h}(\theta)$ is taken as a new parameter, i.e., if $\eta = \mathbf{h}(\theta)$ can be solved for θ in the form $\theta = \mathbf{t}(\eta)$, the distribution belongs to the exponential family with parameter η.

[22] Beside the Gaussian distribution, the exponential family includes such distributions as the *Poisson distribution*, the *binomial distribution*, the *gamma distribution*, the *beta distribution*, and the *Poisson-gamma distribution*.

defined by eq. (3.121) does not depend on $\boldsymbol{x}$ and hence is equal to the Fisher information matrix $\boldsymbol{J}$ (see eq. (3.122)). From eq. (3.145), we see that the covariance matrix $V[\hat{\boldsymbol{\theta}}]$ of the maximum likelihood estimator $\hat{\boldsymbol{\theta}}$ is given as follows (see eqs. (3.127) and (3.144)):

$$V[\hat{\boldsymbol{\theta}}] = \bar{\boldsymbol{L}}^{-} E[\bar{\boldsymbol{l}}\bar{\boldsymbol{l}}^{\top}]\bar{\boldsymbol{L}}^{-} = \boldsymbol{J}^{-}\boldsymbol{J}\boldsymbol{J}^{-} = \boldsymbol{J}^{-}. \tag{3.147}$$

This means that the Cramer-Rao lower bound is attained (see eq. (3.134)). Thus, *the maximum likelihood estimator is efficient in the first order if the distribution belongs to the exponential family.*

3.6.2 Asymptotic behavior

If N independent samples $\boldsymbol{x}_1$, ..., $\boldsymbol{x}_N$ are observed from a distribution whose probability density is $p(\boldsymbol{x};\boldsymbol{\theta})$, the maximum likelihood estimator $\hat{\boldsymbol{\theta}}$ is the value of $\boldsymbol{\theta}$ that maximizes the likelihood $p(\boldsymbol{x}_1;\boldsymbol{\theta})\cdots p(\boldsymbol{x}_N;\boldsymbol{\theta})$. Namely, $\boldsymbol{\theta}$ is the solution of

$$J = -2\sum_{\alpha=1}^{N} \log p(\boldsymbol{x}_\alpha) \to \min. \tag{3.148}$$

In this case, eq. (3.145) is replaced by

$$\Delta\boldsymbol{\theta} = -\left(\sum_{\beta=1}^{N} \bar{\boldsymbol{L}}_\alpha\right)^{-} \sum_{\alpha=1}^{N} \bar{\boldsymbol{l}}_\alpha, \tag{3.149}$$

where $\bar{\boldsymbol{l}}_\alpha$ and $\bar{\boldsymbol{L}}_\alpha$ are, respectively, the score $\boldsymbol{l}$ and the matrix $\boldsymbol{L}$ for $\boldsymbol{x}_\alpha$ evaluated at $\bar{\boldsymbol{\theta}}$. Hence, the covariance matrix $V[\hat{\boldsymbol{\theta}}]$ of the maximum likelihood estimator $\hat{\boldsymbol{\theta}} = \bar{\boldsymbol{\theta}} + \Delta\boldsymbol{\theta}$ is written in the following form:

$$V[\hat{\boldsymbol{\theta}}] = E[\left(\frac{1}{N}\sum_{\alpha=1}^{N} \bar{\boldsymbol{L}}_\alpha\right)^{-} \left(\frac{1}{N}\sum_{\beta=1}^{N} \bar{\boldsymbol{l}}_\beta\right) \left(\frac{1}{N}\sum_{\gamma=1}^{N} \bar{\boldsymbol{l}}_\gamma\right)^{\top} \left(\frac{1}{N}\sum_{\delta=1}^{N} \bar{\boldsymbol{L}}_\delta\right)^{-}]. \tag{3.150}$$

Matrices $\bar{\boldsymbol{L}}_1$, ..., $\bar{\boldsymbol{L}}_N$ are random variables that belong to the same distribution, and their common expectation is the Fisher information matrix $\boldsymbol{J}$. Hence, the following *law of large numbers* holds for a sufficiently large N (see eq. (3.9)):

$$\frac{1}{N}\sum_{\alpha=1}^{N} \bar{\boldsymbol{L}}_\alpha \sim \boldsymbol{J}. \tag{3.151}$$

From this and the independence of each $\boldsymbol{x}_\alpha$, we can write eq. (3.150) as follows (see eqs. (2.81)):

$$V[\hat{\boldsymbol{\theta}}] \sim \frac{1}{N^2}\sum_{\alpha,\beta=1}^{N} \boldsymbol{J}^{-}E[\bar{\boldsymbol{l}}_\alpha\bar{\boldsymbol{l}}_\beta^{\top}]\boldsymbol{J}^{-} = \frac{1}{N^2}\sum_{\alpha=1}^{N} \boldsymbol{J}^{-}\boldsymbol{J}\boldsymbol{J}^{-} = \frac{1}{N}\boldsymbol{J}^{-}. \tag{3.152}$$

This means that the Cramer-Rao lower bound is attained in the asymptotic limit (see eq. (3.139)). Thus, *the maximum likelihood estimator is asymptotically efficient*[23]. We also observe the following (we omit the details):

- An estimator $\hat{\boldsymbol{\theta}}$ of $\boldsymbol{\theta}$ is said to be *consistent* if $\hat{\boldsymbol{\theta}} \sim \boldsymbol{\theta}$ as $N \to \infty$. Since eq. (3.152) implies that $V[\hat{\boldsymbol{\theta}}] \sim O(1/N)$, *the maximum likelihood estimator consistent.*

- Since $\bar{\boldsymbol{l}}_1$, ..., $\bar{\boldsymbol{l}}_N$ are independent random variables of mean $\mathbf{0}$ (see eq. (3.117)) and have the same distribution, the *central limit theorem* (see Section 3.3.1) states that $\sum_{\alpha=1}^{N} \bar{\boldsymbol{l}}_\alpha / N$ is asymptotically Gaussian. It follows from eq. (3.149) that *the maximum likelihood estimator is asymptotically a Gaussian random variable of mean* $\mathbf{0}$ *and covariance matrix* $\boldsymbol{J}^{-}/N$.

3.7 Akaike Information Criterion

3.7.1 Model selection

Suppose N data $\boldsymbol{x}_1$, ..., $\boldsymbol{x}_N$ are observed. A *statistical test* is a procedure for judging if they can be regarded as independent random samples from a particular distribution with probability density $p(\boldsymbol{x})$; the χ^2 *test* is a typical example (see Section 3.3.4). If the data are known to be independent random samples from a distribution with probability density $p(\boldsymbol{x}; \boldsymbol{\theta})$ parameterized by $\boldsymbol{\theta}$, the procedure for determining the parameter $\boldsymbol{\theta}$ that best explains the data is called *statistical estimation*; *maximum likelihood estimation* is a typical example (see Section 3.6.1). But how can we guess a parameterized probability density $p(\boldsymbol{x}; \boldsymbol{\theta})$? In other words, how can we judge if the data can be regarded as independent random samples from a distribution with probability density $p_1(\boldsymbol{x}; \boldsymbol{\theta})$, or with probability density $p_2(\boldsymbol{x}; \boldsymbol{\theta})$, or with other probability densities? A parameterized probability density $p(\boldsymbol{x}; \boldsymbol{\theta})$ is called a *(statistical) model* of the distribution. In order to select a best model, we need a criterion that measures the "goodness" of a particular model.

If we adopt a particular model $p(\boldsymbol{x}; \boldsymbol{\theta})$ and apply maximum likelihood estimation, the parameter $\boldsymbol{\theta}$ is determined by maximizing $\prod_{\alpha=1}^{N} p(\boldsymbol{x}_\alpha; \boldsymbol{\theta})$ or equivalently minimizing $-2\sum_{\alpha=1}^{N} \log p(\boldsymbol{x}_\alpha; \boldsymbol{\theta})$. Let $\hat{\boldsymbol{\theta}}$ be the resulting maximum likelihood estimator of $\boldsymbol{\theta}$. Let us call

$$J[\{\boldsymbol{x}_\alpha\}, \hat{\boldsymbol{\theta}}] = -2\sum_{\alpha=1}^{n} \log p(\boldsymbol{x}_\alpha; \hat{\boldsymbol{\theta}}) \tag{3.153}$$

[23]The effect of the neglected high order terms in $\Delta\theta$ converges to 0 as $N \to \infty$ under mild regularity conditions. Hence, the proviso "in the first order" can be dropped. We omit the details.

simply the *residual*. A good model is expected to have a large likelihood $\prod_{\alpha=1}^{N} p(\boldsymbol{x}_\alpha; \hat{\boldsymbol{\theta}})$, thereby a small residual $J[\{\boldsymbol{x}_\alpha\}, \hat{\boldsymbol{\theta}}]$. Hence, the residual appears to be a good criterion for model selection. However, since $\hat{\boldsymbol{\theta}}$ is determined so as to *minimize the residual for the current data* $\{\boldsymbol{x}_\alpha\}$, the residual can be made arbitrarily small, say, by assuming that $\boldsymbol{x}$ can take only N values $\boldsymbol{x}_1$, ..., $\boldsymbol{x}_N$. Such an artificial model may *explain the current data* very well but may be unable to *predict the data to be observed in the future.* This observation leads to the following idea. Let $\boldsymbol{x}_1^*$, ..., $\boldsymbol{x}_N^*$ be independent random samples to be observed in the future; they are assumed to have the same distribution as the current data $\boldsymbol{x}_1$, ..., $\boldsymbol{x}_N$. For a good model, the residual

$$J[\{\boldsymbol{x}_\alpha^*\}, \hat{\boldsymbol{\theta}}] = -2\sum_{\alpha=1}^{n} \log p(\boldsymbol{x}_\alpha^*; \hat{\boldsymbol{\theta}}) \tag{3.154}$$

for the future data $\{\boldsymbol{x}_\alpha^*\}$ should be small. Since the future data $\{\boldsymbol{x}_\alpha^*\}$ and the maximum likelihood estimator $\hat{\boldsymbol{\theta}}$, which is a function of the current data $\{\boldsymbol{x}_\alpha\}$, are both random variables, the above residual is also a random variable. In order to define a definitive value for the model, we take expectation and consider

$$I = E^*[E[J[\{\boldsymbol{x}_\alpha^*\}, \hat{\boldsymbol{\theta}}]]] \tag{3.155}$$

where $E^*[\,\cdot\,]$ and $E[\,\cdot\,]$ denote expectation with respect to the future data $\{\boldsymbol{x}_\alpha^*\}$ and the current data $\{\boldsymbol{x}_\alpha\}$, respectively. We call I simply the *expected residual* and regard a model as better if the expected residual I is smaller.

3.7.2 Asymptotic expression for the expected residual

As in Sections 3.5 and 3.6, we assume that the data $\boldsymbol{x}_1$, ..., $\boldsymbol{x}_N$ are n-vectors sampled from an n'-dimensional manifold $\mathcal{X} \subset \mathcal{R}^n$. The model parameter $\boldsymbol{\theta}$ is assumed to be an m-vector constrained to be in an m'-dimensional manifold $\mathcal{S} \subset \mathcal{R}^m$. Hence, the model $p(\boldsymbol{x}; \boldsymbol{\theta})$ has m' degrees of freedom. We also assume that the model $p(\boldsymbol{x}; \boldsymbol{\theta})$ contains the true probability density. Suppose the true model is $p(\boldsymbol{x}; \bar{\boldsymbol{\theta}})$, and let $\hat{\boldsymbol{\theta}}$ be the maximum likelihood estimator of $\boldsymbol{\theta}$. Writing $\hat{\boldsymbol{\theta}} = \bar{\boldsymbol{\theta}} + \Delta\hat{\boldsymbol{\theta}}$ and expanding $\log p(\boldsymbol{x}_\alpha^*; \hat{\boldsymbol{\theta}})$ in the neighborhood of $\bar{\boldsymbol{\theta}}$, we obtain

$$\log p(\boldsymbol{x}_\alpha^*; \hat{\boldsymbol{\theta}}) = \log p(\boldsymbol{x}_\alpha^*; \bar{\boldsymbol{\theta}}) + (\bar{\boldsymbol{l}}_\alpha^*, \Delta\hat{\boldsymbol{\theta}}) - \frac{1}{2}(\Delta\hat{\boldsymbol{\theta}}, \bar{\boldsymbol{L}}_\alpha^*\Delta\hat{\boldsymbol{\theta}}) + O(\Delta\hat{\boldsymbol{\theta}})^3, \tag{3.156}$$

where the score $\bar{\boldsymbol{l}}_\alpha$ and the matrix $\bar{\boldsymbol{L}}_\alpha$ are defined as follows (see eqs. (3.113) and (3.121)):

$$\bar{\boldsymbol{l}}_\alpha^* = \nabla_{\boldsymbol{\theta}} \log p(\boldsymbol{x}_\alpha^*; \bar{\boldsymbol{\theta}}), \qquad \bar{\boldsymbol{L}}_\alpha^* = -\nabla_{\boldsymbol{\theta}}^2 \log p(\boldsymbol{x}_\alpha^*; \bar{\boldsymbol{\theta}}). \tag{3.157}$$

Ignoring higher order terms and noting that $\Delta\hat{\boldsymbol{\theta}}$ does not depend on $\{\boldsymbol{x}_\alpha^*\}$, we obtain from eqs. (3.155) and (3.156)

$$I = \sum_{\alpha=1}^{N} \left(-2E^*[\log p(\boldsymbol{x}_\alpha^*; \bar{\boldsymbol{\theta}})] - 2(E^*[\bar{\boldsymbol{l}}_\alpha^*], E[\Delta\hat{\boldsymbol{\theta}}]) + E[(\Delta\hat{\boldsymbol{\theta}}, E^*[\bar{\boldsymbol{L}}_\alpha^*]\Delta\hat{\boldsymbol{\theta}})] \right). \tag{3.158}$$

Since $\{\boldsymbol{x}_\alpha^*\}$ and $\{\boldsymbol{x}_\alpha\}$ have the same distribution, we have

$$E^*[\log p(\boldsymbol{x}_\alpha^*; \bar{\boldsymbol{\theta}})] = E[\log p(\boldsymbol{x}_\alpha; \bar{\boldsymbol{\theta}})]. \tag{3.159}$$

Recall that

$$E^*[\bar{\boldsymbol{l}}_\alpha^*] = \boldsymbol{0}, \qquad E^*[\bar{\boldsymbol{L}}_\alpha^*] = \boldsymbol{J}, \tag{3.160}$$

where $\boldsymbol{J}$ is the Fisher information matrix (see eqs. (3.117), (3.118), and (3.122)). It follows that eq. (3.158) can be rewritten as

$$I = -2E[\sum_{\alpha=1}^{N} \log p(\boldsymbol{x}_\alpha; \bar{\boldsymbol{\theta}})] + E[N(\Delta\hat{\boldsymbol{\theta}}, \boldsymbol{J}\Delta\hat{\boldsymbol{\theta}})]. \tag{3.161}$$

Expanding $\log p(\boldsymbol{x}_\alpha; \bar{\boldsymbol{\theta}})$ in the neighborhood of $\hat{\boldsymbol{\theta}}$, we obtain

$$\log p(\boldsymbol{x}_\alpha; \bar{\boldsymbol{\theta}}) = \log p(\boldsymbol{x}_\alpha; \hat{\boldsymbol{\theta}}) - (\hat{\boldsymbol{l}}_\alpha, \Delta\hat{\boldsymbol{\theta}}) - \frac{1}{2}(\Delta\hat{\boldsymbol{\theta}}, \hat{\boldsymbol{L}}_\alpha \Delta\hat{\boldsymbol{\theta}}) + O(\Delta\hat{\boldsymbol{\theta}})^3, \tag{3.162}$$

where $\hat{\boldsymbol{l}}_\alpha$ and $\hat{\boldsymbol{L}}_\alpha$ are defined by

$$\hat{\boldsymbol{l}}_\alpha = \nabla_{\boldsymbol{\theta}} \log p(\boldsymbol{x}_\alpha; \hat{\boldsymbol{\theta}}), \qquad \hat{\boldsymbol{L}}_\alpha = -\nabla_{\boldsymbol{\theta}}^2 \log p(\boldsymbol{x}_\alpha; \hat{\boldsymbol{\theta}}). \tag{3.163}$$

If we put

$$\bar{\boldsymbol{L}}_\alpha = -\nabla_{\boldsymbol{\theta}}^2 \log p(\boldsymbol{x}_\alpha; \bar{\boldsymbol{\theta}}), \tag{3.164}$$

we have

$$\hat{\boldsymbol{L}}_\alpha = \bar{\boldsymbol{L}}_\alpha + O(\Delta\hat{\boldsymbol{\theta}}). \tag{3.165}$$

Substituting this into eq. (3.162) and summing it over $\alpha = 1, ..., N$, we obtain

$$\sum_{\alpha=1}^{N} \log p(\boldsymbol{x}_\alpha; \bar{\boldsymbol{\theta}}) = \sum_{\alpha=1}^{N} \log p(\boldsymbol{x}_\alpha; \hat{\boldsymbol{\theta}}) - (\sum_{\alpha=1}^{N} \hat{\boldsymbol{l}}_\alpha, \Delta\hat{\boldsymbol{\theta}})$$
$$-\frac{N}{2}(\Delta\hat{\boldsymbol{\theta}}, \left(\frac{1}{N}\sum_{\alpha=1}^{N} \bar{\boldsymbol{L}}_\alpha\right) \Delta\hat{\boldsymbol{\theta}}) + O(\Delta\hat{\boldsymbol{\theta}})^3. \tag{3.166}$$

Since the maximum likelihood estimator $\hat{\boldsymbol{\theta}}$ maximizes $\sum_{\alpha=1}^{N} \log p(\boldsymbol{x}_\alpha; \boldsymbol{\theta})$, we see that

$$(\sum_{\alpha=1}^{N} \hat{\boldsymbol{l}}_\alpha, \Delta\hat{\boldsymbol{\theta}}) = (\nabla_{\boldsymbol{\theta}} \left(\sum_{\alpha=1}^{N} \log p(\boldsymbol{x}_\alpha; \hat{\boldsymbol{\theta}})\right), \Delta\hat{\boldsymbol{\theta}}) = 0 \tag{3.167}$$

for any admissible variation $\Delta\hat{\boldsymbol{\theta}} \in T_{\hat{\boldsymbol{\theta}}}(\mathcal{S})$. If we recall that eq. (3.151) holds for $N \sim \infty$ (the *law of large numbers*) and ignore higher order terms in $\Delta\hat{\boldsymbol{\theta}}$ in eq. (3.166), we have

$$\sum_{\alpha=1}^{N} \log p(\boldsymbol{x}_\alpha; \bar{\boldsymbol{\theta}}) \sim \sum_{\alpha=1}^{N} \log p(\boldsymbol{x}_\alpha; \hat{\boldsymbol{\theta}}) - \frac{N}{2} N(\Delta\hat{\boldsymbol{\theta}}, \boldsymbol{J}\Delta\hat{\boldsymbol{\theta}}) \tag{3.168}$$

for $N \sim \infty$. Substituting this into eq. (3.161), we obtain

$$I \sim -2E[\sum_{\alpha=1}^{N} \log p(\boldsymbol{x}_\alpha; \hat{\boldsymbol{\theta}})] + 2NE[(\Delta\hat{\boldsymbol{\theta}}, \boldsymbol{J}\Delta\hat{\boldsymbol{\theta}})]. \tag{3.169}$$

As shown in Section 3.6.2, the deviation $\Delta\hat{\boldsymbol{\theta}}$ of the maximum likelihood estimator is asymptotically a Gaussian random variable of mean $\boldsymbol{0}$ and covariance matrix $\boldsymbol{J}^-/N$ (the *central limit theorem*). We assume that the distribution is regular, so the Fisher information matrix $\boldsymbol{J}$ has rank m' (see Section 3.5.1). It follows that $(\Delta\hat{\boldsymbol{\theta}}, (\boldsymbol{J}^-/N)^-\Delta\hat{\boldsymbol{\theta}}) = N(\Delta\hat{\boldsymbol{\theta}}, \boldsymbol{J}\Delta\hat{\boldsymbol{\theta}})$ is asymptotically a χ^2 variable of m' degrees of freedom (see eq. (3.61)). Hence, we have

$$E[N(\Delta\hat{\boldsymbol{\theta}}, \boldsymbol{J}\Delta\hat{\boldsymbol{\theta}})] \sim m' \tag{3.170}$$

for $N \sim \infty$. Thus, eq. (3.169) can be expressed in the form

$$I \sim E[J[\{\boldsymbol{x}_\alpha\}, \hat{\boldsymbol{\theta}}]] + 2m'. \tag{3.171}$$

This means that if we define

$$AIC = J[\{\boldsymbol{x}_\alpha\}, \hat{\boldsymbol{\theta}}] + 2m', \tag{3.172}$$

this is an unbiased estimator of the expected residual I for $N \sim \infty$. This estimator is called the *Akaike information criterion*, or *AIC* for short, and can be used as a measure of the goodness of the model; the *predictive capacity* of the model is expected to be large if AIC is small. According to this criterion, a good model should not only have a small residual but at the same time have a *small degree of freedom*; otherwise, one can define a model that fits the current data arbitrarily well by increasing the number of the model parameters.

Chapter 4

Representation of Geometric Objects

This chapter discusses representations of geometric objects in two and three dimensions. First, representations of points and lines in two dimensions are described, and their error behavior is characterized in terms of their covariance matrices. Similar analysis is done for points, lines, and planes in three dimensions. Then, incidence relations are described in terms of the representations of the objects involved, and the error behavior of intersections and joins is analyzed. Geometric properties of conics and quadrics are also discussed. Finally, the coordinate transformation of three dimensional objects and their perspective projection onto the image plane are summarized.

4.1 Image Points and Image Lines

4.1.1 *Representation of image points*

We call points and lines defined in two-dimensional images *image points* and *image lines*, respectively, to distinguish them from points and lines in three-dimensional scenes, which we call *space points* and *space lines*, respectively. An image point is represented by its image coordinates (x, y) with respect to a fixed Cartesian coordinate system. An alternative representation is the use of the 3-vector

$$\boldsymbol{x} = \begin{pmatrix} x \\ y \\ 1 \end{pmatrix}. \tag{4.1}$$

If the image plane is viewed as a two-dimensional *projective space*[1], the three components of the 3-vector $\boldsymbol{x}$ can be interpreted as the *homogeneous coordinates* of the image point it represents. By definition, a 3-vector $\boldsymbol{x}$ represents an image point if and only if

$$(\boldsymbol{k}, \boldsymbol{x}) = 1, \tag{4.2}$$

where $\boldsymbol{k} = (0, 0, 1)^\top$. The distance of the image point $\boldsymbol{x}$ from the image origin o is

$$d = \sqrt{x^2 + y^2} = \sqrt{\|\boldsymbol{x}\|^2 - 1}. \tag{4.3}$$

Consider an XYZ Cartesian coordinate system with origin O in the scene. We call the plane placed parallel to the X and Y axes in distance f from the

[1]A two-dimensional Euclidean space can be identified with a two-dimensional *projective space* if *ideal image points* ("image points at infinity") are added. Ideal image points are represented by homogeneous coordinates whose third components are 0.

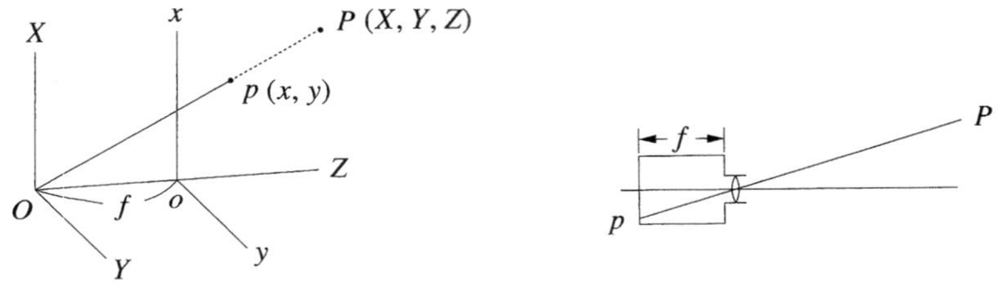

(a) (b)

Fig. 4.1. (a) Perspective projection. (b) Idealized camera model.

XY plane the *image plane*. Define an xy coordinate system on the image plane in such a way that the x- and y-axes are parallel to the X- and Y-axes, respectively, and the origin o is on the Z-axis. If viewed from the origin O, a space point P with coordinates (X, Y, Z) is seen at p on the image plane $Z = f$ with image coordinates (x, y) given by

$$x = f\frac{X}{Z}, \qquad y = f\frac{Y}{Z}. \tag{4.4}$$

We call the origin O and the constant f the *viewpoint* and the *focal length*, respectively. Eqs. (4.4) define a mapping, called *perspective projection*, from the three-dimensional XYZ scene to the two-dimensional xy image plane (Fig. 4.1a). This is an idealized model of camera imaging geometry (Fig. 4.1b): the viewpoint O corresponds to the center of the lens; the Z-axis corresponds to the optical axis of the lens; the focal length f corresponds to the distance from the center of the lens to the surface of the film[2] (or the photo-cells for video cameras). In the following, we use f as the unit of length, so the image plane can be written as $Z = 1$.

If the above geometry of perspective projection is assumed, the vector representation (4.1) can be thought of as identifying an image point p with its position $\boldsymbol{x}$ in three dimensions (Fig. 4.2a). Let us call the space line that starts from the viewpoint O and passes through image point p the *line of sight* of p. The vector $\boldsymbol{x}$ that represents image point p indicates the orientation of the line of sight of p.

Suppose observation of an image point is susceptible to image noise. Let (x, y) be the observed position. If image noise $(\Delta x, \Delta y)$ is randomly added, this position is perturbed into $(x + \Delta x, y + \Delta y)$. The image noise is assumed to be very small[3]: $|\Delta x| \ll 1$, $|\Delta y| \ll 1$. In the vector representation, the observed value $\boldsymbol{x}$ is randomly perturbed into $\boldsymbol{x} + \Delta\boldsymbol{x}$ by image noise $\Delta\boldsymbol{x}$. If $\Delta\boldsymbol{x}$

[2]The focal length f thus defined depends on the position of the object on which the camera is focused. Hence, it is generally different from the *optical* focal length of the lens; the two values coincide only when the object in focus is infinitely far away.

[3]Since we take f as the unit of length, this means that image noise is very small *as*

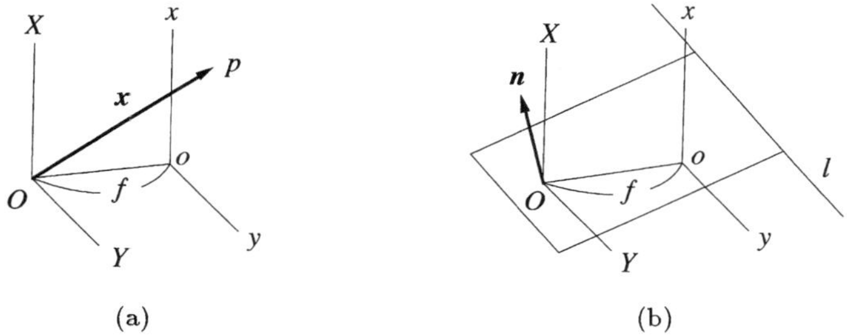

Fig. 4.2. (a) Representation of an image point. (b) Representation of an image line.

is regarded as a random variable of mean $\mathbf{0}$, the uncertainty of the value $\boldsymbol{x}$ is characterized by the covariance matrix $V[\boldsymbol{x}] = E[\Delta\boldsymbol{x}\Delta\boldsymbol{x}^\top]$. Since the image noise $\Delta\boldsymbol{x}$ is orthogonal to $\boldsymbol{k}$, it is constrained to be in the two-dimensional subspace $\{\boldsymbol{k}\}_L^\perp$, the set of all vectors orthogonal to $\boldsymbol{k}$. Consequently, the covariance matrix $V[\boldsymbol{x}]$ is singular; its null space is the one-dimensional subspace $\{\boldsymbol{k}\}_L$ generated by $\boldsymbol{k}$. The fact that rank$V[\boldsymbol{x}] = 2$ simply states that an image point is represented by a 3-vector but has only *two* degrees of freedom. In other words, *the rank of the covariance matrix indicates the degrees of freedom of the representation.*

Example 4.1 If Δx and Δy are independent Gaussian random variables of mean 0 and variance ϵ^2, the covariance matrix of $\boldsymbol{x}$ is given as follows (see eqs. (3.27)):

$$V[\boldsymbol{x}] = \begin{pmatrix} \epsilon^2 & 0 & 0 \\ 0 & \epsilon^2 & 0 \\ 0 & 0 & 0 \end{pmatrix} = \epsilon^2 \boldsymbol{P}_{\boldsymbol{k}}. \tag{4.5}$$

Here, $\boldsymbol{P}_{\boldsymbol{k}} = \boldsymbol{I} - \boldsymbol{k}\boldsymbol{k}^\top$ is the projection matrix onto the XY plane (see eq. (2.46)).

4.1.2 Representation of image lines

An image line l is represented by its equation

$$Ax + By + C = 0. \tag{4.6}$$

Since the coefficients A, B, and C are determined only up to scale, we impose the normalization $A^2 + B^2 + C^2 = 1$. Then, an image line l is represented by

---

compared with the focal length f. The focal length f is usually comparable to or much larger than the physical size of the image. So, this assumption is satisfied if the image noise is much smaller than the size of the image. For an image of 512×512 pixels, for example, this assumption is well satisfied if the image noise is less than, say, five pixels.

a unit 3-vector

$$\boldsymbol{n} = \begin{pmatrix} A \\ B \\ C \end{pmatrix}. \tag{4.7}$$

This representation is not unique: $\boldsymbol{n}$ and $-\boldsymbol{n}$ represent the same image line. If the image plane is viewed as a two-dimensional projective space, the three components of the 3-vector $\boldsymbol{n}$ can be interpreted as the *homogeneous coordinates* of the image line it represents.

Since A and B in eq. (4.6) cannot be zero at the same time[4], an arbitrary 3-vector $\boldsymbol{n}$ represents an image line if and only if

$$\|\boldsymbol{n}\| = 1, \qquad \boldsymbol{n} \neq \pm\boldsymbol{k}. \tag{4.8}$$

The distance of the image line represented by $\boldsymbol{n}$ from the image origin o is

$$d = \frac{|C|}{\sqrt{A^2+B^2}} = \frac{|(\boldsymbol{k},\boldsymbol{n})|}{\sqrt{1-(\boldsymbol{k},\boldsymbol{n})^2}}. \tag{4.9}$$

Recall the geometry of perspective projection described in Fig. 4.1a. The lines of sight of all the image points on an image line l define a space plane. The vector representation (4.7) can be thought of as identifying the image line l with the unit surface normal $\boldsymbol{n}$ to that space plane. In fact, the space plane defined by an image line $Ax + By + C = 0$ is $AX + BY + CZ = 0$, which has surface normal $\boldsymbol{n} = (A, B, C)^\top$ (Fig. 4.2b).

As in the case of an image point, observation of an image line also has uncertainty. Suppose an image line represented by $\boldsymbol{n}$ is randomly perturbed into a position represented by $\boldsymbol{n} = \boldsymbol{n} + \Delta\boldsymbol{n}$. If the error $\Delta\boldsymbol{n}$ is regarded as a random variable of mean $\boldsymbol{0}$, the uncertainty of the value $\boldsymbol{n}$ is characterized by the covariance matrix $V[\boldsymbol{n}] = E[\Delta\boldsymbol{n}\Delta\boldsymbol{n}^\top]$. Since $\boldsymbol{n}$ is normalized into a unit vector, the error $\Delta\boldsymbol{n}$ is orthogonal to $\boldsymbol{n}$ to a first approximation. Hence, the covariance matrix $V[\boldsymbol{n}]$ is singular; its null space is $\{\boldsymbol{n}\}_L$. Thus, although an image line is represented by a 3-vector $\boldsymbol{n}$, it has only *two* degrees of freedom: $\mathrm{rank}V[\boldsymbol{n}] = 2$.

This description of error behavior implies that we are viewing $\boldsymbol{n}$ as a random variable which has a *local distribution* (see Section 3.2.2). Since $\boldsymbol{n}$ is a unit vector, it has a distribution over S^2, a unit sphere centered at the origin. We are assuming that the distribution is sufficiently localized around $\boldsymbol{n}$ and hence its domain can be identified with the tangent plane $T_{\boldsymbol{n}}(S^2)$ at $\boldsymbol{n}$ (see Fig. 3.3).

[4] The set of all ideal image points in a two-dimensional projective space is called the *ideal image line* ("image line at infinity"). In *projective geometry*, the ideal image line can be treated just like an ordinary image line and represented by homogeneous coordinates whose first and second components are both 0.

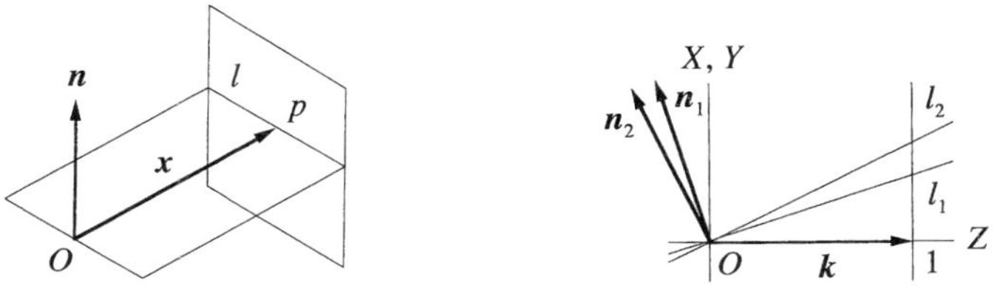

(a) (b)

Fig. 4.3. (a) Incidence of image point p and image line l. (b) Parallel image lines l_1 and l_2.

4.1.3 *Incidence, intersections, and joins*

A. Incidence

Image point p and image line l are *incident* to each other if p is on l, or l passes through p (Fig. 4.3a). Let (x, y) be the image coordinates of p. If image line l is represented by A, B, and C, image point p and image line l are incident to each other if and only if $Ax + By + C = 0$. In other words, an image point p represented by $\boldsymbol{x}$ and an image line l represented by $\boldsymbol{n}$ are incident to each other if and only if

$$(\boldsymbol{n}, \boldsymbol{x}) = 0, \tag{4.10}$$

which is simply the *equation* of image line l if $\boldsymbol{x}$ is regarded as a variable. The distance $D(p, l)$ between an image point p represented by $\boldsymbol{x}$ and an image line l represented by $\boldsymbol{n}$ is

$$D(p, l) = \frac{|(\boldsymbol{n}, \boldsymbol{x})|}{\sqrt{1 - (\boldsymbol{k}, \boldsymbol{n})^2}}. \tag{4.11}$$

B. Intersections

Two image lines $(\boldsymbol{n}_1, \boldsymbol{x}) = 0$ and $(\boldsymbol{n}_2, \boldsymbol{x}) = 0$ are parallel to each other if and only if

$$|\boldsymbol{n}_1, \boldsymbol{n}_2, \boldsymbol{k}| = 0. \tag{4.12}$$

(Fig. 4.3b.) If image lines $(\boldsymbol{n}_1, \boldsymbol{x}) = 0$ and $(\boldsymbol{n}_2, \boldsymbol{x}) = 0$ are not parallel[5], they intersect at a single image point. From Fig. 4.4a, we see that the vector $\boldsymbol{x}$ that represents the intersection must be orthogonal to both $\boldsymbol{n}_1$ and $\boldsymbol{n}_2$. Hence, $\boldsymbol{x} \propto \boldsymbol{n}_1 \times \boldsymbol{n}_2$. Since $(\boldsymbol{k}, \boldsymbol{x}) = 1$, the intersection $\boldsymbol{x}$ is obtained in the following form (see eq. (2.32)):

$$\boldsymbol{x} = \frac{\boldsymbol{n}_1 \times \boldsymbol{n}_2}{|\boldsymbol{n}_1, \boldsymbol{n}_2, \boldsymbol{k}|}. \tag{4.13}$$

[5] Parallel distinct image lines can be thought of as intersecting at an *ideal image point* ("image point at infinity").

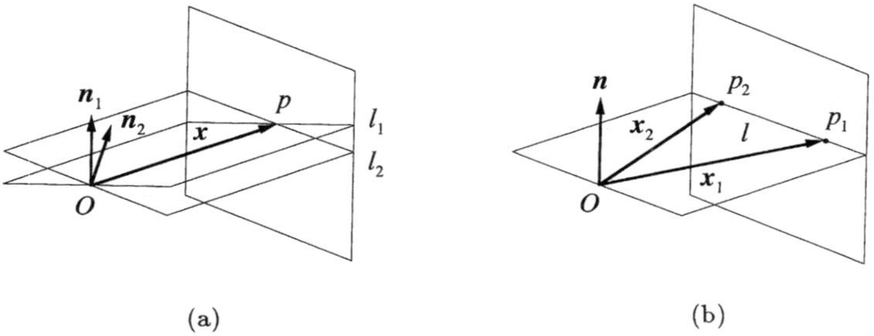

Fig. 4.4. (a) The intersection p of image lines l_1 and l_2. (b) The join l of image points p_1 and p_2.

Let $V[\boldsymbol{n}_1]$ and $V[\boldsymbol{n}_2]$ be the covariance matrices of $\boldsymbol{n}_1$ and $\boldsymbol{n}_2$, respectively. If $\boldsymbol{n}_1$ and $\boldsymbol{n}_2$ are perturbed into $\boldsymbol{n}_1 + \Delta\boldsymbol{n}_1$ and $\boldsymbol{n}_2 + \Delta\boldsymbol{n}_2$, respectively, the intersection $\boldsymbol{x}$ computed by eq. (4.13) is perturbed into $\boldsymbol{x} + \Delta\boldsymbol{x}$ accordingly. The error is to a first approximation

$$\Delta\boldsymbol{x} = \frac{\Delta\boldsymbol{n}_1 \times \boldsymbol{n}_2 + \boldsymbol{n}_1 \times \Delta\boldsymbol{n}_2 - (|\Delta\boldsymbol{n}_1, \boldsymbol{n}_2, \boldsymbol{k}| + |\boldsymbol{n}_1, \Delta\boldsymbol{n}_2, \boldsymbol{k}|)\boldsymbol{x}}{|\boldsymbol{n}_1, \boldsymbol{n}_2, \boldsymbol{k}|}. \tag{4.14}$$

If $\boldsymbol{n}_1$ and $\boldsymbol{n}_2$ are statistically independent[6], the covariance matrix $V[\boldsymbol{x}] = E[\Delta\boldsymbol{x}\Delta\boldsymbol{x}^\top]$ of the intersection $\boldsymbol{x}$ is given by

$$\begin{aligned} V[\boldsymbol{x}] = \frac{1}{|\boldsymbol{n}_1, \boldsymbol{n}_2, \boldsymbol{k}|}\Big(&\boldsymbol{n}_2 \times V[\boldsymbol{n}_1] \times \boldsymbol{n}_2 + 2S[\boldsymbol{n}_2 \times V[\boldsymbol{n}_1](\boldsymbol{n}_2 \times \boldsymbol{k})\boldsymbol{x}^\top] \\ &+(\boldsymbol{n}_2 \times \boldsymbol{k}, V[\boldsymbol{n}_1](\boldsymbol{n}_2 \times \boldsymbol{k}))\boldsymbol{x}\boldsymbol{x}^\top + \boldsymbol{n}_1 \times V[\boldsymbol{n}_2] \times \boldsymbol{n}_1 \\ &+2S[\boldsymbol{n}_1 \times V[\boldsymbol{n}_2](\boldsymbol{n}_1 \times \boldsymbol{k})\boldsymbol{x}^\top] + (\boldsymbol{n}_1 \times \boldsymbol{k}, V[\boldsymbol{n}_2](\boldsymbol{n}_1 \times \boldsymbol{k}))\boldsymbol{x}\boldsymbol{x}^\top\Big), \end{aligned} \tag{4.15}$$

where $S[\,\cdot\,]$ denotes the symmetrization operator (see eqs. (2.205)). In deriving the above expression, the identity $(\Delta\boldsymbol{n}_i \times \boldsymbol{n}_j)(\Delta\boldsymbol{n}_k \times \boldsymbol{n}_l)^\top = \boldsymbol{n}_j \times (\Delta\boldsymbol{n}_i \Delta\boldsymbol{n}_k^\top) \times \boldsymbol{n}_l$ has been used (see eq. (2.43)).

C. Joins

An image line $(\boldsymbol{n}, \boldsymbol{x}) = 0$ that passes through two distinct image points $\boldsymbol{x}_1$ and $\boldsymbol{x}_2$ is called the *join* of $\boldsymbol{x}_1$ and $\boldsymbol{x}_2$. Since $(\boldsymbol{n}, \boldsymbol{x}_1) = 0$ and $(\boldsymbol{n}, \boldsymbol{x}_2) = 0$, the vector $\boldsymbol{n}$ must be orthogonal to both $\boldsymbol{x}_1$ and $\boldsymbol{x}_2$ (Fig. 4.4b). Also, $\boldsymbol{n}$ is normalized into a unit vector. Hence,

$$\boldsymbol{n} = \pm N[\boldsymbol{x}_1 \times \boldsymbol{x}_2], \tag{4.16}$$

[6]We use an informal expression like this, instead of saying that the noise that arises in the observation of $\mathbf{n}_1$ and the noise that arises in the observation of $\mathbf{n}_2$ are independent random variables.

which defines an image line as long as $\boldsymbol{x}_1 \neq \boldsymbol{x}_2$.

Let $V[\boldsymbol{x}_1]$ and $V[\boldsymbol{x}_2]$ be the covariance matrices of image points $\boldsymbol{x}_1$ and $\boldsymbol{x}_2$, respectively. If $\boldsymbol{x}_1$ and $\boldsymbol{x}_2$ are perturbed into $\boldsymbol{x}_1 + \Delta\boldsymbol{x}_1$ and $\boldsymbol{x}_2 + \Delta\boldsymbol{x}_2$, respectively, the vector $\boldsymbol{n}$ computed by (4.16) is perturbed into $\boldsymbol{n} + \Delta\boldsymbol{n}$ accordingly. The error is to a first approximation

$$\Delta\boldsymbol{n} = \pm\frac{\boldsymbol{P_n}(\Delta\boldsymbol{x}_1 \times \boldsymbol{x}_2 + \boldsymbol{x}_1 \times \Delta\boldsymbol{x}_2)}{\|\boldsymbol{x}_1 \times \boldsymbol{x}_2\|}. \tag{4.17}$$

Here, $\boldsymbol{P_n} = \boldsymbol{I} - \boldsymbol{n}\boldsymbol{n}^\top$ is the projection matrix along $\boldsymbol{n}$. If $\boldsymbol{x}_1$ and $\boldsymbol{x}_2$ are statistically independent, the covariance matrix $V[\boldsymbol{n}] = E[\Delta\boldsymbol{n}\Delta\boldsymbol{n}^\top]$ of $\boldsymbol{n}$ is given by

$$V[\boldsymbol{n}] = \frac{\boldsymbol{P_n}(\boldsymbol{x}_1 \times V[\boldsymbol{x}_2] \times \boldsymbol{x}_1 + \boldsymbol{x}_2 \times V[\boldsymbol{x}_1] \times \boldsymbol{x}_2)\boldsymbol{P_n}}{\|\boldsymbol{x}_1 \times \boldsymbol{x}_2\|^2}. \tag{4.18}$$

Example 4.2 As shown in Example 4.1, if each coordinate is perturbed by Gaussian noise independently of mean 0 and variance ϵ^2, the covariance matrices of $\boldsymbol{x}_1$ and $\boldsymbol{x}_2$ are $V[\boldsymbol{x}_1] = V[\boldsymbol{x}_2] = \epsilon^2\boldsymbol{P_k}$. Note the following identity (see eq. (2.42)):

$$\begin{aligned}\boldsymbol{x}_i \times \boldsymbol{P_k} \times \boldsymbol{x}_i &= \boldsymbol{x}_i \times (\boldsymbol{I} - \boldsymbol{k}\boldsymbol{k}^\top) \times \boldsymbol{x}_i \\ &= \|\boldsymbol{x}_i\|^2\boldsymbol{I} - \boldsymbol{x}_i\boldsymbol{x}_i^\top - (\boldsymbol{x}_i \times \boldsymbol{k})(\boldsymbol{x}_i \times \boldsymbol{k})^\top. \end{aligned} \tag{4.19}$$

Let $\boldsymbol{u}$ be the unit vector that indicates the orientation of the join $(\boldsymbol{n}, \boldsymbol{x}) = 0$, and w the separation between the two image points. Their midpoint is $\boldsymbol{x}_C = (\boldsymbol{x}_1 + \boldsymbol{x}_2)/2$, so we can write $\boldsymbol{x}_1 = \boldsymbol{x}_C - w\boldsymbol{u}/2$ and $\boldsymbol{x}_2 = \boldsymbol{x}_C + w\boldsymbol{u}/2$. If the two image points are close to the image origin o, we have $\|\boldsymbol{x}_i\| \approx 1$, $\boldsymbol{x}_i \times \boldsymbol{k} \approx \boldsymbol{0}$, $i = 1, 2$, and $\|\boldsymbol{x}_1 \times \boldsymbol{x}_2\| \approx w$. Using eq. (4.19), we can approximate eq. (4.18) in the form

$$V[\boldsymbol{n}] \approx \frac{2\epsilon^2}{w^2}\left(\boldsymbol{P_n} - \boldsymbol{x}_C\boldsymbol{x}_C^\top - \frac{w^2}{4}\boldsymbol{u}\boldsymbol{u}^\top\right), \tag{4.20}$$

where the identities $\boldsymbol{P_n}\boldsymbol{u} = \boldsymbol{u}$ and $\boldsymbol{P_n}\boldsymbol{x}_C = \boldsymbol{x}_C$ are used. If $\boldsymbol{x}_C$ is approximately perpendicular to $\boldsymbol{u}$, we obtain the approximation $\boldsymbol{P_n} \approx \boldsymbol{x}_C\boldsymbol{x}_C^\top + \boldsymbol{u}\boldsymbol{u}^\top$. Since the distance w is usually very small as compared with the focal length (which we take as the unit of length), we obtain to the following approximation:

$$V[\boldsymbol{n}] \approx \frac{2\epsilon^2}{w^2}\boldsymbol{u}\boldsymbol{u}^\top. \tag{4.21}$$

This means that the vector $\boldsymbol{n}$ is most likely to deviate in the direction $\boldsymbol{u}$ (Fig. 4.5a); the *primary deviation pair*[7] is given as follows (see eq. (3.48)):

$$\boldsymbol{n}^+ \approx N[\boldsymbol{n} + \frac{\sqrt{2}\epsilon}{w}\boldsymbol{u}], \qquad \boldsymbol{n}^- \approx N[\boldsymbol{n} - \frac{\sqrt{2}\epsilon}{w}\boldsymbol{u}]. \tag{4.22}$$

---

[7]The geometric meaning of the primary deviation pair will be discussed in detail in Section 4.5.3.

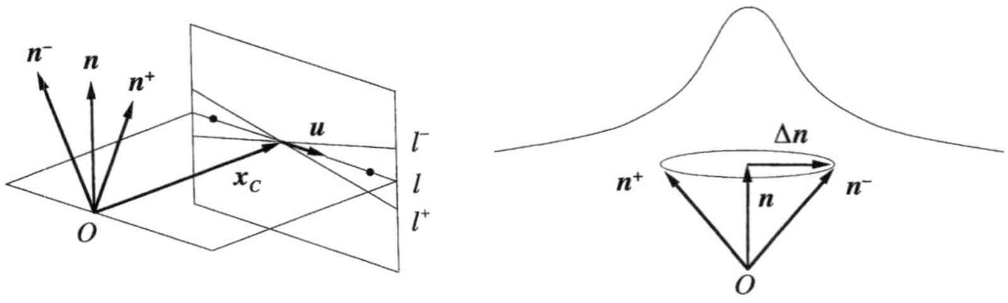

(a) (b)

Fig. 4.5. (a) Primary deviation pair. (b) Deviation of the join.

The two image lines l^+ and l^- defined by these two vectors intersect at the midpoint of the two points, indicating that the join is very likely to pass near the midpoint in the presence of noise (Fig. 4.5b). It is also seen that the error is approximately proportional to the reciprocal of the separation w between the two image points.

4.2 Space Points and Space Lines

4.2.1 Representation of space points

A space point (X, Y, Z) is represented by a 3-vector $\boldsymbol{r} = (X, Y, Z)^\top$. Alternatively, it is represented by a 4-vector

$$\boldsymbol{\rho} = \begin{pmatrix} X \\ Y \\ Z \\ 1 \end{pmatrix} = \boldsymbol{r} \oplus 1. \tag{4.23}$$

The distance from the coordinate origin O is

$$d = \sqrt{X^2 + Y^2 + Z^2} = \|\boldsymbol{r}\| = \sqrt{\|\boldsymbol{\rho}\|^2 - 1}. \tag{4.24}$$

A 4-vector $\boldsymbol{\rho}$ represents a space point if and only if

$$(\boldsymbol{\kappa}, \boldsymbol{\rho}) = 1, \tag{4.25}$$

where $\boldsymbol{\kappa} = (0, 0, 0, 1)^\top$ $(= \boldsymbol{0} \oplus 1)$. If the three-dimensional space is viewed as a three-dimensional *projective space*, the four components of the 4-vector $\boldsymbol{\rho}$ can be interpreted as the *homogeneous coordinates* of the space point[8]. The

[8] As in the two-dimensional case, a three-dimensional Euclidean space can be identified with a three-dimensional *projective space* if *ideal space points* ("space points at infinity") are added. Ideal space points are represented by homogeneous coordinates whose fourth components are zero.

ρ-representation is very convenient for various computations, but we must be careful about the rounding effect in fixed precision computation: if X, Y, and Z are very large, the fourth component of ρ may be treated as effectively 0. Hence, an appropriate scaling must be applied to keep the space coordinates in a reasonable range of magnitude.

Observation of a space point is susceptible to noise. Suppose the observed value $\boldsymbol{r}$ is randomly perturbed into $\boldsymbol{r} + \Delta\boldsymbol{r}$. If $\Delta\boldsymbol{r}$ is regarded as a random variable of mean $\mathbf{0}$, the uncertainty of $\boldsymbol{r}$ is characterized by the covariance matrix $V[\boldsymbol{r}] = E[\Delta\boldsymbol{r}\Delta\boldsymbol{r}^\top]$, which we assume is positive definite: $\text{rank}V[\boldsymbol{r}] = 3$. In the 4-vector representation $\boldsymbol{\rho} = \boldsymbol{r} \oplus 1$, the covariance matrix $V[\boldsymbol{\rho}]$ is singular and has the form

$$V[\boldsymbol{\rho}] = \begin{pmatrix} V[\boldsymbol{r}] & \\ & 0 \end{pmatrix} = V[\boldsymbol{r}] \oplus 0. \tag{4.26}$$

Since the error $\Delta\boldsymbol{\rho}$ is constrained to be in the three-dimensional subspace $\{\boldsymbol{\kappa}\}_L^\perp \subset \mathcal{R}^4$, the null space of $V[\boldsymbol{\rho}]$ is $\{\boldsymbol{\kappa}\}_L$. The fact that $\text{rank}V[\boldsymbol{\rho}] = 3$ states that a space point has *three* degrees of freedom even if it is represented by a 4-vector $\boldsymbol{\rho}$.

Example 4.3 If each component of (X, Y, Z) is perturbed independently by Gaussian noise of mean 0 and variance ϵ^2, the covariance matrix $V[\boldsymbol{r}]$ has the following form (see eq. (3.25)):

$$V[\boldsymbol{r}] = \epsilon^2 \boldsymbol{I}. \tag{4.27}$$

In the 4-vector representation, the covariance matrix $V[\boldsymbol{\rho}]$ has the form

$$V[\boldsymbol{\rho}] = \epsilon^2 \boldsymbol{I} \oplus 0 = \epsilon^2 \boldsymbol{P}_{\mathcal{K}}, \tag{4.28}$$

where $\boldsymbol{P}_{\mathcal{K}} = \boldsymbol{I} - \boldsymbol{\kappa}\boldsymbol{\kappa}^\top$ is the four-dimensional projection matrix along $\boldsymbol{\kappa}$.

4.2.2 Representation of space lines

Consider a space line L. Let H be the point on L closest to the coordinate origin O, and put $\boldsymbol{r}_H = \vec{OH}$. Let $\boldsymbol{m}$ be the unit vector that indicates the orientation of L (Fig. 4.6a). The space line L is represented by two 3-vectors $\{\boldsymbol{m}, \boldsymbol{r}_H\}$; two representations $\{\boldsymbol{m}, \boldsymbol{r}_H\}$ and $\{-\boldsymbol{m}, \boldsymbol{r}_H\}$ define the same space line. A space point $\boldsymbol{r}$ is on L if and only if vector $\boldsymbol{r} - \boldsymbol{r}_H$ is parallel to $\boldsymbol{m}$, or

$$(\boldsymbol{r} - \boldsymbol{r}_H) \times \boldsymbol{m} = \mathbf{0}, \tag{4.29}$$

which is simply the *equation* of space line L if $\boldsymbol{r}$ is regarded as a variable. Two 3-vectors $\{\boldsymbol{m}, \boldsymbol{r}_H\}$ represent a space line if and only if

$$\|\boldsymbol{m}\| = 1, \qquad (\boldsymbol{m}, \boldsymbol{r}_H) = 0. \tag{4.30}$$

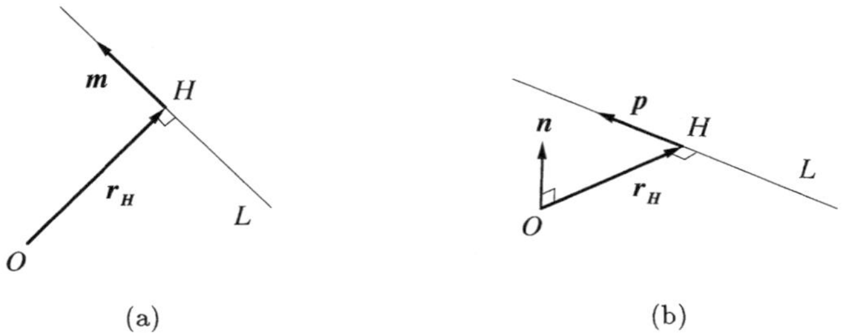

Fig. 4.6. (a) The $\{\mathbf{m}, \mathbf{r}_H\}$-representation. (b) The $\{\mathbf{p}, \mathbf{n}\}$-representation.

Eq. (4.29) can also be written as $\boldsymbol{r} \times \boldsymbol{m} = \boldsymbol{r}_H \times \boldsymbol{m}$. Hence, the equation of a space line can alternatively be written in the form

$$\boldsymbol{r} \times \boldsymbol{p} = \boldsymbol{n}. \tag{4.31}$$

Since both sides can be multiplied by an arbitrary nonzero constant, we normalize the 3-vectors $\boldsymbol{p}$ and $\boldsymbol{n}$ into $\|\boldsymbol{p} \oplus \boldsymbol{n}\| = 1$ (i.e., $\|\boldsymbol{p}\|^2 + \|\boldsymbol{n}\|^2 = 1$). The signs of $\boldsymbol{p}$ and $\boldsymbol{n}$ are chosen in such a way that $\{\boldsymbol{p}, \boldsymbol{n}, \boldsymbol{r}_H\}$ constitute a right-handed orthogonal system (Fig. 4.6b). With this convention, a space line L is also represented by two 3-vectors $\{\boldsymbol{p}, \boldsymbol{n}\}$; two representations $\{\boldsymbol{p}, \boldsymbol{n}\}$ and $\{-\boldsymbol{p}, -\boldsymbol{n}\}$ define the same space line. Geometrically, the vector $\boldsymbol{p}$ indicates the orientation of L, and the vector $\boldsymbol{n}$ indicates the surface normal to the space plane defined by L and the coordinate origin O. The distance of L from O is

$$d = \frac{\|\boldsymbol{n}\|}{\|\boldsymbol{p}\|} = \frac{\sqrt{1 - \|\boldsymbol{p}\|^2}}{\|\boldsymbol{p}\|} = \frac{\|\boldsymbol{n}\|}{\sqrt{1 - \|\boldsymbol{n}\|^2}}. \tag{4.32}$$

It follows that

$$\|\boldsymbol{p}\| = \frac{1}{\sqrt{1 + d^2}}, \qquad \|\boldsymbol{n}\| = \frac{d}{\sqrt{1 + d^2}}. \tag{4.33}$$

The 3-vectors $\boldsymbol{p}$ and $\boldsymbol{n}$ can be interpreted as the *homogeneous coordinates* of a space line, known as the *Plücker* (or *Grassmann*) *coordinates*. Two 3-vectors $\{\boldsymbol{p}, \boldsymbol{n}\}$ represent a space line if and only if

$$\|\boldsymbol{p}\|^2 + \|\boldsymbol{n}\|^2 = 1, \qquad (\boldsymbol{p}, \boldsymbol{n}) = 0. \tag{4.34}$$

The $\{\boldsymbol{m}, \boldsymbol{r}_H\}$-representation and the $\{\boldsymbol{p}, \boldsymbol{n}\}$-representation are related as follows:

$$\boldsymbol{m} = N[\boldsymbol{p}], \qquad \boldsymbol{r}_H = \frac{\boldsymbol{p} \times \boldsymbol{n}}{\|\boldsymbol{p}\|^2}, \tag{4.35}$$

$$\begin{pmatrix} \boldsymbol{p} \\ \boldsymbol{n} \end{pmatrix} = N[\begin{pmatrix} \boldsymbol{m} \\ \boldsymbol{r}_H \times \boldsymbol{m} \end{pmatrix}]. \tag{4.36}$$

Suppose the values $\boldsymbol{m}$ and $\boldsymbol{r}_H$ are randomly perturbed into $\boldsymbol{m} + \Delta\boldsymbol{m}$ and $\boldsymbol{r}_H + \Delta\boldsymbol{r}_H$, respectively. If the errors $\Delta\boldsymbol{m}$ and $\Delta\boldsymbol{r}_H$ are regarded as random variables of mean $\boldsymbol{0}$, the uncertainty of the values $\{\boldsymbol{m}, \boldsymbol{r}_H\}$ is characterized by the six-dimensional covariance matrix $V[\boldsymbol{m} \oplus \boldsymbol{r}_H] = E[(\Delta\boldsymbol{m} \oplus \Delta\boldsymbol{r}_H)(\Delta\boldsymbol{m} \oplus \Delta\boldsymbol{r}_H)^\top]$, which has the following submatrices (see eqs. (3.11) and (3.12)):

$$V[\boldsymbol{m} \oplus \boldsymbol{r}_H] = \begin{pmatrix} V[\boldsymbol{m}] & V[\boldsymbol{m}, \boldsymbol{r}_H] \\ V[\boldsymbol{r}_H, \boldsymbol{m}] & V[\boldsymbol{r}_H] \end{pmatrix}. \tag{4.37}$$

The constraint (4.30) implies that the errors $\Delta\boldsymbol{m}$ and $\Delta\boldsymbol{r}_H$ are constrained by

$$(\boldsymbol{m}, \Delta\boldsymbol{m}) = 0, \qquad (\Delta\boldsymbol{m}, \boldsymbol{r}_H) + (\boldsymbol{m}, \Delta\boldsymbol{r}_H) = 0, \tag{4.38}$$

to a first approximation. Hence, the null space of $V[\boldsymbol{m} \oplus \boldsymbol{r}_H]$ is the two-dimensional subspace

$$\mathcal{N}_{\boldsymbol{m}\oplus\boldsymbol{r}_H} = \{\boldsymbol{m} \oplus \boldsymbol{0}, \boldsymbol{r}_H \oplus \boldsymbol{m}\}_L \subset \mathcal{R}^6. \tag{4.39}$$

It follows that rank$V[\boldsymbol{m} \oplus \boldsymbol{r}_H] = 4$; a space line has *four* degrees of freedom. The six-dimensional projection matrix onto $\mathcal{N}^\perp_{\boldsymbol{m}\oplus\boldsymbol{r}_H}$ is given by

$$\boldsymbol{P}_{\mathcal{N}_{\boldsymbol{m}\oplus\boldsymbol{r}_H}} = \begin{pmatrix} \boldsymbol{P}_{\boldsymbol{m}} & \\ & \boldsymbol{I} \end{pmatrix} - \frac{1}{1 + \|\boldsymbol{r}_H\|^2} \begin{pmatrix} \boldsymbol{r}_H\boldsymbol{r}_H^\top & \boldsymbol{r}_H\boldsymbol{m}^\top \\ \boldsymbol{m}\boldsymbol{r}_H^\top & \boldsymbol{m}\boldsymbol{m}^\top \end{pmatrix}, \tag{4.40}$$

where $\boldsymbol{P}_{\boldsymbol{m}} = \boldsymbol{I} - \boldsymbol{m}\boldsymbol{m}^\top$ is the projection matrix along $\boldsymbol{m}$.

In the $\{\boldsymbol{p}, \boldsymbol{n}\}$-representation, the error behavior is characterized by the six-dimensional covariance matrix

$$V[\boldsymbol{p} \oplus \boldsymbol{n}] = \begin{pmatrix} V[\boldsymbol{p}] & V[\boldsymbol{p}, \boldsymbol{n}] \\ V[\boldsymbol{n}, \boldsymbol{p}] & V[\boldsymbol{n}] \end{pmatrix}. \tag{4.41}$$

The constraint (4.34) implies that the errors $\Delta\boldsymbol{p}$ and $\Delta\boldsymbol{n}$ are constrained by

$$(\boldsymbol{p}, \Delta\boldsymbol{p}) + (\boldsymbol{n}, \Delta\boldsymbol{n}) = 1, \qquad (\Delta\boldsymbol{p}, \boldsymbol{n}) + (\boldsymbol{p}, \Delta\boldsymbol{n}) = 0, \tag{4.42}$$

to a first approximation. Hence, the null space of $V[\boldsymbol{p} \oplus \boldsymbol{n}]$ is the two-dimensional subspace

$$\mathcal{N}_{\boldsymbol{p}\oplus\boldsymbol{n}} = \{\boldsymbol{p} \oplus \boldsymbol{n}, \boldsymbol{n} \oplus \boldsymbol{p}\}_L \subset \mathcal{R}^6. \tag{4.43}$$

Again, rank$V[\boldsymbol{p} \oplus \boldsymbol{n}] = 4$, confirming that a space line indeed has four degrees of freedom. The six-dimensional projection matrix onto $\mathcal{N}^\perp_{\boldsymbol{p}\oplus\boldsymbol{n}}$ is given by

$$\boldsymbol{P}_{\mathcal{N}_{\boldsymbol{p}\oplus\boldsymbol{n}}} = \begin{pmatrix} \boldsymbol{I} - \boldsymbol{p}\boldsymbol{p}^\top - \boldsymbol{n}\boldsymbol{n}^\top & -2S[\boldsymbol{p}\boldsymbol{n}^\top] \\ -2S[\boldsymbol{p}\boldsymbol{n}^\top] & \boldsymbol{I} - \boldsymbol{p}\boldsymbol{p}^\top - \boldsymbol{n}\boldsymbol{n}^\top \end{pmatrix}. \tag{4.44}$$

As in the case of image lines, we are considering a *local distribution* (see Section 3.2.2). Since a space line is represented by two 3-vectors $\{\boldsymbol{m}, \boldsymbol{r}_H\}$

with two constraints given by eqs. (4.30), the set of all space lines defines a *four-dimensional manifold* $\mathcal{S}$ in $\mathcal{R}^6$. We are assuming that the distribution is sufficiently localized around the value $\boldsymbol{m} \oplus \boldsymbol{r}_H$ in $\mathcal{S}$ and hence the domain of the distribution can be identified with the tangent space $T_{\boldsymbol{m}\oplus\boldsymbol{r}_H}(\mathcal{S})$ to $\mathcal{S}$ at $\boldsymbol{m}\oplus\boldsymbol{r}_H$. It follows that the null space $\mathcal{N}_{\boldsymbol{m}\oplus\boldsymbol{r}_H}$ and its orthogonal complement $\mathcal{N}^{\perp}_{\boldsymbol{m}\oplus\boldsymbol{r}_H}$ coincide, respectively, with the normal space $N_{\boldsymbol{m}\oplus\boldsymbol{r}_H}(\mathcal{S})$ and the tangent space $T_{\boldsymbol{m}\oplus\boldsymbol{r}_H}(\mathcal{S})$ (see Section 3.2.1). The same can be said for the $\{\boldsymbol{p}, \boldsymbol{n}\}$-representation.

The covariance matrix $V[\boldsymbol{m} \oplus \boldsymbol{r}_H]$ defined by eq. (4.37) is expressed in terms of the covariance matrix $V[\boldsymbol{p} \oplus \boldsymbol{n}]$ defined by eq. (4.41) as follows (see eqs. (3.18)):

$$V[\boldsymbol{m}] = \frac{1}{\|\boldsymbol{p}\|^2}\boldsymbol{P_m} V[\boldsymbol{p}]\boldsymbol{P_m},$$

$$\begin{aligned} V[\boldsymbol{m}, \boldsymbol{r}_H] &= -\frac{1}{\|\boldsymbol{p}\|^3}\left(\boldsymbol{P_m} V[\boldsymbol{p}] \times \boldsymbol{n} + 2\boldsymbol{P_m} V[\boldsymbol{p}]\boldsymbol{p}\boldsymbol{r}_H^\top - \boldsymbol{P_m} V[\boldsymbol{p}, \boldsymbol{n}] \times \boldsymbol{p}\right) \\ &= V[\boldsymbol{r}_H, \boldsymbol{m}]^\top, \end{aligned}$$

$$\begin{aligned} V[\boldsymbol{r}_H] = \frac{1}{\|\boldsymbol{p}\|^4}\Big(&\boldsymbol{n} \times V[\boldsymbol{p}] \times \boldsymbol{n} - 4S[(\boldsymbol{n} \times V[\boldsymbol{p}]\boldsymbol{p})\boldsymbol{r}_H^\top] \\ &+4(\boldsymbol{p}, V[\boldsymbol{p}]\boldsymbol{p})\boldsymbol{r}_H\boldsymbol{r}_H^\top - 2S[\boldsymbol{n} \times V[\boldsymbol{p}, \boldsymbol{n}] \times \boldsymbol{p}] \\ &- 4S[(\boldsymbol{p} \times V[\boldsymbol{p}, \boldsymbol{n}]^\top \boldsymbol{p})\boldsymbol{r}_H^\top] + \boldsymbol{p} \times V[\boldsymbol{n}] \times \boldsymbol{p}\Big). \end{aligned} \tag{4.45}$$

Here, $\boldsymbol{P_m} = \boldsymbol{I} - \boldsymbol{m}\boldsymbol{m}^\top$ is the projection matrix along $\boldsymbol{m}$. Conversely, the covariance matrix $V[\boldsymbol{p} \oplus \boldsymbol{n}]$ is expressed in terms of the covariance matrix $V[\boldsymbol{m} \oplus \boldsymbol{r}_H]$ in the form

$$V[\boldsymbol{p} \oplus \boldsymbol{n}] = \frac{1}{1 + \|\boldsymbol{r}\|^2}\boldsymbol{P}_{\boldsymbol{p}\oplus\boldsymbol{n}}\begin{pmatrix} V[\boldsymbol{m}] & V[\boldsymbol{m}, \boldsymbol{r} \times \boldsymbol{m}] \\ V[\boldsymbol{r} \times \boldsymbol{m}, \boldsymbol{m}] & V[\boldsymbol{r} \times \boldsymbol{m}] \end{pmatrix}\boldsymbol{P}_{\boldsymbol{p}\oplus\boldsymbol{n}}, \tag{4.46}$$

where

$$V[\boldsymbol{m}, \boldsymbol{r} \times \boldsymbol{m}] = V[\boldsymbol{m}] \times \boldsymbol{r} - V[\boldsymbol{m}, \boldsymbol{r}] \times \boldsymbol{m} = V[\boldsymbol{r} \times \boldsymbol{m}, \boldsymbol{m}]^\top,$$

$$V[\boldsymbol{m} \times \boldsymbol{r}] = \boldsymbol{r} \times V[\boldsymbol{m}] \times \boldsymbol{r} - 2S[\boldsymbol{r} \times V[\boldsymbol{m}, \boldsymbol{r}] \times \boldsymbol{m}] + \boldsymbol{m} \times V[\boldsymbol{r}] \times \boldsymbol{m}. \tag{4.47}$$

Here, $\boldsymbol{P}_{\boldsymbol{p}\oplus\boldsymbol{n}}$ is the six-dimensional projection matrix along $\boldsymbol{p} \oplus \boldsymbol{n}$, which has the form

$$\boldsymbol{P}_{\boldsymbol{p}\oplus\boldsymbol{n}} = \begin{pmatrix} \boldsymbol{I} - \boldsymbol{p}\boldsymbol{p}^\top & -\boldsymbol{p}\boldsymbol{n}^\top \\ -\boldsymbol{n}\boldsymbol{p}^\top & \boldsymbol{I} - \boldsymbol{n}\boldsymbol{n}^\top \end{pmatrix}. \tag{4.48}$$

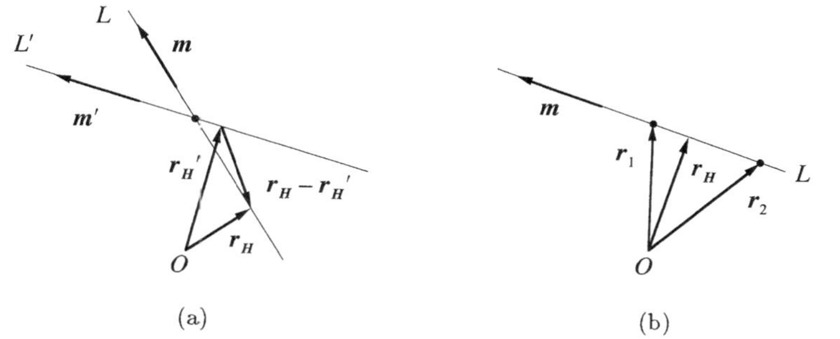

Fig. 4.7. (a) Two intersecting space lines. (b) The join of two space points.

4.2.3 Incidence, intersections, and joins

A. Incidence

A space point P and a space line L are *incident* to each other if P is on L, or L passes through P. If space point P is represented by $\boldsymbol{r}$ and space line L is represented by $\{\boldsymbol{m}, \boldsymbol{r}_H\}$ (or $\{\boldsymbol{p}, \boldsymbol{n}\}$), they are incident to each other if and only if $(\boldsymbol{r} - \boldsymbol{r}_H) \times \boldsymbol{m} = \boldsymbol{0}$ (or $\boldsymbol{r} \times \boldsymbol{p} = \boldsymbol{n}$). The distance $D(P, L)$ between a space point P represented by $\boldsymbol{r}$ and a space line L represented by $(\boldsymbol{r} - \boldsymbol{r}_H) \times \boldsymbol{m} = \boldsymbol{0}$ (or $\boldsymbol{r} \times \boldsymbol{p} = \boldsymbol{n}$) is

$$D(P, L) = \frac{\|\boldsymbol{r} \times \boldsymbol{p} - \boldsymbol{n}\|}{\|\boldsymbol{p}\|} = \|\boldsymbol{P_m}\boldsymbol{r} - \boldsymbol{r}_H\|. \tag{4.49}$$

B. Intersections

The distance $D(L, L')$ between a space line L represented by $\{\boldsymbol{m}, \boldsymbol{r}_H\}$ and a space line L' represented by $\{\boldsymbol{m}', \boldsymbol{r}_H'\}$ is given by

$$D(L, L') = \begin{cases} \dfrac{|\boldsymbol{m}, \boldsymbol{m}', \boldsymbol{r}_H - \boldsymbol{r}_H'|}{\|\boldsymbol{m} \times \boldsymbol{m}'\|}, & \text{if } \boldsymbol{m} \times \boldsymbol{m}' \neq \boldsymbol{0}, \\ \|\boldsymbol{r}_H - \boldsymbol{r}_H'\|, & \text{if } \boldsymbol{m} \times \boldsymbol{m}' = \boldsymbol{0}. \end{cases} \tag{4.50}$$

It follows that the space lines L and L' intersect if and only if the three vectors $\boldsymbol{m}$, $\boldsymbol{m}'$, and $\boldsymbol{r}_H' - \boldsymbol{r}_H$ are coplanar and[9] $\boldsymbol{m} \neq \pm\boldsymbol{m}'$ (Fig. 4.7a), i.e.,

$$|\boldsymbol{m}, \boldsymbol{m}', \boldsymbol{r}_H - \boldsymbol{r}_H'| = 0, \qquad \boldsymbol{m} \times \boldsymbol{m}' \neq \boldsymbol{0}. \tag{4.51}$$

[9] Parallel distinct space lines ($\mathbf{m} = \pm\mathbf{m}'$ or $\mathbf{p} \times \mathbf{p}' = \mathbf{0}$) can be thought of as intersecting at an *ideal space point* ("space point at infinity").

In the $\{\boldsymbol{p},\ \boldsymbol{n}\}$-representation, eqs. (4.50) are replaced by

$$D(L, L') = \begin{cases} \dfrac{(\boldsymbol{p}, \boldsymbol{n}') + (\boldsymbol{p}', \boldsymbol{n})}{\|\boldsymbol{p} \times \boldsymbol{p}'\|}, & \text{if } \boldsymbol{p} \times \boldsymbol{p}' \neq \boldsymbol{0}, \\ \left\| \dfrac{\boldsymbol{n}}{\|\boldsymbol{p}\|} - \dfrac{\boldsymbol{n}'}{\|\boldsymbol{p}'\|} \right\|, & \text{if } \boldsymbol{p} \times \boldsymbol{p}' = \boldsymbol{0}. \end{cases} \tag{4.52}$$

Eqs. (4.51) are replaced by

$$(\boldsymbol{p}, \boldsymbol{n}') + (\boldsymbol{p}', \boldsymbol{n}) = 0, \qquad \boldsymbol{p} \times \boldsymbol{p}' \neq \boldsymbol{0}. \tag{4.53}$$

The intersection $\boldsymbol{r}$ of the space lines L and L' is given by

$$\boldsymbol{r} = \frac{(\boldsymbol{m} \times \boldsymbol{r}_H) \times (\boldsymbol{m}' \times \boldsymbol{r}'_H)}{|\boldsymbol{m}, \boldsymbol{m}', \boldsymbol{r}_H|} = \frac{\boldsymbol{n} \times \boldsymbol{n}'}{(\boldsymbol{p}', \boldsymbol{n})}. \tag{4.54}$$

This expression is convenient for theoretical analysis but not suitable for actual computation, since both the numerator and the denominator vanish when $|\boldsymbol{m}, \boldsymbol{m}', \boldsymbol{r}_H| = 0$ or $(\boldsymbol{p}', \boldsymbol{n}) = 0$. A more convenient expression is

$$\begin{aligned} \boldsymbol{r} &= \boldsymbol{r}_H + \frac{(\boldsymbol{m}, \boldsymbol{r}'_H) + (\boldsymbol{m}, \boldsymbol{m}')(\boldsymbol{m}', \boldsymbol{r}_H)}{\|\boldsymbol{m} \times \boldsymbol{m}'\|^2} \boldsymbol{m} \\ &= \frac{1}{\|\boldsymbol{p}\|^2} \left(\frac{\|\boldsymbol{p}\|^2 |\boldsymbol{p}, \boldsymbol{p}', \boldsymbol{n}'| - (\boldsymbol{p}, \boldsymbol{p}') |\boldsymbol{p}, \boldsymbol{p}', \boldsymbol{n}|}{\|\boldsymbol{p} \times \boldsymbol{p}'\|^2} \boldsymbol{p} + \boldsymbol{p} \times \boldsymbol{n} \right). \end{aligned} \tag{4.55}$$

C. Joins

The *join* of two distinct space points $\boldsymbol{r}_1$ and $\boldsymbol{r}_2$ is a space line that passes through both $\boldsymbol{r}_1$ and $\boldsymbol{r}_2$ (Fig. 4.7b). In the $\{\boldsymbol{m},\ \boldsymbol{r}_H\}$-representation, the join is given by

$$\boldsymbol{m} = N[\boldsymbol{r}_1 - \boldsymbol{r}_2], \qquad \boldsymbol{r}_H = \frac{(\boldsymbol{m}, \boldsymbol{r}_1)\boldsymbol{r}_2 - (\boldsymbol{m}, \boldsymbol{r}_2)\boldsymbol{r}_1}{\|\boldsymbol{r}_1 - \boldsymbol{r}_2\|}. \tag{4.56}$$

In the $\{\boldsymbol{p},\ \boldsymbol{n}\}$-representation,

$$\begin{pmatrix} \boldsymbol{p} \\ \boldsymbol{n} \end{pmatrix} = N[\begin{pmatrix} \boldsymbol{r}_1 - \boldsymbol{r}_2 \\ \boldsymbol{r}_2 \times \boldsymbol{r}_1 \end{pmatrix}]. \tag{4.57}$$

If the space points $\boldsymbol{r}_1$ and $\boldsymbol{r}_2$ are statistically independent and have respective covariance matrices $V[\boldsymbol{r}_1]$ and $V[\boldsymbol{r}_2]$, the covariance matrix of their join is computed from eq. (4.57) in the following form (see eqs. (3.18)):

$$\begin{aligned} V[\boldsymbol{p} \oplus \boldsymbol{n}] = & \frac{1}{\|\boldsymbol{r}_1 - \boldsymbol{r}_2\|^2 + \|\boldsymbol{r}_2 \times \boldsymbol{r}_1\|^2} \boldsymbol{P}_{\boldsymbol{p} \oplus \boldsymbol{n}} \\ & \begin{pmatrix} V[\boldsymbol{r}_1] + V[\boldsymbol{r}_2] & V[\boldsymbol{r}_1] \times \boldsymbol{r}_2 + V[\boldsymbol{r}_2] \times \boldsymbol{r}_1 \\ \boldsymbol{r}_2 \times V[\boldsymbol{r}_1] + \boldsymbol{r}_1 \times V[\boldsymbol{r}_2] & \boldsymbol{r}_2 \times V[\boldsymbol{r}_1] \times \boldsymbol{r}_2 + \boldsymbol{r}_1 \times V[\boldsymbol{r}_2] \times \boldsymbol{r}_1 \end{pmatrix} \\ & \boldsymbol{P}_{\boldsymbol{p} \oplus \boldsymbol{n}}. \end{aligned} \tag{4.58}$$

Here, $\boldsymbol{P}_{\boldsymbol{p} \oplus \boldsymbol{n}}$ is the six-dimensional projection matrix defined by eq. (4.48).

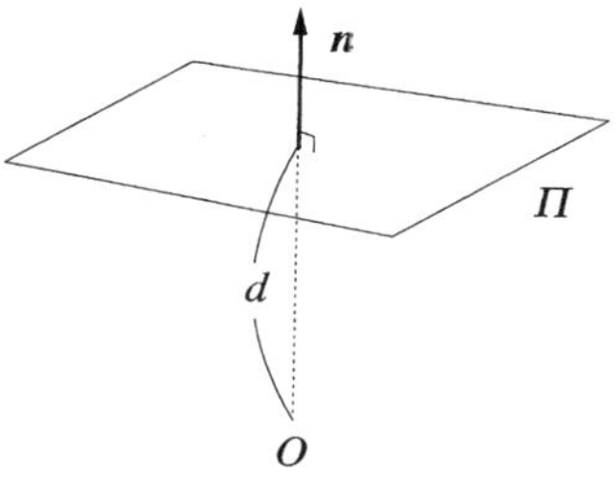

Fig. 4.8. Representation of a space plane.

4.3 Space Planes

4.3.1 Representation of space planes

A space plane Π is represented by its unit surface normal $\boldsymbol{n}$ and its (signed) distance d from the coordinate origin O (Fig. 4.8): the distance d is positive in the direction $\boldsymbol{n}$ and negative in the direction $-\boldsymbol{n}$. Hence, $\{\boldsymbol{n}, d\}$ and $\{-\boldsymbol{n}, -d\}$ represent the same space plane. The equation of a space plane that has unit surface normal $\boldsymbol{n} = (A, B, C)^\top$ and distance d is

$$AX + BY + CZ = d, \tag{4.59}$$

which can be written as

$$(\boldsymbol{n}, \boldsymbol{r}) = d. \tag{4.60}$$

If the 4-vector representation $\boldsymbol{\rho} = \boldsymbol{r} \oplus 1$ is used for space point $\boldsymbol{r}$, eq. (4.60) is expressed in the form

$$(\boldsymbol{\nu}, \boldsymbol{\rho}) = 0, \tag{4.61}$$

where $\boldsymbol{\nu} \propto \boldsymbol{n} \oplus (-1)$. The scale indeterminacy can be removed by imposing the normalization $\|\boldsymbol{\nu}\| = 1$:

$$\boldsymbol{\nu} = \frac{1}{\sqrt{1+d^2}} \begin{pmatrix} \boldsymbol{n} \\ -d \end{pmatrix}. \tag{4.62}$$

Conversely, the unit 4-vector $\boldsymbol{\nu} = (\nu_1, \nu_2, \nu_3, \nu_4)^\top$ can be decomposed into the unit surface normal $\boldsymbol{n}$ and the (signed) distance from the coordinate origin in the form

$$\boldsymbol{n} = N[\begin{pmatrix} \nu_1 \\ \nu_2 \\ \nu_3 \end{pmatrix}], \qquad d = -\frac{(\boldsymbol{\kappa}, \boldsymbol{\nu})}{\sqrt{1 - (\boldsymbol{\kappa}, \boldsymbol{\nu})^2}}, \tag{4.63}$$

where $\boldsymbol{\kappa} = (0, 0, 0, 1)^\top$. As in the case of space points, we must be careful about the rounding effect in fixed precision computation: if the distance d is very large, the first three components of $\boldsymbol{\nu}$ may be treated as effectively 0.

Hence, an appropriate scaling must be applied to keep the distance d in a reasonable range of magnitude.

As we did for image lines and space lines, we assume local distributions for all variables; the error behavior of a space plane is characterized by the four-dimensional covariance matrix

$$V[\boldsymbol{n} \oplus d] = \begin{pmatrix} V[\boldsymbol{n}] & V[\boldsymbol{n}, d] \\ V[\boldsymbol{n}, d] & V[d] \end{pmatrix}, \tag{4.64}$$

where $V[\boldsymbol{n}, d]$ $(= V[d, \boldsymbol{n}]^\top)$ is a 3-vector and $V[d]$ is a scalar. Since $\boldsymbol{n}$ is a unit vector, the null space of $V[\boldsymbol{n} \oplus d]$ is $\{\boldsymbol{n} \oplus 0\}_L$. If the $\boldsymbol{\nu}$-representation, the covariance matrix $V[\boldsymbol{\nu}]$ has the form

$$V[\boldsymbol{\nu}] = \frac{1}{1+d^2}\boldsymbol{P}_{\boldsymbol{\nu}} \begin{pmatrix} V[\boldsymbol{n}] & -V[\boldsymbol{n}, d] \\ -V[\boldsymbol{n}, d] & V[d] \end{pmatrix} \boldsymbol{P}_{\boldsymbol{\nu}}, \tag{4.65}$$

where $\boldsymbol{P}_{\boldsymbol{\nu}} = \boldsymbol{I} - \boldsymbol{\nu}\boldsymbol{\nu}^\top$ is the four-dimensional projection matrix along $\boldsymbol{\nu}$. Since $\boldsymbol{\nu}$ is a unit vector, the null space of $V[\boldsymbol{\nu}]$ is $\{\boldsymbol{\nu}\}_L$. Whichever representation is used, a space plane has *three* degrees of freedom: $\mathrm{rank}V[\boldsymbol{n} \oplus d] = \mathrm{rank}V[\boldsymbol{\nu}] = 3$.

If let $\boldsymbol{\nu} \to \boldsymbol{\nu} + \Delta\boldsymbol{\nu}$ in eqs. (4.63), the unit surface normal $\boldsymbol{n}$ and the distance d are respectively perturbed to a first approximation by

$$\Delta\boldsymbol{n} = \sqrt{1+d^2}\boldsymbol{P}_{\boldsymbol{\nu}} \begin{pmatrix} \Delta\nu_1 \\ \Delta\nu_2 \\ \Delta\nu_3 \end{pmatrix}, \qquad \Delta d = -\sqrt{(1+d^2)^3}\Delta\nu_4, \tag{4.66}$$

where we have used eqs. (3.17). It follows that for a given covariance matrix $V[\boldsymbol{\nu}]$ the the corresponding covariance matrix $V[\boldsymbol{n} \oplus d]$ can be computed as follows:

$$V[\boldsymbol{n}] = (1+d^2)\boldsymbol{P}_{\boldsymbol{\nu}} \begin{pmatrix} V[\boldsymbol{\nu}]_{11} & V[\boldsymbol{\nu}]_{12} & V[\boldsymbol{\nu}]_{13} \\ V[\boldsymbol{\nu}]_{21} & V[\boldsymbol{\nu}]_{22} & V[\boldsymbol{\nu}]_{23} \\ V[\boldsymbol{\nu}]_{31} & V[\boldsymbol{\nu}]_{32} & V[\boldsymbol{\nu}]_{33} \end{pmatrix} \boldsymbol{P}_{\boldsymbol{\nu}},$$

$$V[\boldsymbol{n}, d] = -(1+d^2)^2\boldsymbol{P}_{\boldsymbol{\nu}} \begin{pmatrix} V[\boldsymbol{\nu}]_{14} \\ V[\boldsymbol{\nu}]_{24} \\ V[\boldsymbol{\nu}]_{34} \end{pmatrix}, \quad V[d] = (1+d^2)^3 V[\boldsymbol{\nu}]_{44}. \tag{4.67}$$

4.3.2 Incidence, intersections, and joins

A. Incidence

1. A space point P and a space plane Π are *incident* to each other if P is on Π, or Π passes through P. A space point $\boldsymbol{r}$ (or $\boldsymbol{\rho}$) and a space plane represented by $\{\boldsymbol{n},\ d\}$ (or $\boldsymbol{\nu}$) are incident to each other if and only if

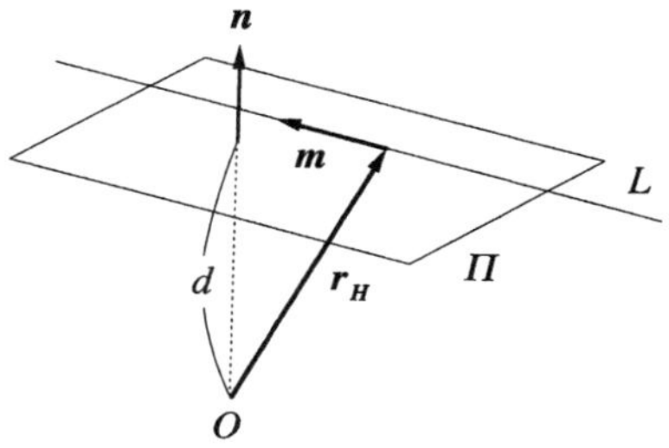

Fig. 4.9. Incidence of space plane Π and space line L.

$(\boldsymbol{n}, \boldsymbol{r}) = d$ (or $(\boldsymbol{\nu}, \boldsymbol{\rho}) = 0$). The distance $D(P, \Pi)$ between a space point P represented by $\boldsymbol{r}$ (or $\boldsymbol{\rho}$) and a space plane represented by $\{\boldsymbol{n},\ d\}$ (or $\boldsymbol{\nu}$) is

$$D(P, \Pi) = |(\boldsymbol{n}, \boldsymbol{r}) - d| = \frac{|(\boldsymbol{\nu}, \boldsymbol{\rho})|}{\sqrt{1 - (\boldsymbol{\kappa}, \boldsymbol{\nu})^2}}. \tag{4.68}$$

2. A space line L and a space plane Π are *incident* to each other if L is on Π, or Π passes through L (Fig. 4.9). A space line $(\boldsymbol{r} - \boldsymbol{r}_H) \times \boldsymbol{m} = 0$ and a space plane $(\boldsymbol{n}, \boldsymbol{r}) = d$ (or $(\boldsymbol{\nu}, \boldsymbol{\rho}) = 0$) are incident to each other if and only if

$$(\boldsymbol{n}, \boldsymbol{m}) = 0, \qquad (\boldsymbol{n}, \boldsymbol{r}_H) = d, \tag{4.69}$$

or equivalently[10]

$$(\boldsymbol{\nu}, \boldsymbol{m} \oplus 0) = 0, \qquad (\boldsymbol{\nu}, \boldsymbol{r}_H \oplus 1) = 0. \tag{4.70}$$

B. Intersections

1. Two distinct space planes $(\boldsymbol{n}_1, \boldsymbol{r}) = d_1$ and $(\boldsymbol{n}_2, \boldsymbol{r}) = d_2$ intersect along a space line as long as[11] $\boldsymbol{n}_1 \neq \pm\boldsymbol{n}_2$ (Fig. 4.10a). In the $\{\boldsymbol{m},\ \boldsymbol{r}_H\}$-representation, the intersection is

$$\boldsymbol{m} = N[\boldsymbol{n}_1 \times \boldsymbol{n}_2],$$

$$\boldsymbol{r}_H = \frac{(d_1 - (\boldsymbol{n}_1, \boldsymbol{n}_2)d_2)\boldsymbol{n}_1 + (d_2 - (\boldsymbol{n}_1, \boldsymbol{n}_2)d_1)\boldsymbol{n}_2}{\|\boldsymbol{n}_1 \times \boldsymbol{n}_2\|^2}. \tag{4.71}$$

[10] The first of eqs. (4.70) can be given the interpretation that the *ideal space point* ("space point at infinity") $\mathbf{m} \oplus \mathbf{0}$ of the space line $(\mathbf{r} - \mathbf{r}_H) \times \mathbf{m} = \mathbf{0}$ is on the space plane $(\nu, \rho) = 0$.

[11] Parallel distinct space planes can be thought of as intersecting along an *ideal space line* ("space line at infinity").

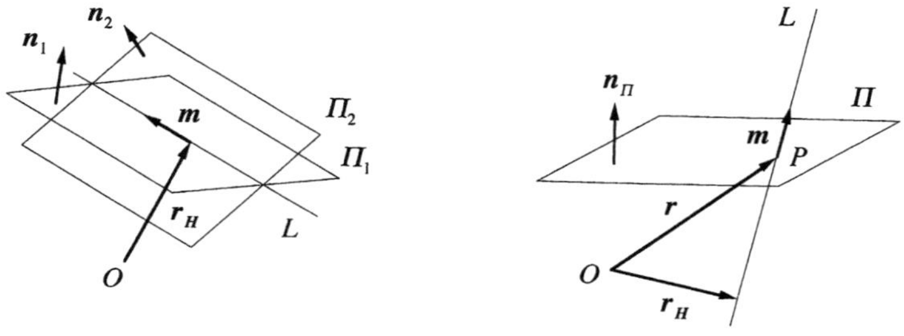

(a) (b)

Fig. 4.10. (a) The intersection L of space planes Π_1 and Π_2. (b) The intersection P of space line L and space plane Π.

In the $\{\boldsymbol{p}, \boldsymbol{n}\}$-representation,

$$\begin{pmatrix} \boldsymbol{p} \\ \boldsymbol{n} \end{pmatrix} = N[\begin{pmatrix} \boldsymbol{n}_1 \times \boldsymbol{n}_2 \\ d_2\boldsymbol{n}_1 - d_1\boldsymbol{n}_2 \end{pmatrix}]. \tag{4.72}$$

Suppose the two space planes $(\boldsymbol{n}_1, \boldsymbol{r}) = d_1$ and $(\boldsymbol{n}_2, \boldsymbol{r}) = d_2$ are statistically independent. Let $V[\boldsymbol{n}_1 \oplus d_1]$ and $V[\boldsymbol{n}_2 \oplus d_2]$ be their respective covariance matrices. The covariance matrix of the intersection is computed from eq. (4.72) in the form

$$V[\boldsymbol{p} \oplus \boldsymbol{n}] = \boldsymbol{P}_{\boldsymbol{p}\oplus\boldsymbol{n}} \begin{pmatrix} \boldsymbol{V}_{\boldsymbol{pp}} & \boldsymbol{V}_{\boldsymbol{pn}} \\ \boldsymbol{V}_{\boldsymbol{np}} & \boldsymbol{V}_{\boldsymbol{nn}} \end{pmatrix} \boldsymbol{P}_{\boldsymbol{p}\oplus\boldsymbol{n}}, \tag{4.73}$$

where

$$\boldsymbol{V}_{\boldsymbol{pp}} = \boldsymbol{n}_2 \times V[\boldsymbol{n}_1] \times \boldsymbol{n}_2 + \boldsymbol{n}_1 \times V[\boldsymbol{n}_2] \times \boldsymbol{n}_1,$$

$$\begin{aligned} \boldsymbol{V}_{\boldsymbol{pn}} &= -d_2\boldsymbol{n}_2 \times V[\boldsymbol{n}_1] - d_1\boldsymbol{n}_1 \times V[\boldsymbol{n}_2] \\ &\quad + \boldsymbol{n}_2 \times V[\boldsymbol{n}_1, d_1]\boldsymbol{n}_2^\top + \boldsymbol{n}_1 \times V[\boldsymbol{n}_2, d_2]\boldsymbol{n}_1^\top = \boldsymbol{V}_{\boldsymbol{np}}^\top, \end{aligned}$$

$$\begin{aligned} \boldsymbol{V}_{\boldsymbol{nn}} &= {d_2}^2 V[\boldsymbol{n}_1] + {d_1}^2 V[\boldsymbol{n}_2] - 2d_2 S[V[\boldsymbol{n}_1, d_1]\boldsymbol{n}_2^\top] \\ &\quad - 2d_1 S[V[\boldsymbol{n}_2, d_2]\boldsymbol{n}_1^\top] + V[d_1]\boldsymbol{n}_2\boldsymbol{n}_2^\top + V[d_2]\boldsymbol{n}_1\boldsymbol{n}_1^\top. \end{aligned} \tag{4.74}$$

Here, $\boldsymbol{P}_{\boldsymbol{p}\oplus\boldsymbol{n}}$ is the six-dimensional projection matrix defined by eq. (4.48).

2. A space plane $(\boldsymbol{n}_\Pi, \boldsymbol{r}) = d$ and a space line $(\boldsymbol{r} - \boldsymbol{r}_H) \times \boldsymbol{m} = \boldsymbol{0}$ (or $\boldsymbol{r} \times \boldsymbol{p} = \boldsymbol{n}_L$) intersect at a space point unless[12] $(\boldsymbol{n}_\Pi, \boldsymbol{m}) = 0$ (or $(\boldsymbol{n}_\Pi, \boldsymbol{p}) = 0$)

[12] A space plane and a space line that are parallel and not incident to each other can be thought of as intersecting at an *ideal space point* ("space point at infinity").

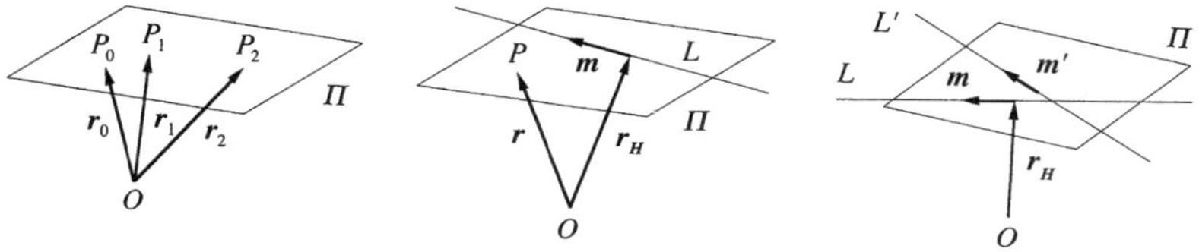

(a) (b) (c)

Fig. 4.11. (a) The join Π of three space points P_0, P_1, and P_2. (b) The join Π of space line L and space point P. (c) The join Π of space lines L and L'.

(Fig. 4.10b). Their intersection $\boldsymbol{r}$ is given by

$$\boldsymbol{r}=\boldsymbol{r}_H+\frac{d-(\boldsymbol{n}_\Pi,\boldsymbol{r}_H)}{(\boldsymbol{n}_\Pi,\boldsymbol{m})}\boldsymbol{m}=\frac{1}{\|\boldsymbol{p}\|^2}\left(\frac{|\boldsymbol{n}_\Pi,\boldsymbol{n}_L,\boldsymbol{p}|+d\|\boldsymbol{p}\|^2}{(\boldsymbol{n}_\Pi,\boldsymbol{p})}\boldsymbol{p}+\boldsymbol{p}\times\boldsymbol{n}_L\right). \tag{4.75}$$

C. Joins

1. Three distinct space points $\boldsymbol{r}_0$, $\boldsymbol{r}_1$, and $\boldsymbol{r}_2$ define a space plane, called their *join* (Fig. 4.11a), represented by

$$\boldsymbol{n} = N[(\boldsymbol{r}_1 - \boldsymbol{r}_0) \times (\boldsymbol{r}_2 - \boldsymbol{r}_0)], \qquad d = (\boldsymbol{n}, \boldsymbol{r}_0). \tag{4.76}$$

2. A space line $(\boldsymbol{r}-\boldsymbol{r}_H)\times\boldsymbol{m} = \boldsymbol{0}$ and a space point $\boldsymbol{r}_P$ define a space plane, called their *join* (Fig. 4.11b), if they are not incident to each other. The join is represented by

$$\boldsymbol{n} = N[\boldsymbol{m} \times (\boldsymbol{r}_P - \boldsymbol{r}_H)], \qquad d = (\boldsymbol{n}, \boldsymbol{r}_P). \tag{4.77}$$

3. Two intersecting space lines $(\boldsymbol{r} - \boldsymbol{r}_H) \times \boldsymbol{m} = \boldsymbol{0}$ and $(\boldsymbol{r} - \boldsymbol{r}'_H) \times \boldsymbol{m}' = \boldsymbol{0}$ define a space plane, called their *join* (Fig. 4.11c), represented by

$$\boldsymbol{n} = N[\boldsymbol{m} \times \boldsymbol{m}'], \qquad d = (\boldsymbol{n}, \boldsymbol{r}_H). \tag{4.78}$$

4.4 Conics

4.4.1 Classification of conics

A *conic* (sometimes referred to as a *conic locus* or *conic section*) is a curve on a two-dimensional plane whose equation has the form

$$Ax^2 + 2Bxy + Cy^2 + 2(Dx + Ey) + F = 0. \tag{4.79}$$

If an image point (x, y) is represented by 3-vector $\boldsymbol{x} = (x, y, 1)^\top$, eq. (4.79) can be written in the form

$$(\boldsymbol{x}, \boldsymbol{Q}\boldsymbol{x}) = 0, \tag{4.80}$$

where $\boldsymbol{Q}$ is a (33)-matrix defined by

$$\boldsymbol{Q} = \begin{pmatrix} A & B & D \\ B & C & E \\ D & E & F \end{pmatrix}. \tag{4.81}$$

A conic does not necessarily define a curve. A conic is *singular* if it defines two (real or imaginary) lines[13] (including one degenerate line); otherwise, the conic is *nonsingular*. It is easy to prove the following:

- Conic $(\boldsymbol{x}, \boldsymbol{Q}\boldsymbol{x}) = 0$ is nonsingular if and only if the matrix $\boldsymbol{Q}$ is nonsingular.
- A nonsingular conic $(\boldsymbol{x}, \boldsymbol{Q}\boldsymbol{x}) = 0$ defines a *real conic* (an *ellipse*, a *parabola*, or a *hyperbola*) if and only if the *signature* of $\boldsymbol{Q}$ is (2,1) or (1,2) (see Section 2.3.4).

If the sign of $\boldsymbol{Q}$ are chosen so that $\det \boldsymbol{Q} \leq 0$, the type of conic $(\boldsymbol{x}, \boldsymbol{Q}\boldsymbol{x}) = 0$ is classified as follows:

1. If $\det \boldsymbol{Q} = 0$, the conic defines two (real or imaginary) lines (including one degenerate line).
2. If $\det \boldsymbol{Q} \neq 0$, then
   (a) if $AC - B^2 > 0$, then
      i. if $A + C > 0$, the conic is an ellipse,
      ii. if $A + C < 0$, the conic is an empty set (an *imaginary ellipse*),
   (b) if $AC - B^2 = 0$, the conic is a parabola,
   (c) if $AC - B^2 < 0$, the conic is a hyperbola.

Given a nonsingular conic $(\boldsymbol{x}, \boldsymbol{Q}\boldsymbol{x}) = 0$ and an image point $\boldsymbol{x}_p$, the image line $(\boldsymbol{n}_p, \boldsymbol{x}) = 0$ for

$$\boldsymbol{n}_p = \pm N[\boldsymbol{Q}\boldsymbol{x}_p] \tag{4.82}$$

is called the *polar* of the image point $\boldsymbol{x}_p$ with respect to the conic $(\boldsymbol{x}, \boldsymbol{Q}\boldsymbol{x}) = 0$. Conversely, given a nonsingular conic $(\boldsymbol{x}, \boldsymbol{Q}\boldsymbol{x}) = 0$ and an image line $(\boldsymbol{n}_p, \boldsymbol{x}) = 0$, the image point

$$\boldsymbol{x}_p = \frac{\boldsymbol{Q}^{-1}\boldsymbol{n}_p}{(\boldsymbol{k}, \boldsymbol{Q}^{-1}\boldsymbol{n}_p)} \tag{4.83}$$

is called the *pole* of the image line $(\boldsymbol{n}_p, \boldsymbol{x}) = 0$ with respect to the conic $(\boldsymbol{x}, \boldsymbol{Q}\boldsymbol{x}) = 0$. We observe the following:

[13] If two imaginary lines intersect, their intersection is an isolated real point.

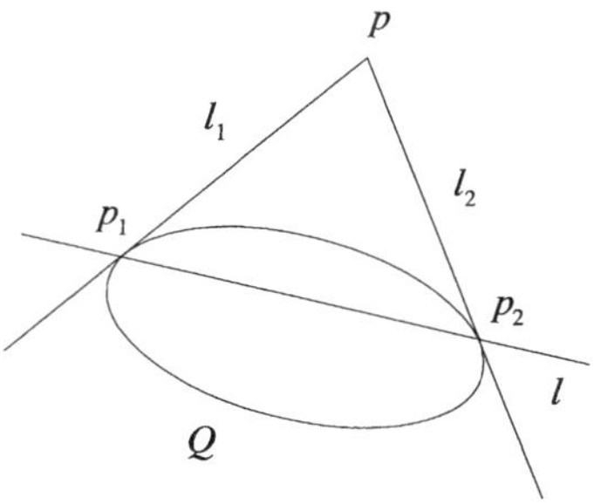

Fig. 4.12. Image point p is the pole of image line l, and image line l is the polar of image pont p. The tangent points p_1 and p_2 are the poles of the tangents l_1 and l_2, respectively, and the tangents l_1 and l_2 are the polars of the tangent points p_1 and p_2, respectively.

- A point $\boldsymbol{x}_p$ is on conic $(\boldsymbol{x}, \boldsymbol{Q}\boldsymbol{x}) = 0$ if and only if its polar $(\boldsymbol{n}_p, \boldsymbol{x}) = 0$ is *tangent* to the conic $(\boldsymbol{x}, \boldsymbol{Q}\boldsymbol{x}) = 0$ at $\boldsymbol{x}_p$.

- Let $\boldsymbol{x}_p$ be the pole of an image line $(\boldsymbol{n}_p, \boldsymbol{x}) = 0$ with respect to conic $(\boldsymbol{x}, \boldsymbol{Q}\boldsymbol{x}) = 0$. If the image line $(\boldsymbol{n}_p, \boldsymbol{x}) = 0$ has two intersections $\boldsymbol{x}_1$ and $\boldsymbol{x}_2$ with the conic $(\boldsymbol{x}, \boldsymbol{Q}\boldsymbol{x}) = 0$, the join of $\boldsymbol{x}_p$ and $\boldsymbol{x}_i$ is tangent to the conic $(\boldsymbol{x}, \boldsymbol{Q}\boldsymbol{x}) = 0$ at $\boldsymbol{x}_i$, $i = 1, 2$ (Fig. 4.12).

4.4.2 Canonical forms of conics

A real conic can be reduced to its *canonical form* by an appropriate translation and rotation of the xy coordinate system. If the coordinate system is rotated by angle θ and translated by (a, b), the transformation of coordinates can be written in the following form[14]:

$$\boldsymbol{x}' = \boldsymbol{A}\boldsymbol{x}, \qquad \boldsymbol{A} = \begin{pmatrix} \cos\theta & \sin\theta & -a\cos\theta - b\sin\theta \\ -\sin\theta & \cos\theta & a\sin\theta - b\cos\theta \\ 0 & 0 & 1 \end{pmatrix}. \tag{4.84}$$

Hence,

$$\boldsymbol{x} = \boldsymbol{A}^{-1}\boldsymbol{x}', \qquad \boldsymbol{A}^{-1} = \begin{pmatrix} \cos\theta & -\sin\theta & a \\ \sin\theta & \cos\theta & b \\ 0 & 0 & 1 \end{pmatrix}. \tag{4.85}$$

By this coordinate transformation, conic $(\boldsymbol{x}, \boldsymbol{Q}\boldsymbol{x}) = 0$ is transformed to conic $(\boldsymbol{x}', \boldsymbol{Q}'\boldsymbol{x}') = 0$ for

$$\boldsymbol{Q}' = (\boldsymbol{A}^{-1})^\top \boldsymbol{Q}\boldsymbol{A}^{-1}. \tag{4.86}$$

---

[14]The set of matrices that have the form shown in eqs. (4.84) is closed under matrix multiplication and called the group of two-dimensional *Euclidean motions*.

This is the *congruence transformation* of $\boldsymbol{Q}$ by matrix $\boldsymbol{A}^{-1}$ (see eq. (2.154)). Since the signature is preserved by a congruence transformation (*Sylvester's law of inertia*; see Section 2.3.4), a real conic is always transformed to a real conic[15].

Consider a nonsingular conic $(\boldsymbol{x}, \boldsymbol{Q}\boldsymbol{x}) = 0$. If the scale and the sign of $\boldsymbol{Q}$ are chosen so that $\det \boldsymbol{Q} = -1$, the conic has the following canonical form. Let

$$\lambda_1, \lambda_2 = \frac{(A+C) \pm \sqrt{(A+C)^2 - 4(AC - B^2)}}{2}. \tag{4.87}$$

1. If $AC - B^2 \neq 0$, then λ_1 and λ_2 are both nonzero. Let

$$\mu = \frac{1}{AC - B^2}, \qquad a = \sqrt{\left|\frac{\mu}{\lambda_1}\right|}, \qquad b = \sqrt{\left|\frac{\mu}{\lambda_2}\right|}. \tag{4.88}$$

   (a) If $\mu\lambda_1 > 0$ and $\mu\lambda_2 > 0$, the conic is an ellipse (Fig. 4.13a) with canonical form

   $$\frac{x^2}{a^2} + \frac{y^2}{b^2} = 1. \tag{4.89}$$

   This ellipse has radii a and b in the x and y directions, respectively.

   (b) If $\mu\lambda_1$ and $\mu\lambda_2$ have opposite signs, the conic is a hyperbola (Fig. 4.13b) with canonical form

   $$\pm \frac{x^2}{a^2} \mp \frac{y^2}{b^2} = 1, \tag{4.90}$$

   where the upper signs are for $\mu\lambda_1 > 0$ and $\mu\lambda_2 < 0$, and the lower signs are for $\mu\lambda_1 < 0$ and $\mu\lambda_2 > 0$. In the former (latter) case, this hyperbola intersects with the x-axis (y-axis) at $x = a$ ($y = b$) and has two *asymptotes* $y = \pm(b/a)x$.

   (c) If $\mu\lambda_1 < 0$ and $\mu\lambda_2 < 0$, the conic is an imaginary ellipse with canonical form

   $$-\frac{x^2}{a^2} - \frac{y^2}{b^2} = 1. \tag{4.91}$$

2. If $AC - B^2 = 0$, either λ_1 or λ_2 is zero. The conic is a parabola (Fig. 4.13c).

   (a) If $\lambda_1 \neq 0$ and $\lambda_2 = 0$, the canonical form is

   $$y = \left|\frac{(A+C)\sqrt{A^2 + B^2}}{2(BD - AE)}\right| x^2. \tag{4.92}$$

[15] This property holds not only for two-dimensional Euclidean motions but also for all two-dimensional *projective transformations*. A two-dimensional projective transformation has the form $\mathbf{x}' \propto \mathbf{A}\mathbf{x}$, where $\mathbf{A}$ is an arbitrary nonsingular 33-matrix. Conics undergo the congruence transformation in the form of eq. (4.86).

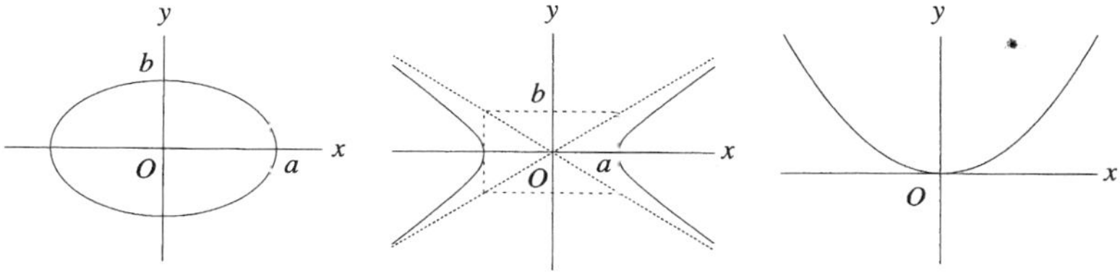

(a) (b) (c)

Fig. 4.13. (a) Canonical form of an ellipse. (b) Canonical form of a hyperbola. (c) Canonical form of a parabola.

(b) If $\lambda_1 = 0$ and $\lambda_2 \neq 0$, the canonical form is

$$y = \left| \frac{(A+C)\sqrt{B^2+C^2}}{2(BE-CD)} \right| x^2. \tag{4.93}$$

The shape of a real conic is characterized by its axes of symmetry and *eccentricity* as follows:

1. An ellipse has two axes of symmetry, called the *principal axes*, that are orthogonal to each other. In the canonical form of eq. (4.89), the x-axis is called the *major* (*minor*) axis if $a < b$ ($a > b$). Its area is

$$S = \pi ab. \tag{4.94}$$

Its eccentricity e is defined to be

$$e = \begin{cases} \frac{\sqrt{a^2-b^2}}{a}, & a \geq b, \\ \frac{\sqrt{b^2-a^2}}{b}, & a < b. \end{cases} \tag{4.95}$$

The eccentricity of a circle is 0.

2. For a hyperbola in the canonical form $x^2/a^2 - y^2/b^2 = 1$, the x- and y-axes are called the *transverse axis* and the *conjugate* axis, respectively. Its eccentricity is defined to be

$$e = \frac{\sqrt{a^2+b^2}}{a}. \tag{4.96}$$

If $e = \sqrt{2}$ (i.e., $a = b$), the curve is said to be a *rectangular hyperbola*, for which the two asymptotes are mutually orthogonal.

3. A parabola has one axis of symmetry. In the canonical form of eq. (4.92) or (4.93), the origin and the y-axis are called the *axis* and the *vertex*, respectively. The eccentricity of a parabola is defined to be $e = 1$.

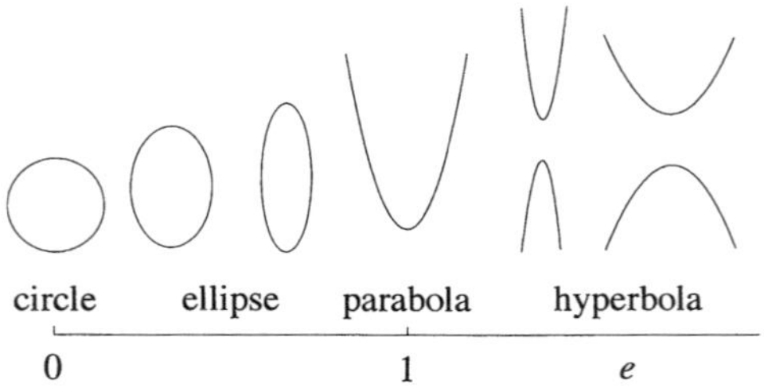

Fig. 4.14. Eccentricity e of conics: $0 \leq e < 1$ for ellipses, $e = 1$ for parabolas, and $e > 1$ for hyperbolas.

Thus, $0 \leq e < 1$ for ellipses, $e = 1$ for parabolas, and $e > 1$ for hyperbolas (Fig. 4.14).

4.5 Space Conics and Quadrics

4.5.1 Representation in three dimensions

A. Space conics

A *space conic* is a conic defined on a space plane. We represent it by *back projection*: we regard a space conic as the intersection of a space plane with a "cone" with vertex at the viewpoint O generated by the lines of sight of image points on a conic (Fig. 4.15). It follows that a space conic is represented by the space plane $(\boldsymbol{n}, \boldsymbol{r}) = d$ on which the space conic lies and its projection $(\boldsymbol{x}, \boldsymbol{Q}\boldsymbol{x}) = 0$ onto the image plane; a space conic represented by $\{\boldsymbol{n}, d, \boldsymbol{Q}\}$ is a set of space points that satisfy

$$(\boldsymbol{n}, \boldsymbol{r}) = d, \qquad (\boldsymbol{r}, \boldsymbol{Q}\boldsymbol{r}) = 0. \tag{4.97}$$

B. Quadrics

A *quadric* is a surface defined by a quadratic equation in the coordinates (X, Y, Z). Quadrics are classified into *ellipsoids*, *paraboloids*, *hyperboloids*, and their degeneracies (e.g., cylinders, pairs of space planes, space points, and empty sets). Paraboloids, hyperboloids, and their degeneracies are called *centered quadrics*, because they have centers of symmetry. Let $\boldsymbol{r}_C$ be the center of symmetry, which we simply call the *center*. Then, a centered quadric is represented in the form

$$(\boldsymbol{r} - \boldsymbol{r}_C, \boldsymbol{S}(\boldsymbol{r} - \boldsymbol{r}_C)) = 1, \tag{4.98}$$

where $\boldsymbol{S}$ is a (33)-matrix. It is easy to see the following:

1. If $\boldsymbol{S}$ is positive definite, the quadric is an ellipsoid.

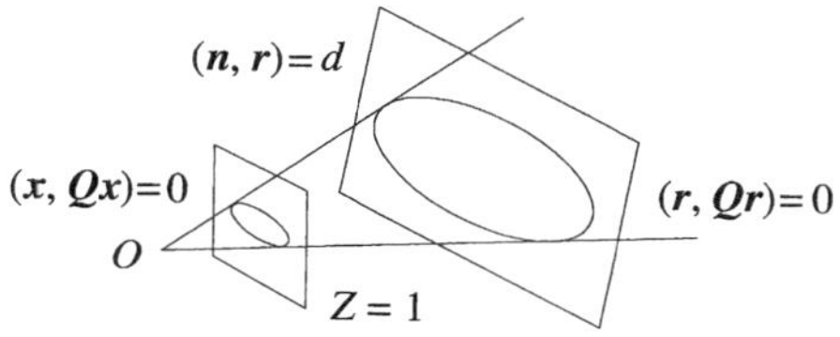

Fig. 4.15. Representation of a space conic.

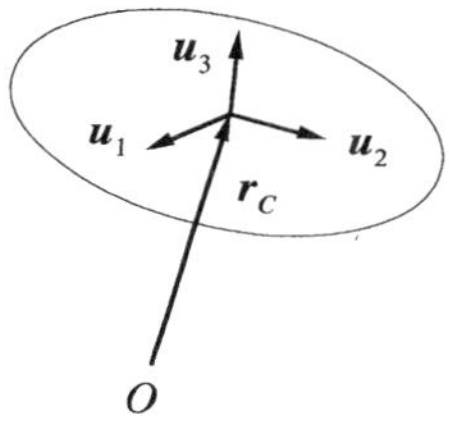

Fig. 4.16. Representation of a quadric.

2. If $\boldsymbol{S}$ is negative definite, the quadric is an empty set.

3. If $\boldsymbol{S}$ has signature (2,1), the quadric is a hyperboloid of one sheet.

4. If $\boldsymbol{S}$ has signature (1,2), the quadric is a hyperboloid of two sheets.

5. If $\boldsymbol{S}$ is singular, various types of degeneracy occur.

The quadric given by eq. (4.98) is said to be *singular* or *nonsingular* depending on the matrix $\boldsymbol{S}$ is singular or nonsingular. The eigenvectors of $\boldsymbol{S}$ are called its *principal axes*. The reciprocal of the square root of its positive eigenvalue is the radius of the quadric for the corresponding principal axis (Fig. 4.16). For an ellipsoid that has an orthonormal system $\{\boldsymbol{u}_1, \boldsymbol{u}_2, \boldsymbol{u}_3\}$ as its principal axes and σ_1, σ_2, and σ_2 as the corresponding radii, the matrix $\boldsymbol{S}$ has the following form (see eq. (2.62)):

$$\boldsymbol{S} = \frac{\boldsymbol{u}_1\boldsymbol{u}_1^\top}{\sigma_1^2} + \frac{\boldsymbol{u}_2\boldsymbol{u}_2^\top}{\sigma_2^2} + \frac{\boldsymbol{u}_3\boldsymbol{u}_3^\top}{\sigma_3^2}. \tag{4.99}$$

Example 4.4 An ellipsoid centered at (a, b, c) with principal axes in the the coordinate axis orientations and the corresponding radii σ_1, σ_2, and σ_2 has the form

$$\frac{(X-a)^2}{\sigma_1^2} + \frac{(Y-b)^2}{\sigma_2^2} + \frac{(Z-c)^2}{\sigma_3^2} = 1. \tag{4.100}$$

The center $\boldsymbol{r}_C$ and the matrix $\boldsymbol{S}$ are given as follows:

$$\boldsymbol{r}_C = \begin{pmatrix} a \\ b \\ c \end{pmatrix}, \quad \boldsymbol{S} = \begin{pmatrix} 1/\sigma_1^2 & & \\ & 1/\sigma_2^2 & \\ & & 1/\sigma_3^2 \end{pmatrix}. \tag{4.101}$$

4.5.2 Polarity and conjugate direction

A. Poles and polars

Consider a nonsingular quadric given by eq. (4.98). The *polar* of a space point $\boldsymbol{r}_p$ with respect to this quadric is a space plane defined by

$$(\boldsymbol{r} - \boldsymbol{r}_C, \boldsymbol{S}(\boldsymbol{r}_p - \boldsymbol{r}_C)) = 1. \tag{4.102}$$

Its unit surface normal $\boldsymbol{n}_p$ and distance d from O are

$$\boldsymbol{n}_p = N[\boldsymbol{S}(\boldsymbol{r}_p - \boldsymbol{r}_C)], \qquad d_p = \frac{1}{\|\boldsymbol{S}(\boldsymbol{r}_p - \boldsymbol{r}_C)\|} + (\boldsymbol{r}_C, \boldsymbol{n}_p). \tag{4.103}$$

Conversely, the space point $\boldsymbol{r}_p$ is called the *pole* of the space plane $(\boldsymbol{n}_p, \boldsymbol{r}) = d_p$ with respect to the same quadric. From eqs. (4.103), we see that

$$\boldsymbol{r}_p = \frac{\boldsymbol{S}^{-1}\boldsymbol{n}}{d_p - (\boldsymbol{r}_C, \boldsymbol{n}_p)}. \tag{4.104}$$

Eq. (4.102) implies the following:

- The polar of $\boldsymbol{r}_p$ passes through $\boldsymbol{r}_p$ *if and only if* $\boldsymbol{r}_p$ *is on the quadric* (4.98).
- If $\boldsymbol{r}_p$ is on the quadric (4.98), the polar of $\boldsymbol{r}_p$ is *tangent* to the quadric at $\boldsymbol{r}_p$ (Fig. 4.17a).
- The surface normal $\boldsymbol{n}$ to the tangent plane to the quadric (4.98) at $\boldsymbol{r}$ has orientation $\boldsymbol{S}(\boldsymbol{r} - \boldsymbol{r}_C)$:

$$\boldsymbol{n} \propto \boldsymbol{S}(\boldsymbol{r} - \boldsymbol{r}_C). \tag{4.105}$$

B. Conjugate direction

Consider a space plane Π that passes through the center $\boldsymbol{r}_C$ of the quadric (4.98). Let $\boldsymbol{n}$ be its surface normal. The *conjugate direction*[16] of the space

[16] If the space undergoes a general linear transformation, vectors $\mathbf{n}$ and $\mathbf{n}^\dagger$ transform differently as a *covariant vector* and a *contravariant vector*, respectively. In this sense, the transformations of $\mathbf{n}$ and $\mathbf{n}^\dagger$ are *contragradient* to each other.

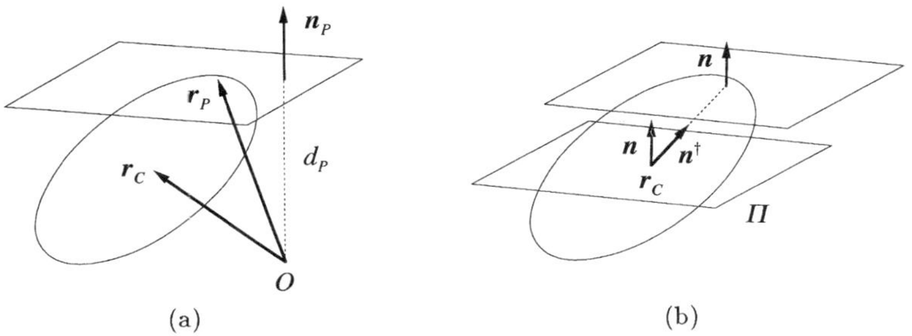

Fig. 4.17. (a) The polar of $\mathbf{r}_p$ passes through $\mathbf{r}_p$ if and only if it is tangent to the quadric at $\mathbf{r}_p$. (b) The conjugate direction $\mathbf{n}^\dagger$ of space plane Π.

plane Π with respect to the quadric (4.98) is the orientation of the vector $\boldsymbol{n}^\dagger$ that starts from the center $\boldsymbol{r}_C$ and points toward a point on the quadric at which the tangent plane has surface normal $\pm\boldsymbol{n}$ (Fig. 4.17b). It follows from eq. (4.105) that

$$\boldsymbol{n}^\dagger \propto \boldsymbol{S}^{-1}\boldsymbol{n}. \tag{4.106}$$

If the quadric is a sphere, the conjugate direction of a space plane coincides with its surface normal. In this sense, the conjugate direction is a generalization of the surface normal to a space plane.

C. Generalization

Quadrics can also be defined in higher dimensions: eq. (4.98) defines a quadric in n dimensions if $\boldsymbol{r}$ and $\boldsymbol{r}_C$ are regarded as n-vectors and $\boldsymbol{S}$ is regarded as an (nn) matrix. If $\boldsymbol{S}$ is nonsingular, the polarity and the conjugate direction are defined as in three dimensions straightforwardly. Suppose $\boldsymbol{S}$ is a positive semi-definite symmetric matrix of rank r ($< n$), and let $\{\boldsymbol{u}_i\}$ be its eigensystem for eigenvalues $\{\lambda_i\}$, where $\lambda_1 \geq ... \geq \lambda_r > \lambda_{r+1} = ... = \lambda_n = 0$. Then, the center $\boldsymbol{r}_C$ and the vectors $\boldsymbol{u}_1$, ..., $\boldsymbol{u}_r$ define an r-dimensional *affine subspace*[17] $\mathcal{S} \subset \mathcal{R}^n$. If the quadric is restricted to this affine subspace (Fig. 4.18), it defines a nonsingular quadric in it, so the polarity and the conjugate direction are defined in $\mathcal{S}$.

4.5.3 Visualization of covariance matrices

Conics are important image features because many man-made objects have circular and spherical parts, and circles and spheres are projected onto ellipses on the image plane. As compared with conics, the role of quadrics appears to be minor, since ellipsoidal objects are rare in real environments. However, quadrics have an important role: a three-dimensional positive semi-definite

[17] See Footnote 17 in Section 3.4.2.

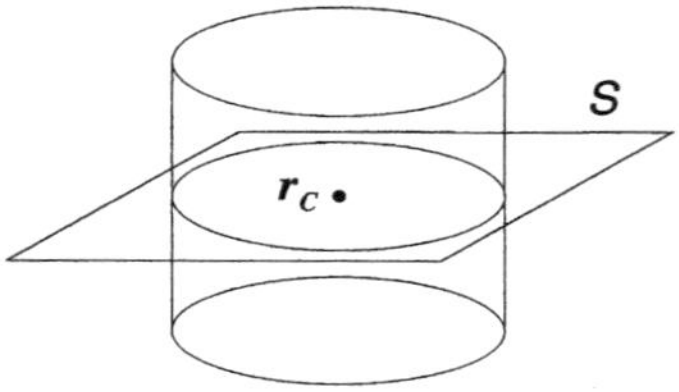

Fig. 4.18. A singular quadric defines a nonsingular quadric if restricted to the affine subspace spanned by eigenvectors for positive eigenvalues.

symmetric matrix is represented by an ellipsoid. This fact allows us to visualize three-dimensional covariance matrices. The following two techniques serve this purpose.

A. Standard confidence region

Let $V[\hat{\boldsymbol{r}}]$ be the covariance matrix of a space point $\hat{\boldsymbol{r}}$. If it has eigenvalues λ_1, λ_2, and λ_3 and the corresponding eigensystem $\{\boldsymbol{u}_1, \boldsymbol{u}_2, \boldsymbol{u}_3\}$, it has the spectral decomposition

$$V[\hat{\boldsymbol{r}}] = \sigma_1^2 \boldsymbol{u}_1 \boldsymbol{u}_1^\top + \sigma_2^2 \boldsymbol{u}_2 \boldsymbol{u}_2^\top + \sigma_3^2 \boldsymbol{u}_3 \boldsymbol{u}_3^\top, \tag{4.107}$$

where $\sigma_i = \sqrt{\lambda_i}$, $i = 1, 2, 3$. We can regard σ_i as the standard deviation in direction $\boldsymbol{u}_i$ (see eq. (3.21)).

1. Suppose $V[\hat{\boldsymbol{r}}]$ is of full rank. If the distribution is Gaussian, the surface on which the probability density is constant is an ellipsoid centered at $\hat{\boldsymbol{r}}$ in the form $(\boldsymbol{r} - \hat{\boldsymbol{r}}, V[\hat{\boldsymbol{r}}]^{-1}(\boldsymbol{r} - \hat{\boldsymbol{r}})) = \text{constant}$ (see eq. (3.38)). If we choose the constant to be $1/\sqrt{(2\pi)^3 |V[\hat{\boldsymbol{r}}]|} e^{-1/2}$, the ellipsoid is

$$(\boldsymbol{r} - \hat{\boldsymbol{r}}, V[\hat{\boldsymbol{r}}]^{-1}(\boldsymbol{r} - \hat{\boldsymbol{r}})) = 1, \tag{4.108}$$

which has principal axes $\{\boldsymbol{u}_1, \boldsymbol{u}_2, \boldsymbol{u}_3\}$ and the corresponding radii σ_1, σ_2, and σ_3. Let us call the region inside this ellipsoid the *standard confidence region* of $\hat{\boldsymbol{r}}$. This is a natural extension of the standard confidence interval (3.42). The probability that the true value of $\boldsymbol{r}$ falls inside the standard confidence region is 19.87% (see eq. (3.62)).

2. If $V[\hat{\boldsymbol{r}}]$ is of rank 2 with $\sigma_3 = 0$, the standard confidence region degenerates into a space conic with principal axes $\{\boldsymbol{u}_1, \boldsymbol{u}_2\}$ and the corresponding radii σ_1 and σ_2. It is represented by

$$(\boldsymbol{u}_3, \boldsymbol{r}) = (\boldsymbol{u}_3, \hat{\boldsymbol{r}}), \qquad (\boldsymbol{x}, \boldsymbol{Q}\boldsymbol{x}) = 0, \tag{4.109}$$

where

$$\boldsymbol{Q} = N[(\boldsymbol{u}_1, \boldsymbol{u}_2, \hat{\boldsymbol{r}})^{\dagger\top} \text{diag}(\sigma_1^2, \sigma_2^2, -1)(\boldsymbol{u}_1, \boldsymbol{u}_2, \hat{\boldsymbol{r}})^{\dagger}]. \tag{4.110}$$

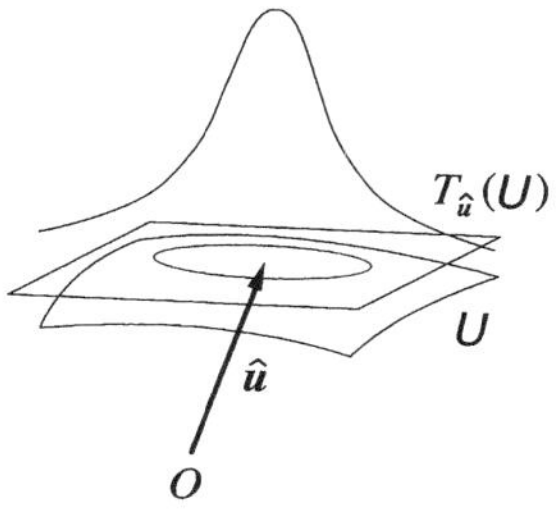

Fig. 4.19. The standard confidence region is defined in the tangent space $T_{\hat{\mathbf{u}}}(\mathcal{U})$ to the manifold $\mathcal{U}$ defined by the constraint.

The symbol $N[\cdot]$ denotes the normalization for a matrix (see eq. (2.145)), and $(\boldsymbol{u}_1, \boldsymbol{u}_2, \hat{\boldsymbol{r}})^\dagger$ is the *cofactor* of the matrix whose columns are $\boldsymbol{u}_1$, $\boldsymbol{u}_2$, and $\hat{\boldsymbol{r}}$ in that order (see eq. (2.18)). The probability that the true value of $\boldsymbol{r}$ falls inside this space conic is 39.35%.

3. If $V[\hat{\boldsymbol{r}}]$ is of rank 1 with $\sigma_2 = \sigma_3 = 0$, the standard confidence region degenerates into the line segment connecting $\hat{\boldsymbol{r}} - \sigma_1 \boldsymbol{u}_1$ and $\hat{\boldsymbol{r}} + \sigma_1 \boldsymbol{u}_1$. The probability that the true value of $\boldsymbol{r}$ falls inside this segment is 68.27%.

The standard confidence region can be defined in higher dimensions in the same way. Let $\hat{\boldsymbol{u}}$ be an n-vector that represents some object, and $V[\hat{\boldsymbol{u}}]$ its covariance matrix. Suppose $\hat{\boldsymbol{u}}$ is constrained to be in an r-dimensional manifold $\mathcal{U}$. Then, $V[\hat{\boldsymbol{u}}]$ is generally a singular matrix of rank r, so the equation

$$(\boldsymbol{u} - \hat{\boldsymbol{u}}, V[\hat{\boldsymbol{u}}]^-(\boldsymbol{u} - \hat{\boldsymbol{u}})) = 1 \tag{4.111}$$

generally defines a singular quadric. However, it defines a nonsingular quadric if it is restricted to the tangent space $T_{\hat{\boldsymbol{u}}}(\mathcal{U})$ to the manifold $\mathcal{U}$ at $\hat{\boldsymbol{u}}$ (Fig. 4.19). We define the standard confidence region of $V[\hat{\boldsymbol{u}}]$ to be inside this quadric.

B. Primary deviation pair

Covariance matrices can be visualized by means of their standard confidence regions only in three dimensions. In higher dimensions, we use an alternative technique. Note that if σ_1 is sufficiently large as compared with σ_2 in eq. (4.107), the error distribution is localized along the *major axis* (i.e., the principal axis with the largest eigenvalue). This suggests the following visualization technique. Let $\hat{\boldsymbol{u}}$ be an n-vector that represents a geometric object, and $V[\hat{\boldsymbol{n}}]$ its covariance matrix. Let

$$V[\hat{\boldsymbol{u}}] = \sum_{i=1}^{n} \sigma_i^2 \boldsymbol{u}_i \boldsymbol{u}_i^\top, \qquad \sigma_1^2 \geq \sigma_2^2 \geq \cdots \geq \sigma_n^2, \tag{4.112}$$

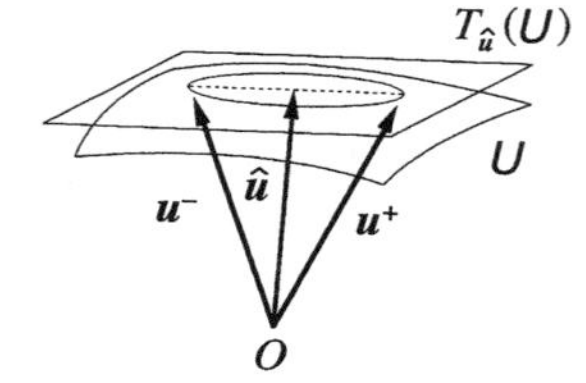

Fig. 4.20. The primary deviation pair.

be its spectral decomposition. The vector $\boldsymbol{u}_1$ can be interpreted as the orientation of the most likely deviation, and σ_1 as the standard deviation in that orientation. Hence, the reliability of $\hat{\boldsymbol{u}}$ can be visualized by displaying the two objects represented by the *primary deviation pair* $\{\boldsymbol{u}^+, \boldsymbol{u}^-\}$ defined as follows (see eq. (3.48)):

$$\boldsymbol{u}^+ = \mathcal{C}[\hat{\boldsymbol{u}} + \sigma_1 \boldsymbol{u}_1], \qquad \boldsymbol{u}^- = \mathcal{C}[\hat{\boldsymbol{u}} - \sigma_1 \boldsymbol{u}_1]. \tag{4.113}$$

Here, the operation $\mathcal{C}[\,\cdot\,]$ designates high order correction, such as normalization, to ensure that $\boldsymbol{u}^+ \in \mathcal{U}$ and $\boldsymbol{u}^- \in \mathcal{U}$. To a first approximation, the primary deviation pair $\{\boldsymbol{u}^+, \boldsymbol{u}^-\}$ indicates diametrically located endpoints in the standard confidence region[18] in the direction of its major axis (Fig. 4.20).

4.6 Coordinate Transformation and Projection

4.6.1 Coordinate transformation

Since objects in three dimensions are represented with respect to a fixed XYZ coordinate system, the same object has different representations if described with respect to different coordinate systems. Suppose a new $X'Y'Z'$ coordinate system is defined in such a way that its axis orientation is obtained by rotating the original XYZ coordinate system by $\boldsymbol{R}$ and its origin O' is translated from O by $\boldsymbol{h}$; the vector $\boldsymbol{h}$ and the rotation matrix $\boldsymbol{R}$ are defined with respect to the original XYZ coordinate system. We call $\{\boldsymbol{h}, \boldsymbol{R}\}$ the *motion parameters* of the coordinate system (Fig. 4.21).

This definition implies that if $\{\boldsymbol{h}, \boldsymbol{R}\}$ are the motion parameters of the $X'Y'Z'$ coordinate system with respect to the XYZ coordinate system, the motion parameters $\{\boldsymbol{h}', \boldsymbol{R}'\}$ of the XYZ coordinate system with respect to the $X'Y'Z'$ coordinate system are given by

$$\boldsymbol{h}' = -\boldsymbol{R}^\top \boldsymbol{h}, \qquad \boldsymbol{R}' = \boldsymbol{R}^\top. \tag{4.114}$$

[18] Note that the standard confidence region of $V[\hat{\mathbf{u}}]$ is defined in the tangent space $T_{\hat{\mathbf{u}}}(\mathcal{U})$, which does not coincide with the manifold $\mathcal{U}$ if $\mathcal{U}$ is "curved" (Fig. 4.19).

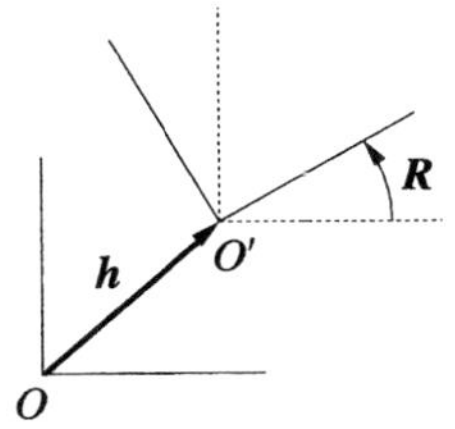

Fig. 4.21. Coordinate transformation and the motion parameters {h, R}.

A. Space points

1. A space point $\boldsymbol{r}$ defined with respect to the XYZ coordinate system is represented with respect to the $X'Y'Z'$ coordinate system in the form

$$\boldsymbol{r}' = \boldsymbol{R}^\top(\boldsymbol{r} - \boldsymbol{h}). \tag{4.115}$$

Hence,

$$\boldsymbol{r} = \boldsymbol{h} + \boldsymbol{R}\boldsymbol{r}'. \tag{4.116}$$

2. In the $\boldsymbol{\rho}$-representation[19],

$$\boldsymbol{\rho}' = \begin{pmatrix} \boldsymbol{R}^\top & -\boldsymbol{R}^\top\boldsymbol{h} \\ \boldsymbol{0} & 1 \end{pmatrix}\boldsymbol{\rho}, \quad \boldsymbol{\rho} = \begin{pmatrix} \boldsymbol{R} & \boldsymbol{h} \\ \boldsymbol{0} & 1 \end{pmatrix}\boldsymbol{\rho}'. \tag{4.117}$$

B. Space lines

1. In the $\{\boldsymbol{m}, \boldsymbol{r}_H\}$-representation, a space line $(\boldsymbol{r}-\boldsymbol{r}_H)\times\boldsymbol{m} = \boldsymbol{0}$ defined with respect to the XYZ coordinate system is represented by $(\boldsymbol{r}' - \boldsymbol{r}'_H) \times \boldsymbol{m}' = \boldsymbol{0}$ with respect to the $X'Y'Z'$ coordinate system, where

$$\boldsymbol{m}' = \boldsymbol{R}^\top\boldsymbol{m}, \quad \boldsymbol{r}'_H = \boldsymbol{R}^\top\boldsymbol{P}_{\boldsymbol{m}}(\boldsymbol{r}_H - \boldsymbol{h}). \tag{4.118}$$

Hence,

$$\boldsymbol{m} = \boldsymbol{R}\boldsymbol{m}', \quad \boldsymbol{r}_H = \boldsymbol{R}\boldsymbol{P}_{\mathbf{m}'}(\boldsymbol{r}'_H + \boldsymbol{R}^\top\boldsymbol{h}). \tag{4.119}$$

2. In the $\{\boldsymbol{p}, \boldsymbol{n}\}$-representation, a space line $\boldsymbol{r}\times\boldsymbol{p} = \boldsymbol{n}$ defined with respect to the XYZ coordinate system is represented by $\boldsymbol{r}'\times\boldsymbol{p}' = \boldsymbol{n}'$ with respect to the $X'Y'Z'$ coordinate system, where

$$\begin{pmatrix} \boldsymbol{p}' \\ \boldsymbol{n}' \end{pmatrix} = N[\begin{pmatrix} \boldsymbol{R}^\top\boldsymbol{p} \\ \boldsymbol{R}^\top(\boldsymbol{n} - \boldsymbol{h}\times\boldsymbol{p}) \end{pmatrix}]. \tag{4.120}$$

[19] Eqs. (4.117) define a transformation of $\mathcal{R}^3$ and its inverse. The set of all such transformations forms the group of three-dimensional *Euclidean motions*. If the rotation matrix $\mathbf{R}$ in eqs. (4.117) is replaced by an arbitrary nonsingular 33-matrix, the resulting group is the group of three-dimensional *affine transformations*. If the 44-matrices in eqs. (4.117) are replaced by an arbitrary nonsingular 44-matrix and its inverse, the resulting group is the group of three-dimensional *projective transformations* of the four-dimensional space $\mathcal{R}^4$, which can be identified with a three-dimensional projective space.

Hence,

$$\begin{pmatrix} \boldsymbol{p} \\ \boldsymbol{n} \end{pmatrix} = N[\begin{pmatrix} \boldsymbol{R}\boldsymbol{p}' \\ \boldsymbol{R}(\boldsymbol{n}' + (\boldsymbol{R}^\top \boldsymbol{h}) \times \boldsymbol{p}') \end{pmatrix}]. \tag{4.121}$$

C. Space plane

1. In the $\{\boldsymbol{n},\ d\}$-representation, a space plane $(\boldsymbol{n}, \boldsymbol{r}) = d$ defined with respect to the XYZ coordinate system is represented by $(\boldsymbol{n}', \boldsymbol{r}') = d'$ with respect to the $X'Y'Z'$ coordinate system, where

$$\boldsymbol{n}' = \boldsymbol{R}^\top \boldsymbol{n}, \qquad d' = d - (\boldsymbol{n}, \boldsymbol{h}). \tag{4.122}$$

Hence,

$$\boldsymbol{n} = \boldsymbol{R}\boldsymbol{n}', \qquad d = d' + (\boldsymbol{n}, \boldsymbol{R}^\top \boldsymbol{h}). \tag{4.123}$$

2. In the $\boldsymbol{\nu}$-representation,

$$\boldsymbol{\nu}' = N[\begin{pmatrix} \boldsymbol{R} & \boldsymbol{0} \\ -\boldsymbol{h}^\top \boldsymbol{R} & 1 \end{pmatrix} \boldsymbol{\nu}], \qquad \boldsymbol{\nu} = N[\begin{pmatrix} \boldsymbol{R}^\top & \boldsymbol{0} \\ \boldsymbol{h}^\top & 1 \end{pmatrix} \boldsymbol{\nu}']. \tag{4.124}$$

D. Space conics and quadrics

1. A space conic $\{\boldsymbol{n},\ d,\ \boldsymbol{Q}\}$ defined with respect to the XYZ coordinate system is represented by $\{\boldsymbol{n}',\ d',\ \boldsymbol{Q}'\}$ with respect to the $X'Y'Z'$ coordinate system, where

$$\boldsymbol{n}' = \boldsymbol{R}^\top \boldsymbol{n}, \qquad d' = d - (\boldsymbol{n}, \boldsymbol{h}),$$

$$\boldsymbol{Q}' = \boldsymbol{R}^\top \left(\boldsymbol{I} + \frac{\boldsymbol{n}\boldsymbol{h}^\top}{d - (\boldsymbol{n}, \boldsymbol{h})} \right) \boldsymbol{Q} \left(\boldsymbol{I} + \frac{\boldsymbol{h}\boldsymbol{n}^\top}{d - (\boldsymbol{n}, \boldsymbol{h})} \right) \boldsymbol{R}. \tag{4.125}$$

Hence[20],

$$\boldsymbol{n} = \boldsymbol{R}\boldsymbol{n}', \qquad d = d' + (\boldsymbol{n}, \boldsymbol{R}^\top \boldsymbol{h}),$$

$$\boldsymbol{Q} = \left(\boldsymbol{I} - \frac{\boldsymbol{n}\boldsymbol{h}^\top}{d} \right) \boldsymbol{R}\boldsymbol{Q}'\boldsymbol{R}^\top \left(\boldsymbol{I} - \frac{\boldsymbol{h}\boldsymbol{n}^\top}{d} \right). \tag{4.126}$$

2. A quadric $(\boldsymbol{r} - \boldsymbol{r}_C, \boldsymbol{S}(\boldsymbol{r} - \boldsymbol{r}_C)) = 1$ defined with respect to the XYZ coordinate system is represented by $(\boldsymbol{r}' - \boldsymbol{r}_C', \boldsymbol{S}'(\boldsymbol{r}' - \boldsymbol{r}_C')) = 1$ with respect to the $X'Y'Z'$ coordinate system, where

$$\boldsymbol{r}_C' = \boldsymbol{R}^\top (\boldsymbol{r}_C - \boldsymbol{h}), \qquad \boldsymbol{S}' = \boldsymbol{R}^\top \boldsymbol{S}\boldsymbol{R}. \tag{4.127}$$

Hence,

$$\boldsymbol{r}_C = \boldsymbol{h} + \boldsymbol{R}\boldsymbol{r}_C', \qquad \boldsymbol{S} = \boldsymbol{R}\boldsymbol{S}'\boldsymbol{R}^\top. \tag{4.128}$$

---

[20] Eqs. (4.125) and (4.126) define a *congruence transformation* (see eq. (2.154)) and its inverse in the form $\mathbf{Q} = \mathbf{A}^\top \mathbf{Q}' \mathbf{A}$ and $\mathbf{Q}' = (\mathbf{A}^{-1})^\top \mathbf{Q} \mathbf{A}^{-1}$, where $\mathbf{A}$ is the matrix that defines the *projective transformation* of the image plane (see eq. (4.86) and the matrix inversion formula (2.24)).

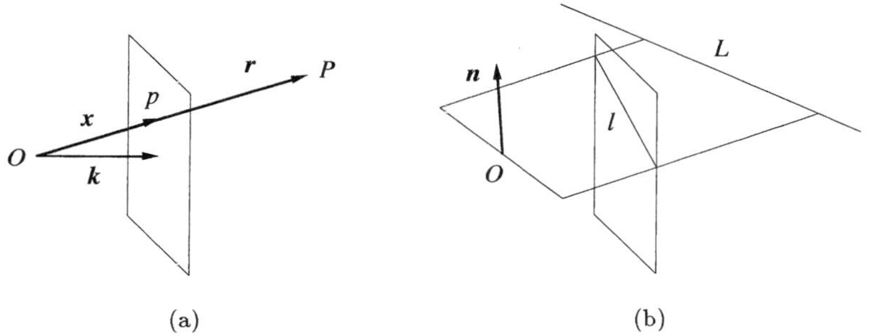

Fig. 4.22. (a) Perspective projection of a space point. (b) Perspective projection of a space line.

4.6.2 *Perspective projection*

A. Space points

1. A space point P not on the XY plane represented by $\boldsymbol{r}$ is perspectively projected onto the intersection p of the image plane $Z = 1$ with the line of sight of P (Fig. 4.22a). It is represented by

$$\boldsymbol{x} = \frac{\boldsymbol{r}}{(\boldsymbol{k}, \boldsymbol{r})}. \tag{4.129}$$

2. Space points on the XY plane produce no images on the image plane[21].

B. Space lines

1. A space line L that does not pass through the viewpoint O is projected onto the intersection l of the image plane $Z = 1$ with the space plane that passes through O and L (Fig. 4.22b). The surface normal to that space plane is $\boldsymbol{n}$ in the $\{\boldsymbol{p}, \boldsymbol{n}\}$-representation, and $\boldsymbol{r}_H \times \boldsymbol{m}$ in the $\{\boldsymbol{m}, \boldsymbol{r}_H\}$-representation (see eq. (4.36)). Hence, the projected image line is represented by $(\boldsymbol{n}_l, \boldsymbol{x}) = 0$, where

$$\boldsymbol{n}_l = N[\boldsymbol{n}] = N[\boldsymbol{r}_H \times \boldsymbol{m}]. \tag{4.130}$$

   If a space point on space line L moves in one direction indefinitely, its projection converges to an image point called the *vanishing point* of L (Fig. 4.23a); the same vanishing point is defined if the space point moves in the opposite direction. The vanishing point is represented by

$$\boldsymbol{x} = \frac{\boldsymbol{m}}{(\boldsymbol{k}, \boldsymbol{m})} = \frac{\boldsymbol{p} \times \boldsymbol{n}}{|\boldsymbol{k}, \boldsymbol{p}, \boldsymbol{n}|}. \tag{4.131}$$

---

[21] A space point $\mathbf{r}$ on the XY plane different from O can be thought of as projected onto the *ideal image point* ("image point at infinity") in the direction of $\mathbf{r}$ on the image plane; perspective projection of the viewpoint O is not defined.

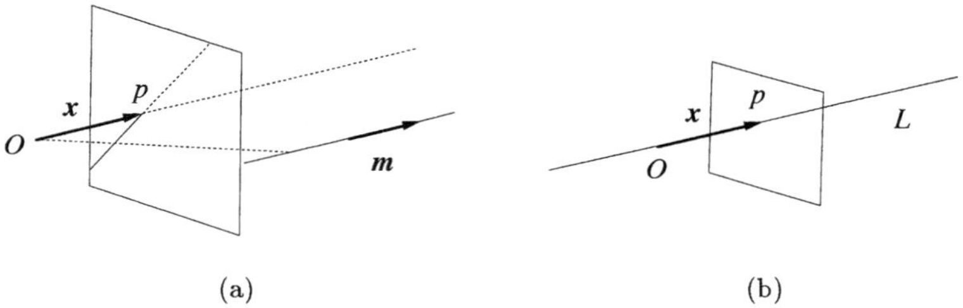

Fig. 4.23. (a) Vanishing point of a space line. (b) Perspective projection of a space line that passes through the viewpoint O.

The vanishing point separates the projected image line into two half lines. One side is the projection of the part of the space line for which $Z > 0$; the other side is the projection of the part for which $Z < 0$. The part for which $Z = 0$ is not projected onto the image plane[22].

2. A space line L that passes through the viewpoint O is projected onto the intersection p of the image plane with L if L is not on the XY plane (Fig. 4.23b). Hence, space line $\boldsymbol{r} \times \boldsymbol{m} = \boldsymbol{0}$ for $(\boldsymbol{k}, \boldsymbol{m}) \neq 0$, or space line $\boldsymbol{r} \times \boldsymbol{p} = \boldsymbol{0}$ for $(\boldsymbol{k}, \boldsymbol{p}) \neq 0$, is projected onto image point

$$\boldsymbol{x} = \frac{\boldsymbol{m}}{(\boldsymbol{k}, \boldsymbol{m})} = \frac{\boldsymbol{p}}{(\boldsymbol{k}, \boldsymbol{p})}. \tag{4.132}$$

3. A space line on the XY plane is not projected onto the image plane[23].

C. Space planes

1. A space plane Π that does not pass through the viewpoint O is projected onto the entire image plane. If the space plane Π is not parallel to the image plane, those space points which are on Π and infinitely far away[24] from the viewpoint O are projected onto an image line, known as the *vanishing line* of Π (Fig. 4.24a). It is easily seen that the vanishing line of space plane $(\boldsymbol{n}_\Pi, \boldsymbol{r}) = d \ (\neq 0)$ is

$$(\boldsymbol{n}_\Pi, \boldsymbol{x}) = 0. \tag{4.133}$$

[22]The part of a space line for which $Z = 0$ can be thought of as projected onto the *ideal image point* ("image point at infinity") of the projected image line.

[23]A space line on the XY plane can be thought of as projected onto the *ideal image line* ("image line at infinity") of the image plane if it does not pass through the viewpoint O; perspective projection of a space line on the XY plane that passes through O is not defined.

[24]The set of all space points on a space plane that are infinitely far away from the viewpoint O is called an *ideal space line* ("space line at infinity"). The ideal space lines defined by mutually parallel space planes are regarded as the same ideal space line.

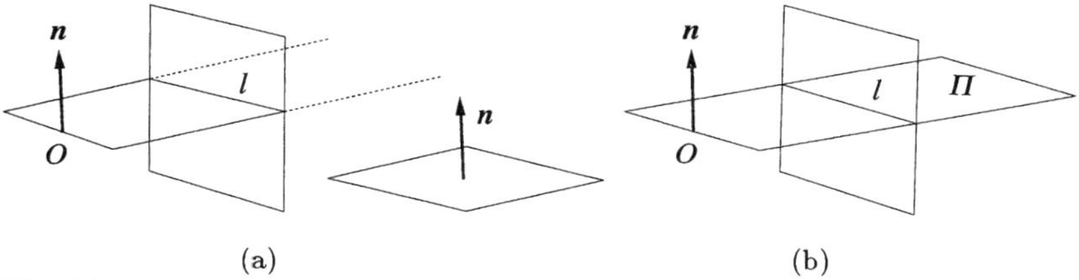

Fig. 4.24. (a) Vanishing line of a space plane. (b) Perspective projection of a space plane that passes through the viewpoint O.

One of the two regions on the image plane separated by the vanishing line is the projection of the part for which $Z > 0$; the other is the projection of the part for which $Z < 0$. The part for which $Z = 0$ is not projected[25]. If $\boldsymbol{n} = \pm\boldsymbol{k}$, no vanishing line appears[26].

2. A space plane Π that passes through the viewpoint O is projected onto the intersection l of the image plane with Π if it does not coincide with the XY plane itself (Fig. 4.24b). Namely, space plane $(\boldsymbol{n}_\Pi, \boldsymbol{r}) = 0$ for $\boldsymbol{n}_\Pi \neq \pm\boldsymbol{k}$ is projected onto image line $(\boldsymbol{n}_\Pi, \boldsymbol{x}) = 0$.

3. The XY plane is not projected onto the the image plane[27].

D. Space conics and quadrics

1. By definition, a space conic $\{\boldsymbol{n}, d, \boldsymbol{Q}\}$ is projected onto conic $(\boldsymbol{x}, \boldsymbol{Q}\boldsymbol{x}) = 0$ (see Fig. 4.15).

2. If space point $\boldsymbol{r} = Z\boldsymbol{x}$ is on quadric $(\boldsymbol{r} - \boldsymbol{r}_C, \boldsymbol{S}(\boldsymbol{r} - \boldsymbol{r}_C)) = 1$, we have

$$(Z\boldsymbol{x} - \boldsymbol{r}_C, \boldsymbol{S}(Z\boldsymbol{x} - \boldsymbol{r}_C)) = Z^2(\boldsymbol{x}, \boldsymbol{S}\boldsymbol{x}) - 2Z(\boldsymbol{x}, \boldsymbol{S}\boldsymbol{r}_C) + (\boldsymbol{r}_C, \boldsymbol{S}\boldsymbol{r}_C) = 1. \tag{4.134}$$

This equation yields two real solutions for Z if the line of sight of $\boldsymbol{x}$ intersects the quadric at two space points; no real solution exists if it does not meet the quadric. The set of those image points whose lines of sight are tangent to an object in the scene is called the (*occluding*) *contour* of the object (Fig. 4.25). It follows that an image pont $\boldsymbol{x}$ is on the contour of quadric $(\boldsymbol{r} - \boldsymbol{r}_C, \boldsymbol{S}(\boldsymbol{r} - \boldsymbol{r}_C)) = 1$ if and only if eq. (4.134) has one multiple root, i.e.,

$$(\boldsymbol{x}, \boldsymbol{S}\boldsymbol{r}_C)^2 - (\boldsymbol{x}, \boldsymbol{S}\boldsymbol{x})((\boldsymbol{r}_C, \boldsymbol{S}\boldsymbol{r}_C) - 1) = 0, \tag{4.135}$$

[25] The part of a space plane for which $Z = 0$ can be thought of as projected onto the *ideal image line* ("image line at infinity").
[26] The vanishing line can be thought of as the *ideal image line* ("image line at infinity").
[27] Space plane $(\mathbf{k}, \mathbf{r}) = 0$ can be thought of as projected onto the *ideal image line* ("image line at infinity").

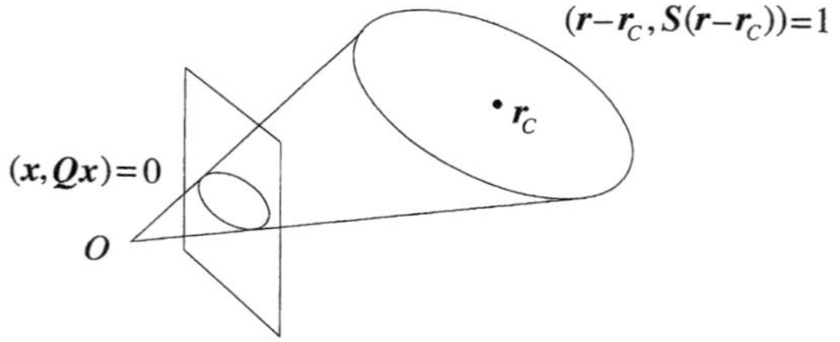

Fig. 4.25. Perspective projection of a quadric.

which defines a conic

$$(\boldsymbol{x}, \boldsymbol{Q}\boldsymbol{x}) = 0, \tag{4.136}$$

where

$$\boldsymbol{Q} = ((\boldsymbol{r}_C, \boldsymbol{S}\boldsymbol{r}_C) - 1)\boldsymbol{S} - \boldsymbol{S}\boldsymbol{r}_C\boldsymbol{r}_C^\top \boldsymbol{S}. \tag{4.137}$$

In other words, quadric $(\boldsymbol{r} - \boldsymbol{r}_C, \boldsymbol{S}(\boldsymbol{r} - \boldsymbol{r}_C)) = 1$ is projected onto the conic given by eq. (4.136). This conic may not be a real conic[28]; if it is a real conic, it may be degenerate.

[28] For example, an ellipsoid centered at the viewpoint O is projected onto an "imaginary conic" (an empty set in the real space).

Chapter 5

Geometric Correction

Multiple geometric objects that are supposedly interrelated by a constraint may not satisfy it if each object is separately observed in the presence of noise. This chapter presents a statistically optimal way to correct the positions of geometric objects so that they satisfy a required constraint. The fundamental principle is the minimization of the *Mahalanobis distance* defined in terms of the covariance matrices of the objects. First, a general theory is formulated in abstract terms. Then, it is applied to typical geometric problems in two and three dimensions—optimally imposing coincidence and incidence on image points, image lines, conics, space points, space lines, and space planes. For each problem, explicit expressions for the correction and the a posteriori covariance matrices are derived. Optimal correction for orthogonality constraints is also studied.

5.1 General Theory

5.1.1 Basic formulation

Consider N geometric objects in two or three dimensions, the αth object being represented by an n_α-vector $\boldsymbol{u}_\alpha$, $\alpha = 1, ..., N$. Let n_α be the dimension of vector $\boldsymbol{u}_\alpha$. The N objects are assumed to be statistically independent, but the components of each $\boldsymbol{u}_\alpha$ may be correlated. We assume that each $\boldsymbol{u}_\alpha$ is constrained to be in an n'_α-dimensional manifold $\mathcal{U}_\alpha \subset \mathcal{R}^{n_\alpha}$, which we call the *data space* of $\boldsymbol{u}_\alpha$. Let $\bar{\boldsymbol{u}}_\alpha$ be the true value we should observe in the absence of noise, and write $\boldsymbol{u}_\alpha = \bar{\boldsymbol{u}}_\alpha + \Delta\boldsymbol{u}_\alpha$. The error $\Delta\boldsymbol{u}_\alpha$ is, to a first approximation, constrained to be in the *tangent space* $T_{\bar{\boldsymbol{u}}_\alpha}(\mathcal{U}_\alpha)$ to the manifold $\mathcal{U}_\alpha$ at $\bar{\boldsymbol{u}}_\alpha$. Let $\bar{V}[\boldsymbol{u}_\alpha]$ be the covariance matrix of the error $\Delta\boldsymbol{u}_\alpha$. We assume that no constraint exists on $\Delta\boldsymbol{u}_\alpha$ other than $\Delta\boldsymbol{u}_\alpha \in T_{\bar{\boldsymbol{u}}_\alpha}(\mathcal{U}_\alpha)$ and hence the range of $\bar{V}[\boldsymbol{u}_\alpha]$ coincides with $T_{\bar{\boldsymbol{u}}_\alpha}(\mathcal{U}_\alpha)$. It follows that

$$\boldsymbol{P}^{\mathcal{U}_\alpha}_{\bar{\boldsymbol{u}}_\alpha}\bar{V}[\boldsymbol{u}_\alpha] = \bar{V}[\boldsymbol{u}_\alpha]\boldsymbol{P}^{\mathcal{U}_\alpha}_{\bar{\boldsymbol{u}}_\alpha} = \bar{V}[\boldsymbol{u}_\alpha],$$

$$\bar{V}[\boldsymbol{u}_\alpha]\bar{V}[\boldsymbol{u}_\alpha]^- = \bar{V}[\boldsymbol{u}_\alpha]^-\bar{V}[\boldsymbol{u}_\alpha] = \boldsymbol{P}^{\mathcal{U}_\alpha}_{\bar{\boldsymbol{u}}_\alpha}, \tag{5.1}$$

where $\boldsymbol{P}^{\mathcal{U}_\alpha}_{\bar{\boldsymbol{u}}_\alpha}$ is the n_α-dimensional projection matrix onto $T_{\bar{\boldsymbol{u}}_\alpha}(\mathcal{U}_\alpha)$.

Suppose L smooth functions $F^{(k)}(\,\cdot\,, ..., \,\cdot\,)$: $\mathcal{R}^{n_1} \times \cdots \mathcal{R}^{n_N} \to \mathcal{R}$ exist and the true values $\bar{\boldsymbol{u}}_1$, ..., $\bar{\boldsymbol{u}}_N$ are known to satisfy

$$F^{(k)}(\bar{\boldsymbol{u}}_1, ..., \bar{\boldsymbol{u}}_N) = 0, \qquad k = 1, ..., L. \tag{5.2}$$

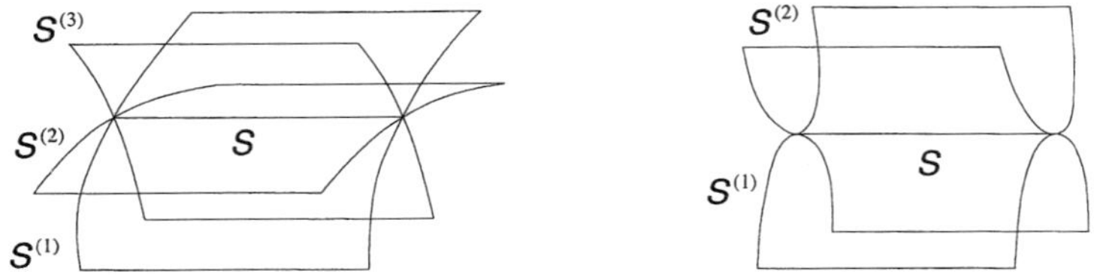

(a) (b)

Fig. 5.1. (a) Nonsingular constraint imposed by three equations that are not independent. (b) Singular constraint imposed by two independent equations.

We call eq. (5.2) simply the *constraint*, whereas we refer to the constraint $\boldsymbol{u}_\alpha \in \mathcal{U}_\alpha$ as the *inherent constraint* on $\boldsymbol{u}_\alpha$.

We now consider the problem of optimally correcting $\boldsymbol{u}_1$, ..., $\boldsymbol{u}_N$ so that these L equations are satisfied. Since each $\boldsymbol{u}_\alpha$ is constrained to be in the data space $\mathcal{U}_\alpha \subset \mathcal{R}^{n_\alpha}$, the direct sum $\bigoplus_{\alpha=1}^N \boldsymbol{u}_\alpha$ is constrained to be in its data space $\mathcal{U} = \bigoplus_{\alpha=1}^N \mathcal{U}_\alpha \subset \mathcal{R}^{\Sigma_{\alpha=1}^N n_\alpha}$. We say that the constraint imposed by eq. (5.2) is *nonsingular* if each of the L equations defines a manifold $\mathcal{S}^{(k)}$ of *codimension 1* in $\mathcal{U}$ and the L manifolds $\mathcal{S}^{(k)}$, $k = 1$, ..., L, meet each other *transversally* in $\mathcal{U}$ (Fig. 5.1a; see Section 3.2.1); otherwise, the constraint is said to be *singular*[1] (Fig. 5.1b). In this chapter, we assume that the constraint (5.2) is nonsingular, although the following theory can also be applied to singular constraints if appropriately modified[2].

The L equations (5.2) may not necessarily be independent (Fig. 5.1a). We call the number r of independent equations the *rank* of the constraint. It follows that the constraint (5.2) defines a manifold $\mathcal{S} = \bigcap_{k=1}^L \mathcal{S}^{(k)}$ of *codimension* r in $\mathcal{U}$; we call $\mathcal{S}$ the (*geometric*) *model* of the constraint (5.2). From the above definition, we see that the rank r equals the dimension of the linear subspace

$$\bar{\mathcal{V}} = \{\bigoplus_{\alpha=1}^N \boldsymbol{P}^{\mathcal{U}_\alpha}_{\bar{\boldsymbol{u}}_\alpha} \nabla_{\boldsymbol{u}_\alpha} \bar{F}^{(1)}, ..., \bigoplus_{\alpha=1}^N \boldsymbol{P}^{\mathcal{U}_\alpha}_{\bar{\boldsymbol{u}}_\alpha} \nabla_{\boldsymbol{u}_\alpha} \bar{F}^{(L)}\}_L \subset \mathcal{R}^{\Sigma_{\alpha=1}^N n_\alpha}, \tag{5.3}$$

where $\nabla_{\boldsymbol{u}_\alpha} \bar{F}^{(k)}$ is the abbreviation of $\nabla_{\boldsymbol{u}_\alpha} F^{(k)}(\bar{\boldsymbol{u}}_1, ..., \bar{\boldsymbol{u}}_N)$. Note that the dimension of this subspace may not be equal to the dimension of the linear subspace

$$\mathcal{V} = \{\bigoplus_{\alpha=1}^N \boldsymbol{P}^{\mathcal{U}_\alpha}_{\boldsymbol{u}_\alpha} \nabla_{\boldsymbol{u}_\alpha} F^{(1)}, ..., \bigoplus_{\alpha=1}^N \boldsymbol{P}^{\mathcal{U}_\alpha}_{\boldsymbol{u}_\alpha} \nabla_{\boldsymbol{u}_\alpha} F^{(L)}\}_L \subset \mathcal{R}^{\Sigma_{\alpha=1}^N n_\alpha} \tag{5.4}$$

---

[1] For example, if the L equations (5.2) are expressed as one equation, say, in the form $\Sigma_{\alpha=1}^N F^{(k)}(\bar{\mathbf{u}}_1, ..., \bar{\mathbf{u}}_N)^2 = 0$, the constraint is singular.

[2] We will see an example of a singular constraint in the motion analysis in Chapter 11.

for $\boldsymbol{u}_\alpha \neq \bar{\boldsymbol{u}}_\alpha$, where $\nabla_{\boldsymbol{u}_\alpha} F^{(k)}$ is the abbreviation of $\nabla_{\boldsymbol{u}_\alpha} F^{(k)}(\boldsymbol{u}_1, ..., \boldsymbol{u}_N)$. We say that the constraint (5.2) is *degenerate*[3] if the dimension of $\mathcal{V}$ is larger that the dimension of the subspace $\bar{\mathcal{V}}$.

Substituting $\boldsymbol{u}_\alpha = \bar{\boldsymbol{u}}_\alpha + \Delta \boldsymbol{u}_\alpha$ into $F^{(k)}(\boldsymbol{u}_1, ..., \boldsymbol{u}_L)$ and taking a linear approximation, we can replace eq. (5.2) to a first approximation by

$$F^{(k)} = \sum_{\alpha=1}^{N} (\nabla_{\boldsymbol{u}_\alpha} \bar{F}^{(k)}, \Delta \boldsymbol{u}_\alpha), \qquad k = 1, ..., L, \tag{5.5}$$

where $F^{(k)}$ is the abbreviation of $F^{(k)}(\boldsymbol{u}_1, ..., \boldsymbol{u}_L)$. This linearized constraint is assumed to be *satisfiable*, i.e., there exists at least one set of solutions $\Delta \boldsymbol{u}_\alpha \in T_{\bar{\boldsymbol{u}}_\alpha}(\mathcal{U}_\alpha)$, $\alpha = 1, ..., N$, that satisfies eq. (5.5).

If $\Delta \boldsymbol{u}_\alpha$ is a solution of eq. (5.5), the correction takes the form $\hat{\boldsymbol{u}}_\alpha = \boldsymbol{u}_\alpha - \Delta \boldsymbol{u}_\alpha$ to a first approximation. However, infinitely many solutions may exist for $\Delta \boldsymbol{u}_\alpha$, $\alpha = 1, ..., N$. From among them, we choose the one which minimizes the square sum of the *Mahalanobis distance* $\|\Delta \boldsymbol{u}_\alpha\|_{\bar{V}[\boldsymbol{u}_\alpha]}$ (see eq. (3.66)), i.e.,

$$J = \sum_{\alpha=1}^{N} (\Delta \boldsymbol{u}_\alpha, \bar{V}[\boldsymbol{u}_\alpha]^- \Delta \boldsymbol{u}_\alpha) \to \min \tag{5.6}$$

under the inherent constraint $\Delta \boldsymbol{u}_\alpha \in T_{\bar{\boldsymbol{u}}_\alpha}(\mathcal{U}_\alpha)$, $\alpha = 1, ..., N$.

Geometrically, we are projecting the direct sum $\bigoplus_{\alpha=1}^{N} \boldsymbol{u}_\alpha$ onto the "closest point" in the model $\mathcal{S}$ determined by eq. (5.2), where the "closeness" is measured in the Mahalanobis distance with respect to the total covariance matrix $\bigoplus_{\alpha=1}^{N} \bar{V}[\boldsymbol{u}_\alpha]$ (Fig. 5.2). This criterion can be justified as *maximum likelihood estimation* for Gaussian noise. Namely, if the errors $\Delta \boldsymbol{u}_1$, ..., $\Delta \boldsymbol{u}_N$ have the probability density

$$p(\Delta \boldsymbol{u}_1, ..., \Delta \boldsymbol{u}_N) = \left(\prod_{\beta=1}^{N} \frac{1}{\sqrt{(2\pi)^{n'_\beta} |\bar{V}[\boldsymbol{u}_\beta]|_+}} \right) e^{-\sum_{\alpha=1}^{N} (\Delta \boldsymbol{u}_\alpha, \bar{V}[\boldsymbol{u}_\alpha]^- \Delta \boldsymbol{u}_\alpha)/2}, \tag{5.7}$$

maximizing the likelihood is equivalent to minimizing the function J given in (5.6) (see eqs. (3.46), (3.81) and (3.82)).

5.1.2 Optimal solution

Let $\{\bar{\boldsymbol{v}}_j^{(\alpha)}\}$, $j = 1, ..., n_\alpha - n'_\alpha$, be an orthonormal basis of $T_{\bar{\boldsymbol{u}}_\alpha}(\mathcal{U}_\alpha)^\perp$. The inherent constraint $\Delta \boldsymbol{u}_\alpha \in T_{\bar{\boldsymbol{u}}_\alpha}(\mathcal{U}_\alpha)$ can be written as

$$(\bar{\boldsymbol{v}}_j^{(\alpha)}, \Delta \boldsymbol{u}_\alpha) = 0, \qquad j = 1, ..., n_\alpha - n'_\alpha. \tag{5.8}$$

[3] A more rigorous argument will be given in Chapter 14.

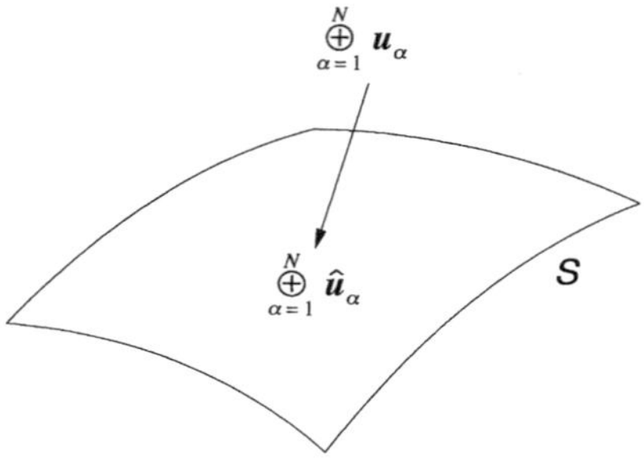

Fig. 5.2. Projecting $\bigoplus_{\alpha=1}^{N} \mathbf{u}_\alpha$ onto the closest point in the model $\mathcal{S}$ measured in the Mahalanobis distance.

Introducing Lagrange multipliers λ_i and $\mu_i^{(\alpha)}$, differentiating

$$J - 2\sum_{k=1}^{L} \lambda_k \sum_{\alpha=1}^{N} (\nabla_{\boldsymbol{u}_\alpha} \bar{F}^{(k)}, \Delta \boldsymbol{u}_\alpha) - 2\sum_{\alpha=1}^{N} \sum_{j=1}^{n_\alpha - n'_\alpha} \mu_j^{(\alpha)} (\bar{\boldsymbol{v}}_j^{(\alpha)}, \Delta \boldsymbol{u}_\alpha) \tag{5.9}$$

with respect to each $\Delta \boldsymbol{u}_\alpha$, and setting the result zero, we obtain

$$\bar{V}[\boldsymbol{u}_\alpha]^- \Delta \boldsymbol{u}_\alpha = \sum_{k=1}^{L} \lambda_k \nabla_{\boldsymbol{u}_\alpha} \bar{F}^{(k)} + \sum_{j=1}^{n_\alpha - n'_\alpha} \mu_{j(\alpha)} \bar{\boldsymbol{v}}_j^{(\alpha)}. \tag{5.10}$$

Multiplying this by $\bar{V}[\boldsymbol{u}_\alpha]$ on both sides and noting that $\bar{\boldsymbol{v}}_j^{(\alpha)} \in T_{\bar{\boldsymbol{u}}_\alpha}(\mathcal{U}_\alpha)^\perp$, we obtain

$$\boldsymbol{P}_{\bar{\boldsymbol{u}}_\alpha}^{\mathcal{U}_\alpha} \Delta \boldsymbol{u}_\alpha = \sum_{k=1}^{L} \lambda_k \bar{V}[\boldsymbol{u}_\alpha] \nabla_{\boldsymbol{u}_\alpha} \bar{F}^{(k)}, \tag{5.11}$$

where eqs. (5.1) have been used. Since $\Delta \boldsymbol{u}_\alpha \in T_{\bar{\boldsymbol{u}}}(\mathcal{U}_\alpha)$, the solution is given by

$$\Delta \boldsymbol{u}_\alpha = \sum_{k=1}^{L} \lambda_k \bar{V}[\boldsymbol{u}_\alpha] \nabla_{\boldsymbol{u}_\alpha} \bar{F}^{(k)}. \tag{5.12}$$

Substitution of this into eq. (5.5) yields

$$\sum_{l=1}^{L} \left(\sum_{\alpha=1}^{N} (\nabla_{\boldsymbol{u}_\alpha} \bar{F}^{(k)}, \bar{V}[\boldsymbol{u}_\alpha] \nabla_{\boldsymbol{u}_\alpha} \bar{F}^{(l)}) \right) \lambda_l = F^{(k)}. \tag{5.13}$$

Since eq. (5.5) is assumed to be satisfiable, this equation is solvable (see Section 2.3.2); the solution is given in the following form:

$$\lambda_k = \sum_{l=1}^{L} \bar{W}^{(kl)} F^{(l)}. \tag{5.14}$$

Here, $\bar{W}^{(kl)}$ is the (kl) element of the (LL)-matrix $\bar{\boldsymbol{W}} = (\bar{W}^{(kl)})$ defined by $\bar{\boldsymbol{W}} = \bar{\boldsymbol{V}}^{-}$, where $\bar{\boldsymbol{V}} = (\bar{V}^{(kl)})$ is the (LL)-matrix defined by

$$(\bar{V}^{(kl)}) = \left(\sum_{\alpha=1}^{N}(\nabla_{\boldsymbol{u}_\alpha}\bar{F}^{(k)}, \bar{V}[\boldsymbol{u}_\alpha]\nabla_{\boldsymbol{u}_\alpha}\bar{F}^{(l)})\right). \tag{5.15}$$

In the following, we use the following abbreviation to denote the (LL)-matrix $\bar{\boldsymbol{W}} = (\bar{W}^{(kl)})$:

$$(\bar{W}^{(kl)}) = \left(\sum_{\alpha=1}^{N}(\nabla_{\boldsymbol{u}_\alpha}\bar{F}^{(k)}, \bar{V}[\boldsymbol{u}_\alpha]\nabla_{\boldsymbol{u}_\alpha}\bar{F}^{(l)})\right)^{-}. \tag{5.16}$$

It can be shown[4] that the rank of the matrix $\bar{\boldsymbol{V}}$ (hence of $\bar{\boldsymbol{W}}$) equals the rank r of the constraint (5.2).

It follows that the optimal correction is given in the following form (see eqs. (2.140) and (3.85)):

$$\Delta\boldsymbol{u}_\alpha = \bar{V}[\boldsymbol{u}_\alpha]\sum_{k,l=1}^{L}\bar{W}^{(kl)}F^{(k)}\nabla_{\boldsymbol{u}_\alpha}\bar{F}^{(l)}. \tag{5.17}$$

This equation has the following geometric interpretation. If the noise is Gaussian, the equiprobability surface for $\boldsymbol{u}_\alpha$ has the form

$$(\boldsymbol{u}_\alpha - \bar{\boldsymbol{u}}_\alpha, \bar{V}[\boldsymbol{u}_\alpha](\boldsymbol{u}_\alpha - \bar{\boldsymbol{u}}_\alpha)) = \text{constant}. \tag{5.18}$$

As discussed in Section 4.5, this equation defines a nonsingular quadric in the tangent space $T_{\bar{\boldsymbol{u}}_\alpha}(\mathcal{U}_\alpha)$. Let $\mathcal{S}_\alpha$ be the restriction of the model $\mathcal{S}$ to $\mathcal{U}_\alpha$ obtained by fixing $\boldsymbol{u}_\beta = \bar{\boldsymbol{u}}_\beta$ for $\beta \neq \alpha$. We now show that the optimal correction $\Delta\boldsymbol{u}_\alpha$ given by eq. (5.17) is in the *conjugate* direction of the tangent space $T_{\bar{\boldsymbol{u}}_\alpha}(\mathcal{S}_\alpha)$ to the model $\mathcal{S}_\alpha$ at $\bar{\boldsymbol{u}}_\alpha$ (see Section 4.5.2).

In $T_{\bar{\boldsymbol{u}}_\alpha}(\mathcal{U}_\alpha)$, the tangent hyperplane to the quadric defined by eq. (5.18) at $\bar{\boldsymbol{u}}_\alpha + \Delta\boldsymbol{u}_\alpha$ has the following surface normal (see eq. (4.105)):

$$\boldsymbol{n}_\alpha \propto \bar{V}[\boldsymbol{u}_\alpha]^{-}\Delta\boldsymbol{u}_\alpha = \sum_{k,l=1}^{L}\bar{W}^{(kl)}F^{(k)}\boldsymbol{P}^{\mathcal{U}_\alpha}_{\bar{\boldsymbol{u}}_\alpha}\nabla_{\boldsymbol{u}_\alpha}\bar{F}^{(l)}. \tag{5.19}$$

Let $\boldsymbol{v}$ be an arbitrary tangent vector to the manifold $\mathcal{S}_\alpha$ at $\bar{\boldsymbol{u}}_\alpha$ (Fig. 5.3). Since the orthogonal complement of $T_{\bar{\boldsymbol{u}}_\alpha}(\mathcal{S}_\alpha)$ with respect to $T_{\bar{\boldsymbol{u}}_\alpha}(\mathcal{U}_\alpha)$ is generated by $\boldsymbol{P}^{\mathcal{U}_\alpha}_{\bar{\boldsymbol{u}}_\alpha}\nabla_{\boldsymbol{u}_\alpha}\bar{F}^{(k)}$, $k = 1, ..., L$, eq. (5.19) implies

$$(\boldsymbol{v}, \boldsymbol{n}_\alpha) \propto \sum_{k,l=1}^{L}\bar{W}^{(kl)}F^{(k)}(\boldsymbol{v}, \boldsymbol{P}^{\mathcal{U}_\alpha}_{\bar{\boldsymbol{u}}_\alpha}\nabla_{\boldsymbol{u}_\alpha}\bar{F}^{(l)}) = 0. \tag{5.20}$$

Thus, $\Delta\boldsymbol{u}_\alpha$ is in the conjugate direction of $T_{\bar{\boldsymbol{u}}_\alpha}(\mathcal{S}_\alpha)$.

[4]The proof will be given in Chapter 14 in a more general framework.

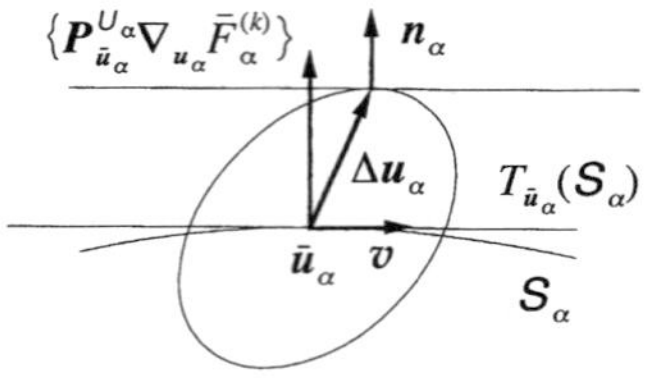

Fig. 5.3. Geometric interpretation of optimal correction.

5.1.3 *Practical considerations*

Eq. (5.17) is merely a theoretical expression, because the right-hand side involves the covariance matrices $\bar{V}[\boldsymbol{u}_\alpha]$ evaluated at the true values $\bar{\boldsymbol{u}}_\alpha$, $\alpha = 1, ..., N$, which we want to compute. It appears that they can be approximated by the covariance matrices $V[\boldsymbol{u}_\alpha]$ evaluated at the observed values $\boldsymbol{u}_\alpha$, $\alpha = 1, ..., N$. However, if the matrix $\bar{\boldsymbol{V}} = (\bar{V}^{(kl)})$ defined by eq. (5.15) is approximated by $\boldsymbol{V} = (V^{(kl)})$ in the form

$$(V^{(kl)}) = \left(\sum_{\alpha=1}^{N}(\nabla_{\boldsymbol{u}_\alpha}F^{(k)}, V[\boldsymbol{u}_\alpha]\nabla_{\boldsymbol{u}_\alpha}F^{(l)})\right), \tag{5.21}$$

matrices $\bar{\boldsymbol{V}}$ and $\boldsymbol{V}$ may have *different ranks*: the rank of $\boldsymbol{V}$ is larger than that of $\bar{\boldsymbol{V}}$ *if the constraint (5.2) is degenerate* (see eqs. (5.3) and (5.4)). Hence, even if $\boldsymbol{V}$ is a good approximation to $\bar{\boldsymbol{V}}$, its generalized inverse $\boldsymbol{W} = \boldsymbol{V}^-$ may be very different from $\bar{\boldsymbol{W}} = \bar{\boldsymbol{V}}^-$.

A practical solution to this difficulty is to compute the *rank-constrained generalized inverse* (see eq. (2.82)). Namely, if the rank of the constraint (5.2) is r, eq. (5.17) is approximated by

$$\Delta\boldsymbol{u}_\alpha = V[\boldsymbol{u}_\alpha]\sum_{k,l=1}^{L} W^{(kl)}F^{(k)}\nabla_{\boldsymbol{u}_\alpha}F^{(l)}, \tag{5.22}$$

where $\boldsymbol{W} = (W^{(kl)})$ is an (LL) matrix defined by $\boldsymbol{W} = (\boldsymbol{V})_r^-$, which we write as

$$(W^{(kl)}) = \left(\sum_{\alpha=1}^{N}(\nabla_{\boldsymbol{u}_\alpha}F^{(k)}, V[\boldsymbol{u}_\alpha]\nabla_{\boldsymbol{u}_\alpha}F^{(l)})\right)_r^-. \tag{5.23}$$

The use of $V[\boldsymbol{u}_\alpha]$ instead of $\bar{V}[\boldsymbol{u}_\alpha]$ has the following geometric interpretation. The quadric defined by eq. (5.18) is centered at the true value $\bar{\boldsymbol{u}}_\alpha$ in the tangent space $T_{\bar{\boldsymbol{u}}_\alpha}(\mathcal{U}_\alpha)$ at $\bar{\boldsymbol{u}}_\alpha$; the correction $\Delta\boldsymbol{u}$ given by eq. (5.17) is an element of $T_{\bar{\boldsymbol{u}}_\alpha}(\mathcal{U}_\alpha)$ (Fig. 5.4). Using $V[\boldsymbol{u}_\alpha]$ instead of $\bar{V}[\boldsymbol{u}_\alpha]$ means replacing eq. (5.18) by

$$(\bar{\boldsymbol{u}}_\alpha - \boldsymbol{u}_\alpha, V[\boldsymbol{u}_\alpha](\bar{\boldsymbol{u}}_\alpha - \boldsymbol{u}_\alpha)) = \text{constant}. \tag{5.24}$$

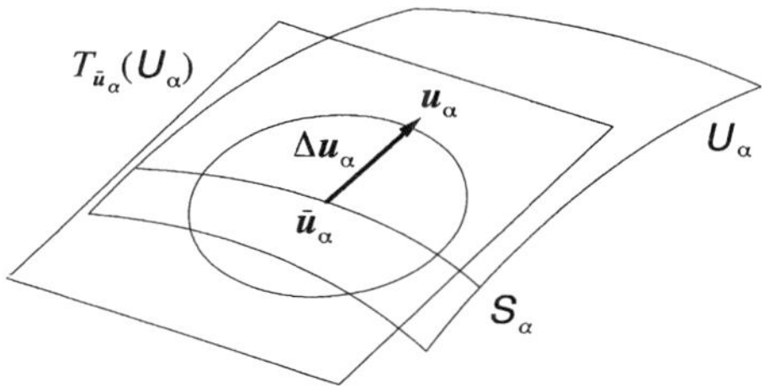

Fig. 5.4. Theoretical analysis is done in the tangent space $T_{\bar{\mathbf{u}}_\alpha}(\mathcal{U}_\alpha)$ at $\bar{\mathbf{u}}_\alpha$.

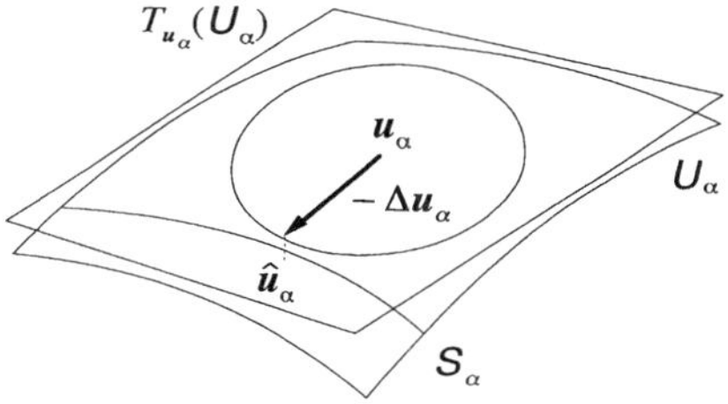

Fig. 5.5. Actual correction is done in the tangent space $T_{\mathbf{u}_\alpha}(\mathcal{U}_\alpha)$ at $\mathbf{u}_\alpha$.

If we regard $\bar{\boldsymbol{u}}_\alpha$ as a variable, this equation defines a quadric centered at the data value $\boldsymbol{u}_\alpha$ in the tangent space $T_{\boldsymbol{u}_\alpha}(\mathcal{U}_\alpha)$ at $\boldsymbol{u}_\alpha$. Hence, the correction $\Delta\boldsymbol{u}$ given by eq. (5.22) is an element of $T_{\boldsymbol{u}_\alpha}(\mathcal{U}_\alpha)$ (Fig. 5.5). This means that the data value $\boldsymbol{u}_\alpha$ is corrected within $T_{\boldsymbol{u}_\alpha}(\mathcal{U}_\alpha)$ in such a way that the Mahalanobis distance $\|\Delta\boldsymbol{u}_\alpha\|_{V[\boldsymbol{u}_\alpha]}$ is minimized.

This observation implies that as long as $\Delta\boldsymbol{u}_\alpha \in T_{\boldsymbol{u}_\alpha}(\mathcal{U}_\alpha)$, the inherent constraint $\hat{\boldsymbol{u}}_\alpha \in \mathcal{U}_\alpha$ on the corrected value $\hat{\boldsymbol{u}}_\alpha = \boldsymbol{u}_\alpha - \Delta\boldsymbol{u}_\alpha$ is satisfied to a first approximation *but may be violated if higher order terms are considered* (Fig. 5.5). It follows that if we want to impose the inherent constraint $\hat{\boldsymbol{u}}_\alpha \in \mathcal{U}_\alpha$ exactly, we need a higher order correction, which we denote by $\mathcal{C}[\,\cdot\,]$ (see eqs. (4.113)):

$$\hat{\boldsymbol{u}}_\alpha = \mathcal{C}[\boldsymbol{u}_\alpha - \Delta\boldsymbol{u}_\alpha]. \tag{5.25}$$

This higher order correction can be made rather arbitrarily, since the correction is optimal in the first order.

Because the correction given by eq. (5.22) is based on the linear approximation (5.5), the values $\{\hat{\boldsymbol{u}}_\alpha\}$ corrected by eq. (5.25) may not exactly satisfy the constraint (5.2) (Fig. 5.5). In order to impose it exactly, the computation is iterated by replacing the original values $\{\boldsymbol{u}_\alpha\}$ by the corrected values $\{\hat{\boldsymbol{u}}_\alpha\}$. This process is essentially the Newton iterations, so the convergence is quadratic; usually two or three iterations are sufficient.

In these iterations, *the covariance matrix $V[\boldsymbol{u}_\alpha]$ must also be updated*, be-

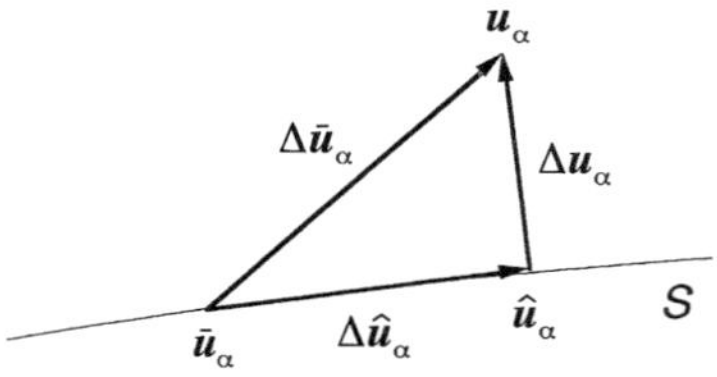

Fig. 5.6. The error of optimal correction.

cause the range of $V[\boldsymbol{u}_\alpha]$ at $\boldsymbol{u}_\alpha$ is $T_{\boldsymbol{u}_\alpha}(\mathcal{U}_\alpha)$ and is generally different from the range at $\hat{\boldsymbol{u}}_\alpha$, which should be $T_{\hat{\boldsymbol{u}}_\alpha}(\mathcal{U}_\alpha)$ (Fig. 5.5). If the covariance matrix $V[\boldsymbol{u}_\alpha]$ is given as a function of $\boldsymbol{u}_\alpha$, it only needs to be re-evaluated at the updated value $\hat{\boldsymbol{u}}_\alpha$. In many practical problems, however, the covariance matrix $V[\boldsymbol{u}_\alpha]$ is given only at the initial value $\boldsymbol{u}_\alpha$. In such a case, a practical compromise is to "project" the covariance matrix $V[\boldsymbol{u}_\alpha]$ onto the tangent space $T_{\hat{\boldsymbol{u}}_\alpha}(\mathcal{U}_\alpha)$ at $\hat{\boldsymbol{u}}_\alpha$ in the form

$$\hat{V}[\boldsymbol{u}_\alpha] = \boldsymbol{P}^{\mathcal{U}_\alpha}_{\hat{\boldsymbol{u}}_\alpha} V[\boldsymbol{u}_\alpha] \boldsymbol{P}^{\mathcal{U}_\alpha}_{\hat{\boldsymbol{u}}_\alpha}, \tag{5.26}$$

where $\boldsymbol{P}^{\mathcal{U}_\alpha}_{\hat{\boldsymbol{u}}_\alpha}$ is the projection matrix onto the the tangent space $T_{\hat{\boldsymbol{u}}_\alpha}(\mathcal{U}_\alpha)$ at $\hat{\boldsymbol{u}}_\alpha$.

5.1.4 A posteriori covariance matrices

Even if the constraint (5.2) is exactly imposed on the corrected values $\{\hat{\boldsymbol{u}}_\alpha\}$, they are random variables because they are computed from the data $\{\boldsymbol{u}_\alpha\}$. Let $\boldsymbol{u}_\alpha = \bar{\boldsymbol{u}}_\alpha + \Delta\bar{\boldsymbol{u}}_\alpha$, $\alpha = 1, ..., N$, be the observed values, where $\Delta\bar{\boldsymbol{u}}_\alpha$ is the actual error in $\boldsymbol{u}_\alpha$. After the correction $\Delta\boldsymbol{u}_\alpha$ given by eq. (5.17) is subtracted, the data value $\boldsymbol{u}_\alpha$ is modified into

$$\begin{aligned}\hat{\boldsymbol{u}}_\alpha &= (\bar{\boldsymbol{u}}_\alpha + \Delta\bar{\boldsymbol{u}}_\alpha) - \bar{V}[\boldsymbol{u}_\alpha] \sum_{k,l=1}^{L} \bar{W}^{(kl)} F(\bar{\boldsymbol{u}}_1 + \Delta\bar{\boldsymbol{u}}_1, ..., \bar{\boldsymbol{u}}_N + \Delta\bar{\boldsymbol{u}}_N) \nabla_{\boldsymbol{u}_\alpha} \bar{F}^{(l)} \\ &= \bar{\boldsymbol{u}}_\alpha + \left(\Delta\bar{\boldsymbol{u}}_\alpha - \bar{V}[\boldsymbol{u}_\alpha] \sum_{k,l=1}^{L} \bar{W}^{(kl)} \sum_{\beta=1}^{N} (\nabla_{\boldsymbol{u}_\beta} \bar{F}^{(k)}, \Delta\bar{\boldsymbol{u}}_\beta) \nabla_{\boldsymbol{u}_\alpha} \bar{F}^{(l)} \right)\end{aligned} \tag{5.27}$$

to a first approximation. Let $\Delta\hat{\boldsymbol{u}}_\alpha = \hat{\boldsymbol{u}}_\alpha - \bar{\boldsymbol{u}}_\alpha$ be the error in the corrected value $\hat{\boldsymbol{u}}_\alpha$ (Fig. 5.6). The covariance matrix $\bar{V}[\hat{\boldsymbol{u}}_\alpha, \hat{\boldsymbol{u}}_\beta] = E[\Delta\hat{\boldsymbol{u}}_\alpha \Delta\hat{\boldsymbol{u}}_\beta^\top]$ of the corrected values $\{\hat{\boldsymbol{u}}_\alpha\}$ is computed as follows:

$$\begin{aligned}&\bar{V}[\hat{\boldsymbol{u}}_\alpha, \hat{\boldsymbol{u}}_\beta] \\ &\quad = E[\Delta\bar{\boldsymbol{u}}_\alpha \Delta\bar{\boldsymbol{u}}_\beta^\top]\end{aligned}$$

$$- \left(\sum_{\delta=1}^{N} \sum_{m,n=1}^{L} \bar{W}^{(mn)} (\bar{V}[\boldsymbol{u}_\beta] \nabla_{\boldsymbol{u}_\beta} \bar{F}^{(n)}) (\nabla_{\boldsymbol{u}_\delta} \bar{F}^{(m)})^\top E[\Delta\bar{\boldsymbol{u}}_\delta \Delta\bar{\boldsymbol{u}}_\alpha^\top] \right)^\top$$

$$- \sum_{\gamma=1}^{N} \sum_{k,l=1}^{L} \bar{W}^{(kl)} (\bar{V}[\boldsymbol{u}_\alpha] \nabla_{\boldsymbol{u}_\alpha} \bar{F}^{(l)}) (\nabla_{\boldsymbol{u}_\gamma} \bar{F}^{(k)})^\top E[\Delta\bar{\boldsymbol{u}}_\gamma \Delta\bar{\boldsymbol{u}}_\beta^\top]$$

$$+ \sum_{\gamma,\delta=1}^{N} \sum_{k,l,m,n=1}^{L} \bar{W}^{(kl)} \bar{W}^{(mn)} (\bar{V}[\boldsymbol{u}_\alpha] \nabla_{\boldsymbol{u}_\alpha} \bar{F}^{(l)})$$

$$(\nabla_{\boldsymbol{u}_\gamma} \bar{F}^{(k)})^\top E[\Delta\bar{\boldsymbol{u}}_\gamma \Delta\bar{\boldsymbol{u}}_\delta^\top] \nabla_{\boldsymbol{u}_\delta} \bar{F}^{(m)} (\bar{V}[\boldsymbol{u}_\beta] \nabla_{\boldsymbol{u}_\beta} \bar{F}^{(n)})^\top$$

$$= \bar{V}[\boldsymbol{u}_\alpha] \delta_{\alpha\beta} - \left(\sum_{m,n=1}^{L} \bar{W}^{(mn)} (\bar{V}[\boldsymbol{u}_\beta] \nabla_{\boldsymbol{u}_\beta} \bar{F}^{(n)}) (\nabla_{\boldsymbol{u}_\alpha} \bar{F}^{(m)})^\top \bar{V}[\boldsymbol{u}_\alpha] \right)^\top$$

$$- \sum_{k,l=1}^{L} \bar{W}^{(kl)} (\bar{V}[\boldsymbol{u}_\alpha] \nabla_{\boldsymbol{u}_\alpha} \bar{F}^{(l)}) (\nabla_{\boldsymbol{u}_\beta} \bar{F}^{(k)})^\top \bar{V}[\boldsymbol{u}_\beta]$$

$$+ \sum_{k,l,m,n=1}^{L} \bar{W}^{(kl)} \bar{W}^{(mn)} (\bar{V}[\boldsymbol{u}_\alpha] \nabla_{\boldsymbol{u}_\alpha} \bar{F}^{(l)})$$

$$\sum_{\gamma=1}^{N} (\nabla_{\boldsymbol{u}_\gamma} \bar{F}^{(k)}, \bar{V}[\boldsymbol{u}_\gamma] \nabla_{\boldsymbol{u}_\gamma} \bar{F}^{(m)}) (\bar{V}[\boldsymbol{u}_\beta] \nabla_{\boldsymbol{u}_\beta} \bar{F}^{(n)})^\top$$

$$= \bar{V}[\boldsymbol{u}_\alpha] \delta_{\alpha\beta} - \sum_{m,n=1}^{L} \bar{W}^{(mn)} (\bar{V}[\boldsymbol{u}_\beta] \nabla_{\boldsymbol{u}_\beta} \bar{F}^{(n)}) (\bar{V}[\boldsymbol{u}_\alpha] \nabla_{\boldsymbol{u}_\alpha} \bar{F}^{(m)})^\top$$

$$- \sum_{k,l=1}^{L} \bar{W}^{(kl)} (\bar{V}[\boldsymbol{u}_\alpha] \nabla_{\boldsymbol{u}_\alpha} \bar{F}^{(l)}) (\bar{V}[\boldsymbol{u}_\beta] \nabla_{\boldsymbol{u}_\beta} \bar{F}^{(k)})^\top$$

$$+ \sum_{k,l,m,n=1}^{L} \bar{W}^{(kl)} \bar{W}^{(mn)} (\bar{V}[\boldsymbol{u}_\alpha] \nabla_{\boldsymbol{u}_\alpha} \bar{F}^{(l)}) \bar{V}^{(km)} (\bar{V}[\boldsymbol{u}_\beta] \nabla_{\boldsymbol{u}_\beta} \bar{F}^{(n)})^\top$$

$$= \bar{V}[\bar{\boldsymbol{u}}_\alpha] \delta_{\alpha\beta} - \sum_{k,l=1}^{L} \bar{W}^{(kl)} (\bar{V}[\boldsymbol{u}_\alpha] \nabla_{\boldsymbol{u}_\alpha} \bar{F}^{(k)}) (\bar{V}[\boldsymbol{u}_\beta] \nabla_{\boldsymbol{u}_\beta} \bar{F}^{(l)})^\top. \quad (5.28)$$

Here, we have invoked the assumption that each $\boldsymbol{u}_\alpha$ is independent and hence $E[\Delta\bar{\boldsymbol{u}}_\alpha \Delta\bar{\boldsymbol{u}}_\beta] = \bar{V}[\boldsymbol{u}_\alpha]\delta_{\alpha\beta}$. We have also used the identity $\bar{\boldsymbol{W}}\bar{\boldsymbol{V}}\bar{\boldsymbol{W}} = \bar{\boldsymbol{W}}\bar{\boldsymbol{W}}^{-}\bar{\boldsymbol{W}} = \bar{\boldsymbol{W}}$ (see eqs. (2.123)).

Letting $\alpha = \beta$ in eq. (5.28), we obtain

$$\bar{V}[\hat{\boldsymbol{u}}_\alpha] = \bar{V}[\boldsymbol{u}_\alpha] - \sum_{k,l=1}^{L} \bar{W}^{(kl)} (\bar{V}[\boldsymbol{u}_\alpha] \nabla_{\boldsymbol{u}_\alpha} \bar{F}^{(k)}) (\bar{V}[\boldsymbol{u}_\alpha] \nabla_{\boldsymbol{u}_\alpha} \bar{F}^{(l)})^\top. \quad (5.29)$$

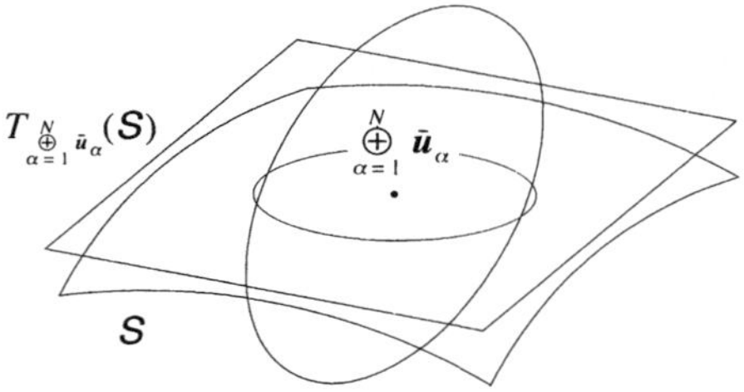

Fig. 5.7. The a priori and a posteriori standard confidence regions.

For $\alpha \neq \beta$, we obtain

$$\bar{V}[\hat{\boldsymbol{u}}_\alpha, \hat{\boldsymbol{u}}_\beta] = -\sum_{k,l=1}^{L} \bar{W}^{(kl)}(\bar{V}[\boldsymbol{u}_\alpha]\nabla_{\boldsymbol{u}_\alpha}\bar{F}^{(k)})(\bar{V}[\boldsymbol{u}_\beta]\nabla_{\boldsymbol{u}_\beta}\bar{F}^{(l)})^\top. \tag{5.30}$$

Thus, *the corrected values* $\{\hat{\boldsymbol{u}}_\alpha\}$ *are correlated* even though the original values $\{\boldsymbol{u}_\alpha\}$ are independent. It can be confirmed[5] that the ranges of $\bar{V}[\hat{\boldsymbol{u}}_\alpha]$ and $\bar{V}[\hat{\boldsymbol{u}}_\alpha, \hat{\boldsymbol{u}}_\beta]$ coincide with the tangent space $T_{\hat{\boldsymbol{u}}_\alpha}(\mathcal{U}_\alpha)$ to the manifold $\mathcal{U}_\alpha$ at $\hat{\boldsymbol{u}}_\alpha$.

Eqs. (5.29) and (5.30) can be given the following geometric interpretation. The a priori covariance matrix $\bigoplus_{\alpha=1}^{N} \bar{V}[\boldsymbol{u}_\alpha]$ defines the *a priori standard confidence region* in the tangent space $T_{\oplus_{\alpha=1}^{N}\bar{\boldsymbol{u}}_\alpha}(\mathcal{U})$ to the manifold $\mathcal{U} = \bigoplus_{\alpha=1}^{N} \mathcal{U}_\alpha$ (Fig. 5.7). The a posteriori covariance matrix given by eq. (5.28) defines the *a posteriori standard confidence region* in the tangent space $T_{\oplus_{\alpha=1}^{N}\bar{\boldsymbol{u}}_\alpha}(\mathcal{S})$ to the model $\mathcal{S}$ defined by the constraint (5.2). This confidence region is the "projection" of the a priori standard confidence region onto $T_{\oplus_{\alpha=1}^{N}\bar{\boldsymbol{u}}_\alpha}(\mathcal{S})$ along the "conjugate direction". It will be shown in Chapter 14 that eq. (5.28) coincides with the *Cramer-Rao lower bound* on the covariance matrix of the corrected values, meaning that the correction given by eq. (5.17) is indeed theoretically optimal.

Eqs. (5.29) and (5.30) are mere theoretical expressions, since they involve values evaluated at $\bar{\boldsymbol{u}}_\alpha$, $\alpha = 1, ..., N$. Furthermore, eqs. (5.29) and (5.30) define the covariance matrices evaluated at $\bar{\boldsymbol{u}}_\alpha$, $\alpha = 1, ..., N$. In actual computation, a consistent approximation is identifying $\bar{\boldsymbol{u}}_\alpha$ with $\hat{\boldsymbol{u}}_\alpha$, i.e., we compute

$$V[\hat{\boldsymbol{u}}_\alpha] = \hat{V}[\boldsymbol{u}_\alpha] - \sum_{k,l=1}^{L} \hat{W}^{(kl)}(\hat{V}[\boldsymbol{u}_\alpha]\nabla_{\boldsymbol{u}_\alpha}\hat{F}^{(k)})(\hat{V}[\boldsymbol{u}_\alpha]\nabla_{\boldsymbol{u}_\alpha}\hat{F}^{(l)})^\top, \tag{5.31}$$

[5]The proof will be given in Chapter 14 in a more general framework.

$$V[\hat{\boldsymbol{u}}_\alpha, \hat{\boldsymbol{u}}_\beta] = -\sum_{k,l=1}^{L} \hat{W}^{(kl)}(\hat{V}[\boldsymbol{u}_\alpha]\nabla_{\boldsymbol{u}_\alpha}\hat{F}^{(k)})(\hat{V}[\boldsymbol{u}_\beta]\nabla_{\boldsymbol{u}_\beta}\hat{F}^{(l)})^\top. \tag{5.32}$$

Here, $\hat{V}[\boldsymbol{u}_\alpha]$ is the matrix defined by eq. (5.26), and $\nabla_{\boldsymbol{u}_\alpha}\hat{F}^{(k)}$ is the abbreviation of $\nabla_{\boldsymbol{u}_\alpha}F^{(k)}(\hat{\boldsymbol{u}}_1, ..., \hat{\boldsymbol{u}}_N)$. The matrix $\hat{\boldsymbol{W}} = (\hat{W}^{(kl)})$ is defined by

$$(\hat{W}^{(kl)}) = \left(\sum_{\alpha=1}^{N}(\nabla_{\boldsymbol{u}_\alpha}\hat{F}^{(k)}, \hat{V}[\boldsymbol{u}_\alpha]\nabla_{\boldsymbol{u}_\alpha}\hat{F}^{(l)})\right)_r^-. \tag{5.33}$$

The ranges of $V[\hat{\boldsymbol{u}}_\alpha]$ and $V[\hat{\boldsymbol{u}}_\alpha, \hat{\boldsymbol{u}}_\beta]$ thus defined coincide with the tangent space $T_{\hat{\boldsymbol{u}}_\alpha}(\mathcal{U}_\alpha)$.

5.1.5 *Hypothesis testing and noise level estimation*

So far, the constraint (5.2) has been assumed given. However, it can be *hypothetical*, and the above correction procedure can provide a means of *testing* this hypothesis. Let us hypothesize that the true values $\bar{\boldsymbol{u}}_\alpha$, $\alpha = 1, ..., N$, satisfy eq. (5.2). Let $\boldsymbol{u}_\alpha = \bar{\boldsymbol{u}}_\alpha + \Delta\boldsymbol{u}_\alpha$, $\alpha = 1, ..., N$, be the observed values, and regard each $\Delta\boldsymbol{u}_\alpha$ as an independent Gaussian random variable with mean $\boldsymbol{0}$ and covariance matrix $\bar{V}[\boldsymbol{u}_\alpha]$.

Since $\nabla_{\boldsymbol{u}_\alpha}\bar{F}^{(k)}$ are deterministic values, eq. (5.5) implies that to a first approximation $F^{(k)}$ is a Gaussian random variable of mean 0. Noting that each $\Delta\boldsymbol{u}_\alpha$ is independent, we can compute the covariance of $F^{(k)}$ and $F^{(l)}$ in the following form:

$$\begin{aligned} V[F^{(k)}, F^{(l)}] &= E[\sum_{\alpha=1}^{N}(\nabla_{\boldsymbol{u}_\alpha}\bar{F}^{(k)}, \Delta\boldsymbol{u}_\alpha)\sum_{\beta=1}^{N}(\nabla_{\boldsymbol{u}_\beta}\bar{F}^{(l)}, \Delta\boldsymbol{u}_\beta)] \\ &= \sum_{\alpha,\beta=1}^{N}(\nabla_{\boldsymbol{u}_\alpha}\bar{F}^{(k)}, E[\Delta\boldsymbol{u}_\alpha\Delta\boldsymbol{u}_\beta^\top]\nabla_{\boldsymbol{u}_\beta}\bar{F}^{(l)}) \\ &= \sum_{\alpha=1}^{N}(\nabla_{\boldsymbol{u}_\alpha}\bar{F}^{(k)}, \bar{V}[\boldsymbol{u}_\alpha]\nabla_{\boldsymbol{u}_\alpha}\bar{F}^{(l)}) = \bar{V}^{(kl)}. \end{aligned} \tag{5.34}$$

The matrix $\bar{\boldsymbol{W}} = (W^{(kl)})$ defined by eq. (5.16) has rank r (= the rank of eq. (5.2)), so the quadratic form

$$\hat{J} = \sum_{k,l=1}^{L} \bar{W}^{(kl)}F^{(k)}F^{(l)} \tag{5.35}$$

is a χ^2 variable with r degrees of freedom (see eq. (3.61)).

It is easily confirmed that the right-hand side of eq. (5.35) coincides with the value obtained by substituting eq. (5.17) into eq. (5.6) (see eqs. (2.141) and

(3.87)). In other words, $\hat{J}$ is the *residual* of the optimization (5.6). If this value is much larger than can be accounted for by the statistical behavior of the noise, the hypothesis (5.2) should be rejected. It follows that the hypothesis (5.2) can be tested by the standard χ^2 *test*: the hypothesis is rejected with significance level $a\%$ if

$$\hat{J} > \chi^2_{r,a}, \tag{5.36}$$

where $\chi^2_{r,a}$ is the $a\%$ significance value of χ^2 with r degrees of freedom (see Section 3.3.4). Intuitively, the hypothesis that eq. (5.2) holds is rejected if the Mahalanobis distance over which the data $\{\boldsymbol{u}_\alpha\}$ must be displaced for imposing eq. (5.2) is too large (see Fig. 5.2).

Eq. (5.35) is merely a theoretical expression, since it involves $\bar{\boldsymbol{W}} = (\bar{W}^{(kl)})$. A simple approximation is using $\boldsymbol{W} = (W^{(kl)})$ defined by eq. (5.23), but now that the optimal estimate $\hat{\boldsymbol{u}}_\alpha$ has been obtained, we can alternatively use $\hat{\boldsymbol{W}} = (\hat{W}^{(kl)})$ defined by eq. (5.33). However, the use of $\hat{\boldsymbol{W}}$ rather than $\boldsymbol{W}$ makes only a second order difference.

In many application problems, the geometric characteristics of noise (e.g., the degree of homogeneity/inhomogeneity and isotropy/anisotropy) can be relatively easily predicted but the absolute magnitude of noise is very difficult to estimate a priori. In such a case, we can write the covariance matrix $V[\boldsymbol{u}_\alpha]$ in the form

$$V[\boldsymbol{u}_\alpha] = \epsilon^2 V_0[\boldsymbol{u}_\alpha], \qquad \alpha = 1, ..., N, \tag{5.37}$$

where $V_0[\boldsymbol{u}_\alpha]$ has a known form while ϵ is unknown. Let us call $V_0[\boldsymbol{u}_\alpha]$ the *normalized covariance matrix* of $\boldsymbol{u}_\alpha$, and ϵ the *noise level*. It is easily seen from eqs. (5.22) and (5.23) that the optimal correction is *not affected by multiplication of* $V[\boldsymbol{u}_\alpha]$ *by an arbitrary positive constant*. Hence, $V[\boldsymbol{u}_\alpha]$ can be replaced by the normalized covariance matrix $V_0[\boldsymbol{u}_\alpha]$. In other words, *we need not know the absolute noise level for the optimal correction.*

Once the optimal solution is computed, the noise level ϵ can be estimated *a posteriori* as follows. If the normalized covariance matrix $V_0[\boldsymbol{u}_\alpha]$ is used for $V[\boldsymbol{u}_\alpha]$ in eq. (5.35), the resulting residual $\hat{J}_0$ equals $\epsilon^2\hat{J}$. Since $\hat{J}$ is a χ^2 variable with r degrees of freedom, its expectation and variance are r and $2r$, respectively (see eqs. (3.59)). Hence, an unbiased estimator $\hat{\epsilon}^2$ of ϵ^2 is obtained in the form

$$\hat{\epsilon}^2 = \frac{\hat{J}_0}{r}. \tag{5.38}$$

Its expectation and variance are respectively given by

$$E[\hat{\epsilon}^2] = \epsilon^2, \qquad V[\hat{\epsilon}^2] = \frac{2\epsilon^4}{r}. \tag{5.39}$$

In geometric terms, we are estimating the noise level from the Mahalanobis distance with respect to $V_0[\boldsymbol{u}_\alpha]$ over which the data $\{\boldsymbol{u}_\alpha\}$ must be displaced for imposing eq. (5.2) (see Fig. 5.2). It follows that the hypothesis test (5.36) can be interpreted as comparing the a priori value ϵ with the a posteriori estimate $\hat{\epsilon}$ computed on the assumption that the hypothesis is true. In fact, eq. (5.36) is equivalently rewritten as[6]

$$\frac{\hat{\epsilon}^2}{\epsilon^2} > \frac{\chi^2_{r,a}}{r}. \tag{5.40}$$

5.1.6 *Linear constraint*

In many problems, the constraint is linear in the form

$$\sum_{\alpha=1}^{N} \boldsymbol{A}_\alpha \bar{\boldsymbol{u}}_\alpha = \boldsymbol{b}, \tag{5.41}$$

where $\boldsymbol{A}_\alpha$ is an Ln_α-matrix and $\boldsymbol{b}$ is an L-vector. The rank r of this constraint equals the rank of the $L(\sum_{\alpha=1}^{N} n_\alpha)$-matrix

$$\boldsymbol{A} = (\boldsymbol{A}_1 \boldsymbol{P}^{\mathcal{U}_1}_{\bar{\boldsymbol{u}}_1}, ..., \boldsymbol{A}_N \boldsymbol{P}^{\mathcal{U}_N}_{\bar{\boldsymbol{u}}_N}), \tag{5.42}$$

in terms of which eq. (5.41) can be written as $\boldsymbol{A} \bigoplus_{\alpha=1}^{N} \bar{\boldsymbol{u}}_\alpha = \boldsymbol{b}$.

Suppose the observed values $\boldsymbol{u}_\alpha$, $\alpha = 1, ..., N$, do not satisfy the constraint (5.41). If we write $\boldsymbol{u}_\alpha = \bar{\boldsymbol{u}}_\alpha + \Delta \boldsymbol{u}_\alpha$, eq. (5.41) can be written as

$$\sum_{\alpha=1}^{N} \boldsymbol{A}_\alpha \Delta \boldsymbol{u}_\alpha = \sum_{\alpha=1}^{N} \boldsymbol{A}_\alpha \boldsymbol{u}_\alpha - \boldsymbol{b}. \tag{5.43}$$

The correction $\Delta \boldsymbol{u}_\alpha$ is determined by the optimization (5.6) under the inherent constraint $\Delta \boldsymbol{u}_\alpha \in T_{\bar{\boldsymbol{u}}_\alpha}(\mathcal{U}_\alpha)$. The solution given by eq. (5.17) reduces to

$$\Delta \boldsymbol{u}_\alpha = V[\boldsymbol{u}_\alpha] \boldsymbol{A}_\alpha^\top \boldsymbol{W} (\sum_{\beta=1}^{N} \boldsymbol{A}_\beta \boldsymbol{u}_\beta - \boldsymbol{b}), \tag{5.44}$$

where the (LL)-matrix $\boldsymbol{W}$ is given as follows (see eq. (5.23)):

$$\boldsymbol{W} = \left(\sum_{\alpha=1}^{N} \boldsymbol{A}_\alpha V[\boldsymbol{u}_\alpha] \boldsymbol{A}_\alpha^\top \right)_r^- . \tag{5.45}$$

Since the constraint (5.41) is linear, no approximation has been made to obtain eq. (5.43). Hence, no iterations are necessary.

The a posteriori covariance matrices given by eqs. (5.31) and (5.32) reduce to

$$V[\hat{\boldsymbol{u}}_\alpha] = \hat{V}[\boldsymbol{u}_\alpha] - \hat{V}[\boldsymbol{u}_\alpha] \boldsymbol{A}_\alpha^\top \boldsymbol{W} \boldsymbol{A}_\alpha \hat{V}[\boldsymbol{u}_\alpha], \tag{5.46}$$

[6] This is a consequence of the fact that $\hat{\epsilon}^2/\epsilon^2$ is a *modified* χ^2 *variable* with r degrees of freedom if the hypothesis is true (see eq. (3.72)).

$$V[\hat{\boldsymbol{u}}_\alpha, \hat{\boldsymbol{u}}_\beta] = \hat{V}[\boldsymbol{u}_\alpha]\boldsymbol{A}_\alpha^\top \boldsymbol{W}\boldsymbol{A}_\beta \hat{V}[\boldsymbol{u}_\beta]. \tag{5.47}$$

The residual of the optimization (5.6) can be written in the form

$$\hat{J} = (\sum_{\alpha=1}^{N} \boldsymbol{A}_\alpha \hat{\boldsymbol{u}}_\alpha - \boldsymbol{b}, \boldsymbol{W}(\sum_{\beta=1}^{N} \boldsymbol{A}_\beta \hat{\boldsymbol{u}}_\beta - \boldsymbol{b})), \tag{5.48}$$

which is a χ^2 variable with r degrees of freedom if the noise is Gaussian.

5.2 Correction of Image Points and Image Lines

5.2.1 Optimal correction for coincidence

A. Image points

Let $\boldsymbol{x}_1$ and $\boldsymbol{x}_2$ be two image points, and $V[\boldsymbol{x}_1]$ and $V[\boldsymbol{x}_2]$ their respective a priori covariance matrices. Suppose $\boldsymbol{x}_1$ and $\boldsymbol{x}_2$ are two different estimates of the same image point. Consider the problem of estimating the true position. Let $\bar{\boldsymbol{x}}_1$ and $\bar{\boldsymbol{x}}_2$ be the true values of $\boldsymbol{x}_1$ and $\boldsymbol{x}_2$, respectively. The constraint to be imposed is

$$\bar{\boldsymbol{x}}_1 = \bar{\boldsymbol{x}}_2, \tag{5.49}$$

which has rank 2 because both sides are orthogonal to $\boldsymbol{k} = (0, 0, 1)^\top$. If the two image points are statistically independent, the optimal estimate $\hat{\boldsymbol{x}} = \boldsymbol{x}_1 - \Delta\boldsymbol{x}_1 = \boldsymbol{x}_2 - \Delta\boldsymbol{x}_2$ is obtained by finding $\Delta\boldsymbol{x}_1$ and $\Delta\boldsymbol{x}_2$ such that[7]

$$J = (\Delta\boldsymbol{x}_1, \bar{V}[\boldsymbol{x}_1]^- \Delta\boldsymbol{x}_1) + (\Delta\boldsymbol{x}_2, \bar{V}[\boldsymbol{x}_2]^- \Delta\boldsymbol{x}_2) \to \min \tag{5.50}$$

under the linearized constraint

$$\Delta\boldsymbol{x}_2 - \Delta\boldsymbol{x}_1 = \boldsymbol{x}_2 - \boldsymbol{x}_1, \qquad \Delta\boldsymbol{x}_1, \Delta\boldsymbol{x}_2 \in \{\boldsymbol{k}\}_L^\perp. \tag{5.51}$$

The first order solution is given by[8]

$$\Delta\boldsymbol{x}_1 = V[\boldsymbol{x}_1]\boldsymbol{W}(\boldsymbol{x}_1 - \boldsymbol{x}_2),$$
$$\Delta\boldsymbol{x}_2 = V[\boldsymbol{x}_2]\boldsymbol{W}(\boldsymbol{x}_2 - \boldsymbol{x}_1), \tag{5.52}$$

where $\boldsymbol{W}$ is a (33)-matrix defined by

$$\boldsymbol{W} = \Big(V[\boldsymbol{x}_1] + V[\boldsymbol{x}_2]\Big)^-. \tag{5.53}$$

---

[7]We adopt the convention that $\bar{V}[\cdot]$ denotes the value of the covariance matrix $V[\cdot]$ evaluated at the true value of the variable.

[8]We mean by "first order solution" the approximation expressed in terms of the data and the covariance matrices evaluated at the data values (see eq. (5.22)).

The a posteriori covariance matrix of the estimate $\hat{\boldsymbol{x}}$ is

$$V[\hat{\boldsymbol{x}}] = V[\boldsymbol{x}_1] - V[\boldsymbol{x}_1]\boldsymbol{W}V[\boldsymbol{x}_1] = V[\boldsymbol{x}_1]\boldsymbol{W}V[\boldsymbol{x}_2]$$
$$= V[\boldsymbol{x}_2] - V[\boldsymbol{x}_2]\boldsymbol{W}V[\boldsymbol{x}_2]. \tag{5.54}$$

The residual of J can be written as

$$\hat{J} = (\boldsymbol{x}_2 - \boldsymbol{x}_1, \boldsymbol{W}(\boldsymbol{x}_2 - \boldsymbol{x}_1)), \tag{5.55}$$

which is a χ^2 variable with two degrees of freedom[9]. This fact provides a *coincidence test* for image points: the hypothesis that image points $\boldsymbol{x}_1$ and $\boldsymbol{x}_2$ coincide with each other is rejected with significance level $a\%$ if

$$\hat{J} > \chi^2_{2,a}. \tag{5.56}$$

Example 5.1 If each coordinate is perturbed independently by Gaussian noise of mean 0 and variance ϵ^2, the covariance matrices of $\Delta\boldsymbol{x}_1$ and $\Delta\boldsymbol{x}_2$ are $V[\boldsymbol{x}_1] = V[\boldsymbol{x}_2] = \epsilon^2\boldsymbol{P}_{\boldsymbol{k}}$ (see Example 4.1). The optimal estimate $\hat{\boldsymbol{x}}$ is

$$\hat{\boldsymbol{x}} = \boldsymbol{x}_1 - \Delta\boldsymbol{x}_1 = \boldsymbol{x}_2 - \Delta\boldsymbol{x}_2 = \frac{1}{2}(\boldsymbol{x}_1 + \boldsymbol{x}_2). \tag{5.57}$$

The a posteriori covariance matrix of $\hat{\boldsymbol{x}}$ is

$$V[\hat{\boldsymbol{x}}] = \frac{\epsilon^2}{2}\boldsymbol{P}_{\boldsymbol{k}}. \tag{5.58}$$

The residual (5.55) can be written as

$$\hat{J} = \frac{1}{2\epsilon^2}\|\boldsymbol{x}_2 - \boldsymbol{x}_1\|^2. \tag{5.59}$$

Hence, an unbiased estimator of the variance ϵ^2 is obtained in the form

$$\hat{\epsilon}^2 = \frac{1}{4}\|\boldsymbol{x}_2 - \boldsymbol{x}_1\|^2. \tag{5.60}$$

The value $\hat{\epsilon}$ thus estimated equals the half-distance between the two image points. If the value ϵ is given a priori, the coincidence test takes the form

$$\frac{\hat{\epsilon}^2}{\epsilon^2} > \frac{\chi^2_{2,a}}{2}. \tag{5.61}$$

[9]We assume Gaussian noise and do first order analysis whenever we refer to χ^2 distributions and χ^2 tests.

B. Image lines

The same analysis can be done for two image lines $(\boldsymbol{n}_1, \boldsymbol{x}) = 0$ and $(\boldsymbol{n}_2, \boldsymbol{x}) = 0$. Let $V[\boldsymbol{n}_1]$ and $V[\boldsymbol{n}_2]$ be their respective a priori covariance matrices. The signs of $\boldsymbol{n}_1$ and $\boldsymbol{n}_2$ are chosen in such a way that $\boldsymbol{n}_1 \approx \boldsymbol{n}_2$. Let $(\bar{\boldsymbol{n}}_1, \boldsymbol{x}) = 0$ and $(\bar{\boldsymbol{n}}_2, \boldsymbol{x}) = 0$ be the true image lines. The constraint to be imposed is

$$\bar{\boldsymbol{n}}_1 = \bar{\boldsymbol{n}}_2, \tag{5.62}$$

which has rank 2 because both sides are unit vectors[10]. If the two image lines are statistically independent, the first order solution of the optimization

$$J = (\Delta\boldsymbol{n}_1, \bar{V}[\boldsymbol{n}_1]^- \Delta\boldsymbol{n}_1) + (\Delta\boldsymbol{n}_2, \bar{V}[\boldsymbol{n}_2]^- \Delta\boldsymbol{n}_2) \rightarrow \min \tag{5.63}$$

under the linearized constraint

$$\begin{gathered} \Delta\boldsymbol{n}_2 - \Delta\boldsymbol{n}_1 = \boldsymbol{n}_2 - \boldsymbol{n}_1, \\ \Delta\boldsymbol{n}_1 \in \{\bar{\boldsymbol{n}}_1\}_L^\perp, \qquad \Delta\boldsymbol{n}_2 \in \{\bar{\boldsymbol{n}}_2\}_L^\perp, \end{gathered} \tag{5.64}$$

is given by

$$\begin{aligned} \Delta\boldsymbol{n}_1 &= V[\boldsymbol{n}_1]\boldsymbol{W}(\boldsymbol{n}_1 - \boldsymbol{n}_2), \\ \Delta\boldsymbol{n}_2 &= V[\boldsymbol{n}_2]\boldsymbol{W}(\boldsymbol{n}_2 - \boldsymbol{n}_1), \end{aligned} \tag{5.65}$$

where $\boldsymbol{W}$ is a (33)-matrix defined by[11]

$$\boldsymbol{W} = \Big(V[\boldsymbol{n}_1] + V[\boldsymbol{n}_2]\Big)_2^- . \tag{5.66}$$

A realistic form of the correction is

$$\hat{\boldsymbol{n}} = N[\boldsymbol{n}_1 - \Delta\boldsymbol{n}_1] = N[\boldsymbol{n}_2 - \Delta\boldsymbol{n}_2]. \tag{5.67}$$

The a posteriori covariance matrix of the estimate $\hat{\boldsymbol{n}}$ is

$$\begin{aligned} V[\hat{\boldsymbol{n}}] &= \hat{V}[\boldsymbol{n}_1] - \hat{V}[\boldsymbol{n}_1]\hat{\boldsymbol{W}}\hat{V}[\boldsymbol{n}_1] = \hat{V}[\boldsymbol{n}_1]\hat{\boldsymbol{W}}\hat{V}[\boldsymbol{n}_2] \\ &= \hat{V}[\boldsymbol{n}_2] - \hat{V}[\boldsymbol{n}_2]\hat{\boldsymbol{W}}\hat{V}[\boldsymbol{n}_2]. \end{aligned} \tag{5.68}$$

Here,

$$\hat{V}[\boldsymbol{n}_i] = \boldsymbol{P}_{\hat{\boldsymbol{n}}} V[\boldsymbol{n}_i] \boldsymbol{P}_{\hat{\boldsymbol{n}}}, \qquad i = 1, 2, \tag{5.69}$$

where $\boldsymbol{P}_{\hat{\boldsymbol{n}}}$ is the projection matrix along $\hat{\boldsymbol{n}}$. The matrix $\hat{\boldsymbol{W}}$ is obtained by replacing $V[\boldsymbol{n}_i]$ by $\hat{V}[\boldsymbol{n}_i]$, $i = 1, 2$, in eq. (5.66). The residual of J can be written as

$$\hat{J} = (\boldsymbol{n}_2 - \boldsymbol{n}_1, \hat{\boldsymbol{W}}(\boldsymbol{n}_2 - \boldsymbol{n}_1)), \tag{5.70}$$

---

[10]This constraint is degenerate.

[11]The rank-constrained generalized inverse $(\cdot)_2^-$ is used because the ranges of $V[\mathbf{n}_1]$ and $V[\mathbf{n}_2]$ are different from the ranges of $\bar{V}[\mathbf{n}_1]$ and $\bar{V}[\mathbf{n}_2]$. Consequently, although $\bar{V}[\mathbf{n}_1] + \bar{V}[\mathbf{n}_2]$ is a singular matrix of rank 2, the matrix $V[\mathbf{n}_1] + V[\mathbf{n}_2]$ is generally nonsingular.

which is a χ^2 variable with two degrees of freedom. This fact provides a *coincidence test* for image lines: the hypothesis that image lines $(\boldsymbol{n}_1, \boldsymbol{x}) = 0$ and $(\boldsymbol{n}_2, \boldsymbol{x}) = 0$ coincide with each other is rejected with significance level a% if

$$\hat{J} > \chi^2_{2,a}. \tag{5.71}$$

5.2.2 *Optimal correction for incidence*

A. Simultaneous correction

Suppose image point $\boldsymbol{x}$ and image line $(\boldsymbol{n}, \boldsymbol{x}) = 0$ are, respectively, estimates of an image point p and an image line l that should be incident to each other in the absence of noise. Consider the problem of optimally correcting them so as to make them incident. In other words, we want to find $\Delta\boldsymbol{x}$ and $\Delta\boldsymbol{n}$ such that $\bar{\boldsymbol{x}} = \boldsymbol{x} - \Delta\boldsymbol{x}$ and $\bar{\boldsymbol{n}} = \boldsymbol{n} - \Delta\boldsymbol{n}$ satisfy

$$(\bar{\boldsymbol{n}}, \bar{\boldsymbol{x}}) = 0. \tag{5.72}$$

The rank of this constraint is 1. Let $V[\boldsymbol{x}]$ and $V[\boldsymbol{n}]$ be the a priori covariance matrices of $\boldsymbol{x}$ and $\boldsymbol{n}$, respectively. If the image point and the image line are statistically independent, the problem can be written as the optimization

$$J = (\Delta\boldsymbol{x}, \bar{V}[\boldsymbol{x}]^- \Delta\boldsymbol{x}) + (\Delta\boldsymbol{n}, \bar{V}[\boldsymbol{n}]^- \Delta\boldsymbol{n}) \to \min \tag{5.73}$$

under the linearized constraint

$$(\bar{\boldsymbol{n}}, \Delta\boldsymbol{x}) + (\bar{\boldsymbol{x}}, \Delta\boldsymbol{n}) = (\boldsymbol{n}, \boldsymbol{x}),$$
$$\Delta\boldsymbol{x} \in \{\boldsymbol{k}\}^\perp_L, \qquad \Delta\boldsymbol{n} \in \{\bar{\boldsymbol{n}}\}^\perp_L. \tag{5.74}$$

The first order solution is given by

$$\Delta\boldsymbol{x} = \frac{(\boldsymbol{n}, \boldsymbol{x}) V[\boldsymbol{x}]\boldsymbol{n}}{(\boldsymbol{n}, V[\boldsymbol{x}]\boldsymbol{n}) + (\boldsymbol{x}, V[\boldsymbol{n}]\boldsymbol{x})},$$
$$\Delta\boldsymbol{n} = \frac{(\boldsymbol{n}, \boldsymbol{x}) V[\boldsymbol{n}]\boldsymbol{x}}{(\boldsymbol{n}, V[\boldsymbol{x}]\boldsymbol{n}) + (\boldsymbol{x}, V[\boldsymbol{n}]\boldsymbol{x})}. \tag{5.75}$$

A realistic form of the correction is

$$\hat{\boldsymbol{x}} = \boldsymbol{x} - \Delta\boldsymbol{x}, \qquad \hat{\boldsymbol{n}} = N[\boldsymbol{n} - \Delta\boldsymbol{n}]. \tag{5.76}$$

The a posteriori covariance matrices of the corrected values $\hat{\boldsymbol{x}}$ and $\hat{\boldsymbol{n}}$ are

$$V[\hat{\boldsymbol{x}}] = V[\boldsymbol{x}] - \frac{(V[\boldsymbol{x}]\hat{\boldsymbol{n}})(V[\boldsymbol{x}]\hat{\boldsymbol{n}})^\top}{(\hat{\boldsymbol{n}}, V[\boldsymbol{x}]\hat{\boldsymbol{n}}) + (\hat{\boldsymbol{x}}, \hat{V}[\boldsymbol{n}]\hat{\boldsymbol{x}})},$$
$$V[\hat{\boldsymbol{n}}] = \hat{V}[\boldsymbol{n}] - \frac{(\hat{V}[\boldsymbol{n}]\hat{\boldsymbol{x}})(\hat{V}[\boldsymbol{n}]\hat{\boldsymbol{x}})^\top}{(\hat{\boldsymbol{n}}, V[\boldsymbol{x}]\hat{\boldsymbol{n}}) + (\hat{\boldsymbol{x}}, \hat{V}[\boldsymbol{n}]\hat{\boldsymbol{x}})},$$

$$V[\hat{\boldsymbol{x}}, \hat{\boldsymbol{n}}] = -\frac{(V[\boldsymbol{x}]\hat{\boldsymbol{n}})(\hat{V}[\boldsymbol{n}]\hat{\boldsymbol{x}})^\top}{(\hat{\boldsymbol{n}}, V[\boldsymbol{x}]\hat{\boldsymbol{n}}) + (\hat{\boldsymbol{x}}, \hat{V}[\boldsymbol{n}]\hat{\boldsymbol{x}})} = V[\hat{\boldsymbol{n}}, \hat{\boldsymbol{x}}]^\top, \tag{5.77}$$

where

$$\hat{V}[\boldsymbol{n}] = \boldsymbol{P}_{\hat{\boldsymbol{n}}} V[\boldsymbol{n}] \boldsymbol{P}_{\hat{\boldsymbol{n}}}. \tag{5.78}$$

The residual of J can be written as

$$\hat{J} = \frac{(\boldsymbol{n}, \boldsymbol{x})^2}{(\hat{\boldsymbol{n}}, V[\boldsymbol{x}]\hat{\boldsymbol{n}}) + (\hat{\boldsymbol{x}}, \hat{V}[\boldsymbol{n}]\hat{\boldsymbol{x}})}, \tag{5.79}$$

which is a χ^2 variable with one degree of freedom. This fact provides an *incidence test* for an image point and an image line: the hypothesis that image point $\boldsymbol{x}$ and image line $(\boldsymbol{n}, \boldsymbol{x}) = 0$ are incident to each other is rejected with significance level $a\%$ if

$$\hat{J} > \chi^2_{1,a}. \tag{5.80}$$

B. Image point correction

If the image line $(\boldsymbol{n}, \boldsymbol{x}) = 0$ is fixed, the linearized constraint is

$$(\boldsymbol{n}, \Delta\boldsymbol{x}) = (\boldsymbol{n}, \boldsymbol{x}), \qquad \Delta\boldsymbol{x} \in \{\boldsymbol{k}\}^\perp_L. \tag{5.81}$$

The first-order correction of the image point $\boldsymbol{x}$ is obtained by letting $V[\boldsymbol{n}] = \boldsymbol{O}$ in eqs. (5.75):

$$\Delta\boldsymbol{x} = \frac{(\boldsymbol{n}, \boldsymbol{x}) V[\boldsymbol{x}]\boldsymbol{n}}{(\boldsymbol{n}, V[\boldsymbol{x}]\boldsymbol{n})}. \tag{5.82}$$

The a posteriori covariance matrix of the corrected value $\hat{\boldsymbol{x}}$ is

$$V[\hat{\boldsymbol{x}}] = V[\boldsymbol{x}] - \frac{(V[\boldsymbol{x}]\boldsymbol{n})(V[\boldsymbol{x}]\boldsymbol{n})^\top}{(\boldsymbol{n}, V[\boldsymbol{x}]\boldsymbol{n})}. \tag{5.83}$$

Since $V[\hat{\boldsymbol{x}}]\boldsymbol{n} = \boldsymbol{0}$, the rank of $V[\hat{\boldsymbol{x}}]$ is 1; its null space is $\{\boldsymbol{k}, \boldsymbol{n}\}_L$, which is orthogonal to the orientation $\boldsymbol{m} = N[\boldsymbol{k} \times \boldsymbol{n}]$ of the space line $(\boldsymbol{n}, \boldsymbol{x}) = 0$. The residual

$$\hat{J} = \frac{(\boldsymbol{n}, \boldsymbol{x})^2}{(\boldsymbol{n}, V[\boldsymbol{x}]\boldsymbol{n})} \tag{5.84}$$

is a χ^2 variable with one degree of freedom. Hence, and the incidence test given by eq. (5.80) can be applied.

Example 5.2 If each coordinate is perturbed independently by Gaussian noise of mean 0 and variance ϵ^2, the covariance matrix of $\Delta\boldsymbol{x}$ is $V[\boldsymbol{x}] = \epsilon^2 \boldsymbol{P}_{\boldsymbol{k}}$. The optimal correction (5.82) reduces to

$$\Delta\boldsymbol{x} = -\frac{(\boldsymbol{n}, \boldsymbol{x})\boldsymbol{P}_{\boldsymbol{k}}\boldsymbol{n}}{1 - (\boldsymbol{k}, \boldsymbol{n})^2}. \tag{5.85}$$

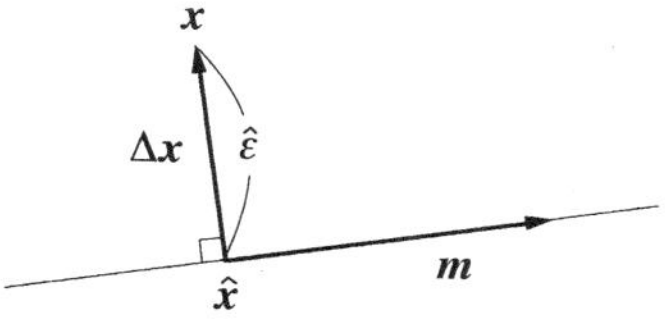

Fig. 5.8. Optimal incidence correction for an image point.

Consequently, $\boldsymbol{x}$ is displaced onto the image line $(\boldsymbol{n}, \boldsymbol{x}) = 0$ perpendicularly (Fig. 5.8). The a posteriori covariance matrix of the corrected value $\hat{\boldsymbol{x}}$ can be written as

$$V[\hat{\boldsymbol{x}}] = \epsilon^2 \boldsymbol{m}\boldsymbol{m}^\top, \tag{5.86}$$

where $\boldsymbol{m} = N[\boldsymbol{k} \times \boldsymbol{n}]$ is the orientation of the image line $(\boldsymbol{n}, \boldsymbol{x}) = 0$. The residual (5.84) can be written as

$$\hat{J} = \frac{1}{\epsilon^2} \frac{(\boldsymbol{n}, \boldsymbol{x})^2}{1 - (\boldsymbol{k}, \boldsymbol{n})^2}. \tag{5.87}$$

Hence, an unbiased estimator of the variance ϵ^2 is obtained in the form

$$\hat{\epsilon}^2 = \frac{(\boldsymbol{n}, \boldsymbol{x})^2}{1 - (\boldsymbol{k}, \boldsymbol{n})^2}. \tag{5.88}$$

The value $\hat{\epsilon}$ thus estimated equals the distance between the image point and the image line (see eq. (4.11)). If the value ϵ is given a priori, the incidence test takes the form

$$\frac{\hat{\epsilon}^2}{\epsilon^2} > \chi^2_{1,a}. \tag{5.89}$$

C. Image line correction

If the image point $\boldsymbol{x} = 0$ is fixed, the linearized constraint is

$$(\Delta \boldsymbol{n}, \boldsymbol{x}) = (\boldsymbol{n}, \boldsymbol{x}), \qquad \Delta \boldsymbol{n} \in \{\bar{\boldsymbol{n}}\}^\perp_L. \tag{5.90}$$

The first-order correction of the image line $(\boldsymbol{n}, \boldsymbol{x}) = 0$ is obtained by letting $V[\boldsymbol{x}] = \boldsymbol{O}$ in eqs. (5.75):

$$\Delta \boldsymbol{n} = \frac{(\boldsymbol{n}, \boldsymbol{x}) V[\boldsymbol{n}]\boldsymbol{x}}{(\boldsymbol{x}, V[\boldsymbol{n}]\boldsymbol{x})}. \tag{5.91}$$

The a posteriori covariance matrix of the corrected value $\hat{\boldsymbol{n}}$ is

$$V[\hat{\boldsymbol{n}}] = \hat{V}[\boldsymbol{n}] - \frac{(\hat{V}[\boldsymbol{n}]\boldsymbol{x})(\hat{V}[\boldsymbol{n}]\boldsymbol{x})^\top}{(\boldsymbol{x}, \hat{V}[\boldsymbol{n}]\boldsymbol{x})}. \tag{5.92}$$

The rank of $V[\hat{\boldsymbol{n}}]$ is 1; its null space is $\{\hat{\boldsymbol{n}}, \boldsymbol{x}\}_L$. The residual

$$\hat{J} = \frac{(\boldsymbol{n}, \boldsymbol{x})^2}{(\boldsymbol{x}, \hat{V}[\boldsymbol{n}]\boldsymbol{x})} \tag{5.93}$$

is a χ^2 variable with one degree of freedom. Hence, the incidence test given by eq. (5.80) can be applied.

5.3 Correction of Space Points and Space Lines

5.3.1 Optimal correction for coincidence

A. Space points

Two space points $\boldsymbol{r}_1$ and $\boldsymbol{r}_2$ that are supposed to coincide can be optimally corrected in the same way as in the case of image points. Let $V[\boldsymbol{r}_1]$ and $V[\boldsymbol{r}_2]$ be their respective a priori covariance matrices. Let $\bar{\boldsymbol{r}}_1$ and $\bar{\boldsymbol{r}}_2$ be the true positions of $\boldsymbol{r}_1$ and $\boldsymbol{r}_2$, respectively. The constraint to be imposed is

$$\bar{\boldsymbol{r}}_1 = \bar{\boldsymbol{r}}_2, \tag{5.94}$$

which has rank 3. If the two space points are statistically independent, the problem is finding $\Delta\boldsymbol{r}_1$ and $\Delta\boldsymbol{r}_2$ such that

$$J = (\Delta\boldsymbol{r}_1, \bar{V}[\boldsymbol{r}_1]^{-1}\Delta\boldsymbol{r}_1) + (\Delta\boldsymbol{r}_2, \bar{V}[\boldsymbol{r}_2]^{-1}\Delta\boldsymbol{r}_2) \to \min \tag{5.95}$$

under the linearized constraint

$$\Delta\boldsymbol{r}_2 - \Delta\boldsymbol{r}_1 = \boldsymbol{r}_2 - \boldsymbol{r}_1. \tag{5.96}$$

The first order solution is given by

$$\Delta\boldsymbol{r}_1 = V[\boldsymbol{r}_1]\boldsymbol{W}(\boldsymbol{r}_1 - \boldsymbol{r}_2),$$
$$\Delta\boldsymbol{r}_2 = V[\boldsymbol{r}_2]\boldsymbol{W}(\boldsymbol{r}_2 - \boldsymbol{r}_1), \tag{5.97}$$

where $\boldsymbol{W}$ is a (33)-matrix defined by

$$\boldsymbol{W} = \Big(V[\boldsymbol{r}_1] + V[\boldsymbol{r}_2]\Big)^{-1}. \tag{5.98}$$

The a posteriori covariance matrix of the estimate $\hat{\boldsymbol{r}}$ is

$$\begin{aligned} V[\hat{\boldsymbol{r}}] &= V[\boldsymbol{r}_1] - V[\boldsymbol{r}_1]\boldsymbol{W}V[\boldsymbol{r}_1] = V[\boldsymbol{r}_1]\boldsymbol{W}V[\boldsymbol{r}_2] \\ &= V[\boldsymbol{r}_2] - V[\boldsymbol{r}_2]\boldsymbol{W}V[\boldsymbol{r}_2]. \end{aligned} \tag{5.99}$$

The residual of J can be written as

$$\hat{J} = (\boldsymbol{r}_2 - \boldsymbol{r}_1, \boldsymbol{W}(\boldsymbol{r}_2 - \boldsymbol{r}_1)), \tag{5.100}$$

which is a χ^2 variable with three degrees of freedom. This fact provides a *coincidence test* for space points: the hypothesis that space points $\boldsymbol{r}_1$ and $\boldsymbol{r}_2$ coincide with each other is rejected with significance level $a\%$ if

$$\hat{J} > \chi^2_{3,a}. \tag{5.101}$$

Example 5.3 If each coordinate is perturbed independently by Gaussian noise of mean 0 and variance ϵ^2, the covariance matrices of $\Delta\boldsymbol{r}_1$ and $\Delta\boldsymbol{r}_2$ are $V[\boldsymbol{r}_1] = V[\boldsymbol{r}_2] = \epsilon^2\boldsymbol{I}$. The optimal estimate $\hat{\boldsymbol{r}}$ is

$$\hat{\boldsymbol{r}} = \boldsymbol{r}_1 - \Delta\boldsymbol{r}_1 = \boldsymbol{r}_2 - \Delta\boldsymbol{r}_2 = \frac{1}{2}(\boldsymbol{r}_1 + \boldsymbol{r}_2). \tag{5.102}$$

The a posteriori covariance matrix of $\hat{\boldsymbol{r}}$ is

$$V[\hat{\boldsymbol{r}}] = \frac{\epsilon^2}{2}\boldsymbol{I}. \tag{5.103}$$

The residual (5.100) can be written as

$$\hat{J} = \frac{1}{2\epsilon^2}\|\boldsymbol{r}_2 - \boldsymbol{r}_1\|^2. \tag{5.104}$$

Hence, an unbiased estimator of the variance ϵ^2 is obtained in the form

$$\hat{\epsilon}^2 = \frac{1}{6}\|\boldsymbol{r}_2 - \boldsymbol{r}_1\|^2. \tag{5.105}$$

The value $\hat{\epsilon}$ thus estimated equals $1/\sqrt{3}$ times the half-distance between the two space points. If the value ϵ is given a priori, the coincidence test takes the form

$$\frac{\hat{\epsilon}^2}{\epsilon^2} > \frac{\chi^2_{3,a}}{3}. \tag{5.106}$$

B. Space lines

The same analysis can be done for two space lines $\boldsymbol{r} \times \boldsymbol{p}_1 = \boldsymbol{n}_1$ and $\boldsymbol{r} \times \boldsymbol{p}_2 = \boldsymbol{n}_2$. Let $V[\boldsymbol{p}_1 \oplus \boldsymbol{n}_1]$ and $V[\boldsymbol{p}_2 \oplus \boldsymbol{n}_2]$ be their a priori covariance matrices. The signs of $\{\boldsymbol{p}_1, \boldsymbol{n}_1\}$ and $\{\boldsymbol{p}_2, \boldsymbol{n}_2\}$ are chosen so that $\boldsymbol{p}_1 \approx \boldsymbol{p}_2$ and $\boldsymbol{n}_1 \approx \boldsymbol{n}_2$. Let $\boldsymbol{r} \times \bar{\boldsymbol{p}}_1 = \bar{\boldsymbol{n}}_1$ and $\boldsymbol{r} \times \bar{\boldsymbol{p}}_2 = \bar{\boldsymbol{n}}_2$ be the true space lines. The constraint to be imposed is

$$\bar{\boldsymbol{p}}_1 = \bar{\boldsymbol{p}}_2, \qquad \bar{\boldsymbol{n}}_1 = \bar{\boldsymbol{n}}_2, \tag{5.107}$$

which has rank 4 because the representations $\{\bar{\boldsymbol{p}}_1, \bar{\boldsymbol{n}}_1\}$ and $\{\bar{\boldsymbol{p}}_2, \bar{\boldsymbol{n}}_2\}$ have four degrees of freedom[12] (see Section 4.2.2). If the two space lines are statistically independent, the problem is finding $\Delta\boldsymbol{p}_1$, $\Delta\boldsymbol{n}_1$, $\Delta\boldsymbol{p}_2$, and $\Delta\boldsymbol{n}_2$ such that

$$\begin{aligned} J = {} & (\Delta\boldsymbol{p}_1 \oplus \Delta\boldsymbol{n}_1, \bar{V}[\boldsymbol{p}_1 \oplus \boldsymbol{n}_1]^-(\Delta\boldsymbol{p}_1 \oplus \Delta\boldsymbol{n}_1)) \\ & + (\Delta\boldsymbol{p}_2 \oplus \Delta\boldsymbol{n}_2, \bar{V}[\boldsymbol{p}_2 \oplus \boldsymbol{n}_2]^-(\Delta\boldsymbol{p}_2 \oplus \Delta\boldsymbol{n}_2)) \to \min \end{aligned} \tag{5.108}$$

[12]This constraint is degenerate.

under the linearized constraint

$$\begin{aligned}\Delta\boldsymbol{p}_2 \oplus \Delta\boldsymbol{n}_2 - \Delta\boldsymbol{p}_1 \oplus \Delta\boldsymbol{n}_1 = \boldsymbol{p}_2 \oplus \boldsymbol{n}_2 - \boldsymbol{p}_1 \oplus \boldsymbol{n}_1,\\ \Delta\boldsymbol{p}_1 \oplus \Delta\boldsymbol{n}_1 \in \{\bar{\boldsymbol{p}}_1 \oplus \bar{\boldsymbol{n}}_1, \bar{\boldsymbol{n}}_1 \oplus \bar{\boldsymbol{p}}_1\}_L^\perp,\\ \Delta\boldsymbol{p}_2 \oplus \Delta\boldsymbol{n}_2 \in \{\bar{\boldsymbol{p}}_2 \oplus \bar{\boldsymbol{n}}_2, \bar{\boldsymbol{n}}_2 \oplus \bar{\boldsymbol{p}}_2\}_L^\perp.\end{aligned} \tag{5.109}$$

The first order solution is

$$\begin{aligned}\Delta\boldsymbol{p}_1 \oplus \Delta\boldsymbol{n}_1 = V[\boldsymbol{p}_1 \oplus \boldsymbol{n}_1]\boldsymbol{W}(\boldsymbol{p}_1 \oplus \boldsymbol{n}_1 - \boldsymbol{p}_2 \oplus \boldsymbol{n}_2),\\ \Delta\boldsymbol{p}_2 \oplus \Delta\boldsymbol{n}_2 = V[\boldsymbol{p}_2 \oplus \boldsymbol{n}_2]\boldsymbol{W}(\boldsymbol{p}_2 \oplus \boldsymbol{n}_2 - \boldsymbol{p}_1 \oplus \boldsymbol{n}_1),\end{aligned} \tag{5.110}$$

where $\boldsymbol{W}$ is a (66)-matrix defined by[13]

$$\boldsymbol{W} = \Big(V[\boldsymbol{p}_1 \oplus \boldsymbol{n}_1] + V[\boldsymbol{p}_2 \oplus \boldsymbol{n}_2]\Big)_4^-. \tag{5.111}$$

A realistic form of the correction is

$$\begin{pmatrix} \hat{\boldsymbol{p}} \\ \hat{\boldsymbol{n}} \end{pmatrix} = N_\perp[\begin{pmatrix} \boldsymbol{p}_1 - \Delta\boldsymbol{p}_1 \\ \boldsymbol{n}_1 - \Delta\boldsymbol{n}_1 \end{pmatrix}] = N_\perp[\begin{pmatrix} \boldsymbol{p}_2 - \Delta\boldsymbol{p}_2 \\ \boldsymbol{n}_2 - \Delta\boldsymbol{n}_2 \end{pmatrix}], \tag{5.112}$$

where the operation $N_\perp[\,\cdot\,]$ is defined by

$$N_\perp[\begin{pmatrix} \boldsymbol{a} \\ \boldsymbol{b} \end{pmatrix}] = \begin{cases} N[\boldsymbol{a} \oplus \boldsymbol{P}_{N[\boldsymbol{a}]}\boldsymbol{b}] & \text{if } \|\boldsymbol{a}\| \geq \|\boldsymbol{b}\|, \\ N[\boldsymbol{P}_{N[\boldsymbol{b}]}\boldsymbol{a} \oplus \boldsymbol{b}] & \text{otherwise.} \end{cases} \tag{5.113}$$

The a posteriori covariance matrix of the estimate $\hat{\boldsymbol{p}} \oplus \hat{\boldsymbol{n}}$ is

$$\begin{aligned}V[\hat{\boldsymbol{p}} \oplus \hat{\boldsymbol{n}}] &= \hat{V}[\boldsymbol{p}_1 \oplus \boldsymbol{n}_1] - \hat{V}[\boldsymbol{p}_1 \oplus \boldsymbol{n}_1]\hat{\boldsymbol{W}}\hat{V}[\boldsymbol{p}_1 \oplus \boldsymbol{n}_1]\\ &= \hat{V}[\boldsymbol{p}_1 \oplus \boldsymbol{n}_1]\hat{\boldsymbol{W}}\hat{V}[\boldsymbol{p}_2 \oplus \boldsymbol{n}_2]\\ &= \hat{V}[\boldsymbol{p}_2 \oplus \boldsymbol{n}_2] - \hat{V}[\boldsymbol{p}_2 \oplus \boldsymbol{n}_2]\hat{\boldsymbol{W}}\hat{V}[\boldsymbol{p}_2 \oplus \boldsymbol{n}_2].\end{aligned} \tag{5.114}$$

Here,

$$\hat{V}[\boldsymbol{p}_i \oplus \boldsymbol{n}_i] = \boldsymbol{P}_{\mathcal{N}_{\hat{p}\oplus\hat{n}}} V[\boldsymbol{p}_i \oplus \boldsymbol{n}_i]\boldsymbol{P}_{\mathcal{N}_{\hat{p}\oplus\hat{n}}}, \quad i = 1, 2, \tag{5.115}$$

where $\boldsymbol{P}_{\mathcal{N}_{\hat{p}\oplus\hat{n}}}$ is the six-dimensional projection matrix onto $\mathcal{N}^\perp_{\hat{\boldsymbol{p}}\oplus\hat{\boldsymbol{n}}}$ (see eq. (4.44)). The matrix $\hat{\boldsymbol{W}}$ is obtained by replacing $V[\boldsymbol{p}_i \oplus \boldsymbol{n}_i]$ by $\hat{V}[\boldsymbol{p}_i \oplus \boldsymbol{n}_i]$, $i = 1, 2$, in eq. (5.111). The residual of J can be written as

$$\hat{J} = (\boldsymbol{p}_2 \oplus \boldsymbol{n}_2 - \boldsymbol{p}_1 \oplus \boldsymbol{n}_1, \hat{\boldsymbol{W}}(\boldsymbol{p}_2 \oplus \boldsymbol{n}_2 - \boldsymbol{p}_1 \oplus \boldsymbol{n}_1)), \tag{5.116}$$

which is a χ^2 variable with four degrees of freedom. This fact provides a *coincidence test* for space lines: the hypothesis that space lines $\boldsymbol{r} \times \boldsymbol{p}_1 = \boldsymbol{n}_1$ and $\boldsymbol{r} \times \boldsymbol{p}_2 = \boldsymbol{n}_2$ coincide with each other is rejected with significance level a% if

$$\hat{J} > \chi^2_{3,a}. \tag{5.117}$$

[13] The ranges of $V[\mathbf{p}_1 \oplus \mathbf{n}_1]$ and $V[\mathbf{p}_2 \oplus \mathbf{n}_2]$ are different from the ranges of $\bar{V}[\mathbf{p}_1 \oplus \mathbf{n}_1]$ and $\bar{V}[\mathbf{p}_2 \oplus \mathbf{n}_2]$. Consequently, although $\bar{V}[\mathbf{p}_1 \oplus \mathbf{n}_1] + \bar{V}[\mathbf{p}_2 \oplus \mathbf{n}_2]$ is a singular matrix of rank 4, the matrix $V[\mathbf{p}_1 \oplus \mathbf{n}_1] + V[\mathbf{p}_2 \oplus \mathbf{n}_2]$ is generally nonsingular.

5.3.2 *Optimal correction for incidence*

A. Simultaneous correction

As in two dimensions, a space point $\boldsymbol{r}$ and a space line $\boldsymbol{r} \times \boldsymbol{p} = \boldsymbol{n}$ can be optimally corrected so as to make them incident. Let $V[\boldsymbol{r}]$ and $V[\boldsymbol{p} \oplus \boldsymbol{n}]$ be their a priori covariance matrices. The problem is finding $\Delta\boldsymbol{r}$, $\Delta\boldsymbol{p}$, and $\Delta\boldsymbol{n}$ such that $\bar{\boldsymbol{r}} = \boldsymbol{r} - \Delta\boldsymbol{r}$, $\bar{\boldsymbol{p}} = \boldsymbol{p} - \Delta\boldsymbol{p}$, and $\bar{\boldsymbol{n}} = \boldsymbol{n} - \Delta\boldsymbol{n}$ satisfy

$$\bar{\boldsymbol{r}} \times \bar{\boldsymbol{p}} = \bar{\boldsymbol{n}}. \tag{5.118}$$

The rank of this constraint is 2 because the three component equations are algebraically dependent[14]. If the space point and the space line are statistically independent, the problem can be written as the optimization

$$J = (\Delta\boldsymbol{r}, \bar{V}[\boldsymbol{r}]^{-1}\Delta\boldsymbol{r}) + (\Delta\boldsymbol{p} \oplus \Delta\boldsymbol{n}, \bar{V}[\boldsymbol{p} \oplus \boldsymbol{n}]^{-}(\Delta\boldsymbol{p} \oplus \Delta\boldsymbol{n})) \to \min \tag{5.119}$$

under the linearized constraint

$$\Delta\boldsymbol{r} \times \bar{\boldsymbol{p}} + \bar{\boldsymbol{r}} \times \Delta\boldsymbol{p} - \Delta\boldsymbol{n} = \boldsymbol{r} \times \boldsymbol{p} - \boldsymbol{n},$$
$$\Delta\boldsymbol{p} \oplus \Delta\boldsymbol{n} \in \{\bar{\boldsymbol{p}} \oplus \bar{\boldsymbol{n}}, \bar{\boldsymbol{n}} \oplus \bar{\boldsymbol{p}}\}_L^{\perp}. \tag{5.120}$$

The first order solution is given by

$$\Delta\boldsymbol{r} = -(V[\boldsymbol{r}] \times \boldsymbol{p})\boldsymbol{W}(\boldsymbol{r} \times \boldsymbol{p} - \boldsymbol{n}),$$
$$\Delta\boldsymbol{p} = (V[\boldsymbol{p}] \times \boldsymbol{r} - V[\boldsymbol{p}, \boldsymbol{n}])\boldsymbol{W}(\boldsymbol{r} \times \boldsymbol{p} - \boldsymbol{n}),$$
$$\Delta\boldsymbol{n} = (V[\boldsymbol{n}, \boldsymbol{p}] \times \boldsymbol{r} - V[\boldsymbol{n}])\boldsymbol{W}(\boldsymbol{r} \times \boldsymbol{p} - \boldsymbol{n}), \tag{5.121}$$

where $\boldsymbol{W}$ is a (33)-matrix defined by[15]

$$\boldsymbol{W} = \Big(\boldsymbol{p} \times V[\boldsymbol{r}] \times \boldsymbol{p} + \boldsymbol{r} \times V[\boldsymbol{p}] \times \boldsymbol{r} - 2S[\boldsymbol{r} \times V[\boldsymbol{p}, \boldsymbol{n}]] + V[\boldsymbol{n}]\Big)_2^{-}. \tag{5.122}$$

The symbol $S[\,\cdot\,]$ denotes the symmetrization operator (see eqs. (2.205)). A realistic form of the correction is

$$\hat{\boldsymbol{r}} = \boldsymbol{r} - \Delta\boldsymbol{r}, \qquad \begin{pmatrix} \hat{\boldsymbol{p}} \\ \hat{\boldsymbol{n}} \end{pmatrix} = N_{\perp}[\begin{pmatrix} \boldsymbol{p} - \Delta\boldsymbol{p} \\ \boldsymbol{n} - \Delta\boldsymbol{n} \end{pmatrix}], \tag{5.123}$$

where the operator $N_{\perp}[\,\cdot\,]$ is defined by eq. (5.113). The a posteriori covariance matrices of the corrected values $\hat{\boldsymbol{r}}$, $\hat{\boldsymbol{p}}$, and $\hat{\boldsymbol{n}}$ are

$$V[\hat{\boldsymbol{r}}] = V[\boldsymbol{r}] - (V[\boldsymbol{r}] \times \hat{\boldsymbol{p}})\hat{\boldsymbol{W}}(\hat{\boldsymbol{p}} \times V[\boldsymbol{r}]),$$

[14] This constraint is degenerate.

[15] The rank-constrained generalized inverse $(\,\cdot\,)_2^-$ is used because $(\,\cdot\,)$ is generally nonsingular if evaluated at the data values; it should be a singular matrix of rank 2 if evaluated at the true values.

$$V[\hat{\boldsymbol{p}}] = \hat{V}[\boldsymbol{p}] - (\hat{V}[\boldsymbol{p}] \times \hat{\boldsymbol{r}} - \hat{V}[\boldsymbol{p},\boldsymbol{n}])\hat{\boldsymbol{W}}(\hat{\boldsymbol{r}} \times \hat{V}[\boldsymbol{p}] - \hat{V}[\boldsymbol{n},\boldsymbol{p}]),$$

$$V[\hat{\boldsymbol{p}},\hat{\boldsymbol{n}}] = \hat{V}[\boldsymbol{p},\boldsymbol{n}] - (\hat{V}[\boldsymbol{p}] \times \hat{\boldsymbol{r}} - \hat{V}[\boldsymbol{p},\boldsymbol{n}])\hat{\boldsymbol{W}}(\hat{\boldsymbol{r}} \times \hat{V}[\boldsymbol{p},\boldsymbol{n}] - \hat{V}[\boldsymbol{n}]) = V[\hat{\boldsymbol{n}},\hat{\boldsymbol{p}}]^\top,$$

$$V[\hat{\boldsymbol{n}}] = \hat{V}[\boldsymbol{n}] - (\hat{V}[\boldsymbol{n},\boldsymbol{p}] \times \hat{\boldsymbol{r}} - \hat{V}[\boldsymbol{n}])\hat{\boldsymbol{W}}(\hat{\boldsymbol{r}} \times \hat{V}[\boldsymbol{p},\boldsymbol{n}] - \hat{V}[\boldsymbol{n}]),$$

$$V[\hat{\boldsymbol{r}},\hat{\boldsymbol{p}}] = (V[\boldsymbol{r}] \times \hat{\boldsymbol{p}})\hat{\boldsymbol{W}}(\hat{\boldsymbol{r}} \times \hat{V}[\boldsymbol{p}] - \hat{V}[\boldsymbol{n},\boldsymbol{p}]) = V[\hat{\boldsymbol{p}},\hat{\boldsymbol{r}}]^\top,$$

$$V[\hat{\boldsymbol{r}},\hat{\boldsymbol{n}}] = (V[\boldsymbol{r}] \times \hat{\boldsymbol{p}})\hat{\boldsymbol{W}}(\hat{\boldsymbol{r}} \times \hat{V}[\boldsymbol{p},\boldsymbol{n}] - \hat{V}[\boldsymbol{n}]) = V[\hat{\boldsymbol{n}},\hat{\boldsymbol{r}}]^\top. \tag{5.124}$$

The matrices $\hat{V}[\boldsymbol{p}]$, $\hat{V}[\boldsymbol{p},\boldsymbol{n}]$, and $\hat{V}[\boldsymbol{n}]$ are obtained as submatrices of

$$\hat{V}[\boldsymbol{p} \oplus \boldsymbol{n}] = \boldsymbol{P}_{\mathcal{N}_{\hat{p}\oplus\hat{n}}} V[\boldsymbol{p} \oplus \boldsymbol{n}] \boldsymbol{P}_{\mathcal{N}_{\hat{p}\oplus\hat{n}}}. \tag{5.125}$$

The matrix $\hat{\boldsymbol{W}}$ is obtained by replacing $\boldsymbol{r}$, $\boldsymbol{p}$, $V[\boldsymbol{p}]$, $V[\boldsymbol{p},\boldsymbol{n}]$, and $V[\boldsymbol{n}]$ by $\hat{\boldsymbol{r}}$, $\hat{\boldsymbol{p}}$, $\hat{V}[\boldsymbol{p}]$, $\hat{V}[\boldsymbol{p},\boldsymbol{n}]$, and $\hat{V}[\boldsymbol{n}]$, respectively, in eq. (5.122). The residual of J can be written as

$$\hat{J} = (\boldsymbol{r} \times \boldsymbol{p} - \boldsymbol{n}, \hat{\boldsymbol{W}}(\boldsymbol{r} \times \boldsymbol{p} - \boldsymbol{n})), \tag{5.126}$$

which is a χ^2 variable with two degrees of freedom. This fact provides an *incidence test* for a space point and a space line: the hypothesis that space point $\boldsymbol{r}$ and space line $\boldsymbol{r} \times \boldsymbol{p} = \boldsymbol{n}$ are incident to each other is rejected with significance level a% if

$$\hat{J} > \chi^2_{2,a}. \tag{5.127}$$

B. Space point correction

If the space line $\boldsymbol{r} \times \boldsymbol{p} = \boldsymbol{n}$ is fixed, the linearized constraint is

$$\Delta\boldsymbol{r} \times \boldsymbol{p} = \boldsymbol{r} \times \boldsymbol{p} - \boldsymbol{n}. \tag{5.128}$$

The first order correction of the space point $\boldsymbol{r}$ is

$$\Delta\boldsymbol{r} = -(V[\boldsymbol{r}] \times \boldsymbol{p})\boldsymbol{W}(\boldsymbol{r} \times \boldsymbol{p} - \boldsymbol{n}), \tag{5.129}$$

where $\boldsymbol{W}$ is a (33)-matrix given by

$$\boldsymbol{W} = \left(\boldsymbol{p} \times V[\boldsymbol{r}] \times \boldsymbol{p}\right)_2^-. \tag{5.130}$$

The a posteriori covariance matrix $V[\hat{\boldsymbol{r}}]$ of the corrected value $\hat{\boldsymbol{r}}$ is given in the form shown in eqs. (5.124), where $\hat{\boldsymbol{p}}$ and $\hat{\boldsymbol{n}}$ are replaced by $\boldsymbol{p}$ and $\boldsymbol{n}$, respectively. Matrix $V[\hat{\boldsymbol{r}}]$ has rank 1; its null space is $\{\boldsymbol{n}, \boldsymbol{r}_H\}_L$ in the $\{\boldsymbol{m}, \boldsymbol{r}_H\}$-representation. The residual $\hat{J}$ is given in the form of eq. (5.126) and is a χ^2 variable with two degrees of freedom. Hence, the incidence test given by eq. (5.127) can be applied.

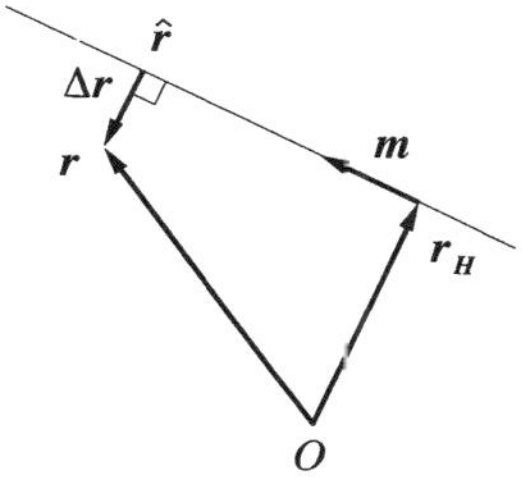

Fig. 5.9. Optimal incidence correction for a space point.

Example 5.4 If each coordinate is perturbed independently by Gaussian noise of mean 0 and variance ϵ^2, the covariance matrix of $\Delta \boldsymbol{r}$ is $V[\boldsymbol{r}] = \epsilon^2 \boldsymbol{I}$. The optimal correction (5.129) reduces to

$$\Delta \boldsymbol{r} = \frac{\boldsymbol{p} \times (\boldsymbol{r} \times \boldsymbol{p} - \boldsymbol{n})}{\|\boldsymbol{p}\|^2} = \boldsymbol{P}_{\boldsymbol{m}} \boldsymbol{r} - \boldsymbol{r}_H. \tag{5.131}$$

Consequently, $\boldsymbol{r}$ is displaced onto the space line $\boldsymbol{r} \times \boldsymbol{p} = \boldsymbol{n}$ perpendicularly (Fig. 5.9). The a posteriori covariance matrix of the corrected value $\hat{\boldsymbol{r}}$ has the form

$$V[\hat{\boldsymbol{r}}] = \epsilon^2 \boldsymbol{m}\boldsymbol{m}^\top, \tag{5.132}$$

where $\boldsymbol{m} = N[\boldsymbol{p}]$ is the orientation of the space line $\boldsymbol{r} \times \boldsymbol{p} = \boldsymbol{n}$. The residual (5.126) can be written as

$$\hat{J} = \frac{1}{\epsilon^2} \frac{\|\boldsymbol{r} \times \boldsymbol{p} - \boldsymbol{n}\|^2}{\|\boldsymbol{p}\|^2} = \frac{1}{\epsilon^2} \|\boldsymbol{P}_{\boldsymbol{m}} \boldsymbol{r} - \boldsymbol{r}_H\|^2. \tag{5.133}$$

Hence, an unbiased estimator of the variance ϵ^2 is obtained in the form

$$\hat{\epsilon}^2 = \frac{\|\boldsymbol{r} \times \boldsymbol{p} - \boldsymbol{n}\|^2}{2\|\boldsymbol{p}\|^2} = \frac{1}{2} \|\boldsymbol{P}_{\boldsymbol{m}} \boldsymbol{r} - \boldsymbol{r}_H\|^2. \tag{5.134}$$

The value $\hat{\epsilon}$ thus estimated equals $1/\sqrt{2}$ times the distance between the space point and the space line (see eq. 4.49)). If the value ϵ is given a priori, the incidence test takes the form

$$\frac{\hat{\epsilon}^2}{\epsilon^2} > \frac{\chi^2_{2,a}}{2}. \tag{5.135}$$

C. Space line correction

If the space point $\boldsymbol{r}$ is fixed, the linearized constraint is

$$\boldsymbol{r} \times \Delta \boldsymbol{p} - \Delta \boldsymbol{n} = \boldsymbol{r} \times \boldsymbol{p} - \boldsymbol{n},$$

$$\Delta \boldsymbol{p} \oplus \boldsymbol{n} \in \{\bar{\boldsymbol{p}} \oplus \bar{\boldsymbol{n}}, \bar{\boldsymbol{n}} \oplus \bar{\boldsymbol{p}}\}_L^{\perp}. \tag{5.136}$$

The optimal correction of the space line $\boldsymbol{r} \times \boldsymbol{p} = \boldsymbol{n}$ is

$$\Delta \boldsymbol{p} = (V[\boldsymbol{p}] \times \boldsymbol{r} - V[\boldsymbol{p}, \boldsymbol{n}])\boldsymbol{W}(\boldsymbol{r} \times \boldsymbol{p} - \boldsymbol{n}),$$

$$\Delta \boldsymbol{n} = ((\boldsymbol{r} \times V[\boldsymbol{p}, \boldsymbol{n}])^{\top} - V[\boldsymbol{n}])\boldsymbol{W}(\boldsymbol{r} \times \boldsymbol{p} - \boldsymbol{n}), \tag{5.137}$$

where $\boldsymbol{W}$ is a (33)-matrix given by

$$\boldsymbol{W} = \left(\boldsymbol{r} \times V[\boldsymbol{p}] \times \boldsymbol{r} - \boldsymbol{r} \times V[\boldsymbol{p}, \boldsymbol{n}] - (\boldsymbol{r} \times V[\boldsymbol{p}, \boldsymbol{n}])^{\top} + V[\boldsymbol{n}]\right)_2^{-}. \tag{5.138}$$

The a posteriori covariance matrices of the corrected values $\hat{\boldsymbol{p}}$ and $\hat{\boldsymbol{n}}$ are given in the form shown in eqs. (5.124), where $\hat{\boldsymbol{r}}$ is replaced by $\boldsymbol{r}$. The residual is

$$\hat{J} = (\boldsymbol{r} \times \boldsymbol{p} - \boldsymbol{n}, \hat{\boldsymbol{W}}(\boldsymbol{r} \times \boldsymbol{p} - \boldsymbol{n})). \tag{5.139}$$

This is a χ^2 variable with two degrees of freedom. Hence, the incidence test given by eq. (5.127) can be applied.

5.4 Correction of Space Planes

5.4.1 Optimal correction for coincidence

Two space planes $(\boldsymbol{\nu}_1, \boldsymbol{\rho}) = 0$ and $(\boldsymbol{\nu}_2, \boldsymbol{\rho}) = 0$ that are supposed to coincide can also be optimally corrected. Let $V[\boldsymbol{\nu}_1]$ and $V[\boldsymbol{\nu}_2]$ be their respective a priori covariance matrices. The signs of the 4-vectors $\boldsymbol{\nu}_1$ and $\boldsymbol{\nu}_2$ are chosen so that $\boldsymbol{\nu}_1 \approx \boldsymbol{\nu}_2$. Let $(\bar{\boldsymbol{\nu}}_1, \boldsymbol{\rho}) = 0$ and $(\bar{\boldsymbol{\nu}}_2, \boldsymbol{\rho}) = 0$ be the true space planes. The constraint to be imposed is

$$\bar{\boldsymbol{\nu}}_1 = \bar{\boldsymbol{\nu}}_2, \tag{5.140}$$

which has rank 3 because both sides are unit vectors[16]. If the two space planes are statistically independent, the problem is finding $\Delta\boldsymbol{\nu}_1$ and $\Delta\boldsymbol{\nu}_2$ such that

$$J = (\Delta\boldsymbol{\nu}_1, \bar{V}[\boldsymbol{\nu}_1]^{-1}\Delta\boldsymbol{\nu}_1) + (\Delta\boldsymbol{\nu}_2, \bar{V}[\boldsymbol{\nu}_2]^{-1}\Delta\boldsymbol{\nu}_2) \to \min \tag{5.141}$$

under the linearized constraint

$$\Delta\boldsymbol{\nu}_2 - \Delta\boldsymbol{\nu}_1 = \boldsymbol{\nu}_2 - \boldsymbol{\nu}_1,$$

$$\Delta\boldsymbol{\nu}_1 \in \{\bar{\boldsymbol{\nu}}_1\}_L^{\perp}, \qquad \Delta\boldsymbol{\nu}_2 \in \{\bar{\boldsymbol{\nu}}_2\}_L^{\perp}. \tag{5.142}$$

The first order solution is given by

$$\Delta\boldsymbol{\nu}_1 = V[\boldsymbol{\nu}_1]\boldsymbol{W}(\boldsymbol{\nu}_1 - \boldsymbol{\nu}_2),$$

[16] This constraint is degenerate.

$$\Delta\boldsymbol{\nu}_2 = V[\boldsymbol{\nu}_2]\boldsymbol{W}(\boldsymbol{\nu}_2 - \boldsymbol{\nu}_1), \tag{5.143}$$

where $\boldsymbol{W}$ is a (44)-matrix defined by[17]

$$\boldsymbol{W} = \left(V[\boldsymbol{\nu}_1] + V[\boldsymbol{\nu}_2]\right)^-_3. \tag{5.144}$$

A realistic form of the correction is

$$\hat{\boldsymbol{\nu}} = N[\boldsymbol{\nu}_1 - \Delta\boldsymbol{\nu}_1] = N[\boldsymbol{\nu}_2 - \Delta\boldsymbol{\nu}_1]. \tag{5.145}$$

The a posteriori covariance matrix of the estimate $\hat{\boldsymbol{\nu}}$ is

$$\begin{aligned} V[\hat{\boldsymbol{\nu}}] &= \hat{V}[\boldsymbol{\nu}_1] - \hat{V}[\boldsymbol{\nu}_1]\hat{\boldsymbol{W}}\hat{V}[\boldsymbol{\nu}_1] = \hat{V}[\boldsymbol{\nu}_1]\hat{\boldsymbol{W}}\hat{V}[\boldsymbol{\nu}_2] \\ &= \hat{V}[\boldsymbol{\nu}_2] - \hat{V}[\boldsymbol{\nu}_2]\hat{\boldsymbol{W}}\hat{V}[\boldsymbol{\nu}_2]. \end{aligned} \tag{5.146}$$

Here,

$$\hat{V}[\boldsymbol{\nu}_i] = \boldsymbol{P}_{\hat{\nu}_i} V[\boldsymbol{\nu}_i] \boldsymbol{P}_{\hat{\nu}_i}, \qquad i = 1, 2, \tag{5.147}$$

where $\boldsymbol{P}_{\hat{\nu}_i}$ is the four-dimensional projection matrix along $\boldsymbol{\nu}_i$. The matrix $\hat{\boldsymbol{W}}$ is obtained by replacing $V[\boldsymbol{\nu}_i]$ by $\hat{V}[\boldsymbol{\nu}_i]$, i = 1, 2, in eq. (5.144). The residual of J can be written as

$$\hat{J} = (\boldsymbol{\nu}_2 - \boldsymbol{\nu}_1, \bar{\boldsymbol{W}}(\boldsymbol{\nu}_2 - \boldsymbol{\nu}_1)), \tag{5.148}$$

which is a χ^2 variable with three degrees of freedom. This fact provides a *coincidence test* for space planes: the hypothesis that two space planes $(\boldsymbol{\nu}_1, \boldsymbol{\rho}) = 0$ and $(\boldsymbol{\nu}_2, \boldsymbol{\rho}) = 0$ coincide with each other is rejected with significance level a% if

$$\hat{J} > \chi^2_{3,a}. \tag{5.149}$$

5.4.2 Optimal incidence with space points

A. Simultaneous correction

A space point $\boldsymbol{\rho}$ and a space plane $(\boldsymbol{\nu}, \boldsymbol{\rho}) = 0$ can be optimally corrected so as to make them incident. Let $V[\boldsymbol{\rho}]$ and $V[\boldsymbol{\nu}]$ be their respective a priori covariance matrices. The problem is finding $\Delta\boldsymbol{\rho}$ and $\Delta\boldsymbol{\nu}$ such that $\bar{\boldsymbol{\rho}} = \boldsymbol{\rho} - \Delta\boldsymbol{\rho}$ and $\bar{\boldsymbol{\nu}} = \boldsymbol{\nu} - \Delta\boldsymbol{\nu}$ satisfy

$$(\bar{\boldsymbol{\nu}}, \bar{\boldsymbol{\rho}}) = 0. \tag{5.150}$$

The rank of this constraint is 1. If the space point and the space plane are statistically independent, the problem can be written as the optimization

$$J = (\Delta\boldsymbol{\rho}, \bar{V}[\boldsymbol{\rho}]^- \Delta\boldsymbol{\rho}) + (\Delta\boldsymbol{\nu}, \bar{V}[\boldsymbol{\nu}]^- \Delta\boldsymbol{\nu}) \to \min \tag{5.151}$$

[17]The the ranges of $V[\nu_1]$ and $V[\nu_2]$ are different from the ranges of $\bar{V}[\nu_1]$ and $\bar{V}[\nu_2]$. Consequently, although $\bar{V}[\nu_1]+\bar{V}[\nu_2]$ is a singular matrix of rank 3, the matrix $V[\nu_1]+V[\nu_2]$ is generally nonsingular.

under the linearized constraint

$$(\Delta\boldsymbol{\nu}, \bar{\boldsymbol{\rho}}) + (\bar{\boldsymbol{\nu}}, \Delta\boldsymbol{\rho}) = (\boldsymbol{\nu}, \boldsymbol{\rho}),$$

$$\Delta\boldsymbol{\rho} \in \{\boldsymbol{\kappa}\}_L^\perp, \qquad \Delta\boldsymbol{\nu} \in \{\bar{\boldsymbol{\nu}}\}_L^\perp, \tag{5.152}$$

where $\boldsymbol{\kappa} = (0,0,0,1)^\top$. The first order solution is given by

$$\Delta\boldsymbol{\rho} = \frac{(\boldsymbol{\nu}, \boldsymbol{\rho})V[\boldsymbol{\rho}]\boldsymbol{\nu}}{(\boldsymbol{\nu}, V[\boldsymbol{\rho}]\boldsymbol{\nu}) + (\boldsymbol{\rho}, V[\boldsymbol{\nu}]\boldsymbol{\rho})},$$

$$\Delta\boldsymbol{\nu} = \frac{(\boldsymbol{\nu}, \boldsymbol{\rho})V[\boldsymbol{\nu}]\boldsymbol{\rho}}{(\boldsymbol{\nu}, V[\boldsymbol{\rho}]\boldsymbol{\nu}) + (\boldsymbol{\rho}, V[\boldsymbol{\nu}]\boldsymbol{\rho})}. \tag{5.153}$$

A realistic form of the correction is

$$\hat{\boldsymbol{\rho}} = \boldsymbol{\rho} - \Delta\boldsymbol{\rho}, \qquad \hat{\boldsymbol{\nu}} = N[\boldsymbol{\nu} - \Delta\boldsymbol{\nu}]. \tag{5.154}$$

The a posteriori covariance matrices of the corrected values $\hat{\boldsymbol{\rho}}$ and $\hat{\boldsymbol{\nu}}$ are

$$V[\hat{\boldsymbol{\rho}}] = V[\boldsymbol{\rho}] - \frac{(V[\boldsymbol{\rho}]\hat{\boldsymbol{\nu}})(V[\boldsymbol{\rho}]\hat{\boldsymbol{\nu}})^\top}{(\hat{\boldsymbol{\nu}}, V[\boldsymbol{\rho}]\hat{\boldsymbol{\nu}}) + (\hat{\boldsymbol{\rho}}, \hat{V}[\boldsymbol{\nu}]\hat{\boldsymbol{\rho}})},$$

$$V[\hat{\boldsymbol{\nu}}] = \hat{V}[\boldsymbol{\nu}] - \frac{(\hat{V}[\boldsymbol{\nu}]\hat{\boldsymbol{\rho}})(\hat{V}[\boldsymbol{\nu}]\hat{\boldsymbol{\rho}})^\top}{(\hat{\boldsymbol{\nu}}, V[\boldsymbol{\rho}]\hat{\boldsymbol{\nu}}) + (\hat{\boldsymbol{\rho}}, \hat{V}[\boldsymbol{\nu}]\hat{\boldsymbol{\rho}})},$$

$$V[\hat{\boldsymbol{\rho}}, \hat{\boldsymbol{\nu}}] = -\frac{(V[\boldsymbol{\rho}]\hat{\boldsymbol{\nu}})(\hat{V}[\boldsymbol{\nu}]\hat{\boldsymbol{\rho}})^\top}{(\hat{\boldsymbol{\nu}}, V[\boldsymbol{\rho}]\hat{\boldsymbol{\nu}}) + (\hat{\boldsymbol{\rho}}, \hat{V}[\boldsymbol{\nu}]\hat{\boldsymbol{\rho}})} = V[\hat{\boldsymbol{\nu}}, \hat{\boldsymbol{\rho}}]^\top. \tag{5.155}$$

Here,

$$\hat{V}[\boldsymbol{\nu}] = \boldsymbol{P}_{\hat{\boldsymbol{\nu}}} V[\boldsymbol{\nu}] \boldsymbol{P}_{\hat{\boldsymbol{\nu}}}, \tag{5.156}$$

where $\boldsymbol{P}_{\hat{\boldsymbol{\nu}}}$ is the four-dimensional projection matrix along $\hat{\boldsymbol{\nu}}$. The residual of J can be written as

$$\hat{J} = \frac{(\boldsymbol{\nu}, \boldsymbol{\rho})^2}{(\hat{\boldsymbol{\nu}}, V[\boldsymbol{\rho}]\hat{\boldsymbol{\nu}}) + (\hat{\boldsymbol{\rho}}, \hat{V}[\boldsymbol{\nu}]\hat{\boldsymbol{\rho}})}, \tag{5.157}$$

which is a χ^2 variable with one degree of freedom. This fact provides an *incidence test* for a space point and a space plane: the hypothesis that space point $\boldsymbol{\rho}$ and space plane $(\boldsymbol{\nu}, \boldsymbol{\rho}) = 0$ are incident to each other is rejected with significance level $a\%$ if

$$\hat{J} > \chi^2_{1,a}. \tag{5.158}$$

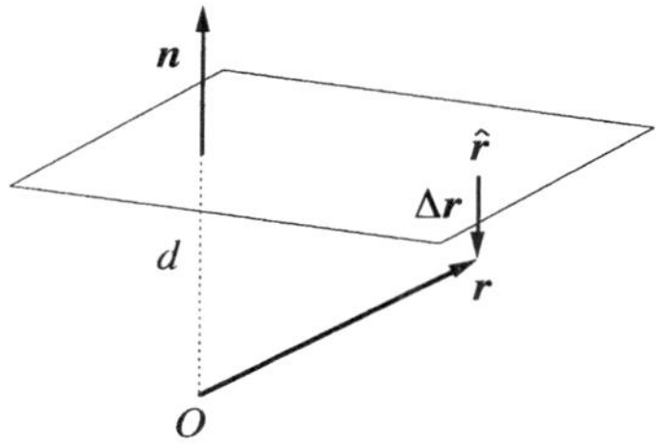

Fig. 5.10. Optimal incidence correction for a space point.

B. Space point correction

If the space plane $(\boldsymbol{\nu}, \boldsymbol{\rho}) = 0$ is fixed, the linearized constraint is

$$(\boldsymbol{\nu}, \Delta\boldsymbol{\rho}) = (\boldsymbol{\nu}, \boldsymbol{\rho}), \qquad \Delta\boldsymbol{\rho} \in \{\boldsymbol{\kappa}\}_L^{\perp}. \tag{5.159}$$

The optimal correction of $\boldsymbol{\rho}$ is

$$\Delta\boldsymbol{\rho} = \frac{(\boldsymbol{\nu}, \boldsymbol{\rho})V[\boldsymbol{\rho}]\boldsymbol{\nu}}{(\boldsymbol{\nu}, V[\boldsymbol{\rho}]\boldsymbol{\nu})}. \tag{5.160}$$

The a posteriori covariance matrix of the corrected value $\hat{\boldsymbol{\rho}}$ is

$$V[\hat{\boldsymbol{\rho}}] = V[\boldsymbol{\rho}] - \frac{(V[\boldsymbol{\rho}]\boldsymbol{\nu})(V[\boldsymbol{\rho}]\boldsymbol{\nu})^\top}{(\boldsymbol{\nu}, V[\boldsymbol{\rho}]\boldsymbol{\nu})}. \tag{5.161}$$

Since $V[\hat{\boldsymbol{\rho}}]\boldsymbol{\nu} = \mathbf{0}$, the rank of $V[\hat{\boldsymbol{\rho}}]$ is 2; its null space is $\{\boldsymbol{\kappa}, \boldsymbol{\nu}\}_L$. The residual

$$\hat{J} = \frac{(\boldsymbol{\nu}, \boldsymbol{\rho})^2}{(\boldsymbol{\nu}, V[\boldsymbol{\rho}]\boldsymbol{\nu})} \tag{5.162}$$

is a χ^2 variable with one degree of freedom. Hence, the incidence test given by eq. (5.158) can be applied.

Example 5.5 If each coordinate is perturbed independently by Gaussian noise of mean 0 and variance ϵ^2, the covariance matrix of $\boldsymbol{\rho}$ is $V[\boldsymbol{\rho}] = \epsilon^2 \boldsymbol{P}_{\boldsymbol{\kappa}}$ ($= \epsilon^2 \boldsymbol{I} \oplus 0$). In the $\{\boldsymbol{n}, d\}$-representation, the optimal correction (5.160) reduces to

$$\Delta\boldsymbol{r} = ((\boldsymbol{n}, \boldsymbol{r}) - d)\boldsymbol{n}. \tag{5.163}$$

Consequently, $\boldsymbol{r}$ is displaced onto the space plane $(\boldsymbol{n}, \boldsymbol{r}) = d$ perpendicularly (Fig. 5.10). The a posteriori covariance matrix of the corrected value $\hat{\boldsymbol{r}}$ is

$$V[\hat{\boldsymbol{r}}] = \epsilon^2 \boldsymbol{P}_{\boldsymbol{n}}. \tag{5.164}$$

The residual (5.162) can be written as

$$\hat{J} = \frac{1}{\epsilon^2}((\boldsymbol{n}, \boldsymbol{r}) - d)^2. \tag{5.165}$$

Hence, an unbiased estimator of the variance ϵ^2 is obtained in the form

$$\hat{\epsilon}^2 = ((\boldsymbol{n}, \boldsymbol{r}) - d)^2. \tag{5.166}$$

The value $\hat{\epsilon}$ thus estimated equals the distance between the space plane and the space point (see eq. (4.68)). If the value ϵ is given a priori, the incidence test takes the form

$$\frac{\hat{\epsilon}^2}{\epsilon^2} > \chi^2_{1,a}. \tag{5.167}$$

C. Space plane correction

If the space point $\boldsymbol{\rho}$ is fixed, the linearized constraint is

$$(\Delta\boldsymbol{\nu}, \boldsymbol{\rho}) = (\boldsymbol{\nu}, \boldsymbol{\rho}), \qquad \Delta\boldsymbol{\nu} \in \{\bar{\boldsymbol{\nu}}\}_L^\perp. \tag{5.168}$$

The optimal correction of space plane $(\boldsymbol{\nu}, \boldsymbol{\rho}) = 0$ is

$$\Delta\boldsymbol{\nu} = \frac{(\boldsymbol{\nu}, \boldsymbol{\rho})V[\boldsymbol{\nu}]\boldsymbol{\rho}}{(\boldsymbol{\rho}, V[\boldsymbol{\nu}]\boldsymbol{\rho})}. \tag{5.169}$$

The a posteriori covariance matrix the corrected value $\hat{\boldsymbol{\nu}}$ is

$$V[\hat{\boldsymbol{\nu}}] = \hat{V}[\boldsymbol{\nu}] - \frac{(\hat{V}[\boldsymbol{\nu}]\boldsymbol{\rho})(\hat{V}[\boldsymbol{\nu}]\boldsymbol{\rho})^\top}{(\boldsymbol{\rho}, \hat{V}[\boldsymbol{\nu}]\boldsymbol{\rho})}. \tag{5.170}$$

The rank of $V[\hat{\boldsymbol{\nu}}]$ is 2; its null space is $\{\hat{\boldsymbol{n}}, \boldsymbol{\rho}\}_L$. The residual

$$\hat{J} = \frac{(\boldsymbol{\nu}, \boldsymbol{\rho})^2}{(\boldsymbol{\rho}, \hat{V}[\boldsymbol{\nu}]\boldsymbol{\rho})} \tag{5.171}$$

is a χ^2 variable with one degree of freedom. Hence, the incidence test given by eq. (5.158) can be applied.

5.4.3 Optimal incidence with space lines

A. Simultaneous correction

A space line $(\boldsymbol{r} - \boldsymbol{r}_H) \times \boldsymbol{m} = \boldsymbol{0}$ and a space plane $(\boldsymbol{n}, \boldsymbol{r}) = d$ can be optimally corrected so as to make them incident. Let $V[\boldsymbol{m} \oplus \boldsymbol{r}_H]$ and $V[\boldsymbol{n} \oplus d]$ be their respective a priori covariance matrices, and $(\boldsymbol{r} - \bar{\boldsymbol{r}}_H) \times \bar{\boldsymbol{m}} = \boldsymbol{0}$ and $(\bar{\boldsymbol{n}}, \boldsymbol{r}) = \bar{d}$ their true equations. The constraint is

$$(\bar{\boldsymbol{n}}, \bar{\boldsymbol{m}}) = 0, \qquad (\bar{\boldsymbol{n}}, \bar{\boldsymbol{r}}_H) = \bar{d}, \tag{5.172}$$

which has rank 2. If the space line and the space plane are statistically independent, the problem is finding $\Delta\boldsymbol{m}$, $\Delta\boldsymbol{r}_H$, $\Delta\boldsymbol{n}$, and Δd such that

$$\begin{aligned} J = {} & (\Delta\boldsymbol{m} \oplus \Delta\boldsymbol{r}_H, \bar{V}[\boldsymbol{m} \oplus \boldsymbol{r}_H]^-(\Delta\boldsymbol{m} \oplus \Delta\boldsymbol{r}_H)) \\ & +(\Delta\boldsymbol{n} \oplus \Delta d, \bar{V}[\boldsymbol{n} \oplus d]^-(\Delta\boldsymbol{n} \oplus \Delta d)) \to \min \end{aligned} \tag{5.173}$$

under the linearized constraint

$$(\Delta \boldsymbol{n}, \bar{\boldsymbol{m}}) + (\bar{\boldsymbol{n}}, \Delta \boldsymbol{m}) = (\boldsymbol{n}, \boldsymbol{m}),$$

$$(\Delta \boldsymbol{n}, \bar{\boldsymbol{r}}_H) + (\bar{\boldsymbol{n}}, \Delta \boldsymbol{r}_H) - \Delta d = (\boldsymbol{n}, \boldsymbol{r}_H) - d,$$

$$\Delta \boldsymbol{m} \oplus \Delta \boldsymbol{r}_H \in \{\bar{\boldsymbol{m}} \oplus 0, \bar{\boldsymbol{r}}_H \oplus \bar{\boldsymbol{m}}\}_L^{\perp}, \qquad \Delta \boldsymbol{n} \in \{\bar{\boldsymbol{n}}\}_L^{\perp}. \tag{5.174}$$

The first order solution is given by

$$\begin{pmatrix} \Delta \boldsymbol{m} \\ \Delta \boldsymbol{r}_H \end{pmatrix} = \begin{pmatrix} V[\boldsymbol{m}]\boldsymbol{n} & V[\boldsymbol{m}, \boldsymbol{r}_H]\boldsymbol{n} \\ V[\boldsymbol{r}_H, \boldsymbol{m}]^\top \boldsymbol{n} & V[\boldsymbol{r}_H]\boldsymbol{n} \end{pmatrix} \boldsymbol{W} \begin{pmatrix} (\boldsymbol{n}, \boldsymbol{m}) \\ (\boldsymbol{n}, \boldsymbol{r}_H) - d \end{pmatrix},$$

$$\begin{pmatrix} \Delta \boldsymbol{n} \\ \Delta d \end{pmatrix} = \begin{pmatrix} V[\boldsymbol{n}]\boldsymbol{m} & V[\boldsymbol{n}]\boldsymbol{r}_H - V[\boldsymbol{n}, d] \\ (\boldsymbol{m}, V[\boldsymbol{n}, d]) & (\boldsymbol{r}_H, V[\boldsymbol{n}, d]) - V[d] \end{pmatrix} \boldsymbol{W} \begin{pmatrix} (\boldsymbol{n}, \boldsymbol{m}) \\ (\boldsymbol{n}, \boldsymbol{r}_H) - d \end{pmatrix}, \tag{5.175}$$

where $\boldsymbol{W}$ is a (22)-matrix defined by

$$\boldsymbol{W} = \begin{pmatrix} (\boldsymbol{n}, V[\boldsymbol{m}]\boldsymbol{n}) + (\boldsymbol{m}, V[\boldsymbol{n}]\boldsymbol{m}) & (\boldsymbol{n}, V[\boldsymbol{m}, \boldsymbol{r}_H]\boldsymbol{n}) + (\boldsymbol{m}, V[\boldsymbol{n}]\boldsymbol{r}_H) - (\boldsymbol{m}, V[\boldsymbol{n}, d]) \\ (\boldsymbol{n}, V[\boldsymbol{m}, \boldsymbol{r}_H]\boldsymbol{n}) + (\boldsymbol{r}_H, V[\boldsymbol{n}]\boldsymbol{m}) & (\boldsymbol{n}, V[\boldsymbol{r}_H]\boldsymbol{n}) + (\boldsymbol{r}_H, V[\boldsymbol{n}]\boldsymbol{r}_H) - (\boldsymbol{r}_H, V[\boldsymbol{n}, d]) \end{pmatrix}^{-1}. \tag{5.176}$$

A realistic form of the correction is

$$\hat{\boldsymbol{m}} = N[\boldsymbol{m} - \Delta \boldsymbol{m}], \qquad \hat{\boldsymbol{r}}_H = \boldsymbol{P}_{\hat{\boldsymbol{m}}}(\boldsymbol{r}_H - \Delta \boldsymbol{r}_H),$$

$$\hat{\boldsymbol{n}} = N[\boldsymbol{n} - \Delta \boldsymbol{n}], \qquad \hat{d} = d - \Delta d. \tag{5.177}$$

The a posteriori covariance matrices of the corrected values $\hat{\boldsymbol{m}}$, $\hat{\boldsymbol{r}}_H$, $\hat{\boldsymbol{n}}$, and $\hat{d}$ are

$$\begin{pmatrix} V[\hat{\boldsymbol{m}}] & V[\hat{\boldsymbol{m}}, \hat{\boldsymbol{r}}_H] \\ V[\hat{\boldsymbol{r}}_H, \hat{\boldsymbol{m}}] & V[\hat{\boldsymbol{r}}_H] \end{pmatrix} = \begin{pmatrix} \hat{V}[\boldsymbol{m}] & \hat{V}[\boldsymbol{m}, \boldsymbol{r}_H] \\ \hat{V}[\boldsymbol{r}_H, \boldsymbol{m}] & \hat{V}[\boldsymbol{r}_H] \end{pmatrix}$$
$$- \begin{pmatrix} \hat{V}[\boldsymbol{m}]\hat{\boldsymbol{n}} & \hat{V}[\boldsymbol{m}, \boldsymbol{r}_H]\hat{\boldsymbol{n}} \\ \hat{V}[\boldsymbol{r}, \boldsymbol{m}]\hat{\boldsymbol{n}} & \hat{V}[\boldsymbol{r}_H]\hat{\boldsymbol{n}} \end{pmatrix} \hat{\boldsymbol{W}} \begin{pmatrix} \hat{V}[\boldsymbol{m}]\hat{\boldsymbol{n}} & \hat{V}[\boldsymbol{m}, \boldsymbol{r}_H]\hat{\boldsymbol{n}} \\ \hat{V}[\boldsymbol{r}, \boldsymbol{m}]\hat{\boldsymbol{n}} & \hat{V}[\boldsymbol{r}_H]\hat{\boldsymbol{n}} \end{pmatrix}^\top,$$

$$\begin{pmatrix} V[\hat{\boldsymbol{n}}] & V[\hat{\boldsymbol{n}}, \hat{d}] \\ V[\hat{d}, \hat{\boldsymbol{n}}] & V[\hat{d}] \end{pmatrix} = \begin{pmatrix} \hat{V}[\boldsymbol{n}] & \hat{V}[\boldsymbol{n}, d] \\ \hat{V}[d, \boldsymbol{n}] & \hat{V}[d] \end{pmatrix}$$
$$- \begin{pmatrix} \hat{V}[\boldsymbol{n}]\hat{\boldsymbol{m}} & \hat{V}[\boldsymbol{n}]\hat{\boldsymbol{r}}_H - \hat{V}[\boldsymbol{n}, d] \\ (\hat{\boldsymbol{m}}, \hat{V}[\boldsymbol{n}, d]) & (\hat{\boldsymbol{r}}_H, \hat{V}[\boldsymbol{n}, d]) - \hat{V}[d] \end{pmatrix}$$
$$\hat{\boldsymbol{W}} \begin{pmatrix} \hat{V}[\boldsymbol{n}]\hat{\boldsymbol{m}} & \hat{V}[\boldsymbol{n}]\hat{\boldsymbol{r}}_H - \hat{V}[\boldsymbol{n}, d] \\ (\hat{\boldsymbol{m}}, \hat{V}[\boldsymbol{n}, d]) & (\hat{\boldsymbol{r}}_H, \hat{V}[\boldsymbol{n}, d]) - \hat{V}[d] \end{pmatrix}^\top,$$

$$\begin{pmatrix} V[\hat{m},\hat{n}] & V[\hat{m},\hat{d}] \\ V[\hat{r}_H,\hat{n}] & V[\hat{r}_H,\hat{d}] \end{pmatrix} = -\begin{pmatrix} \hat{V}[m]\hat{n} & \hat{V}[m,r_H]\hat{n} \\ \hat{V}[r_H,m]\hat{n} & \hat{V}[r_H]\hat{n} \end{pmatrix}$$

$$\hat{W}\begin{pmatrix} \hat{V}[n]\hat{m} & \hat{V}[n]\hat{r}_H - \hat{V}[n,d] \\ (\hat{m},\hat{V}[n,d]) & (\hat{r}_H,\hat{V}[n,d]) - \hat{V}[d] \end{pmatrix}^{\top}, \quad (5.178)$$

where $\hat{V}[m]$, $\hat{V}[m,r_H]$, etc. are computed as submatrices of

$$\hat{V}[m\oplus r_H] = P_{\mathcal{N}_{\hat{m}\oplus\hat{r}_H}} V[m\oplus r_H] P_{\mathcal{N}_{\hat{m}\oplus\hat{r}_H}},$$

$$\hat{V}[n\oplus d] = (P_{\hat{n}}\oplus 1)V[n\oplus d](P_{\hat{n}}\oplus 1). \quad (5.179)$$

Here, $P_{\mathcal{N}_{\hat{m}\oplus\hat{r}_H}}$ is the six-dimensional projection matrix onto $\mathcal{N}^{\perp}_{\hat{m}\oplus\hat{r}_H}$ (see eq. (4.40)). The matrix $\hat{W}$ is obtained by replacing m, r_H, n, $V[m]$, $V[m,r_H]$, etc. by $\hat{m}$, $\hat{r}_H$, $\hat{n}$, $\hat{V}[m]$, $\hat{V}[m,r_H]$, etc., respectively, in eq. (5.176). The residual of J can be written as

$$\hat{J} = \hat{W}^{(11)}(n,m)^2 + 2\hat{W}^{(12)}(n,m)((n,r_H)-d) + \hat{W}^{(22)}((n,r_H)-d)^2, \quad (5.180)$$

which is a χ^2 variable with two degrees of freedom. This fact provides an *incidence test* for a space line and a space plane: the hypothesis that space line $(r - r_H)\times m = 0$ and space plane are incident to each other is rejected with significance level a% if

$$\hat{J} > \chi^2_{2,a}. \quad (5.181)$$

B. Space line correction

If the space plane $(n, r) = d$ is fixed, the linearized constraint is

$$(n,\Delta m) = (n,m), \qquad (n,\Delta r_H) = (n,r_H) - d,$$

$$\Delta m\oplus\Delta r_H \in \{\bar{m}\oplus 0, \bar{r}_H\oplus\bar{m}\}^{\perp}_{L}. \quad (5.182)$$

The optimal correction of the space line $(r - r_H)\times m = \mathbf{0}$ is

$$\begin{pmatrix} \Delta m \\ \Delta r_H \end{pmatrix} = \begin{pmatrix} V[m]n & V[m,r_H]n \\ V[r_H,m]^{\top}n & V[r_H]n \end{pmatrix} W \begin{pmatrix} (n,m) \\ (n,r_H) - d \end{pmatrix}, \quad (5.183)$$

where W is a (22)-matrix defined by

$$W = \begin{pmatrix} (n,V[m]n) & (n,V[m,r_H]n) \\ (n,V[m,r_H]n) & (n,V[r_H]n) \end{pmatrix}^{-1}. \quad (5.184)$$

The a posteriori covariance matrices of the corrected values $\hat{m}$ and $\hat{r}_H$ are given in the form shown in eqs. (5.178), where $\hat{n}$ and $\hat{d}$ are replaced by n and d, respectively. The residual $\hat{J}$ is given in the form of eq. (5.178) and is a χ^2 variable with two degrees of freedom. Hence, the incidence test given by eq. (5.181) can be applied.

C. Space plane correction

If the space line $(\boldsymbol{r}-\boldsymbol{r}_H)\times\boldsymbol{m}=\boldsymbol{0}$ is fixed, the linearized constraint is

$$(\Delta\boldsymbol{n},\boldsymbol{m})=(\boldsymbol{n},\boldsymbol{m}),\qquad (\Delta\boldsymbol{n},\boldsymbol{r}_H)-\Delta d=(\boldsymbol{n},\boldsymbol{r}_H)-d,$$

$$\Delta\boldsymbol{n}\in\{\bar{\boldsymbol{n}}\}_L^{\perp}. \tag{5.185}$$

The optimal correction of the space plane $(\boldsymbol{n},\boldsymbol{r})=d$ is

$$\begin{pmatrix}\Delta\boldsymbol{n}\\ \Delta d\end{pmatrix}=\begin{pmatrix}V[\boldsymbol{n}]\boldsymbol{m} & V[\boldsymbol{n}]\boldsymbol{r}_H-V[\boldsymbol{n},d]\\ (\boldsymbol{m},V[\boldsymbol{n},d]) & (\boldsymbol{r},V[\boldsymbol{n},d])-V[d]\end{pmatrix}\boldsymbol{W}\begin{pmatrix}(\boldsymbol{n},\boldsymbol{m})\\ (\boldsymbol{n},\boldsymbol{r}_H)-d\end{pmatrix}, \tag{5.186}$$

where $\boldsymbol{W}$ is a (22)-matrix defined by

$$\boldsymbol{W}=\begin{pmatrix}(\boldsymbol{m},V[\boldsymbol{n}]\boldsymbol{m}) & (\boldsymbol{m},V[\boldsymbol{n}]\boldsymbol{r}_H)-(\boldsymbol{m},V[\boldsymbol{n},d])\\ (\boldsymbol{r}_H,V[\boldsymbol{n}]\boldsymbol{m}) & (\boldsymbol{r}_H,V[\boldsymbol{n}]\boldsymbol{r}_H)-(\boldsymbol{r},V[\boldsymbol{n},d])\end{pmatrix}^{-1}. \tag{5.187}$$

The a posteriori covariance matrices of the corrected values $\hat{\boldsymbol{n}}$ and $\hat{d}$ are given in the form shown in eqs. (5.178), where $\hat{\boldsymbol{r}}_H$ and $\hat{\boldsymbol{m}}$ are replaced by $\boldsymbol{r}_H$ and $\boldsymbol{m}$, respectively. The residual $\hat{J}$ is given in the form of eq. (5.178) and is a χ^2 variable with two degrees of freedom. Hence, the incidence test given by eq. (5.181) can be applied.

5.5 Orthogonality Correction

5.5.1 *Correction of two orientations*

A. Simultaneous correction

Let $\boldsymbol{m}_1$ and $\boldsymbol{m}_2$ be unit 3-vectors that indicate orientations supposedly orthogonal. Let $V[\boldsymbol{m}_1]$ and $V[\boldsymbol{m}_2]$ be their respective a priori covariance matrices. In the presence of noise, $\boldsymbol{m}_1$ and $\boldsymbol{m}_2$ are not exactly orthogonal. Consider the problem of optimally correcting them so as to make them orthogonal (Fig. 5.11). In other words, we want to find $\Delta\boldsymbol{m}_1$ and $\Delta\boldsymbol{m}_2$ such that $\bar{\boldsymbol{m}}_1=\boldsymbol{m}_1-\Delta\boldsymbol{m}_1$ and $\bar{\boldsymbol{m}}_2=\boldsymbol{m}_2-\Delta\boldsymbol{m}_2$ satisfy

$$(\bar{\boldsymbol{m}}_1,\bar{\boldsymbol{m}}_2)=0. \tag{5.188}$$

The rank of this constraint is 1. If the two orientations are statistically independent, the problem can be written as the optimization

$$J=(\Delta\boldsymbol{m}_1,\bar{V}[\boldsymbol{m}_1]^-\Delta\boldsymbol{m}_1)+(\Delta\boldsymbol{m}_2,\bar{V}[\boldsymbol{m}_2]^-\Delta\boldsymbol{m}_2)\to\min \tag{5.189}$$

under the linearized constraint

$$(\Delta\boldsymbol{m}_1,\bar{\boldsymbol{m}}_2)+(\bar{\boldsymbol{m}}_1,\Delta\boldsymbol{m}_2)=(\boldsymbol{m}_1,\boldsymbol{m}_2),$$

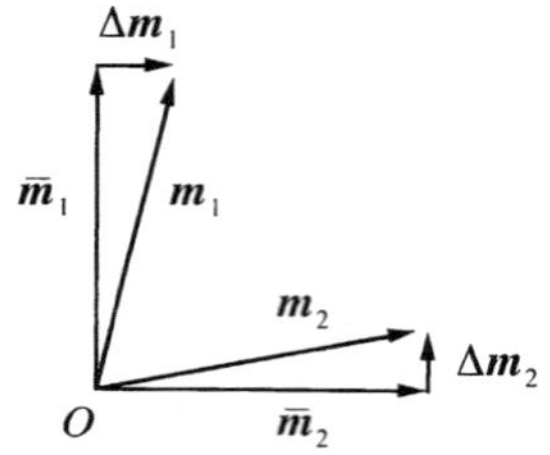

Fig. 5.11. Orthogonality correction for two orientations.

$$\Delta \boldsymbol{m}_1 \in \{\bar{\boldsymbol{m}}_1\}_L^\perp, \qquad \Delta \boldsymbol{m}_2 \in \{\bar{\boldsymbol{m}}_2\}_L^\perp. \tag{5.190}$$

The first order solution is given by

$$\Delta \boldsymbol{m}_1 = \frac{(\boldsymbol{m}_1, \boldsymbol{m}_2) V[\boldsymbol{m}_1] \boldsymbol{m}_2}{(\boldsymbol{m}_2, V[\boldsymbol{m}_1]\boldsymbol{m}_2) + (\boldsymbol{m}_1, V[\boldsymbol{m}_2]\boldsymbol{m}_1)},$$

$$\Delta \boldsymbol{m}_2 = \frac{(\boldsymbol{m}_1, \boldsymbol{m}_2) V[\boldsymbol{m}_2] \boldsymbol{m}_1}{(\boldsymbol{m}_2, V[\boldsymbol{m}_1]\boldsymbol{m}_2) + (\boldsymbol{m}_1, V[\boldsymbol{m}_2]\boldsymbol{m}_1)}. \tag{5.191}$$

A realistic form of the correction is

$$\hat{\boldsymbol{m}}_1 = N[\boldsymbol{m}_1 - \Delta \boldsymbol{m}_1], \qquad \hat{\boldsymbol{m}}_2 = N[\boldsymbol{m}_2 - \Delta \boldsymbol{m}_2]. \tag{5.192}$$

The a posteriori covariance matrices of the corrected values $\hat{\boldsymbol{m}}_1$ and $\hat{\boldsymbol{m}}_2$ are

$$V[\hat{\boldsymbol{m}}_1] = \hat{V}[\boldsymbol{m}_1] - \frac{(\hat{V}[\boldsymbol{m}_1]\hat{\boldsymbol{m}}_2)(\hat{V}[\boldsymbol{m}_1]\hat{\boldsymbol{m}}_2)^\top}{(\hat{\boldsymbol{m}}_2, V[\boldsymbol{m}_1]\hat{\boldsymbol{m}}_2) + (\hat{\boldsymbol{m}}_1, V[\boldsymbol{m}_2]\hat{\boldsymbol{m}}_1)},$$

$$V[\hat{\boldsymbol{m}}_2] = \hat{V}[\boldsymbol{m}_2] - \frac{(\hat{V}[\boldsymbol{m}_2]\hat{\boldsymbol{m}}_1)(\hat{V}[\boldsymbol{m}_2]\hat{\boldsymbol{m}}_1)^\top}{(\hat{\boldsymbol{m}}_2, V[\boldsymbol{m}_1]\hat{\boldsymbol{m}}_2) + (\hat{\boldsymbol{m}}_1, V[\boldsymbol{m}_2]\hat{\boldsymbol{m}}_1)},$$

$$V[\hat{\boldsymbol{m}}_1, \hat{\boldsymbol{m}}_2] = -\frac{(\hat{V}[\boldsymbol{m}_1]\hat{\boldsymbol{m}}_2)(\hat{V}[\boldsymbol{m}_2]\hat{\boldsymbol{m}}_1)^\top}{(\hat{\boldsymbol{m}}_2, V[\boldsymbol{m}_1]\hat{\boldsymbol{m}}_2) + (\hat{\boldsymbol{m}}_1, V[\boldsymbol{m}_2]\hat{\boldsymbol{m}}_1)}. \tag{5.193}$$

Here,

$$\hat{V}[\boldsymbol{m}_i] = \boldsymbol{P}_{\hat{\boldsymbol{m}}_i} V[\boldsymbol{m}_i] \boldsymbol{P}_{\hat{\boldsymbol{m}}_i}, \qquad i = 1, 2, \tag{5.194}$$

where $\boldsymbol{P}_{\hat{\boldsymbol{m}}_i}$ is the projection matrix along $\hat{\boldsymbol{m}}_i$. The residual of J can be written as

$$\hat{J} = \frac{(\boldsymbol{m}_1, \boldsymbol{m}_2)^2}{(\hat{\boldsymbol{m}}_2, \hat{V}[\boldsymbol{m}_1]\hat{\boldsymbol{m}}_2) + (\hat{\boldsymbol{m}}_1, \hat{V}[\boldsymbol{m}_2]\hat{\boldsymbol{m}}_1)}, \tag{5.195}$$

which is a χ^2 variable with one degree of freedom. This fact provides an *orthogonality test* for two orientations: the hypothesis that the two orientations $\boldsymbol{m}_1$ and $\boldsymbol{m}_2$ are orthogonal to each other is rejected with significance level $a\%$ if

$$\hat{J} > \chi^2_{1,a}. \tag{5.196}$$

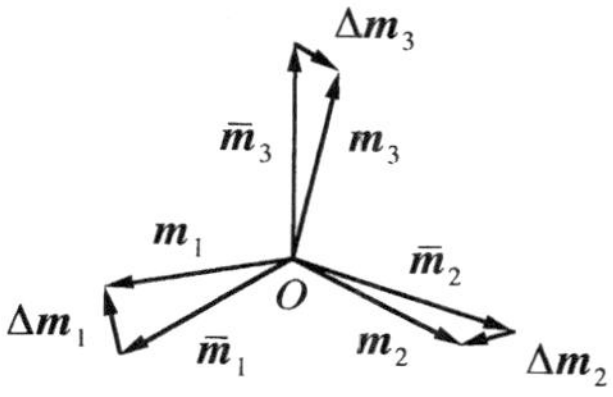

Fig. 5.12. Orthogonality correction for three orientations.

B. Correction of one orientation

If $\boldsymbol{m}_2$ is fixed, the linearized constraint is

$$(\Delta \boldsymbol{m}_1, \boldsymbol{m}_2) = (\boldsymbol{m}_1, \boldsymbol{m}_2), \qquad \Delta \boldsymbol{m}_1 \in \{\bar{\boldsymbol{m}}_1\}_L^{\perp}. \tag{5.197}$$

The optimal correction of $\boldsymbol{m}_1$ is

$$\Delta \boldsymbol{m}_1 = \frac{(\boldsymbol{m}_1, \boldsymbol{m}_2) V[\boldsymbol{m}_1] \boldsymbol{m}_2}{(\boldsymbol{m}_2, V[\boldsymbol{m}_1] \boldsymbol{m}_2)}. \tag{5.198}$$

The a posteriori covariance matrix of the corrected value $\hat{\boldsymbol{m}}_1$ is

$$V[\hat{\boldsymbol{m}}_1] = \hat{V}[\boldsymbol{m}_1] - \frac{(\hat{V}[\boldsymbol{m}_1]\boldsymbol{m}_2)(\hat{V}[\boldsymbol{m}_1]\boldsymbol{m}_2)^\top}{(\boldsymbol{m}_2, \hat{V}[\boldsymbol{m}_1]\boldsymbol{m}_2)}. \tag{5.199}$$

Since $V[\hat{\boldsymbol{m}}_1]\boldsymbol{m}_2 = \boldsymbol{0}$, the rank of $V[\hat{\boldsymbol{m}}_1]$ is 1; its null space is $\{\hat{\boldsymbol{m}}_1, \boldsymbol{m}_2\}_L$. The residual

$$\hat{J} = \frac{(\boldsymbol{m}_1, \boldsymbol{m}_2)^2}{(\boldsymbol{m}_2, V[\boldsymbol{m}_1]\boldsymbol{m}_2)} \tag{5.200}$$

is a χ^2 variable with one degree of freedom. Hence, the orthogonality test given by eq. (5.196) can be applied.

5.5.2 *Correction of three orientations*

A. Simultaneous correction

The same procedure can be applied to three orientations. Let $\boldsymbol{m}_1$, $\boldsymbol{m}_2$, and $\boldsymbol{m}_3$ be unit 3-vectors that indicate three orientations supposedly orthogonal (Fig. 5.12). The problem is finding $\Delta \boldsymbol{m}_i$ such that $\bar{\boldsymbol{m}}_i = \boldsymbol{m}_i - \Delta \boldsymbol{m}_i$ satisfies

$$(\bar{\boldsymbol{m}}_i, \bar{\boldsymbol{m}}_j) = \delta_{ij}, \qquad i, j = 1, 2, 3. \tag{5.201}$$

The rank of this constraint is 3. Let $V[\boldsymbol{m}_i]$ be the a priori covariance matrix of $\boldsymbol{m}_i$. If the three orientations are statistically independent, the problem can be written as the optimization

$$J = \sum_{i=1}^{3} (\Delta \boldsymbol{m}_i, \bar{V}[\boldsymbol{m}_i]^- \Delta \boldsymbol{m}_i) \to \min \tag{5.202}$$

under the linearized constraint

$$(\bar{\boldsymbol{m}}_3, \Delta\boldsymbol{m}_2) + (\bar{\boldsymbol{m}}_2, \Delta\boldsymbol{m}_3) = (\boldsymbol{m}_2, \boldsymbol{m}_3),$$

$$(\bar{\boldsymbol{m}}_1, \Delta\boldsymbol{m}_3) + (\bar{\boldsymbol{m}}_3, \Delta\boldsymbol{m}_1) = (\boldsymbol{m}_3, \boldsymbol{m}_1),$$

$$(\bar{\boldsymbol{m}}_2, \Delta\boldsymbol{m}_1) + (\bar{\boldsymbol{m}}_1, \Delta\boldsymbol{m}_2) = (\boldsymbol{m}_1, \boldsymbol{m}_2),$$

$$\boldsymbol{m}_i \in \{\bar{\boldsymbol{m}}_i\}_L^\perp, \qquad i = 1, 2, 3. \tag{5.203}$$

The first order solution is given by

$$\begin{pmatrix} \Delta\boldsymbol{m}_1 \\ \Delta\boldsymbol{m}_2 \\ \Delta\boldsymbol{m}_3 \end{pmatrix} = \begin{pmatrix} \boldsymbol{0} & V[\boldsymbol{m}_1]\boldsymbol{m}_3 & V[\boldsymbol{m}_1]\boldsymbol{m}_2 \\ V[\boldsymbol{m}_2]\boldsymbol{m}_3 & \boldsymbol{0} & V[\boldsymbol{m}_2]\boldsymbol{m}_1 \\ V[\boldsymbol{m}_3]\boldsymbol{m}_2 & V[\boldsymbol{m}_3]\boldsymbol{m}_1 & \boldsymbol{0} \end{pmatrix} \boldsymbol{W} \begin{pmatrix} (\boldsymbol{m}_2, \boldsymbol{m}_3) \\ (\boldsymbol{m}_3, \boldsymbol{m}_1) \\ (\boldsymbol{m}_1, \boldsymbol{m}_2) \end{pmatrix}, \tag{5.204}$$

where $\boldsymbol{W}$ is a (33)-matrix defined by

$$\boldsymbol{W} = \left(\begin{pmatrix} \boldsymbol{0} & \boldsymbol{m}_3 & \boldsymbol{m}_2 \\ \boldsymbol{m}_3 & \boldsymbol{0} & \boldsymbol{m}_1 \\ \boldsymbol{m}_2 & \boldsymbol{m}_1 & \boldsymbol{0} \end{pmatrix}^\top \begin{pmatrix} V[\boldsymbol{m}_1] & & \\ & V[\boldsymbol{m}_2] & \\ & & V[\boldsymbol{m}_3] \end{pmatrix} \begin{pmatrix} \boldsymbol{0} & \boldsymbol{m}_3 & \boldsymbol{m}_2 \\ \boldsymbol{m}_3 & \boldsymbol{0} & \boldsymbol{m}_1 \\ \boldsymbol{m}_2 & \boldsymbol{m}_1 & \boldsymbol{0} \end{pmatrix} \right)^-. \tag{5.205}$$

A realistic form of the correction is

$$\hat{\boldsymbol{m}}_i = N[\boldsymbol{m}_i - \Delta\boldsymbol{m}_i], \qquad i = 1, 2, 3. \tag{5.206}$$

The a posteriori covariance matrices of the corrected values $\hat{\boldsymbol{m}}_i$ are

$$\begin{pmatrix} V[\hat{\boldsymbol{m}}_1] & V[\hat{\boldsymbol{m}}_1, \hat{\boldsymbol{m}}_2] & V[\hat{\boldsymbol{m}}_1, \hat{\boldsymbol{m}}_3] \\ V[\hat{\boldsymbol{m}}_2, \hat{\boldsymbol{m}}_1] & V[\hat{\boldsymbol{m}}_2] & V[\hat{\boldsymbol{m}}_2, \hat{\boldsymbol{m}}_3] \\ V[\hat{\boldsymbol{m}}_3, \hat{\boldsymbol{m}}_1] & V[\hat{\boldsymbol{m}}_3, \hat{\boldsymbol{m}}_2] & V[\hat{\boldsymbol{m}}_3] \end{pmatrix} = \begin{pmatrix} \hat{V}[\boldsymbol{m}_1] & & \\ & \hat{V}[\boldsymbol{m}_2] & \\ & & \hat{V}[\boldsymbol{m}_3] \end{pmatrix}$$
$$- \begin{pmatrix} \boldsymbol{0} & \hat{V}[\boldsymbol{m}_1]\hat{\boldsymbol{m}}_3 & \hat{V}[\boldsymbol{m}_1]\hat{\boldsymbol{m}}_2 \\ \hat{V}[\boldsymbol{m}_2]\hat{\boldsymbol{m}}_3 & \boldsymbol{0} & \hat{V}[\boldsymbol{m}_2]\hat{\boldsymbol{m}}_1 \\ \hat{V}[\boldsymbol{m}_3]\hat{\boldsymbol{m}}_2 & \hat{V}[\boldsymbol{m}_3]\hat{\boldsymbol{m}}_1 & \boldsymbol{0} \end{pmatrix}$$
$$\hat{\boldsymbol{W}} \begin{pmatrix} \boldsymbol{0} & \hat{V}[\boldsymbol{m}_1]\hat{\boldsymbol{m}}_3 & \hat{V}[\boldsymbol{m}_1]\hat{\boldsymbol{m}}_2 \\ \hat{V}[\boldsymbol{m}_2]\hat{\boldsymbol{m}}_3 & \boldsymbol{0} & \hat{V}[\boldsymbol{m}_2]\hat{\boldsymbol{m}}_1 \\ \hat{V}[\boldsymbol{m}_3]\hat{\boldsymbol{m}}_2 & \hat{V}[\boldsymbol{m}_3]\hat{\boldsymbol{m}}_1 & \boldsymbol{0} \end{pmatrix}^\top, \tag{5.207}$$

where

$$\hat{V}[\boldsymbol{m}_i] = \boldsymbol{P}_{\hat{\boldsymbol{m}}_i} V[\boldsymbol{m}_i] \boldsymbol{P}_{\hat{\boldsymbol{m}}_i}, \qquad i = 1, 2, 3. \tag{5.208}$$

The matrix $\hat{\boldsymbol{W}}$ is obtained by replacing $\boldsymbol{m}_i$ and $V[\boldsymbol{m}_i]$ by $\hat{\boldsymbol{m}}_i$ and $\hat{V}[\boldsymbol{m}_i]$, respectively, in eq. (5.207). The residual of J can be written as

$$\hat{J} = (\begin{pmatrix} (\boldsymbol{m}_2, \boldsymbol{m}_3) \\ (\boldsymbol{m}_3, \boldsymbol{m}_1) \\ (\boldsymbol{m}_1, \boldsymbol{m}_2) \end{pmatrix}, \hat{\boldsymbol{W}} \begin{pmatrix} (\boldsymbol{m}_2, \boldsymbol{m}_3) \\ (\boldsymbol{m}_3, \boldsymbol{m}_1) \\ (\boldsymbol{m}_1, \boldsymbol{m}_2) \end{pmatrix}), \tag{5.209}$$

which is a χ^2 variable with three degrees of freedom. This fact provides an *orthogonality test* for three orientations: the hypothesis that the three orientations $\boldsymbol{m}_i$, $i = 1, 2, 3$, are orthogonal to each other is rejected with significance level $a\%$ if

$$\hat{J} > \chi^2_{3,a}. \tag{5.210}$$

B. Correction of one orientation

If $\boldsymbol{m}_1$ and $\boldsymbol{m}_2$ are fixed in such a way that $(\boldsymbol{m}_1, \boldsymbol{m}_2) = 0$, the rank of the constraint decreases to 2, and the linearized constraint is

$$(\boldsymbol{m}_1, \Delta\boldsymbol{m}_3) = (\boldsymbol{m}_1, \boldsymbol{m}_3), \qquad (\boldsymbol{m}_2, \Delta\boldsymbol{m}_3) = (\boldsymbol{m}_2, \boldsymbol{m}_3),$$
$$\Delta\boldsymbol{m}_3 \in \{\bar{\boldsymbol{m}}_3\}^\perp_L. \tag{5.211}$$

The optimal correction of $\boldsymbol{m}_3$ is

$$\Delta\boldsymbol{m}_3 = V[\boldsymbol{m}_3](\boldsymbol{m}_1, \boldsymbol{m}_2)\boldsymbol{W} \begin{pmatrix} (\boldsymbol{m}_1, \boldsymbol{m}_3) \\ (\boldsymbol{m}_2, \boldsymbol{m}_3) \end{pmatrix}, \tag{5.212}$$

where $\boldsymbol{W}$ is a (22)-matrix defined by

$$\boldsymbol{W} = \begin{pmatrix} (\boldsymbol{m}_1, V[\boldsymbol{m}_3]\boldsymbol{m}_1) & (\boldsymbol{m}_1, V[\boldsymbol{m}_3]\boldsymbol{m}_2) \\ (\boldsymbol{m}_2, V[\boldsymbol{m}_3]\boldsymbol{m}_1) & (\boldsymbol{m}_2, V[\boldsymbol{m}_3]\boldsymbol{m}_2) \end{pmatrix}^{-1}. \tag{5.213}$$

It is evident from the underlying geometry that

$$\hat{\boldsymbol{m}}_3 = \pm\boldsymbol{m}_1 \times \boldsymbol{m}_2 \tag{5.214}$$

is the exact solution if the sign is appropriately chosen. Hence, its covariance is

$$V[\hat{\boldsymbol{m}}_1] = \boldsymbol{O}. \tag{5.215}$$

The residual can be written as

$$\hat{J} = \hat{W}^{(11)}(\boldsymbol{m}_1, \boldsymbol{m}_3)^2 + 2\hat{W}^{(12)}(\boldsymbol{m}_1, \boldsymbol{m}_3)(\boldsymbol{m}_2, \boldsymbol{m}_3) + \hat{W}^{(22)}(\boldsymbol{m}_2, \boldsymbol{m}_3)^2, \tag{5.216}$$

which is a χ^2 variable with two degrees of freedom. Here, the matrix $\hat{\boldsymbol{W}} = (\hat{W}^{(kl)})$ is obtained by replacing $\boldsymbol{m}_1$, $\boldsymbol{m}_2$, and $V[\boldsymbol{m}_3]$ by $\hat{\boldsymbol{m}}_1$, $\hat{\boldsymbol{m}}_2$, and $\hat{V}[\boldsymbol{m}_3]$, respectively, in eq. (5.213). The orthogonality test takes the form

$$\hat{J} > \chi^2_{2,a}. \tag{5.217}$$

C. Correction of two orientations

If $\boldsymbol{m}_3$ is fixed, the rank of the constraint is 3, and the linearized constraint is

$$(\boldsymbol{m}_3, \Delta\boldsymbol{m}_1) = (\boldsymbol{m}_3, \boldsymbol{m}_1), \qquad (\boldsymbol{m}_3, \Delta\boldsymbol{m}_2) = (\boldsymbol{m}_3, \boldsymbol{m}_2),$$
$$(\Delta\boldsymbol{m}_1, \bar{\boldsymbol{m}}_2) + (\bar{\boldsymbol{m}}_1, \Delta\boldsymbol{m}_2) = (\boldsymbol{m}_1, \boldsymbol{m}_2),$$
$$\Delta\boldsymbol{m}_1 \in \{\bar{\boldsymbol{m}}_1\}_L^\perp, \qquad \Delta\boldsymbol{m}_2 \in \{\bar{\boldsymbol{m}}_2\}_L^\perp. \tag{5.218}$$

The optimal correction of $\boldsymbol{m}_1$ and $\boldsymbol{m}_2$ is given by

$$\begin{pmatrix} \Delta\boldsymbol{m}_1 \\ \Delta\boldsymbol{m}_2 \end{pmatrix} = \begin{pmatrix} V[\boldsymbol{m}_1]\boldsymbol{m}_3 & \boldsymbol{0} & V[\boldsymbol{m}_1]\boldsymbol{m}_2 \\ \boldsymbol{0} & V[\boldsymbol{m}_2]\boldsymbol{m}_3 & V[\boldsymbol{m}_2]\boldsymbol{m}_1 \end{pmatrix} \boldsymbol{W} \begin{pmatrix} (\boldsymbol{m}_3, \boldsymbol{m}_1) \\ (\boldsymbol{m}_3, \boldsymbol{m}_2) \end{pmatrix}, \tag{5.219}$$

where $\boldsymbol{W}$ is a (33)-matrix defined by

$$\boldsymbol{W} = \begin{pmatrix} (\boldsymbol{m}_3, V[\boldsymbol{m}_1]\boldsymbol{m}_3) & & (\boldsymbol{m}_3, V[\boldsymbol{m}_1]\boldsymbol{m}_2) \\ & (\boldsymbol{m}_3, V[\boldsymbol{m}_2]\boldsymbol{m}_3) & (\boldsymbol{m}_3, V[\boldsymbol{m}_2]\boldsymbol{m}_1) \\ (\boldsymbol{m}_2, V[\boldsymbol{m}_1]\boldsymbol{m}_3) & (\boldsymbol{m}_1, V[\boldsymbol{m}_2]\boldsymbol{m}_3) & (\boldsymbol{m}_2, V[\boldsymbol{m}_1]\boldsymbol{m}_2) + (\boldsymbol{m}_1, V[\boldsymbol{m}_2]\boldsymbol{m}_1) \end{pmatrix}^{-1}. \tag{5.220}$$

The a posteriori covariance matrices of the corrected values $\hat{\boldsymbol{m}}_1$ and $\hat{\boldsymbol{m}}_2$ are

$$V[\hat{\boldsymbol{m}}_1] = \hat{V}[\boldsymbol{m}_1] - \hat{W}^{(11)}(\hat{V}[\boldsymbol{m}_1]\boldsymbol{m}_3)(\hat{V}[\boldsymbol{m}_1]\boldsymbol{m}_3)^\top,$$
$$V[\hat{\boldsymbol{m}}_2] = \hat{V}[\boldsymbol{m}_2] - \hat{W}^{(22)}(\hat{V}[\boldsymbol{m}_2]\boldsymbol{m}_3)(\hat{V}[\boldsymbol{m}_2]\boldsymbol{m}_3)^\top,$$
$$V[\hat{\boldsymbol{m}}_1, \hat{\boldsymbol{m}}_2] = -\hat{W}^{(12)}(\hat{V}[\boldsymbol{m}_1]\boldsymbol{m}_3)(\hat{V}[\boldsymbol{m}_2]\boldsymbol{m}_3)^\top = V[\hat{\boldsymbol{m}}_2, \hat{\boldsymbol{m}}_1]^\top, \tag{5.221}$$

where the matrix $\hat{\boldsymbol{W}}$ is obtained by replacing $\boldsymbol{m}_i$ and $V[\boldsymbol{m}_i]$ by $\hat{\boldsymbol{m}}_i$ and $\hat{V}[\boldsymbol{m}_i]$, respectively, in eq. (5.220). The residual can be written as

$$\hat{J} = \hat{W}^{(11)}(\boldsymbol{m}_3, \boldsymbol{m}_1)^2 + 2\hat{W}^{(12)}(\boldsymbol{m}_3, \boldsymbol{m}_1)(\boldsymbol{m}_3, \boldsymbol{m}_2) + W^{\hat{(22)}}(\boldsymbol{m}_3, \boldsymbol{m}_1)^2, \tag{5.222}$$

which is a χ^2 variable with two degrees of freedom. Hence, the orthogonality test given by eq. (5.217) can be applied.

5.6 Conic Incidence Correction

Consider a conic $(\boldsymbol{x}, \boldsymbol{Q}\boldsymbol{x}) = 0$ (see eq. (4.80)). Let $\boldsymbol{x}$ be an image point not on conic $(\boldsymbol{x}, \boldsymbol{Q}\boldsymbol{x}) = 0$. We consider the problem of optimally correcting $\boldsymbol{x}$ so as to make it incident to the conic $(\boldsymbol{x}, \boldsymbol{Q}\boldsymbol{x}) = 0$. In other words, we want to find $\Delta\boldsymbol{x}$ such that $\bar{\boldsymbol{x}} = \boldsymbol{x} - \Delta\boldsymbol{x}$ satisfies

$$(\bar{\boldsymbol{x}}, \boldsymbol{Q}\bar{\boldsymbol{x}}) = 0. \tag{5.223}$$

The rank of this constraint is 1. Let $V[\boldsymbol{x}]$ be the a priori covariance matrix of $\boldsymbol{x}$. The problem can be written as the optimization

$$J = (\Delta\boldsymbol{x}, \bar{V}[\boldsymbol{x}]^- \Delta\boldsymbol{x}) \to \min \tag{5.224}$$

under the linearized constraint

$$(\Delta\boldsymbol{x}, \boldsymbol{Q}\bar{\boldsymbol{x}}) = \frac{1}{2}(\boldsymbol{x}, \boldsymbol{Q}\boldsymbol{x}), \qquad \Delta\boldsymbol{x} \in \{\boldsymbol{k}\}_L^\perp. \tag{5.225}$$

The first order solution is given by

$$\Delta\boldsymbol{x} = \frac{(\boldsymbol{x}, \boldsymbol{Q}\boldsymbol{x})V[\boldsymbol{x}]\boldsymbol{Q}\boldsymbol{x}}{2(\boldsymbol{x}, \boldsymbol{Q}V[\boldsymbol{x}]\boldsymbol{Q}\boldsymbol{x})}. \tag{5.226}$$

If we put $\boldsymbol{n} = N[\boldsymbol{Q}\boldsymbol{x}]$, eq. (5.226) can be written as

$$\Delta\boldsymbol{x} = \frac{(\boldsymbol{n}, \boldsymbol{x})V[\boldsymbol{x}]\boldsymbol{n}}{2(\boldsymbol{n}, V[\boldsymbol{x}]\boldsymbol{n})}. \tag{5.227}$$

This problem can be viewed as imposing the incidence constraint on the image point $\boldsymbol{x}$ and its *polar* $(\boldsymbol{n}, \boldsymbol{x}) = 0$ with respect to the conic $(\boldsymbol{x}, \boldsymbol{Q}\boldsymbol{x}) = 0$ (see eq. (4.82)). The difference in the factor 2 (see eq. (5.82)) is due to the fact that as the image point $\boldsymbol{x}$ approaches, its polar $(\boldsymbol{n}, \boldsymbol{x}) = 0$ also approaches its *pole* $\boldsymbol{x}$ by the same distance.

The a posteriori covariance matrix of the corrected position $\hat{\boldsymbol{x}}$ is

$$V[\hat{\boldsymbol{x}}] = V[\boldsymbol{x}] - \frac{(V[\boldsymbol{x}]\hat{\boldsymbol{n}})(V[\boldsymbol{x}]\hat{\boldsymbol{n}})^\top}{(\hat{\boldsymbol{n}}, V[\boldsymbol{x}]\hat{\boldsymbol{n}})}, \tag{5.228}$$

where

$$\hat{\boldsymbol{n}} = N[\boldsymbol{Q}\hat{\boldsymbol{x}}]. \tag{5.229}$$

Eq. (5.228) has the same form as eq. (5.83). Hence, the rank of $V[\hat{\boldsymbol{x}}]$ is 1; its null space is $\{\boldsymbol{k}, \hat{\boldsymbol{n}}\}_L$, which is orthogonal to the orientation $\hat{\boldsymbol{m}} = N[\boldsymbol{k} \times \hat{\boldsymbol{n}}]$ of the polar $(\hat{\boldsymbol{n}}, \boldsymbol{x}) = 0$. The residual of J can be written as

$$\hat{J} = \frac{(\boldsymbol{x}, \boldsymbol{Q}\boldsymbol{x})^2}{4(\hat{\boldsymbol{x}}, \boldsymbol{Q}V[\boldsymbol{x}]\boldsymbol{Q}\hat{\boldsymbol{x}})}, \tag{5.230}$$

which is a χ^2 variable with one degree of freedom. This fact provides a *conic incidence test*: the hypothesis that image point $\boldsymbol{x}$ is on conic $(\boldsymbol{x}, \boldsymbol{Q}\boldsymbol{x}) = 0$ is rejected with significance level a% if

$$\hat{J} > \chi^2_{1,a}. \tag{5.231}$$

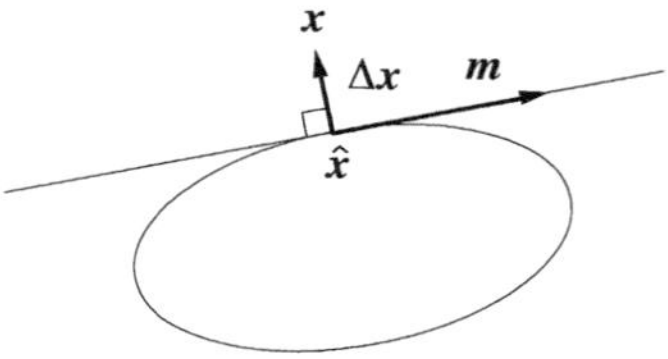

Fig. 5.13. Optimal incidence correction for an image point.

Example 5.6 If each coordinate is perturbed independently by Gaussian noise of mean 0 and variance ϵ^2, the covariance matrix of $\boldsymbol{x}$ is $V[\boldsymbol{x}] = \epsilon^2 \boldsymbol{P}_{\boldsymbol{k}}$. The optimal correction (5.226) reduces to

$$\Delta \boldsymbol{x} = \frac{(\boldsymbol{x}, \boldsymbol{Q}\boldsymbol{x})\boldsymbol{P}_{\boldsymbol{k}}\boldsymbol{Q}\boldsymbol{x}}{2\|\boldsymbol{P}_{\boldsymbol{k}}\boldsymbol{Q}\boldsymbol{x}\|^2}. \tag{5.232}$$

Consequently, $\boldsymbol{x}$ is displaced onto the conic perpendicularly (Fig. 5.13). The a posteriori covariance matrix of the corrected value $\hat{\boldsymbol{x}}$ is

$$V[\hat{\boldsymbol{x}}] = \epsilon^2 \hat{\boldsymbol{m}}\hat{\boldsymbol{m}}^\top, \tag{5.233}$$

where $\hat{\boldsymbol{m}} = N[\boldsymbol{k} \times \hat{\boldsymbol{n}}]$ is the orientation of the polar $(\hat{\boldsymbol{n}}, \boldsymbol{x}) = 0$. The residual (5.230) can be written as

$$\hat{J} = \frac{1}{4\epsilon^2} \frac{(\boldsymbol{x}, \boldsymbol{Q}\boldsymbol{x})^2}{\|\boldsymbol{P}_{\boldsymbol{k}}\boldsymbol{Q}\hat{\boldsymbol{x}}\|^2}. \tag{5.234}$$

Hence, an unbiased estimator of the variance ϵ^2 is obtained in the form

$$\hat{\epsilon}^2 = \frac{1}{4} \frac{(\boldsymbol{x}, \boldsymbol{Q}\boldsymbol{x})^2}{\|\boldsymbol{P}_{\boldsymbol{k}}\boldsymbol{Q}\hat{\boldsymbol{x}}\|^2}. \tag{5.235}$$

The value $\hat{\epsilon}$ thus estimated equals half the distance between the image point and its polar with respect to the conic. If the value ϵ is given a priori, the conic incidence test takes the form

$$\frac{\hat{\epsilon}^2}{\epsilon^2} > \chi^2_{1,a}. \tag{5.236}$$

Chapter 6

3-D Computation by Stereo Vision

As seen in the preceding chapter, the covariance matrix plays a fundamental role in any type of optimization. In the beginning, the a priori covariance matrices of raw data are determined by the characteristics of the sensing device. Then, the raw data define geometric objects, and the resulting geometric objects in turn define another class of geometric objects, which define geometric objects in a higher level and so on. In this process, the error characteristics can be traced bottom up: the covariance matrix of an object is computed from the covariance matrices of the objects in the lower hierarchy. In this chapter, we do such an analysis for *stereo vision*. We first study the *epipolar constraint* of a stereo system and derive explicit expressions for optimal 3-D reconstruction of points and lines. Then, the error behavior of the reconstructed space points and space lines is analyzed. We also derive expressions for optimal *back projection* of image points and image lines onto a space plane and analyze the behavior of the errors involved. Finally, we evaluate the effect of camera calibration errors.

6.1 Epipolar Constraint

6.1.1 Camera imaging geometry

Stereo vision is a means of reconstructing 3-D structures from two-dimensional images by triangulation using two cameras[1]. To do this, the camera geometry must be known, since 3-D structures are reconstructed by inverting the imaging process. As discussed in Section 4.1.1, the camera geometry is modeled as *perspective projection*: a space point P is projected onto the intersection p of the image plane with the line of sight that starts from the center of the lens and passes through P (see Fig. 4.1). The center of the lens is called the *viewpoint*, and the distance f from the viewpoint to the image plane is called the *focal length*. Define an XYZ coordinate system by identifying the origin O with the viewpoint and taking the Z-axis along the optical axis of the lens. The unit of length is scaled so that the focal length f is unity. First-order analysis based on covariance matrices is justified if image noise is very small as compared with the focal length, which is usually the case.

Given two cameras, define an XYZ coordinate system for the first camera, and an $X'Y'Z'$ coordinate system for the second (Fig. 6.1). The two cameras

[1]More than two cameras can be used to enhance the reliability of the reconstruction. If the number of cameras needs to be specified, such terms as *binocular stereo vision* and *trinocular stereo vision* are used.

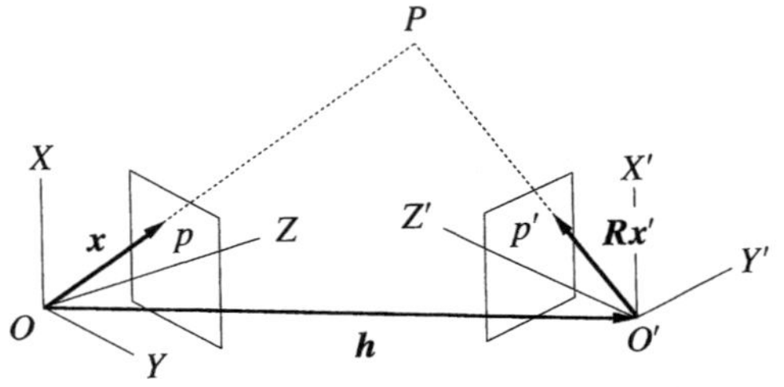

Fig. 6.1. Geometry of stereo vision.

are assumed to have the same focal length. Let $\boldsymbol{h}$ represent the origin O' of the $X'Y'Z'$ coordinate system with respect to the XYZ coordinate system; we call it the *base-line vector*. Let $\boldsymbol{R}$ be the rotation matrix that describes the orientation of the $X'Y'Z'$ coordinate system relative to the XYZ coordinate system. The relative configuration of the two cameras is specified by the pair $\{\boldsymbol{h}, \boldsymbol{R}\}$, which we call the *motion parameters* (see Section 4.6.1).

Consider a space point which has coordinates (X, Y, Z) with respect to the XYZ coordinate system. Let (X', Y', Z') be the coordinates of the same space point with respect to the $X'Y'Z'$ coordinate system. If we put $\boldsymbol{r} = (X, Y, Z)^\top$ and $\boldsymbol{r}' = (X', Y', Z')^\top$, the following relationship holds (see eq. (4.116)):

$$\boldsymbol{r} = \boldsymbol{h} + \boldsymbol{R}\boldsymbol{r}'. \tag{6.1}$$

6.1.2 Epipolar equation

Let p be the perspective projection of a space point P onto the image plane of the first camera, and p' that for the second camera. Evidently, the three vectors $\vec{Op}$, $\vec{OO'}$, and $\vec{O'p'}$ must be coplanar (Fig. 6.1). Hence,

$$|\vec{Op}, \vec{OO'}, \vec{O'p'}| = 0, \tag{6.2}$$

where $|\cdot, \cdot, \cdot|$ denotes the scalar triple product (see eq. (2.32)). Eq. (6.2) is known as the *epipolar constraint*; the plane defined by $\vec{Op}$, $\vec{OO'}$, and $\vec{O'p'}$ is called the *epipolar plane*.

Let (x, y) be the image coordinates of p, and (x', y') those of p'. The image points p and p' are respectively represented by the following 3-vectors (see eq. (4.1)):

$$\boldsymbol{x} = \begin{pmatrix} x \\ y \\ 1 \end{pmatrix}, \qquad \boldsymbol{x}' = \begin{pmatrix} x' \\ y' \\ 1 \end{pmatrix}. \tag{6.3}$$

Since the $X'Y'Z'$ coordinate system is rotated by $\boldsymbol{R}$ relative to the XYZ coordinate system, vector $\vec{O'p'}$ is represented by $\boldsymbol{R}\boldsymbol{x}'$ with respect to the

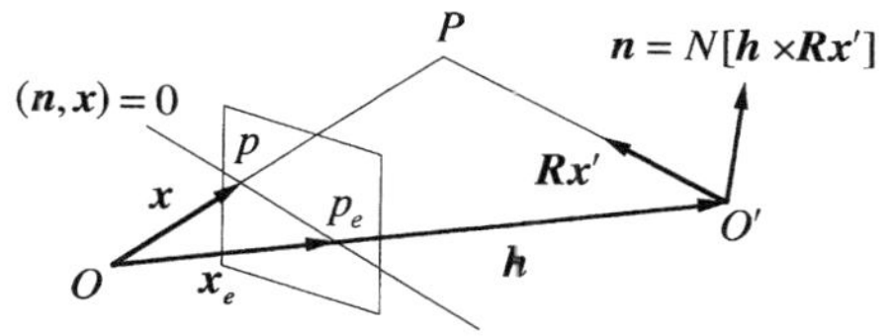

Fig. 6.2. Epipole and epipolar.

XYZ coordinate system. Noting that $\vec{Op} = \boldsymbol{x}$ and $\vec{OO'} = \boldsymbol{h}$, we can express eq. (6.3) with respect to the XYZ coordinate system in the form

$$|\boldsymbol{x}, \boldsymbol{h}, \boldsymbol{R}\boldsymbol{x}'| = 0, \tag{6.4}$$

which we call the *epipolar equation.*

The epipolar equation provides a strong clue to automatically detecting point-to-point correspondences between the two images. Consider the image plane of the first camera. For a fixed value of $\boldsymbol{x}'$, eq. (6.4) defines an image line if $\boldsymbol{x}$ is regarded as a variable. This line is called the *epipolar* of $\boldsymbol{x}'$. Eq. (6.4) states that the epipolar of $\boldsymbol{x}'$ passes through $\boldsymbol{x}$. Let p_e be the image point represented by

$$\boldsymbol{x}_e = \frac{\boldsymbol{h}}{(\boldsymbol{k}, \boldsymbol{h})}, \tag{6.5}$$

where $\boldsymbol{k} = (0, 0, 1)^\top$. This point is called the *epipole*[2] of the first image plane (Fig. 6.2). Since $|\boldsymbol{x}_e, \boldsymbol{h}, \boldsymbol{R}\boldsymbol{x}'| = 0$ holds irrespective of the value of $\boldsymbol{x}'$, *all epipolars pass through the epipole.*

Consider the image plane of the second camera. For a fixed value of $\boldsymbol{x}$, eq. (6.4) defines an image line if $\boldsymbol{x}'$ is regarded as a variable. This line is called the *epipolar* of $\boldsymbol{x}$. Eq. (6.4) states that the epipolar of $\boldsymbol{x}$ passes through $\boldsymbol{x}'$. Let p'_e be the image point represented by

$$\boldsymbol{x}'_e = \frac{\boldsymbol{R}^\top \boldsymbol{h}}{(\boldsymbol{k}, \boldsymbol{R}^\top \boldsymbol{h})}. \tag{6.6}$$

This point is also called the *epipole* of the second image plane. Since $|\boldsymbol{x}, \boldsymbol{h}, \boldsymbol{R}\boldsymbol{x}'_e| = 0$ holds irrespective of the value of $\boldsymbol{x}$, *all epipolars passes through the epipole.*

The above observations are summarized as follows:

- The image point p that corresponds to an image point p' in the other image is located on the epipolar of p', and vice versa.

[2]In a real camera, the size of the image frame is finite, so the epipole p_e may be located outside the image frame, or it can be an *ideal image point* located at infinity. In the following analysis, the image plane is assumed to be infinitely extended.

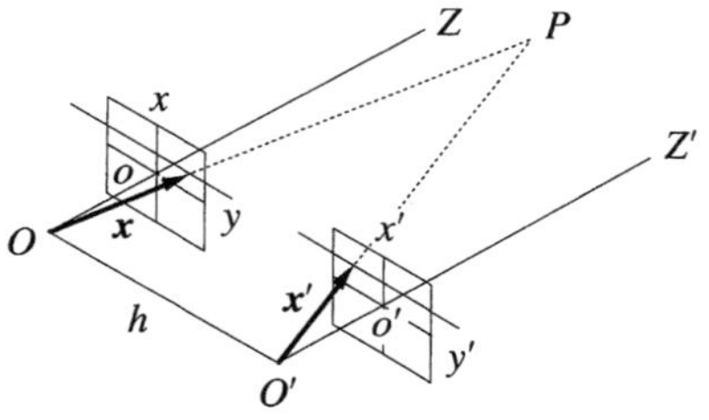

Fig. 6.3. Parallel stereo system.

- The epipolar in each image is determined by the location of the corresponding point in the other image and the motion parameters $\{\boldsymbol{h}, \boldsymbol{R}\}$.
- In each image, epipolars are *concurrent*, all passing through the epipole.

If we define the *essential matrix*

$$\boldsymbol{G} = \boldsymbol{h} \times \boldsymbol{R}, \tag{6.7}$$

the epipolar equation (6.4) can be written as

$$(\boldsymbol{x}, \boldsymbol{G}\boldsymbol{x}') = 0. \tag{6.8}$$

Hence, the epipolar in the first image is an image line represented by

$$(\boldsymbol{n}, \boldsymbol{x}) = 0, \qquad \boldsymbol{n} = \boldsymbol{G}\boldsymbol{x}'. \tag{6.9}$$

The epipolar in the second image is an image line represented by

$$(\boldsymbol{n}', \boldsymbol{x}') = 0, \qquad \boldsymbol{n}' = \boldsymbol{G}^\top \boldsymbol{x}. \tag{6.10}$$

6.1.3 Parallel stereo system

Let us call a stereo system *parallel* if the optical axes of the two cameras are parallel and the base-line vector is perpendicular to them (Fig. 6.3). If the Y-axis is taken in the direction of the base-line vector, the motion parameters have the form

$$\boldsymbol{h} = \begin{pmatrix} 0 \\ h \\ 0 \end{pmatrix}, \qquad \boldsymbol{R} = \boldsymbol{I}. \tag{6.11}$$

The essential matrix (6.7) reduces to

$$\boldsymbol{G} = h \begin{pmatrix} 0 & 0 & 1 \\ 0 & 0 & 0 \\ -1 & 0 & 0 \end{pmatrix}. \tag{6.12}$$

The epipolar equation (6.4) is simply

$$x - x' = 0, \tag{6.13}$$

which defines a line parallel to the y-axis on each image plane. The epipole is an ideal image point located at infinity in the direction of the base-line vector. The epipolar constraint simply states that *corresponding points must have the same x coordinate.*

We now show that *any stereo system can be regarded as a parallel stereo system by changing the camera coordinate systems.* Consider a stereo system with motion parameters $\{\boldsymbol{h}, \boldsymbol{R}\}$. Let $\tilde{\boldsymbol{j}} = N[\boldsymbol{h}]$, and let $\tilde{\boldsymbol{k}}$ be an arbitrary unit vector orthogonal to $\tilde{\boldsymbol{j}}$. If we define $\tilde{\boldsymbol{i}} = \tilde{\boldsymbol{j}} \times \tilde{\boldsymbol{k}}$, the set $\{\tilde{\boldsymbol{i}}, \tilde{\boldsymbol{j}}, \tilde{\boldsymbol{k}}\}$ is an orthonormal system. Define a new $\tilde{X}\tilde{Y}\tilde{Z}$ coordinate system for the first camera by taking $\{\tilde{\boldsymbol{i}}, \tilde{\boldsymbol{j}}, \tilde{\boldsymbol{k}}\}$ as the axis orientations. An image point represented by $\boldsymbol{x}$ with respect to the XYZ coordinate system is now represented by[3]

$$\tilde{\boldsymbol{x}} = \frac{\tilde{\boldsymbol{R}}^\top \boldsymbol{x}}{(\boldsymbol{k}, \tilde{\boldsymbol{R}}^\top \boldsymbol{x})}, \tag{6.14}$$

where

$$\tilde{\boldsymbol{R}} = (\tilde{\boldsymbol{i}}, \tilde{\boldsymbol{j}}, \tilde{\boldsymbol{k}}). \tag{6.15}$$

Similarly, define a new $\tilde{X}'\tilde{Y}'\tilde{Z}'$ coordinate system for the second camera by taking the same axis orientations. An image point represented by $\boldsymbol{x}'$ with respect to the $X'Y'Z'$ coordinate system is now represented by

$$\tilde{\boldsymbol{x}}' = \frac{\tilde{\boldsymbol{R}}^\top \boldsymbol{R}\boldsymbol{x}'}{(\boldsymbol{k}, \tilde{\boldsymbol{R}}^\top \boldsymbol{R}\boldsymbol{x}')}. \tag{6.16}$$

With respect to the $\tilde{X}\tilde{Y}\tilde{Z}$ and $\tilde{X}'\tilde{Y}'\tilde{Z}'$ coordinate systems, the stereo system is parallel with motion parameters $\{\|\boldsymbol{h}\|\boldsymbol{j}, \boldsymbol{I}\}$.

6.2 Optimal Correction of Correspondence

6.2.1 *Correspondence detection and optimal correction*

In order to do error analysis based on corresponding image points, we need to consider the way they are detected. Basically, there are two possibilities:

- *Edge-based method.* We first detect corresponding *edge segments.* Then, point-to-point correspondence is established by computing the intersection of one edge segment with the epipolars of the pixels of the other edge segment.

[3] Eq. (6.14) defines a two-dimensional *projective transformation* if the image plane is identified with the two-dimensional *projective space* by adding ideal image points. If $\tilde{\mathbf{R}}$ runs over all rotations and $\tilde{\mathrm{k}}$ is its third column, the set of the corresponding transformations is a subgroup of the two-dimensional projective transformations.

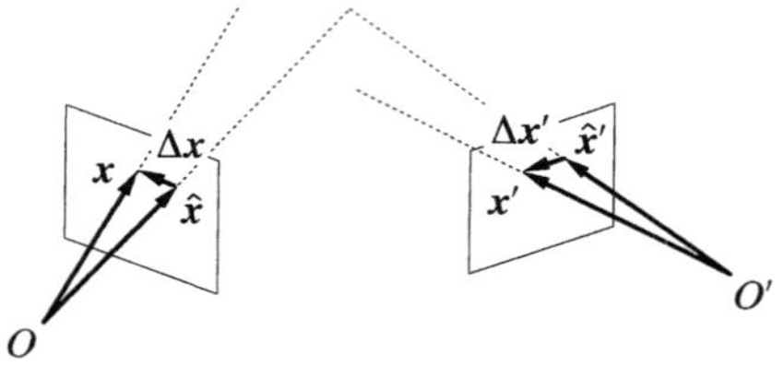

Fig. 6.4. Corresponding image points are corrected so that they define a space point.

- *Feature matching method.* We match feature points directly: for a small neighborhood of each feature point in one image, the other image is searched for a portion that has a maximum correlation with that neighborhood. Many types of correlation measure and searching strategy are conceivable.

The edge-based method has the advantage that a large number of corresponding pairs can be obtained easily. However, computational difficulties may arise for those edge segments which are nearly parallel to the epipolars. Moreover, the motion parameters $\{\boldsymbol{h}, \boldsymbol{R}\}$ must be known accurately for computing the epipolars. In contrast, the feature matching method does not require any knowledge of the motion parameters. If they are known, the epipolar constraint can be used for limiting the search region—only a small neighborhood of each epipolar needs to be searched. In Chapter 11, we will show that the motion parameters $\{\boldsymbol{h}, \boldsymbol{R}\}$ can be *computed* if a sufficient number of corresponding feature points are detected. The feature matching method is also suitable if natural or artificial markers are involved or humans intervene by using cursors and mice. In this section, we assume that corresponding points are detected by the feature matching method. The case in which correspondence is given between straight edge segments will be studied in Section 6.4.

If corresponding image points $\boldsymbol{x}$ and $\boldsymbol{x}'$ are detected by the feature matching method in the presence of image noise, they may not strictly satisfy the epipolar equation (6.4). This means that their lines of sight do not intersect in the scene. In order that a unique space point be determined, image points $\boldsymbol{x}$ and $\boldsymbol{x}'$ must be corrected so as to satisfy the epipolar equation exactly. In geometric terms, this means modifying $\boldsymbol{x}$ and $\boldsymbol{x}'$ so that their lines of sight meet in the scene (Fig. 6.4). This problem is equivalent to finding $\Delta\boldsymbol{x}$ and $\Delta\boldsymbol{x}'$ such that $\bar{\boldsymbol{x}} = \boldsymbol{x} - \Delta\boldsymbol{x}$ and $\bar{\boldsymbol{x}}' = \boldsymbol{x}' - \Delta\boldsymbol{x}'$ satisfy the epipolar equation

$$(\bar{\boldsymbol{x}}, \boldsymbol{G}\bar{\boldsymbol{x}}') = 0. \tag{6.17}$$

The rank of this constraint is 1. Let $V[\boldsymbol{x}]$ and $V[\boldsymbol{x}']$ be the a priori covariance matrices of $\boldsymbol{x}$ and $\boldsymbol{x}'$, respectively. If $\boldsymbol{x}$ and $\boldsymbol{x}'$ are statistically independent, the problem can be written as the optimization

$$J = (\Delta \boldsymbol{x}, V[\boldsymbol{x}]^{-} \Delta \boldsymbol{x}) + (\Delta \boldsymbol{x}', V[\boldsymbol{x}']^{-} \Delta \boldsymbol{x}') \rightarrow \min \tag{6.18}$$

under the linearized constraint

$$(\Delta \boldsymbol{x}, \boldsymbol{G}\boldsymbol{x}') + (\boldsymbol{x}, \boldsymbol{G}\Delta \boldsymbol{x}') = (\boldsymbol{x}, \boldsymbol{G}\boldsymbol{x}'),$$
$$\Delta \boldsymbol{x}, \Delta \boldsymbol{x}' \in \{\boldsymbol{k}\}_L^{\perp}. \tag{6.19}$$

The first order solution is given as follows (see eq. (5.17)):

$$\Delta \boldsymbol{x} = \frac{(\boldsymbol{x}, \boldsymbol{G}\boldsymbol{x}')V[\boldsymbol{x}]\boldsymbol{G}\boldsymbol{x}'}{(\boldsymbol{x}', \boldsymbol{G}^{\top} V[\boldsymbol{x}]\boldsymbol{G}\boldsymbol{x}') + (\boldsymbol{x}, \boldsymbol{G}V[\boldsymbol{x}']\boldsymbol{G}^{\top}\boldsymbol{x})},$$

$$\Delta \boldsymbol{x}' = \frac{(\boldsymbol{x}, \boldsymbol{G}\boldsymbol{x}')V[\boldsymbol{x}']\boldsymbol{G}^{\top}\boldsymbol{x}}{(\boldsymbol{x}', \boldsymbol{G}^{\top} V[\boldsymbol{x}]\boldsymbol{G}\boldsymbol{x}') + (\boldsymbol{x}, \boldsymbol{G}V[\boldsymbol{x}']\boldsymbol{G}^{\top}\boldsymbol{x})}. \tag{6.20}$$

Since the constraint (6.19) is obtained by a linear approximation, corrections $\boldsymbol{x} \leftarrow \boldsymbol{x} - \Delta \boldsymbol{x}$ and $\boldsymbol{x}' \leftarrow \boldsymbol{x}' - \Delta \boldsymbol{x}'$ need to be iterated until $(\boldsymbol{x}, \boldsymbol{G}\boldsymbol{x}') = 0$ is sufficiently satisfied (see Section 5.1.3). The a posteriori covariance matrices of the corrected positions $\hat{\boldsymbol{x}} = \boldsymbol{x} - \Delta \boldsymbol{x}$ and $\hat{\boldsymbol{x}}' = \boldsymbol{x}' - \Delta \boldsymbol{x}'$ are computed as follows (see eqs. (5.31) and (5.32)):

$$V[\hat{\boldsymbol{x}}] = V[\boldsymbol{x}] - \frac{(V[\boldsymbol{x}]\boldsymbol{G}\hat{\boldsymbol{x}}')(V[\boldsymbol{x}]\boldsymbol{G}\hat{\boldsymbol{x}}')^{\top}}{(\hat{\boldsymbol{x}}', \boldsymbol{G}^{\top} V[\boldsymbol{x}]\boldsymbol{G}\hat{\boldsymbol{x}}') + (\hat{\boldsymbol{x}}, \boldsymbol{G}V[\boldsymbol{x}']\boldsymbol{G}^{\top}\hat{\boldsymbol{x}})},$$

$$V[\hat{\boldsymbol{x}}'] = V[\boldsymbol{x}'] - \frac{(V[\boldsymbol{x}']\boldsymbol{G}^{\top}\hat{\boldsymbol{x}})(V[\boldsymbol{x}']\boldsymbol{G}^{\top}\hat{\boldsymbol{x}})^{\top}}{(\hat{\boldsymbol{x}}', \boldsymbol{G}^{\top} V[\boldsymbol{x}]\boldsymbol{G}\hat{\boldsymbol{x}}') + (\hat{\boldsymbol{x}}, \boldsymbol{G}V[\boldsymbol{x}']\boldsymbol{G}^{\top}\hat{\boldsymbol{x}})},$$

$$V[\hat{\boldsymbol{x}}, \hat{\boldsymbol{x}}'] = -\frac{(V[\boldsymbol{x}]\boldsymbol{G}\hat{\boldsymbol{x}}')(V[\boldsymbol{x}']\boldsymbol{G}^{\top}\hat{\boldsymbol{x}})^{\top}}{(\hat{\boldsymbol{x}}', \boldsymbol{G}^{\top} V[\boldsymbol{x}]\boldsymbol{G}\hat{\boldsymbol{x}}') + (\hat{\boldsymbol{x}}, \boldsymbol{G}V[\boldsymbol{x}']\boldsymbol{G}^{\top}\hat{\boldsymbol{x}})} = V[\hat{\boldsymbol{x}}', \hat{\boldsymbol{x}}]^{\top}. \tag{6.21}$$

6.2.2 Correspondence test and noise level estimation

The residual of the function J given in eq. (6.18) can be written as follows (see eq. (5.34)):

$$\hat{J} = \frac{(\boldsymbol{x}, \boldsymbol{G}\boldsymbol{x}')^2}{(\hat{\boldsymbol{x}}', \boldsymbol{G}^{\top} V[\boldsymbol{x}]\boldsymbol{G}\hat{\boldsymbol{x}}') + (\hat{\boldsymbol{x}}, \boldsymbol{G}V[\boldsymbol{x}']\boldsymbol{G}^{\top}\hat{\boldsymbol{x}})}. \tag{6.22}$$

This is a χ^2 variable with one degree of freedom if the noise is Gaussian (see Section 5.1.5). This fact provides a *correspondence test*: the hypothesis that image points $\boldsymbol{x}$ and $\boldsymbol{x}'$ correspond to each other is rejected with significance level a% if

$$\hat{J} > \chi^2_{1,a}. \tag{6.23}$$

As discussed in Section 5.1.5, it is usually very difficult to estimate the absolute magnitude of image noise, but often its geometric characteristics (e.g., the degree of homogeneity/inhomogeneity and isotropy/anisotropy) can be relatively easily predicted. Let us assume that the covariance matrices $V[\boldsymbol{x}]$ and $V[\boldsymbol{x}']$ can be written in the form

$$V[\boldsymbol{x}] = \epsilon^2 V_0[\boldsymbol{x}], \qquad V[\boldsymbol{x}'] = \epsilon^2 V_0[\boldsymbol{x}'], \tag{6.24}$$

where matrices $V_0[\boldsymbol{x}]$ and $V_0[\boldsymbol{x}']$ are known but the constant ϵ^2 is unknown. We call $V_0[\boldsymbol{x}]$ and $V_0[\boldsymbol{x}']$ the *normalized covariance matrices* and ϵ the *noise level*. From eq. (6.20), we see that the optimal correction is not affected if the covariance matrices $V[\boldsymbol{x}]$ and $V[\boldsymbol{x}']$ are replaced by the normalized covariance matrices $V_0[\boldsymbol{x}]$ and $V_0[\boldsymbol{x}']$, respectively. The unknown noise level ϵ can be estimated *a posteriori* as follows.

If $V[\boldsymbol{x}]$ and $V[\boldsymbol{x}']$ in eq. (6.22) are replaced by $V_0[\boldsymbol{x}]$ and $V_0[\boldsymbol{x}']$, respectively, the right-hand side is multiplied by $1/\epsilon^2$. Since $\hat{J}$ is a χ^2 variable with one degree of freedom, an unbiased estimator of ϵ^2 is obtained in the following form (see eq. (5.38)):

$$\hat{\epsilon}^2 = \frac{(\boldsymbol{x}, \boldsymbol{G}\boldsymbol{x}')^2}{(\hat{\boldsymbol{x}}', \boldsymbol{G}^\top V_0[\boldsymbol{x}]\boldsymbol{G}\hat{\boldsymbol{x}}') + (\hat{\boldsymbol{x}}, \boldsymbol{G}V_0[\boldsymbol{x}']\boldsymbol{G}^\top \hat{\boldsymbol{x}})}. \tag{6.25}$$

Its expectation and variance are given as follows (see eqs. (5.39)):

$$E[\hat{\epsilon}^2] = \epsilon^2, \qquad V[\hat{\epsilon}^2] = 2\epsilon^4. \tag{6.26}$$

If the value ϵ is given a priori, the χ^2 test (6.23) takes the following form (see eq. (5.40)):

$$\frac{\hat{\epsilon}^2}{\epsilon^2} > \chi^2_{1,a}. \tag{6.27}$$

Example 6.1 If each coordinate is perturbed independently by Gaussian noise of mean 0 and variance ϵ^2, the covariance matrices of $\boldsymbol{x}$ and $\boldsymbol{x}'$ are $V[\boldsymbol{x}] = V[\boldsymbol{x}'] = \epsilon^2 \boldsymbol{P}_{\boldsymbol{k}}$. The optimal correction (6.20) reduces to

$$\Delta\boldsymbol{x} = \frac{(\boldsymbol{x}, \boldsymbol{G}\boldsymbol{x}')\boldsymbol{P}_{\boldsymbol{k}}\boldsymbol{G}\boldsymbol{x}'}{\|\boldsymbol{P}_{\boldsymbol{k}}\boldsymbol{G}^\top\boldsymbol{x}\|^2 + \|\boldsymbol{P}_{\boldsymbol{k}}\boldsymbol{G}\boldsymbol{x}'\|^2},$$

$$\Delta\boldsymbol{x}' = \frac{(\boldsymbol{x}, \boldsymbol{G}\boldsymbol{x}')\boldsymbol{P}_{\boldsymbol{k}}\boldsymbol{G}^\top\boldsymbol{x}}{\|\boldsymbol{P}_{\boldsymbol{k}}\boldsymbol{G}^\top\boldsymbol{x}\|^2 + \|\boldsymbol{P}_{\boldsymbol{k}}\boldsymbol{G}\boldsymbol{x}'\|^2}. \tag{6.28}$$

The a posteriori covariance matrices (6.21) become

$$V[\hat{\boldsymbol{x}}] = \epsilon^2 \left(\boldsymbol{P}_{\boldsymbol{k}} - \frac{(\boldsymbol{P}_{\boldsymbol{k}}\boldsymbol{G}\hat{\boldsymbol{x}}')(\boldsymbol{P}_{\boldsymbol{k}}\boldsymbol{G}\hat{\boldsymbol{x}}')^\top}{\|\boldsymbol{P}_{\boldsymbol{k}}\boldsymbol{G}^\top\hat{\boldsymbol{x}}\|^2 + \|\boldsymbol{P}_{\boldsymbol{k}}\boldsymbol{G}\hat{\boldsymbol{x}}'\|^2} \right),$$

$$V[\hat{\boldsymbol{x}}'] = \epsilon^2 \left(\boldsymbol{P_k} - \frac{(\boldsymbol{P_k}\boldsymbol{G}^\top \hat{\boldsymbol{x}})(\boldsymbol{P_k}\boldsymbol{G}^\top \hat{\boldsymbol{x}})^\top}{\|\boldsymbol{P_k}\boldsymbol{G}^\top \hat{\boldsymbol{x}}\|^2 + \|\boldsymbol{P_k}\boldsymbol{G}\hat{\boldsymbol{x}}'\|^2} \right),$$

$$V[\hat{\boldsymbol{x}}, \hat{\boldsymbol{x}}'] = -\frac{\epsilon^2 (\boldsymbol{P_k}\boldsymbol{G}\hat{\boldsymbol{x}}')(\boldsymbol{P_k}\boldsymbol{G}^\top \hat{\boldsymbol{x}})^\top}{\|\boldsymbol{P_k}\boldsymbol{G}^\top \hat{\boldsymbol{x}}\|^2 + \|\boldsymbol{P_k}\boldsymbol{G}\hat{\boldsymbol{x}}'\|^2} = V[\hat{\boldsymbol{x}}', \hat{\boldsymbol{x}}]^\top. \tag{6.29}$$

An unbiased estimator of the variance ϵ^2 is obtained in the form

$$\hat{\epsilon}^2 = \frac{(\boldsymbol{x}, \boldsymbol{G}\boldsymbol{x}')^2}{\|\boldsymbol{P_k}\boldsymbol{G}^\top \hat{\boldsymbol{x}}\|^2 + \|\boldsymbol{P_k}\boldsymbol{G}\hat{\boldsymbol{x}}'\|^2}. \tag{6.30}$$

Example 6.2 Consider the parallel stereo system described in Section 6.1.3 with the noise characteristics given in Example 6.1. Let (x, y) and (x', y') be the corresponding image points on the first and the second image planes, respectively. The optimal correction (6.28) reduces to

$$\hat{\boldsymbol{x}} = \begin{pmatrix} (x + x')/2 \\ y \\ 1 \end{pmatrix}, \qquad \hat{\boldsymbol{x}}' = \begin{pmatrix} (x + x')/2 \\ y' \\ 1 \end{pmatrix}. \tag{6.31}$$

The residual (6.22) is simply

$$\hat{J} = \frac{1}{2\epsilon^2}(x - x')^2, \tag{6.32}$$

which gives an unbiased estimator of ϵ^2 in the form

$$\hat{\epsilon}^2 = \frac{1}{2}(x - x')^2. \tag{6.33}$$

The a posteriori covariance matrices (6.29) reduce to

$$V[\hat{\boldsymbol{x}}] = V[\hat{\boldsymbol{x}}'] = \epsilon^2 \begin{pmatrix} 1/2 & & \\ & 1 & \\ & & 0 \end{pmatrix}, \quad V[\hat{\boldsymbol{x}}, \hat{\boldsymbol{x}}'] = \epsilon^2 \begin{pmatrix} 1/2 & & \\ & 0 & \\ & & 0 \end{pmatrix}. \tag{6.34}$$

6.3 3-D Reconstruction of Points

6.3.1 Depth reconstruction

Let $\hat{\boldsymbol{x}}$ and $\hat{\boldsymbol{x}}'$ be the corrected positions of the corresponding image points on the first and the second image planes, respectively. Since the epipolar equations $(\hat{\boldsymbol{x}}, \boldsymbol{G}\hat{\boldsymbol{x}}') = 0$ is satisfied, there exist constants Z and Z' such that

$$Z\hat{\boldsymbol{x}} = \boldsymbol{h} + Z'\boldsymbol{R}\hat{\boldsymbol{x}}'. \tag{6.35}$$

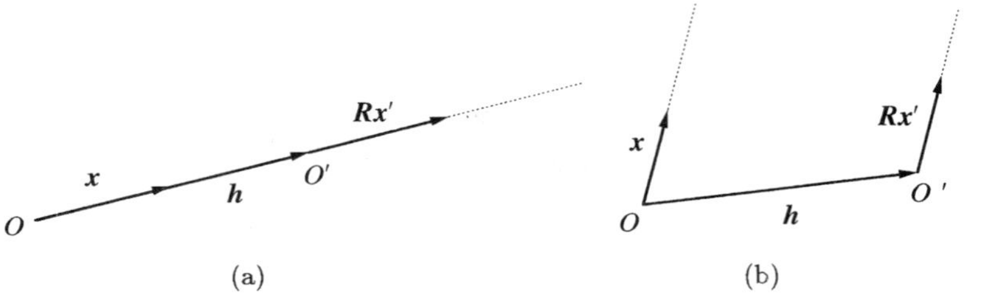

Fig. 6.5. (a) The feature point is infinitely far away. (b) The feature point is in the direction of the base-line vector **h**.

We call Z and Z' the *depths* of $\boldsymbol{x}$ and $\boldsymbol{x}'$, respectively; they are the distances of the reconstructed space point P from the XY plane of the first camera and from the $X'Y'$ plane of the second camera, respectively (see eq. (6.1)). Taking the vector product with $\boldsymbol{R}\hat{\boldsymbol{x}}'$ on both sides of eq. (6.35), we obtain

$$Z\hat{\boldsymbol{x}} \times \boldsymbol{R}\hat{\boldsymbol{x}}' = \boldsymbol{h} \times \boldsymbol{R}\hat{\boldsymbol{x}}'. \tag{6.36}$$

Taking the vector product with $\hat{\boldsymbol{x}}$ on both sides of eq. (6.35), we obtain

$$\boldsymbol{0} = \hat{\boldsymbol{x}} \times \boldsymbol{h} + Z'\hat{\boldsymbol{x}} \times \boldsymbol{R}\hat{\boldsymbol{x}}'. \tag{6.37}$$

From these, we obtain the depths Z and Z' in the following form[4]:

$$Z = \frac{(\boldsymbol{h} \times \boldsymbol{R}\hat{\boldsymbol{x}}', \hat{\boldsymbol{x}} \times \boldsymbol{R}\hat{\boldsymbol{x}}')}{\|\hat{\boldsymbol{x}} \times \boldsymbol{R}\hat{\boldsymbol{x}}'\|^2}, \qquad Z' = \frac{(\boldsymbol{h} \times \hat{\boldsymbol{x}}, \hat{\boldsymbol{x}} \times \boldsymbol{R}\hat{\boldsymbol{x}}')}{\|\hat{\boldsymbol{x}} \times \boldsymbol{R}\hat{\boldsymbol{x}}'\|^2}. \tag{6.38}$$

The corresponding space point is given by

$$\boldsymbol{r} = Z\hat{\boldsymbol{x}}. \tag{6.39}$$

Eqs. (6.38) have a *singularity*: the depths Z and Z' cannot be determined when (and only when) $\hat{\boldsymbol{x}} \times \boldsymbol{R}\hat{\boldsymbol{x}}' = \boldsymbol{0}$. This configuration occurs in two cases:

- The space point P is infinitely far away from the two cameras (Fig. 6.5a).
- The space point P and the two viewpoints O and O' are *collinear*, i.e., the space point P is projected onto the *epipoles* of the two image planes (Fig. 6.5b).

---

[4]The depths Z and Z' can be equivalently written in various different forms such as $Z = \|\mathbf{h} \times \mathbf{R}\hat{\mathbf{x}}'\|/\|\hat{\mathbf{x}} \times \mathbf{R}\hat{\mathbf{x}}'\|$ and $Z' = \|\mathbf{h} \times \hat{\mathbf{x}}\|/\|\hat{\mathbf{x}} \times \mathbf{R}\hat{\mathbf{x}}'\|$. Among them, eqs. (6.38) are the most convenient. For one thing, this form makes sense even if the depth is negative (i.e., the object is *behind* the camera).

Example 6.3 For the parallel stereo system described in Section 6.1.3, the space point $\boldsymbol{r}$ reconstructed from corresponding image points (x, y) and (x', y') has the depths

$$Z = Z' = \frac{h}{D}, \tag{6.40}$$

where $D = y - y'$ is called the *disparity* of the two image points.

6.3.2 Error behavior of reconstructed space points

The depths Z and Z' are defined in such a way that eq. (6.35) holds. If $\hat{\boldsymbol{x}}$ and $\hat{\boldsymbol{x}}'$ are perturbed by $\Delta\hat{\boldsymbol{x}}$ and $\Delta\hat{\boldsymbol{x}}'$, respectively, the resulting perturbations ΔZ and $\Delta Z'$ of Z and Z' satisfy to a first approximation

$$\Delta Z\hat{\boldsymbol{x}} = \Delta Z'\boldsymbol{R}\hat{\boldsymbol{x}}' - Z\Delta\hat{\boldsymbol{x}} + Z'\boldsymbol{R}\Delta\hat{\boldsymbol{x}}'. \tag{6.41}$$

Taking the vector product with $\boldsymbol{R}\hat{\boldsymbol{x}}'$ on both sides, we obtain

$$\Delta Z(\boldsymbol{R}\hat{\boldsymbol{x}}') \times \hat{\boldsymbol{x}} = -Z(\boldsymbol{R}\hat{\boldsymbol{x}}') \times \Delta\hat{\boldsymbol{x}} + Z'(\boldsymbol{R}\hat{\boldsymbol{x}}') \times \boldsymbol{R}\Delta\hat{\boldsymbol{x}}'. \tag{6.42}$$

The unit surface normal to the epipolar plane, on which $\hat{\boldsymbol{x}}$, $\boldsymbol{h}$, and $\boldsymbol{R}\hat{\boldsymbol{x}}'$ lie, is given by

$$\hat{\boldsymbol{n}} = N[\boldsymbol{h} \times \hat{\boldsymbol{x}}]. \tag{6.43}$$

Taking the inner product with $\hat{\boldsymbol{n}}$ on both sides of eq. (6.42), we obtain

$$\Delta Z(\hat{\boldsymbol{n}}, (\boldsymbol{R}\hat{\boldsymbol{x}}') \times \hat{\boldsymbol{x}}) = -Z(\hat{\boldsymbol{n}}, (\boldsymbol{R}\hat{\boldsymbol{x}}') \times \Delta\hat{\boldsymbol{x}}) + Z'(\hat{\boldsymbol{n}}, (\boldsymbol{R}\hat{\boldsymbol{x}}') \times \boldsymbol{R}\Delta\hat{\boldsymbol{x}}'). \tag{6.44}$$

If we define

$$\hat{\boldsymbol{m}} = \hat{\boldsymbol{n}} \times \boldsymbol{R}\hat{\boldsymbol{x}}', \tag{6.45}$$

eq. (6.44) can be written as

$$\Delta Z = -\frac{(\hat{\boldsymbol{m}}, Z\Delta\hat{\boldsymbol{x}} - Z'\boldsymbol{R}\Delta\hat{\boldsymbol{x}}')}{(\hat{\boldsymbol{m}}, \hat{\boldsymbol{x}})}. \tag{6.46}$$

Using the relationship

$$|(\hat{\boldsymbol{m}}, \boldsymbol{x})| = |(\boldsymbol{n}, \hat{\boldsymbol{x}} \times \boldsymbol{R}\hat{\boldsymbol{x}}')| = \|\hat{\boldsymbol{x}} \times \boldsymbol{R}\hat{\boldsymbol{x}}'\|, \tag{6.47}$$

and noting that $E[\Delta\hat{\boldsymbol{x}}\Delta\hat{\boldsymbol{x}}^\top] = V[\hat{\boldsymbol{x}}]$, $E[\Delta\hat{\boldsymbol{x}}\Delta\hat{\boldsymbol{x}}'^\top] = V[\hat{\boldsymbol{x}}, \hat{\boldsymbol{x}}']$, and $E[\Delta\hat{\boldsymbol{x}}'\Delta\hat{\boldsymbol{x}}'^\top] = V[\hat{\boldsymbol{x}}']$ (see eqs. (6.21)), we obtain the variance $V[Z] = E[(\Delta Z)^2]$ of Z from eq. (6.46) in the following form:

$$V[Z] = \frac{Z^2(\hat{\boldsymbol{m}}, V[\hat{\boldsymbol{x}}]\hat{\boldsymbol{m}}) - 2ZZ'(\hat{\boldsymbol{m}}, V[\hat{\boldsymbol{x}}, \hat{\boldsymbol{x}}']\boldsymbol{R}^\top\hat{\boldsymbol{m}}) + Z'^2(\hat{\boldsymbol{m}}, \boldsymbol{R}V[\hat{\boldsymbol{x}}']\boldsymbol{R}^\top\hat{\boldsymbol{m}})}{\|\hat{\boldsymbol{x}} \times \boldsymbol{R}\hat{\boldsymbol{x}}'\|^2}. \tag{6.48}$$

The covariance vector $V[\hat{\boldsymbol{x}}, Z] = E[\Delta\hat{\boldsymbol{x}}\Delta Z]$ has the form

$$V[\hat{\boldsymbol{x}}, Z] = -\frac{(ZV[\hat{\boldsymbol{x}}] - Z'V[\hat{\boldsymbol{x}}, \hat{\boldsymbol{x}}']\boldsymbol{R}^\top)\hat{\boldsymbol{m}}}{(\hat{\boldsymbol{m}}, \hat{\boldsymbol{x}})}. \tag{6.49}$$

The covariance matrix of the reconstructed space point $\boldsymbol{r} = Z\hat{\boldsymbol{x}}$ is given by

$$V[\boldsymbol{r}] = Z^2 V[\hat{\boldsymbol{x}}] + 2ZS[V[\hat{\boldsymbol{x}}, Z]\hat{\boldsymbol{x}}^\top] + V[Z]\hat{\boldsymbol{x}}\hat{\boldsymbol{x}}^\top, \tag{6.50}$$

where $S[\,\cdot\,]$ is the symmetrization operator (see eqs. (2.205)).

Example 6.4 Consider the parallel stereo system described in Section 6.1.3 with the noise characteristics given in Example 6.1. Eqs. (6.48) and (6.49) reduce to

$$V[Z] = \frac{2\epsilon^2 Z^4}{h^2}, \qquad V[\hat{\boldsymbol{x}}, Z] = -\frac{\epsilon^2 Z^2}{h}\boldsymbol{j}, \tag{6.51}$$

where $\boldsymbol{j} = (0, 1, 0)^\top$. Put

$$\boldsymbol{x}^* = \frac{1}{2}(\hat{\boldsymbol{x}} + \hat{\boldsymbol{x}}'), \tag{6.52}$$

which represents the midpoint of $\hat{\boldsymbol{x}}$ and $\hat{\boldsymbol{x}}'$ if the two image planes are identified. The covariance matrix $V[\boldsymbol{r}]$ of the reconstructed space point $\boldsymbol{r}$ given by eq. (6.50) reduces to

$$V[\boldsymbol{r}] = \frac{\epsilon^2 Z^2}{2}\left(\boldsymbol{P}_{\boldsymbol{k}} + \frac{4}{D^2}\boldsymbol{x}^*\boldsymbol{x}^{*\top}\right), \tag{6.53}$$

where $D = y - y'$ is the disparity. It is geometrically evident that the error in $\boldsymbol{r}$ is proportional to ϵ and Z. Eq. (6.53) also implies that the error in orientation $\boldsymbol{x}^*$ is very large. If $\boldsymbol{x}^* \approx \boldsymbol{k}$, for instance, the error is approximately isotropic around the Z-axis but its magnitude along the Z-axis is about $2/D$ times as large as in the direction orthogonal to it. If the focal length of the camera is 700 pixels and the disparity is 10 pixels, for example, we have $2/D = 140$. If the disparity is 1 pixel, we have $2/D = 1400$. Thus, the uncertainty of the Z coordinate is very large as compared with that of the X and Y coordinates.

Example 6.5 Fig. 6.6 are simulated stereo images (480×680 pixels with focal length $f = 600$ (pixels)) of a cylindrical grid. Gaussian noise of mean 0 and standard deviation $\sigma = 2$ (pixels) is independently added to the x- and y-coordinates of each grid point, so the noise level is $\epsilon = \sigma/f = 1/300$. However, the value of ϵ is treated as unknown in the 3-D reconstruction computation and estimated a posteriori by using eq. (6.30) for each grid point. Fig. 6.7 shows the *standard confidence regions* (see Section 4.5.3) of the grid points computed from eq. (6.50).

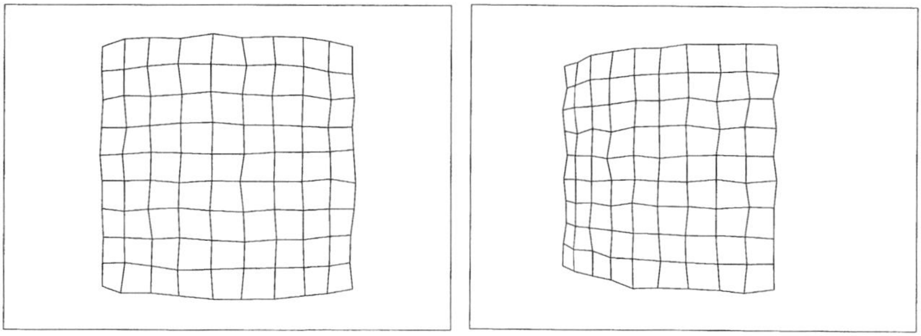

Fig. 6.6. Simulated stereo images.

Example 6.6 Fig. 6.8 are real stereo images (512 × 512 pixels with focal length $f = 1870$ (pixels)) of a surface of a building. Fig. 6.9a shows a grid pattern defined by feature points (corners of the windows) extracted from the left image in Fig. 6.8. Assuming that the image noise is isotropic, we estimate the noise level by eq. (6.30). Fig. 6.9b displays the reconstructed grid and the two shapes that envelop the *primary deviation pairs* of the grid points (see Section 4.5.3). In this example, the base-line length $\|\boldsymbol{h}\|$ is very short as compared with the distance to the surface (approximately 1/16), so a very small error in camera calibration sensitively affects the computation, thereby reducing the reliability of 3-D reconstruction. Since the noise level is estimated from "the degree to which the epipolar equation is not satisfied", the error in the motion parameters is also treated as "image noise".

6.3.3 Mahalanobis distance in the scene

If a particular stereo system with known motion parameters $\{\boldsymbol{h}, \boldsymbol{R}\}$ is fixed and particular characteristics of the image noise are assumed, eq. (6.50) can be viewed as defining an *uncertainty field* over the 3-D scene. Given an arbitrary space point $\boldsymbol{r}$, its depths Z and Z' are computed by

$$Z = (\boldsymbol{k}, \boldsymbol{r}), \qquad Z' = (\boldsymbol{R}\boldsymbol{k}, \boldsymbol{r} - \boldsymbol{h}). \tag{6.54}$$

The corresponding image points are given by

$$\boldsymbol{x} = \frac{1}{Z}\boldsymbol{r}, \qquad \boldsymbol{x}' = \frac{1}{Z'}\boldsymbol{R}^\top(\boldsymbol{r} - \boldsymbol{h}). \tag{6.55}$$

By construction, the epipolar equation $|\boldsymbol{x}, \boldsymbol{h}, \boldsymbol{R}\boldsymbol{x}'| = 0$ is satisfied. Identifying $\boldsymbol{x}$ and $\boldsymbol{x}'$ with $\hat{\boldsymbol{x}}$ and $\hat{\boldsymbol{x}}'$, we can compute the covariance matrix $V[\boldsymbol{r}]$ from eqs. (6.21), (6.48), (6.49), and (6.50). Hence, $V[\boldsymbol{r}]$ is a function[5] of $\boldsymbol{r}$, and

[5]As we noted earlier, singularities exist along the base line OO'. In reality, 3-D cannot be reconstructed in the part outside either of the views of the two cameras, including the

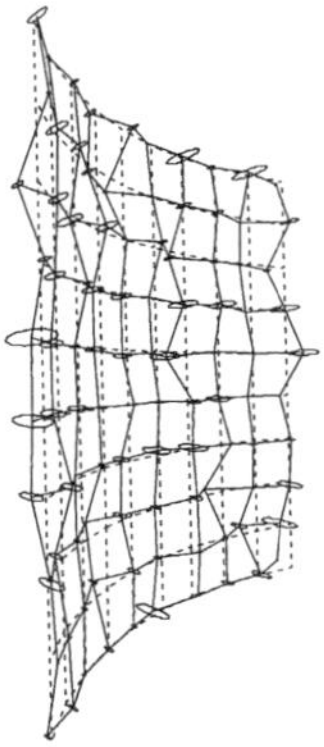

Fig. 6.7. 3-D reconstruction and standard confidence regions of grid points.

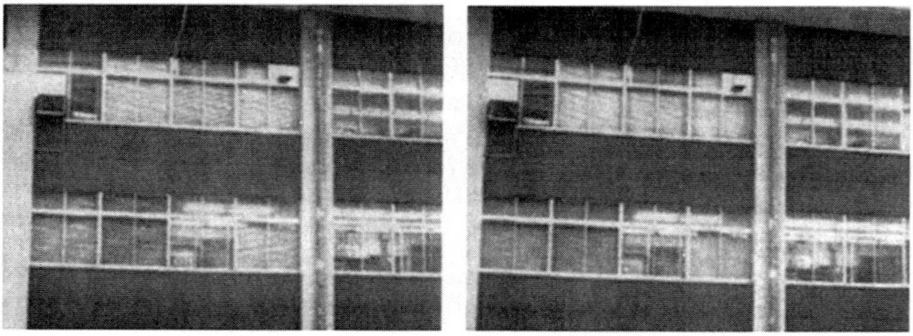

Fig. 6.8. Real stereo images.

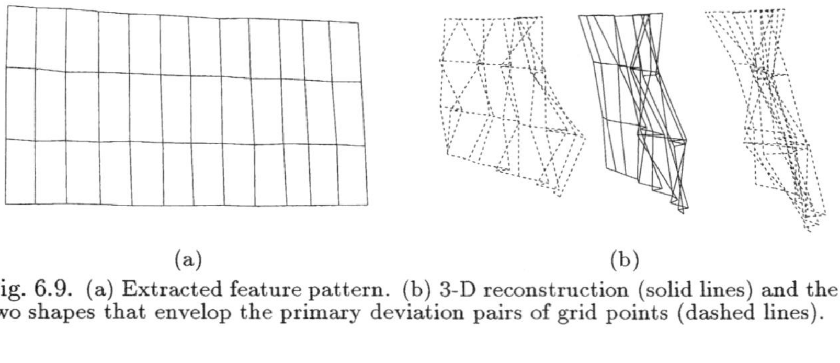

(a) (b)

Fig. 6.9. (a) Extracted feature pattern. (b) 3-D reconstruction (solid lines) and the two shapes that envelop the primary deviation pairs of grid points (dashed lines).

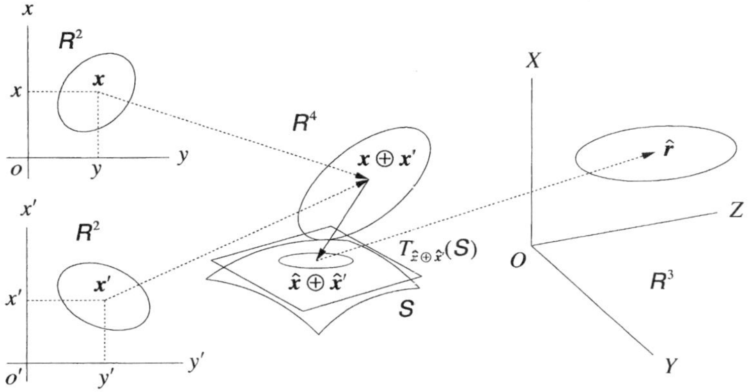

Fig. 6.10. Geometry of 3-D reconstruction by stereo.

it defines the *Mahalanobis distance*[6] that measures the degree of uncertainty in the scene; a perturbation in the same Mahalanobis distance has the same probability density (see Section 3.3.4). The following geometric interpretation is given to this distance.

An image plane can be identified with a two-dimensional Cartesian space $\mathcal{R}^2$, and the direct sum $\boldsymbol{x} \oplus \boldsymbol{x}'$ can be viewed as a point in the *four*-dimensional direct sum Cartesian space $\mathcal{R}^4 = \mathcal{R}^2 \oplus \mathcal{R}^2$. The set of all $\boldsymbol{x} \oplus \boldsymbol{x}'$ that satisfy the epipolar equation (6.4) defines a three-dimensional manifold $\mathcal{S}$ in $\mathcal{R}^4$. The 3-D reconstruction by eqs. (6.38) can be thought of as establishing a one-to-one and onto mapping from the manifold $\mathcal{S}$ to the scene $\mathcal{R}^3$. Fig. 6.10 schematically illustrates these relations, where the ellipses indicate the standard confidence regions defined by the covariance matrices.

The covariance matrices $V[\boldsymbol{x}]$ and $V[\boldsymbol{x}']$ define the Mahalanobis distances that measure the uncertainty of $\boldsymbol{x}$ and $\boldsymbol{x}'$ in the two images. The optimization (6.18) means projecting each direct sum point $\boldsymbol{x} \oplus \boldsymbol{x}' \in \mathcal{R}^4 = \mathcal{R}^2 \times \mathcal{R}^2$ onto the "nearest point" $\hat{\boldsymbol{x}} \oplus \hat{\boldsymbol{x}}'$ in $\mathcal{S}$ measured in the Mahalanobis distance defined by the direct sum covariance matrix $V[\boldsymbol{x}] \oplus V[\boldsymbol{x}']$. Eqs. (6.20) describe this projection, and eqs. (6.21) define the standard confidence region of $\hat{\boldsymbol{x}} \oplus \hat{\boldsymbol{x}}'$ in the *tangent space* $T_{\hat{\boldsymbol{x}} \oplus \hat{\boldsymbol{x}}'}(\mathcal{S})$ to $\mathcal{S}$ at $\hat{\boldsymbol{x}} \oplus \hat{\boldsymbol{x}}'$. Eq. (6.50) can be viewed as defining the Mahalanobis distance in $\mathcal{R}^3$ by mapping[7] the Mahalanobis distance in $\mathcal{S}$ in accordance with the 3-D reconstruction equation $\boldsymbol{r} = Z\hat{\boldsymbol{x}}$;

part behind the cameras, but mathematically the value of $V[\mathbf{r}]$ is defined everywhere except at singularities.

[6]This distance is generally *Riemannian* (*non-Euclidean*) and defines a nonzero *Riemannian curvature* in the scene. Hence, the 3-D scene can be regarded as a *Riemannian space*.

[7]Mathematically, this process of projecting the distance in $\mathcal{R}^4$ onto $\mathcal{S}$ and mapping the distance in $\mathcal{S}$ to $\mathcal{R}^3$ is defined by a procedure called *pull-back* of a tensor field.

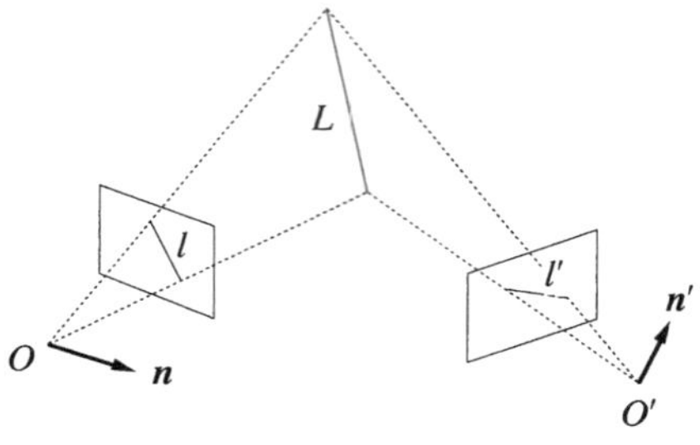

Fig. 6.11. 3-D reconstruction of a space line.

the standard confidence region of $\hat{x} \oplus \hat{x}'$ in $T_{\hat{x} \oplus \hat{x}'}(\mathcal{S})$ is mapped to define the standard confidence region of r in $\mathcal{R}^3$.

6.4 3-D Reconstruction of Lines

6.4.1 Line reconstruction

Image lines are also very important image features. As we noted in Section 6.2.1, they are usually detected as *edge segments*, i.e., sequences of pixels. If straight lines are fitted to them (the fitting procedure will be discussed in subsequent chapters), point-to-point correspondence is easily established once line-to-line correspondence is obtained: for a given image point p on one image line l, the corresponding image point p' on the other image line l' is at the intersection[8] of the epipolar of p with l'. By construction, the corresponding image points p and p' satisfy the epipolar equation. Hence, 3-D can be reconstructed pointwise.

However, if the two image lines are represented by $(\boldsymbol{n}, \boldsymbol{x}) = 0$ and $(\boldsymbol{n}', \boldsymbol{x}') = 0$, a space line is directly reconstructed from $\boldsymbol{n}$ and $\boldsymbol{n}'$: all we need to do is compute the intersection of the space plane defined by the viewpoint O and the image line l with the space plane defined by the viewpoint O' and the image line l' (Fig. 6.11). In other words, *the point-to-point correspondence need not be computed.*

Note that a space line $\boldsymbol{r} \times \boldsymbol{p} = \boldsymbol{n}$ is perspectively projected onto the image plane of the first camera as an image line $(\boldsymbol{n}, \boldsymbol{x}) = 0$ (see eq. (4.130)). In order to invert this projection, we must be careful about the scaling of the vector $\boldsymbol{n}$. Recall that we adopted the scale $\|\boldsymbol{n}\| = 1$ in representing an image line (see eqs. (4.8)) while we adopted the scale $\|\boldsymbol{p}\|^2 + \|\boldsymbol{n}\|^2 = 1$ in representing a space line (see eqs. (4.34)). For the convenience of computation, let us temporarily

[8]As pointed out in Section 6.2.1, computational difficulties may arise in the intersection computation if the epipolar is nearly parallel to l'.

adopt the scale $\|\boldsymbol{n}\| = 1$ for the $\{\boldsymbol{p}, \boldsymbol{n}\}$-representation of space lines[9]. Let the two corresponding image lines l and l' be represented by

$$(\boldsymbol{n}, \boldsymbol{x}) = 0, \qquad (\boldsymbol{n}', \boldsymbol{x}') = 0. \tag{6.56}$$

The space plane defined by the first viewpoint O and image line l is

$$(\boldsymbol{n}, \boldsymbol{r}) = 0. \tag{6.57}$$

The space plane defined by the second viewpoint O' and image line l' is

$$(\boldsymbol{R}\boldsymbol{n}', \boldsymbol{r} - \boldsymbol{h}) = 0. \tag{6.58}$$

The intersection of these two space planes define a space line L. Let its equation be $\boldsymbol{r} \times \boldsymbol{p} = \boldsymbol{n}$. The vector $\boldsymbol{p}$ is given as follows (see eq. (4.72)):

$$\boldsymbol{p} = \frac{\boldsymbol{n} \times \boldsymbol{R}\boldsymbol{n}'}{(\boldsymbol{h}, \boldsymbol{R}\boldsymbol{n}')}. \tag{6.59}$$

After $\boldsymbol{p}$ is computed, the vectors $\{\boldsymbol{p}, \boldsymbol{n}\}$ are normalized to $\|\boldsymbol{p}\|^2 + \|\boldsymbol{n}\|^2 = 1$, if necessary.

6.4.2 Error behavior of reconstructed space lines

First, assume the scale $\|\boldsymbol{n}\| = \|\boldsymbol{n}'\| = 1$. If $\boldsymbol{n}$ and $\boldsymbol{n}'$ are perturbed by $\Delta\boldsymbol{n}$ and $\Delta\boldsymbol{n}'$, respectively, the vector $\boldsymbol{p}$ computed by eq. (6.59) is perturbed to a first approximation by

$$\begin{aligned}\Delta\boldsymbol{p} &= \frac{\Delta\boldsymbol{n} \times \boldsymbol{R}\boldsymbol{n}' + \boldsymbol{n} \times \boldsymbol{R}\Delta\boldsymbol{n}'}{(\boldsymbol{h}, \boldsymbol{R}\boldsymbol{n}')} - \frac{(\boldsymbol{h}, \boldsymbol{R}\Delta\boldsymbol{n}')\boldsymbol{n} \times \boldsymbol{R}\boldsymbol{n}'}{(\boldsymbol{h}, \boldsymbol{R}\boldsymbol{n}')^2} \\ &= -\frac{(\boldsymbol{R}\boldsymbol{n}' \times \boldsymbol{I})\Delta\boldsymbol{n} + (\boldsymbol{p}\boldsymbol{h}^\top \boldsymbol{R} - \boldsymbol{n} \times \boldsymbol{R})\Delta\boldsymbol{n}'}{(\boldsymbol{h}, \boldsymbol{R}\boldsymbol{n}')}.\end{aligned} \tag{6.60}$$

Let $V[\boldsymbol{n}]$ and $V[\boldsymbol{n}']$ be the covariance matrices of $\boldsymbol{n}$ and $\boldsymbol{n}'$, respectively. Assuming that the two image lines are obtained by separately processing the two images, we regard $\boldsymbol{n}$ and $\boldsymbol{n}'$ as independent random variables. The covariance matrices $V[\boldsymbol{p}] = E[\Delta\boldsymbol{p}\Delta\boldsymbol{p}^\top]$ and $V[\boldsymbol{p}, \boldsymbol{n}] = E[\Delta\boldsymbol{p}\Delta\boldsymbol{n}^\top]$ are given as follows (see eq. (2.43)):

$$\begin{aligned}V[\boldsymbol{p}] = \frac{1}{(\boldsymbol{h}, \boldsymbol{R}\boldsymbol{n}')^2}\Big((\boldsymbol{R}\boldsymbol{n}') \times V[\boldsymbol{n}] \times (\boldsymbol{R}\boldsymbol{n}') + (\boldsymbol{h}, \boldsymbol{R}V[\boldsymbol{n}']\boldsymbol{R}^\top\boldsymbol{h})\boldsymbol{p}\boldsymbol{p}^\top \\ - 2S[\boldsymbol{n} \times \boldsymbol{R}V[\boldsymbol{n}']\boldsymbol{R}^\top\boldsymbol{h}\boldsymbol{p}^\top] + \boldsymbol{n} \times \boldsymbol{R}V[\boldsymbol{n}']\boldsymbol{R}^\top \times \boldsymbol{n}\Big),\end{aligned}$$

[9]If $\mathbf{n}$ is scaled to a unit vector, space lines that pass through the coordinate origin O cannot be represented in the form $\mathbf{r} \times \mathbf{p} = \mathbf{n}$. However, such space lines are "invisible" when viewed from O, so we need not consider them for the purpose of 3-D reconstruction.

$$V[\boldsymbol{p}, \boldsymbol{n}] = -\frac{(\boldsymbol{R}\boldsymbol{n}') \times V[\boldsymbol{n}]}{(\boldsymbol{h}, \boldsymbol{R}\boldsymbol{n}')}. \tag{6.61}$$

Once the covariance matrices $V[\boldsymbol{p}]$, $V[\boldsymbol{p}, \boldsymbol{n}]$ $(= V[\boldsymbol{n}, \boldsymbol{p}]^\top)$, and $V[\boldsymbol{n}]$ are obtained for the vectors $\{\boldsymbol{p}, \boldsymbol{n}\}$, $\|\boldsymbol{n}\| = 1$, the corresponding covariance matrices for the rescaled vectors $\{\tilde{\boldsymbol{p}}, \tilde{\boldsymbol{n}}\}$, $\|\tilde{\boldsymbol{p}}\|^2 + \|\tilde{\boldsymbol{n}}\|^2 = 1$, are computed as follows (see eqs. (3.18)):

$$\begin{pmatrix} V[\tilde{\boldsymbol{p}}] & V[\tilde{\boldsymbol{p}}, \tilde{\boldsymbol{n}}] \\ V[\tilde{\boldsymbol{n}}, \tilde{\boldsymbol{p}}] & V[\tilde{\boldsymbol{n}}] \end{pmatrix} = \frac{1}{\|\boldsymbol{p}\|^2 + \|\boldsymbol{n}\|^2} \boldsymbol{P}_{\tilde{\boldsymbol{p}}\oplus\tilde{\boldsymbol{n}}} \begin{pmatrix} V[\boldsymbol{p}] & V[\boldsymbol{p}, \boldsymbol{n}] \\ V[\boldsymbol{n}, \boldsymbol{p}] & V[\boldsymbol{n}] \end{pmatrix} \boldsymbol{P}_{\tilde{\boldsymbol{p}}\oplus\tilde{\boldsymbol{n}}}. \tag{6.62}$$

Here, $\boldsymbol{P}_{\tilde{\boldsymbol{p}}\oplus\tilde{\boldsymbol{n}}}$ is the (66)-projection matrix onto $\{\tilde{\boldsymbol{p}} \oplus \tilde{\boldsymbol{n}}\}_L^\perp$ (see eq. (4.48)).

6.5 Optimal Back Projection onto a Space Plane

6.5.1 *Back projection of a point*

A. Image transformation between the two images

If a feature point is known to be on a space plane Π whose equation is $(\boldsymbol{n}_\Pi, \boldsymbol{r}) = d$, a single image is sufficient to compute their 3D positions: all we need to do is *back project* the image point onto the space plane Π along its line of sight (see Section 4.5.1). In fact, let $\boldsymbol{x}$ be an image point on the first image plane. If the reconstructed space point $\boldsymbol{r} = Z\boldsymbol{x}$ is on space plane $(\boldsymbol{n}_\Pi, \boldsymbol{r}) = d$, we have $(\boldsymbol{n}_\Pi, Z\boldsymbol{x}) = d$. Hence, the depth Z is given by

$$Z = \frac{d}{(\boldsymbol{n}_\Pi, \boldsymbol{x})}, \tag{6.63}$$

and the space point is reconstructed in the position

$$\boldsymbol{r} = \frac{d\boldsymbol{x}}{(\boldsymbol{n}_\Pi, \boldsymbol{x})}. \tag{6.64}$$

It follows that a pair of stereo images have redundant information. We can take advantage of this fact: the reliability of 3D reconstruction can be enhanced by *optimizing this redundancy.*

Suppose there exists no image noise, and consider a stereo system with motion parameters $\{\boldsymbol{h}, \boldsymbol{R}\}$. Let $\boldsymbol{x}$ be an image point in the first image. The corresponding image point $\boldsymbol{x}'$ in the second image satisfies $Z\boldsymbol{x} = \boldsymbol{h} + Z'\boldsymbol{R}\boldsymbol{x}'$ (see eqs. (6.1) and (6.35)). Hence,

$$\begin{aligned} \boldsymbol{x}' &= \frac{Z}{Z'}\boldsymbol{R}^\top\left(\boldsymbol{x} - \frac{1}{Z}\boldsymbol{h}\right) = \frac{Z}{Z'}\boldsymbol{R}^\top\left(\boldsymbol{x} - \frac{(\boldsymbol{n}_\Pi, \boldsymbol{x})\boldsymbol{h}}{d}\right) \\ &= -\frac{Z}{Z'd}\boldsymbol{R}^\top(\boldsymbol{h}\boldsymbol{n}_\Pi^\top - d\boldsymbol{I})\boldsymbol{x}. \end{aligned} \tag{6.65}$$

It follows that the transformation between the two image planes can be written in the form

$$\boldsymbol{x}' = k\boldsymbol{A}\boldsymbol{x}, \tag{6.66}$$

where

$$\boldsymbol{A} = \boldsymbol{R}^\top(\boldsymbol{h}\boldsymbol{n}_\Pi^\top - d\boldsymbol{I}). \tag{6.67}$$

The constant k is chosen so that $(\boldsymbol{k}, \boldsymbol{x}') = 1$ holds. An image transformation in the form of eq. (6.66) is called a (two-dimensional) *projective transformation*[10] or *collineation* (see Sections 4.4.2 and 4.6.1). Note that eq. (6.66) implies the epipolar equation (6.4) and hence is a *stronger* condition than eq. (6.4). In fact, we see from eqs. (6.66) and (6.67) that

$$\begin{aligned}|\boldsymbol{x}, \boldsymbol{h}, \boldsymbol{R}\boldsymbol{x}'| &= k(\boldsymbol{x} \times \boldsymbol{h}, \boldsymbol{R}\boldsymbol{A}\boldsymbol{x}) = k(\boldsymbol{x} \times \boldsymbol{h}, (\boldsymbol{h}\boldsymbol{n}_\Pi^\top - d\boldsymbol{I})\boldsymbol{x}) \\ &= k(\boldsymbol{x} \times \boldsymbol{h}, (\boldsymbol{n}_\Pi, \boldsymbol{x})\boldsymbol{h} - d\boldsymbol{x}) = 0.\end{aligned} \tag{6.68}$$

Example 6.7 Consider the parallel stereo system described in Section 6.1.3. If we write $\boldsymbol{n}_\Pi = (n_{\Pi(1)}, n_{\Pi(2)}, n_{\Pi(3)})^\top$, the transformation matrix $\boldsymbol{A}$ has the form

$$\boldsymbol{A} = \begin{pmatrix} -d & 0 & 0 \\ hn_{\Pi(1)} & hn_{\Pi(2)} - d & hn_{\Pi(3)} \\ 0 & 0 & -d \end{pmatrix}. \tag{6.69}$$

It follows that the constant k in eq. (6.66) should be chosen to be $k = -1/d$. In image coordinates, the transformation given by eq. (6.66) has the following form:

$$\begin{pmatrix} x' \\ y' \end{pmatrix} = \begin{pmatrix} 1 & 0 \\ -hn_{\Pi(1)}/d & 1 - hn_{\Pi(2)}/d \end{pmatrix}\begin{pmatrix} x \\ y \end{pmatrix} + \begin{pmatrix} 0 \\ -hn_{\Pi(3)}/d \end{pmatrix}. \tag{6.70}$$

This is an *affine transformation*. It is easily seen that although the transformation between two images of a space plane is a projective transformation in general, it reduces to an affine transformation *if and only if the two optical axes are parallel and the base-line vector is orthogonal to them*, i.e., $(\boldsymbol{k}, \boldsymbol{h}) = 0$ and $\boldsymbol{R}\boldsymbol{k} = \boldsymbol{k}$. As pointed out in Section 6.1.3, any stereo system can be equivalently treated as a parallel stereo system by applying the projective transformations given by eqs. (6.14) and (6.16) to the two images. It follows that the transformation between two images of a space plane reduces to an affine transformation for any stereo system if an appropriate transformation is applied to each image.

---

[10]The set of all image transformations in the form of eq. (6.66) for an arbitrary nonsingular matrix $\mathbf{A}$ forms a group of two-dimensional *projective transformations* with respect to the composition operation. Note that eq. (4.126) can be written as $\mathbf{Q} = \mathbf{A}^\top\mathbf{Q}'\mathbf{A}/d^2$ in terms of the matrix $\mathbf{A}$ defined in eq. (6.67).

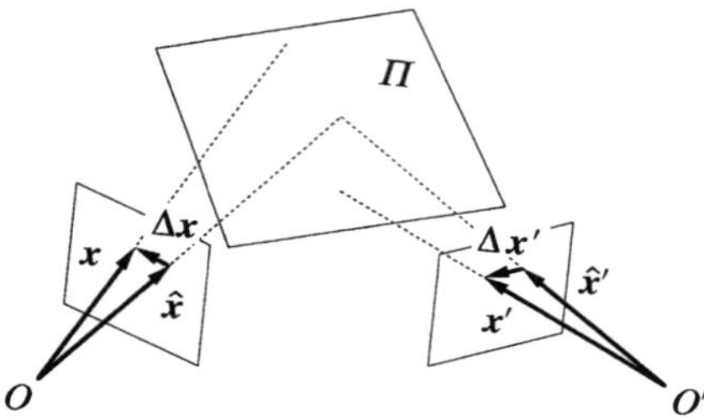

Fig. 6.12. Two corresponding image points are back projected onto a space plane.

B. Optimal correction

In the presence of noise, corresponding image points $\boldsymbol{x}$ and $\boldsymbol{x}'$ do not necessarily satisfy eq. (6.66). Hence, we optimally correct them. In geometric terms, this means modifying $\boldsymbol{x}$ and $\boldsymbol{x}$ so that their lines of sight meet exactly on the space plane Π (Fig. 6.12). This problem is equivalent to finding $\Delta\boldsymbol{x}$ and $\Delta\boldsymbol{x}'$ such that $\bar{\boldsymbol{x}} = \boldsymbol{x} - \Delta\boldsymbol{x}$ and $\bar{\boldsymbol{x}}' = \boldsymbol{x}' - \Delta\boldsymbol{x}'$ satisfy eq. (6.66) or equivalently[11]

$$\bar{\boldsymbol{x}}' \times \boldsymbol{A}\bar{\boldsymbol{x}} = \boldsymbol{0}. \tag{6.71}$$

The rank of this constraint[12] is 2, since only two of the three component equations are algebraically independent[13].

Let $V[\boldsymbol{x}]$ and $V[\boldsymbol{x}']$ be the a priori covariance matrices of $\boldsymbol{x}$ and $\boldsymbol{x}'$, respectively. As discussed in Section 5.1.1, the optimal correction is determined by the optimization

$$J = (\Delta\boldsymbol{x}, V[\boldsymbol{x}]^-\Delta\boldsymbol{x}) + (\Delta\boldsymbol{x}', V[\boldsymbol{x}']^-\Delta\boldsymbol{x}') \to \min \tag{6.72}$$

under the linearized constraint

$$\boldsymbol{x}' \times \boldsymbol{A}\Delta\boldsymbol{x} - (\boldsymbol{A}\boldsymbol{x}) \times \Delta\boldsymbol{x}' = \boldsymbol{x}' \times \boldsymbol{A}\boldsymbol{x},$$

$$\Delta\boldsymbol{x}, \Delta\boldsymbol{x}' \in \{\boldsymbol{k}\}_L^\perp. \tag{6.73}$$

The first order solution is given as follows (see eq. (5.17)):

$$\Delta\boldsymbol{x} = (V[\boldsymbol{x}]\boldsymbol{A}^\top \times \boldsymbol{x}')\boldsymbol{W}(\boldsymbol{x}' \times \boldsymbol{A}\boldsymbol{x}),$$

$$\Delta\boldsymbol{x}' = -(V[\boldsymbol{x}'] \times (\boldsymbol{A}\boldsymbol{x}))\boldsymbol{W}(\boldsymbol{x}' \times \boldsymbol{A}\boldsymbol{x}). \tag{6.74}$$

---

[11] If eq. (6.71) satisfied, there exists a value k such that $\bar{\mathbf{x}}' = k\mathbf{A}\bar{\mathbf{x}}$; it is chosen so that $(\mathbf{k}, \mathbf{x}') = 1$ holds.

[12] This constraint is degenerate (see Section 5.1.3).

[13] The third one can be obtained by multiplying the first and the second ones by $-\bar{x}'$ and $-\bar{y}'$, respectively, and adding them together.

Here, $\boldsymbol{W}$ is a (33)-matrix defined as follows (see Section 5.1.3):

$$\boldsymbol{W} = \left(\boldsymbol{x}' \times \boldsymbol{A}V[\boldsymbol{x}]\boldsymbol{A}^\top \times \boldsymbol{x}' + (\boldsymbol{A}\boldsymbol{x}) \times V[\boldsymbol{x}'] \times (\boldsymbol{A}\boldsymbol{x})\right)_2^- . \tag{6.75}$$

Since the constraint (6.73) is obtained by a linear approximation, corrections $\boldsymbol{x} \leftarrow \boldsymbol{x} - \Delta\boldsymbol{x}$ and $\boldsymbol{x}' \leftarrow \boldsymbol{x}' - \Delta\boldsymbol{x}'$ need to be iterated until the constraint (6.71) is sufficiently satisfied (see Section 5.1.3). The a posteriori covariance matrices of the corrected positions $\hat{\boldsymbol{x}}$ and $\hat{\boldsymbol{x}}'$ are given as follows (see eqs. (5.31) and (5.32)):

$$V[\hat{\boldsymbol{x}}] = V[\boldsymbol{x}] - (V[\boldsymbol{x}]\boldsymbol{A}^\top \times \hat{\boldsymbol{x}}')\hat{\boldsymbol{W}}(V[\boldsymbol{x}]\boldsymbol{A}^\top \times \hat{\boldsymbol{x}}')^\top,$$

$$V[\hat{\boldsymbol{x}}'] = V[\boldsymbol{x}'] - (V[\boldsymbol{x}'] \times (\boldsymbol{A}\hat{\boldsymbol{x}}))\hat{\boldsymbol{W}}(V[\boldsymbol{x}'] \times (\boldsymbol{A}\hat{\boldsymbol{x}}))^\top,$$

$$V[\hat{\boldsymbol{x}}, \hat{\boldsymbol{x}}'] = (V[\boldsymbol{x}]\boldsymbol{A}^\top \times \hat{\boldsymbol{x}}')\hat{\boldsymbol{W}}(V[\boldsymbol{x}'] \times (\boldsymbol{A}\hat{\boldsymbol{x}}))^\top = V[\hat{\boldsymbol{x}}', \hat{\boldsymbol{x}}]^\top. \tag{6.76}$$

Here, the matrix $\hat{\boldsymbol{W}}$ is obtained by replacing $\boldsymbol{x}$ and $\boldsymbol{x}'$ by $\hat{\boldsymbol{x}}$ and $\hat{\boldsymbol{x}}'$, respectively, in eq. (6.75).

The 3-D position $\boldsymbol{r}$ of the back projected point is determined by the depth Z given by eq. (6.63). Its covariance matrix $V[\boldsymbol{r}]$ is computed from eqs. (6.48), (6.49), and (6.50). Since $\boldsymbol{r}$ is constrained to be on Π, its covariance matrix $V[\boldsymbol{r}]$ is singular; the surface normal $\boldsymbol{n}_\Pi$ is its eigenvector for eigenvalue 0, and the remaining eigenvectors lie on Π.

C. Incidence test and noise level estimation

The residual of J given in eq. (6.72) can be written as follows (see eq. (5.34)):

$$\hat{J} = (\boldsymbol{x}' \times \boldsymbol{A}\boldsymbol{x}, \hat{\boldsymbol{W}}(\boldsymbol{x}' \times \boldsymbol{A}\boldsymbol{x})). \tag{6.77}$$

If the noise is Gaussian, the residual $\hat{J}$ is a χ^2 variable with two degrees of freedom (see Section 5.1.5). This fact provides an *incidence test*: the hypothesis that the space point defined by image points $\boldsymbol{x}$ and $\boldsymbol{x}'$ is on the space plane $(\boldsymbol{n}_\Pi, \boldsymbol{r}) = d$ is rejected with significance level $a\%$ if

$$\hat{J} > \chi^2_{2,a}. \tag{6.78}$$

Suppose the covariance matrices $V[\boldsymbol{x}]$ and $V[\boldsymbol{x}']$ are expressed in terms of the normalized covariance matrices $V_0[\boldsymbol{x}]$ and $V_0[\boldsymbol{x}']$ and the noise level ϵ in the form of eqs. (6.24). From eqs. (6.74) and (6.75), we see that the optimal correction is not affected if the covariance matrices $V[\boldsymbol{x}]$ and $V[\boldsymbol{x}']$ are replaced by the normalized covariance matrices $V_0[\boldsymbol{x}]$ and $V_0[\boldsymbol{x}']$, respectively. The unknown noise level ϵ can be estimated *a posteriori* in the form

$$\hat{\epsilon}^2 = \frac{1}{2}(\boldsymbol{x}' \times \boldsymbol{A}\boldsymbol{x}, \hat{\boldsymbol{W}}_0(\boldsymbol{x}' \times \boldsymbol{A}\boldsymbol{x})), \tag{6.79}$$

where

$$\hat{\boldsymbol{W}}_0 = \left(\hat{\boldsymbol{x}}' \times \boldsymbol{A}V_0[\boldsymbol{x}]\boldsymbol{A}^\top \times \hat{\boldsymbol{x}}' + (\boldsymbol{A}\hat{\boldsymbol{x}}) \times V_0[\boldsymbol{x}'] \times (\boldsymbol{A}\hat{\boldsymbol{x}})\right)_2^- . \tag{6.80}$$

The expectation and variance of $\hat{\epsilon}^2$ are given as follows (see eqs. (5.39)):

$$E[\hat{\epsilon}^2] = \epsilon^2, \qquad V[\hat{\epsilon}^2] = \epsilon^4. \tag{6.81}$$

If the value ϵ is given a priori, the χ^2 test (6.78) takes the following form (see eq. (5.40)):

$$\frac{\hat{\epsilon}^2}{\epsilon^2} > \frac{\chi^2_{2,a}}{2}. \tag{6.82}$$

Example 6.8 Consider the parallel stereo system described in Section 6.1.3 with the noise characteristics given in Example 6.1. Suppose a feature point is known to be on space plane $Z = d$. Let (x, y) and (x', y') be its image coordinates on the first and the second image planes, respectively. According to eqs. (6.74) and (6.75), the two image points are respectively corrected into

$$\hat{\boldsymbol{x}} = \begin{pmatrix} (x + x')/2 \\ (y + y' + h/d)/2 \\ 1 \end{pmatrix}, \qquad \hat{\boldsymbol{x}}' = \begin{pmatrix} (x + x')/2 \\ (y + y' - h/d)/2 \\ 1 \end{pmatrix}. \tag{6.83}$$

The a posteriori covariance matrices (6.76) reduce to

$$V[\hat{\boldsymbol{x}}] = V[\hat{\boldsymbol{x}}'] = V[\hat{\boldsymbol{x}}, \hat{\boldsymbol{x}}'] = \frac{\epsilon^2}{2}\boldsymbol{P}_{\boldsymbol{k}}. \tag{6.84}$$

The residual (6.77) is

$$\hat{J} = \frac{1}{2\epsilon^2}\left((x - x')^2 + (y - y' - \frac{h}{d})^2\right). \tag{6.85}$$

Hence, an unbiased estimator of ϵ^2 is obtained in the form

$$\hat{\epsilon}^2 = \frac{1}{4}\left((x - x')^2 + (y - y' - \frac{h}{d})^2\right). \tag{6.86}$$

The depths Z and Z' computed by eqs. (6.38) are simply

$$Z = Z' = d. \tag{6.87}$$

Eq. (6.84) reduces eqs. (6.48) and (6.49) to

$$V[Z] = 0, \qquad V[\hat{\boldsymbol{x}}, Z] = \boldsymbol{0}. \tag{6.88}$$

From eq. (6.50), we see that the covariance matrix of the reconstructed space point $\boldsymbol{r}$ is

$$V[\boldsymbol{r}] = \frac{\epsilon^2 d^2}{2}\boldsymbol{P}_{\boldsymbol{k}}, \tag{6.89}$$

meaning that errors in $\boldsymbol{r}$ are constrained to be on the space plane $Z = 1$, on which the distribution is isotropic with variance $\epsilon d/2$ in each orientation.

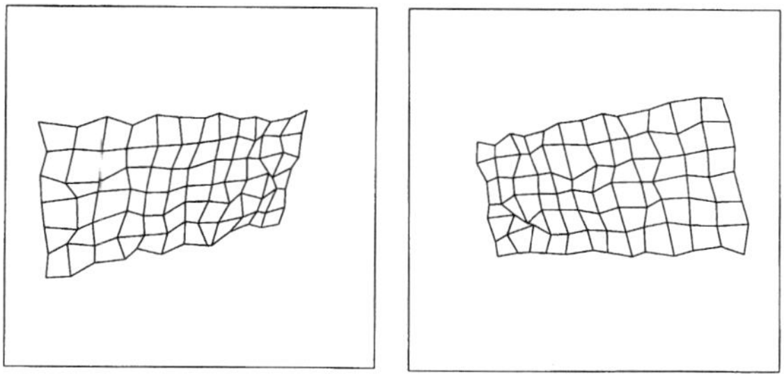

Fig. 6.13. Stereo images of a planar grid.

(a) (b)

Fig. 6.14. (a) Simple back projection. (b) Optimal back projection.

Example 6.9 Fig. 6.13 shows two simulated stereo images (512×512 pixels with focal length $f = 600$ (pixels)) of a planar grid in the scene. The x and y coordinates of each grid point are independently perturbed by Gaussian noise of standard deviation $\sigma = 5$ (pixels), so the noise level is $\epsilon = \sigma/f = 1/120$, which is treated as unknown in the subsequent computation and estimated a posteriori by using eq. (6.79). The equation of the space plane on which the grid lies is assume to be known. Fig. 6.14a shows the back projected pattern computed by eq. (6.64) viewed from a different angle; Fig. 6.14b is the corresponding result obtained after the feature points are optimally corrected. We can clearly see that the correction enhances the accuracy of 3-D reconstruction.

6.5.2 *Back projection of a line*

Suppose we observe a space line which is known to lie on a space plane $(\boldsymbol{n}_\Pi, \boldsymbol{r}) = d$. Let l and l' be its stereo images; let $(\boldsymbol{n}, \boldsymbol{x}) = 0$ and $(\boldsymbol{n}', \boldsymbol{x}') = 0$ be their respective equations. The image point $\boldsymbol{x}'$ that corresponds to an image point $\boldsymbol{x}$ on l must be on l'. If there is no image noise, the corresponding image points $\boldsymbol{x}$ and $\boldsymbol{x}'$ are related by eq. (6.66). Hence, every $\boldsymbol{x}$ that satisfies $(\boldsymbol{n}, \boldsymbol{x})$

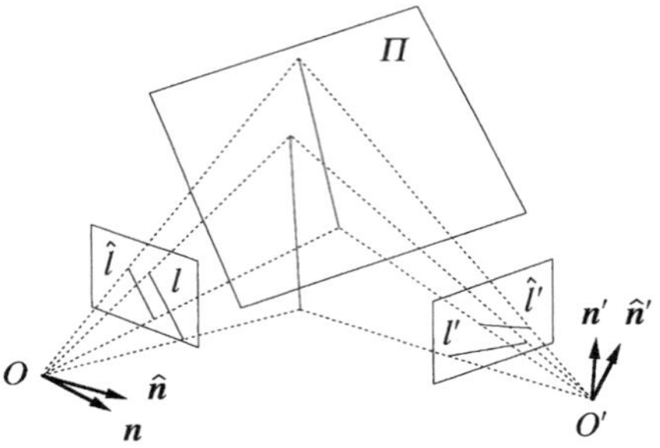

Fig. 6.15. Two corresponding image lines are back projected onto a space plane.

= 0 must satisfy $(\boldsymbol{n}', k\boldsymbol{A}\boldsymbol{x}) = 0$ or $(\boldsymbol{A}^\top \boldsymbol{n}', \boldsymbol{x}) = 0$, which implies[14]

$$\boldsymbol{n} = k'\boldsymbol{A}^\top \boldsymbol{n}', \tag{6.90}$$

where the constant k' is chosen so that $\|k'\boldsymbol{A}^\top \boldsymbol{n}'\| = 1$ holds.

In the presence of noise, corresponding image lines $(\boldsymbol{n}, \boldsymbol{x}) = 0$ and $(\boldsymbol{n}', \boldsymbol{x}')$ $= 0$ do not necessarily satisfy eq. (6.90). Hence, we optimally correct $\boldsymbol{n}$ and $\boldsymbol{n}'$ so that eq. (6.90) is strictly satisfied. In geometric terms, this means modifying the image lines l and l' so that the space plane defined by the viewpoint O and l meet the space plane defined by the viewpoint O' and l' exactly on the space plane Π (Fig. 6.15). This problem is equivalent to finding $\Delta\boldsymbol{n}$ and $\Delta\boldsymbol{n}'$ such that $\bar{\boldsymbol{n}} = \boldsymbol{n} - \Delta\boldsymbol{n}$ and $\bar{\boldsymbol{n}}' = \boldsymbol{n}' - \Delta\boldsymbol{n}'$ satisfy eq. (6.90) or equivalently

$$\bar{\boldsymbol{n}} \times \boldsymbol{A}^\top \bar{\boldsymbol{n}}' = \boldsymbol{0}. \tag{6.91}$$

As in the case of image points, this constraint[15] has rank is 2.

Let $V[\boldsymbol{n}]$ and $V[\boldsymbol{n}']$ be the a priori covariance matrices of $\boldsymbol{n}$ and $\boldsymbol{n}'$, respectively. The problem can be written as the optimization

$$J = (\Delta\boldsymbol{n}, V[\boldsymbol{n}]^- \Delta\boldsymbol{n}) + (\Delta\boldsymbol{n}', V[\boldsymbol{n}']^- \Delta\boldsymbol{n}') \to \min \tag{6.92}$$

under the linearized constraint

$$-\boldsymbol{A}^\top \boldsymbol{n}' \times \Delta\boldsymbol{n} + \boldsymbol{n} \times \boldsymbol{A}^\top \Delta\boldsymbol{n}' = \boldsymbol{n} \times \boldsymbol{A}^\top \boldsymbol{n}',$$

$$\Delta\boldsymbol{n} \in \{\boldsymbol{n}\}_L^\perp, \qquad \Delta\boldsymbol{n}' \in \{\boldsymbol{n}'\}_L^\perp. \tag{6.93}$$

The first order solution is given as follows (see eq. (5.17)):

$$\Delta\boldsymbol{n} = -(V[\boldsymbol{n}] \times (\boldsymbol{A}^\top \boldsymbol{n}'))\boldsymbol{W}(\boldsymbol{n} \times \boldsymbol{A}^\top \boldsymbol{n}'),$$

[14]If we note that $(\mathbf{A}^\top)^{-1} = (\mathbf{A}^{-1})^\top$ (see eqs. (2.21)), we see from eq. (6.90) that $\mathbf{n}'$ $= (\mathbf{A}^{-1})^\top \mathbf{n}/k'$. This is also a *projective transformation* (or *collineation*) of image lines. This transformation is *contragradient* to eq. (6.66). This *duality* between image points and image lines is a fundamental property of *projective geometry*.

[15]This constraint is degenerate (see Section 5.1.3).

$$\Delta \boldsymbol{n}' = (V[\boldsymbol{n}']\boldsymbol{A} \times \boldsymbol{n})\boldsymbol{W}(\boldsymbol{n} \times \boldsymbol{A}^\top \boldsymbol{n}'). \tag{6.94}$$

Here, $\boldsymbol{W}$ is a (33)-matrix defined as follows (see Section 5.1.3):

$$\boldsymbol{W} = \Big((\boldsymbol{A}^\top \boldsymbol{n}') \times V[\boldsymbol{n}] \times (\boldsymbol{A}^\top \boldsymbol{n}') + \boldsymbol{n} \times \boldsymbol{A}^\top V[\boldsymbol{n}']\boldsymbol{A} \times \boldsymbol{n}\Big)_2^- . \tag{6.95}$$

Since the constraint (6.93) is obtained by a linear approximation, corrections $\boldsymbol{n} \leftarrow \boldsymbol{n} + \Delta \boldsymbol{n}$ and $\boldsymbol{n}' \leftarrow \boldsymbol{n}' + \Delta \boldsymbol{n}'$ need to be iterated until the constraint (6.91) is sufficiently satisfied (see Section 5.1.3). The a posteriori covariance matrices of the corrected values $\hat{\boldsymbol{n}}$ and $\hat{\boldsymbol{n}}'$ are given as follows (see eqs. (5.31) and (5.32)):

$$V[\hat{\boldsymbol{n}}] = \hat{V}[\boldsymbol{n}] - (\hat{V}[\boldsymbol{n}] \times (\boldsymbol{A}^\top \hat{\boldsymbol{n}}'))\hat{\boldsymbol{W}}(\hat{V}[\boldsymbol{n}] \times (\boldsymbol{A}^\top \hat{\boldsymbol{n}}'))^\top,$$

$$V[\hat{\boldsymbol{n}}'] = \hat{V}[\boldsymbol{n}'] - (\hat{V}[\boldsymbol{n}']\boldsymbol{A} \times \hat{\boldsymbol{n}})\hat{\boldsymbol{W}}(\hat{V}[\boldsymbol{n}']\boldsymbol{A} \times \hat{\boldsymbol{n}})^\top,$$

$$V[\hat{\boldsymbol{n}}, \hat{\boldsymbol{n}}'] = (\hat{V}[\boldsymbol{n}] \times (\boldsymbol{A}^\top \hat{\boldsymbol{n}}'))\hat{\boldsymbol{W}}(\hat{V}[\boldsymbol{n}']\boldsymbol{A} \times \hat{\boldsymbol{n}})^\top = V[\hat{\boldsymbol{n}}', \hat{\boldsymbol{n}}]^\top. \tag{6.96}$$

Here, we define

$$\hat{V}[\boldsymbol{n}] = \boldsymbol{P}_{\hat{\boldsymbol{n}}} V[\boldsymbol{n}] \boldsymbol{P}_{\hat{\boldsymbol{n}}}, \qquad \hat{V}[\boldsymbol{n}'] = \boldsymbol{P}_{\hat{\boldsymbol{n}}} V[\boldsymbol{n}'] \boldsymbol{P}_{\hat{\boldsymbol{n}}}. \tag{6.97}$$

The matrix $\hat{\boldsymbol{W}}$ is obtained by replacing $\boldsymbol{n}$ and $\boldsymbol{n}'$ by $\hat{\boldsymbol{n}}$ and $\hat{\boldsymbol{n}}'$, respectively, in eq. (6.95). The space line reconstructed by eq. (6.59) lies exactly on the space plane Π. Its covariance matrix $V[\boldsymbol{p} \oplus \boldsymbol{n}]$ is computed from eqs. (6.61).

The residual of J can be written as follows (see eq. (5.34)):

$$\hat{J} = (\hat{\boldsymbol{n}} \times \boldsymbol{A}^\top \hat{\boldsymbol{n}}', \hat{\boldsymbol{W}}(\hat{\boldsymbol{n}} \times \boldsymbol{A}^\top \hat{\boldsymbol{n}}')). \tag{6.98}$$

If noise is Gaussian, the residual $\hat{J}$ is a χ^2 variable with two degrees of freedom (see Section 5.1.5). This fact provides an *incidence test*: the hypothesis that the space line defined by image lines $(\boldsymbol{n}_1, \boldsymbol{x}) = 0$ and $(\boldsymbol{n}_2, \boldsymbol{x}) = 0$ is on the space plane $(\boldsymbol{n}_\Pi, \boldsymbol{r}) = d$ is rejected with significance level $a\%$ if

$$\hat{J} > \chi^2_{2,a}. \tag{6.99}$$

If the noise level in the images is unknown, it can be estimated a posteriori as in the case of image point back projection.

6.6 Scenes Infinitely Far Away

6.6.1 Space points infinitely far away

If $\boldsymbol{x}$ and $\boldsymbol{x}'$ are stereo images of a feature point that belongs to an object located practically infinitely far away (e.g., a mountain, a boat on the sea,

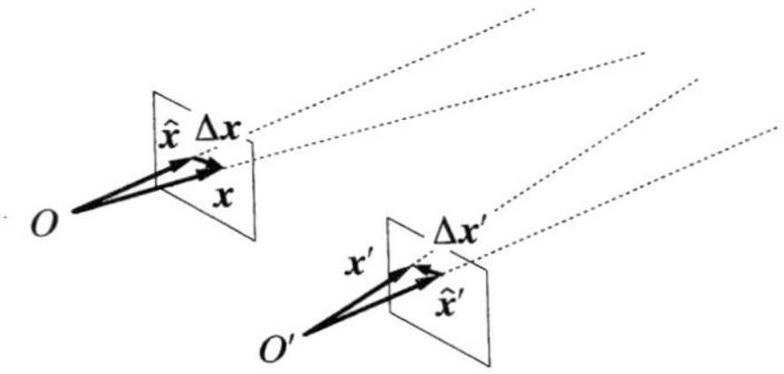

Fig. 6.16. Two corresponding image points are corrected so that their lines of sight become parallel.

or an airplane in the sky), then lines of sight of $\boldsymbol{x}$ and $\boldsymbol{x}'$ should be parallel. This condition can be written as

$$\boldsymbol{x} = k\boldsymbol{R}\boldsymbol{x}' \tag{6.100}$$

for some constant k. This equation may not necessarily be satisfied exactly in the presence of noise (Fig. 6.16). Consider the problem of optimally correcting them: we want to find $\Delta\boldsymbol{x}$ and $\Delta\boldsymbol{x}'$ such that $\bar{\boldsymbol{x}} = \boldsymbol{x} - \Delta\boldsymbol{x}$ and $\bar{\boldsymbol{x}}' = \boldsymbol{x}' - \Delta\boldsymbol{x}'$ satisfy eq. (6.100) or equivalently

$$\bar{\boldsymbol{x}} \times \boldsymbol{R}\bar{\boldsymbol{x}}' = \boldsymbol{0}. \tag{6.101}$$

Like eq. (6.71), this constraint[16] has rank 2.

Let $V[\boldsymbol{x}]$ and $V[\boldsymbol{x}']$ be the a priori covariance matrices of $\boldsymbol{x}$ and $\boldsymbol{x}'$, respectively. As discussed in Section 5.1.1, the optimal correction is determined by the optimization

$$J = (\Delta\boldsymbol{x}, V[\boldsymbol{x}]^{-}\Delta\boldsymbol{x}) + (\Delta\boldsymbol{x}', V[\boldsymbol{x}']^{-}\Delta\boldsymbol{x}') \to \min \tag{6.102}$$

under the linearized constraint

$$\boldsymbol{x} \times \boldsymbol{R}\Delta\boldsymbol{x}' - (\boldsymbol{R}\boldsymbol{x}') \times \Delta\boldsymbol{x} = \boldsymbol{x} \times \boldsymbol{R}\boldsymbol{x}',$$

$$\Delta\boldsymbol{x}, \Delta\boldsymbol{x}' \in \{\boldsymbol{k}\}_L^{\perp}. \tag{6.103}$$

The first order solution is given as follows (see eq. (5.17)):

$$\Delta\boldsymbol{x} = -(V[\boldsymbol{x}] \times (\boldsymbol{R}\boldsymbol{x}'))\boldsymbol{W}(\boldsymbol{x} \times \boldsymbol{R}\boldsymbol{x}'),$$

$$\Delta\boldsymbol{x}' = (V[\boldsymbol{x}']\boldsymbol{R}^{\top} \times \boldsymbol{x})\boldsymbol{W}(\boldsymbol{x} \times \boldsymbol{R}\boldsymbol{x}'). \tag{6.104}$$

Here, $\boldsymbol{W}$ is a (33)-matrix defined as follows:

$$\boldsymbol{W} = \Big((\boldsymbol{R}\boldsymbol{x}') \times V[\boldsymbol{x}] \times (\boldsymbol{R}\boldsymbol{x}') + \boldsymbol{x} \times \boldsymbol{R}V[\boldsymbol{x}']\boldsymbol{R}^{\top} \times \boldsymbol{x}\Big)_2^{-}. \tag{6.105}$$

[16]This constraint is degenerate (see Section 5.1.3).

Since the constraint (6.103) is obtained by a linear approximation, corrections $\boldsymbol{x} \leftarrow \boldsymbol{x} - \Delta\boldsymbol{x}'$ and $\boldsymbol{x}' \leftarrow \boldsymbol{x}' - \Delta\boldsymbol{x}$ need to be iterated until the constraint (6.101) is sufficiently satisfied (see Section 5.1.3).

The a posteriori covariance matrices of the corrected positions $\hat{\boldsymbol{x}}$ and $\hat{\boldsymbol{x}}'$ are given as follows (see eqs. (5.31) and (5.32)):

$$V[\hat{\boldsymbol{x}}] = V[\boldsymbol{x}] - (V[\boldsymbol{x}] \times (\boldsymbol{R}\hat{\boldsymbol{x}}'))\hat{\boldsymbol{W}}(V[\boldsymbol{x}] \times (\boldsymbol{R}\hat{\boldsymbol{x}}'))^\top,$$

$$V[\hat{\boldsymbol{x}}'] = V[\boldsymbol{x}'] - (V[\boldsymbol{x}']\boldsymbol{R}^\top \times \hat{\boldsymbol{x}})\hat{\boldsymbol{W}}(V[\boldsymbol{x}']\boldsymbol{R}^\top \times \hat{\boldsymbol{x}})^\top,$$

$$V[\hat{\boldsymbol{x}}, \hat{\boldsymbol{x}}'] = (V[\boldsymbol{x}] \times (\boldsymbol{R}\hat{\boldsymbol{x}}'))\hat{\boldsymbol{W}}(V[\boldsymbol{x}']\boldsymbol{R}^\top \times \hat{\boldsymbol{x}})^\top = V[\hat{\boldsymbol{x}}', \hat{\boldsymbol{x}}]^\top. \tag{6.106}$$

The matrix $\hat{\boldsymbol{W}}$ is obtained by replacing $\boldsymbol{x}$ and $\boldsymbol{x}'$ by $\hat{\boldsymbol{x}}$ and $\hat{\boldsymbol{x}}'$, respectively, in eq. (6.105).

The residual of J given in eq. (6.102) can be written as follows (see eq. (5.34)):

$$\hat{J} = (\boldsymbol{x} \times \boldsymbol{R}\boldsymbol{x}', \hat{\boldsymbol{W}}(\boldsymbol{x} \times \boldsymbol{R}\boldsymbol{x}')). \tag{6.107}$$

If the noise is Gaussian, the residual $\hat{J}$ is a χ^2 variable with two degrees of freedom (see Section 5.1.5). This fact provides an *infinity test*: the hypothesis that the space point defined by image points $\boldsymbol{x}$ and $\boldsymbol{x}'$ is infinitely far away is rejected with significance level a% if

$$\hat{J} > \chi^2_{2,a}. \tag{6.108}$$

In actual applications of stereo vision, feature points reconstructed to be very far away from the cameras are often disregarded as meaningless, but the judgement as to how far is far enough is usually done ad hoc. The above procedure provides a rigorous statistical criterion.

Suppose the covariance matrices $V[\boldsymbol{x}]$ and $V[\boldsymbol{x}']$ are expressed in terms of the noise level ϵ and the normalized covariance matrices $V_0[\boldsymbol{x}]$ and $V_0[\boldsymbol{x}']$ in the form of eqs. (6.24). From eqs. (6.104) and (6.105), we see that the optimal correction is not affected if the covariance matrices $V[\boldsymbol{x}]$ and $V[\boldsymbol{x}']$ are replaced by the normalized covariance matrices $V_0[\boldsymbol{x}]$ and $V_0[\boldsymbol{x}']$, respectively. The unknown noise level ϵ can be estimated *a posteriori* in the form

$$\hat{\epsilon}^2 = \frac{1}{2}(\boldsymbol{x} \times \boldsymbol{R}\boldsymbol{x}', \hat{\boldsymbol{W}}_0(\boldsymbol{x} \times \boldsymbol{R}\boldsymbol{x}')), \tag{6.109}$$

where

$$\hat{\boldsymbol{W}}_0 = \Big((\boldsymbol{R}\hat{\boldsymbol{x}}') \times V_0[\boldsymbol{x}] \times (\boldsymbol{R}\hat{\boldsymbol{x}}') + \hat{\boldsymbol{x}} \times \boldsymbol{R}V_0[\boldsymbol{x}']\boldsymbol{R}^\top \times \hat{\boldsymbol{x}}\Big)^-_2. \tag{6.110}$$

The expectation and variance of $\hat{\epsilon}^2$ are given as follows (see eqs. (5.39)):

$$E[\hat{\epsilon}^2] = \epsilon^2, \qquad V[\hat{\epsilon}^2] = \epsilon^4. \tag{6.111}$$

If the value ϵ is given a priori, the χ^2 test (6.108) takes the following form (see eq. (5.40)):

$$\frac{\hat{\epsilon}^2}{\epsilon^2} > \frac{\chi^2_{2,a}}{2}. \tag{6.112}$$

Example 6.10 Consider the parallel stereo system described in Section 6.1.3 with the noise characteristics given in Example 6.1. Suppose a feature point is known to be infinitely far away. Let (x, y) and (x', y') be its image coordinates on the first and the second image planes, respectively. Eqs. (6.104) and (6.105) lead to

$$\hat{\boldsymbol{x}} = \hat{\boldsymbol{x}}' = \begin{pmatrix} (x+x')/2 \\ (y+y')/2 \\ 1 \end{pmatrix}. \tag{6.113}$$

Namely, they are corrected into their "midpoint" if the two image planes are identified. The a posteriori covariance matrices (6.106) reduce to

$$V[\hat{\boldsymbol{x}}] = V[\hat{\boldsymbol{x}}'] = V[\hat{\boldsymbol{x}}, \hat{\boldsymbol{x}}'] = \frac{\epsilon^2}{2}\boldsymbol{P_k}. \tag{6.114}$$

The residual (6.107) is

$$\hat{J} = \frac{1}{2\epsilon^2}\left((x-x')^2 + (y-y')^2\right). \tag{6.115}$$

An unbiased estimator of ϵ^2 is obtained in the form

$$\hat{\epsilon}^2 = \frac{1}{4}\left((x-x')^2 + (y-y')^2\right). \tag{6.116}$$

6.6.2 Space lines infinitely far away

Let l and l' be stereo images of a space line located infinitely far away. If $(\boldsymbol{n}, \boldsymbol{x}) = 0$ and $(\boldsymbol{n}', \boldsymbol{x}') = 0$ are their respective equations, the space planes $(\boldsymbol{n}, \boldsymbol{r}) = 0$ and $(\boldsymbol{Rn}', \boldsymbol{r}) = 0$ should be parallel to each other. This condition can be written as

$$\boldsymbol{n} = k\boldsymbol{Rn}' \tag{6.117}$$

for some constant k. This equation may not necessarily be satisfied exactly in the presence of noise (Fig. 6.17). Consider the problem of optimally correcting $\boldsymbol{n}$ and $\boldsymbol{n}'$: we want to find $\Delta\boldsymbol{n}$ and $\Delta\boldsymbol{n}'$ such that $\bar{\boldsymbol{n}} = \boldsymbol{n} - \Delta\boldsymbol{n}$ and $\bar{\boldsymbol{n}}' = \boldsymbol{n}' - \Delta\boldsymbol{n}'$ satisfy eq. (6.117) or equivalently

$$\bar{\boldsymbol{n}} \times \boldsymbol{R}\bar{\boldsymbol{n}}' = \boldsymbol{0}. \tag{6.118}$$

As in the case of image points, this constraint[17] has rank 2.

[17]This constraint is degenerate (see Section 5.1.3).

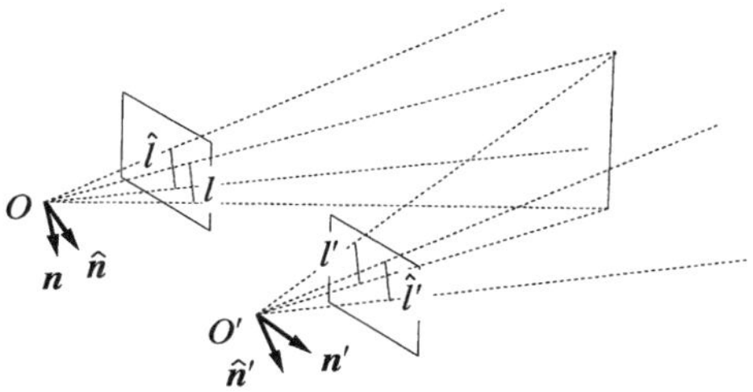

Fig. 6.17. Two corresponding image lines are corrected so as to define a space line located infinitely far away.

Let $V[\boldsymbol{n}]$ and $V[\boldsymbol{n}']$ be the a priori covariance matrices of $\boldsymbol{n}$ and $\boldsymbol{n}'$, respectively. The problem can be written as the optimization

$$J = (\Delta\boldsymbol{n}, V[\boldsymbol{n}]^{-}\Delta\boldsymbol{n}) + (\Delta\boldsymbol{n}', V[\boldsymbol{n}']^{-}\Delta\boldsymbol{n}') \to \min \tag{6.119}$$

under the linearized constraint

$$-\boldsymbol{R}\boldsymbol{n}' \times \Delta\boldsymbol{n} + \boldsymbol{n} \times \boldsymbol{R}\Delta\boldsymbol{n}' = \boldsymbol{n} \times \boldsymbol{R}\boldsymbol{n}',$$
$$\Delta\boldsymbol{n} \in \{\boldsymbol{n}\}_L^{\perp}, \qquad \Delta\boldsymbol{n}' \in \{\boldsymbol{n}'\}_L^{\perp}. \tag{6.120}$$

The first order solution is given as follows (see eq. (5.17)):

$$\Delta\boldsymbol{n} = -(V[\boldsymbol{n}] \times (\boldsymbol{R}\boldsymbol{n}'))\boldsymbol{W}(\boldsymbol{n} \times \boldsymbol{R}\boldsymbol{n}'),$$
$$\Delta\boldsymbol{n}' = (V[\boldsymbol{n}']\boldsymbol{R}^{\top} \times \boldsymbol{n})\boldsymbol{W}(\boldsymbol{n} \times \boldsymbol{R}\boldsymbol{n}'). \tag{6.121}$$

Here, $\boldsymbol{W}$ is a (33)-matrix defined as follows (see Section 5.1.3):

$$\boldsymbol{W} = \Big((\boldsymbol{R}\boldsymbol{n}') \times V[\boldsymbol{n}] \times (\boldsymbol{R}\boldsymbol{n}') + \boldsymbol{n} \times \boldsymbol{R}V[\boldsymbol{n}']\boldsymbol{R}^{\top} \times \boldsymbol{n}\Big)_2^{-}. \tag{6.122}$$

Since the constraint (6.120) is obtained by a linear approximation, corrections $\boldsymbol{n} \leftarrow \boldsymbol{n} + \Delta\boldsymbol{n}$ and $\boldsymbol{n}' \leftarrow \boldsymbol{n}' + \Delta\boldsymbol{n}'$ need to be iterated until the constraint (6.118) is sufficiently satisfied (see Section 5.1.3).

The a posteriori covariance matrices of the corrected values $\hat{\boldsymbol{n}}$ and $\hat{\boldsymbol{n}}'$ are given as follows (see eqs. (5.31) and (5.32)):

$$V[\hat{\boldsymbol{n}}] = \hat{V}[\boldsymbol{n}] - (\hat{V}[\boldsymbol{n}] \times (\boldsymbol{R}\hat{\boldsymbol{n}}'))\hat{\boldsymbol{W}}(\hat{V}[\boldsymbol{n}] \times (\boldsymbol{R}\hat{\boldsymbol{n}}'))^{\top},$$
$$V[\hat{\boldsymbol{n}}'] = \hat{V}[\boldsymbol{n}'] - (\hat{V}[\boldsymbol{n}']\boldsymbol{R}^{\top} \times \hat{\boldsymbol{n}})\hat{\boldsymbol{W}}(\hat{V}[\boldsymbol{n}']\boldsymbol{R}^{\top} \times \hat{\boldsymbol{n}})^{\top},$$
$$V[\hat{\boldsymbol{n}}, \hat{\boldsymbol{n}}'] = (\hat{V}[\boldsymbol{n}] \times (\boldsymbol{R}\hat{\boldsymbol{n}}'))\hat{\boldsymbol{W}}(\hat{V}[\boldsymbol{n}']\boldsymbol{R}^{\top} \times \hat{\boldsymbol{n}})^{\top} = V[\hat{\boldsymbol{n}}', \hat{\boldsymbol{n}}]^{\top}. \tag{6.123}$$

Here, we define

$$\hat{V}[\boldsymbol{n}] = \boldsymbol{P}_{\hat{\boldsymbol{n}}}V[\boldsymbol{n}]\boldsymbol{P}_{\hat{\boldsymbol{n}}}, \qquad \hat{V}[\boldsymbol{n}'] = \boldsymbol{P}_{\hat{\boldsymbol{n}}}V[\boldsymbol{n}']\boldsymbol{P}_{\hat{\boldsymbol{n}}}. \tag{6.124}$$

The matrix $\hat{\boldsymbol{W}}$ is obtained by replacing $\boldsymbol{n}$ and $\boldsymbol{n}'$ by $\hat{\boldsymbol{n}}$ and $\hat{\boldsymbol{n}}'$, respectively, in eq. (6.122).

The residual of J can be written as follows (see eq. (5.34)):

$$\hat{J} = (\hat{\boldsymbol{n}} \times \boldsymbol{R}\hat{\boldsymbol{n}}', \hat{\boldsymbol{W}}(\hat{\boldsymbol{n}} \times \boldsymbol{R}\hat{\boldsymbol{n}}')). \tag{6.125}$$

If noise is Gaussian, the residual $\hat{J}$ is a χ^2 variable with two degrees of freedom (see Section 5.1.5). This fact provides an *infinity test*: the hypothesis that the space line defined by image lines $(\boldsymbol{n}_1, \boldsymbol{x}) = 0$ and $(\boldsymbol{n}_2, \boldsymbol{x}) = 0$ is infinitely far away is rejected with significance level $a\%$ if

$$\hat{J} > \chi^2_{2,a}. \tag{6.126}$$

If the noise level is unknown, it can be estimated a posteriori as in the case of image points.

6.7 Camera Calibration Errors

6.7.1 *Errors in base-line*

So far, our analysis has been based on the assumption that the motion parameters and the focal length are accurately calibrated beforehand. In reality, however, accurately estimating such parameters is very difficult. We now analyze how the accuracy of the camera parameters affects the accuracy of 3-D reconstruction. The effect of calibration errors is very different from the effect of image noise: the statistical characteristics of all space points reconstructed from perturbed camera parameters are *correlated*, while the effect of image noise is generally *independent* from point to point.

For simplicity, errors in the camera parameters are assumed to be independent of image noise. This means that when we analyze the effect of errors in the camera parameters, we can assume that image noise does not exist, since the combined effect of image noise and calibration errors is, to a first approximation, a superimposition of the effect of image noise with accurate calibration and the effect of calibration errors with no image noise.

First, assume that errors in the base-line vector $\boldsymbol{h}$ are independent of errors in other parameters and image noise. For the reason stated above, other parameters and image points can be assumed accurate. Let $V[\boldsymbol{h}] = E[\Delta\boldsymbol{h}\Delta\boldsymbol{h}^\top]$ be the covariance matrix of $\boldsymbol{h}$. Since the image points are assumed correct, we have $\hat{\boldsymbol{x}} = \boldsymbol{x}$ and $\hat{\boldsymbol{x}}' = \boldsymbol{x}'$. If $\boldsymbol{h}$ is perturbed into $\boldsymbol{h} + \Delta\boldsymbol{h}$, the image points $\hat{\boldsymbol{x}}$ and $\hat{\boldsymbol{x}}'$ are corrected into $\hat{\boldsymbol{x}} + \Delta\hat{\boldsymbol{x}}$ and $\hat{\boldsymbol{x}}' + \Delta\hat{\boldsymbol{x}}'$, respectively, so as to satisfy the perturbed epipolar equation for $\boldsymbol{G}' = (\boldsymbol{h} + \Delta\boldsymbol{h}) \times \boldsymbol{R}$. Noting that $(\hat{\boldsymbol{x}}, \boldsymbol{G}\hat{\boldsymbol{x}}') = 0$, we put

$$\Delta\hat{e} = (\hat{\boldsymbol{x}}, \boldsymbol{G}'\hat{\boldsymbol{x}}) = (\hat{\boldsymbol{x}}, \Delta\boldsymbol{h} \times \boldsymbol{R}\hat{\boldsymbol{x}}) = -(\hat{\boldsymbol{x}} \times \boldsymbol{R}\hat{\boldsymbol{x}}, \Delta\boldsymbol{h}) = -(\hat{\boldsymbol{a}}, \Delta\boldsymbol{h}), \tag{6.127}$$

where

$$\hat{\boldsymbol{a}} = \hat{\boldsymbol{x}} \times \boldsymbol{R}\hat{\boldsymbol{x}}'. \tag{6.128}$$

According to eqs. (6.20), the correction of the image points has the form

$$\Delta\hat{\boldsymbol{x}} = -\frac{\Delta\hat{e}V[\boldsymbol{x}]\boldsymbol{G}\hat{\boldsymbol{x}}'}{(\hat{\boldsymbol{x}}', \boldsymbol{G}^\top V[\boldsymbol{x}]\boldsymbol{G}\hat{\boldsymbol{x}}') + (\hat{\boldsymbol{x}}, \boldsymbol{G}V[\boldsymbol{x}']\boldsymbol{G}^\top\hat{\boldsymbol{x}})},$$

$$\Delta\hat{\boldsymbol{x}}' = -\frac{\Delta\hat{e}V[\boldsymbol{x}']\boldsymbol{G}^\top\hat{\boldsymbol{x}}}{(\hat{\boldsymbol{x}}', \boldsymbol{G}^\top V[\boldsymbol{x}]\boldsymbol{G}\hat{\boldsymbol{x}}') + (\hat{\boldsymbol{x}}, \boldsymbol{G}V[\boldsymbol{x}']\boldsymbol{G}^\top\hat{\boldsymbol{x}})}. \tag{6.129}$$

Hence, the a posteriori covariance matrices of the corrected points are given by

$$V[\hat{\boldsymbol{x}}] = \frac{V[\hat{e}](V[\boldsymbol{x}]\boldsymbol{G}\hat{\boldsymbol{x}}')(V[\boldsymbol{x}]\boldsymbol{G}\hat{\boldsymbol{x}}')^\top}{\left((\hat{\boldsymbol{x}}', \boldsymbol{G}^\top V[\boldsymbol{x}]\boldsymbol{G}\hat{\boldsymbol{x}}') + (\hat{\boldsymbol{x}}, \boldsymbol{G}V[\boldsymbol{x}']\boldsymbol{G}^\top\hat{\boldsymbol{x}})\right)^2},$$

$$V[\hat{\boldsymbol{x}}'] = \frac{V[\hat{e}](V[\boldsymbol{x}']\boldsymbol{G}^\top\hat{\boldsymbol{x}})(V[\boldsymbol{x}']\boldsymbol{G}^\top\hat{\boldsymbol{x}})^\top}{\left((\hat{\boldsymbol{x}}', \boldsymbol{G}^\top V[\boldsymbol{x}]\boldsymbol{G}\hat{\boldsymbol{x}}') + (\hat{\boldsymbol{x}}, \boldsymbol{G}V[\boldsymbol{x}']\boldsymbol{G}^\top\hat{\boldsymbol{x}})\right)^2},$$

$$V[\hat{\boldsymbol{x}}, \hat{\boldsymbol{x}}'] = \frac{V[\hat{e}](V[\boldsymbol{x}]\boldsymbol{G}\hat{\boldsymbol{x}}')(V[\boldsymbol{x}']\boldsymbol{G}^\top\hat{\boldsymbol{x}})^\top}{\left((\hat{\boldsymbol{x}}', \boldsymbol{G}^\top V[\boldsymbol{x}]\boldsymbol{G}\hat{\boldsymbol{x}}') + (\hat{\boldsymbol{x}}, \boldsymbol{G}V[\boldsymbol{x}']\boldsymbol{G}^\top\hat{\boldsymbol{x}})\right)^2}, \tag{6.130}$$

where

$$V[\hat{e}] = E[\Delta\hat{e}^2] = (\hat{\boldsymbol{a}}, V[\boldsymbol{h}]\hat{\boldsymbol{a}}). \tag{6.131}$$

We also obtain

$$V[\boldsymbol{h}, \hat{\boldsymbol{x}}] = -\frac{V[\boldsymbol{h}, \hat{e}](V[\boldsymbol{x}]\boldsymbol{G}\hat{\boldsymbol{x}}')^\top}{(\hat{\boldsymbol{x}}', \boldsymbol{G}^\top V[\boldsymbol{x}]\boldsymbol{G}\hat{\boldsymbol{x}}') + (\hat{\boldsymbol{x}}, \boldsymbol{G}V[\boldsymbol{x}']\boldsymbol{G}^\top\hat{\boldsymbol{x}})},$$

$$V[\boldsymbol{h}, \hat{\boldsymbol{x}}'] = -\frac{V[\boldsymbol{h}, \hat{e}](V[\boldsymbol{x}']\boldsymbol{G}^\top\hat{\boldsymbol{x}})^\top}{(\hat{\boldsymbol{x}}', \boldsymbol{G}^\top V[\boldsymbol{x}]\boldsymbol{G}\hat{\boldsymbol{x}}') + (\hat{\boldsymbol{x}}, \boldsymbol{G}V[\boldsymbol{x}']\boldsymbol{G}^\top\hat{\boldsymbol{x}})}, \tag{6.132}$$

where

$$V[\boldsymbol{h}, \hat{e}] = E[\Delta\boldsymbol{h}\Delta\hat{e}] = -V[\boldsymbol{h}]\hat{\boldsymbol{a}}. \tag{6.133}$$

If $\boldsymbol{h}$ in eq. (6.35) is perturbed by $\Delta\boldsymbol{h}$, eq. (6.41) is replaced by

$$\Delta Z\hat{\boldsymbol{x}} = \Delta Z'\boldsymbol{R}\hat{\boldsymbol{x}}' + \Delta\boldsymbol{h} - Z\Delta\hat{\boldsymbol{x}} + Z'\boldsymbol{R}\Delta\hat{\boldsymbol{x}}', \tag{6.134}$$

from which we obtain

$$\Delta Z = -\frac{(\hat{\boldsymbol{m}}, Z\Delta\hat{\boldsymbol{x}} - Z'\boldsymbol{R}\Delta\hat{\boldsymbol{x}}' - \Delta\boldsymbol{h})}{(\hat{\boldsymbol{m}}, \hat{\boldsymbol{x}})}, \tag{6.135}$$

where $\hat{\boldsymbol{m}}$ is defined by eq. (6.45). The variance $V[Z] = E[(\Delta Z)^2]$ of Z is given by

$$V[Z] = \frac{1}{\|\hat{\boldsymbol{x}} \times \boldsymbol{R}\hat{\boldsymbol{x}}'\|^2} \Big(Z^2(\hat{\boldsymbol{m}}, V[\hat{\boldsymbol{x}}]\hat{\boldsymbol{m}}) - 2ZZ'(\hat{\boldsymbol{m}}, V[\hat{\boldsymbol{x}}, \hat{\boldsymbol{x}}']\boldsymbol{R}^\top \hat{\boldsymbol{m}}) + Z'^2(\hat{\boldsymbol{m}}, \boldsymbol{R}V[\hat{\boldsymbol{x}}']\boldsymbol{R}^\top \hat{\boldsymbol{m}}) - 2Z(\hat{\boldsymbol{m}}, V[\hat{\boldsymbol{x}}, \boldsymbol{h}]\hat{\boldsymbol{m}}) + 2Z'(\hat{\boldsymbol{m}}, \boldsymbol{R}V[\hat{\boldsymbol{x}}', \boldsymbol{h}]\hat{\boldsymbol{m}}) + (\hat{\boldsymbol{m}}, V[\boldsymbol{h}]\hat{\boldsymbol{m}}) \Big). \qquad (6.136)$$

The covariance vector $V[\hat{\boldsymbol{x}}, Z] = E[\Delta\hat{\boldsymbol{x}}\Delta Z]$ has the form

$$V[\hat{\boldsymbol{x}}, Z] = -\frac{ZV[\hat{\boldsymbol{x}}]\hat{\boldsymbol{m}} - Z'V[\hat{\boldsymbol{x}}, \hat{\boldsymbol{x}}']\boldsymbol{R}^\top \hat{\boldsymbol{m}} - V[\hat{\boldsymbol{x}}, \boldsymbol{h}]\hat{\boldsymbol{m}}}{(\hat{\boldsymbol{m}}, \hat{\boldsymbol{x}})}. \qquad (6.137)$$

The covariance matrix of the reconstructed space point $\boldsymbol{r} = Z\hat{\boldsymbol{x}}$ is given by eq. (6.50).

Although the above computation involves the a priori covariance matrices $V[\boldsymbol{x}]$ and $V[\boldsymbol{x}']$, the result is invariant to multiplication of them by any positive constant. Hence, the covariance matrices $V[\boldsymbol{x}]$ and $V[\boldsymbol{x}']$ need to be given only up to scale for this analysis.

6.7.2 Errors in camera orientation

Suppose the rotation matrix $\boldsymbol{R}$ that describes the relative orientation of the two cameras is not accurate. If $\boldsymbol{R}$ is perturbed into $\boldsymbol{R} + \Delta\boldsymbol{R}$, both $\boldsymbol{R}$ and $\boldsymbol{R} + \Delta\boldsymbol{R}$ are rotation matrices, so the difference is also a small rotation. Since a small rotation is given in the form of eq. (2.57), there exists a small vector $\Delta\boldsymbol{\Omega}$ such that to a first approximation

$$\Delta\boldsymbol{R} = \Delta\boldsymbol{\Omega} \times \boldsymbol{R}. \qquad (6.138)$$

This means that the second camera is further rotated around axis $\Delta\boldsymbol{\Omega}$ by angle $\|\Delta\boldsymbol{\Omega}\|$, where the vector $\Delta\boldsymbol{\Omega}$ is defined with respect to the first camera coordinate system (see eqs. (2.58) and (3.32)). The covariance matrix of $\boldsymbol{R}$ is defined by $V[\boldsymbol{R}] = E[\Delta\boldsymbol{\Omega}\Delta\boldsymbol{\Omega}^\top]$ (see eq. (3.33)).

We assume that errors in $\boldsymbol{R}$ are independent of errors in other parameters and the image noise. It follows that other parameters and image points can be assumed accurate, so we let $\hat{\boldsymbol{x}} = \boldsymbol{x}$ and $\hat{\boldsymbol{x}}' = \boldsymbol{x}'$. If $\boldsymbol{R}$ is perturbed into $\boldsymbol{R}+\Delta\boldsymbol{R}$, the image points $\hat{\boldsymbol{x}}$ and $\hat{\boldsymbol{x}}'$ are corrected into $\hat{\boldsymbol{x}}+\Delta\hat{\boldsymbol{x}}$ and $\hat{\boldsymbol{x}}'+\Delta\hat{\boldsymbol{x}}'$, respectively, so as to satisfy the perturbed epipolar equation for $\boldsymbol{G}' = \boldsymbol{h} \times (\boldsymbol{R} + \Delta\boldsymbol{\Omega} \times \boldsymbol{R})$. Put

$$\Delta\hat{e} = (\hat{\boldsymbol{x}}, \boldsymbol{G}'\hat{\boldsymbol{x}}) = (\hat{\boldsymbol{x}}, \boldsymbol{h} \times (\Delta\boldsymbol{\Omega} \times \boldsymbol{R}\hat{\boldsymbol{x}})) = (\boldsymbol{h}, \boldsymbol{R}\hat{\boldsymbol{x}}')(\hat{\boldsymbol{x}}, \Delta\boldsymbol{\Omega}) - (\hat{\boldsymbol{x}}, \boldsymbol{R}\hat{\boldsymbol{x}}')(\boldsymbol{h}, \Delta\boldsymbol{\Omega}) = -(\hat{\boldsymbol{b}}, \Delta\boldsymbol{\Omega}), \qquad (6.139)$$

where

$$\hat{\boldsymbol{b}} = (\hat{\boldsymbol{x}}, \boldsymbol{R}\hat{\boldsymbol{x}}')\boldsymbol{h} - (\boldsymbol{h}, \boldsymbol{R}\hat{\boldsymbol{x}}')\hat{\boldsymbol{x}}. \tag{6.140}$$

The corrections $\Delta\hat{\boldsymbol{x}}$ and $\Delta\hat{\boldsymbol{x}}'$ are given in the form of eqs. (6.129). Hence, the a posteriori covariance matrices $V[\hat{\boldsymbol{x}}]$, $V[\hat{\boldsymbol{x}}']$, and $V[\hat{\boldsymbol{x}}, \hat{\boldsymbol{x}}']$ are given in the form of eqs. (6.130), where eq. (6.131) is replaced by

$$V[\hat{e}] = E[\Delta\hat{e}^2] = (\hat{\boldsymbol{b}}, V[\boldsymbol{h}]\hat{\boldsymbol{b}}). \tag{6.141}$$

Similarly, the covariance matrices $V[\boldsymbol{R}, \hat{\boldsymbol{x}}] = E[\Delta\boldsymbol{\Omega}\Delta\hat{\boldsymbol{x}}^\top]$ and $V[\boldsymbol{R}, \hat{\boldsymbol{x}}'] = E[\Delta\boldsymbol{\Omega}\Delta\hat{\boldsymbol{x}}'^\top]$ are given in the form

$$V[\boldsymbol{R}, \hat{\boldsymbol{x}}] = -\frac{V[\boldsymbol{R}, \hat{e}](V[\boldsymbol{x}]\boldsymbol{G}\hat{\boldsymbol{x}}')^\top}{(\hat{\boldsymbol{x}}', \boldsymbol{G}^\top V[\boldsymbol{x}]\boldsymbol{G}\hat{\boldsymbol{x}}') + (\hat{\boldsymbol{x}}, \boldsymbol{G}V[\boldsymbol{x}']\boldsymbol{G}^\top\hat{\boldsymbol{x}})},$$

$$V[\boldsymbol{R}, \hat{\boldsymbol{x}}'] = -\frac{V[\boldsymbol{R}, \hat{e}](V[\boldsymbol{x}']\boldsymbol{G}^\top\hat{\boldsymbol{x}})^\top}{(\hat{\boldsymbol{x}}', \boldsymbol{G}^\top V[\boldsymbol{x}]\boldsymbol{G}\hat{\boldsymbol{x}}') + (\hat{\boldsymbol{x}}, \boldsymbol{G}V[\boldsymbol{x}']\boldsymbol{G}^\top\hat{\boldsymbol{x}})}, \tag{6.142}$$

where

$$V[\boldsymbol{R}, \hat{e}] = E[\Delta\boldsymbol{\Omega}\Delta\hat{e}] = -V[\boldsymbol{R}]\hat{\boldsymbol{b}}. \tag{6.143}$$

If $\boldsymbol{R}$ in eq. (6.35) is perturbed by $\Delta\boldsymbol{R} = \boldsymbol{\Omega} \times \boldsymbol{R}$, eq. (6.41) is replaced by

$$\Delta Z\hat{\boldsymbol{x}} = \Delta Z'\boldsymbol{R}\hat{\boldsymbol{x}}' - Z'(\boldsymbol{R}\hat{\boldsymbol{x}}') \times \Delta\boldsymbol{\Omega} - Z\Delta\hat{\boldsymbol{x}} + Z'\boldsymbol{R}\Delta\hat{\boldsymbol{x}}', \tag{6.144}$$

from which we obtain

$$\Delta Z = -\frac{(\hat{\boldsymbol{m}}, Z\Delta\hat{\boldsymbol{x}} - Z'\boldsymbol{R}\Delta\hat{\boldsymbol{x}}') - Z'\|\hat{\boldsymbol{x}}'\|^2(\hat{\boldsymbol{n}}, \Delta\boldsymbol{\Omega})}{(\hat{\boldsymbol{m}}, \hat{\boldsymbol{x}})}. \tag{6.145}$$

Hence, the variance $V[Z] = E[(\Delta Z)^2]$ of Z is given by

$$V[Z] = \frac{1}{\|\hat{\boldsymbol{x}} \times \boldsymbol{R}\hat{\boldsymbol{x}}'\|^2}\Big(Z^2(\hat{\boldsymbol{m}}, V[\hat{\boldsymbol{x}}]\hat{\boldsymbol{m}}) - 2ZZ'(\hat{\boldsymbol{m}}, V[\hat{\boldsymbol{x}}, \hat{\boldsymbol{x}}']\boldsymbol{R}^\top\hat{\boldsymbol{m}})$$
$$+ Z'^2(\hat{\boldsymbol{m}}, \boldsymbol{R}V[\hat{\boldsymbol{x}}']\boldsymbol{R}^\top\hat{\boldsymbol{m}}) - 2Z'\|\hat{\boldsymbol{x}}'\|^2(Z(\hat{\boldsymbol{m}}, V[\hat{\boldsymbol{x}}, \boldsymbol{R}]\hat{\boldsymbol{n}})$$
$$- Z'(\hat{\boldsymbol{m}}, \boldsymbol{R}V[\hat{\boldsymbol{x}}', \boldsymbol{R}]\hat{\boldsymbol{n}})) + Z'^2\|\hat{\boldsymbol{x}}'\|^4(\hat{\boldsymbol{n}}, V[\boldsymbol{R}]\hat{\boldsymbol{n}})\Big). \tag{6.146}$$

The covariance vector $V[\hat{\boldsymbol{x}}, Z]$ has the form

$$V[\hat{\boldsymbol{x}}, Z] = -\frac{ZV[\hat{\boldsymbol{x}}]\hat{\boldsymbol{m}} - Z'V[\hat{\boldsymbol{x}}, \hat{\boldsymbol{x}}']\boldsymbol{R}^\top\hat{\boldsymbol{m}} - Z'\|\hat{\boldsymbol{x}}'\|^2V[\hat{\boldsymbol{x}}, \boldsymbol{R}]\hat{\boldsymbol{n}}}{(\hat{\boldsymbol{m}}, \hat{\boldsymbol{x}})}. \tag{6.147}$$

The covariance matrix of the reconstructed space point $\boldsymbol{r} = Z\hat{\boldsymbol{x}}$ is given by eq. (6.50). As in the case of base-line vector errors, the covariance matrices $V[\boldsymbol{x}]$ and $V[\boldsymbol{x}']$ need to be given only up to scale for this analysis.

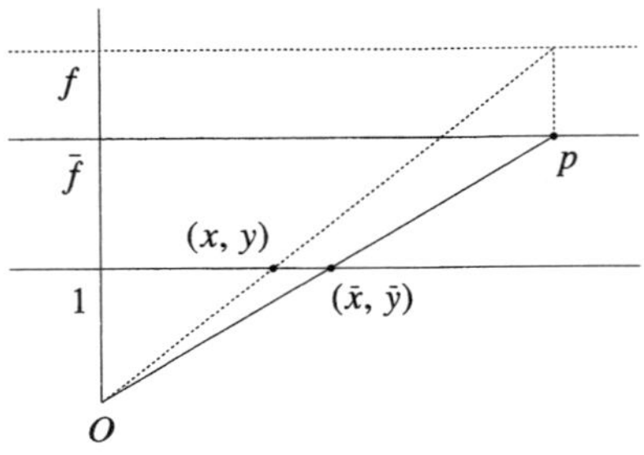

Fig. 6.18. Image coordinates (x, y) are computed with respect to a perturbed focal length f.

6.7.3 Errors in focal length

As mentioned in Section 6.1.1, the *focal length* f is the distance between the image plane and the viewpoint O. In 3-D analysis, we take f as the unit of length (see also Section 4.1.1). However, the focal length f is also difficult to calibrate accurately. Let $\bar{f}$ be the correct focal length. Consider an image point p whose correct image coordinates are $(\bar{x}, \bar{y})$. The physical location of p measured in pixels is $(\bar{f}\bar{x}, \bar{f}\bar{y})$. If (x, y) are the image coordinates of p computed by using a perturbed value $f = \bar{f} + \Delta f$, we have

$$((\bar{f} + \Delta f)x, (\bar{f} + \Delta f)y) = (\bar{f}\bar{x}, \bar{f}\bar{y}), \tag{6.148}$$

since the physical location of the image point should be the same (Fig. 6.18). From eq. (6.148), we obtain to a first approximation

$$\bar{x} = x + \frac{\Delta f}{f}x, \qquad \bar{y} = y + \frac{\Delta f}{f}y. \tag{6.149}$$

In other words, the true location of an image point which appears to be at $\boldsymbol{x}$ is

$$\bar{\boldsymbol{x}} = \boldsymbol{x} + \frac{\Delta f}{f}\boldsymbol{P_k}\boldsymbol{x}. \tag{6.150}$$

A. Independent cameras

Consider the case in which the two cameras have different focal lengths f and f', which are separately calibrated. Then, errors in f and errors in f' are regarded as independent. Suppose f is not accurate but f' is accurate. Other parameters and image points are also assumed accurate. Let $\boldsymbol{x}$ and $\boldsymbol{x}'$ be corresponding image points. Because f is not accurate, $\boldsymbol{x}$ and $\boldsymbol{x}'$ do not exactly satisfy the epipolar equation, but their true locations $\bar{\boldsymbol{x}}$ and $\bar{\boldsymbol{x}}'$ ($= \boldsymbol{x}'$) do. From eq. (6.150), we see that to a first approximation

$$0 = (\bar{\boldsymbol{x}}, \boldsymbol{G}\bar{\boldsymbol{x}}') = (\boldsymbol{x} + \frac{\Delta f}{f}\boldsymbol{P_k}\boldsymbol{x}, \boldsymbol{G}\boldsymbol{x}') = ((\boldsymbol{x}, \boldsymbol{G}\boldsymbol{x}') + \frac{\Delta f}{f}(\boldsymbol{P_k}\boldsymbol{x}, \boldsymbol{G}\boldsymbol{x}'). \quad (6.151)$$

Hence,

$$(\boldsymbol{x}, \boldsymbol{G}\boldsymbol{x}') = -\frac{\Delta f}{f}(\boldsymbol{x}, \boldsymbol{P_k}\boldsymbol{G}\boldsymbol{x}'). \quad (6.152)$$

The image points $\boldsymbol{x}$ and $\boldsymbol{x}'$ are corrected into $\hat{\boldsymbol{x}} = \boldsymbol{x} + \Delta\boldsymbol{x}$ and $\hat{\boldsymbol{x}}' = \boldsymbol{x}' + \Delta\boldsymbol{x}$ so as to satisfy $(\hat{\boldsymbol{x}}, \boldsymbol{G}\hat{\boldsymbol{x}}') = 0$. The corrections $\Delta\boldsymbol{x}$ and $\Delta\boldsymbol{x}'$ are given by eqs. (6.20). Let $\Delta\hat{\boldsymbol{x}} = \hat{\boldsymbol{x}} - \bar{\boldsymbol{x}}$. Then,

$$\Delta\hat{\boldsymbol{x}} = \boldsymbol{x} + \Delta\boldsymbol{x} - \bar{\boldsymbol{x}} = \frac{\Delta f}{f}\boldsymbol{P_k}\boldsymbol{x} + \Delta\boldsymbol{x}$$
$$= \frac{\Delta f}{f}\left(\boldsymbol{P_k}\boldsymbol{x} + \frac{(\boldsymbol{x}, \boldsymbol{P_k}\boldsymbol{G}\boldsymbol{x}')V[\boldsymbol{x}]\boldsymbol{G}\boldsymbol{x}'}{(\boldsymbol{x}', \boldsymbol{G}^\top V[\boldsymbol{x}]\boldsymbol{G}\boldsymbol{x}') + (\boldsymbol{x}, \boldsymbol{G}V[\boldsymbol{x}']\boldsymbol{G}^\top\boldsymbol{x})}\right). \quad (6.153)$$

Similarly,

$$\Delta\hat{\boldsymbol{x}}' = \frac{\Delta f}{f}\frac{(\boldsymbol{x}, \boldsymbol{P_k}\boldsymbol{G}\boldsymbol{x}')V[\boldsymbol{x}']\boldsymbol{G}^\top\boldsymbol{x}}{(\boldsymbol{x}', \boldsymbol{G}^\top V[\boldsymbol{x}]\boldsymbol{G}\boldsymbol{x}') + (\boldsymbol{x}, \boldsymbol{G}V[\boldsymbol{x}']\boldsymbol{G}^\top\boldsymbol{x})}. \quad (6.154)$$

Let $V[f] = E[\Delta f^2]$ be the variance of f. The covariance matrices of the corrected position $\hat{\boldsymbol{x}}$ and $\hat{\boldsymbol{x}}'$ are given as follows:

$$V[\hat{\boldsymbol{x}}] = \frac{V[f]}{f^2}\left(\boldsymbol{P_k}\boldsymbol{x} + \frac{(\boldsymbol{x}, \boldsymbol{P_k}\boldsymbol{G}\boldsymbol{x}')V[\boldsymbol{x}]\boldsymbol{G}\boldsymbol{x}'}{(\boldsymbol{x}', \boldsymbol{G}^\top V[\boldsymbol{x}]\boldsymbol{G}\boldsymbol{x}') + (\boldsymbol{x}, \boldsymbol{G}V[\boldsymbol{x}']\boldsymbol{G}^\top\boldsymbol{x})}\right)$$
$$\left(\boldsymbol{P_k}\boldsymbol{x} + \frac{(\boldsymbol{x}, \boldsymbol{P_k}\boldsymbol{G}\boldsymbol{x}')V[\boldsymbol{x}]\boldsymbol{G}\boldsymbol{x}'}{(\boldsymbol{x}', \boldsymbol{G}^\top V[\boldsymbol{x}]\boldsymbol{G}\boldsymbol{x}') + (\boldsymbol{x}, \boldsymbol{G}V[\boldsymbol{x}']\boldsymbol{G}^\top\boldsymbol{x})}\right)^\top,$$

$$V[\hat{\boldsymbol{x}}'] = \frac{V[f]}{f^2}\frac{(\boldsymbol{x}, \boldsymbol{P_k}\boldsymbol{G}\boldsymbol{x}')^2(V[\boldsymbol{x}']\boldsymbol{G}^\top\boldsymbol{x})(V[\boldsymbol{x}']\boldsymbol{G}^\top\boldsymbol{x})^\top}{\left((\boldsymbol{x}', \boldsymbol{G}^\top V[\boldsymbol{x}]\boldsymbol{G}\boldsymbol{x}') + (\boldsymbol{x}, \boldsymbol{G}V[\boldsymbol{x}']\boldsymbol{G}^\top\boldsymbol{x})\right)},$$

$$V[\hat{\boldsymbol{x}}, \hat{\boldsymbol{x}}'] = \frac{V[f]}{f^2}\left(\boldsymbol{P_k}\boldsymbol{x} + \frac{(\boldsymbol{x}, \boldsymbol{P_k}\boldsymbol{G}\boldsymbol{x}')V[\boldsymbol{x}]\boldsymbol{G}\boldsymbol{x}'}{(\boldsymbol{x}', \boldsymbol{G}^\top V[\boldsymbol{x}]\boldsymbol{G}\boldsymbol{x}') + (\boldsymbol{x}, \boldsymbol{G}V[\boldsymbol{x}']\boldsymbol{G}^\top\boldsymbol{x})}\right)$$
$$\frac{(\boldsymbol{x}, \boldsymbol{P_k}\boldsymbol{G}\boldsymbol{x}')(V[\boldsymbol{x}']\boldsymbol{G}^\top\boldsymbol{x})^\top}{(\boldsymbol{x}', \boldsymbol{G}^\top V[\boldsymbol{x}]\boldsymbol{G}\boldsymbol{x}') + (\boldsymbol{x}, \boldsymbol{G}V[\boldsymbol{x}']\boldsymbol{G}^\top\boldsymbol{x})}. \quad (6.155)$$

The subsequent analysis is the same as in Section 6.3.2. Namely, the covariance matrix of the reconstructed space point $\boldsymbol{r} = Z\hat{\boldsymbol{x}}$ is computed by eqs. (6.48), (6.49), and (6.50). The covariance matrices $V[\boldsymbol{x}]$ and $V[\boldsymbol{x}']$ need to be given only up to scale.

If f is accurate but f' is not, $V[f']$ being its variance, a similar analysis leads to

$$V[\hat{\boldsymbol{x}}] = \frac{V[f']}{f'^2} \frac{(\boldsymbol{x}, \boldsymbol{G}\boldsymbol{P_k}\boldsymbol{x}')^2 (V[\boldsymbol{x}]\boldsymbol{G}\boldsymbol{x}')(V[\boldsymbol{x}]\boldsymbol{G}\boldsymbol{x}')^\top}{\left((\boldsymbol{x}', \boldsymbol{G}^\top V[\boldsymbol{x}]\boldsymbol{G}\boldsymbol{x}') + (\boldsymbol{x}, \boldsymbol{G}V[\boldsymbol{x}']\boldsymbol{G}^\top \boldsymbol{x})\right)^2},$$

$$V[\hat{\boldsymbol{x}}'] = \frac{V[f']}{f'^2} \left(\boldsymbol{P_k}\boldsymbol{x}' + \frac{(\boldsymbol{x}, \boldsymbol{G}\boldsymbol{P_k}\boldsymbol{x}')V[\boldsymbol{x}']\boldsymbol{G}^\top \boldsymbol{x}}{(\boldsymbol{x}', \boldsymbol{G}^\top V[\boldsymbol{x}]\boldsymbol{G}\boldsymbol{x}') + (\boldsymbol{x}, \boldsymbol{G}V[\boldsymbol{x}']\boldsymbol{G}^\top \boldsymbol{x})}\right) \left(\boldsymbol{P_k}\boldsymbol{x}' + \frac{(\boldsymbol{x}, \boldsymbol{G}\boldsymbol{P_k}\boldsymbol{x}')V[\boldsymbol{x}']\boldsymbol{G}^\top \boldsymbol{x}}{(\boldsymbol{x}', \boldsymbol{G}^\top V[\boldsymbol{x}]\boldsymbol{G}\boldsymbol{x}') + (\boldsymbol{x}, \boldsymbol{G}V[\boldsymbol{x}']\boldsymbol{G}^\top \boldsymbol{x})}\right)^\top.$$

$$V[\hat{\boldsymbol{x}}, \hat{\boldsymbol{x}}'] = \frac{V[f']}{f'^2} \frac{(\boldsymbol{x}, \boldsymbol{G}\boldsymbol{P_k}\boldsymbol{x}')V[\boldsymbol{x}]\boldsymbol{G}\boldsymbol{x}'}{(\boldsymbol{x}', \boldsymbol{G}^\top V[\boldsymbol{x}]\boldsymbol{G}\boldsymbol{x}') + (\boldsymbol{x}, \boldsymbol{G}V[\boldsymbol{x}']\boldsymbol{G}^\top \boldsymbol{x})} \left(\boldsymbol{P_k}\boldsymbol{x}' + \frac{(\boldsymbol{x}, \boldsymbol{G}\boldsymbol{P_k}\boldsymbol{x}')V[\boldsymbol{x}']\boldsymbol{G}^\top \boldsymbol{x}}{(\boldsymbol{x}', \boldsymbol{G}^\top V[\boldsymbol{x}]\boldsymbol{G}\boldsymbol{x}') + (\boldsymbol{x}, \boldsymbol{G}V[\boldsymbol{x}']\boldsymbol{G}^\top \boldsymbol{x})}\right)^\top. \quad (6.156)$$

If f and f' are both inaccurate but independent, the error behavior of 3-D reconstruction is described to a first approximation as the superimposition of that for inaccurate f and accurate f' and that for accurate f and inaccurate f'.

B. Identical cameras

Consider the case in which the two cameras are identical and have the same focal length but the value f we use is not necessarily correct. Let $V[f]$ be its variance. In this case, eq. (6.152) is modified to

$$(\boldsymbol{x}, \boldsymbol{G}\boldsymbol{x}') = -\frac{\Delta f}{f}\left((\boldsymbol{x}\boldsymbol{P_k}, \boldsymbol{G}\boldsymbol{x}') + (\boldsymbol{x}, \boldsymbol{G}\boldsymbol{P_k}\boldsymbol{x}')\right) = -\frac{\Delta f}{f}(\boldsymbol{x}, (\boldsymbol{P_k}\boldsymbol{G} + \boldsymbol{G}\boldsymbol{P_k})\boldsymbol{x}'). \quad (6.157)$$

The corrections $\Delta\hat{\boldsymbol{x}}$ and $\Delta\hat{\boldsymbol{x}}'$ are given as follows:

$$\Delta\hat{\boldsymbol{x}} = \frac{\Delta f}{f}\left(\boldsymbol{P_k}\boldsymbol{x} + \frac{(\boldsymbol{x}, (\boldsymbol{P_k}\boldsymbol{G} + \boldsymbol{G}\boldsymbol{P_k})\boldsymbol{x}')V[\boldsymbol{x}]\boldsymbol{G}\boldsymbol{x}'}{(\boldsymbol{x}', \boldsymbol{G}^\top V[\boldsymbol{x}]\boldsymbol{G}\boldsymbol{x}') + (\boldsymbol{x}, \boldsymbol{G}V[\boldsymbol{x}']\boldsymbol{G}^\top \boldsymbol{x})}\right),$$

$$\Delta\hat{\boldsymbol{x}}' = \frac{\Delta f}{f}\left(\boldsymbol{P_k}\boldsymbol{x}' + \frac{(\boldsymbol{x}, (\boldsymbol{P_k}\boldsymbol{G} + \boldsymbol{G}\boldsymbol{P_k})\boldsymbol{x}')V[\boldsymbol{x}']\boldsymbol{G}^\top \boldsymbol{x}}{(\boldsymbol{x}', \boldsymbol{G}^\top V[\boldsymbol{x}]\boldsymbol{G}\boldsymbol{x}') + (\boldsymbol{x}, \boldsymbol{G}V[\boldsymbol{x}']\boldsymbol{G}^\top \boldsymbol{x})}\right). \quad (6.158)$$

Hence,

$$V[\hat{\boldsymbol{x}}] = \frac{V[f]}{f^2}\left(\boldsymbol{P_k x} + \frac{(\boldsymbol{x}, (\boldsymbol{P_k G} + \boldsymbol{G P_k})\boldsymbol{x}')V[\boldsymbol{x}]\boldsymbol{G x}'}{(\boldsymbol{x}', \boldsymbol{G}^\top V[\boldsymbol{x}]\boldsymbol{G x}') + (\boldsymbol{x}, \boldsymbol{G}V[\boldsymbol{x}']\boldsymbol{G}^\top \boldsymbol{x})}\right)$$
$$\left(\boldsymbol{P_k x} + \frac{(\boldsymbol{x}, (\boldsymbol{P_k G} + \boldsymbol{G P_k})\boldsymbol{x}')V[\boldsymbol{x}]\boldsymbol{G x}'}{(\boldsymbol{x}', \boldsymbol{G}^\top V[\boldsymbol{x}]\boldsymbol{G x}') + (\boldsymbol{x}, \boldsymbol{G}V[\boldsymbol{x}']\boldsymbol{G}^\top \boldsymbol{x})}\right)^\top,$$

$$V[\hat{\boldsymbol{x}}'] = \frac{V[f]}{f^2}\left(\boldsymbol{P_k x}' + \frac{(\boldsymbol{x}, (\boldsymbol{P_k G} + \boldsymbol{G P_k})\boldsymbol{x}')V[\boldsymbol{x}']\boldsymbol{G}^\top \boldsymbol{x}}{(\boldsymbol{x}', \boldsymbol{G}^\top V[\boldsymbol{x}]\boldsymbol{G x}') + (\boldsymbol{x}, \boldsymbol{G}V[\boldsymbol{x}']\boldsymbol{G}^\top \boldsymbol{x})}\right)$$
$$\left(\boldsymbol{P_k x}' + \frac{(\boldsymbol{x}, (\boldsymbol{P_k G} + \boldsymbol{G P_k})\boldsymbol{x}')V[\boldsymbol{x}']\boldsymbol{G}^\top \boldsymbol{x}}{(\boldsymbol{x}', \boldsymbol{G}^\top V[\boldsymbol{x}]\boldsymbol{G x}') + (\boldsymbol{x}, \boldsymbol{G}V[\boldsymbol{x}']\boldsymbol{G}^\top \boldsymbol{x})}\right)^\top,$$

$$V[\hat{\boldsymbol{x}}, \hat{\boldsymbol{x}}'] = \frac{V[f]}{f^2}\left(\boldsymbol{P_k x} + \frac{(\boldsymbol{x}, (\boldsymbol{P_k G} + \boldsymbol{G P_k})\boldsymbol{x}')V[\boldsymbol{x}]\boldsymbol{G x}'}{(\boldsymbol{x}', \boldsymbol{G}^\top V[\boldsymbol{x}]\boldsymbol{G x}') + (\boldsymbol{x}, \boldsymbol{G}V[\boldsymbol{x}']\boldsymbol{G}^\top \boldsymbol{x})}\right)$$
$$\left(\boldsymbol{P_k x}' + \frac{(\boldsymbol{x}, (\boldsymbol{P_k G} + \boldsymbol{G P_k})\boldsymbol{x}')V[\boldsymbol{x}']\boldsymbol{G}^\top \boldsymbol{x}}{(\boldsymbol{x}', \boldsymbol{G}^\top V[\boldsymbol{x}]\boldsymbol{G x}') + (\boldsymbol{x}, \boldsymbol{G}V[\boldsymbol{x}']\boldsymbol{G}^\top \boldsymbol{x})}\right)^\top. \quad (6.159)$$

Again, the covariance matrices $V[\boldsymbol{x}]$ and $V[\boldsymbol{x}']$ need to be given only up to scale. The subsequent analysis is the same as given in Section 6.3.2.

Chapter 7

Parametric Fitting

This chapter studies methods for fitting a geometric object to multiple instances of another geometric object in an optimal manner in the presence of noise. First, the criterion for optimal fitting is derived in general terms as maximum likelihood estimation. The covariance matrix of the resulting estimate is explicitly obtained, and the statistical behavior of the residual of optimization is analyzed. This analysis provides a criterion for testing the hypothesis that the observed objects are in a special configuration. Then, we study various types of fitting problems in two and three dimensions such as finding an optimal average, estimating a common intersection, and fitting a common line or plane. In two dimensions, this theory predicts the statistical behavior of an image line fitted to an edge segment.

7.1 General Theory

7.1.1 *Parametric fitting*

We consider the problem of defining a geometric object from a number of data. Let $\boldsymbol{u}$ be the vector that represents the object to be determined, and let $\boldsymbol{a}_1$, ..., $\boldsymbol{a}_N$ be the data. Suppose the data $\boldsymbol{a}_1$, ..., $\boldsymbol{a}_N$ are vectors that represent N instances of the same type of object (e.g., image point). They are supposed to satisfy the same geometric relationship (e.g., incidence) with the object in question in the absence of noise. We want to determine the value of $\boldsymbol{u}$ by optimally imposing the relationship on the data. This type of problem frequently arises in computer vision and robotics applications.

Let n be the dimension of the vector $\boldsymbol{u}$. Its domain is assumed to be an n'-dimensional manifold $\mathcal{U} \subset \mathcal{R}^n$, which we call the *parameter space*. Consider an object represented by an m-vector $\boldsymbol{a}$. Its domain is assumed to be an m'-dimensional manifold $\mathcal{A} \subset \mathcal{R}^m$, which we call the *data space*. Suppose N different instances $\boldsymbol{a}_1$, ..., $\boldsymbol{a}_N$ of vector $\boldsymbol{a}$ are observed in the presence of noise. Let $\bar{\boldsymbol{a}}_\alpha$ be the true value of $\boldsymbol{a}_\alpha$, and write

$$\boldsymbol{a}_\alpha = \bar{\boldsymbol{a}}_\alpha + \Delta\boldsymbol{a}_\alpha, \qquad \alpha = 1, ..., N. \tag{7.1}$$

Assuming that each $\boldsymbol{a}_\alpha$ has been obtained by a separate process, we regard the noise $\Delta\boldsymbol{a}_\alpha$ as a statistically independent random variable of mean $\boldsymbol{0}$ and covariance matrix $\bar{V}[\boldsymbol{a}_\alpha]$. Since the data are constrained to be in the data space $\mathcal{A}$, the noise $\Delta\boldsymbol{a}_\alpha$ is, to a first approximation, constrained to be in the tangent space $T_{\bar{\boldsymbol{a}}_\alpha}(\mathcal{A})$ to the manifold $\mathcal{A}$ at $\bar{\boldsymbol{a}}_\alpha$. Hence, the domain and the null space of the covariance matrix $\bar{V}[\boldsymbol{a}_\alpha]$ are $T_{\bar{\boldsymbol{a}}_\alpha}(\mathcal{A})$ and $T_{\bar{\boldsymbol{a}}_\alpha}(\mathcal{A})^\perp$, respectively.

Suppose L smooth functions $F^{(k)}(\,\cdot\,,\,\cdot\,)$: $\mathcal{R}^m \times \mathcal{R}^n \to \mathcal{R}$ exist and the true values $\bar{\boldsymbol{a}}_1$, ..., $\bar{\boldsymbol{a}}_N$ satisfy

$$F^{(k)}(\bar{\boldsymbol{a}}_\alpha, \boldsymbol{u}) = 0, \qquad k = 1, ..., L, \tag{7.2}$$

for $\alpha = 1$, ..., N for *some* value $\boldsymbol{u}$. Eq. (7.2) imposes a parameterized constraint on $\{\bar{\boldsymbol{a}}_\alpha\}$; we call this the *hypothesis*, because the value of the parameter $\boldsymbol{u}$ is unknown. Our goal is to optimally estimate the parameter $\boldsymbol{u}$ from the data $\boldsymbol{a}_1$, ..., $\boldsymbol{a}_N$ and their covariance matrices $V[\boldsymbol{a}_1]$, ..., $V[\boldsymbol{a}_N]$.

We assume that the hypothesis (7.2) is *nonsingular* in the sense discussed in Section 5.1. The L equations (7.2) may not be independent; we call the number of independent equations the *rank* of the hypothesis (7.2). To be specific, the rank is the *codimension* of the manifold $\mathcal{S}$ defined by the L equations $F^{(k)}(\,\cdot\,, \boldsymbol{u}) = 0$, $k = 1$, ..., L, in the data space $\mathcal{A}$ (see Section 3.2.1). We call $\mathcal{S}$ the *(geometric) model* of the hypothesis (7.2). It can be shown[1] that the rank of the hypothesis (7.2), or the codimension of the model $\mathcal{S}$, generally coincides with the dimension of the linear subspace

$$\bar{\mathcal{V}}_\alpha = \{\boldsymbol{P}^{\mathcal{A}}_{\bar{\boldsymbol{a}}_\alpha} \nabla_{\boldsymbol{a}} F^{(1)}(\bar{\boldsymbol{a}}_\alpha, \boldsymbol{u}), ..., \boldsymbol{P}^{\mathcal{A}}_{\bar{\boldsymbol{a}}_\alpha} \nabla_{\boldsymbol{a}} F^{(L)}(\bar{\boldsymbol{a}}_\alpha, \boldsymbol{u})\}_L \subset \mathcal{R}^m \tag{7.3}$$

for all α, where $\boldsymbol{P}^{\mathcal{A}}_{\bar{\boldsymbol{a}}_\alpha}$ is the m-dimensional projection matrix onto the tangent space $T_{\bar{\boldsymbol{a}}_\alpha}(\mathcal{A})$ at $\bar{\boldsymbol{a}}_\alpha$. If the dimension of this linear subspace is smaller than the rank of the hypothesis (7.2), we call $\boldsymbol{a}_\alpha$ a *singular datum*; otherwise, it is a *nonsingular datum*. In the following, we assume that singular data are removed from the fitting data. However, the dimension of the linear subspace defined by eq. (7.3) may not be equal to the dimension of the linear subspace

$$\mathcal{V}_\alpha = \{\boldsymbol{P}^{\mathcal{A}}_{\boldsymbol{a}_\alpha} \nabla_{\boldsymbol{a}} F^{(1)}(\boldsymbol{a}_\alpha, \boldsymbol{u}), ..., \boldsymbol{P}^{\mathcal{A}}_{\boldsymbol{a}_\alpha} \nabla_{\boldsymbol{a}} F^{(L)}(\boldsymbol{a}_\alpha, \boldsymbol{u})\}_L \subset \mathcal{R}^m \tag{7.4}$$

for $\boldsymbol{a}_\alpha \neq \bar{\boldsymbol{a}}_\alpha$ (see Section 5.1.1); we say that the hypothesis (7.2) is *degenerate* if the dimension of the subspace $\mathcal{V}_\alpha$ is larger than the dimension of $\bar{\mathcal{V}}_\alpha$ (see Section 5.1.3).

7.1.2 Maximum likelihood estimation

Our approach consists of two stages[2]: the correction stage and the estimation stage.

A. Correction stage

We first assume a particular value of the parameter $\boldsymbol{u}$. If the data $\{\boldsymbol{a}_\alpha\}$ do not satisfy the hypothesis for the assumed value $\boldsymbol{u}$, we optimally corrected them

[1] Detailed discussions will be given in Chapter 14.

[2] The proof that the following approach is indeed optimal will be given in Chapter 14.

by applying the theory given in Chapter 5. Namely, we find $\Delta\boldsymbol{a}_\alpha$ such that $\bar{\boldsymbol{a}}_\alpha = \boldsymbol{a}_\alpha - \Delta\boldsymbol{a}_\alpha$ satisfies eq. (7.2). As discussed in Section 5.1.1, this correction is done for each α by minimizing the Mahalanobis distance $\|\Delta\boldsymbol{a}_\alpha\|_{\bar{V}[\boldsymbol{a}_\alpha]}$, i.e., by the optimization

$$J_\alpha = (\Delta\boldsymbol{a}_\alpha, \bar{V}[\boldsymbol{a}_\alpha]^- \Delta\boldsymbol{a}_\alpha) \to \min. \tag{7.5}$$

This can be justified as maximum likelihood estimation for Gaussian noise (see Section 5.1.1). Eq. (7.2) imposes the following linearized constraint on $\Delta\boldsymbol{a}_\alpha \in T_{\bar{\boldsymbol{a}}_\alpha}(\mathcal{A})$:

$$(\nabla_{\boldsymbol{a}} F^{(k)}(\bar{\boldsymbol{a}}_\alpha, \boldsymbol{u}), \Delta\boldsymbol{a}_\alpha) = F^{(k)}(\boldsymbol{a}_\alpha, \boldsymbol{u}), \qquad k = 1, ..., L. \tag{7.6}$$

The solution of the optimization (7.5) is given as follows (see eq. (5.17)):

$$\Delta\boldsymbol{a}_\alpha = \bar{V}[\boldsymbol{a}_\alpha] \sum_{k,l=1}^{L} \bar{W}_\alpha^{(kl)}(\boldsymbol{u}) F^{(k)}(\boldsymbol{a}_\alpha, \boldsymbol{u}) \nabla_{\boldsymbol{a}} F^{(l)}(\bar{\boldsymbol{a}}_\alpha, \boldsymbol{u}). \tag{7.7}$$

Here, $\bar{\boldsymbol{W}}_\alpha(\boldsymbol{u}) = (\bar{W}_\alpha^{(kl)}(\boldsymbol{u}))$ is the (LL)-matrix defined by

$$(\bar{W}_\alpha^{(kl)}) = \left((\nabla_{\boldsymbol{a}} F^{(k)}(\bar{\boldsymbol{a}}_\alpha, \boldsymbol{u}), \bar{V}[\boldsymbol{a}_\alpha] \nabla_{\boldsymbol{a}} F_\alpha^{(l)}(\bar{\boldsymbol{a}}_\alpha, \boldsymbol{u}))\right)^-. \tag{7.8}$$

As in Section 5.1.2, this expression is the abbreviation of $\bar{\boldsymbol{W}}_\alpha(\boldsymbol{u}) = \bar{\boldsymbol{V}}_\alpha(\boldsymbol{u})^-$ for the (LL)-matrix $\bar{\boldsymbol{V}}_\alpha(\boldsymbol{u}) = (\bar{V}_\alpha^{(kl)}(\boldsymbol{u}))$ defined as follows (see eqs. (5.15) and (5.16)):

$$(\bar{V}_\alpha^{(kl)}(\boldsymbol{u})) = \left((\nabla_{\boldsymbol{a}} F^{(k)}(\bar{\boldsymbol{a}}_\alpha, \boldsymbol{u}), \bar{V}[\boldsymbol{a}_\alpha] \nabla_{\boldsymbol{a}} F_\alpha^{(l)}(\bar{\boldsymbol{a}}_\alpha, \boldsymbol{u}))\right). \tag{7.9}$$

It can be proved[3] that the rank of matrix $\bar{\boldsymbol{V}}_\alpha(\boldsymbol{u})$ (hence of matrix $\bar{\boldsymbol{W}}_\alpha(\boldsymbol{u})$) equals the rank r of the hypothesis (7.2).

B. Estimation stage

If $\Delta\boldsymbol{a}_\alpha$ is an independent Gaussian random variable of mean $\boldsymbol{0}$ and covariance matrix $\bar{V}[\boldsymbol{a}_\alpha]$, the joint probability density of all $\{\Delta\boldsymbol{a}_\alpha\}$, $\alpha = 1, ..., N$, has the following form (see eq. (3.46)):

$$\left(\prod_{\beta=1}^{N} \frac{1}{\sqrt{(2\pi)^{r_\beta} |\bar{V}[\boldsymbol{a}_\beta]|_+}}\right) e^{-\sum_{\alpha=1}^{N} (\Delta\boldsymbol{a}_\alpha, \bar{V}[\boldsymbol{a}_\alpha]^- \Delta\boldsymbol{a}_\alpha)/2}. \tag{7.10}$$

This probability density can be viewed as the *likelihood* of the observed values $\{\Delta\boldsymbol{a}_\alpha\}$, $\alpha = 1, ..., N$. Let $\hat{J}_\alpha$ be the residual of J_α obtained by substituting

[3]The proof will be given in Chapter 14.

eq. (7.7) into eq. (7.5). The likelihood takes the form

$$\left(\prod_{\beta=1}^{N} \frac{1}{\sqrt{(2\pi)^{r_\beta}|\bar{V}[\boldsymbol{a}_\beta]|_+}}\right) e^{-\sum_{\alpha=1}^{N} \hat{J}_\alpha/2}, \tag{7.11}$$

which is a function of $\boldsymbol{u}$ alone. We now seek the value $\boldsymbol{u}$ that maximizes this likelihood. This is equivalent to minimizing the sum of the residuals $\hat{J}_\alpha$, which we write as $\bar{J}[\boldsymbol{u}]$:

$$\bar{J}[\boldsymbol{u}] = \sum_{\alpha=1}^{N} \hat{J}_\alpha \to \min. \tag{7.12}$$

Substituting eq. (7.7) into eq. (7.5), we see that this minimization takes the following form (see eq. (5.35)):

$$\bar{J}[\boldsymbol{u}] = \sum_{\alpha=1}^{N} \sum_{k,l=1}^{L} \bar{W}_\alpha^{(kl)}(\boldsymbol{u}) F^{(k)}(\boldsymbol{a}_\alpha, \boldsymbol{u}) F^{(l)}(\boldsymbol{a}_\alpha, \boldsymbol{u}) \to \min. \tag{7.13}$$

C. Practical considerations

In practice, the function $\bar{J}[\boldsymbol{u}]$ cannot be computed from the data alone, since $\bar{W}_\alpha^{(kl)}(\boldsymbol{u})$ involves the true value $\bar{\boldsymbol{a}}_\alpha$. Hence, as we did in Section 5.1.3, the (LL)-matrix $\bar{\boldsymbol{W}}_\alpha(\boldsymbol{u}) = (\bar{W}_\alpha^{(kl)}(\boldsymbol{u}))$ is approximated by the (LL)-matrix $\boldsymbol{W}_\alpha(\boldsymbol{u}) = (W_\alpha^{(kl)}(\boldsymbol{u}))$ obtained by using the rank-constrained generalized inverse (see eq. (2.82)):

$$(W_\alpha^{(kl)}(\boldsymbol{u})) = \left((\nabla_{\boldsymbol{a}} F^{(k)}(\boldsymbol{a}_\alpha, \boldsymbol{u}), V[\boldsymbol{a}_\alpha]\nabla_{\boldsymbol{a}} F^{(l)}(\boldsymbol{a}_\alpha, \boldsymbol{u}))\right)_r^-. \tag{7.14}$$

Here, r is the rank of the hypothesis (7.2). Eq. (7.14) is the abbreviation of $\boldsymbol{W}_\alpha(\boldsymbol{u}) = (\boldsymbol{V}_\alpha(\boldsymbol{u}))_r^-$ for the (LL)-matrix $\boldsymbol{V}_\alpha(\boldsymbol{u}) = (V_\alpha^{(kl)}(\boldsymbol{u}))$ defined by

$$(V_\alpha^{(kl)}(\boldsymbol{u})) = \left((\nabla_{\boldsymbol{a}} F^{(k)}(\boldsymbol{a}_\alpha, \boldsymbol{u}), V[\boldsymbol{a}_\alpha]\nabla_{\boldsymbol{a}} F^{(l)}(\boldsymbol{a}_\alpha, \boldsymbol{u}))\right). \tag{7.15}$$

The rank-constrained generalized inverse is used because matrix $\boldsymbol{V}_\alpha(\boldsymbol{u})$ has a larger rank than matrix $\bar{\boldsymbol{V}}_\alpha(\boldsymbol{u})$ if the hypothesis (7.2) is degenerate (see Section 5.1.3).

Thus, the actual computation is

$$J[\boldsymbol{u}] = \sum_{\alpha=1}^{N} \sum_{k,l=1}^{L} W_\alpha^{(kl)}(\boldsymbol{u}) F^{(k)}(\boldsymbol{a}_\alpha, \boldsymbol{u}) F^{(l)}(\boldsymbol{a}_\alpha, \boldsymbol{u}) \to \min. \tag{7.16}$$

Since $\boldsymbol{u}$ is constrained to be in the parameter space $\mathcal{U} \subset \mathcal{R}^n$, the minimization search is done in $\mathcal{U}$. In principle, the solution $\hat{\boldsymbol{u}}$ can be obtained by numerical computation, and many kinds of numerical software are available for that purpose (e.g., the quasi-Newton method and the conjugate gradient method). In the rest of this chapter, we assume that the optimal solution $\hat{\boldsymbol{u}}$ has been obtained by some numerical means; actual computational schemes for this optimization will be given in subsequent chapters.

7.1.3 Covariance matrix of the optimal fit

The optimal estimate $\hat{\boldsymbol{u}}$ is a random variable, because it is computed from the data $\{\boldsymbol{a}_\alpha\}$. We now study its statistical behavior. To do this, we need to distinguish $\boldsymbol{u}$ regarded as a variable from its true value. Let $\bar{\boldsymbol{u}}$ be the true value that satisfies the hypothesis (7.2), and put

$$\boldsymbol{u} = \bar{\boldsymbol{u}} + \Delta\boldsymbol{u}. \tag{7.17}$$

Since $\boldsymbol{u}$ is constrained to be in the parameter space $\mathcal{U} \subset \mathcal{R}^n$, the variation $\Delta\boldsymbol{u}$ is constrained, to a first approximation, to be in the tangent space $T_{\bar{\boldsymbol{u}}}(\mathcal{U})$ to the manifold $\mathcal{U}$ at $\bar{\boldsymbol{u}}$.

If we substitute $\boldsymbol{a}_\alpha = \bar{\boldsymbol{a}}_\alpha + \Delta\boldsymbol{a}_\alpha$ and $\boldsymbol{u} = \bar{\boldsymbol{u}} + \Delta\boldsymbol{u}$ into $F^{(k)}(\boldsymbol{a}_\alpha, \boldsymbol{u})$ and expand it in the neighborhood of $\bar{\boldsymbol{a}}_\alpha$ and $\bar{\boldsymbol{u}}$, we obtain

$$F^{(k)}(\boldsymbol{a}_\alpha, \boldsymbol{u}) = (\nabla_{\boldsymbol{a}}\bar{F}^{(k)}_\alpha, \Delta\boldsymbol{a}_\alpha) + (\nabla_{\boldsymbol{u}}\bar{F}^{(k)}_\alpha, \Delta\boldsymbol{u}_\alpha) + O(\Delta\boldsymbol{a}_\alpha, \Delta\boldsymbol{u})^2, \tag{7.18}$$

where $\nabla_{\boldsymbol{a}}\bar{F}^{(k)}_\alpha$ and $\nabla_{\boldsymbol{u}}\bar{F}^{(k)}_\alpha$ are the abbreviations of $\nabla_{\boldsymbol{a}}F^{(k)}(\bar{\boldsymbol{a}}_\alpha, \bar{\boldsymbol{u}})$ and $\nabla_{\boldsymbol{u}}F^{(k)}(\bar{\boldsymbol{a}}_\alpha, \bar{\boldsymbol{u}})$, respectively. The symbol $O(\cdots)^p$ denotes terms of order p or higher in $\cdots$. Noting that $\bar{W}^{(kl)}_\alpha(\boldsymbol{u}) = \bar{W}^{(kl)}_\alpha(\bar{\boldsymbol{u}}) + O(\Delta\boldsymbol{u})$, we see from eq. (7.13) that

$$\bar{J}[\boldsymbol{u}] = \sum_{\alpha=1}^{N}\sum_{k,l=1}^{L} \bar{W}^{(kl)}_\alpha(\bar{\boldsymbol{u}})\Big((\nabla_{\boldsymbol{u}}\bar{F}^{(k)}_\alpha, \Delta\boldsymbol{u}) + (\nabla_{\boldsymbol{a}}\bar{F}^{(k)}_\alpha, \Delta\boldsymbol{a}_\alpha)\Big)$$
$$\Big((\nabla_{\boldsymbol{u}}\bar{F}^{(l)}_\alpha, \Delta\boldsymbol{u}) + (\nabla_{\boldsymbol{a}}\bar{F}^{(l)}_\alpha, \Delta\boldsymbol{a}_\alpha)\Big) + O(\Delta\boldsymbol{a}_\alpha, \Delta\boldsymbol{u})^3. \tag{7.19}$$

Let $\{\bar{\boldsymbol{v}}_i\}$, $i = 1, ..., n - n'$, be an orthonormal basis of $T_{\bar{\boldsymbol{u}}}(\mathcal{U})^\perp$. The constraint $\Delta\boldsymbol{u} \in T_{\bar{\boldsymbol{u}}}(\mathcal{U})$ can be written as

$$(\bar{\boldsymbol{v}}_i, \Delta\boldsymbol{u}) = 0, \qquad i = 1, ..., n - n'. \tag{7.20}$$

The minimum of eq. (7.19) is obtained by introducing Lagrange multipliers λ_i and differentiating $\bar{J}[\boldsymbol{u}] - \sum_{i=1}^{n-n'} \lambda_i(\bar{\boldsymbol{v}}_i, \Delta\boldsymbol{u})$ with respect to $\Delta\boldsymbol{u}$. Ignoring higher order terms, we obtain

$$\sum_{\alpha=1}^{N}\sum_{k,l=1}^{L} \bar{W}^{(kl)}_\alpha(\bar{\boldsymbol{u}})\Big((\nabla_{\boldsymbol{u}}\bar{F}^{(k)}_\alpha, \Delta\boldsymbol{u}) + (\nabla_{\boldsymbol{a}}\bar{F}^{(k)}, \Delta\boldsymbol{a}_\alpha)\Big)\nabla_{\boldsymbol{u}}\bar{F}^{(l)}_\alpha = \sum_{i=1}^{n-n'} \lambda_i\bar{\boldsymbol{v}}_i. \tag{7.21}$$

Let $\boldsymbol{P}^{\mathcal{U}}_{\bar{\boldsymbol{u}}}$ be the n-dimensional projection matrix onto the tangent space $T_{\bar{\boldsymbol{u}}}(\mathcal{U})$ at $\bar{\boldsymbol{u}}$. Multiplying $\boldsymbol{P}^{\mathcal{U}}_{\bar{\boldsymbol{u}}}$ on both sides and noting that $\boldsymbol{P}^{\mathcal{U}}_{\bar{\boldsymbol{u}}}\Delta\boldsymbol{u} = \Delta\boldsymbol{u}$ and $\boldsymbol{P}^{\mathcal{U}}_{\bar{\boldsymbol{u}}}\bar{\boldsymbol{v}}_i = 0$, $i = 1, ..., n - n'$, we obtain

$$\left(\sum_{\alpha=1}^{N}\sum_{k,l=1}^{L}\bar{W}_\alpha^{(kl)}(\bar{\boldsymbol{u}})\boldsymbol{P}^{\mathcal{U}}_{\bar{\boldsymbol{u}}}\nabla_{\boldsymbol{u}}\bar{F}_\alpha^{(k)}(\nabla_{\boldsymbol{u}}\bar{F}_\alpha^{(l)})^\top\boldsymbol{P}^{\mathcal{U}}_{\bar{\boldsymbol{u}}}\right)\Delta\boldsymbol{u}$$
$$+\sum_{\alpha=1}^{N}\left(\sum_{k,l=1}^{L}W_\alpha^{(kl)}(\bar{\boldsymbol{u}})\boldsymbol{P}^{\mathcal{U}}_{\bar{\boldsymbol{u}}}\nabla_{\boldsymbol{u}}\bar{F}_\alpha^{(k)}(\nabla_{\boldsymbol{a}}\bar{F}_\alpha^{(l)})^\top\right)\Delta\boldsymbol{a}_\alpha = \boldsymbol{0}. \tag{7.22}$$

If we define the *moment matrix* $\bar{\boldsymbol{M}}$ by

$$\bar{\boldsymbol{M}} = \sum_{\alpha=1}^{N}\sum_{k,l=1}^{L}\bar{W}_\alpha^{(kl)}(\bar{\boldsymbol{u}})(\boldsymbol{P}^{\mathcal{U}}_{\bar{\boldsymbol{u}}}\nabla_{\boldsymbol{u}}\bar{F}_\alpha^{(k)})(\boldsymbol{P}^{\mathcal{U}}_{\bar{\boldsymbol{u}}}\nabla_{\boldsymbol{u}}\bar{F}_\alpha^{(l)})^\top, \tag{7.23}$$

eq. (7.22) can be written in the following form:

$$\bar{\boldsymbol{M}}\Delta\boldsymbol{u} = -\sum_{\alpha=1}^{N}\left(\sum_{k,l=1}^{L}\bar{W}_\alpha^{(kl)}(\bar{\boldsymbol{u}})(\boldsymbol{P}^{\mathcal{U}}_{\bar{\boldsymbol{u}}}\nabla_{\boldsymbol{u}}\bar{F}_\alpha^{(k)})(\nabla_{\boldsymbol{a}}\bar{F}_\alpha^{(l)})^\top\right)\Delta\boldsymbol{a}_\alpha. \tag{7.24}$$

The moment matrix $\bar{\boldsymbol{M}}$ is an (nn)-matrix, and its range is contained in $T_{\bar{\boldsymbol{u}}}(\mathcal{U})$. Hence, the rank of $\bar{\boldsymbol{M}}$ is at most n'. Here, we assume that the number of data $\{\boldsymbol{a}_\alpha\}$ is sufficiently large and no special relationship exists among $\bar{\boldsymbol{a}}_1, ..., \bar{\boldsymbol{a}}_N$ other than eq. (7.2). Then, the range of $\bar{\boldsymbol{M}}$ generally coincides with $T_{\bar{\boldsymbol{u}}}(\mathcal{U})$. It follows that

$$\bar{\boldsymbol{M}}\bar{\boldsymbol{M}}^- = \bar{\boldsymbol{M}}^-\bar{\boldsymbol{M}} = \boldsymbol{P}^{\mathcal{U}}_{\bar{\boldsymbol{u}}}. \tag{7.25}$$

Multiplying $\bar{\boldsymbol{M}}^-$ on both sides of eq. (7.24) and noting that $\bar{\boldsymbol{M}}^-\bar{\boldsymbol{M}}\Delta\boldsymbol{u} = \boldsymbol{P}^{\mathcal{U}}_{\bar{\boldsymbol{u}}}\Delta\boldsymbol{u} = \Delta\boldsymbol{u}$, we obtain

$$\Delta\boldsymbol{u} = -\bar{\boldsymbol{M}}^-\sum_{\alpha=1}^{N}\left(\sum_{k,l=1}^{L}\bar{W}_\alpha^{(kl)}(\bar{\boldsymbol{u}})(\boldsymbol{P}^{\mathcal{U}}_{\bar{\boldsymbol{u}}}\nabla_{\boldsymbol{u}}\bar{F}_\alpha^{(k)})(\nabla_{\boldsymbol{a}}\bar{F}_\alpha^{(l)})^\top\right)\Delta\boldsymbol{a}_\alpha. \tag{7.26}$$

The covariance matrix $\bar{V}[\hat{\boldsymbol{u}}] = E[\Delta\boldsymbol{u}\Delta\boldsymbol{u}^\top]$ of the optimal estimate $\hat{\boldsymbol{u}} = \bar{\boldsymbol{u}}+\Delta\boldsymbol{u}$ is computed as follows:

$$\bar{V}[\hat{\boldsymbol{u}}] = \bar{\boldsymbol{M}}^-\sum_{\alpha,\beta=1}^{N}\sum_{k,l,m,n=1}^{L}\bar{W}_\alpha^{(kl)}(\bar{\boldsymbol{u}})\bar{W}_\beta^{(mn)}(\bar{\boldsymbol{u}})(\boldsymbol{P}^{\mathcal{U}}_{\bar{\boldsymbol{u}}}\nabla_{\boldsymbol{u}}\bar{F}_\alpha^{(k)})$$
$$(\nabla_{\boldsymbol{a}}\bar{F}_\alpha^{(l)})^\top E[\Delta\boldsymbol{a}_\alpha\Delta\boldsymbol{a}_\beta^\top](\nabla_{\boldsymbol{a}}\bar{F}_\alpha^{(n)})(\boldsymbol{P}^{\mathcal{U}}_{\bar{\boldsymbol{u}}}\nabla_{\boldsymbol{u}}\bar{F}_\beta^{(m)})^\top\bar{\boldsymbol{M}}^-$$
$$= \bar{\boldsymbol{M}}^-\sum_{\alpha,\beta=1}^{N}\sum_{k,l,m,n=1}^{L}\bar{W}_\alpha^{(kl)}(\bar{\boldsymbol{u}})\bar{W}_\beta^{(mn)}(\bar{\boldsymbol{u}})(\boldsymbol{P}^{\mathcal{U}}_{\bar{\boldsymbol{u}}}\nabla_{\boldsymbol{u}}\bar{F}_\alpha^{(k)})$$

$$(\nabla_{\boldsymbol{a}}\bar{F}_\alpha^{(l)})^\top \delta_{\alpha\beta}\bar{V}[\boldsymbol{a}_\alpha](\nabla_{\boldsymbol{a}}\bar{F}_\alpha^{(n)})(\boldsymbol{P}_{\bar{\boldsymbol{u}}}^{\mathcal{U}}\nabla_{\boldsymbol{u}}\bar{F}_\beta^{(m)})^\top \bar{\boldsymbol{M}}^-$$

$$= \bar{\boldsymbol{M}}^- \sum_{\alpha=1}^N \sum_{k,l,m,n=1}^L \bar{W}_\alpha^{(kl)}(\bar{\boldsymbol{u}})\bar{W}_\alpha^{(mn)}(\bar{\boldsymbol{u}})(\boldsymbol{P}_{\bar{\boldsymbol{u}}}^{\mathcal{U}}\nabla_{\boldsymbol{u}}\bar{F}_\alpha^{(k)})$$

$$(\nabla_{\boldsymbol{a}}\bar{F}_\alpha^{(l)}, \bar{V}[\boldsymbol{a}_\alpha]\nabla_{\boldsymbol{a}}\bar{F}_\alpha^{(n)})(\boldsymbol{P}_{\bar{\boldsymbol{u}}}^{\mathcal{U}}\nabla_{\boldsymbol{u}}\bar{F}_\alpha^{(m)})^\top \bar{\boldsymbol{M}}^-$$

$$= \bar{\boldsymbol{M}}^- \sum_{\alpha=1}^N \sum_{k,m=1}^L \left(\sum_{l,n=1}^L \bar{W}_\alpha^{(kl)}(\bar{\boldsymbol{u}})\bar{V}_\alpha^{(ln)}(\bar{\boldsymbol{u}})\bar{W}_\alpha^{(mn)}(\bar{\boldsymbol{u}})\right)$$

$$(\boldsymbol{P}_{\bar{\boldsymbol{u}}}^{\mathcal{U}}\nabla_{\boldsymbol{u}}\bar{F}_\alpha^{(k)})(\boldsymbol{P}_{\bar{\boldsymbol{u}}}^{\mathcal{U}}\nabla_{\boldsymbol{u}}\bar{F}_\alpha^{(m)})^\top \bar{\boldsymbol{M}}^-. \tag{7.27}$$

Here, we have used eq. (7.9). Since $\sum_{l,m=1}^L \bar{W}_\alpha^{(kl)}(\bar{\boldsymbol{u}})\bar{V}_\alpha^{(ln)}(\bar{\boldsymbol{u}})\bar{W}_\alpha^{(mn)}(\bar{\boldsymbol{u}})$ equals the (kn) element of matrix $\bar{\boldsymbol{W}}_\alpha(\bar{\boldsymbol{u}})\bar{\boldsymbol{W}}_\alpha(\bar{\boldsymbol{u}})^-\bar{\boldsymbol{W}}_\alpha(\bar{\boldsymbol{u}}) = \bar{\boldsymbol{W}}_\alpha(\bar{\boldsymbol{u}})$ (see eq. (2.81)), we obtain

$$\bar{V}[\hat{\boldsymbol{u}}] = \bar{\boldsymbol{M}}^- \sum_{\alpha=1}^N \sum_{k,l=1}^L \bar{W}_\alpha^{(kl)}(\bar{\boldsymbol{u}})(\boldsymbol{P}_{\bar{\boldsymbol{u}}}^{\mathcal{U}}\nabla_{\boldsymbol{u}}\bar{F}_\alpha^{(k)})(\boldsymbol{P}_{\bar{\boldsymbol{u}}}^{\mathcal{U}}\nabla_{\boldsymbol{u}}\bar{F}_\alpha^{(l)})^\top \bar{\boldsymbol{M}}^-$$

$$= \bar{\boldsymbol{M}}^-\bar{\boldsymbol{M}}\bar{\boldsymbol{M}}^- = \bar{\boldsymbol{M}}^-. \tag{7.28}$$

Thus, the covariance matrix of the optimal estimate $\hat{\boldsymbol{u}}$ is given by

$$\bar{V}[\hat{\boldsymbol{u}}] = \left(\sum_{\alpha=1}^N \sum_{k,l=1}^L \bar{W}_\alpha^{(kl)}(\bar{\boldsymbol{u}})(\boldsymbol{P}_{\bar{\boldsymbol{u}}}^{\mathcal{U}}\nabla_{\boldsymbol{u}}\bar{F}_\alpha^{(k)})(\boldsymbol{P}_{\bar{\boldsymbol{u}}}^{\mathcal{U}}\nabla_{\boldsymbol{u}}\bar{F}_\alpha^{(l)})^\top\right)^-, \tag{7.29}$$

which has rank n'; its null space is $T_{\bar{\boldsymbol{u}}}(\mathcal{U})^\perp$. Eq. (7.29) coincides with the *Cramer-Rao lower bound*[4] on the attainable accuracy of parametric fitting.

Eq. (7.29) is purely a theoretical expression because it involves the true values of $\boldsymbol{u}$ and $\boldsymbol{a}_\alpha$. A simple approximation to compute it form the data alone is replacing the true value $\bar{\boldsymbol{u}}$ by the optimal estimate $\hat{\boldsymbol{u}}$ and the true value $\bar{\boldsymbol{a}}_\alpha$ by the optimally corrected value $\hat{\boldsymbol{a}}_\alpha = \boldsymbol{a}_\alpha - \Delta\boldsymbol{a}_\alpha$, where $\Delta\boldsymbol{a}_\alpha$ is given by eq. (7.7). Then, the covariance matrix $V[\hat{\boldsymbol{x}}]$ takes the form

$$V[\hat{\boldsymbol{u}}] = \left(\sum_{\alpha=1}^N \sum_{k,l=1}^L W_\alpha^{(kl)}(\hat{\boldsymbol{u}})(\boldsymbol{P}_{\hat{\boldsymbol{u}}}^{\mathcal{U}}\nabla_{\boldsymbol{u}}\hat{F}_\alpha^{(k)})(\boldsymbol{P}_{\hat{\boldsymbol{u}}}^{\mathcal{U}}\nabla_{\boldsymbol{u}}\hat{F}_\alpha^{(l)})^\top\right)^-, \tag{7.30}$$

where $\nabla_{\boldsymbol{u}}\hat{F}_\alpha^{(k)}$ is the abbreviation of $\nabla_{\boldsymbol{u}}F^{(k)}(\hat{\boldsymbol{a}}_\alpha, \hat{\boldsymbol{u}})$. However, the data $\{\boldsymbol{a}_\alpha\}$ themselves can be used instead of the corrected values $\{\hat{\boldsymbol{a}}_\alpha\}$ to a first approximation. In whichever form, the resulting covariance matrix $V[\hat{\boldsymbol{u}}]$ also has rank n'; its null space is $T_{\hat{\boldsymbol{u}}}(\mathcal{U})^\perp$.

[4] Detailed discussions will be given in Chapter 14.

7.1.4 Hypothesis testing and noise level estimation

The above estimation procedure is based on the hypothesis that the data $\{\boldsymbol{a}_\alpha\}$ are random deviations from the values $\{\bar{\boldsymbol{a}}_\alpha\}$ that satisfy eq. (7.2); minimizing the function $\bar{J}[\boldsymbol{u}]$ defined by eq. (7.13) can be interpreted as choosing the value $\hat{\boldsymbol{u}}$ that make the hypothesis the most likely. If the hypothesis is correct, the residual $\bar{J}[\hat{\boldsymbol{u}}]$ should be 0 for the true values $\{\bar{\boldsymbol{a}}_\alpha\}$. However, the residual is generally positive for the data $\{\boldsymbol{a}_\alpha\}$. This suggests that if the residual $\bar{J}[\hat{\boldsymbol{u}}]$ is much larger than can be accounted for by the statistical behavior of the noise in $\{\boldsymbol{a}_\alpha\}$, the hypothesis should be rejected. In order to formulate this process as a statistical test, we need to derive the probability distribution of the residual $\bar{J}[\hat{\boldsymbol{u}}]$. We do this by assuming that the noise is Gaussian.

A. Testing of a strong hypothesis

As a preliminary step, consider the residual $\bar{J}[\bar{\boldsymbol{u}}]$ *for the true value* $\bar{\boldsymbol{u}}$. Letting $\Delta\boldsymbol{u} = \boldsymbol{0}$ in eq. (7.19) and neglecting higher order terms, we observe that

$$\bar{J}[\bar{\boldsymbol{u}}] = \sum_{\alpha=1}^{N}\sum_{k,l=1}^{L} \bar{W}_\alpha^{(kl)}(\bar{\boldsymbol{u}})(\nabla_{\boldsymbol{a}}\bar{F}_\alpha^{(k)}, \Delta\boldsymbol{a}_\alpha)(\nabla_{\boldsymbol{a}}\bar{F}_\alpha^{(l)}, \Delta\boldsymbol{a}_\alpha). \tag{7.31}$$

If we put

$$e_\alpha^{(k)} = (\nabla_{\boldsymbol{a}}\bar{F}_\alpha^{(k)}, \Delta\boldsymbol{a}_\alpha), \tag{7.32}$$

the L-vector $\boldsymbol{e}_\alpha = (e_\alpha^{(1)}, ..., e_\alpha^{(L)})^\top$ is a Gaussian random variable of mean $\boldsymbol{0}$. Its covariance matrix is

$$\begin{aligned} V[\boldsymbol{e}_\alpha] &= \left(E[e_\alpha^{(k)}e_\alpha^{(l)}]\right) = \left((\nabla_{\boldsymbol{a}}\bar{F}_\alpha^{(k)}, E[\Delta\boldsymbol{a}_\alpha\Delta\boldsymbol{a}_\alpha^\top]\nabla_{\boldsymbol{a}}\bar{F}_\alpha^{(l)})\right) \\ &= \left((\nabla_{\boldsymbol{a}}\bar{F}_\alpha^{(k)}, \bar{V}[\boldsymbol{a}_\alpha]\nabla_{\boldsymbol{a}}\bar{F}_\alpha^{(l)})\right) = \left(\bar{V}_\alpha^{(kl)}(\bar{\boldsymbol{u}})\right) = \left(\bar{W}_\alpha^{(kl)}(\bar{\boldsymbol{u}})\right)^-, \end{aligned} \tag{7.33}$$

where eqs. (7.9) and (7.8) are used. Hence, eq. (7.31) can be written as

$$\bar{J}[\bar{\boldsymbol{u}}] = \sum_{\alpha=1}^{N}(\boldsymbol{e}_\alpha, V[\boldsymbol{e}_\alpha]^-\boldsymbol{e}_\alpha). \tag{7.34}$$

The rank of $V[\boldsymbol{e}_\alpha]$ $(= \bar{\boldsymbol{W}}_\alpha(\bar{\boldsymbol{u}})^-)$ equals the rank r of the hypothesis (7.2), and each $\boldsymbol{e}_\alpha$ is an independent Gaussian random variable. Hence, the residual $\bar{J}[\bar{\boldsymbol{u}}]$ is a χ^2 variable with rN degrees of freedom (see eq. (3.63)). Its expectation and variance are given as follows (see eq. (3.59)):

$$E[\bar{J}[\bar{\boldsymbol{u}}]] = rN, \qquad V[\bar{J}[\bar{\boldsymbol{u}}]] = 2rN. \tag{7.35}$$

It follows that if a *particular* value $\bar{\boldsymbol{u}}$ is given *independently of the data* $\{\boldsymbol{a}_\alpha\}$, the strong hypothesis that *the true value is* $\bar{\boldsymbol{u}}$ can be tested by the standard the χ^2 test (see Section 3.3.4). Namely, the strong hypothesis is rejected with $a\%$ significance level if

$$\bar{J}[\bar{\boldsymbol{u}}] > \chi^2_{rN}. \tag{7.36}$$

In practice, however, the function $\bar{J}[\boldsymbol{u}]$ defined by eq. (7.13) cannot be computed, since $\bar{W}_\alpha^{(kl)}$ involves true value $\bar{\boldsymbol{a}}_\alpha$. Hence, $\bar{J}[\boldsymbol{u}]$ is approximated by the function $J[\boldsymbol{u}]$ given by eq. (7.16).

B. Testing of a weak hypothesis

Next, consider the residual $\bar{J}[\hat{\boldsymbol{u}}]$ *for the optimal estimate* $\hat{\boldsymbol{u}}$. Substituting eq. (7.26) into eq. (7.19) and using eqs. (7.24) and (7.29), we obtain to a first approximation

$$\begin{aligned}
\bar{J}[\hat{\boldsymbol{u}}] &= \sum_{\alpha=1}^{N}\sum_{k,l=1}^{L} \bar{W}_\alpha^{(kl)}(\bar{\boldsymbol{u}}) \Big((\nabla_{\boldsymbol{a}}\bar{F}_\alpha^{(k)}, \Delta\boldsymbol{a}_\alpha)(\nabla_{\boldsymbol{a}}\bar{F}_\alpha^{(l)}, \Delta\boldsymbol{a}_\alpha) \\
&\quad + (\nabla_{\boldsymbol{u}}\bar{F}_\alpha^{(k)}, \Delta\boldsymbol{u})(\nabla_{\boldsymbol{a}}\bar{F}_\alpha^{(l)}, \Delta\boldsymbol{a}_\alpha) + (\nabla_{\boldsymbol{a}}\bar{F}_\alpha^{(k)}, \Delta\boldsymbol{a}_\alpha)(\nabla_{\boldsymbol{u}}\bar{F}_\alpha^{(l)}, \Delta\boldsymbol{u}) \\
&\quad + (\nabla_{\boldsymbol{u}}\bar{F}_\alpha^{(k)}, \Delta\boldsymbol{u})(\nabla_{\boldsymbol{u}}\bar{F}_\alpha^{(l)}, \Delta\boldsymbol{u})\Big) \\
&= \bar{J}[\bar{\boldsymbol{u}}] + (\Delta\boldsymbol{u}, \sum_{\alpha=1}^{N}\left(\sum_{k,l=1}^{L} \bar{W}_\alpha^{(kl)}(\bar{\boldsymbol{u}})(\boldsymbol{P}^{\mathcal{U}}_{\bar{\boldsymbol{u}}}\nabla_{\boldsymbol{u}}\bar{F}_\alpha^{(k)})(\nabla_{\boldsymbol{a}}\bar{F}_\alpha^{(l)})^\top\right)\Delta\boldsymbol{a}_\alpha) \\
&\quad +(\Delta\boldsymbol{u}, \sum_{\alpha=1}^{N}\left(\sum_{k,l=1}^{L} \bar{W}_\alpha^{(kl)}(\bar{\boldsymbol{u}})(\boldsymbol{P}^{\mathcal{U}}_{\bar{\boldsymbol{u}}}\nabla_{\boldsymbol{u}}\bar{F}_\alpha^{(l)})(\nabla_{\boldsymbol{a}}\bar{F}_\alpha^{(k)})^\top\right)\Delta\boldsymbol{a}_\alpha) \\
&\quad +(\Delta\boldsymbol{u}, \left(\sum_{\alpha=1}^{N}\sum_{k,l=1}^{L} \bar{W}_\alpha^{(kl)}(\bar{\boldsymbol{u}})(\boldsymbol{P}^{\mathcal{U}}_{\bar{\boldsymbol{u}}}\nabla_{\boldsymbol{u}}\bar{F}_\alpha^{(k)})(\boldsymbol{P}^{\mathcal{U}}_{\bar{\boldsymbol{u}}}\nabla_{\boldsymbol{u}}\bar{F}_\alpha^{(l)})^\top\right)\Delta\boldsymbol{u}) \\
&= \bar{J}[\bar{\boldsymbol{u}}] - (\Delta\boldsymbol{u}, \bar{V}[\hat{\boldsymbol{u}}]^-\Delta\boldsymbol{u}).
\end{aligned} \tag{7.37}$$

Since $\bar{V}[\hat{\boldsymbol{u}}]$ is the covariance matrix of $\Delta\boldsymbol{u}$ and has rank n', the quadratic form $(\Delta\boldsymbol{u}, \bar{V}[\hat{\boldsymbol{u}}]^-\Delta\boldsymbol{u})$ is a χ^2 variable with n' degrees of freedom (see eq. (3.61)). From eq. (7.32), we see that eq. (7.26) is expressed in terms of $e_\alpha^{(k)}$ in the form

$$\Delta\boldsymbol{u} = -\bar{\boldsymbol{V}}\sum_{\alpha=1}^{N}\sum_{k,l=1}^{L} \bar{W}_\alpha^{(kl)}(\bar{\boldsymbol{u}})(\boldsymbol{P}^{\mathcal{U}}_{\bar{\boldsymbol{u}}}\nabla_{\boldsymbol{u}}\bar{F}_\alpha^{(k)})e_\alpha^{(l)}, \tag{7.38}$$

meaning that $\Delta\boldsymbol{u}$ is obtained by a linear mapping from $\{\boldsymbol{e}_\alpha\}$. Hence, the residual $\bar{J}[\hat{\boldsymbol{u}}]$ is a χ^2 variable with $rN - n'$ degrees of freedom (see eq. (3.64)). Its expectation and variance are given as follows (see eq. (3.59)):

$$E[\bar{J}[\hat{\boldsymbol{u}}]] = rN - n', \qquad V[\bar{J}[\hat{\boldsymbol{u}}]] = 2(rN - n'). \tag{7.39}$$

It follows that the number N of data must be such that

$$N \geq \frac{n'}{r}. \tag{7.40}$$

From eqs. (7.39), we see that with a high probability *the residual* $\bar{J}[\hat{\boldsymbol{u}}]$ *is smaller than the residual* $\bar{J}[\bar{\boldsymbol{u}}]$. This is because the estimate $\hat{\boldsymbol{u}}$ is defined as the value that minimizes $\bar{J}[\boldsymbol{u}]$; the estimate $\hat{\boldsymbol{u}}$ is a function of the data $\{\boldsymbol{a}_\alpha\}$ and hence is *correlated* with them.

The above analysis can be used to test the weak hypothesis that eq. (7.2) is satisfied by *some* value $\boldsymbol{u}$. Namely, the weak hypothesis is rejected with a% significance level if

$$\bar{J}[\hat{\boldsymbol{u}}] > \chi^2_{rN-n',a}. \tag{7.41}$$

In practice, the function $\bar{J}[\boldsymbol{u}]$ is approximated by the function $J[\boldsymbol{u}]$ given by eq. (7.16).

C. Noise level estimation

Note that although the covariance matrices $V[\boldsymbol{a}_\alpha]$, $\alpha = 1, ..., N$, are involved in the optimization (7.16), *we need not know their absolute scale*: it suffices to know them only *up to scale*. In fact, if the covariance matrices $V[\boldsymbol{a}_\alpha]$ are multiplied by a positive constant c, the matrix $(W_\alpha^{(kl)}(\boldsymbol{u}))$ is multiplied by $1/c$ (see eq. (7.14)), and multiplication of $J[\boldsymbol{u}]$ by a positive constant does not affect the value that minimizes it.

The covariance matrix $V[\boldsymbol{a}_\alpha]$ can be decomposed into the *noise level* ϵ and the *normalized covariance matrix* $V_0[\boldsymbol{a}_\alpha]$ in the form

$$V[\boldsymbol{a}_\alpha] = \epsilon^2 V_0[\boldsymbol{a}_\alpha], \qquad \alpha = 1, ..., N. \tag{7.42}$$

As discussed in Section 5.1.5, the normalized covariance matrices $V_0[\boldsymbol{a}_\alpha]$ can be relatively easily predicted in many practical problems while the noise level ϵ is very difficult to estimate a priori. Since the optimal solution is invariant to the scale of $V[\boldsymbol{a}_\alpha]$, the normalized covariance matrices $V_0[\boldsymbol{a}_\alpha]$ can be used in the computation.

Once the optimal estimate $\hat{\boldsymbol{u}}$ is obtained, the noise level ϵ is estimated *a posteriori* (see Section 5.1.5). In fact, let $J_0[\hat{\boldsymbol{u}}]$ be the residual computed by using the normalized covariance matrices $V_0[\boldsymbol{a}_\alpha]$. An unbiased estimator $\hat{\epsilon}^2$ of ϵ^2 is obtained in the form

$$\hat{\epsilon}^2 = \frac{J_0[\hat{\boldsymbol{u}}]}{rN - n'}. \tag{7.43}$$

Its expectation and variance are respectively given by

$$E[\hat{\epsilon}^2] = \epsilon^2, \qquad V[\hat{\epsilon}^2] = \frac{2\epsilon^4}{rN - n'}. \tag{7.44}$$

On the other hand, testing of hypotheses requires a priori knowledge of the noise level ϵ, because a hypothesis is tested by comparing the a priori value ϵ with the a posteriori estimate $\hat{\epsilon}$ computed *on the assumption that the hypothesis is true*. The χ^2 test (7.41) can be rewritten in terms of the estimate $\hat{\epsilon}$ given by eq. (7.43) in the following form[5]:

$$\frac{\hat{\epsilon}^2}{\epsilon^2} > \frac{\chi^2_{rN-n',a}}{rN-n'}. \tag{7.45}$$

7.1.5 Linear hypothesis

In many computer vision and robotics applications, eq. (7.2) takes the following linear form:

$$(\bar{\boldsymbol{a}}_\alpha^{(k)}, \boldsymbol{u}) = 0, \qquad k = 1, ..., L. \tag{7.46}$$

For each k, vectors $\bar{\boldsymbol{a}}_1^{(k)}$, ..., $\bar{\boldsymbol{a}}_N^{(k)}$ are the true values of the data $\boldsymbol{a}_1^{(k)}$, ..., $\boldsymbol{a}_N^{(k)}$, which are N instances of an n-vector variable $\boldsymbol{a}^{(k)}$. If we define the direct sum vector

$$\boldsymbol{a} = \boldsymbol{a}^{(1)} \oplus \cdots \oplus \boldsymbol{a}^{(L)}, \tag{7.47}$$

and define k functions

$$F^{(k)}(\boldsymbol{a}, \boldsymbol{u}) = (\boldsymbol{a}^{(k)}, \boldsymbol{u}), \qquad k = 1, ..., L, \tag{7.48}$$

eq. (7.46) can be viewed as a special case of eq. (7.2). From the above definition, we see that

$$\nabla_{\boldsymbol{a}} F^{(k)}(\boldsymbol{a}, \boldsymbol{u}) = \boldsymbol{0} \oplus \cdots \oplus \overset{(k)}{\boldsymbol{u}} \oplus \cdots \oplus \boldsymbol{0}, \tag{7.49}$$

where the stacked symbol (k) means that $\boldsymbol{u}$ is the kth component of the direct sum.

Write

$$\boldsymbol{a}_\alpha^{(k)} = \bar{\boldsymbol{a}}_\alpha^{(k)} + \Delta \boldsymbol{a}_\alpha^{(k)}, \qquad \alpha = 1, ..., N, \tag{7.50}$$

and assume that the noise $\Delta \boldsymbol{a}_\alpha^{(k)}$ is a random variable of mean $\boldsymbol{0}$, independent for each α. However, $\Delta \boldsymbol{a}_\alpha^{(1)}$, ..., $\Delta \boldsymbol{a}_\alpha^{(L)}$ may be correlated. We write their covariance matrices as

$$\bar{V}[\boldsymbol{a}_\alpha^{(k)}, \boldsymbol{a}_\alpha^{(l)}] = E[\Delta \boldsymbol{a}_\alpha^{(k)} \Delta \boldsymbol{a}_\alpha^{(l)\top}]. \tag{7.51}$$

If we abbreviate $\bar{V}[\boldsymbol{a}_\alpha^{(k)}, \boldsymbol{a}_\alpha^{(k)}]$ as $\bar{V}[\boldsymbol{a}_\alpha^{(k)}]$ $(= E[\Delta \boldsymbol{a}_\alpha^{(k)} \Delta \boldsymbol{a}_\alpha^{(k)\top}])$, the covariance matrix of the direct sum vector $\boldsymbol{a}_\alpha = \boldsymbol{a}_\alpha^{(1)} \oplus \cdots \oplus \boldsymbol{a}_\alpha^{(L)}$ has the following

[5] This is a consequence of the fact that $\hat{\epsilon}^2/\epsilon^2$ is a *modified* χ^2 *variable* with $rN - n'$ degrees of freedom if the hypothesis is true (see eq. (3.72) and Footnote 6 in Section 5.1.5).

submatrices (see eq. (3.11)):

$$\bar{V}[\boldsymbol{a}_\alpha] = \begin{pmatrix} \bar{V}[\boldsymbol{a}_\alpha^{(1)}] & \cdots & \bar{V}[\boldsymbol{a}_\alpha^{(1)}, \boldsymbol{a}_\alpha^{(L)}] \\ \vdots & \cdots & \vdots \\ \bar{V}[\boldsymbol{a}_\alpha^{(L)}, \boldsymbol{a}_\alpha^{(1)}] & \cdots & \bar{V}[\boldsymbol{a}_\alpha^{(L)}] \end{pmatrix}. \tag{7.52}$$

Eq. (7.7) now reads

$$\Delta \boldsymbol{a}_\alpha^{(k)} = \sum_{l,m=1}^{L} \bar{W}_\alpha^{(lm)}(\boldsymbol{u})(\boldsymbol{a}_\alpha^{(m)}, \boldsymbol{u})\bar{V}[\boldsymbol{a}_\alpha^{(k)}, \boldsymbol{a}_\alpha^{(l)}]\boldsymbol{u}. \tag{7.53}$$

Eq. (7.8) is replaced by

$$(\bar{W}_\alpha^{(kl)}) = \left((\boldsymbol{u}, \bar{V}[\boldsymbol{a}_\alpha^{(k)}, \boldsymbol{a}_\alpha^{(l)}]\boldsymbol{u})\right)^{-}. \tag{7.54}$$

The optimization (7.16) reads

$$J[\boldsymbol{u}] = \sum_{\alpha=1}^{N} \sum_{k,l=1}^{L} W_\alpha^{(kl)}(\boldsymbol{u})(\boldsymbol{a}_\alpha^{(k)}, \boldsymbol{u})(\boldsymbol{a}_\alpha^{(l)}, \boldsymbol{u}) \to \min. \tag{7.55}$$

Eq. (7.14) is replaced by

$$(W_\alpha^{(kl)}(\boldsymbol{u})) = \left((\boldsymbol{u}, V[\boldsymbol{a}_\alpha^{(k)}, \boldsymbol{a}_\alpha^{(l)}]\boldsymbol{u})\right)_r^{-}, \tag{7.56}$$

where r is the rank of the hypothesis (7.46) and $V[\boldsymbol{a}_\alpha^{(k)}, \boldsymbol{a}_\alpha^{(l)}]$ is the value of the covariance matrix $\bar{V}[\boldsymbol{a}_\alpha^{(k)}, \boldsymbol{a}_\alpha^{(l)}]$ evaluated at $\boldsymbol{a}_\alpha^{(k)}$ and $\boldsymbol{a}_\alpha^{(l)}$.

From eq. (7.30), the covariance matrix of the optimal estimate $\hat{\boldsymbol{u}}$ is obtained in the following form:

$$V[\hat{\boldsymbol{u}}] = \left(\sum_{\alpha=1}^{N} \sum_{k,l=1}^{L} W_\alpha^{(kl)}(\hat{\boldsymbol{u}})(\boldsymbol{P}_{\hat{\boldsymbol{u}}}^{\mathcal{U}} \hat{\boldsymbol{a}}_\alpha^{(k)})(\boldsymbol{P}_{\hat{\boldsymbol{u}}}^{\mathcal{U}} \hat{\boldsymbol{a}}_\alpha^{(l)})^\top\right)^{-}. \tag{7.57}$$

Here, $\hat{\boldsymbol{a}}_\alpha^{(k)} = \boldsymbol{a}_\alpha^{(k)} - \Delta \boldsymbol{a}_\alpha^{(k)}$ is the value of $\boldsymbol{a}_\alpha^{(k)}$ corrected by using eq. (7.53), in which $\bar{W}_\alpha^{(lm)}$ and $\bar{V}[\boldsymbol{a}_\alpha^{(k)}, \boldsymbol{a}_\alpha^{(k)}]$ are approximated by $W_\alpha^{(lm)}$ and $V[\boldsymbol{a}_\alpha^{(k)}, \boldsymbol{a}_\alpha^{(k)}]$, respectively. To a first approximation, however, we may use the data $\{\boldsymbol{a}_\alpha^{(k)}\}$ themselves instead of $\{\hat{\boldsymbol{a}}_\alpha^{(k)}\}$.

7.2 Optimal Fitting for Image Points

7.2.1 Image point fitting

Given N image points $\{\boldsymbol{x}_\alpha\}$, $\alpha = 1, ..., N$, consider the problem of fitting an image point $\boldsymbol{x}$ to them. This means finding an *optimal average* $\boldsymbol{x}$ of $\{\boldsymbol{x}_\alpha\}$.

We write

$$\boldsymbol{x}_\alpha = \bar{\boldsymbol{x}}_\alpha + \Delta\boldsymbol{x}_\alpha, \tag{7.58}$$

and regard each $\Delta\boldsymbol{x}_\alpha$ as an independent random variable of mean $\boldsymbol{0}$ and covariance matrix $V[\boldsymbol{x}_\alpha]$. The hypothesis is

$$\bar{\boldsymbol{x}}_\alpha = \boldsymbol{x}, \qquad \alpha = 1, ..., N, \tag{7.59}$$

which has rank 2 because both sides are orthogonal to $\boldsymbol{k} = (0,0,1)^\top$.

For each $\boldsymbol{x}_\alpha$, the optimal estimate of its true value $\bar{\boldsymbol{x}}_\alpha$ is evidently $\boldsymbol{x}$, so the optimal average $\hat{\boldsymbol{x}}$ can be obtained by the optimization

$$J[\boldsymbol{x}] = \sum_{\alpha=1}^{N} (\boldsymbol{x}_\alpha - \boldsymbol{x}, V[\boldsymbol{x}_\alpha]^- (\boldsymbol{x}_\alpha - \boldsymbol{x})) \to \min \tag{7.60}$$

under the constraint $\boldsymbol{x} \in \{\boldsymbol{k}\}_L^\perp$. The solution is obtained in the following form:

$$\hat{\boldsymbol{x}} = \left(\sum_{\alpha=1}^{N} V[\boldsymbol{x}_\alpha]^-\right)^- \sum_{\alpha=1}^{N} V[\boldsymbol{x}_\alpha]^- \boldsymbol{x}_\alpha + \boldsymbol{k}. \tag{7.61}$$

Its covariance matrix is

$$V[\hat{\boldsymbol{x}}] = \left(\sum_{\alpha=1}^{N} V[\boldsymbol{x}_\alpha]^-\right)^-, \tag{7.62}$$

which has rank 2; its null space is $\{\boldsymbol{k}\}_L$.

The residual $J[\hat{\boldsymbol{x}}]$ is a χ^2 variable with $2(N-1)$ degrees of freedom[6]. Evidently, the number N of image points must be such that

$$N \geq 1. \tag{7.63}$$

The residual is 0 when $N = 1$. For $N \geq 2$, we obtain a *coincidence test* for image points: the hypothesis that image points $\{\boldsymbol{x}_\alpha\}$ coincide is rejected with significance level $a\%$ if

$$J[\hat{\boldsymbol{x}}] > \chi^2_{2(N-1),a}. \tag{7.64}$$

The $2(N-1)$ degrees of freedom can be intuitively interpreted as follows. An image point has two degrees of freedom, so N image points have $2N$ degrees of freedom for their deviations from the true position $\bar{\boldsymbol{x}}$. But if we consider deviations from the *average* $\hat{\boldsymbol{x}}$, no deviation occurs when $N = 1$. Hence, the degrees of freedom decrease by two.

---

[6] As in Chapter 5, we assume Gaussian noise and do first order analysis whenever we refer to χ^2 distributions and χ^2 tests.

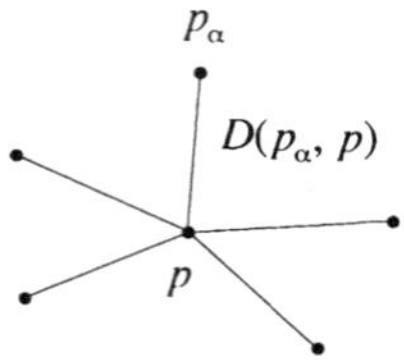

Fig. 7.1. Optimally averaging image points by least squares.

Example 7.1 If each coordinate is perturbed independently by Gaussian noise of mean 0 and variance ϵ^2, the covariance matrix of each image point $\boldsymbol{x}_\alpha$ is $V[\boldsymbol{x}_\alpha] = \epsilon^2 \boldsymbol{P_k}$, so eq. (7.60) reduces to the usual least-squares optimization

$$J_0[\boldsymbol{x}] = \sum_{\alpha=1}^{N} \|\boldsymbol{x}_\alpha - \boldsymbol{x}\|^2 = \sum_{\alpha=1}^{N} D(p_\alpha, p)^2 \to \min, \tag{7.65}$$

where $D(p_\alpha, p)$ is the distance from the αth image point p_α to the image point p to be fitted (Fig. 7.1). The optimal average given by eq. (7.61) reduces to the sample average

$$\hat{\boldsymbol{x}} = \frac{1}{N} \sum_{\alpha=1}^{N} \boldsymbol{x}_\alpha. \tag{7.66}$$

Its covariance matrix is

$$V[\hat{\boldsymbol{x}}] = \frac{\epsilon^2}{N} \boldsymbol{P_k}. \tag{7.67}$$

An unbiased estimator of the variance ϵ^2 is obtained in the form

$$\hat{\epsilon}^2 = \frac{1}{2(N-1)} \sum_{\alpha=1}^{N} \|\boldsymbol{x}_\alpha - \hat{\boldsymbol{x}}\|^2. \tag{7.68}$$

If the value ϵ is given a priori, the coincidence test takes the form

$$\frac{\hat{\epsilon}^2}{\epsilon^2} > \frac{\chi^2_{2(N-1),a}}{2(N-1)}. \tag{7.69}$$

7.2.2 Image line fitting

Given N image points $\{\boldsymbol{x}_\alpha\}$, $\alpha = 1, ..., N$, consider the problem of fitting an image line $(\boldsymbol{n}, \boldsymbol{x}) = 0$ to them. We write

$$\boldsymbol{x}_\alpha = \bar{\boldsymbol{x}}_\alpha + \Delta \boldsymbol{x}_\alpha, \tag{7.70}$$

and regard each $\Delta \boldsymbol{x}_\alpha$ as an independent random variable with mean $\boldsymbol{0}$ and covariance matrix $V[\boldsymbol{x}_\alpha]$. The hypothesis is

$$(\boldsymbol{n}, \bar{\boldsymbol{x}}_\alpha) = 0, \qquad \alpha = 1, ..., N, \tag{7.71}$$

which has rank 1. An optimal estimate of $\boldsymbol{n}$ can be obtained by the optimization

$$J[\boldsymbol{n}] = \sum_{\alpha=1}^{N} \frac{(\boldsymbol{n}, \boldsymbol{x}_\alpha)^2}{(\boldsymbol{n}, V[\boldsymbol{x}_\alpha]\boldsymbol{n})} \to \min \tag{7.72}$$

under the constraint $\|\boldsymbol{n}\| = 1$. The covariance matrix of the solution $\hat{\boldsymbol{n}}$ is

$$V[\hat{\boldsymbol{n}}] = \left(\sum_{\alpha=1}^{N} \frac{(\boldsymbol{P}_{\hat{\boldsymbol{n}}}\hat{\boldsymbol{x}}_\alpha)(\boldsymbol{P}_{\hat{\boldsymbol{n}}}\hat{\boldsymbol{x}}_\alpha)^\top}{(\hat{\boldsymbol{n}}, V[\boldsymbol{x}_\alpha]\hat{\boldsymbol{n}})} \right)^-, \tag{7.73}$$

where $\boldsymbol{P}_{\hat{\boldsymbol{n}}}$ and $\hat{\boldsymbol{x}}_\alpha$ are the projection matrix along $\hat{\boldsymbol{n}}$ and the optimally corrected value of $\boldsymbol{x}_\alpha$, respectively (see eq. (5.82)). The rank of $V[\hat{\boldsymbol{n}}]$ is 2; its null space is $\{\hat{\boldsymbol{n}}\}_L$.

The residual $J[\hat{\boldsymbol{n}}]$ is a χ^2 variable with $N-2$ degrees of freedom. Evidently, the number N of image points must be such that

$$N \geq 2. \tag{7.74}$$

The residual is 0 when $N = 2$. For $N \geq 3$, we obtain a *collinearity test* for image points: the hypothesis that image points $\{\boldsymbol{x}_\alpha\}$ are collinear is rejected with significance level $a\%$ if

$$J[\hat{\boldsymbol{n}}] > \chi^2_{N-2,a}. \tag{7.75}$$

The $N-2$ degrees of freedom can be intuitively interpreted as follows. A free image point has two degrees of freedom. An image point constrained to be on an image line keeps the incidence if it moves along that image line, so there remains one degree of freedom to break the incidence. Consequently, N image points have N degrees of freedom for their deviations from the true image line $(\bar{\boldsymbol{n}}, \boldsymbol{x}) = 0$. But if we consider their deviations from the *fitted* image line $(\hat{\boldsymbol{n}}, \boldsymbol{x}) = 0$, no deviations occur when $N = 2$. Hence, the degrees of freedom decrease by two.

Example 7.2 If each coordinate is perturbed independently by Gaussian noise of mean 0 and variance ϵ^2, image point $\boldsymbol{x}_\alpha$ has covariance matrix $V[\boldsymbol{x}_\alpha] = \epsilon^2 \boldsymbol{P}_{\boldsymbol{k}}$, so eq. (7.72) reduces to the least-squares optimization

$$J_0[\boldsymbol{n}] = \sum_{\alpha=1}^{N} \frac{(\boldsymbol{n}, \boldsymbol{x}_\alpha)^2}{1 - (\boldsymbol{k}, \boldsymbol{n})^2} = \sum_{\alpha=1}^{N} D(p_\alpha, l)^2 \to \min, \tag{7.76}$$

where $D(p_\alpha, l)$ is the distance from the αth image point p_α to the image line l to be fitted (Fig. 7.2; see eq. (4.11)). The covariance matrix of the solution $\hat{\boldsymbol{n}}$ for image points $\{(x_\alpha, y_\alpha)\}$ is given by

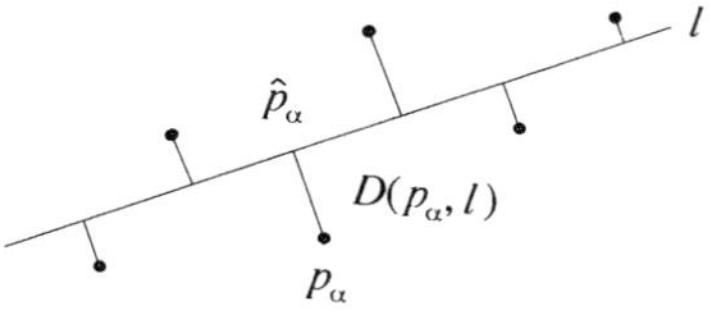

Fig. 7.2. Line fitting by least squares.

$$V[\hat{\boldsymbol{n}}] = \frac{\epsilon^2}{1+\hat{d}^2} \begin{pmatrix} \sum_{\alpha=1}^N \hat{x}_\alpha^2 & \sum_{\alpha=1}^N \hat{x}_\alpha \hat{y}_\alpha & \sum_{\alpha=1}^N \hat{x}_\alpha \\ \sum_{\alpha=1}^N \hat{y}_\alpha \hat{x}_\alpha & \sum_{\alpha=1}^N \hat{y}_\alpha^2 & \sum_{\alpha=1}^N \hat{y}_\alpha \\ \sum_{\alpha=1}^N \hat{x}_\alpha & \sum_{\alpha=1}^N \hat{y}_\alpha & N \end{pmatrix}^-, \tag{7.77}$$

where $\hat{d}$ is the distance of the image line $(\hat{\boldsymbol{n}}, \boldsymbol{x}) = 0$ from the image origin o (see eq. (4.9)). Here, $(\hat{x}_\alpha, \hat{y}_\alpha)$ is the optimally corrected position of (x_α, y_α), i.e., orthogonal projection onto the fitted image line (see eq. (5.85)). If we write $\hat{\boldsymbol{n}} = (\hat{A}, \hat{B}, \hat{C})^\top$, the residual of the minimization (7.76) can be written as

$$J_0[\hat{\boldsymbol{n}}] = (1+\hat{d}^2) \sum_{\alpha=1}^N (\hat{A}x_\alpha + \hat{B}y_\alpha + \hat{C})^2. \tag{7.78}$$

Hence, an unbiased estimator of the variance ϵ^2 is obtained in the form

$$\hat{\epsilon}^2 = \frac{1+\hat{d}^2}{N-2} \sum_{\alpha=1}^N (\hat{A}x_\alpha + \hat{B}y_\alpha + \hat{C})^2. \tag{7.79}$$

If the value ϵ is given a priori, the collinearity test takes the form

$$\frac{\hat{\epsilon}^2}{\epsilon^2} > \frac{\chi^2_{N-2,a}}{N-2}. \tag{7.80}$$

The solution of the least-squares optimization (7.76) can be obtained analytically. In fact, let $x \cos\theta + y \sin\theta = d$ be the image line to be fitted. The function to minimize can be written in the following form:

$$J(\theta, d) = \sum_{\alpha=1}^N (x_\alpha \cos\theta + y_\alpha \sin\theta - d)^2. \tag{7.81}$$

Differentiating this with respect to d and setting the result 0, we obtain d in the form

$$d = \bar{x}\cos\theta + \bar{y}\sin\theta, \tag{7.82}$$

where

$$\bar{x} = \frac{1}{N}\sum_{\alpha=1}^N x_\alpha, \qquad \bar{y} = \frac{1}{N}\sum_{\alpha=1}^N y_\alpha. \tag{7.83}$$

Eq. (7.82) states that the image line should pass through the *centroid* $(\bar{x}, \bar{y})$ of the data $\{(x_\alpha, y_\alpha)\}$. Substituting eq. (7.82) into eq. (7.81), we obtain a function of θ to minimize in the form

$$J(\theta) = \sum_{\alpha=1}^{N} \Big((x_\alpha - \bar{x}) \cos\theta + (y_\alpha - \bar{y}) \sin\theta \Big)^2. \tag{7.84}$$

If we put $\vec{n} = (\cos\theta, \sin\theta)^\top$, this equation can be rewritten as

$$J[\vec{n}] = (\vec{n}, M\vec{n}), \tag{7.85}$$

where M is the two-dimensional *moment matrix*

$$M = \begin{pmatrix} \sum_{\alpha=1}^N (x_\alpha - \bar{x})^2 & \sum_{\alpha=1}^N (x_\alpha - \bar{x})(y_\alpha - \bar{y}) \\ \sum_{\alpha=1}^N (y_\alpha - \bar{y})(x_\alpha - \bar{x}) & \sum_{\alpha=1}^N (y_\alpha - \bar{y})^2 \end{pmatrix}. \tag{7.86}$$

Since $\vec{n}$ is a unit vector, $J[\vec{n}]$ is minimized by the unit eigenvector of the moment matrix M for the smallest eigenvalue (see eqs. (2.86)). The value of d is given by eq. (7.82).

Example 7.3 Suppose N is an odd number in Example 7.2. Let w be the distance between (x_1, y_1) and (x_N, y_N). Put

$$\boldsymbol{x}_C = \begin{pmatrix} x_{(N+1)/2} \\ y_{(N+1)/2} \\ 1 \end{pmatrix}, \tag{7.87}$$

which represents the midpoint. Let $\boldsymbol{u}$ be the unit vector that indicates the orientation of the image line $(\hat{\boldsymbol{n}}, \boldsymbol{x}) = 0$. If $\{(x_\alpha, y_\alpha)\}$, $\alpha = 1, ..., N$, are approximately equidistant, we have

$$\boldsymbol{x}_\alpha \approx \boldsymbol{x}_C + \frac{w}{N-1}(\alpha - \frac{N+1}{2})\boldsymbol{u}. \tag{7.88}$$

Then, eq. (7.77) is approximated by

$$V[\hat{\boldsymbol{n}}] \approx \frac{\epsilon^2}{N(1+\hat{d}^2)} \left(\boldsymbol{x}_C \boldsymbol{x}_C^\top + \frac{w^2(N+1)}{12(N-1)} \boldsymbol{u}\boldsymbol{u}^\top \right)^-. \tag{7.89}$$

If the number N of the data points is very large, this expression gives the covariance matrix of the image line fitted to an *edge segment* having length w, orientation $\boldsymbol{u}$, and midpoint $\boldsymbol{x}_C$ (Fig. 7.3). Define the *edge density* (the number of edge pixels per unit length) ρ by

$$\rho = \frac{N}{w}. \tag{7.90}$$

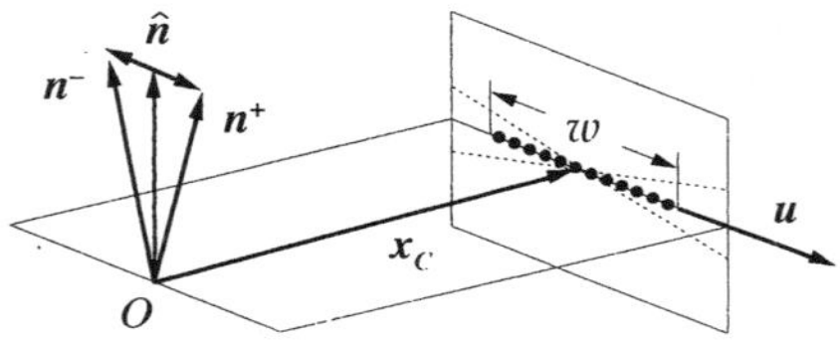

Fig. 7.3. Line fitting to an edge segment.

In the limit $N \to \infty$, we have the following asymptotic expression:

$$V[\hat{\boldsymbol{n}}] = \frac{12\epsilon^2}{\rho w^3(1+\hat{d}^2)}\left(\boldsymbol{u}\boldsymbol{u}^\top + \frac{12}{w^2}\boldsymbol{x}_C\boldsymbol{x}_C^\top\right)^- . \tag{7.91}$$

The distance $\hat{d}$ of the fitted line from the image origin o is usually very small as compared to the focal length (which we take as the unit of length). If the edge segment is near the image origin o, we can assume that $(\boldsymbol{x}_C, \boldsymbol{u}) \approx 0$ and $\|\boldsymbol{x}_C\| \approx 1$. The length w of the edge segment is usually very small as compared with the focal length, so $1 \ll 12/w^2$. Then, eq. (7.91) has the following approximation:

$$V[\hat{\boldsymbol{n}}] \approx \frac{12\epsilon^2}{\rho w^3}\boldsymbol{u}\boldsymbol{u}^\top. \tag{7.92}$$

It follows that the primary deviation pair is approximated as follows (see Section 4.5.3):

$$\boldsymbol{n}^+ \approx N[\hat{\boldsymbol{n}} + \frac{2\sqrt{3}\epsilon}{\rho^{1/2}w^{3/2}}\boldsymbol{u}], \qquad \boldsymbol{n}^- \approx N[\hat{\boldsymbol{n}} - \frac{2\sqrt{3}\epsilon}{\rho^{1/2}w^{3/2}}\boldsymbol{u}]. \tag{7.93}$$

We see from this that the fitted line is very likely to pass near the midpoint of the edge segment. We also see that error is approximately proportional to $\rho^{-1/2}$ and $w^{-3/2}$ (cf. Example 4.2 in Section 4.1.3).

7.3 Optimal Fitting for Image Lines

7.3.1 Image point fitting

Given N image lines $\{(\boldsymbol{n}_\alpha, \boldsymbol{x}) = 0\}$, $\alpha = 1, ..., N$, consider the problem of fitting an image point $\boldsymbol{x}$ to them. This means estimating their *common intersection*. We write

$$\boldsymbol{n}_\alpha = \bar{\boldsymbol{n}}_\alpha + \Delta\boldsymbol{n}_\alpha, \tag{7.94}$$

and regard each $\Delta\boldsymbol{n}_\alpha$ as an independent random variable with mean $\mathbf{0}$ and covariance matrix $V[\boldsymbol{n}_\alpha]$. The hypothesis is

$$(\bar{\boldsymbol{n}}_\alpha, \boldsymbol{x}) = 0, \qquad \alpha = 1, ..., N, \tag{7.95}$$

which has rank 1. An optimal estimate of $\boldsymbol{x}$ can be obtained by the optimization

$$J[\boldsymbol{x}] = \sum_{\alpha=1}^{N} \frac{(\boldsymbol{n}_\alpha, \boldsymbol{x})^2}{(\boldsymbol{x}, V[\boldsymbol{n}_\alpha]\boldsymbol{x})} \to \min \tag{7.96}$$

under the constraint $\boldsymbol{x} \in \{\boldsymbol{k}\}_L^\perp$. The covariance matrix of the solution $\hat{\boldsymbol{x}}$ is given by

$$V[\hat{\boldsymbol{x}}] = \left(\sum_{\alpha=1}^{N} \frac{(\boldsymbol{P_k}\hat{\boldsymbol{n}}_\alpha)(\boldsymbol{P_k}\hat{\boldsymbol{n}}_\alpha)^\top}{(\hat{\boldsymbol{x}}, V[\boldsymbol{n}_\alpha]\hat{\boldsymbol{x}})} \right)^-, \tag{7.97}$$

where $\hat{\boldsymbol{n}}_\alpha$ is the optimally corrected value of $\boldsymbol{n}_\alpha$ (see eq. (5.91)). The rank of $V[\hat{\boldsymbol{x}}]$ is 2; its null space is $\{\boldsymbol{k}\}_L$.

The residual $J[\hat{\boldsymbol{x}}]$ is a χ^2 variable with $N-2$ degrees of freedom. Evidently, the number N of image lines must be such that

$$N \geq 2. \tag{7.98}$$

The residual is 0 when $N = 2$. For $N \geq 3$, we obtain a *concurrency test* for image lines: the hypothesis that image lines $\{(\boldsymbol{n}_\alpha, \boldsymbol{x}) = 0\}$ are concurrent is rejected with significance level $a\%$ if

$$J[\hat{\boldsymbol{x}}] > \chi^2_{N-2,a}. \tag{7.99}$$

The $N-2$ degrees of freedom can be intuitively interpreted as follows. A free image line has two degrees of freedom. An image line passing through an image point keeps the incidence if it changes its orientation around that image point, so there remains one degree of freedom to break the incidence. Consequently, N image lines have N degrees of freedom for their deviations from the true intersection $\bar{\boldsymbol{x}}$. But if we consider their deviations from the *estimated* intersection $\hat{\boldsymbol{x}}$, no deviations occur when $N = 2$. Hence, the degrees of freedom decrease by two.

Example 7.4 Suppose each image line l_α is likely to be translated by noise into a position parallel to l_α, and suppose the distance of such a parallel translation is an independent random variable for each image line with mean 0 and standard deviation ϵ, which is assumed to be very small. If image line l_α is represented by $(\boldsymbol{n}_\alpha, \boldsymbol{x}) = 0$, the covariance matrix of $\boldsymbol{n}_\alpha$ for this statistical model is

$$V[\boldsymbol{n}_\alpha] = \epsilon^2(1 - (\boldsymbol{k}, \boldsymbol{n}_\alpha)^2)(\boldsymbol{P}_{\boldsymbol{n}_\alpha}\boldsymbol{k})(\boldsymbol{P}_{\boldsymbol{n}_\alpha}\boldsymbol{k})^\top. \tag{7.100}$$

If image point $\boldsymbol{x}$ is very close to each image line $(\boldsymbol{n}_\alpha, \boldsymbol{x}) = 0$, we see that

$$(\boldsymbol{x}, V[\boldsymbol{n}_\alpha]\boldsymbol{x}) = \epsilon^2(1 - (\boldsymbol{k}, \boldsymbol{n}_\alpha)^2)(\boldsymbol{x}, \boldsymbol{P}_{\boldsymbol{n}_\alpha}\boldsymbol{k})^2 = \epsilon^2(1 - (\boldsymbol{k}, \boldsymbol{n}_\alpha)^2)(\boldsymbol{P}_{\boldsymbol{n}_\alpha}\boldsymbol{x}, \boldsymbol{k})^2$$
$$\approx \epsilon^2(1 - (\boldsymbol{k}, \boldsymbol{n}_\alpha)^2)(\boldsymbol{x}, \boldsymbol{k})^2 = \epsilon^2(1 - (\boldsymbol{k}, \boldsymbol{n}_\alpha)^2). \tag{7.101}$$

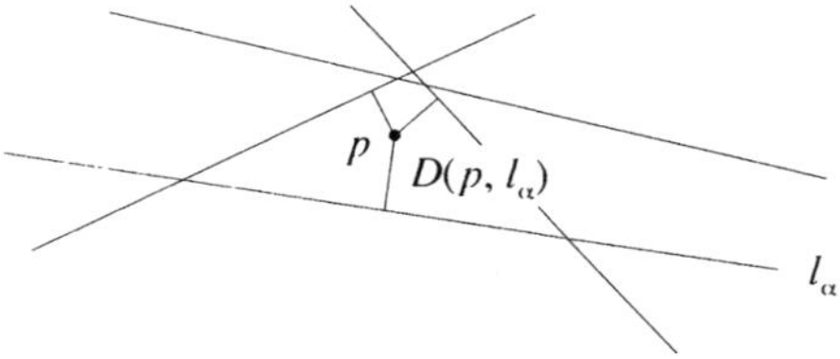

Fig. 7.4. Intersection estimation by least squares.

Hence, eq. (7.96) reduces to the following least-squares optimization (Fig. 7.4; see eq. (4.11)):

$$J_0[\boldsymbol{x}] \approx \sum_{\alpha=1}^{N} \frac{(\boldsymbol{n}_\alpha, \boldsymbol{x})^2}{1-(\boldsymbol{k}, \boldsymbol{n}_\alpha)^2} = \sum_{\alpha=1}^{N} D(p, l_\alpha)^2 \to \min. \tag{7.102}$$

The solution can be obtained analytically. In fact, let $x\cos\theta_\alpha + y\sin\theta_\alpha = d_\alpha$ be the αth image line. The function to minimize can be written in the following form:

$$J(x,y) = \sum_{\alpha=1}^{N} (x\cos\theta_\alpha + y\sin\theta_\alpha - d_\alpha)^2. \tag{7.103}$$

This is a quadratic polynomial in x and y, so the solution is given by solving the following linear equation:

$$\begin{pmatrix} \sum_{\alpha=1}^{N} \cos^2\theta_\alpha & \sum_{\alpha=1}^{N} \cos\theta_\alpha \sin\theta_\alpha \\ \sum_{\alpha=1}^{N} \sin\theta_\alpha \cos\theta_\alpha & \sum_{\alpha=1}^{N} \sin^2\theta_\alpha \end{pmatrix} \begin{pmatrix} x \\ y \end{pmatrix} = \begin{pmatrix} \sum_{\alpha=1}^{N} d_\alpha \cos\theta_\alpha \\ \sum_{\alpha=1}^{N} d_\alpha \sin\theta_\alpha \end{pmatrix}. \tag{7.104}$$

However, the statistical model of noise given by eq. (7.100) is very artificial and unnatural (see Examples 4.2 and 7.3). Hence, the least-squares optimization (7.102) for intersection estimation is not appropriate even though the least-squares optimization (7.77) is appropriate for line fitting.

7.3.2 *Image line fitting*

Given N image lines $\{(\boldsymbol{n}_\alpha, \boldsymbol{x}) = 0\}$, $\alpha = 1, ..., N$, consider the problem of fitting an image line $(\boldsymbol{n}, \boldsymbol{x}) = 0$ to them. This means finding an optimal average $\boldsymbol{n}$ of $\{\boldsymbol{n}_\alpha\}$. We write

$$\boldsymbol{n}_\alpha = \bar{\boldsymbol{n}}_\alpha + \Delta\boldsymbol{n}_\alpha, \tag{7.105}$$

and regard each $\Delta\boldsymbol{n}_\alpha$ as an independent random variable of mean $\boldsymbol{0}$ and covariance matrix $V[\boldsymbol{n}_\alpha]$. The hypothesis is

$$\bar{\boldsymbol{n}}_\alpha = \boldsymbol{n}, \qquad \alpha = 1, ..., N, \tag{7.106}$$

which has rank 2 because both sides are unit vectors[7].

For each $\boldsymbol{n}_\alpha$, the optimal estimate of its true value $\bar{\boldsymbol{n}}_\alpha$ is evidently $\boldsymbol{n}$, so the optimal average $\hat{\boldsymbol{n}}$ can be obtained by the optimization

$$J[\boldsymbol{n}] = \sum_{\alpha=1}^{N} (\boldsymbol{n}_\alpha - \boldsymbol{n}, V[\boldsymbol{n}_\alpha]^- (\boldsymbol{n}_\alpha - \boldsymbol{n})) \to \min \tag{7.107}$$

under the constraint $\|\boldsymbol{n}\| = 1$. The covariance matrix $V[\boldsymbol{n}_\alpha]$ has the null space $\{\boldsymbol{n}_\alpha\}_L$, so $V[\boldsymbol{n}_\alpha]\boldsymbol{n}_\alpha = \boldsymbol{0}$. Hence, eq. (7.107) can be rewritten as

$$J[\boldsymbol{n}] = (\boldsymbol{n}, \left(\sum_{\alpha=1}^{N} V[\boldsymbol{n}_\alpha]^-\right) \boldsymbol{n}) \to \min. \tag{7.108}$$

The solution $\hat{\boldsymbol{n}}$ is given by the unit eigenvector of the matrix

$$\boldsymbol{M} = \sum_{\alpha=1}^{N} V[\boldsymbol{n}_\alpha]^- \tag{7.109}$$

for the smallest eigenvalue (see eqs. (2.86)). The covariance matrix of the solution $\hat{\boldsymbol{n}}$ is

$$V[\hat{\boldsymbol{n}}] = \left(\sum_{\alpha=1}^{N} \boldsymbol{P}_{\hat{\boldsymbol{n}}} V[\boldsymbol{n}_\alpha]^- \boldsymbol{P}_{\hat{\boldsymbol{n}}}\right)^-, \tag{7.110}$$

which has rank 2; its null space is $\{\hat{\boldsymbol{n}}\}_L$.

The residual $J[\hat{\boldsymbol{n}}]$ is a χ^2 variable with $2(N-1)$ degrees of freedom. Evidently, the number N of image lines must be such that

$$N \geq 1. \tag{7.111}$$

The residual is 0 when $N = 1$. For $N \geq 2$, we obtain a *coincidence test* for image lines: the hypothesis that image lines $\{(\boldsymbol{n}_\alpha, \boldsymbol{x}) = 0\}$ coincide is rejected with significance level a% if

$$J[\hat{\boldsymbol{n}}] > \chi^2_{2(N-1),a}. \tag{7.112}$$

The $2(N-1)$ degrees of freedom can be intuitively interpreted as follows. An image line has two degrees of freedom, so N image lines have $2N$ degrees of freedom for their deviations from the true position $(\bar{\boldsymbol{n}}, \boldsymbol{x}) = 0$. But if we consider deviations from the *average* $(\hat{\boldsymbol{n}}, \boldsymbol{x}) = 0$, no deviation occurs when $N = 1$. Hence, the degrees of freedom decrease by two.

[7] This hypothesis is degenerate.

7.4 Optimal Fitting for Space Points

7.4.1 Space point fitting

Given N space points $\{\boldsymbol{r}_\alpha\}$, $\alpha = 1, ..., N$, consider the problem of fitting a space point $\boldsymbol{r}$ to them. This means finding an optimal average $\boldsymbol{r}$ of $\{\boldsymbol{r}_\alpha\}$. We write

$$\boldsymbol{r}_\alpha = \bar{\boldsymbol{r}}_\alpha + \Delta\boldsymbol{r}_\alpha \tag{7.113}$$

and regard each $\Delta\boldsymbol{r}_\alpha$ as an independent random variable of mean $\boldsymbol{0}$ and covariance matrix $V[\boldsymbol{r}_\alpha]$. The hypothesis is

$$\bar{\boldsymbol{r}}_\alpha = \boldsymbol{r}, \qquad \alpha = 1, ..., N, \tag{7.114}$$

which has rank 3.

For each $\boldsymbol{r}_\alpha$, the optimal estimate of its true value $\bar{\boldsymbol{r}}_\alpha$ is evidently $\boldsymbol{r}$, so the optimal average can be obtained by the optimization

$$J[\boldsymbol{r}] = \sum_{\alpha=1}^{N}(\boldsymbol{r}_\alpha - \boldsymbol{r}, V[\boldsymbol{r}_\alpha]^{-1}(\boldsymbol{r}_\alpha - \boldsymbol{r})) \to \min. \tag{7.115}$$

The solution is given by

$$\hat{\boldsymbol{r}} = \left(\sum_{\alpha=1}^{N} V[\boldsymbol{r}_\alpha]^{-1}\right)^{-1} \sum_{\alpha=1}^{N} V[\boldsymbol{r}_\alpha]^{-1}\boldsymbol{r}_\alpha. \tag{7.116}$$

Its covariance matrix is

$$V[\hat{\boldsymbol{r}}] = \left(\sum_{\alpha=1}^{N} V[\boldsymbol{r}_\alpha]^{-1}\right)^{-1}, \tag{7.117}$$

which has rank 3.

The residual $J[\hat{\boldsymbol{r}}]$ is a χ^2 variable with $3(N-1)$ degrees of freedom. Evidently, the number N of space points must be such that

$$N \geq 1. \tag{7.118}$$

The residual is 0 when $N = 1$. For $N \geq 2$, we obtain a *coincidence test* for space points: the hypothesis that space points $\{\boldsymbol{r}_\alpha\}$ coincide is rejected with significance level $a\%$ if

$$J[\hat{\boldsymbol{r}}] > \chi^2_{3(N-1),a}. \tag{7.119}$$

The $3(N-1)$ degrees of freedom can be intuitively interpreted as follows. A space point has three degrees of freedom, so N space points have $3N$ degrees of freedom for their deviations from the true position $\bar{\boldsymbol{r}}$. But if we consider deviations from the *average* $\hat{\boldsymbol{r}}$, no deviation occurs when $N = 1$. Hence, the degrees of freedom decrease by two.

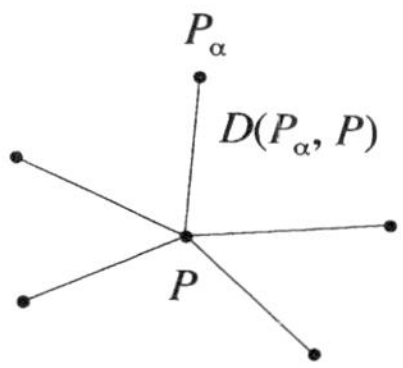

Fig. 7.5. Optimally averaging space points by least squares.

Example 7.5 If each coordinate is perturbed independently by Gaussian noise of mean 0 and variance ϵ^2, the covariance matrix of each space point $\boldsymbol{r}_\alpha$ is $V[\boldsymbol{r}_\alpha] = \epsilon^2 \boldsymbol{I}$, so eq. (7.115) reduces to the least-squares optimization

$$J_0[\boldsymbol{r}] = \sum_{\alpha=1}^{N} \|\boldsymbol{r}_\alpha - \boldsymbol{r}\|^2 = \sum_{\alpha=1}^{N} D(P_\alpha, P)^2 \to \min, \tag{7.120}$$

where $D(P_\alpha, P)$ is the distance from the αth space point P_α to the space point P to be fitted (Fig. 7.5). The optimal average given by eq. (7.116) reduces to

$$\hat{\boldsymbol{r}} = \frac{1}{N}\sum_{\alpha=1}^{N} \boldsymbol{r}_\alpha. \tag{7.121}$$

Its covariance matrix is

$$V[\hat{\boldsymbol{r}}] = \frac{\epsilon^2}{N}\boldsymbol{I}. \tag{7.122}$$

An unbiased estimator of the variance ϵ^2 is obtained in the form

$$\hat{\epsilon}^2 = \frac{1}{3(N-1)}\sum_{\alpha=1}^{N} \|\boldsymbol{r}_\alpha - \hat{\boldsymbol{r}}\|^2. \tag{7.123}$$

If the value ϵ is given a priori, the coincidence test takes the form

$$\frac{\hat{\epsilon}^2}{\epsilon^2} > \frac{\chi^2_{3(N-1),a}}{3(N-1)}. \tag{7.124}$$

7.4.2 Space line fitting

Given N space points $\{\boldsymbol{r}_\alpha\}$, $\alpha = 1, ..., N$, consider the problem of fitting a space line $\boldsymbol{r} \times \boldsymbol{p} = \boldsymbol{n}$ to them. We write

$$\boldsymbol{r}_\alpha = \bar{\boldsymbol{r}}_\alpha + \Delta\boldsymbol{r}_\alpha, \tag{7.125}$$

and regard each $\Delta\boldsymbol{r}_\alpha$ as an independent random variable with mean $\boldsymbol{0}$ and covariance matrix $V[\boldsymbol{r}_\alpha]$. The hypothesis is

$$\bar{\boldsymbol{r}}_\alpha \times \boldsymbol{p} = \boldsymbol{n}, \qquad \alpha = 1, ..., N, \tag{7.126}$$

which has rank 2 because both sides are orthogonal to $\boldsymbol{p}$.

If we let

$$\boldsymbol{e}^{(1)} = \begin{pmatrix} 1 \\ 0 \\ 0 \end{pmatrix}, \quad \boldsymbol{e}^{(2)} = \begin{pmatrix} 0 \\ 1 \\ 0 \end{pmatrix}, \quad \boldsymbol{e}^{(3)} = \begin{pmatrix} 0 \\ 0 \\ 1 \end{pmatrix}, \tag{7.127}$$

and

$$\boldsymbol{r}_\alpha^{(k)} = \boldsymbol{r}_\alpha \times \boldsymbol{e}^{(k)}, \qquad k = 1, 2, 3, \tag{7.128}$$

the hypothesis (7.126) can be equivalently written in the form

$$(\bar{\boldsymbol{r}}_\alpha^{(k)}, \boldsymbol{p}) + (\boldsymbol{e}^{(k)}, \boldsymbol{n}) = 0, \qquad k = 1, 2, 3. \tag{7.129}$$

If we define 6-vectors

$$\boldsymbol{a}_\alpha^{(k)} = \boldsymbol{r}_\alpha^{(k)} \oplus \boldsymbol{e}^{(k)}, \qquad \boldsymbol{u} = \boldsymbol{p} \oplus \boldsymbol{n}, \tag{7.130}$$

eq. (7.129) can be further rewritten in the form

$$(\bar{\boldsymbol{a}}_\alpha^{(k)}, \boldsymbol{u}) = 0, \qquad \alpha = 1, ..., N, \tag{7.131}$$

where the bar refers to the true value. Eq. (7.131) has the same form as eq. (7.46), so the result in Section 7.1.5 can be applied. Since $\boldsymbol{e}^{(k)}$ does not incur noise, the covariance matrix $V[\boldsymbol{a}_\alpha^{(k)}, \boldsymbol{a}_\alpha^{(l)}]$ has the form

$$V[\boldsymbol{a}_\alpha^{(k)}, \boldsymbol{a}_\alpha^{(l)}] = V[\boldsymbol{r}_\alpha^{(k)}, \boldsymbol{r}_\alpha^{(l)}] \oplus \boldsymbol{O}. \tag{7.132}$$

The covariance matrix of $\boldsymbol{r}_\alpha^{(k)}$ is given as follows (see eq. (2.43)):

$$V[\boldsymbol{r}_\alpha^{(k)}, \boldsymbol{r}_\alpha^{(l)}] = \boldsymbol{e}^{(k)} \times V[\boldsymbol{r}_\alpha] \times \boldsymbol{e}^{(l)}. \tag{7.133}$$

From eqs. (7.128) and (7.130), we see that

$$(\boldsymbol{a}_\alpha^{(k)}, \boldsymbol{u}) = -(\boldsymbol{r}_\alpha \times \boldsymbol{p} - \boldsymbol{n}, \boldsymbol{e}^{(k)}). \tag{7.134}$$

Hence, eq. (7.55) reduces to the optimization

$$J[\boldsymbol{p} \oplus \boldsymbol{n}] = \sum_{\alpha=1}^{N} (\boldsymbol{r}_\alpha \times \boldsymbol{p} - \boldsymbol{n}, \boldsymbol{W}_\alpha(\boldsymbol{p})(\boldsymbol{r}_\alpha \times \boldsymbol{p} - \boldsymbol{n})) \to \min \tag{7.135}$$

under the constraints $(\boldsymbol{p}, \boldsymbol{n}) = 0$ and $\|\boldsymbol{p}\|^2 + \|\boldsymbol{n}\|^2 = 1$. The (33)-matrix $\boldsymbol{W}_\alpha(\boldsymbol{p})$ is given by

$$\boldsymbol{W}_\alpha(\boldsymbol{p}) = \left(\boldsymbol{p} \times V[\boldsymbol{r}_\alpha] \times \boldsymbol{p}\right)^{-}. \tag{7.136}$$

The covariance matrix of the solution $\hat{\boldsymbol{p}} \oplus \hat{\boldsymbol{n}}$ is given by eq. (7.57), which reduces to

$$V[\hat{\boldsymbol{p}} \oplus \hat{\boldsymbol{n}}] = \Bigg(\boldsymbol{P}_{\mathcal{N}_{\hat{p}\oplus\hat{n}}} \begin{pmatrix} \sum_{\alpha=1}^N \hat{\boldsymbol{r}}_\alpha \times \boldsymbol{W}_\alpha(\hat{\boldsymbol{p}}) \times \hat{\boldsymbol{r}}_\alpha & \sum_{\alpha=1}^N \hat{\boldsymbol{r}}_\alpha \times \boldsymbol{W}_\alpha(\hat{\boldsymbol{p}}) \\ \sum_{\alpha=1}^N \boldsymbol{W}_\alpha(\hat{\boldsymbol{p}}) \times \hat{\boldsymbol{r}}_\alpha & \sum_{\alpha=1}^N \boldsymbol{W}_\alpha(\hat{\boldsymbol{p}}) \end{pmatrix} \boldsymbol{P}_{\mathcal{N}_{\hat{p}\oplus\hat{n}}}\Bigg)^-, \qquad (7.137)$$

where $\hat{\boldsymbol{r}}_\alpha$ is the optimally corrected value of $\boldsymbol{r}_\alpha$ (see eq. (5.129)). Here, $\boldsymbol{P}_{\mathcal{N}_{\hat{p}\oplus\hat{n}}}$ is the six-dimensional projection matrix onto $\mathcal{N}^\perp_{\hat{\mathbf{p}}\oplus\hat{\boldsymbol{n}}}$ (see eqs. (4.43) and (4.44)). The rank of the covariance matrix $V[\hat{\boldsymbol{p}} \oplus \hat{\boldsymbol{n}}]$ is 4; its null space is $\mathcal{N}_{\hat{\mathbf{p}}\oplus\hat{\mathbf{n}}}$.

The residual $J[\hat{\boldsymbol{p}} \oplus \hat{\boldsymbol{n}}]$ is a χ^2 variable with $2(N-2)$ degrees of freedom. Evidently, the number N of space points must be such that

$$N \geq 2. \qquad (7.138)$$

The residual is 0 when $N = 2$. For $N \geq 3$, we obtain a *collinearity test* for space points: the hypothesis that space points $\{\boldsymbol{r}_\alpha\}$ are collinear is rejected with significance level $a\%$ if

$$J[\hat{\boldsymbol{p}} \oplus \hat{\boldsymbol{n}}] > \chi^2_{2(N-2),a}. \qquad (7.139)$$

The $2(N-2)$ degrees of freedom can be intuitively interpreted as follows. A free space point has three degrees of freedom. A space point constrained to be on a space line keeps the incidence if it moves along that space line, so there remain two degrees of freedom to break the incidence. Consequently, N space points have $2N$ degrees of freedom for their deviations from the true image line $\boldsymbol{r} \times \bar{\boldsymbol{p}} = \bar{\boldsymbol{n}}$. But if we consider their deviations from the *fitted* space line $\boldsymbol{r} \times \hat{\boldsymbol{p}} = \hat{\boldsymbol{n}}$, no deviations occur when $N = 2$. Hence, the degrees of freedom decrease by four.

Example 7.6 If each coordinate is perturbed independently by Gaussian noise of mean 0 and variance ϵ^2, space point $\boldsymbol{r}_\alpha$ has covariance matrix $V[\boldsymbol{r}_\alpha] = \epsilon^2\boldsymbol{I}$, so eq. (7.136) can be rewritten as follows (see eq. (2.42)):

$$\boldsymbol{W}_\alpha(\boldsymbol{p}) = \left(\epsilon^2\boldsymbol{p} \times \boldsymbol{I} \times \boldsymbol{p}\right)^- = \frac{1}{\epsilon^2}\left(\|\boldsymbol{p}\|^2\boldsymbol{I} - \boldsymbol{p}\boldsymbol{p}^\top\right)^-$$
$$= \frac{1}{\epsilon^2\|\boldsymbol{p}\|^2}\left(\boldsymbol{I} - \frac{\boldsymbol{p}\boldsymbol{p}^\top}{\|\boldsymbol{p}\|^2}\right). \qquad (7.140)$$

Then, eq. (7.135) reduces to the least-squares optimization

$$J_0[\boldsymbol{p} \oplus \boldsymbol{n}] = \sum_{\alpha=1}^N \frac{\|\boldsymbol{r}_\alpha \times \boldsymbol{p} - \boldsymbol{n}\|^2}{\|\boldsymbol{p}\|^2} = \sum_{\alpha=1}^N D(P_\alpha, L)^2 \to \min, \qquad (7.141)$$

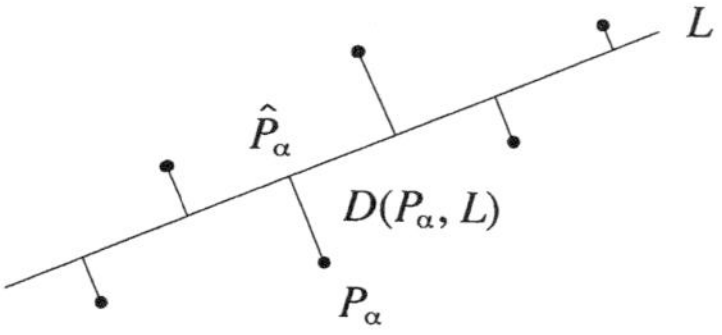

Fig. 7.6. Space line fitting by least squares.

where $D(P_\alpha, L)$ is the distance from the αth space point to the space line L to be fitted (Fig. 7.6; see eq. (4.49)). Eq. (7.137) reduces to

$$V[\hat{\boldsymbol{p}} \oplus \hat{\boldsymbol{n}}] = \epsilon^2 \|\hat{\boldsymbol{p}}\|^2 \Bigg(\boldsymbol{P}_{\mathcal{N}_{\hat{p}\oplus\hat{n}}} \begin{pmatrix} \sum_{\alpha=1}^N \hat{\boldsymbol{r}}_\alpha \times \boldsymbol{I} \times \hat{\boldsymbol{r}}_\alpha & \sum_{\alpha=1}^N \hat{\boldsymbol{r}}_\alpha \times \boldsymbol{I} \\ \sum_{\alpha=1}^N (\hat{\boldsymbol{r}}_\alpha \times \boldsymbol{I})^\top & N\boldsymbol{I} \end{pmatrix} \boldsymbol{P}_{\mathcal{N}_{\hat{p}\oplus\hat{n}}} \Bigg)^-, \tag{7.142}$$

where $\hat{\boldsymbol{r}}_\alpha$ is the orthogonal projection of $\boldsymbol{r}_\alpha$ onto the fitted space line (see eq. (5.131)). An unbiased estimator of ϵ^2 is obtained in the form

$$\hat{\epsilon}^2 = \frac{1}{2(N-2)} \sum_{\alpha=1}^N \frac{\|\boldsymbol{r}_\alpha \times \hat{\boldsymbol{p}} - \hat{\boldsymbol{n}}\|^2}{\|\hat{\boldsymbol{p}}\|^2}. \tag{7.143}$$

If the value ϵ is given a priori, the collinearity test takes the form

$$\frac{\hat{\epsilon}^2}{\epsilon^2} > \frac{\chi^2_{2(N-2),a}}{2(N-2)}. \tag{7.144}$$

The solution of the least-squares optimization (7.141) can be obtained analytically. In fact, let $(\boldsymbol{r} - \boldsymbol{r}_H) \times \boldsymbol{m} = \boldsymbol{0}$ be the $\{\boldsymbol{m}, \boldsymbol{r}_H\}$-representation of the space line to be fitted. The function to minimize can be written in the following form (see eq. (4.49)):

$$J[\boldsymbol{m}, \boldsymbol{r}_H] = \sum_{\alpha=1}^N \|\boldsymbol{P}_{\boldsymbol{m}} \boldsymbol{r}_\alpha - \boldsymbol{r}_H\|^2. \tag{7.145}$$

Differentiating this with respect to $\boldsymbol{r}_H$ and setting the result $\boldsymbol{0}$, we obtain $\boldsymbol{r}_H$ in the form

$$\boldsymbol{r}_H = \boldsymbol{P}_{\boldsymbol{m}} \bar{\boldsymbol{r}}, \qquad \bar{\boldsymbol{r}} = \frac{1}{N} \sum_{\alpha=1}^N \boldsymbol{r}_\alpha. \tag{7.146}$$

This means that the space line should passes through the *centroid* $\bar{\boldsymbol{r}}$ of the data $\{\boldsymbol{r}_\alpha\}$. If we note the identity $\|\boldsymbol{P}_{\boldsymbol{m}} \boldsymbol{a}\|^2 = \|\boldsymbol{a}\|^2 - (\boldsymbol{m}, \boldsymbol{a})^2$ for an arbitrary vector $\boldsymbol{a}$ (see Fig. 2.2), eq. (7.145) reduces to a function of $\boldsymbol{m}$ in the form

$$J[\boldsymbol{m}] = \sum_{\alpha=1}^{N} \|\boldsymbol{P_m}(\boldsymbol{r}_\alpha - \bar{\boldsymbol{r}})\|^2 = \sum_{\alpha=1}^{N} \|\boldsymbol{r}_\alpha - \bar{\boldsymbol{r}}\|^2 - \sum_{\alpha=1}^{N} (\boldsymbol{m}, \boldsymbol{r}_\alpha - \bar{\boldsymbol{r}})^2$$
$$= \sum_{\alpha=1}^{N} \|\boldsymbol{r}_\alpha - \bar{\boldsymbol{r}}\|^2 - (\boldsymbol{m}, \boldsymbol{M}\boldsymbol{m}), \tag{7.147}$$

where $\boldsymbol{M}$ is the *moment matrix*

$$\boldsymbol{M} = \sum_{\alpha=1}^{N} (\boldsymbol{r}_\alpha - \bar{\boldsymbol{r}})(\boldsymbol{r}_\alpha - \bar{\boldsymbol{r}})^\top. \tag{7.148}$$

The function $J[\boldsymbol{m}]$ is minimized if $(\boldsymbol{m}, \boldsymbol{M}\boldsymbol{m})$ is maximized. Since $\boldsymbol{m}$ is a unit vector, the solution is obtained as the unit eigenvector of the moment matrix $\boldsymbol{M}$ for the largest eigenvalue (see eqs. (2.86)). The vector $\boldsymbol{r}_H$ is given by eq. (7.146).

7.4.3 Space plane fitting

Given N space points $\{\boldsymbol{\rho}_\alpha\}$, $\alpha = 1, ..., N$, consider the problem of fitting a space plane $(\boldsymbol{\nu}, \boldsymbol{\rho}) = 0$ to them. We write

$$\boldsymbol{\rho}_\alpha = \bar{\boldsymbol{\rho}}_\alpha + \Delta\boldsymbol{\rho}_\alpha, \tag{7.149}$$

and regard each $\Delta\boldsymbol{\rho}_\alpha$ as an independent random variable with mean $\boldsymbol{0}$ and covariance matrix $V[\boldsymbol{\rho}_\alpha]$. The hypothesis is

$$(\boldsymbol{\nu}, \bar{\boldsymbol{\rho}}_\alpha) = 0, \qquad \alpha = 1, ..., N, \tag{7.150}$$

which has rank 1. An optimal estimate $\hat{\boldsymbol{\nu}}$ can be obtained by the optimization

$$J[\boldsymbol{\nu}] = \sum_{\alpha=1}^{N} \frac{(\boldsymbol{\nu}, \boldsymbol{\rho}_\alpha)^2}{(\boldsymbol{\nu}, V[\boldsymbol{\rho}_\alpha]\boldsymbol{\nu})} \to \min \tag{7.151}$$

under the constraint $\|\boldsymbol{\nu}\| = 1$. The covariance matrix of the solution $\hat{\boldsymbol{\nu}}$ is

$$V[\hat{\boldsymbol{\nu}}] = \left(\sum_{\alpha=1}^{N} \frac{(\boldsymbol{P}_{\hat{\boldsymbol{\nu}}}\hat{\boldsymbol{\rho}}_\alpha)(\boldsymbol{P}_{\hat{\boldsymbol{\nu}}}\hat{\boldsymbol{\rho}}_\alpha)^\top}{(\hat{\boldsymbol{\nu}}, V[\boldsymbol{\rho}_\alpha]\hat{\boldsymbol{\nu}})} \right)^-, \tag{7.152}$$

where $\hat{\boldsymbol{\rho}}_\alpha$ is the optimally corrected value of $\boldsymbol{\rho}_\alpha$ (see eq. (5.160)). The rank of $V[\hat{\boldsymbol{\nu}}]$ is 3; its null space is $\{\hat{\boldsymbol{\nu}}\}_L$.

The residual $J[\hat{\boldsymbol{\nu}}]$ is a χ^2 variable with $N-3$ degrees of freedom. Evidently, the number N of space points must be such that

$$N \geq 3. \tag{7.153}$$

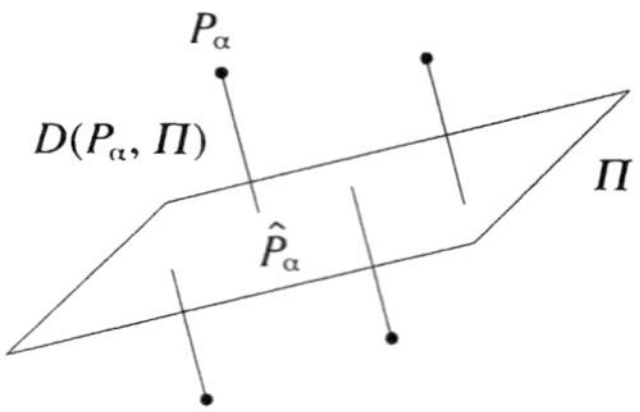

Fig. 7.7. Space plane fitting by least squares.

The residual is 0 when $N = 3$. For $N \geq 4$, we obtain a *coplanarity test* for space points: the hypothesis that space points $\{\boldsymbol{r}_\alpha\}$ are coplanar is rejected with significance level $a\%$ if

$$J[\hat{\boldsymbol{\nu}}] > \chi^2_{N-3,a}. \tag{7.154}$$

The $N - 3$ degrees of freedom can be intuitively interpreted as follows. A free space point has three degrees of freedom. A space point constrained to be on a space plane keeps the incidence if it moves within that space plane, so there remains one degree of freedom to break the incidence. Consequently, N space points have N degrees of freedom for their deviations form the true space plane $(\bar{\boldsymbol{\nu}}, \boldsymbol{\rho}) = 0$. But if we consider their deviations from the *fitted* space plane $(\hat{\boldsymbol{\nu}}, \boldsymbol{\rho}) = 0$, no deviations occur when $N = 3$. Hence, the degrees of freedom decrease by three.

Example 7.7 If each coordinate is perturbed independently by Gaussian noise of mean 0 and variance ϵ^2, space point $\boldsymbol{\rho}_\alpha$ has covariance matrix $V[\boldsymbol{\rho}_\alpha]$ $= \epsilon^2 \boldsymbol{I} \oplus 0 = \epsilon^2(\boldsymbol{I} - \boldsymbol{\kappa}\boldsymbol{\kappa}^\top)$, where $\boldsymbol{\kappa} = (0, 0, 0, 1)^\top$. In this case, eq. (7.151) reduces to the least-squares optimization

$$J_0[\boldsymbol{\nu}] = \sum_{\alpha=1}^{N} \frac{(\boldsymbol{\nu}, \boldsymbol{\rho}_\alpha)^2}{1 - (\boldsymbol{\kappa}, \boldsymbol{\nu})^2} = \sum_{\alpha=1}^{N} D(P_\alpha, \Pi)^2 \to \min, \tag{7.155}$$

where $D(P_\alpha, \Pi)$ is the distance from the αth space point to the space line Π to be fitted (Fig. 7.7; see eq. (4.68)). The covariance matrix of the solution $\hat{\boldsymbol{\nu}}$ for space points $\{(X_\alpha, Y_\alpha, Z_\alpha)\}$ is given in the form

$$V[\hat{\boldsymbol{\nu}}] = \frac{\epsilon^2}{1 + \hat{d}^2} \begin{pmatrix} \sum_{\alpha=1}^N \hat{X}_\alpha^2 & \sum_{\alpha=1}^N \hat{X}_\alpha \hat{Y}_\alpha & \sum_{\alpha=1}^N \hat{X}_\alpha \hat{Z}_\alpha & \sum_{\alpha=1}^N \hat{X}_\alpha \\ \sum_{\alpha=1}^N \hat{Y}_\alpha \hat{X}_\alpha & \sum_{\alpha=1}^N \hat{Y}_\alpha^2 & \sum_{\alpha=1}^N \hat{Y}_\alpha \hat{Z}_\alpha & \sum_{\alpha=1}^N \hat{Y}_\alpha \\ \sum_{\alpha=1}^N \hat{X}_\alpha & \sum_{\alpha=1}^N \hat{Y}_\alpha & \sum_{\alpha=1}^N \hat{Z}_\alpha & N \end{pmatrix}^-, \tag{7.156}$$

where $\hat{d}$ is the distance of the space plane $(\hat{\boldsymbol{\nu}}, \boldsymbol{\rho}) = 0$ from the origin O (see eqs. (4.63)). Here, $(\hat{X}_\alpha, \hat{Y}_\alpha, \hat{Z}_\alpha)$ is the orthogonal projection of $(X_\alpha, Y_\alpha, Z_\alpha)$

onto the fitted space plane (see eq. (5.163)). If we write $\hat{\boldsymbol{\nu}} = (\hat{A}, \hat{B}, \hat{C}, \hat{D})^\top$, the residual of the minimization (7.155) can be written as

$$J_0[\hat{\boldsymbol{\nu}}] = (1 + \hat{d}^2) \sum_{\alpha=1}^{N} (\hat{A}X_\alpha + \hat{B}Y_\alpha + \hat{C}Z_\alpha + \hat{D})^2. \tag{7.157}$$

Hence, an unbiased estimator of the variance ϵ^2 is obtained in the form

$$\hat{\epsilon}^2 = \frac{1 + \hat{d}^2}{N - 3} \sum_{\alpha=1}^{N} (\hat{A}X_\alpha + \hat{B}Y_\alpha + \hat{C}Z_\alpha + \hat{D})^2. \tag{7.158}$$

If the value ϵ is given a priori, the coplanarity test takes the form

$$\frac{\hat{\epsilon}^2}{\epsilon^2} > \frac{\chi^2_{N-3,a}}{N-3}. \tag{7.159}$$

The solution of the least-squares optimization (7.155) can be obtained analytically. In fact, let $(\boldsymbol{n}, \boldsymbol{r}) = d$ be the $\{\boldsymbol{n}, d\}$-representation of the space plane to be fitted. The function function to minimize can be written in the following form (see eq. (4.68)):

$$J[\boldsymbol{n}, d] = \sum_{\alpha=1}^{N} \left((\boldsymbol{n}, \boldsymbol{r}_\alpha)^2 - d \right)^2. \tag{7.160}$$

Differentiating this with respect to d and setting the result 0, we obtain d in the form

$$d = (\boldsymbol{n}, \bar{\boldsymbol{r}}), \qquad \bar{\boldsymbol{r}} = \frac{1}{N} \sum_{\alpha=1}^{N} \boldsymbol{r}_\alpha. \tag{7.161}$$

This means that the space plane should pass through the *centroid* $\bar{\boldsymbol{r}}$ of the data $\{\boldsymbol{r}_\alpha\}$. Substituting eq. (7.161) into eq. (7.160), we obtain a function of $\boldsymbol{n}$ to minimize in the form

$$J[\boldsymbol{n}] = \sum_{\alpha=1}^{N} (\boldsymbol{n}, \boldsymbol{r}_\alpha - \bar{\boldsymbol{r}})^2 = (\boldsymbol{n}, \boldsymbol{M}\boldsymbol{n}), \tag{7.162}$$

where $\boldsymbol{M}$ is the *moment matrix*

$$\boldsymbol{M} = \sum_{\alpha=1}^{N} (\boldsymbol{r}_\alpha - \bar{\boldsymbol{r}})(\boldsymbol{r}_\alpha - \bar{\boldsymbol{r}})^\top. \tag{7.163}$$

Since $\boldsymbol{n}$ is a unit vector, $J[\boldsymbol{n}]$ is minimized by the unit eigenvector of the moment matrix M for the smallest eigenvalue (see eqs. (2.86)). The value of d is given by eq. (7.161).

7.5 Optimal Fitting for Space Lines

7.5.1 Space point fitting

Given N space lines $\{\boldsymbol{r} \times \boldsymbol{p}_\alpha = \boldsymbol{n}_\alpha\}$, $\alpha = 1, ..., N$, consider the problem of fitting a space point $\boldsymbol{r}$ to them. This means optimally estimating their common intersection. We write

$$\boldsymbol{p}_\alpha = \bar{\boldsymbol{p}}_\alpha + \Delta \boldsymbol{p}_\alpha, \qquad \boldsymbol{n}_\alpha = \bar{\boldsymbol{n}}_\alpha + \Delta \boldsymbol{n}_\alpha, \tag{7.164}$$

and regard $\Delta \boldsymbol{p}_\alpha$ and $\Delta \boldsymbol{n}_\alpha$ as random variables of mean $\boldsymbol{0}$ and covariance matrix $V[\boldsymbol{p}_\alpha \oplus \boldsymbol{n}_\alpha]$, independent for each α. The hypothesis[8] is

$$\boldsymbol{r} \times \bar{\boldsymbol{p}}_\alpha = \bar{\boldsymbol{n}}_\alpha, \qquad \alpha = 1, ..., N, \tag{7.165}$$

which has rank 2 because both sides are orthogonal to $\bar{\boldsymbol{p}}_\alpha$.

Let $\boldsymbol{\rho} = (X, Y, Z, 1)^\top$, and define $\boldsymbol{e}^{(1)}$, $\boldsymbol{e}^{(2)}$, and $\boldsymbol{e}^{(3)}$ by eqs. (7.127). If we define 4-vector

$$\boldsymbol{a}_\alpha^{(k)} = (\boldsymbol{e}^{(k)} \times \boldsymbol{p}_\alpha) \oplus (\boldsymbol{e}^{(k)}, \boldsymbol{n}_\alpha), \tag{7.166}$$

the hypothesis (7.165) can be rewritten in the form

$$(\bar{\boldsymbol{a}}_\alpha^{(k)}, \boldsymbol{\rho}) = 0, \qquad k = 1, 2, 3, \tag{7.167}$$

where the bar refers to the true value. Since $\boldsymbol{e}^{(k)}$ does not incur noise, the covariance matrix $V[\boldsymbol{a}_\alpha^{(k)}, \boldsymbol{a}_\alpha^{(l)}]$ is given as follows (see eqs. (2.39) and (2.44)):

$$V[\boldsymbol{a}_\alpha^{(k)}, \boldsymbol{a}_\alpha^{(l)}] = \begin{pmatrix} \boldsymbol{e}^{(k)} \times V[\boldsymbol{p}_\alpha] \times \boldsymbol{e}^{(l)} & \boldsymbol{e}^{(k)} \times V[\boldsymbol{p}_\alpha]\boldsymbol{e}^{(l)} \\ (\boldsymbol{e}^{(l)} \times V[\boldsymbol{p}_\alpha]\boldsymbol{e}^{(k)})^\top & (\boldsymbol{e}^{(k)}, V[\boldsymbol{p}_\alpha]\boldsymbol{e}^{(l)}) \end{pmatrix}. \tag{7.168}$$

From eq. (7.166), we see that

$$(\boldsymbol{a}_\alpha^{(k)}, \boldsymbol{\rho}) = -(\boldsymbol{r} \times \boldsymbol{p}_\alpha - \boldsymbol{n}_\alpha, \boldsymbol{e}^{(k)}). \tag{7.169}$$

Hence, the optimization (7.55) can be rewritten in the form

$$J[\boldsymbol{r}] = \sum_{\alpha=1}^{N} (\boldsymbol{r} \times \boldsymbol{p}_\alpha - \boldsymbol{n}_\alpha, \boldsymbol{W}_\alpha(\boldsymbol{r})(\boldsymbol{r} \times \boldsymbol{p}_\alpha - \boldsymbol{n}_\alpha)) \to \min, \tag{7.170}$$

where the (33)-matrix $\boldsymbol{W}_\alpha(\boldsymbol{r})$ is given by

$$\boldsymbol{W}_\alpha(\boldsymbol{r}) = \Big(\boldsymbol{r} \times V[\boldsymbol{p}_\alpha] \times \boldsymbol{r} - \boldsymbol{r} \times V[\boldsymbol{p}_\alpha, \boldsymbol{n}_\alpha] - V[\boldsymbol{p}_\alpha, \boldsymbol{n}_\alpha] \times \boldsymbol{r} + V[\boldsymbol{n}_\alpha]\Big)_2^- . \tag{7.171}$$

[8] This hypothesis is degenerate.

The covariance matrix of the solution $\hat{\boldsymbol{r}}$ is given by eq. (7.57), which reduces to

$$V[\hat{\boldsymbol{r}}] = \left(\sum_{\alpha=1}^{N} \hat{\boldsymbol{p}}_\alpha \times \hat{\boldsymbol{W}}_\alpha(\hat{\boldsymbol{r}}) \times \hat{\boldsymbol{p}}_\alpha \right)^{-1}, \tag{7.172}$$

where $\hat{\boldsymbol{p}}_\alpha$ is the optimally corrected value of $\boldsymbol{p}_\alpha$ (see eqs. (5.137)).

The residual $J[\hat{\boldsymbol{r}}]$ is a χ^2 variable with $2N-3$ degrees of freedom. Evidently, the number N of space lines must be such that

$$N \geq 2. \tag{7.173}$$

However, the residual is *not* 0 when $N = 2$. For $N \geq 3$, we obtain a *concurrency test* for space lines: the hypothesis that space lines $\{\boldsymbol{r} \times \boldsymbol{p}_\alpha = \boldsymbol{n}_\alpha\}$ are concurrent is rejected with significance level $a\%$ if

$$J[\hat{\boldsymbol{r}}] > \chi^2_{2N-3,a}. \tag{7.174}$$

The $2N-3$ degrees of freedom can be intuitively interpreted as follows. A free space line has four degrees of freedom. If it is constrained to pass through a space point, it can still change its orientation freely around that space point, so there remain two degrees of freedom to break the incidence. Consequently, N space lines have $2N$ degrees of freedom for their deviations from the true intersection $\bar{\boldsymbol{r}}$. But we are considering their deviations form the *estimated* intersection $\hat{\boldsymbol{r}}$. For $N = 2$, there remains only one degree of freedom to break the concurrency: the freedom of translation orthogonal to the two space lines. Hence, the total degrees of freedom are $2(N-2)+1 = 2N-3$.

7.5.2 *Space line fitting*

Given N space lines $\{\boldsymbol{r} \times \boldsymbol{p}_\alpha = \boldsymbol{n}_\alpha\}$, $\alpha = 1, ..., N$, consider the problem of fitting a space line $\boldsymbol{r} \times \boldsymbol{p} = \boldsymbol{n}$ to them. This means finding optimal averages $\boldsymbol{p}$ and $\boldsymbol{n}$ of $\{\boldsymbol{p}_\alpha\}$ and $\{\boldsymbol{n}_\alpha\}$, respectively. We write

$$\boldsymbol{p}_\alpha = \bar{\boldsymbol{p}}_\alpha + \Delta\boldsymbol{p}_\alpha, \qquad \boldsymbol{n}_\alpha = \bar{\boldsymbol{n}}_\alpha + \Delta\boldsymbol{n}_\alpha, \tag{7.175}$$

and regard $\Delta\boldsymbol{p}_\alpha$ and $\Delta\boldsymbol{n}_\alpha$ as random variables of mean $\boldsymbol{0}$ and covariance matrix $V[\boldsymbol{p}_\alpha \oplus \boldsymbol{n}_\alpha]$, independent for each α. The hypothesis is

$$\bar{\boldsymbol{p}}_\alpha = \boldsymbol{p}, \qquad \bar{\boldsymbol{n}}_\alpha = \boldsymbol{n}, \qquad \alpha = 1, ..., N, \tag{7.176}$$

which has rank 4 because $\{\boldsymbol{p}, \boldsymbol{n}\}$ have four degrees of freedom[9].

For each α, the optimal estimates of the true values $\bar{\boldsymbol{p}}_\alpha$ and $\bar{\boldsymbol{n}}_\alpha$ are evidently $\boldsymbol{p}$ and $\boldsymbol{n}$, so the optimal averages can be obtained by the optimization

$$J[\boldsymbol{p}, \boldsymbol{n}] = \sum_{\alpha=1}^{N} (\boldsymbol{p}_\alpha \oplus \boldsymbol{n}_\alpha - \boldsymbol{p} \oplus \boldsymbol{n}, V[\boldsymbol{p}_\alpha \oplus \boldsymbol{n}_\alpha]^- (\boldsymbol{p}_\alpha \oplus \boldsymbol{n}_\alpha - \boldsymbol{p} \oplus \boldsymbol{n})) \to \min \tag{7.177}$$

[9] This hypothesis is degenerate.

under the constraints $(\boldsymbol{p}, \boldsymbol{n}) = 0$ and $\|\boldsymbol{p}\|^2 + \|\boldsymbol{n}\|^2 = 1$. Since the covariance matrix $V[\boldsymbol{p}_\alpha \oplus \boldsymbol{n}_\alpha]$ has null space $\mathcal{N}_{\boldsymbol{p}_\alpha \oplus \boldsymbol{n}_\alpha} = \{\boldsymbol{p}_\alpha \oplus \boldsymbol{n}_\alpha, \boldsymbol{n}_\alpha \oplus \boldsymbol{p}_\alpha\}_L \subset \mathcal{R}^6$ (see eq. (4.43)), we have $V[\boldsymbol{p}_\alpha \oplus \boldsymbol{n}_\alpha](\boldsymbol{p}_\alpha \oplus \boldsymbol{n}_\alpha) = \boldsymbol{0}$. Hence, the optimization (7.177) can be rewritten in the form

$$J[\boldsymbol{p}, \boldsymbol{n}] = \sum_{\alpha=1}^{N} (\boldsymbol{p} \oplus \boldsymbol{n}, V[\boldsymbol{p}_\alpha \oplus \boldsymbol{n}_\alpha]^- (\boldsymbol{p} \oplus \boldsymbol{n})) \to \min. \tag{7.178}$$

The solution $\hat{\boldsymbol{p}} \oplus \hat{\boldsymbol{n}}$ is given by the unit eigenvector of the matrix

$$\boldsymbol{M} = \sum_{\alpha=1}^{N} V[\boldsymbol{p}_\alpha \oplus \boldsymbol{n}_\alpha]^- \tag{7.179}$$

for the smallest eigenvalue (see eqs. (2.86)). The covariance matrix of of the solution $\hat{\boldsymbol{p}} \oplus \hat{\boldsymbol{n}}$ is

$$V[\hat{\boldsymbol{p}} \oplus \hat{\boldsymbol{n}}] = \left(\sum_{\alpha=1}^{N} \boldsymbol{P}_{\mathcal{N}_{\hat{p} \oplus \hat{n}}} V[\boldsymbol{p}_\alpha \oplus \boldsymbol{n}_\alpha]^- \boldsymbol{P}_{\mathcal{N}_{\hat{p} \oplus \hat{n}}} \right)^-, \tag{7.180}$$

where $\boldsymbol{P}_{\mathcal{N}_{\hat{p} \oplus \hat{n}}}$ is the six-dimensional projection matrix onto $\mathcal{N}^\perp_{\hat{\boldsymbol{p}} \oplus \hat{\boldsymbol{n}}}$ (see eq. (4.44)). The covariance matrix $V[\hat{\boldsymbol{p}} \oplus \hat{\boldsymbol{n}}]$ has rank 4; its null space is $\mathcal{N}_{\hat{\boldsymbol{p}} \oplus \hat{\boldsymbol{n}}}$.

The residual

$$J[\hat{\boldsymbol{p}}, \hat{\boldsymbol{n}}] = \sum_{\alpha=1}^{N} (\boldsymbol{p}_\alpha \oplus \boldsymbol{n}_\alpha - \hat{\boldsymbol{p}} \oplus \hat{\boldsymbol{n}}, V[\boldsymbol{p}_\alpha \oplus \boldsymbol{n}_\alpha]^- (\boldsymbol{p}_\alpha \oplus \boldsymbol{n}_\alpha - \hat{\boldsymbol{p}} \oplus \hat{\boldsymbol{n}})) \tag{7.181}$$

is a χ^2 variable with $4(N-1)$ degrees of freedom. Evidently, the number N of space lines must be such that

$$N \geq 1. \tag{7.182}$$

The residual is 0 when $N = 1$. For $N \geq 2$, we obtain a *coincidence test* for space lines: the hypothesis that space lines $\{\boldsymbol{r} \times \boldsymbol{p}_\alpha = \boldsymbol{n}_\alpha\}$ coincide is rejected with significance level a% if

$$J[\hat{\boldsymbol{p}}, \hat{\boldsymbol{n}}] > \chi^2_{4(N-1),a}. \tag{7.183}$$

The $4(N-1)$ degrees of freedom can be intuitively interpreted as follows. A free space line has four degrees of freedom, so N space lines have $4N$ degrees of freedom for their deviations from the true position $\boldsymbol{r} \times \bar{\boldsymbol{p}} = \bar{\boldsymbol{n}}$. But if we consider deviations from the *average* $\boldsymbol{r} \times \hat{\boldsymbol{p}} = \hat{\boldsymbol{n}}$, no deviation occurs when $N = 1$. Hence, the degrees of freedom decrease by four.

7.5.3 Space plane fitting

Given N space lines $\{(\boldsymbol{r}-\boldsymbol{r}_H)\times\boldsymbol{m}_\alpha=\boldsymbol{0}\}$, $\alpha=1,\ldots,N$, consider the problem of fitting a space plane $(\boldsymbol{\nu},\boldsymbol{\rho})=0$ to them. We write

$$\boldsymbol{m}_\alpha=\bar{\boldsymbol{m}}_\alpha+\Delta\boldsymbol{m}_\alpha,\qquad \boldsymbol{r}_{H\alpha}=\bar{\boldsymbol{r}}_{H\alpha}+\Delta\boldsymbol{r}_{H\alpha},\tag{7.184}$$

and regard $\Delta\boldsymbol{m}_\alpha$ and $\Delta\boldsymbol{r}_{H\alpha}$ as random variables of mean $\boldsymbol{0}$ and covariance matrix $V[\boldsymbol{m}_\alpha\oplus\boldsymbol{r}_{H\alpha}]$, independent for each α. The hypothesis can be written as follows (see eq. (4.70)):

$$(\boldsymbol{\nu},\bar{\boldsymbol{m}}_\alpha\oplus 0)=0,\qquad (\boldsymbol{\nu},\bar{\boldsymbol{r}}_{H\alpha}\oplus 1)=0,\qquad \alpha=1,\ldots,N.\tag{7.185}$$

The rank of this hypothesis is 2.

If we define 4-vectors

$$\boldsymbol{a}_\alpha^{(1)}=\boldsymbol{m}_\alpha\oplus 0,\qquad \boldsymbol{a}_\alpha^{(2)}=\boldsymbol{r}_{H\alpha}\oplus 1,\tag{7.186}$$

the hypothesis (7.187) can be rewritten in the form

$$(\bar{\boldsymbol{a}}_\alpha^{(k)},\boldsymbol{\rho})=0,\qquad k=1,2,\tag{7.187}$$

where the bar refers to the true value. The covariance matrices $V[\boldsymbol{a}_\alpha^{(k)},\boldsymbol{a}_\alpha^{(l)}]$, $k,l=1,2$, are given as follows:

$$V[\boldsymbol{a}_\alpha^{(1)},\boldsymbol{a}_\alpha^{(1)}]=V[\boldsymbol{m}_\alpha]\oplus 0,\qquad V[\boldsymbol{a}_\alpha^{(1)},\boldsymbol{a}_\alpha^{(2)}]=V[\boldsymbol{m}_\alpha,\boldsymbol{r}_{H\alpha}]\oplus 0,$$
$$V[\boldsymbol{a}_\alpha^{(2)},\boldsymbol{a}_\alpha^{(1)}]=V[\boldsymbol{m}_\alpha,\boldsymbol{r}_{H\alpha}]^\top\oplus 0,\qquad V[\boldsymbol{a}_\alpha^{(2)},\boldsymbol{a}_\alpha^{(2)}]=V[\boldsymbol{r}_{H\alpha}]\oplus 0.\tag{7.188}$$

Hence, eq (7.55) reduces to the optimization

$$J[\boldsymbol{n},d]=\frac{1}{1+d^2}\sum_{\alpha=1}^N\Big(W_\alpha^{(11)}(\boldsymbol{n},d)(\boldsymbol{n},\boldsymbol{m}_\alpha)^2 + 2W_\alpha^{(12)}(\boldsymbol{n},d)(\boldsymbol{n},\boldsymbol{m}_\alpha)((\boldsymbol{n},\boldsymbol{r}_{H\alpha})-d) + W_\alpha^{(22)}(\boldsymbol{n},d)((\boldsymbol{n},\boldsymbol{r}_{H\alpha})-d)^2\Big)\to\min\tag{7.189}$$

under the constraint $\|\boldsymbol{n}\|=1$, where the (22)-matrix $\boldsymbol{W}_\alpha(\boldsymbol{n},d)$ is given by

$$\boldsymbol{W}_\alpha(\boldsymbol{n},d)=\frac{1}{1+d^2}\begin{pmatrix}(\boldsymbol{n},V[\boldsymbol{m}_\alpha]\boldsymbol{n}) & (\boldsymbol{n},V[\boldsymbol{m}_\alpha,\boldsymbol{r}_{H\alpha}]\boldsymbol{n})\\ (\boldsymbol{n},V[\boldsymbol{r}_{H\alpha},\boldsymbol{m}_\alpha]\boldsymbol{n}) & (\boldsymbol{n},V[\boldsymbol{r}_{H\alpha}]\boldsymbol{n})\end{pmatrix}^{-1}.\tag{7.190}$$

The covariance matrix of the solution $\hat{\boldsymbol{\nu}}$ is given by eq. (7.57), which reduces to

$$V[\hat{\boldsymbol{\nu}}]=\left(\boldsymbol{P}_{\hat{\boldsymbol{\nu}}}\begin{pmatrix}\sum_{\alpha=1}^N\Big(W_\alpha^{(11)}(\hat{\boldsymbol{n}},\hat{d})\hat{\boldsymbol{m}}_\alpha\hat{\boldsymbol{m}}_\alpha^\top+2W_\alpha^{(12)}(\hat{\boldsymbol{n}},\hat{d})S[\hat{\boldsymbol{m}}_\alpha\hat{\boldsymbol{r}}_{H\alpha}^\top]+W_\alpha^{(22)}(\hat{\boldsymbol{n}},\hat{d})\hat{\boldsymbol{r}}_{H\alpha}\hat{\boldsymbol{r}}_{H\alpha}^\top\Big) & \sum_{\alpha=1}^N\Big(W_\alpha^{(12)}(\hat{\boldsymbol{n}},\hat{d})\hat{\boldsymbol{m}}_\alpha+W_\alpha^{(22)}(\hat{\boldsymbol{n}},\hat{d})\hat{\boldsymbol{r}}_{H\alpha}\Big)\\ \sum_{\alpha=1}^N\Big(W_\alpha^{(12)}(\hat{\boldsymbol{n}},\hat{d})\hat{\boldsymbol{m}}_\alpha+W_\alpha^{(22)}(\hat{\boldsymbol{n}},\hat{d})\hat{\boldsymbol{r}}_{H\alpha}\Big)^\top & \sum_{\alpha=1}^N W_\alpha^{(22)}(\hat{\boldsymbol{n}},\hat{d})\end{pmatrix}\boldsymbol{P}_{\hat{\boldsymbol{\nu}}}\right)^-,\tag{7.191}$$

where $\hat{m}_\alpha$ and $\hat{r}_{H\alpha}$ are the optimally corrected values of m_α and $r_{H\alpha}$, respectively (see eq. (5.183)). The symbol $S[\cdot]$ denotes the symmetrization operator (see eqs. (2.205)). The rank of $V[\hat{\nu}]$ is 3; its null space is $\{\nu\}_L$.

The residual $J[\hat{n}, \hat{d}]$ is a χ^2 variable with $2N-3$ degrees of freedom. Evidently, the number N of space lines must be such that

$$N \geq 2. \tag{7.192}$$

However, the residual is *not* 0 when $N = 2$. For $N \geq 3$, we obtain a *coplanarity test* for space lines: the hypothesis that space lines $\{(r - r_{H\alpha}) \times m_\alpha = 0\}$ are coplanar is rejected with significance level $a\%$ if

$$J[\hat{n}, \hat{d}] > \chi^2_{2N-3,a}. \tag{7.193}$$

The $2N-3$ degrees of freedom can be intuitively interpreted as follows. A free space line has four degrees of freedom. If it is constrained to be on a space plane, it can still translate and rotate freely within that space plane, so there remain two degrees of freedom to break the incidence. Consequently, N space lines have $2N$ degrees of freedom for their deviations from the true plane $(\bar{\nu}, \rho) = 0$. But we are considering their deviations from the *fitted* plane $(\hat{\nu}, \rho) = 0$. For $N = 2$, there remains only one degree of freedom to break the coplanarity: the freedom of translation orthogonal to the two space lines. Hence, the total degrees of freedom are $2(N-2)+1 = 2N-3$.

7.6 Optimal Fitting for Space Planes

7.6.1 Space point fitting

Given N space planes $\{(\nu_\alpha, \rho) = 0\}$, $\alpha = 1, ..., N$, consider the problem of fitting a space point ρ to them. This means optimally estimating their common intersection point. We write

$$\nu_\alpha = \bar{\nu}_\alpha + \Delta\nu_\alpha, \tag{7.194}$$

and regard each $\Delta\nu_\alpha$ as an independent random variable of mean $\mathbf{0}$ and covariance matrix $V[\nu_\alpha]$. The hypothesis can be written as follows:

$$(\bar{\nu}_\alpha, \rho) = 0, \qquad \alpha = 1, ..., N. \tag{7.195}$$

The rank of this hypothesis is 1. An optimal estimate $\hat{\nu}$ can be obtained by the optimization

$$J[\rho] = \sum_{\alpha=1}^{N} \frac{(\nu_\alpha, \rho)^2}{(\rho, V[\nu_\alpha]\rho)} \to \min \tag{7.196}$$

under the constraint $(\boldsymbol{\kappa}, \boldsymbol{\rho}) = 1$, where $\boldsymbol{\kappa} = (0,0,0,1)^\top$. The covariance matrix of the solution $\hat{\boldsymbol{\rho}}$ is

$$V[\hat{\boldsymbol{\rho}}] = \left(\sum_{\alpha=1}^{N} \frac{(\boldsymbol{P}_{\boldsymbol{\kappa}}\hat{\boldsymbol{\nu}}_\alpha)(\boldsymbol{P}_{\boldsymbol{\kappa}}\hat{\boldsymbol{\nu}}_\alpha)^\top}{(\hat{\boldsymbol{\rho}}, V[\boldsymbol{\nu}_\alpha]\hat{\boldsymbol{\rho}})} \right)^-, \tag{7.197}$$

where $\hat{\boldsymbol{\nu}}_\alpha$ is the optimally corrected value of $\boldsymbol{\nu}_\alpha$ (see eq. (5.169)). The rank of $V[\hat{\boldsymbol{\rho}}]$ is 3; its null space is $\mathcal{N}_{\hat{\rho}} = \{\boldsymbol{\kappa}\}_L$.

The residual $J[\hat{\boldsymbol{\rho}}]$ is a χ^2 variable with $N-3$ degrees of freedom. Evidently, the number N of space lines must be such that

$$N \geq 3. \tag{7.198}$$

The residual is 0 when $N = 3$. For $N \geq 4$, we can test if space planes have a common intersection: the hypothesis that space planes $\{(\boldsymbol{\nu}_\alpha, \boldsymbol{\rho}) = 0\}$ have a common intersection point is rejected with significance level $a\%$ if

$$J[\hat{\boldsymbol{\rho}}] > \chi^2_{N-3,a}. \tag{7.199}$$

The $N-3$ degrees of freedom can be intuitively interpreted as follows. A free space plane has three degrees of freedom. If it is constrained to pass through a space point, it can still change its orientation freely around that space point, so there remains one degree of freedom to break the incidence. Consequently, N space lines have N degrees of freedom for their deviations from the true space point $\bar{\boldsymbol{\rho}}$. But if we consider their deviations from the *estimated* intersection point $\hat{\boldsymbol{\rho}}$, no deviations occur when $N = 3$. Hence, the degrees of freedom decrease by three.

7.6.2 Space line fitting

Given N space planes $\{(\boldsymbol{\nu}_\alpha, \boldsymbol{\rho}) = 0\}$, $\alpha = 1, ..., N$, consider the problem of fitting a space line $(\boldsymbol{r} - \boldsymbol{r}_H) \times \boldsymbol{m} = \boldsymbol{0}$ to them. This means optimally estimating their common intersection line. We write

$$\boldsymbol{\nu}_\alpha = \bar{\boldsymbol{\nu}}_\alpha + \Delta\boldsymbol{\nu}_\alpha, \tag{7.200}$$

and regard each $\Delta\boldsymbol{\nu}_\alpha$ as an independent random variable of mean $\boldsymbol{0}$ and covariance matrix $V[\boldsymbol{\nu}_\alpha]$. The hypothesis can be written as follows (see eq. (4.70)):

$$(\boldsymbol{\nu}_\alpha, \boldsymbol{m} \oplus 0) = 0, \qquad (\boldsymbol{\nu}_\alpha, \boldsymbol{r}_H \oplus 1) = 0, \qquad \alpha = 1, ..., N. \tag{7.201}$$

The rank of this hypothesis is 2.

If we define 8-vectors

$$\boldsymbol{u} = \boldsymbol{m} \oplus 0 \oplus \boldsymbol{r}_H \oplus 1, \qquad \boldsymbol{a}_\alpha^{(1)} = \boldsymbol{\nu}_\alpha \oplus \boldsymbol{0}, \qquad \boldsymbol{a}_\alpha^{(2)} = \boldsymbol{0} \oplus \boldsymbol{\nu}_\alpha, \tag{7.202}$$

the hypothesis (7.201) can be rewritten in the form

$$(\bar{\boldsymbol{a}}_\alpha^{(k)}, \boldsymbol{u}) = 0, \qquad k = 1, 2, \tag{7.203}$$

where the bar refers to the true value. The covariance matrices $V[\boldsymbol{a}_\alpha^{(k)}, \boldsymbol{a}_\alpha^{(l)}]$, $k, l = 1, 2$, are given as follows:

$$V[\boldsymbol{a}_\alpha^{(1)}, \boldsymbol{a}_\alpha^{(1)}] = \begin{pmatrix} V[\boldsymbol{\nu}_\alpha] & \boldsymbol{O} \\ \boldsymbol{O} & \boldsymbol{O} \end{pmatrix}, \quad V[\boldsymbol{a}_\alpha^{(1)}, \boldsymbol{a}_\alpha^{(2)}] = \begin{pmatrix} \boldsymbol{O} & V[\boldsymbol{\nu}_\alpha] \\ \boldsymbol{O} & \boldsymbol{O} \end{pmatrix},$$

$$V[\boldsymbol{a}_\alpha^{(2)}, \boldsymbol{a}_\alpha^{(1)}] = \begin{pmatrix} \boldsymbol{O} & \boldsymbol{O} \\ V[\boldsymbol{\nu}_\alpha] & \boldsymbol{O} \end{pmatrix}, \quad V[\boldsymbol{a}_\alpha^{(2)}, \boldsymbol{a}_\alpha^{(2)}] = \begin{pmatrix} \boldsymbol{O} & \boldsymbol{O} \\ \boldsymbol{O} & V[\boldsymbol{\nu}_\alpha] \end{pmatrix}. \tag{7.204}$$

From eqs. (7.202), eq. (7.55) reduces to the optimization

$$\begin{aligned} J[\boldsymbol{m}, \boldsymbol{r}_H] = \sum_{\alpha=1}^{N} \frac{1}{1 + d_\alpha^2} \Big(& W_\alpha^{(11)}(\boldsymbol{m}, \boldsymbol{r}_H)(\boldsymbol{n}_\alpha, \boldsymbol{m})^2 \\ & + 2W_\alpha^{(12)}(\boldsymbol{m}, \boldsymbol{r}_H)(\boldsymbol{n}_\alpha, \boldsymbol{m})((\boldsymbol{n}_\alpha, \boldsymbol{r}_H) - d_\alpha) \\ & + W_\alpha^{(22)}(\boldsymbol{m}, \boldsymbol{r}_H)((\boldsymbol{n}_\alpha, \boldsymbol{r}_H) - d_\alpha)^2 \Big) \to \min \end{aligned} \tag{7.205}$$

under the constraints $\|\boldsymbol{m}\| = 1$ and $(\boldsymbol{m}, \boldsymbol{r}_H) = 0$, where the (22)-matrix $\boldsymbol{W}_\alpha(\boldsymbol{m}, \boldsymbol{r}_H)$ is given by

$$\boldsymbol{W}_\alpha(\boldsymbol{m}, \boldsymbol{r}_H) = \begin{pmatrix} (\boldsymbol{m} \oplus 0, V[\boldsymbol{\nu}_\alpha](\boldsymbol{m} \oplus 0)) & (\boldsymbol{m} \oplus 0, V[\boldsymbol{\nu}_\alpha](\boldsymbol{r}_H \oplus 1)) \\ (\boldsymbol{r}_H \oplus 1, V[\boldsymbol{\nu}_\alpha](\boldsymbol{m} \oplus 0)) & (\boldsymbol{r}_H \oplus 1, V[\boldsymbol{\nu}_\alpha](\boldsymbol{r}_H \oplus 1)) \end{pmatrix}^{-1}. \tag{7.206}$$

The covariance matrix of the solution $\hat{\boldsymbol{u}}$ is given by eq. (7.57), which reduces to

$$V[\hat{\boldsymbol{u}}] = \left(\boldsymbol{P}_{\mathcal{N}_{\hat{u}}} \begin{pmatrix} \sum_{\alpha=1}^N W_\alpha^{(11)}(\hat{\boldsymbol{m}}, \hat{\boldsymbol{r}}_H)\hat{\boldsymbol{\nu}}_\alpha\hat{\boldsymbol{\nu}}_\alpha^\top & \sum_{\alpha=1}^N W_\alpha^{(12)}(\hat{\boldsymbol{m}}, \hat{\boldsymbol{r}}_H)\hat{\boldsymbol{\nu}}_\alpha\hat{\boldsymbol{\nu}}_\alpha^\top \\ \sum_{\alpha=1}^N W_\alpha^{(12)}(\hat{\boldsymbol{m}}, \hat{\boldsymbol{r}}_H)\hat{\boldsymbol{\nu}}_\alpha\hat{\boldsymbol{\nu}}_\alpha^\top & \sum_{\alpha=1}^N W_\alpha^{(22)}(\hat{\boldsymbol{m}}, \hat{\boldsymbol{r}}_H)\hat{\boldsymbol{\nu}}_\alpha\hat{\boldsymbol{\nu}}_\alpha^\top \end{pmatrix} \boldsymbol{P}_{\mathcal{N}_{\hat{u}}} \right)^-, \tag{7.207}$$

where $\hat{\boldsymbol{\nu}}_\alpha$ is the optimally corrected value of $\boldsymbol{\nu}_\alpha$ (see eq. (5.186)). Here, $\boldsymbol{P}_{\mathcal{N}_{\hat{u}}}$ is the eight-dimensional projection matrix onto the orthogonal complement of

$$\mathcal{N}_{\hat{\boldsymbol{u}}} = \{ \begin{pmatrix} \hat{\boldsymbol{m}} \\ 0 \\ \boldsymbol{0} \\ 0 \end{pmatrix}, \begin{pmatrix} \hat{\boldsymbol{r}}_H \\ 0 \\ \hat{\boldsymbol{m}} \\ 0 \end{pmatrix}, \begin{pmatrix} \boldsymbol{0} \\ 1 \\ \boldsymbol{0} \\ 0 \end{pmatrix}, \begin{pmatrix} \boldsymbol{0} \\ 0 \\ \boldsymbol{0} \\ 1 \end{pmatrix} \}_L, \tag{7.208}$$

which is the four-dimensional null space of $\hat{\boldsymbol{u}}$; the covariance matrix $V[\hat{\boldsymbol{u}}]$ has rank 4.

The residual $J[\hat{\boldsymbol{m}}, \hat{\boldsymbol{r}}_H]$ is a χ^2 variable with $2(N-2)$ degrees of freedom. Evidently, the number N of space planes must be such that

$$N \geq 2. \tag{7.209}$$

The residual is 0 when $N = 2$. For $N \geq 3$, we can test if space planes have a common intersection line: the hypothesis that space planes $\{(\boldsymbol{n}_\alpha, \boldsymbol{r}) = d_\alpha\}$ have a common intersection line is rejected with significance level $a\%$ if

$$J[\hat{\boldsymbol{m}}, \hat{\boldsymbol{r}}_H] > \chi^2_{2(N-2),a}. \tag{7.210}$$

The $2(N-1)$ degrees of freedom can be intuitively interpreted as follows. A free space plane has three degrees of freedom. If it is constrained to pass through a space line, it can still rotate freely around that space line, so there remain two degrees of freedom to break the incidence. Consequently, N space planes have $2N$ degrees of freedom for their deviations from the true intersection line $(\boldsymbol{r} - \bar{\boldsymbol{r}}_H) \times \bar{\boldsymbol{m}} = \boldsymbol{0}$. But if we consider their deviations from the *estimated* intersection line $(\boldsymbol{r} - \hat{\boldsymbol{r}}_H) \times \hat{\boldsymbol{m}} = \boldsymbol{0}$, no deviations occur when $N = 2$. Hence, the degrees of freedom decrease by two.

7.6.3 Space plane fitting

Given N space planes $\{(\boldsymbol{\nu}_\alpha, \boldsymbol{\rho}) = 0\}$, $\alpha = 1, ..., N$, consider the problem of fitting a space plane $(\boldsymbol{\nu}, \boldsymbol{\rho}) = 0$ to them. This means finding an optimal average $\boldsymbol{\nu}$ of $\{\boldsymbol{\nu}_\alpha\}$. We write

$$\boldsymbol{\nu}_\alpha = \bar{\boldsymbol{\nu}}_\alpha + \Delta\boldsymbol{\nu}_\alpha, \tag{7.211}$$

and regard each $\Delta\boldsymbol{\nu}_\alpha$ as an independent random variable of mean $\boldsymbol{0}$ and covariance matrix $V[\boldsymbol{\nu}_\alpha]$. The hypothesis is

$$\bar{\boldsymbol{\nu}}_\alpha = \boldsymbol{\nu}, \qquad \alpha = 1, ..., N, \tag{7.212}$$

which has rank 3 because both sides are unit 4-vectors[10].

For each $\boldsymbol{\nu}_\alpha$, the optimal estimate of its true value $\bar{\boldsymbol{\nu}}_\alpha$ is evidently $\boldsymbol{\nu}$, so the optimal average can be obtained by the optimization

$$J[\boldsymbol{\nu}] = \sum_{\alpha=1}^{N} (\boldsymbol{\nu}_\alpha - \boldsymbol{\nu}, V[\boldsymbol{\nu}_\alpha]^- (\boldsymbol{\nu}_\alpha - \boldsymbol{\nu})) \to \min \tag{7.213}$$

under the constraint $\|\boldsymbol{\nu}\| = 1$. Since the covariance matrix $V[\boldsymbol{\nu}_\alpha]$ has null space $\mathcal{N}_{\boldsymbol{\nu}_\alpha} = \{\boldsymbol{\nu}_\alpha\}_L$, we have $V[\boldsymbol{\nu}_\alpha]\boldsymbol{\nu}_\alpha = \boldsymbol{0}$. Hence, the optimization (7.213) can be rewritten in the form

$$J[\boldsymbol{\nu}] = (\boldsymbol{\nu}, \left(\sum_{\alpha=1}^{N} V[\boldsymbol{\nu}_\alpha]^-\right) \boldsymbol{\nu})) \to \min. \tag{7.214}$$

[10]This hypothesis is degenerate.

The solution $\hat{\boldsymbol{\nu}}$ is given by the unit eigenvector of the matrix

$$\boldsymbol{M} = \sum_{\alpha=1}^{N} V[\boldsymbol{\nu}_\alpha]^- \tag{7.215}$$

for the smallest eigenvalue (see eqs. (2.86)). The covariance matrix of of the solution $\hat{\boldsymbol{\nu}}$ is given by

$$V[\hat{\boldsymbol{\nu}}] = \left(\sum_{\alpha=1}^{N} \boldsymbol{P}_{\hat{\boldsymbol{\nu}}} V[\boldsymbol{\nu}_\alpha]^- \boldsymbol{P}_{\hat{\boldsymbol{\nu}}} \right)^-, \tag{7.216}$$

which has rank 3; its null space is $\{\hat{\boldsymbol{n}}_\alpha\}_L$.

The residual $J[\hat{\boldsymbol{\nu}}]$ is a χ^2 variable with $3(N-1)$ degrees of freedom. Evidently, the number N of space planes must be such that

$$N \geq 1. \tag{7.217}$$

The residual is 0 when $N = 1$. For $N \geq 2$, we obtain a *coincidence test* for space planes: the hypothesis that space planes $\{(\boldsymbol{\nu}_\alpha, \boldsymbol{\rho}) = 0\}$ coincide is rejected with significance level a% if

$$J[\hat{\boldsymbol{\nu}}] > \chi^2_{3(N-1),a}. \tag{7.218}$$

The $3(N-1)$ degrees of freedom can be intuitively interpreted as follows. A space plane has three degrees of freedom, so N space planes have $3N$ degrees of freedom for their deviations from the true position $(\bar{\boldsymbol{\nu}}, \boldsymbol{\rho}) = 0$. But if we consider deviations from the *average* $(\hat{\boldsymbol{\nu}}, \boldsymbol{\rho}) = 0$, no deviation occurs when $N = 1$. Hence, the degrees of freedom decrease by three.

Chapter 8

Optimal Filter

In the preceding chapter, the statistical characteristics of the solution of the parametric fitting problem were examined by assuming that the solution was computed by some numerical method. This and the next chapters present such numerical methods. In this chapter, we construct a filter that starts from an initial estimate and optimally updates it each time a new datum is read; we call it simply the *optimal filter*. The update rule is derived from the *Bayesian* standpoint: we apply the principle of *maximum a posteriori probability estimation* by assuming that the noise is Gaussian. The assumptions and approximations introduced in the derivation are elucidated, and the philosophical implications of the Bayesian approach are discussed. Then, the update rule is simplified by introducing the *effective gradient approximation*. The resulting expression is compared with the Kalman filter with no internal dynamics. Finally, the update rule for linear hypotheses is derived.

8.1 General Theory

8.1.1 Bayesian approach

In the preceding chapter, it was implicitly assumed that the entire set of data $\{\boldsymbol{a}_\alpha\}$, $\alpha = 1, ..., N$, was available for computing the optimal estimate of parameter $\boldsymbol{u}$. Suppose the data are a time sequence, the αth value $\boldsymbol{a}_\alpha$ read at time $\alpha = 1, 2, 3, ...$ If the sequence is very long, it is realistic to compute an estimate $\boldsymbol{u}_\alpha$ of $\boldsymbol{u}$ at time α in such a way that it is optimal for the data $\boldsymbol{a}_1$, ..., $\boldsymbol{a}_\alpha$ and update it to $\boldsymbol{u}_{\alpha+1}$ at time $\alpha+1$ so that it is optimal for $\boldsymbol{a}_1$, ..., $\boldsymbol{a}_{\alpha+1}$, and so on (Fig. 8.1). If the update computation does not require rereading the past data, the estimation process can be efficiently run in real time. In this chapter, we derive a theoretically optimal update rule for *Gaussian* noise. The resulting filter not only suits real time data processing but also serves as a numerical algorithm for the parametric fitting problem even when the entire set of data is simultaneously available.

If we try to formulate such an update rule, however, we immediately encounter a subtle problem, which, although purely philosophical, has beset many theoreticians. A natural form of update is computing the αth estimate $\boldsymbol{u}_\alpha$ and its covariance matrix $V[\boldsymbol{u}_\alpha]$ at time α and correcting them into $\boldsymbol{u}_{\alpha+1}$ and $V[\boldsymbol{u}_{\alpha+1}]$ by using the $(\alpha+1)$st datum $\boldsymbol{a}_{\alpha+1}$. As the update proceeds, the estimate $\boldsymbol{u}_\alpha$ is expected to approach the true value, and its covariance matrix $V[\boldsymbol{u}_\alpha]$ is expected to decrease. In this process, we are viewing the estimate $\boldsymbol{u}_\alpha$ as a *random variable*.

In the preceding chapter, we derived the covariance matrix $V[\hat{\boldsymbol{u}}]$ of the optimal estimate $\hat{\boldsymbol{u}}$. The optimal estimate $\hat{\boldsymbol{u}}$ is a random variable because it is computed from the data $\{\boldsymbol{a}_\alpha\}$. The statistical characteristics of the data $\{\boldsymbol{a}_\alpha\}$ reflect *the accuracy and reliability of the data acquisition process* (e.g., image processing and 3-D sensing), and the noise distribution can be estimated, at least in principle, by repeating the data acquisition process many times. Thus, the statistical characteristics of the optimal estimate $\hat{\boldsymbol{u}}$ are a mere (and faithful) reflection of the uncertainty of the data acquisition process.

On the other hand, starting such an update procedure requires an initial value $\boldsymbol{u}_0$ and its covariance matrix $V[\boldsymbol{u}_0]$. The question is: *is $\boldsymbol{u}_0$ a random variable?* Since no data exist *yet*, it has nothing to do with the uncertainty of the data acquisition process. If it is a random variable, how can its distribution be defined? This observation necessitates that we think of $\boldsymbol{u}_0$ as our *subjective belief*, and its distribution as the *relative strength of our belief in particular values*. Once this subjective interpretation is accepted, logical consistency dictates that we view the subsequent update process in the same way: each time a new datum is read, the belief is influenced and its subjective distribution is modified. The approach that admits such a subjective interpretation is known as *Bayesian*[1], whereas that which rejects it is called *non-Bayesian*.

The distinction between the Bayesian and the non-Bayesian approaches is purely philosophical and often not so clear in reality. For example, one can compute the initial estimate $\boldsymbol{u}_0$ and its covariance matrix $V[\boldsymbol{u}_0]$ from a small number of data, say $\boldsymbol{a}_1$, $\boldsymbol{a}_2$, and $\boldsymbol{a}_3$, and then start the update process by regarding the subsequent data $\boldsymbol{a}_4$, $\boldsymbol{a}_5$, ... as a new sequence $\boldsymbol{a}'_1$, $\boldsymbol{a}'_2$, ... In this case, the statistical characteristics of $\boldsymbol{u}_0$ reflect the characteristics of the errors in $\boldsymbol{a}_1$, $\boldsymbol{a}_2$, and $\boldsymbol{a}_3$. Hence, the approach is non-Bayesian. On the other hand, if one *guesses* $\boldsymbol{u}_0$ and $V[\boldsymbol{u}_0]$ and applies the same procedure thereafter, the approach is Bayesian. In this chapter, we adopt the Bayesian approach for the sake of formal consistency, but we do not worry about its philosophical foundation any further. The distribution of an estimate before data are observed is called the *a priori* distribution (or simply the *prior*), while the distribution updated by the data is called the *a posteriori* distribution[2] (or simply the *posterior*).

8.1.2 Maximum a posteriori probability estimation

Given an m-vector datum $\boldsymbol{a}$, we write

$$\boldsymbol{a} = \bar{\boldsymbol{a}} + \Delta\boldsymbol{a}, \tag{8.1}$$

[1] This terminology derives from the *Bayes formula* (see eq. (3.95)), which plays an essential role in this approach.

[2] In the preceding chapters, the terms *a priori* and *a posteriori* were used in the non-Bayesian context.

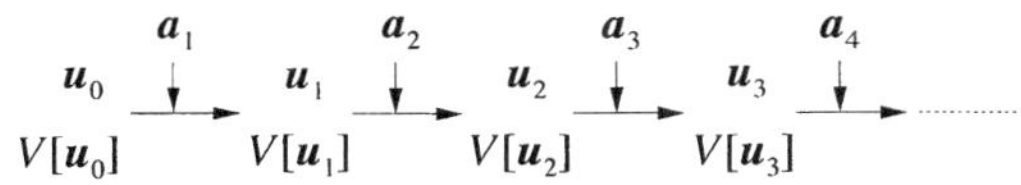

Fig. 8.1. The αth estimate $\hat{\mathbf{u}}_\alpha$ and its covariance matrix $V[\hat{\mathbf{u}}_\alpha]$ are updated by the $(\alpha+1)$th datum $\mathbf{a}_{\alpha+1}$.

and regard $\Delta \boldsymbol{a}$ as a Gaussian random variable of mean $\mathbf{0}$ and covariance matrix $\bar{V}[\boldsymbol{a}]$. We assume that $\boldsymbol{a}$ and $\bar{\boldsymbol{a}}$ are constrained to be in an m'-dimensional manifold $\mathcal{A} \subset \mathcal{R}^m$, which we call the *data space*. It follows that the noise $\Delta \boldsymbol{a}$ is constrained, to a first approximation, to be in the tangent space $T_{\bar{\boldsymbol{a}}}(\mathcal{A})$ to the manifold $\mathcal{A}$ at $\bar{\boldsymbol{a}}$. We also assume that no constraint is imposed on $\Delta \boldsymbol{a}$ other than $\Delta \boldsymbol{a} \in T_{\bar{\boldsymbol{a}}}(\mathcal{A})$, so the range of the covariance matrix $\bar{V}[\boldsymbol{a}]$ coincides with the tangent space $T_{\bar{\boldsymbol{a}}}(\mathcal{A})$. Our task is to estimate the n-vector $\boldsymbol{u}$ that satisfies the hypothesis

$$F^{(k)}(\bar{\boldsymbol{a}}, \boldsymbol{u}) = 0, \qquad k = 1, ..., L, \tag{8.2}$$

given by L smooth functions $F^{(k)}(\,\cdot\,,\,\cdot\,)$: $\mathcal{R}^m \times \mathcal{R}^n \to \mathcal{R}$. We assume that the domain of the parameter $\boldsymbol{u}$ is an n'-dimensional manifold $\mathcal{U} \subset \mathcal{R}^n$, which we call the *parameter space*.

As argued in Section 7.1.2, this problem is solved in two stages: the correction stage and the estimation stage. The difference from Section 7.1.2 is that the parameter $\boldsymbol{u}$ is now a random variable that has an a priori probability density. The formulation in Section 7.1.2 is modified as follows.

A. Correction stage

We estimate the value $\bar{\boldsymbol{a}} = \boldsymbol{a} - \Delta \boldsymbol{a}$ that satisfies the hypothesis (8.2) for a particular value of $\boldsymbol{u}$. As shown in Section 5.1.1, the optimal correction $\Delta \boldsymbol{a}$ is determined by the optimization

$$J = (\Delta \boldsymbol{a}, \bar{V}[\boldsymbol{a}]^- \Delta \boldsymbol{a}) \to \min \tag{8.3}$$

under the linearized constraint

$$(\nabla_{\boldsymbol{a}} F^{(k)}(\bar{\boldsymbol{a}}, \boldsymbol{u}), \Delta \boldsymbol{a}) = F^{(k)}(\boldsymbol{a}, \boldsymbol{u}), \qquad k = 1, ..., L, \tag{8.4}$$

together with $\Delta \boldsymbol{a} \in T_{\bar{\boldsymbol{a}}}(\mathcal{A})$. The optimization (8.3), which minimizes the Mahalanobis distance $\|\Delta \boldsymbol{a}\|_{\bar{V}[\boldsymbol{a}]}$, can be justified as maximum likelihood estimation for Gaussian noise (see Section 5.1.1). The first order solution is given as follows (see eq. (5.17)):

$$\Delta \boldsymbol{a} = \bar{V}[\boldsymbol{a}] \sum_{k,l=1}^{L} \bar{W}^{(kl)}(\boldsymbol{u}) F^{(k)}(\boldsymbol{a}, \boldsymbol{u}) \nabla_{\boldsymbol{a}} F^{(l)}(\boldsymbol{a}, \boldsymbol{u}). \tag{8.5}$$

Here, $\bar{\boldsymbol{W}}(\boldsymbol{u}) = (\bar{W}^{(kl)}(\boldsymbol{u}))$ is the (LL)-matrix defined by

$$(\bar{W}^{(kl)}(\boldsymbol{u})) = \left((\nabla_{\boldsymbol{a}} F^{(k)}(\bar{\boldsymbol{a}}, \boldsymbol{u}), \bar{V}[\boldsymbol{a}]\nabla_{\boldsymbol{a}} F^{(l)}(\bar{\boldsymbol{a}}, \boldsymbol{u}))\right)^-, \tag{8.6}$$

by which we mean $\bar{\boldsymbol{W}}(\boldsymbol{u}) = \bar{\boldsymbol{V}}(\boldsymbol{u})^-$, where $\bar{\boldsymbol{V}}(\boldsymbol{u}) = (\bar{V}^{(kl)}(\boldsymbol{u}))$ is the (LL)-matrix defined by

$$(\bar{V}^{(kl)}(\boldsymbol{u})) = \left((\nabla_{\boldsymbol{a}} F^{(k)}(\bar{\boldsymbol{a}}, \boldsymbol{u}), \bar{V}[\boldsymbol{a}]\nabla_{\boldsymbol{a}} F^{(l)}(\bar{\boldsymbol{a}}, \boldsymbol{u}))\right). \tag{8.7}$$

Hence, for a particular value $\boldsymbol{u}$, the true value of $\boldsymbol{a}$ is estimated to be

$$\bar{\boldsymbol{a}} = \boldsymbol{a} - \bar{V}[\boldsymbol{a}] \sum_{k,l=1}^{L} \bar{W}^{(kl)}(\boldsymbol{u}) F^{(k)}(\boldsymbol{a}, \boldsymbol{u}) \nabla_{\boldsymbol{a}} F^{(l)}(\boldsymbol{a}, \boldsymbol{u}). \tag{8.8}$$

If $\bar{\boldsymbol{a}}$ is the true value, a particular value $\boldsymbol{a}$ has the following probability density (see eq. (3.46)):

$$p(\boldsymbol{a}|\boldsymbol{u}) = \frac{e^{-(\boldsymbol{a}-\bar{\boldsymbol{a}}(\boldsymbol{a},\boldsymbol{u}), \bar{V}[\boldsymbol{a}]^-(\boldsymbol{a}-\bar{\boldsymbol{a}}(\boldsymbol{a},\boldsymbol{u})))/2}}{\sqrt{(2\pi)^{m'}|\bar{V}[\boldsymbol{a}]|_+}}. \tag{8.9}$$

Here, m' is the rank of $\bar{V}[\boldsymbol{a}]$, i.e., the degrees of freedom of the m-vector $\boldsymbol{a}$. Since eq. (8.9) gives a probability density of $\boldsymbol{a}$ for a given value $\boldsymbol{u}$, it defines a *conditional probability density* conditioned on $\boldsymbol{u}$.

B. Estimation stage

Suppose the parameter $\boldsymbol{u}$ has an *a priori probability density* of the form

$$p(\boldsymbol{u}) = \frac{e^{-(\boldsymbol{u}-\boldsymbol{u}_0, V[\boldsymbol{u}_0]^-(\boldsymbol{u}-\boldsymbol{u}_0))/2}}{\sqrt{(2\pi)^{n'}|V[\boldsymbol{u}_0]|_+}}, \tag{8.10}$$

where n' is the rank of $V[\boldsymbol{u}_0]$, i.e., the degrees of freedom of the n-vector $\boldsymbol{u}$. The *joint probability density* of $\boldsymbol{u}$ and $\boldsymbol{a}$ is

$$p(\boldsymbol{a}|\boldsymbol{u})p(\boldsymbol{u}) = \frac{e^{-(\boldsymbol{u}-\boldsymbol{u}_0, V[\boldsymbol{u}_0]^-(\boldsymbol{u}-\boldsymbol{u}_0))/2-(\boldsymbol{a}-\bar{\boldsymbol{a}}, \bar{V}[\boldsymbol{a}]^-(\boldsymbol{a}-\bar{\boldsymbol{a}}))/2}}{\sqrt{(2\pi)^{n'+m'}|V[\boldsymbol{u}_0]|_+|\bar{V}[\boldsymbol{a}]|_+}}. \tag{8.11}$$

According to the *Bayes formula* (3.95), the *a posteriori probability density* of $\boldsymbol{u}$ determined by a particular value $\boldsymbol{a}$ is

$$p(\boldsymbol{u}|\boldsymbol{a}) = \frac{p(\boldsymbol{a}|\boldsymbol{u})p(\boldsymbol{u})}{\int p(\boldsymbol{a}|\boldsymbol{u})p(\boldsymbol{u})d\boldsymbol{u}}. \tag{8.12}$$

The *maximum a posteriori probability estimator*, or simply the *Bayes estimator*[3], is the value $\boldsymbol{u}$ that maximizes this a posteriori probability density for given $\boldsymbol{a}$. Since the denominator of the right-hand side of eq. (8.12) does not depend on $\boldsymbol{u}$, maximizing $p(\boldsymbol{u}|\boldsymbol{a})$ is equivalent to maximizing $p(\boldsymbol{a}|\boldsymbol{u})p(\boldsymbol{u})$, which in turn is equivalent to minimizing

$$J[\boldsymbol{u}] = (\boldsymbol{u} - \boldsymbol{u}_0, V[\boldsymbol{u}_0]^-(\boldsymbol{u} - \boldsymbol{u}_0)) + (\boldsymbol{a} - \bar{\boldsymbol{a}}, \bar{V}[\boldsymbol{a}]^-(\boldsymbol{a} - \bar{\boldsymbol{a}})). \tag{8.13}$$

Substitution of eq. (8.8) into eq. (8.13) yields the following expression (see eq. (7.13)):

$$J[\boldsymbol{u}] = (\boldsymbol{u}-\boldsymbol{u}_0, V[\boldsymbol{u}_0]^-(\boldsymbol{u}-\boldsymbol{u}_0)) + \sum_{k,l=1}^{L} \bar{W}^{(kl)}(\boldsymbol{u})F^{(k)}(\boldsymbol{a},\boldsymbol{u})F^{(l)}(\boldsymbol{a},\boldsymbol{u}). \tag{8.14}$$

The above argument tacitly assumes that the distributions of $\boldsymbol{a}$ and $\boldsymbol{u}$ are *local* (see Section 3.2.2). Namely, eq. (8.9) is based on the assumption that the distribution of $\boldsymbol{a}$ is sufficiently concentrated around $\bar{\boldsymbol{a}}$ in the data space $\mathcal{A}$ and hence the domain of the distribution can be identified with the tangent space $T_{\bar{\boldsymbol{a}}}(\mathcal{A})$ of the manifold $\mathcal{A}$ at $\bar{\boldsymbol{a}}$. Similarly, eq. (8.10) is based on the assumption that the distribution of $\boldsymbol{u}$ is sufficiently concentrated around $\boldsymbol{u}_0$ in the parameter space $\mathcal{U}$ and hence the domain of the distribution can be identified with with the tangent space $T_{\boldsymbol{u}_0}(\mathcal{U})$ of the manifold $\mathcal{U}$ at $\boldsymbol{u}_0$. It follows that $J[\boldsymbol{u}]$ has a meaning only for $\boldsymbol{u} - \boldsymbol{u}_0 \in T_{\boldsymbol{u}_0}(\mathcal{U})$.

8.2 Iterative Estimation Scheme

8.2.1 Optimal update rule

Eq. (8.14) is merely a theoretical expression because $\bar{W}^{(kl)}(\boldsymbol{u})$ involves the true value $\bar{\boldsymbol{a}}$ and the covariance matrix $\bar{V}[\boldsymbol{a}]$ evaluated at $\bar{\boldsymbol{a}}$. In order to compute it from the datum $\boldsymbol{a}$ and its covariance matrix $V[\boldsymbol{a}]$, we apply the approximation introduced in Section 7.1.2. Namely, the (LL)-matrix $\bar{\boldsymbol{W}}_\alpha(\boldsymbol{u})$ is approximated by the rank-constrained generalized inverse

$$(W^{(kl)}(\boldsymbol{u})) = \Big((\nabla_{\boldsymbol{a}} F^{(k)}(\boldsymbol{a},\boldsymbol{u}), V[\boldsymbol{a}]\nabla_{\boldsymbol{a}} F^{(l)}(\boldsymbol{a},\boldsymbol{u}))\Big)_r^-, \tag{8.15}$$

where r is the rank of the hypothesis (8.2) (see eq. (7.14)).

Expanding $W^{(kl)}(\boldsymbol{u})$, $F^{(k)}(\boldsymbol{a},\boldsymbol{u})$, and $F^{(l)}(\boldsymbol{a},\boldsymbol{u})$ in the neighborhood of $\boldsymbol{u}_0$, we obtain

$$W^{(kl)}(\boldsymbol{u}) = W_0^{(kl)} + (\nabla_{\boldsymbol{u}} W_0^{(kl)}, \boldsymbol{u} - \boldsymbol{u}_0) + \frac{1}{2}(\boldsymbol{u} - \boldsymbol{u}_0, \nabla_{\boldsymbol{u}}^2 W_0^{(kl)}(\boldsymbol{u} - \boldsymbol{u}_0))$$
$$+O(\boldsymbol{u} - \boldsymbol{u}_0)^3,$$

[3] To be precise, the *Bayes estimator* is defined as the value that minimizes the expectation of a *cost* (or *loss*) *function* with respect an a posteriori distribution. It can be identified with the maximum a posteriori probability estimator if the cost function is quadratic and the a posteriori distribution is Gaussian.

$$F^{(k)}(\boldsymbol{a}, \boldsymbol{u}) = F_0^{(k)} + (\nabla_{\boldsymbol{u}} F_0^{(k)}, \boldsymbol{u} - \boldsymbol{u}_0) + \frac{1}{2}(\boldsymbol{u} - \boldsymbol{u}_0, \nabla^2_{\boldsymbol{u}} F_0^{(k)}(\boldsymbol{u} - \boldsymbol{u}_0)) + O(\boldsymbol{u} - \boldsymbol{u}_0)^3,$$

$$F^{(l)}(\boldsymbol{a}, \boldsymbol{u}) = F_0^{(l)} + (\nabla_{\boldsymbol{u}} F_0^{(l)}, \boldsymbol{u} - \boldsymbol{u}_0) + \frac{1}{2}(\boldsymbol{u} - \boldsymbol{u}_0, \nabla^2_{\boldsymbol{u}} F_0^{(l)}(\boldsymbol{u} - \boldsymbol{u}_0)) + O(\boldsymbol{u} - \boldsymbol{u}_0)^3. \tag{8.16}$$

Here, the subscript 0 refers to the value evaluated at $\boldsymbol{u}_0$; the symbol $O(\boldsymbol{u}-\boldsymbol{u}_0)^3$ denotes terms of order 3 or higher in $\boldsymbol{u} - \boldsymbol{u}_0$. We use the notation $\nabla^2_{\boldsymbol{u}} f$ to denote the matrix whose (ij) element is $\partial^2 f / \partial u_i \partial u_j$. Substituting eqs. (8.16) into eq. (8.14) and ignoring $O(\boldsymbol{u} - \boldsymbol{u}_0)^3$, we obtain

$$J[\boldsymbol{u}] = (\boldsymbol{u} - \boldsymbol{u}_0, V[\boldsymbol{u}_0]^-(\boldsymbol{u} - \boldsymbol{u}_0)) + (\boldsymbol{u} - \boldsymbol{u}_0, \boldsymbol{S}_0(\boldsymbol{u} - \boldsymbol{u}_0)) + 2(\boldsymbol{t}_0, \boldsymbol{u} - \boldsymbol{u}_0) + E_0, \tag{8.17}$$

$$\boldsymbol{S}_0 = \sum_{k,l=1}^{L} \Big(W_0^{(kl)}((\nabla_{\boldsymbol{u}} F_0^{(k)})(\nabla_{\boldsymbol{u}} F_0^{(l)})^\top + F_0^{(k)} \nabla^2_{\boldsymbol{u}} F_0^{(l)}) + 2F_0^{(k)} S[\nabla_{\boldsymbol{u}} W_0^{(kl)} (\nabla_{\boldsymbol{u}} F_0^{(l)})^\top] + \frac{1}{2} F_0^{(k)} F_0^{(l)} \nabla^2_{\boldsymbol{u}} W_0^{(kl)} \Big), \tag{8.18}$$

$$\boldsymbol{t}_0 = \sum_{k,l=1}^{L} \left(W_0^{(kl)} F_0^{(k)} \nabla_{\boldsymbol{u}} F_0^{(l)} + \frac{1}{2} F_0^{(k)} F_0^{(l)} \nabla_{\boldsymbol{u}} W_0^{(kl)} \right), \tag{8.19}$$

$$E_0 = \sum_{k,l=1}^{L} W_0^{(kl)} F_0^{(k)} F_0^{(l)}, \tag{8.20}$$

where $S[\,\cdot\,]$ is the symmetrization operator (see eqs. (2.205)).

As pointed out in the preceding section, the function $J[\boldsymbol{u}]$ is defined only for $\boldsymbol{u} - \boldsymbol{u}_0 \in T_{\boldsymbol{u}_0}(\mathcal{U})$. Let $\{\boldsymbol{v}_{j(0)}\}$, $j = 1, ..., n - n'$, be an orthonormal basis of the orthogonal complement $T_{\boldsymbol{u}_0}(\mathcal{U})^\perp$. The constraint $\boldsymbol{u} - \boldsymbol{u}_0 \in T_{\boldsymbol{u}_0}(\mathcal{U})$ can be expressed in the form

$$(\boldsymbol{v}_{j(0)}, \boldsymbol{u} - \boldsymbol{u}_0) = 0, \qquad j = 1, ..., n - n'. \tag{8.21}$$

Introducing Lagrange multipliers λ_j, differentiating $J[\boldsymbol{u}] - 2\sum_{j=1}^{n-n'} (\boldsymbol{v}_{j(0)}, \boldsymbol{u} - \boldsymbol{u}_0)$ with respect to $\boldsymbol{u}$, and setting the result zero, we obtain

$$V[\boldsymbol{u}_0]^-(\boldsymbol{u} - \boldsymbol{u}_0) + \boldsymbol{S}_0(\boldsymbol{u} - \boldsymbol{u}_0) + \boldsymbol{t}_0 = \sum_{j=1}^{n-n'} \boldsymbol{v}_{j(0)}. \tag{8.22}$$

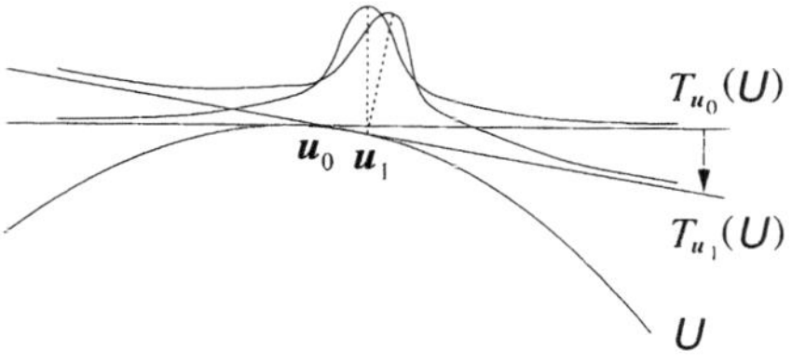

Fig. 8.2. The mapping from $T\mathbf{u}_0(\mathcal{U})$ to $T\mathbf{u}_1(\mathcal{U})$.

Let $\boldsymbol{P}^{\mathcal{U}}_{\boldsymbol{u}_0}$ be the n-dimensional projection matrix onto $T_{\boldsymbol{u}_0}(\mathcal{U})$. Multiplying eq. (8.22) by $\boldsymbol{P}^{\mathcal{U}}_{\boldsymbol{u}_0}$ on both sides and noting that $\boldsymbol{P}^{\mathcal{U}}_{\boldsymbol{u}_0}V[\boldsymbol{u}_0] = V[\boldsymbol{u}_0]$, $\boldsymbol{P}^{\mathcal{U}}_{\boldsymbol{u}_0}(\boldsymbol{u}-\boldsymbol{u}_0) = \boldsymbol{u}-\boldsymbol{u}_0$, and $\boldsymbol{P}^{\mathcal{U}}_{\boldsymbol{u}_0}\boldsymbol{v}_{j(0)} = \boldsymbol{0}$, we obtain

$$V[\boldsymbol{u}_0]^-(\boldsymbol{u}-\boldsymbol{u}_0) + \boldsymbol{P}^{\mathcal{U}}_{\boldsymbol{u}_0}\boldsymbol{S}_0\boldsymbol{P}^{\mathcal{U}}_{\boldsymbol{u}_0}(\boldsymbol{u}-\boldsymbol{u}_0) + \boldsymbol{P}^{\mathcal{U}}_{\boldsymbol{u}_0}\boldsymbol{t}_0 = \boldsymbol{0}. \qquad (8.23)$$

Solving this for $\boldsymbol{u}$, we obtain the Bayes estimator, which we write as $\boldsymbol{u}_1$, in the form

$$\boldsymbol{u}_1 = \boldsymbol{u}_0 - \left(V[\boldsymbol{u}_0]^- + \boldsymbol{P}^{\mathcal{U}}_{\boldsymbol{u}_0}\boldsymbol{S}_0\boldsymbol{P}^{\mathcal{U}}_{\boldsymbol{u}_0}\right)^- \boldsymbol{t}_0. \qquad (8.24)$$

If this is substituted for $\boldsymbol{u}$, eq. (8.17) can be rearranged into the form

$$J[\boldsymbol{u}] = (\boldsymbol{u}-\boldsymbol{u}_1, (V[\boldsymbol{u}_0]^- + \boldsymbol{P}^{\mathcal{U}}_{\boldsymbol{u}_0}\boldsymbol{S}_0\boldsymbol{P}^{\mathcal{U}}_{\boldsymbol{u}_0})(\boldsymbol{u}-\boldsymbol{u}_1)) + E_1, \qquad (8.25)$$

where

$$E_1 = E_0 - (\boldsymbol{t}_0, \left(V[\boldsymbol{u}_0]^- + \boldsymbol{P}^{\mathcal{U}}_{\boldsymbol{u}_0}\boldsymbol{S}_0\boldsymbol{P}^{\mathcal{U}}_{\boldsymbol{u}_0}\right)^- \boldsymbol{t}_0). \qquad (8.26)$$

Since eq. (8.11) has the form $p(\boldsymbol{a}|\boldsymbol{u})p(\boldsymbol{u}) \propto e^{-J[\boldsymbol{u}]/2}$, the Bayes formula (8.12) can be written in the form

$$p(\boldsymbol{u}|\boldsymbol{a}) \propto e^{-(\boldsymbol{u}-\boldsymbol{u}_1, (V[\boldsymbol{u}_0]^- + \boldsymbol{P}^{\mathcal{U}}_{\boldsymbol{u}_0}\boldsymbol{S}_0\boldsymbol{P}^{\mathcal{U}}_{\boldsymbol{u}_0})(\boldsymbol{u}-\boldsymbol{u}_1))/2}, \qquad (8.27)$$

which is a function defined for $\boldsymbol{u}-\boldsymbol{u}_1 \in T_{\boldsymbol{u}_0}(\mathcal{U})$.

In order to carry out the update, we need to view $p(\boldsymbol{u}|\boldsymbol{a})$ as a function defined for $\boldsymbol{u}-\boldsymbol{u}_1 \in T_{\boldsymbol{u}_1}(\mathcal{U})$. From eq. (8.27), we see that $p(\boldsymbol{u}|\boldsymbol{a}) = p(\boldsymbol{u}+\boldsymbol{v}|\boldsymbol{a})$ for an arbitrary $\boldsymbol{v} \in T_{\boldsymbol{u}_0}(\mathcal{U})^\perp$. Hence, we can regard eq. (8.27) as defined for an arbitrary $\boldsymbol{u} \in \mathcal{R}^n$: the value $p(\boldsymbol{u}|\boldsymbol{a})$ is extended "cylindrically" into $T_{\boldsymbol{u}_0}(\mathcal{U})^\perp$ (Fig. 8.2). The function $p(\boldsymbol{u}|\boldsymbol{a})$ thus extended over $\boldsymbol{u} \in \mathcal{R}^n$ is then restricted to the domain $\boldsymbol{u}-\boldsymbol{u}_1 \in T_{\boldsymbol{u}_1}(\mathcal{U})$. Let $\boldsymbol{P}^{\mathcal{U}}_{\boldsymbol{u}_1}$ be the projection matrix onto $T_{\boldsymbol{u}_1}(\mathcal{U})$. Since $\boldsymbol{P}^{\mathcal{U}}_{\boldsymbol{u}_1}(\boldsymbol{u}-\boldsymbol{u}_1) = \boldsymbol{u}-\boldsymbol{u}_1$ for $\boldsymbol{u}-\boldsymbol{u}_1 \in T_{\boldsymbol{u}_1}(\mathcal{U})$, the value of $p(\boldsymbol{u}|\boldsymbol{a})$ for $\boldsymbol{u}-\boldsymbol{u}_1 \in T_{\boldsymbol{u}_1}(\mathcal{U})$ can be written as

$$p(\boldsymbol{u}|\boldsymbol{a}) \propto e^{-(\boldsymbol{u}-\boldsymbol{u}_1, \boldsymbol{P}^{\mathcal{U}}_{\boldsymbol{u}_1}(V[\boldsymbol{u}_0]^- + \boldsymbol{P}^{\mathcal{U}}_{\boldsymbol{u}_0}\boldsymbol{S}_0\boldsymbol{P}^{\mathcal{U}}_{\boldsymbol{u}_0})\boldsymbol{P}^{\mathcal{U}}_{\boldsymbol{u}_1}(\boldsymbol{u}-\boldsymbol{u}_1))/2}, \qquad (8.28)$$

Eq. (8.28) states that *a posteriori* the parameter $\boldsymbol{u}$ has a Gaussian distribution with mean $\boldsymbol{u}_1$ and covariance matrix

$$V[\boldsymbol{u}_1] = \left(\boldsymbol{P}^{\mathcal{U}}_{\boldsymbol{u}_1} V[\boldsymbol{u}_0]^- \boldsymbol{P}^{\mathcal{U}}_{\boldsymbol{u}_1} + \boldsymbol{P}^{\mathcal{U}}_{\boldsymbol{u}_1} \boldsymbol{P}^{\mathcal{U}}_{\boldsymbol{u}_0} \boldsymbol{S}_0 \boldsymbol{P}^{\mathcal{U}}_{\boldsymbol{u}_0} \boldsymbol{P}^{\mathcal{U}}_{\boldsymbol{u}_1}\right)^- . \tag{8.29}$$

Having thus computed $\boldsymbol{u}_1$ and $V[\boldsymbol{u}_1]$, we can compute $\boldsymbol{u}_2$ and $V[\boldsymbol{u}_2]$ by the same procedure when the next datum is read; the update rule has been established.

Since the (generally nonlinear) constraint on $\boldsymbol{u}$ has been linearized in the above derivation, the updated value $\boldsymbol{u}_1$ may not exactly satisfy the inherent constraint on it. Hence, a higher order correction $\mathcal{C}[\,\cdot\,]$ need to be added (see eq. (5.25)). Writing $\boldsymbol{u}_0$ as $\boldsymbol{u}_\alpha$, we obtain the update rule of the *optimal filter* in the following form:

$$\boldsymbol{u}_{\alpha+1} = \mathcal{C}[\boldsymbol{u}_\alpha - \left(V[\boldsymbol{u}_\alpha]^- + \boldsymbol{P}^{\mathcal{U}}_{\boldsymbol{u}_\alpha} \boldsymbol{S}_\alpha \boldsymbol{P}^{\mathcal{U}}_{\boldsymbol{u}_\alpha}\right)^- \boldsymbol{t}_\alpha], \tag{8.30}$$

$$V[\boldsymbol{u}_{\alpha+1}] = \left(\tilde{V}[\boldsymbol{u}_\alpha]^- + \boldsymbol{P}^{\mathcal{U}}_{\boldsymbol{u}_{\alpha+1}} \boldsymbol{P}^{\mathcal{U}}_{\boldsymbol{u}_\alpha} \boldsymbol{S}_\alpha \boldsymbol{P}^{\mathcal{U}}_{\boldsymbol{u}_\alpha} \boldsymbol{P}^{\mathcal{U}}_{\boldsymbol{u}_{\alpha+1}}\right)^- , \tag{8.31}$$

$$\boldsymbol{S}_\alpha = \sum_{k,l=1}^{L} \Big(W^{(kl)}_\alpha ((\nabla_{\boldsymbol{u}} F^{(k)}_\alpha)(\nabla_{\boldsymbol{u}} F^{(l)}_\alpha)^\top + F^{(k)}_\alpha \nabla^2_{\boldsymbol{u}} F^{(l)}_\alpha)$$
$$+ 2F^{(k)}_\alpha S[\nabla_{\boldsymbol{u}} W^{(kl)}_\alpha (\nabla_{\boldsymbol{u}} F^{(l)}_\alpha)^\top] + \frac{1}{2} F^{(k)}_\alpha F^{(l)}_\alpha \nabla^2_{\boldsymbol{u}} W^{(kl)}_\alpha \Big), \tag{8.32}$$

$$\boldsymbol{t}_\alpha = \sum_{k,l=1}^{L} \left(W^{(kl)}_\alpha F^{(k)}_\alpha \nabla_{\boldsymbol{u}} F^{(l)}_\alpha + \frac{1}{2} F^{(k)}_\alpha F^{(l)}_\alpha \nabla_{\boldsymbol{u}} W^{(kl)}_\alpha \right), \tag{8.33}$$

$$\tilde{V}[\boldsymbol{u}_\alpha] = \left(\boldsymbol{P}^{\mathcal{U}}_{\boldsymbol{u}_{\alpha+1}} V[\boldsymbol{u}_\alpha]^- \boldsymbol{P}^{\mathcal{U}}_{\boldsymbol{u}_{\alpha+1}}\right)^- . \tag{8.34}$$

Here, the subscript α refers to the value evaluated at $\boldsymbol{u} = \boldsymbol{u}_\alpha$ and $\boldsymbol{a} = \boldsymbol{a}_{\alpha+1}$. For instance, $F^{(k)}_\alpha$ is the abbreviation of $F^{(k)}(\boldsymbol{a}_{\alpha+1}, \boldsymbol{u}_\alpha)$. The (LL)-matrix $\boldsymbol{W}_\alpha$ is given by

$$(W^{(kl)}_\alpha) = \left((\nabla_{\boldsymbol{a}} F^{(k)}_\alpha, V[\boldsymbol{a}_{\alpha+1}] \nabla_{\boldsymbol{a}} F^{(l)}_\alpha)\right)^-_r . \tag{8.35}$$

8.2.2 Bayesian interpretation

From the Bayesian standpoint, eqs. (8.30)–(8.35) have the following interpretation:

- The matrix $\boldsymbol{S}_\alpha$ is generally positive definite, so eq. (8.31) implies that $V[\boldsymbol{u}_\alpha]$ usually decreases after each update. Hence, the uncertainty of the estimate generally decreases as the update progresses.

- If $V[\boldsymbol{u}_\alpha] \approx \boldsymbol{O}$, then $\left(V[\boldsymbol{u}_\alpha]^- + \boldsymbol{P}^{\mathcal{U}}_{\boldsymbol{u}_\alpha} \boldsymbol{S}_\alpha \boldsymbol{P}^{\mathcal{U}}_{\boldsymbol{u}_\alpha}\right)^- \approx \boldsymbol{O}$. Hence, $\boldsymbol{u}_{\alpha+1} \approx \boldsymbol{u}_\alpha$ and $V[\boldsymbol{u}_{\alpha+1}] \approx \boldsymbol{O}$. Namely, if the estimate $\boldsymbol{u}_\alpha$ is very certain, the amount of update is very small, and the resulting estimate is also very certain.

- If $V[\boldsymbol{a}_{\alpha+1}] \approx \infty$, eq. (8.35) implies that $W_\alpha^{(kl)} \approx 0$, so $\boldsymbol{S}_\alpha \approx \boldsymbol{O}$ and $\boldsymbol{t}_\alpha \approx \boldsymbol{0}$. Hence, $\boldsymbol{u}_{\alpha+1} \approx \boldsymbol{u}_\alpha$ and $V[\boldsymbol{u}_{\alpha+1}] \approx V[\boldsymbol{u}_\alpha]$. Namely, if the datum is very inaccurate, it adds little information, so the estimate is updated little.

- If $F^{(k)}(\boldsymbol{a}_{\alpha+1}, \boldsymbol{u}_\alpha) \approx 0$, eq. (8.33) implies that $\boldsymbol{t}_\alpha \approx \boldsymbol{0}$, so $\boldsymbol{u}_{\alpha+1} \approx \boldsymbol{u}_\alpha$, but $V[\boldsymbol{u}_\alpha]$ generally decreases. Namely, if the observed datum is compatible with the predicted estimate, the estimate is updated little, but the belief in it is enhanced.

In Section 7.1.2, we showed that an optimal estimate of $\boldsymbol{u}$ was given by the optimization

$$J[\boldsymbol{u}] = \sum_{\alpha=1}^{N} \sum_{k,l=1}^{L} W^{(kl)}(\boldsymbol{a}_\alpha, \boldsymbol{u}) F^{(k)}(\boldsymbol{a}_\alpha, \boldsymbol{u}) F^{(l)}(\boldsymbol{a}_\alpha, \boldsymbol{u}) \to \min. \tag{8.36}$$

The following question arises: in what sense is the final Bayes estimator $\boldsymbol{u}_N$ an approximation to the solution of this optimization? We now show that $\boldsymbol{u}_N$ is the solution of the optimization

$$\hat{J}[\boldsymbol{u}] = (\boldsymbol{u} - \boldsymbol{u}_0, V[\boldsymbol{u}_0]^-(\boldsymbol{u} - \boldsymbol{u}_0)) + \sum_{\alpha=1}^{N} \sum_{k,l=1}^{L} W^{(kl)}(\boldsymbol{a}_\alpha, \boldsymbol{u}) F^{(k)}(\boldsymbol{a}_\alpha, \boldsymbol{u}) F^{(l)}(\boldsymbol{a}_\alpha, \boldsymbol{u}) \to \min. \tag{8.37}$$

In fact, rewriting eq. (8.14) in the form of eq. (8.25) is equivalent to rewriting the above $\hat{J}[\boldsymbol{u}]$ in the form

$$\hat{J}[\boldsymbol{u}] = (\boldsymbol{u} - \boldsymbol{u}_1, V[\boldsymbol{u}_1]^-(\boldsymbol{u} - \boldsymbol{u}_1)) + \sum_{\alpha=2}^{N} \sum_{k,l=1}^{L} W^{(kl)}(\boldsymbol{a}_\alpha, \boldsymbol{u}) F^{(k)}(\boldsymbol{a}_\alpha, \boldsymbol{u}) F^{(l)}(\boldsymbol{a}_\alpha, \boldsymbol{u}) + E_1, \tag{8.38}$$

which in turn can be rewritten in the form

$$\hat{J}[\boldsymbol{u}] = (\boldsymbol{u} - \boldsymbol{u}_2, V[\boldsymbol{u}_2]^-(\boldsymbol{u} - \boldsymbol{u}_2)) + \sum_{\alpha=3}^{N} \sum_{k,l=1}^{L} W^{(kl)}(\boldsymbol{a}_\alpha, \boldsymbol{u}) F^{(k)}(\boldsymbol{a}_\alpha, \boldsymbol{u}) F^{(l)}(\boldsymbol{a}_\alpha, \boldsymbol{u}) + E_1 + E_2, \tag{8.39}$$

and so forth. The final expression is

$$\hat{J}[\boldsymbol{u}] = (\boldsymbol{u} - \boldsymbol{u}_N, V[\boldsymbol{u}_N]^-(\boldsymbol{u} - \boldsymbol{u}_N)) + \sum_{\alpha=1}^{N} E_\alpha, \tag{8.40}$$

whose minimum is attained by $\boldsymbol{u} = \boldsymbol{u}_N$, meaning that $\boldsymbol{u}_N$ is the solution of the optimization (8.37).

From eqs. (8.36) and (8.37), we see that

$$\hat{J}[\boldsymbol{u}_N] = (\boldsymbol{u}_N - \boldsymbol{u}_0, V[\boldsymbol{u}_0]^-(\boldsymbol{u}_N - \boldsymbol{u}_0)) + J[\boldsymbol{u}_N]. \tag{8.41}$$

Hence, it is concluded that the Bayes estimator $\boldsymbol{u}_N$ is a good approximation to the solution of the optimization (8.36) if $\boldsymbol{u}_N \approx \boldsymbol{u}_0$ or $V[\boldsymbol{u}_0]^- \approx \boldsymbol{O}$. In other words, $\boldsymbol{u}_N$ is expected to be the optimal solution of (8.36) with a high probability if we start with a good initial guess $\boldsymbol{u}_0$ or assume a large covariance matrix $V[\boldsymbol{u}_0] \approx \infty$. In Bayesian terms, a good estimate is obtained if initially we know *very much* about the true solution or we know *very little*.

Although the controversy over the Bayesian and the non-Bayesian philosophies is very deep, the above observation shows that the distinction is vague in analytical terms: the Bayesian approach simply adds a "starter" $(\boldsymbol{u} - \boldsymbol{u}_0, V[\boldsymbol{u}_0]^-(\boldsymbol{u} - \boldsymbol{u}_0))$ to the optimization criterion so that computation becomes easy.

8.3 Effective Gradient Approximation

Since the function $F^{(k)}(\boldsymbol{a}, \boldsymbol{u})$ takes value 0 if $\boldsymbol{a}$ and $\boldsymbol{u}$ are correctly chosen, we can assume that $F_\alpha^{(k)} = F^{(k)}(\boldsymbol{a}_{\alpha+1}, \boldsymbol{u}_\alpha)$ is very small. If $F^{(k)}(\boldsymbol{a}, \boldsymbol{u})$ is very smooth in $\boldsymbol{u}$, the second derivative $\nabla_{\boldsymbol{u}}^2 F^{(k)}$ is also very small. Hence, we can assume that

$$F_\alpha^{(k)} \nabla_{\boldsymbol{u}}^2 F_\alpha^{(l)} \approx \boldsymbol{0}. \tag{8.42}$$

Consider the (LL)-matrix $\boldsymbol{W}_\alpha(\boldsymbol{u}) = (W_\alpha^{(kl)}(\boldsymbol{u}))$ defined by

$$(W_\alpha^{(kl)}(\boldsymbol{u})) = \Big((\nabla_{\boldsymbol{a}} F^{(k)}(\boldsymbol{a}_{\alpha+1}, \boldsymbol{u}), V[\boldsymbol{a}_{\alpha+1}]\nabla_{\boldsymbol{a}} F^{(l)}(\boldsymbol{a}_\alpha, \boldsymbol{u}))\Big)_r^-. \tag{8.43}$$

If $W_\alpha^{(kl)}(\boldsymbol{u})$ is very smooth in $\boldsymbol{u}$, the second derivative $\nabla_{\boldsymbol{u}}^2 W_\alpha^{(kl)} = \nabla_{\boldsymbol{u}}^2 W_\alpha^{(kl)}(\boldsymbol{u}_\alpha)$ is very small. Hence, we can assume that

$$F_\alpha^{(k)} F_\alpha^{(k)} \nabla_{\boldsymbol{u}}^2 W_\alpha^{(kl)} \approx \boldsymbol{0}. \tag{8.44}$$

Under these approximations, eqs. (8.32) and (8.33) are respectively approximated by

$$\boldsymbol{S}_\alpha \approx \sum_{k,l=1}^{L} \Big(W_\alpha^{(kl)}(\nabla_{\boldsymbol{u}} F_\alpha^{(k)})(\nabla_{\boldsymbol{u}} F_\alpha^{(l)})^\top + 2F_\alpha^{(k)} S[\nabla_{\boldsymbol{u}} W_\alpha^{(kl)}(\nabla_{\boldsymbol{u}} F_\alpha^{(l)})^\top]\Big), \tag{8.45}$$

$$t_\alpha \approx \sum_{k,l=1}^{L} W_\alpha^{(kl)} F_\alpha^{(k)} \nabla_{\boldsymbol{u}} F_\alpha^{(l)}. \tag{8.46}$$

Let $\mathcal{L}_\alpha(\boldsymbol{u})$ be the range of the (LL)-matrix $\boldsymbol{W}_\alpha(\boldsymbol{u})$ defined by eq. (8.43); it is a linear subspace of $\mathcal{R}^L$. Let $\boldsymbol{P}^{\mathcal{L}_\alpha(\boldsymbol{u})}$ be the L-dimensional projection matrix onto the linear subspace $\mathcal{L}_\alpha(\boldsymbol{u})$. Since $\boldsymbol{P}^{\mathcal{L}_\alpha(\boldsymbol{u})}\boldsymbol{W}_\alpha(\boldsymbol{u}) = \boldsymbol{W}_\alpha(\boldsymbol{u})$, we have

$$\frac{\partial \boldsymbol{P}^{\mathcal{L}_\alpha(\boldsymbol{u})}}{\partial u_i}\boldsymbol{W}_\alpha(\boldsymbol{u}) + \boldsymbol{P}^{\mathcal{L}_\alpha(\boldsymbol{u})}\frac{\partial \boldsymbol{W}_\alpha(\boldsymbol{u})}{\partial u_i} = \frac{\partial \boldsymbol{W}_\alpha(\boldsymbol{u})}{\partial u_i}. \tag{8.47}$$

If $\boldsymbol{W}_\alpha(\boldsymbol{u})$ is very smooth in $\boldsymbol{u}$, its range $\mathcal{L}_\alpha(\boldsymbol{u})$ does not depend on $\boldsymbol{u}$ to a first approximation. Hence, $\partial \boldsymbol{P}^{\mathcal{L}_\alpha(\boldsymbol{u})}/\partial u_i \approx \boldsymbol{O}$. Noting that $\boldsymbol{P}^{\mathcal{L}_\alpha(\boldsymbol{u})} = \boldsymbol{W}_\alpha(\boldsymbol{u})\boldsymbol{W}_\alpha(\boldsymbol{u})^-$, we obtain the approximation

$$\sum_{l,m=1}^{L} W_\alpha^{(kl)} (W_\alpha^{(lm)})^- \nabla_{\boldsymbol{u}} W_\alpha^{(mn)} \approx \nabla_{\boldsymbol{u}} W_\alpha^{(kn)}, \tag{8.48}$$

where $(W_\alpha^{(kl)})^-$ denotes the (kl) element of the generalized inverse of the (LL)-matrix $\boldsymbol{W}_\alpha = (W_\alpha^{(kl)})$.

Define the *effective gradient* $\nabla^*_{\boldsymbol{u}} F_\alpha^{(k)}$ by

$$\nabla^*_{\boldsymbol{u}} F_\alpha^{(k)} = \nabla_{\boldsymbol{u}} F_\alpha^{(k)} + \sum_{m,n=1}^{L} (W_\alpha^{(km)})^- F_\alpha^{(n)} \nabla_{\boldsymbol{u}} W_\alpha^{(mn)}. \tag{8.49}$$

Ignoring terms of $O(\sum_{l=1}^L F_\alpha^{(l)} \nabla_{\boldsymbol{u}} W_\alpha^{(kl)})^2$, we obtain the following approximations:

$$\begin{aligned}
&\sum_{k,l=1}^{L} W_\alpha^{(kl)} (\nabla^*_{\boldsymbol{u}} F_\alpha^{(k)})(\nabla^*_{\boldsymbol{u}} F_\alpha^{(l)})^\top \\
&= \sum_{k,l=1}^{L} W_\alpha^{(kl)} \Big(\nabla_{\boldsymbol{u}} F_\alpha^{(k)} + \sum_{m,n=1}^{L} (W_\alpha^{(km)})^- F_\alpha^{(n)} \nabla_{\boldsymbol{u}} W_\alpha^{(mn)}\Big) \\
&\qquad \Big(\nabla_{\boldsymbol{u}} F_\alpha^{(l)} + \sum_{p,q=1}^{L} (W_\alpha^{(lp)})^- F_\alpha^{(q)} \nabla_{\boldsymbol{u}} W_\alpha^{(pq)}\Big)^\top \\
&\approx \sum_{k,l=1}^{L} W_\alpha^{(kl)} (\nabla_{\boldsymbol{u}} F_\alpha^{(k)})(\nabla_{\boldsymbol{u}} F_\alpha^{(l)})^\top \\
&\quad + \sum_{l,n=1}^{L} F_\alpha^{(n)} \Big(\sum_{k,m=1}^{L} W_\alpha^{(lk)} (W_\alpha^{(km)})^- \nabla_{\boldsymbol{u}} W_\alpha^{(mn)}\Big)(\nabla_{\boldsymbol{u}} F_\alpha^{(l)})^\top \\
&\quad + \sum_{k,q=1}^{L} F_\alpha^{(q)} \nabla_{\boldsymbol{u}} F_\alpha^{(k)} \sum_{l,p=1}^{L} (W_\alpha^{(kl)} (W_\alpha^{(lp)})^- \nabla_{\boldsymbol{u}} W_\alpha^{(pq)})^\top
\end{aligned}$$

$$\approx \sum_{k,l=1}^{L} W_\alpha^{(kl)}(\nabla_{\boldsymbol{u}} F_\alpha^{(k)})(\nabla_{\boldsymbol{u}} F_\alpha^{(l)})^\top + \sum_{l,n=1}^{L} F_\alpha^{(n)} \nabla_{\boldsymbol{u}} W_\alpha^{(ln)})(\nabla_{\boldsymbol{u}} F_\alpha^{(l)})^\top$$

$$+ \sum_{k,q=1}^{L} F_\alpha^{(q)} \nabla_{\boldsymbol{u}} F_\alpha^{(k)} (\nabla_{\boldsymbol{u}} W_\alpha^{(pq)})^\top$$

$$\approx \sum_{k,l=1}^{L} \Big(W_\alpha^{(kl)}(\nabla_{\boldsymbol{u}} F_\alpha^{(k)})(\nabla_{\boldsymbol{u}} F_\alpha^{(l)})^\top + 2F_\alpha^{(k)} S[\nabla_{\boldsymbol{u}} W_\alpha^{(kl)} (\nabla_{\boldsymbol{u}} F_\alpha^{(l)})^\top] \Big), \quad (8.50)$$

$$\sum_{k,l=1}^{L} W_\alpha^{(kl)} F_\alpha^{(k)} \nabla_{\boldsymbol{u}}^* F_\alpha^{(l)}$$

$$= \sum_{k,l=1}^{L} W_\alpha^{(kl)} F_\alpha^{(k)} \Big(\nabla_{\boldsymbol{u}} F_\alpha^{(l)} + \sum_{m,n=1}^{L} (W_\alpha^{(lm)})^- F_\alpha^{(n)} \nabla_{\boldsymbol{u}} W_\alpha^{(mn)} \Big)$$

$$= \sum_{k,l=1}^{L} W_\alpha^{(kl)} F_\alpha^{(k)} \nabla_{\boldsymbol{u}} F_\alpha^{(l)}$$

$$+ \sum_{k,n=1}^{L} F_\alpha^{(k)} F_\alpha^{(n)} (\sum_{l,m=1}^{L} W_\alpha^{(kl)} (W_\alpha^{(lm)})^- \nabla_{\boldsymbol{u}} W_\alpha^{(mn)})$$

$$\approx \sum_{k,l=1}^{L} W_\alpha^{(kl)} F_\alpha^{(k)} \nabla_{\boldsymbol{u}} F_\alpha^{(l)} + \sum_{k,n=1}^{L} F_\alpha^{(k)} F_\alpha^{(n)} \nabla_{\boldsymbol{u}} W_\alpha^{(kn)})$$

$$\approx \sum_{k,l=1}^{L} W_\alpha^{(kl)} F_\alpha^{(k)} \nabla_{\boldsymbol{u}} F_\alpha^{(l)}. \quad (8.51)$$

Hence, eqs. (8.32) and (8.33) are approximated by

$$\boldsymbol{S}_\alpha \approx \sum_{k,l=1}^{L} W_\alpha^{(kl)} (\nabla_{\boldsymbol{u}}^* F_\alpha^{(k)})(\nabla_{\boldsymbol{u}}^* F_\alpha^{(l)})^\top, \quad (8.52)$$

$$\boldsymbol{t}_\alpha \approx \sum_{k,l=1}^{L} W_\alpha^{(kl)} F_\alpha^{(k)} \nabla_{\boldsymbol{u}}^* F_\alpha^{(l)}. \quad (8.53)$$

We call these the *effective gradient approximation* and use "=" instead of "≈" in the following.

From the *generalized matrix inversion formula* (2.84), we obtain

$$\Big(V[\boldsymbol{u}_\alpha]^- + \boldsymbol{P}_{\boldsymbol{u}_\alpha}^{\mathcal{U}} \boldsymbol{S}_\alpha \boldsymbol{P}_{\boldsymbol{u}_\alpha}^{\mathcal{U}} \Big)^-$$

$$= V[\boldsymbol{u}_\alpha] - \sum_{k,l=1}^{L} \hat{W}_\alpha^{(kl)} (V[\boldsymbol{u}_\alpha] \nabla_{\boldsymbol{u}}^* F_\alpha^{(k)})(V[\boldsymbol{u}_\alpha] \nabla_{\boldsymbol{u}}^* F_\alpha^{(l)})^\top. \quad (8.54)$$

Here, the (LL)-matrix $\hat{\boldsymbol{W}}_\alpha = (\hat{W}_\alpha^{(kl)})$ is defined by

$$\hat{\boldsymbol{W}}_\alpha = \left(\boldsymbol{W}_\alpha^- + \boldsymbol{P}^{\mathcal{L}_\alpha(\boldsymbol{u}_\alpha)}\hat{\boldsymbol{V}}_\alpha\boldsymbol{P}^{\mathcal{L}_\alpha(\boldsymbol{u}_\alpha)}\right)^-, \tag{8.55}$$

where $\hat{\boldsymbol{V}}_\alpha = (\hat{V}_\alpha^{(kl)})$ is the (LL)-matrix defined by

$$(\hat{V}_\alpha^{(kl)}) = \left((\nabla_{\boldsymbol{u}}^* F_\alpha^{(k)}, V[\boldsymbol{u}_\alpha]\nabla_{\boldsymbol{u}}^* F_\alpha^{(l)})\right). \tag{8.56}$$

Since the matrices $\boldsymbol{W}_\alpha$ and $\hat{\boldsymbol{W}}_\alpha$ share the same range $\mathcal{L}_\alpha(\boldsymbol{u}_\alpha)$, we see that

$$\begin{aligned}
\hat{\boldsymbol{W}}_\alpha\hat{\boldsymbol{V}}_\alpha\boldsymbol{W}_\alpha &= \hat{\boldsymbol{W}}_\alpha\boldsymbol{P}^{\mathcal{L}_\alpha(\boldsymbol{u}_\alpha)}\hat{\boldsymbol{V}}_\alpha\boldsymbol{P}^{\mathcal{L}_\alpha(\boldsymbol{u}_\alpha)}\boldsymbol{W}_\alpha \\
&= \hat{\boldsymbol{W}}_\alpha\left((\boldsymbol{W}_\alpha^- + \boldsymbol{P}^{\mathcal{L}_\alpha(\boldsymbol{u}_\alpha)}\hat{\boldsymbol{V}}_\alpha\boldsymbol{P}^{\mathcal{L}_\alpha(\boldsymbol{u}_\alpha)}) - \boldsymbol{W}_\alpha^-\right)\boldsymbol{W}_\alpha \\
&= \hat{\boldsymbol{W}}_\alpha\hat{\boldsymbol{W}}_\alpha^-\boldsymbol{W}_\alpha - \hat{\boldsymbol{W}}_\alpha\boldsymbol{W}_\alpha^-\boldsymbol{W}_\alpha \\
&= \boldsymbol{P}^{\mathcal{L}_\alpha(\boldsymbol{u}_\alpha)}\boldsymbol{W}_\alpha - \hat{\boldsymbol{W}}_\alpha\boldsymbol{P}^{\mathcal{L}_\alpha(\boldsymbol{u}_\alpha)} \\
&= \boldsymbol{W}_\alpha - \hat{\boldsymbol{W}}_\alpha.
\end{aligned} \tag{8.57}$$

From eqs. (8.52) and (8.54), we obtain

$$\begin{aligned}
&\left(V[\boldsymbol{u}_\alpha]^- + \boldsymbol{P}_{\boldsymbol{u}_\alpha}^{\mathcal{U}}\boldsymbol{S}_\alpha\boldsymbol{P}_{\boldsymbol{u}_\alpha}^{\mathcal{U}}\right)^- \boldsymbol{t}_\alpha \\
&\quad = V[\boldsymbol{u}_\alpha]\sum_{k,l=1}^L W_\alpha^{(kl)}F_\alpha^{(k)}\nabla_{\boldsymbol{u}}^* F_\alpha^{(l)} \\
&\qquad -V[\boldsymbol{u}_\alpha]\sum_{k,l,m,n=1}^L \hat{W}_\alpha^{(kl)}W_\alpha^{(mn)}F_\alpha^{(m)}(\nabla_{\boldsymbol{u}}^* F_\alpha^{(k)})(\nabla_{\boldsymbol{u}}^* F_\alpha^{(l)}, V[\boldsymbol{u}_\alpha]\nabla_{\boldsymbol{u}}^* F_\alpha^{(n)}) \\
&\quad = V[\boldsymbol{u}_\alpha]\sum_{k,l=1}^L W_\alpha^{(kl)}F_\alpha^{(k)}\nabla_{\boldsymbol{u}}^* F_\alpha^{(l)} \\
&\qquad -V[\boldsymbol{u}_\alpha]\sum_{k,m=1}^L \left(\sum_{l,n=1}^L \hat{W}_\alpha^{(kl)}\hat{V}_\alpha^{(ln)}W_\alpha^{(nm)}\right)F_\alpha^{(m)}\nabla_{\boldsymbol{u}}^* F_\alpha^{(k)} \\
&\quad = V[\boldsymbol{u}_\alpha]\sum_{k,l=1}^L W_\alpha^{(kl)}F_\alpha^{(k)}\nabla_{\boldsymbol{u}}^* F_\alpha^{(l)} \\
&\qquad -V[\boldsymbol{u}_\alpha]\sum_{k,m=1}^L (W_\alpha^{(km)} - \hat{W}_\alpha^{(km)})F_\alpha^{(m)}\nabla_{\boldsymbol{u}}^* F_\alpha^{(k)} \\
&\quad = V[\boldsymbol{u}_\alpha]\sum_{k,l=1}^L \hat{W}_\alpha^{(kl)}F_\alpha^{(k)}\nabla_{\boldsymbol{u}}^* F_\alpha^{(l)}.
\end{aligned} \tag{8.58}$$

On the other hand, we obtain the following identity from the generalized matrix inversion formula (2.84) and eq. (8.52):

$$\left(\tilde{V}[\boldsymbol{u}_\alpha]^- + \boldsymbol{P}^{\mathcal{U}}_{\boldsymbol{u}_{\alpha+1}} \boldsymbol{P}^{\mathcal{U}}_{\boldsymbol{u}_\alpha} \boldsymbol{S}_\alpha \boldsymbol{P}^{\mathcal{U}}_{\boldsymbol{u}_\alpha} \boldsymbol{P}^{\mathcal{U}}_{\boldsymbol{u}_{\alpha+1}}\right)^-$$

$$= \tilde{V}[\boldsymbol{u}_\alpha] - \sum_{k,l=1}^{L} \tilde{W}_\alpha^{(kl)} (\tilde{V}[\boldsymbol{u}_\alpha] \boldsymbol{P}^{\mathcal{U}}_{\boldsymbol{u}_\alpha} \nabla^*_{\boldsymbol{u}} F_\alpha^{(k)}) (\tilde{V}[\boldsymbol{u}_\alpha] \boldsymbol{P}^{\mathcal{U}}_{\boldsymbol{u}_\alpha} \nabla^*_{\boldsymbol{u}} F_\alpha^{(l)})^\top. \quad (8.59)$$

Here, the (LL)-matrix $\tilde{\boldsymbol{W}}_\alpha = (\tilde{W}_\alpha^{(kl)})$ is defined by

$$\tilde{\boldsymbol{W}}_\alpha = \left(\boldsymbol{W}_\alpha^- + \boldsymbol{P}^{\mathcal{L}_\alpha(\boldsymbol{u}_\alpha)} \tilde{\boldsymbol{V}}_\alpha \boldsymbol{P}^{\mathcal{L}_\alpha(\boldsymbol{u}_\alpha)}\right)^-, \quad (8.60)$$

where $\tilde{\boldsymbol{V}}_\alpha = (\tilde{V}_\alpha^{(kl)})$ is the (LL)-matrix defined by

$$(\tilde{V}_\alpha^{(kl)}) = \left((\nabla^*_{\boldsymbol{u}} F_\alpha^{(k)}, \boldsymbol{P}^{\mathcal{U}}_{\boldsymbol{u}_\alpha} \tilde{V}[\boldsymbol{u}_\alpha] \boldsymbol{P}^{\mathcal{U}}_{\boldsymbol{u}_\alpha} \nabla^*_{\boldsymbol{u}} F_\alpha^{(l)})\right). \quad (8.61)$$

Hence, we obtain the following effective gradient approximation to the update rule of $\boldsymbol{u}_\alpha$ and $V[\boldsymbol{u}_\alpha]$:

$$\boldsymbol{u}_{\alpha+1} = \mathcal{C}[\boldsymbol{u}_\alpha - V[\boldsymbol{u}_\alpha] \sum_{k,l=1}^{L} \hat{W}_\alpha^{(kl)} F_\alpha^{(k)} \nabla^*_{\boldsymbol{u}} F_\alpha^{(l)}] \quad (8.62)$$

$$V[\boldsymbol{u}_{\alpha+1}] = \tilde{V}[\boldsymbol{u}_\alpha] - \sum_{k,l=1}^{L} \tilde{W}_\alpha^{(kl)} (\tilde{V}[\boldsymbol{u}_\alpha] \boldsymbol{P}^{\mathcal{U}}_{\boldsymbol{u}_\alpha} \nabla^*_{\boldsymbol{u}} F_\alpha^{(k)}) (\tilde{V}[\boldsymbol{u}_\alpha] \boldsymbol{P}^{\mathcal{U}}_{\boldsymbol{u}_\alpha} \nabla^*_{\boldsymbol{u}} F_\alpha^{(l)})^\top. \quad (8.63)$$

8.4 Reduction from the Kalman Filter

We now show that the *Kalman filter* described in Section 3.4.4 can be applied to solve our problem. For simplicity, let us assume that $V[\boldsymbol{a}_\alpha]$ is of full rank. Let $\{\boldsymbol{u}_\alpha\}$ be a sequence of estimates of $\boldsymbol{u}$. For the moment, we simply assume that such a sequence is somehow given and do not question how it is defined. We have

$$\begin{aligned} 0 &= F^{(k)}(\bar{\boldsymbol{a}}_\alpha, \boldsymbol{u}) = F^{(k)}(\boldsymbol{a}_\alpha - \Delta\boldsymbol{a}_\alpha, \boldsymbol{u}_{\alpha-1} + (\boldsymbol{u} - \boldsymbol{u}_{\alpha-1})) \\ &= F^{(k)}(\boldsymbol{a}_\alpha, \boldsymbol{u}_{\alpha-1}) - (\nabla_{\boldsymbol{a}} F^{(k)}(\boldsymbol{a}_\alpha, \boldsymbol{u}_{\alpha-1}), \Delta\boldsymbol{a}_\alpha) \\ &\quad + (\nabla_{\boldsymbol{u}} F^{(k)}(\boldsymbol{a}_\alpha, \boldsymbol{u}_{\alpha-1}), \boldsymbol{u} - \boldsymbol{u}_{\alpha-1}) + O(\Delta\boldsymbol{a}_\alpha, \boldsymbol{u} - \boldsymbol{u}_{\alpha-1})^2. \end{aligned} \quad (8.64)$$

Assuming that $\boldsymbol{u}_{\alpha-1}$ is a good estimate of $\boldsymbol{u}$ and ignoring $O(\Delta\boldsymbol{a}_\alpha, \boldsymbol{u}-\boldsymbol{u}_{\alpha-1})^2$, we obtain

$$F^{(k)}(\boldsymbol{a}_\alpha, \boldsymbol{u}_{\alpha-1}) = (\nabla_{\boldsymbol{a}} F_{\alpha-1}^{(k)}, \Delta\boldsymbol{a}_\alpha) - (\nabla_{\boldsymbol{u}} F_{\alpha-1}^{(k)}, \boldsymbol{u} - \boldsymbol{u}_{\alpha-1}), \quad (8.65)$$

where $\nabla_{\boldsymbol{a}} F_\alpha^{(k)}$ and $\nabla_{\boldsymbol{u}} F_\alpha^{(k)}$ are the abbreviations of $\nabla_{\boldsymbol{a}} F^{(k)}(\boldsymbol{a}_{\alpha+1}, \boldsymbol{u}_\alpha)$ and $\nabla_{\boldsymbol{u}} F^{(k)}(\boldsymbol{a}_{\alpha+1}, \boldsymbol{u}_\alpha)$, respectively.

If we define an n-vector $\boldsymbol{x}_\alpha$, an L-vector $\boldsymbol{y}_\alpha$, an L-vector $\boldsymbol{w}_\alpha$, and an Ln-matrix $\boldsymbol{C}$ by

$$\boldsymbol{x}_\alpha = -(\boldsymbol{u} - \boldsymbol{u}_{\alpha-1}), \qquad \boldsymbol{y}_\alpha = \begin{pmatrix} F^{(1)}(\boldsymbol{a}_\alpha, \boldsymbol{u}_{\alpha-1}) \\ \vdots \\ F^{(L)}(\boldsymbol{a}_\alpha, \boldsymbol{u}_{\alpha-1}) \end{pmatrix}, \tag{8.66}$$

$$\boldsymbol{w}_\alpha = \begin{pmatrix} (\nabla_{\boldsymbol{a}} F_{\alpha-1}^{(1)}, \Delta \boldsymbol{a}_\alpha) \\ \vdots \\ (\nabla_{\boldsymbol{a}} F_{\alpha-1}^{(L)}, \Delta \boldsymbol{a}_\alpha) \end{pmatrix},$$

$$\boldsymbol{C}_\alpha = \begin{pmatrix} \partial F_{\alpha-1}^{(1)}/\partial u_1 & \cdots & \partial F_{\alpha-1}^{(1)}/\partial u_n \\ \vdots & \cdots & \vdots \\ \partial F_{\alpha-1}^{(L)}/\partial u_1 & \cdots & \partial F_{\alpha-1}^{(L)}/\partial u_n \end{pmatrix}, \tag{8.67}$$

eq. (8.65) can be written in the form

$$\boldsymbol{y}_\alpha = \boldsymbol{C}_\alpha \boldsymbol{x}_\alpha + \boldsymbol{w}_\alpha. \tag{8.68}$$

Since $\Delta \boldsymbol{a}_\alpha$ is a random variable of mean $\mathbf{0}$ and covariance matrix $V[\boldsymbol{a}_\alpha]$, the L-vector $\boldsymbol{w}_\alpha$ is also a random variable; its expectation and covariance matrix are given as follows:

$$E[\boldsymbol{w}_\alpha] = \Big((\nabla_{\boldsymbol{a}} F_{\alpha-1}^{(k)}, E[\Delta \boldsymbol{a}_\alpha]) \Big) = \mathbf{0}, \tag{8.69}$$

$$\begin{aligned} V[\boldsymbol{w}_\alpha] &= \Big((\nabla_{\boldsymbol{a}} F_{\alpha-1}^{(k)}, E[\Delta \boldsymbol{a}_\alpha \Delta \boldsymbol{a}_\alpha^\top] \nabla_{\boldsymbol{a}} F_{\alpha-1}^{(l)}) \Big) \\ &= \Big((\nabla_{\boldsymbol{a}} F_{\alpha-1}^{(k)}, V[\boldsymbol{a}_\alpha] \nabla_{\boldsymbol{a}} F_{\alpha-1}^{(l)}) \Big). \end{aligned} \tag{8.70}$$

Now,

$$\begin{aligned} \boldsymbol{x}_\alpha &= -(\boldsymbol{u} - \boldsymbol{u}_{\alpha-1}) = -(\boldsymbol{u} - \boldsymbol{u}_{\alpha-2}) + (\boldsymbol{u}_{\alpha-1} - \boldsymbol{u}_{\alpha-2}) \\ &= \boldsymbol{x}_{\alpha-1} + (\boldsymbol{u}_{\alpha-1} - \boldsymbol{u}_{\alpha-2}). \end{aligned} \tag{8.71}$$

If we put

$$\boldsymbol{A}_\alpha = \boldsymbol{I}, \qquad \boldsymbol{B}_\alpha = \boldsymbol{I}, \qquad \boldsymbol{v}_\alpha = \boldsymbol{u}_\alpha - \boldsymbol{u}_{\alpha-1}, \tag{8.72}$$

eq. (8.71) can be written as

$$\boldsymbol{x}_\alpha = \boldsymbol{A}_{\alpha-1} \boldsymbol{x}_{\alpha-1} + \boldsymbol{B}_{\alpha-1} \boldsymbol{v}_{\alpha-1}. \tag{8.73}$$

This can be viewed as a "linear dynamical system" with input $\boldsymbol{v}_\alpha$, although no internal dynamics exists. Strictly speaking, the input $\boldsymbol{v}_\alpha$ is not a random variable, but it can be regarded as a random variable of *zero covariance*, i.e.,

$$E[\boldsymbol{v}_\alpha] = \boldsymbol{u}_\alpha - \boldsymbol{u}_{\alpha-1}, \qquad V[\boldsymbol{v}_\alpha] = \boldsymbol{O}. \tag{8.74}$$

Eqs. (8.73) and (8.68) have exactly the same form as eqs. (3.104) and (3.105). Hence, the Kalman filter defined by eqs. (3.106)–(3.109) can be applied. Substituting eqs. (8.69), (8.70), (8.72), and (8.74) into eqs. (3.106)–(3.109), the estimate $\hat{\boldsymbol{x}}_\alpha$ and its covariance matrix $V[\hat{\boldsymbol{x}}_\alpha]$ are computed in the following form:

$$\begin{aligned}\hat{\boldsymbol{x}}_\alpha = \ &\hat{\boldsymbol{x}}_{\alpha-1} + \boldsymbol{u}_{\alpha-1} - \boldsymbol{u}_{\alpha-2} \\ &+ V[\hat{\boldsymbol{x}}_\alpha]\boldsymbol{C}_\alpha^\top V[\boldsymbol{w}_\alpha]^{-1}(\boldsymbol{y}_\alpha - \boldsymbol{C}_\alpha(\hat{\boldsymbol{x}}_{\alpha-1} + \boldsymbol{u}_{\alpha-1} - \boldsymbol{u}_{\alpha-2})),\end{aligned} \tag{8.75}$$

$$V[\hat{\boldsymbol{x}}_\alpha] = \left(V[\hat{\boldsymbol{x}}_{\alpha-1}]^{-1} + \boldsymbol{C}_\alpha^\top V[\boldsymbol{w}_\alpha]^{-1}\boldsymbol{C}_\alpha\right)^{-1}. \tag{8.76}$$

Applying the matrix inversion formula (2.22), we can rewrite eq. (8.76) in the form

$$V[\hat{\boldsymbol{x}}_\alpha] = V[\hat{\boldsymbol{x}}_{\alpha-1}] - V[\hat{\boldsymbol{x}}_{\alpha-1}]\boldsymbol{C}_\alpha^\top \tilde{\boldsymbol{V}}_{\alpha-1}^{-1}(V[\hat{\boldsymbol{x}}_{\alpha-1}]\boldsymbol{C}_\alpha^\top)^\top, \tag{8.77}$$

where

$$\tilde{\boldsymbol{V}}_\alpha = V[\boldsymbol{w}_{\alpha+1}] + \boldsymbol{C}_{\alpha+1}V[\hat{\boldsymbol{x}}_\alpha]\boldsymbol{C}_{\alpha+1}^\top. \tag{8.78}$$

From eq. (8.77), we obtain

$$\begin{aligned}&V[\hat{\boldsymbol{x}}_\alpha]\boldsymbol{C}_\alpha^\top \boldsymbol{V}_{\alpha-1}^{-1} \\ &= V[\hat{\boldsymbol{x}}_{\alpha-1}]\boldsymbol{C}_\alpha^\top \boldsymbol{V}_{\alpha-1}^{-1} - V[\hat{\boldsymbol{x}}_{\alpha-1}]\boldsymbol{C}_\alpha^\top \tilde{\boldsymbol{V}}_{\alpha-1}^{-1}\boldsymbol{C}_\alpha V[\hat{\boldsymbol{x}}_{\alpha-1}]\boldsymbol{C}_\alpha^\top \boldsymbol{V}_{\alpha-1}^{-1} \\ &= V[\hat{\boldsymbol{x}}_{\alpha-1}]\boldsymbol{C}_\alpha^\top \boldsymbol{V}_{\alpha-1}^{-1} \\ &\quad - V[\hat{\boldsymbol{x}}_{\alpha-1}]\boldsymbol{C}_\alpha^\top \tilde{\boldsymbol{V}}_{\alpha-1}^{-1}\left((\boldsymbol{V}_{\alpha-1} + \boldsymbol{C}_\alpha V[\hat{\boldsymbol{x}}_{\alpha-1}]\boldsymbol{C}_\alpha^\top) - \boldsymbol{V}_{\alpha-1}\right)\boldsymbol{V}_{\alpha-1}^{-1} \\ &= V[\hat{\boldsymbol{x}}_{\alpha-1}]\boldsymbol{C}_\alpha^\top \boldsymbol{V}_{\alpha-1}^{-1} - V[\hat{\boldsymbol{x}}_{\alpha-1}]\boldsymbol{C}_\alpha^\top \boldsymbol{V}_{\alpha-1}^{-1} + V[\hat{\boldsymbol{x}}_{\alpha-1}]\boldsymbol{C}_\alpha^\top \tilde{\boldsymbol{V}}_{\alpha-1}^{-1} \\ &= V[\hat{\boldsymbol{x}}_{\alpha-1}]\boldsymbol{C}_\alpha^\top \tilde{\boldsymbol{V}}_{\alpha-1}^{-1}.\end{aligned} \tag{8.79}$$

Substituting this into eq. (8.75) and incrementing α by 1, we obtain

$$\begin{aligned}\hat{\boldsymbol{x}}_{\alpha+1} = \ &\hat{\boldsymbol{x}}_\alpha + \boldsymbol{u}_\alpha - \boldsymbol{u}_{\alpha-1} \\ &+ V[\hat{\boldsymbol{x}}_\alpha]\boldsymbol{C}_{\alpha+1}^\top \tilde{\boldsymbol{V}}_\alpha^{-1}(\boldsymbol{y}_{\alpha+1} - \boldsymbol{C}_{\alpha+1}(\hat{\boldsymbol{x}}_\alpha + \boldsymbol{u}_\alpha - \boldsymbol{u}_{\alpha-1})).\end{aligned} \tag{8.80}$$

If $\hat{\boldsymbol{x}}_{\alpha+1}$ is the optimal estimate of $\boldsymbol{x}_{\alpha+1} = -(\boldsymbol{u} - \boldsymbol{u}_\alpha)$, the optimal estimate of $\boldsymbol{u}$ at time $\alpha + 1$ is given by

$$\hat{\boldsymbol{u}}_{\alpha+1} = \boldsymbol{u}_\alpha - \hat{\boldsymbol{x}}_{\alpha+1}. \tag{8.81}$$

Substituting eq. (8.80) into this and noting that $\hat{\boldsymbol{x}}_\alpha + \boldsymbol{u}_\alpha - \boldsymbol{u}_{\alpha-1} = \boldsymbol{u}_\alpha - \hat{\boldsymbol{u}}_\alpha$, we obtain the optimal estimate $\hat{\boldsymbol{u}}_{\alpha+1}$ in the following form:

$$\hat{\boldsymbol{u}}_{\alpha+1} = \hat{\boldsymbol{u}}_\alpha - V[\hat{\boldsymbol{u}}_\alpha]\boldsymbol{C}_{\alpha+1}^\top \tilde{\boldsymbol{V}}_\alpha^{-1}(\boldsymbol{y}_{\alpha+1} - \boldsymbol{C}_{\alpha+1}(\boldsymbol{u}_\alpha - \hat{\boldsymbol{u}}_\alpha)), \tag{8.82}$$

Since $V[\hat{\boldsymbol{u}}_\alpha] = V[\hat{\boldsymbol{x}}_\alpha]$, the covariance matrix of $\hat{\boldsymbol{u}}_{\alpha+1}$ is given by

$$V[\hat{\boldsymbol{u}}_{\alpha+1}] = V[\hat{\boldsymbol{u}}_\alpha] - V[\hat{\boldsymbol{u}}_\alpha]\boldsymbol{C}_{\alpha+1}^\top \tilde{\boldsymbol{V}}_\alpha^{-1}(V[\hat{\boldsymbol{u}}_\alpha]\boldsymbol{C}_{\alpha+1}^\top)^\top. \tag{8.83}$$

Recall that the sequence $\{\boldsymbol{u}_\alpha\}$ is *yet to be defined*. We now define it so that it *happens* to coincides with $\{\hat{\boldsymbol{u}}_\alpha\}$. Using eqs. (8.67) and (8.70), we can now reduce eqs. (8.82) and (8.83) to the following form:

$$\hat{\boldsymbol{u}}_{\alpha+1} = \hat{\boldsymbol{u}}_\alpha - V[\hat{\boldsymbol{u}}_\alpha] \sum_{k,l=1}^{L} \tilde{W}_\alpha^{(kl)} F_\alpha^{(k)} \nabla_{\boldsymbol{u}} F_\alpha^{(l)}, \tag{8.84}$$

$$V[\hat{\boldsymbol{u}}_{\alpha+1}] = V[\hat{\boldsymbol{u}}_\alpha] - \sum_{k,l=1}^{L} \tilde{W}_\alpha^{(kl)} (V[\hat{\boldsymbol{u}}_\alpha]\nabla_{\boldsymbol{u}} F_\alpha^{(k)})(V[\hat{\boldsymbol{u}}_\alpha]\nabla_{\boldsymbol{u}} F_\alpha^{(l)})^\top, \tag{8.85}$$

$$\tilde{W}_\alpha^{(kl)} = \Big((\nabla_{\boldsymbol{a}} F_\alpha^{(k)}, V[\boldsymbol{a}_{\alpha+1}]\nabla_{\boldsymbol{a}} F_\alpha^{(l)}) + (\nabla_{\boldsymbol{u}} F_\alpha^{(k)}, V[\hat{\boldsymbol{u}}_\alpha]\nabla_{\boldsymbol{u}} F_\alpha^{(k)})\Big)^{-1}. \tag{8.86}$$

Applying a linear approximation at assumed estimates, we can make the Kalman filter applicable to nonlinear dynamical systems as well; the resulting rule is called the *extended Kalman filter*. In this sense, eqs. (8.84)—(8.86) can be called the extended Kalman filter, although nonlinearity enters only the observation process (i.e., no internal dynamics exists).

Comparing eqs. (8.84) and (8.85) with eqs. (8.62) and (8.63), we see that eqs. (8.62) and (8.63) reduce to eqs. (8.84) and (8.85) if

1. the effective gradient $\nabla_{\boldsymbol{u}}^* F_\alpha^{(k)}$ in eqs. (8.62) and (8.63) is replaced by the ordinary gradient $\nabla_{\boldsymbol{u}} F_\alpha^{(k)}$, and
2. no constraint is imposed on the parameter space for $\boldsymbol{u}$.

Approximating $\nabla_{\boldsymbol{u}}^* F_\alpha^{(k)}$ by $\nabla_{\boldsymbol{u}} F_\alpha^{(k)}$ is equivalent to assuming that

$$\nabla_{\boldsymbol{u}} W_\alpha^{(kl)} = \boldsymbol{0}. \tag{8.87}$$

However, this approximation introduces *statistical bias* into the solution, as will be discussed in the next chapter. Thus, the update rule derived from the Kalman filter by restricting the underlying linear dynamical system to a linear "statical" system is cruder than the optimal filter directly derived by applying the maximum a posteriori probability principle, on which the Kalman filter is also based (see Section 3.4.4).

8.5 Estimation from Linear Hypotheses

As seen in Chapter 7, many hypotheses involving image and space objects (e.g., their coincidence, incidence, collinearity, concurrency, and coplanarity) are linear in the parameters to be estimated and can be written in the form

$$(\bar{\boldsymbol{a}}_\alpha^{(k)}, \boldsymbol{u}) = 0, \qquad k = 1, ..., L, \tag{8.88}$$

where $\boldsymbol{u}$ and $\bar{\boldsymbol{a}}_\alpha^{(k)}$, $k = 1, ..., L$, $\alpha = 1, ..., N$, are all n-vectors. Let $\{\boldsymbol{a}_\alpha^{(k)}\}$ be the observed data, and let $V[\boldsymbol{a}_\alpha^{(k)}, \boldsymbol{a}_\alpha^{(l)}]$ be the covariance matrix of $\boldsymbol{a}_\alpha^{(k)}$ and $\boldsymbol{a}_\alpha^{(l)}$. As in Section 7.1.5, eq. (8.88) can be written in the form $F^{(k)}(\bar{\boldsymbol{a}}_\alpha, \boldsymbol{u}) = 0$ if we define the direct sum vector $\bar{\boldsymbol{a}}_\alpha$ by

$$\bar{\boldsymbol{a}}_\alpha = \bar{\boldsymbol{a}}_\alpha^{(1)} \oplus \cdots \oplus \bar{\boldsymbol{a}}_\alpha^{(L)}, \tag{8.89}$$

and L functions $F^{(k)}(\,\cdot\,,\,\cdot\,)$: $\mathcal{R}^{Ln} \times \mathcal{R}^n \to \mathcal{R}$ by

$$F^{(k)}(\boldsymbol{a}^{(1)} \oplus \cdots \oplus \boldsymbol{a}^{(L)}, \boldsymbol{u}) = (\boldsymbol{a}^{(k)}, \boldsymbol{u}). \tag{8.90}$$

Hence, the general theory described in Sections 8.2.1 and 8.3 can be applied if we put

$$F_\alpha^{(k)} = (\boldsymbol{a}_{\alpha+1}^{(k)}, \boldsymbol{u}_\alpha), \qquad \nabla_{\boldsymbol{u}} F_\alpha^{(k)} = \boldsymbol{a}_{\alpha+1}^{(k)},$$

$$\nabla_{\boldsymbol{a}} F_\alpha^{(k)} = \boldsymbol{0} \oplus \cdots \oplus \overset{(k)}{\boldsymbol{u}_\alpha} \oplus \cdots \oplus \boldsymbol{0}. \tag{8.91}$$

The (LL)-matrix $\boldsymbol{W}_\alpha(\boldsymbol{u}) = (W_\alpha^{(kl)}(\boldsymbol{u}))$ defined by eq. (8.43) can be rewritten as

$$(W_\alpha^{(kl)}(\boldsymbol{u})) = \Big((\boldsymbol{u}, V[\boldsymbol{a}_\alpha^{(k)}, \boldsymbol{a}_\alpha^{(l)}]\boldsymbol{u})\Big)_r^-. \tag{8.92}$$

Let $\mathcal{L}_\alpha(\boldsymbol{u})$ be the range of the (LL)-matrix $\boldsymbol{W}_\alpha(\boldsymbol{u})$. Differentiating $\boldsymbol{W}_\alpha(\boldsymbol{u})^-\boldsymbol{W}_\alpha(\boldsymbol{u}) = \boldsymbol{P}^{\mathcal{L}_\alpha(\boldsymbol{u})}$ with respect to $\boldsymbol{u}$, we obtain

$$\frac{\partial \boldsymbol{W}_\alpha(\boldsymbol{u})^-}{\partial u_i}\boldsymbol{W}_\alpha(\boldsymbol{u}) + \boldsymbol{W}_\alpha(\boldsymbol{u})^-\frac{\partial \boldsymbol{W}_\alpha(\boldsymbol{u})}{\partial u_i} = \frac{\partial \boldsymbol{P}^{\mathcal{L}_\alpha(\boldsymbol{u})}}{\partial u_i}, \tag{8.93}$$

where $\boldsymbol{P}^{\mathcal{L}_\alpha(\boldsymbol{u})}$ is the L-dimensional projection matrix onto the linear subspace $\mathcal{L}_\alpha(\boldsymbol{u})$. As in Section 8.3, we assume that the range $\mathcal{L}_\alpha(\boldsymbol{u})$ is smooth in $\boldsymbol{u}$ and hence $\partial\boldsymbol{P}^{\mathcal{L}_\alpha(\boldsymbol{u})}/\partial u_i \approx \boldsymbol{O}$. Abbreviating $W_\alpha^{(kl)}(\boldsymbol{u}_\alpha)$ and $\nabla_{\boldsymbol{u}} W_\alpha^{(kl)}(\boldsymbol{u}_\alpha)$ to $W_\alpha^{(kl)}$ and $\nabla_{\boldsymbol{u}} W_\alpha^{(kl)}$, respectively, we obtain

$$\begin{aligned}\sum_{l=1}^{L}(W_\alpha^{(kl)})^-\nabla_{\boldsymbol{u}} W_\alpha^{(lm)} &\approx -\sum_{l=1}^{L} W_\alpha^{(lm)}\nabla_{\boldsymbol{u}}(W_\alpha^{(kl)})^- \\ &\approx -\sum_{l=1}^{L} W_\alpha^{(lm)}\nabla_{\boldsymbol{u}}(\boldsymbol{u}, V[\boldsymbol{a}_{\alpha+1}^{(k)}, \boldsymbol{a}_{\alpha+1}^{(l)}]\boldsymbol{u})|_{\boldsymbol{u}=\boldsymbol{u}_\alpha} \\ &= -2\sum_{l=1}^{L} W_\alpha^{(lm)} S[V[\boldsymbol{a}_{\alpha+1}^{(k)}, \boldsymbol{a}_{\alpha+1}^{(l)}]]\boldsymbol{u}_\alpha.\end{aligned} \tag{8.94}$$

Define the *effective value* $\boldsymbol{a}_\alpha^{*(k)}$ of $\boldsymbol{a}_\alpha^{(k)}$ by the effective gradient $\nabla_{\boldsymbol{u}}^* F_{\alpha-1}^{(k)}$ (see eq. (8.49)):

$$\boldsymbol{a}_\alpha^{*(k)} = \nabla_{\boldsymbol{u}}^* F_{\alpha-1}^{(k)} \approx \boldsymbol{a}_\alpha^{(k)} - 2\sum_{l,m=1}^{L} W_{\alpha-1}^{(lm)}(\boldsymbol{a}_\alpha^{(l)}, \boldsymbol{u}_{\alpha-1}) S[V[\boldsymbol{a}_\alpha^{(k)}, \boldsymbol{a}_\alpha^{(m)}]]\boldsymbol{u}_{\alpha-1}. \tag{8.95}$$

The update rule given by eqs. (8.62) and (8.63) can be written in the form

$$\boldsymbol{u}_{\alpha+1} = N[\boldsymbol{u}_\alpha - V[\boldsymbol{u}_\alpha]\sum_{k,l=1}^{L} \hat{W}_\alpha^{(kl)}(\boldsymbol{a}_{\alpha+1}^{(k)}, \boldsymbol{u}_\alpha)\boldsymbol{a}_{\alpha+1}^{*(l)}], \tag{8.96}$$

$$V[\boldsymbol{u}_{\alpha+1}] = \tilde{V}[\boldsymbol{u}_\alpha] - \sum_{k,l=1}^{L} \tilde{W}_\alpha^{(kl)}(\tilde{V}[\boldsymbol{u}_\alpha]\boldsymbol{P}_{\boldsymbol{u}_\alpha}^{\mathcal{U}}\boldsymbol{a}_{\alpha+1}^{*(k)})(\tilde{V}[\boldsymbol{u}_\alpha]\boldsymbol{P}_{\boldsymbol{u}_\alpha}^{\mathcal{U}}\boldsymbol{a}_{\alpha+1}^{*(l)})^\top, \tag{8.97}$$

where

$$\tilde{V}[\boldsymbol{u}_\alpha] = \left(\boldsymbol{P}_{\boldsymbol{u}_{\alpha+1}}^{\mathcal{U}} V[\boldsymbol{u}_\alpha]^- \boldsymbol{P}_{\boldsymbol{u}_{\alpha+1}}^{\mathcal{U}}\right)^-. \tag{8.98}$$

The (LL)-matrices $\hat{\boldsymbol{W}}_\alpha = (\hat{W}_\alpha^{(kl)})$ and $\tilde{\boldsymbol{W}}_\alpha = (\tilde{W}_\alpha^{(kl)})$ are defined by

$$\hat{\boldsymbol{W}}_\alpha = \left(\boldsymbol{W}_\alpha^- + \boldsymbol{P}^{\mathcal{L}_\alpha(\boldsymbol{u}_\alpha)}\hat{\boldsymbol{V}}_\alpha \boldsymbol{P}^{\mathcal{L}_\alpha(\boldsymbol{u}_\alpha)}\right)^-,$$

$$\tilde{\boldsymbol{W}}_\alpha = \left(\boldsymbol{W}_\alpha^- + \boldsymbol{P}^{\mathcal{L}_\alpha(\boldsymbol{u}_\alpha)}\tilde{\boldsymbol{V}}_\alpha \boldsymbol{P}^{\mathcal{L}_\alpha(\boldsymbol{u}_\alpha)}\right)^-, \tag{8.99}$$

where $\hat{\boldsymbol{V}}_\alpha = (\hat{V}_\alpha^{(kl)})$ and $\tilde{\boldsymbol{V}}_\alpha = (\tilde{V}_\alpha^{(kl)})$ are the (LL)-matrices defined by

$$\hat{V}_\alpha^{(kl)} = \left((\boldsymbol{a}_{\alpha+1}^{*(k)}, V[\boldsymbol{u}_\alpha]\boldsymbol{a}_{\alpha+1}^{*(l)})\right),$$

$$\tilde{V}_\alpha^{(kl)} = \left((\boldsymbol{a}_{\alpha+1}^{*(k)}, \boldsymbol{P}_{\boldsymbol{u}_\alpha}^{\mathcal{U}}\tilde{V}[\boldsymbol{u}_\alpha]\boldsymbol{P}_{\boldsymbol{u}_\alpha}^{\mathcal{U}}\boldsymbol{a}_{\alpha+1}^{*(l)})\right). \tag{8.100}$$

Example 8.1 If the hypothesis consists of a single equation $(\boldsymbol{a}, \boldsymbol{u}) = 0$ and no constraint is imposed on $\boldsymbol{u}$ other than normalization $\|\boldsymbol{u}\| = 1$, the optimal filter takes the following form:

$$\boldsymbol{u}_{\alpha+1} = N[\boldsymbol{u}_\alpha - \frac{(\boldsymbol{a}_{\alpha+1}, \boldsymbol{u}_\alpha)V[\boldsymbol{u}_\alpha]\boldsymbol{a}_{\alpha+1}^*}{(\boldsymbol{u}_\alpha, V[\boldsymbol{a}_{\alpha+1}]\boldsymbol{u}_\alpha) + (\boldsymbol{a}_{\alpha+1}^*, V[\boldsymbol{u}_\alpha]\boldsymbol{a}_{\alpha+1}^*)}], \tag{8.101}$$

$$V[\boldsymbol{u}_{\alpha+1}] = \tilde{V}[\boldsymbol{u}_\alpha] - \frac{(\tilde{V}[\boldsymbol{u}_\alpha]\boldsymbol{P}_{\boldsymbol{u}_\alpha}\boldsymbol{a}_{\alpha+1}^*)(\tilde{V}[\boldsymbol{u}_\alpha]\boldsymbol{P}_{\boldsymbol{u}_\alpha}\boldsymbol{a}_{\alpha+1}^*)^\top}{(\boldsymbol{u}_\alpha, V[\boldsymbol{a}_{\alpha+1}]\boldsymbol{u}_\alpha) + (\boldsymbol{a}_{\alpha+1}^*, \boldsymbol{P}_{\boldsymbol{u}_\alpha}\tilde{V}[\boldsymbol{u}_\alpha]\boldsymbol{P}_{\boldsymbol{u}_\alpha}\boldsymbol{a}_{\alpha+1}^*)}, \tag{8.102}$$

$$\boldsymbol{a}_{\alpha+1}^* = \boldsymbol{a}_{\alpha+1} - \frac{2(\boldsymbol{a}_{\alpha+1}, \boldsymbol{u}_\alpha)V[\boldsymbol{a}_{\alpha+1}]\boldsymbol{u}_\alpha}{(\boldsymbol{u}_\alpha, V[\boldsymbol{a}_{\alpha+1}]\boldsymbol{u}_\alpha)}, \tag{8.103}$$

$$\tilde{V}[\boldsymbol{u}_\alpha] = \left(\boldsymbol{P}_{\boldsymbol{u}_{\alpha+1}} V[\boldsymbol{u}_\alpha]^- \boldsymbol{P}_{\boldsymbol{u}_{\alpha+1}}\right)^-. \tag{8.104}$$

Here, $\boldsymbol{P}_{\boldsymbol{u}_\alpha}$ and $\boldsymbol{P}_{\boldsymbol{u}_{\alpha+1}}$ are the n-dimensional projection matrices along $\boldsymbol{u}_\alpha$ and $\boldsymbol{u}_{\alpha+1}$, respectively.

Chapter 9

Renormalization

This chapter focuses on the parametric fitting problem for a linear hypothesis with no constraints other than normalization. To a first approximation, the problem reduces to least-squares fitting, for which the solution can be obtained analytically by solving the eigenvalue problem. We first show that this least-squares approximation introduces *statistical bias* into the solution whatever weights are used. After analyzing the statistical bias in quantitative terms, we present an iterative procedure, called *renormalization*, for removing the bias by automatically adjusting to the noise. In contrast to the optimal filter we studied in the preceding chapter, renormalization requires no initial estimate and no knowledge of the noise level. We then discuss a procedure called *linearization*, which enables us to apply renormalization to nonlinear constraints. Finally, we define *second order renormalization* which removes statistical bias up to second order terms.

9.1 Eigenvector Fit

9.1.1 Least-squares approximation

As shown in Chapter 7, many hypotheses involving image and space objects (e.g., coincidence, incidence, collinearity, concurrency, and coplanarity) are linear in the parameters to be estimated. If the hypothesis is linear, the problem takes the form of estimating an n-vector $\boldsymbol{u}$ such that

$$(\bar{\boldsymbol{a}}_\alpha^{(k)}, \boldsymbol{u}) = 0, \qquad k = 1, ..., L, \tag{9.1}$$

from n-vector data $\{\boldsymbol{a}_\alpha^{(k)}\}$, $k = 1, ..., L$, $\alpha = 1, ..., N$. We write

$$\boldsymbol{a}_\alpha^{(k)} = \bar{\boldsymbol{a}}_\alpha^{(k)} + \Delta\boldsymbol{a}_\alpha^{(k)}, \tag{9.2}$$

and assume that the noise $\Delta\boldsymbol{a}_\alpha^{(k)}$ is a random variable of mean $\boldsymbol{0}$, independent for each α. Let $V[\boldsymbol{a}_\alpha^{(k)}, \boldsymbol{a}_\alpha^{(l)}]$ $(= E[\Delta\boldsymbol{a}_\alpha^{(k)}\Delta\boldsymbol{a}_\alpha^{(k)\top}])$ be the covariance matrix of $\boldsymbol{a}_\alpha^{(k)}$ and $\boldsymbol{a}_\alpha^{(l)}$. In this chapter, we further assume that each $\Delta\boldsymbol{a}_\alpha^{(k)}$ is $O(\epsilon)$, where ϵ is an appropriately defined constant that measures the average magnitude of the noise. We write

$$V[\boldsymbol{a}_\alpha^{(k)}, \boldsymbol{a}_\alpha^{(l)}] = \epsilon^2 V_0[\boldsymbol{a}_\alpha^{(k)}, \boldsymbol{a}_\alpha^{(l)}], \tag{9.3}$$

and call ϵ and $V_0[\boldsymbol{a}_\alpha^{(k)}, \boldsymbol{a}_\alpha^{(l)}]$ the *noise level* and the *normalized covariance matrix*, respectively (see eqs. (5.37) and (7.42)).

According to the general theory in Section 7.1.5, the optimal estimate of $\boldsymbol{u}$ is obtained by the minimization (7.55). Since multiplication of eq. (7.55) by a positive constant does not affect the solution, we can alternatively minimize

$$J[\boldsymbol{u}] = \frac{1}{N}\sum_{\alpha=1}^{N}\sum_{k,l=1}^{L} W_\alpha^{(kl)}(\boldsymbol{u})(\boldsymbol{a}_\alpha^{(k)}, \boldsymbol{u})(\boldsymbol{a}_\alpha^{(l)}, \boldsymbol{u}). \tag{9.4}$$

In this equation, $W_\alpha^{(kl)}(\boldsymbol{u})$ can be multiplied by an arbitrary positive constant, so we hereafter define the (LL)-matrix $\boldsymbol{W}_\alpha(\boldsymbol{u}) = (W_\alpha^{(kl)}(\boldsymbol{u}))$ by using the normalized covariance matrix $V_0[\boldsymbol{a}_\alpha^{(k)}, \boldsymbol{a}_\alpha^{(l)}]$ instead of $V[\boldsymbol{a}_\alpha^{(k)}, \boldsymbol{a}_\alpha^{(l)}]$ (see eq. (7.55)):

$$\boldsymbol{W}_\alpha(\boldsymbol{u}) = \Big((\boldsymbol{u}, V_0[\boldsymbol{a}_\alpha^{(k)}, \boldsymbol{a}_\alpha^{(l)}]\boldsymbol{u})\Big)^{-}. \tag{9.5}$$

Here, the matrix $V_0[\boldsymbol{a}_\alpha^{(k)}, \boldsymbol{a}_\alpha^{(l)}]$ is evaluated at the true values[1] $\bar{\boldsymbol{a}}_\alpha^{(k)}$ and $\bar{\boldsymbol{a}}_\alpha^{(l)}$. In actual computation, we replace the matrix $V_0[\boldsymbol{a}_\alpha^{(k)}, \boldsymbol{a}_\alpha^{(l)}]$ by the values evaluated at the data values $\boldsymbol{a}_\alpha^{(k)}$ and $\boldsymbol{a}_\alpha^{(l)}$ and approximate the matrix $\boldsymbol{W}_\alpha(\boldsymbol{u})$ by computing the rank-constrained generalized inverse

$$\boldsymbol{W}_\alpha(\boldsymbol{u}) = \Big((\boldsymbol{u}, V_0[\boldsymbol{a}_\alpha^{(k)}, \boldsymbol{a}_\alpha^{(l)}]\boldsymbol{u})\Big)_r^{-}, \tag{9.6}$$

where r is the rank of the hypothesis (9.1) (see eq. (7.56)).

From eqs. (9.4) and (9.5), we immediately see that the scale of $\boldsymbol{u}$ is indeterminate if no constraints are imposed. In order to remove this indeterminacy, we impose normalization $\|\boldsymbol{u}\| = 1$. We further assume that no other constraints exist on $\boldsymbol{u}$. It follows that the parameter space is an $(n-1)$-dimensional unit sphere S^{n-1}.

If the functions $W_\alpha^{(kl)}(\boldsymbol{u})$ are replaced by constants $W_\alpha^{(kl)}$, eq. (9.4) has the form

$$\tilde{J}[\boldsymbol{u}] = \frac{1}{N}\sum_{\alpha=1}^{N}\sum_{k,l=1}^{L} W_\alpha^{(kl)}(\boldsymbol{a}_\alpha^{(k)}, \boldsymbol{u})(\boldsymbol{a}_\alpha^{(l)}, \boldsymbol{u}). \tag{9.7}$$

The constants $W_\alpha^{(kl)}$ are chosen, for example, in the form $W_\alpha^{(kl)} = W_\alpha^{(kl)}(\boldsymbol{u}^*)$, where $\boldsymbol{u}^*$ is an appropriate estimate of $\boldsymbol{u}$. If the hypothesis consists of a single equation $(\bar{\boldsymbol{a}}_\alpha, \boldsymbol{u}) = 0$, minimization of eq. (9.7) is equivalent to

$$\frac{1}{N}\sum_{\alpha=1}^{N} W_\alpha(\boldsymbol{a}_\alpha, \boldsymbol{u})^2 \to \min, \tag{9.8}$$

---

[1]To be consistent with the notation in Chapters 7 and 8, we should write $V_0[\mathbf{a}_\alpha^{(k)}, \mathbf{a}_\alpha^{(l)}]$ and $W_\alpha^{(kl)}(\mathbf{u})$ as $\bar{V}_0[\mathbf{a}_\alpha^{(k)}, \mathbf{a}_\alpha^{(l)}]$ and $\bar{W}_\alpha^{(kl)}(\mathbf{u})$, respectively. In the following, however, we omit the bars to simplify the notation.

which is a special type of least-squares optimization weighted by W_α. Since minimization of eq. (9.7) is a straightforward generalization of eq. (9.8) to multiple hypotheses, we call eq. (9.7) the *least-squares approximation* to eq. (9.4).

The least-squares approximation has a practical significance: *the solution is computed analytically*. Define the *moment matrix*

$$\boldsymbol{M} = \frac{1}{N}\sum_{\alpha=1}^{N}\sum_{k,l=1}^{L} W_\alpha^{(kl)} \boldsymbol{a}_\alpha^{(k)} \boldsymbol{a}_\alpha^{(l)\top}. \tag{9.9}$$

This is a positive semi-definite symmetric matrix. Eq. (9.7) can be rewritten in the form

$$\tilde{J}[\boldsymbol{u}] = (\boldsymbol{u}, \boldsymbol{M}\boldsymbol{u}) \to \min. \tag{9.10}$$

The right-hand side is a quadratic form in $\boldsymbol{u}$. Hence, $\tilde{J}[\boldsymbol{u}]$ is minimized under the constraint $\|\boldsymbol{u}\| = 1$ by the unit eigenvector $\tilde{\boldsymbol{u}}$ of $\boldsymbol{M}$ for the smallest eigenvalue (see eqs. (2.86) and (2.139)); the smallest eigenvalue equals the residual $\tilde{J}[\tilde{\boldsymbol{u}}]$. We call the solution $\tilde{\boldsymbol{u}}$ thus obtained the *eigenvector fit*.

9.1.2 Statistical bias of eigenvector fit

In spite of the computational advantage of the eigenvector fit, it has a drawback. For computing the weights $W_\alpha^{(kl)} = W_\alpha^{(kl)}(\boldsymbol{u}^*)$, we need a good estimate $\boldsymbol{u}^*$. How can we choose it? A naive strategy for this is to substitute an initial guess $\boldsymbol{u}_0$ for $\boldsymbol{u}^*$ and compute the eigenvector fit $\boldsymbol{u}_1$; substituting $\boldsymbol{u}_1$ for $\boldsymbol{u}^*$, we again compute the eigenvector fit $\boldsymbol{u}_2$ and so on. However, such iterations introduce *statistical bias* into the solution. Before going into the details, we must note the following two facts:

1. Statistical bias is a *second order effect*: if the noise is of mean zero and of order $O(\epsilon)$, the bias of the output is of order $O(\epsilon^2)$, since the expectations of noise terms of odd orders vanish.

2. The fact that an estimator X of $\bar{X}$ is unbiased (i.e., $E[X] = \bar{X}$) makes little practical sense unless its variance $V[X]$ is very small. If X is determined from a large number of data and if $\lim_{N\to\infty} E[X] = \bar{X}$ and $\lim_{N\to\infty} V[X] = \boldsymbol{O}$, where N is the number of the data, X is said to be a *consistent estimator* (see Section 3.6.2). A typical example is when X is an average of a large number of independent data of mean zero and of the same order of magnitude (the *law of large numbers*; see eq. (3.9)).

We also need the following preliminary results:

- Consider the identity

$$\boldsymbol{W}_\alpha(\boldsymbol{u})\boldsymbol{W}_\alpha(\boldsymbol{u})^- = \boldsymbol{P}^{\mathcal{L}_\alpha(\boldsymbol{u})}, \tag{9.11}$$

  where $\boldsymbol{P}^{\mathcal{L}_\alpha(\boldsymbol{u})}$ is the L-dimensional projection matrix onto the range $\mathcal{L}_\alpha(\boldsymbol{u})$ of the (LL)-matrix $\boldsymbol{W}_\alpha(\boldsymbol{u})$ (see eqs. (2.80)).

- Taking the trace of eq. (9.11) on both sides, we obtain

$$\mathrm{tr}(\boldsymbol{W}_\alpha(\boldsymbol{u})\boldsymbol{W}_\alpha(\boldsymbol{u})^-) = r, \tag{9.12}$$

where r is the rank of $\boldsymbol{W}_\alpha(\boldsymbol{u})$ (see eqs. (2.51)), which equals the rank of the hypothesis (9.1). If eq. (9.5) is substituted into eq. (9.12), we have

$$\sum_{k,l=1}^{L} W_\alpha^{(kl)}(\boldsymbol{u})(\boldsymbol{u}, V_0[\boldsymbol{a}_\alpha^{(k)}, \boldsymbol{a}_\alpha^{(l)}]\boldsymbol{u}) = r. \tag{9.13}$$

- Differentiating eq. (9.13) with respect to $\boldsymbol{u}$ on both sides, we obtain

$$\sum_{k,l=1}^{L} \nabla_{\boldsymbol{u}} W_\alpha^{(kl)}(\boldsymbol{u})(\boldsymbol{u}, V_0[\boldsymbol{a}_\alpha^{(k)}, \boldsymbol{a}_\alpha^{(l)}]\boldsymbol{u}) + 2\sum_{k,l=1}^{L} W_\alpha^{(kl)}(\boldsymbol{u}) V_0[\boldsymbol{a}_\alpha^{(k)}, \boldsymbol{a}_\alpha^{(l)}]\boldsymbol{u} = \boldsymbol{0}. \tag{9.14}$$

- Substituting eq. (9.5) into the identity $\boldsymbol{W}_\alpha(\boldsymbol{u})\boldsymbol{W}_\alpha(\boldsymbol{u})^-\boldsymbol{W}_\alpha(\boldsymbol{u}) = \boldsymbol{W}_\alpha(\boldsymbol{u})$ (see eqs. (2.81)), we obtain

$$\sum_{l,m=1}^{L} W_\alpha^{(kl)}(\boldsymbol{u}) W_\alpha^{(mn)}(\boldsymbol{u})(\boldsymbol{u}, V_0[\boldsymbol{a}_\alpha^{(l)}, \boldsymbol{a}_\alpha^{(m)}]\boldsymbol{u}) = W_\alpha^{(kn)}. \tag{9.15}$$

- Substituting eq. (9.2) into eq. (9.9), we obtain

$$\boldsymbol{M} = \frac{1}{N}\sum_{\alpha=1}^{N}\sum_{k,l=1}^{L} W_\alpha^{(kl)}(\bar{\boldsymbol{a}}_\alpha^{(k)} + \Delta\boldsymbol{a}_\alpha^{(k)})(\bar{\boldsymbol{a}}_\alpha^{(l)} + \Delta\boldsymbol{a}_\alpha^{(l)})^\top = \bar{\boldsymbol{M}} + \Delta\boldsymbol{M}, \tag{9.16}$$

where $\bar{\boldsymbol{M}}$ is the unperturbed moment matrix defined by

$$\bar{\boldsymbol{M}} = \frac{1}{N}\sum_{\alpha=1}^{N}\sum_{k,l=1}^{L} W_\alpha^{(kl)} \bar{\boldsymbol{a}}_\alpha^{(k)} \bar{\boldsymbol{a}}_\alpha^{(l)\top}. \tag{9.17}$$

The deviation $\Delta\boldsymbol{M}$ is expressed in the following form:

$$\Delta\boldsymbol{M} = \frac{1}{N}\sum_{\alpha=1}^{N}\sum_{k,l=1}^{L} W_\alpha^{(kl)}(\bar{\boldsymbol{a}}_\alpha^{(k)}\Delta\boldsymbol{a}_\alpha^{(l)\top} + \Delta\boldsymbol{a}_\alpha^{(k)}\bar{\boldsymbol{a}}_\alpha^{(l)\top}) + \frac{1}{N}\sum_{\alpha=1}^{N}\sum_{k,l=1}^{L} W_\alpha^{(kl)}\Delta\boldsymbol{a}_\alpha^{(k)}\Delta\boldsymbol{a}_\alpha^{(l)\top}. \tag{9.18}$$

The first term on the right-hand side is $O(\epsilon/\sqrt{N})$ since it is an average of independent random variables of mean zero, while the second term is $O(\epsilon^2)$. Since $E[\Delta\boldsymbol{a}_\alpha^{(k)}] = \boldsymbol{0}$, we see that

$$E[\Delta\boldsymbol{M}] = \frac{\epsilon^2}{N}\sum_{\alpha=1}^{N}\sum_{k,l=1}^{L} W_\alpha^{(kl)} V_0[\boldsymbol{a}_\alpha^{(k)}, \boldsymbol{a}_\alpha^{(l)}]. \tag{9.19}$$

- Let $\bar{\boldsymbol{u}}$ be the true value of $\boldsymbol{u}$ that satisfies the hypothesis $(\bar{\boldsymbol{a}}_\alpha^{(k)}, \boldsymbol{u}) = 0$, $k = 1, ..., L$, $\alpha = 1, ..., N$. From eq. (9.17), it is immediately seen that

$$\bar{\boldsymbol{M}}\bar{\boldsymbol{u}} = \boldsymbol{0}, \tag{9.20}$$

meaning that the true value $\bar{\boldsymbol{u}}$ is the unit eigenvector of the unperturbed moment matrix $\bar{\boldsymbol{M}}$ for eigenvalue 0.

Now, consider the gradient of $\tilde{J}[\boldsymbol{u}]$ at the true value $\bar{\boldsymbol{u}}$. Differentiating eq. (9.10) with respect to $\boldsymbol{u}$, noting $\bar{\boldsymbol{M}}\bar{\boldsymbol{u}} = \boldsymbol{0}$ and $(\bar{\boldsymbol{a}}_\alpha^{(k)}, \bar{\boldsymbol{u}}) = 0$, and using eq. (9.18), we obtain

$$\begin{aligned}\nabla_{\boldsymbol{u}}\tilde{J}[\bar{\boldsymbol{u}}] &= 2\boldsymbol{M}\bar{\boldsymbol{u}} = 2(\bar{\boldsymbol{M}} + \Delta\boldsymbol{M})\bar{\boldsymbol{u}} = 2\Delta\boldsymbol{M}\bar{\boldsymbol{u}} \\ &= \frac{2}{N}\sum_{\alpha=1}^{N}\sum_{k,l=1}^{L} W_\alpha^{(kl)}\left((\Delta\boldsymbol{a}_\alpha^{(l)}, \bar{\boldsymbol{u}})\bar{\boldsymbol{a}}_\alpha^{(k)} + \Delta\boldsymbol{a}_\alpha^{(k)}\Delta\boldsymbol{a}_\alpha^{(l)\top}\bar{\boldsymbol{u}}\right).\end{aligned} \tag{9.21}$$

Let $\boldsymbol{P}_{\bar{\boldsymbol{u}}}$ be the n-dimensional projection matrix along $\bar{\boldsymbol{u}}$. Since $(\bar{\boldsymbol{a}}_\alpha^{(k)}, \bar{\boldsymbol{u}}) = 0$ implies $\boldsymbol{P}_{\bar{\boldsymbol{u}}}\bar{\boldsymbol{a}}_\alpha^{(k)} = \bar{\boldsymbol{a}}_\alpha^{(k)}$, we see from eq. (9.21) that

$$\begin{aligned}\boldsymbol{P}_{\bar{\boldsymbol{u}}}\nabla_{\boldsymbol{u}}\tilde{J}[\bar{\boldsymbol{u}}] = \frac{2}{N}\sum_{\alpha=1}^{N}\sum_{k,l=1}^{L} W_\alpha^{(kl)}\Big((\Delta\boldsymbol{a}_\alpha^{(l)}, \bar{\boldsymbol{u}})\bar{\boldsymbol{a}}_\alpha^{(k)} &+ \Delta\boldsymbol{a}_\alpha^{(k)}\Delta\boldsymbol{a}_\alpha^{(l)\top}\bar{\boldsymbol{u}} \\ &-(\bar{\boldsymbol{u}}, (\Delta\boldsymbol{a}_\alpha^{(k)}\Delta\boldsymbol{a}_\alpha^{(l)\top})\bar{\boldsymbol{u}})\bar{\boldsymbol{u}}\Big).\end{aligned} \tag{9.22}$$

Suppose the estimate $\boldsymbol{u}^*$ for computing the weights $W_\alpha^{(kl)}$ is chosen to be the true value $\bar{\boldsymbol{u}}$. Using eq. (9.13), we observe that

$$\begin{aligned}E[\boldsymbol{P}_{\bar{\boldsymbol{u}}}\nabla_{\boldsymbol{u}}\tilde{J}[\bar{\boldsymbol{u}}]] &= \frac{2\epsilon^2}{N}\sum_{\alpha=1}^{N}\sum_{k,l=1}^{L} W_\alpha^{(kl)}\left(V_0[\boldsymbol{a}_\alpha^{(k)}, \boldsymbol{a}_\alpha^{(l)}]\bar{\boldsymbol{u}} - (\bar{\boldsymbol{u}}, V_0[\boldsymbol{a}_\alpha^{(k)}, \boldsymbol{a}_\alpha^{(l)}])\bar{\boldsymbol{u}})\bar{\boldsymbol{u}}\right) \\ &= 2\epsilon^2\left(\frac{1}{N}\sum_{\alpha=1}^{N}\sum_{k,l=1}^{L} W_\alpha^{(kl)}V_0[\boldsymbol{a}_\alpha^{(k)}, \boldsymbol{a}_\alpha^{(l)}]\bar{\boldsymbol{u}} - r\bar{\boldsymbol{u}}\right).\end{aligned} \tag{9.23}$$

The right-hand side is $O(\epsilon^2)$ but not zero in general even in the limit $N \to \infty$. In other words, $\boldsymbol{P}_{\bar{\boldsymbol{u}}}\nabla_{\boldsymbol{u}}\tilde{J}[\bar{\boldsymbol{u}}]$ is not zero on average however many data we use. This implies that $\tilde{J}[\boldsymbol{u}]$ does not take its minimum at $\bar{\boldsymbol{u}}$ because $\tilde{J}[\boldsymbol{u}]$ can be further reduced without violating the constraint $\|\boldsymbol{u}\| = 1$ if $\boldsymbol{u}$ is infinitesimally incremented in the direction $-\boldsymbol{P}_{\bar{\boldsymbol{u}}}\nabla_{\boldsymbol{u}}\tilde{J}[\bar{\boldsymbol{u}}]$. It follows that if $\tilde{\boldsymbol{u}}$ minimizes $\tilde{J}[\boldsymbol{u}]$, the probability that $(\tilde{\boldsymbol{u}} - \bar{\boldsymbol{u}}, E[\boldsymbol{P}_{\bar{\boldsymbol{u}}}\nabla_{\boldsymbol{u}}\tilde{J}[\bar{\boldsymbol{u}}]]) < 0$ is larger than the probability that $(\tilde{\boldsymbol{u}} - \bar{\boldsymbol{u}}, E[\boldsymbol{P}_{\bar{\boldsymbol{u}}}\nabla_{\boldsymbol{u}}\tilde{J}[\bar{\boldsymbol{u}}]]) > 0$. Hence, the eigenvector fit $\tilde{\boldsymbol{u}}$ is biased (Fig. 9.1).

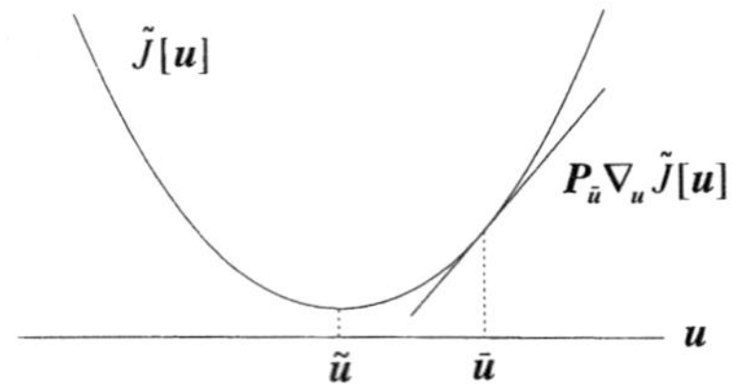

Fig. 9.1. Bias of the value $\tilde{\mathbf{u}}$ that minimizes $\tilde{J}[\mathbf{u}]$.

On the other hand, *no such bias exists if the the original function* $J[\boldsymbol{u}]$ *is minimized*. Differentiating eq. (9.4) with respect to $\boldsymbol{u}$, we obtain

$$\nabla_{\boldsymbol{u}} J[\boldsymbol{u}] = \frac{1}{N}\sum_{\alpha=1}^{N}\sum_{k,l=1}^{L}\Big(\nabla_{\boldsymbol{u}} W_\alpha^{(kl)}(\boldsymbol{u})(\boldsymbol{a}_\alpha^{(k)},\boldsymbol{u})(\boldsymbol{a}_\alpha^{(l)},\boldsymbol{u}) + 2W_\alpha^{(kl)}(\boldsymbol{u})(\boldsymbol{a}_\alpha^{(k)},\boldsymbol{u})\boldsymbol{a}_\alpha^{(l)}\Big). \tag{9.24}$$

Evaluating this at $\bar{\boldsymbol{u}}$ and substituting eq. (9.2) into this, we observe that

$$\begin{aligned}\nabla_{\boldsymbol{u}} J[\bar{\boldsymbol{u}}] &= \frac{1}{N}\sum_{\alpha=1}^{N}\sum_{k,l=1}^{L}\Big(\nabla_{\boldsymbol{u}} W_\alpha^{(kl)}(\bar{\boldsymbol{u}})(\bar{\boldsymbol{a}}_\alpha^{(k)}+\Delta\boldsymbol{a}_\alpha^{(k)},\bar{\boldsymbol{u}})(\bar{\boldsymbol{a}}_\alpha^{(l)}+\Delta\boldsymbol{a}_\alpha^{(l)},\bar{\boldsymbol{u}}) \\ &\qquad + 2W_\alpha^{(kl)}(\bar{\boldsymbol{u}})(\bar{\boldsymbol{a}}_\alpha^{(k)}+\Delta\boldsymbol{a}_\alpha^{(k)},\bar{\boldsymbol{u}})(\bar{\boldsymbol{a}}_\alpha^{(l)}+\Delta\boldsymbol{a}_\alpha^{(l)})\Big) \\ &= \frac{1}{N}\sum_{\alpha=1}^{N}\sum_{k,l=1}^{L}\Big(\nabla_{\boldsymbol{u}} W_\alpha^{(kl)}(\bar{\boldsymbol{u}})(\Delta\boldsymbol{a}_\alpha^{(k)},\bar{\boldsymbol{u}})(\Delta\boldsymbol{a}_\alpha^{(l)},\bar{\boldsymbol{u}}) \\ &\qquad + 2W_\alpha^{(kl)}(\bar{\boldsymbol{u}})((\Delta\boldsymbol{a}_\alpha^{(k)},\bar{\boldsymbol{u}})\bar{\boldsymbol{a}}_\alpha^{(l)} + (\Delta\boldsymbol{a}_\alpha^{(k)},\bar{\boldsymbol{u}})\Delta\boldsymbol{a}_\alpha^{(l)})\Big) \\ &= \frac{1}{N}\sum_{\alpha=1}^{N}\sum_{k,l=1}^{L}\Big(\nabla_{\boldsymbol{u}} W_\alpha^{(kl)}(\bar{\boldsymbol{u}})(\bar{\boldsymbol{u}},(\Delta\boldsymbol{a}_\alpha^{(k)}\Delta\boldsymbol{a}_\alpha^{(l)\top})\bar{\boldsymbol{u}}) \\ &\qquad + 2W_\alpha^{(kl)}(\bar{\boldsymbol{u}})((\bar{\boldsymbol{a}}_\alpha^{(l)}\Delta\boldsymbol{a}_\alpha^{(k)\top})\bar{\boldsymbol{u}} + (\Delta\boldsymbol{a}_\alpha^{(k)}\Delta\boldsymbol{a}_\alpha^{(l)\top})\bar{\boldsymbol{u}})\Big).\end{aligned} \tag{9.25}$$

From eq. (9.14), we see that

$$E[\nabla_{\boldsymbol{u}} J[\bar{\boldsymbol{u}}]] = \frac{\epsilon^2}{N}\sum_{\alpha=1}^{N}\left(\sum_{k,l=1}^{L}\nabla_{\boldsymbol{u}} W_\alpha^{(kl)}(\bar{\boldsymbol{u}})(\bar{\boldsymbol{u}},V_0[\boldsymbol{a}_\alpha^{(k)},\boldsymbol{a}_\alpha^{(l)}]\bar{\boldsymbol{u}}) + 2\sum_{k,l=1}^{L} W_\alpha^{(kl)}(\bar{\boldsymbol{u}})V_0[\boldsymbol{a}_\alpha^{(k)},\boldsymbol{a}_\alpha^{(l)}]\bar{\boldsymbol{u}}\right) = \mathbf{0}. \tag{9.26}$$

Consequently, $E[\boldsymbol{P}_{\bar{\boldsymbol{u}}}\nabla_{\boldsymbol{u}}J[\bar{\boldsymbol{u}}]] = \boldsymbol{0}$. Since $\boldsymbol{P}_{\bar{\boldsymbol{u}}}\nabla_{\boldsymbol{u}}J[\bar{\boldsymbol{u}}]$ is an average of a large number of independent data of mean zero and of order $O(\epsilon)$, we have $\boldsymbol{P}_{\bar{\boldsymbol{u}}}\nabla_{\boldsymbol{u}}J[\bar{\boldsymbol{u}}] = O(\epsilon^2/\sqrt{N})$. This means that the minimum of $J[\boldsymbol{u}]$ occurs at the true value $\bar{\boldsymbol{u}}$ in the limit $N \to \infty$: the optimal solution is a consistent estimator.

From this argument, we also see that the statistical bias of the eigenvector fit is solely due to the least-squares approximation $W_\alpha^{(kl)}(\boldsymbol{u}) \approx W_\alpha^{(kl)}(\boldsymbol{u}^*)$, or equivalently

$$\nabla_{\boldsymbol{u}} W_\alpha^{(kl)} \approx \boldsymbol{0}. \tag{9.27}$$

(See eq. (8.87).) Hence, the bias is unavoidable *however the estimate* $\boldsymbol{u}^*$ *is chosen*—even for $\boldsymbol{u}^* = \bar{\boldsymbol{u}}$.

9.2 Unbiased Eigenvector Fit

9.2.1 Unbiased least-squares approximation

The analysis in the preceding section suggests that an unbiased solution could be obtained if

$$\hat{J}[\boldsymbol{u}] = \frac{1}{N}\sum_{\alpha=1}^{N}\sum_{k,l=1}^{L} W_\alpha^{(kl)}\left((\boldsymbol{a}_\alpha^{(k)}, \boldsymbol{u})(\boldsymbol{a}_\alpha^{(l)}, \boldsymbol{u}) - V[\boldsymbol{a}_\alpha^{(k)}, \boldsymbol{a}_\alpha^{(l)}]\right) \tag{9.28}$$

is minimized instead of eq. (9.7). By the same analysis as shown earlier (see eq. (9.23)), we conclude that $E[\nabla_{\boldsymbol{u}}\hat{J}[\bar{\boldsymbol{u}}]] = \boldsymbol{0}$ and hence $E[\boldsymbol{P}_{\bar{\boldsymbol{u}}}\nabla_{\boldsymbol{u}}\hat{J}[\bar{\boldsymbol{u}}]] = \boldsymbol{0}$. Since $\boldsymbol{P}_{\bar{\boldsymbol{u}}}\nabla_{\boldsymbol{u}}\hat{J}[\bar{\boldsymbol{u}}]$ is an average of a large number of independent data of mean zero and of order $O(\epsilon)$, we have $\boldsymbol{P}_{\bar{\boldsymbol{u}}}\nabla_{\boldsymbol{u}}\hat{J}[\bar{\boldsymbol{u}}] = O(\epsilon^2/\sqrt{N})$. Hence, the solution is a consistent estimator. We call eq. (9.28) the *unbiased least-squares approximation* to eq. (9.4).

The unbiased least-squares approximation is computationally convenient. In fact, if we define the *unbiased moment matrix*

$$\hat{\boldsymbol{M}} = \frac{1}{N}\sum_{\alpha=1}^{N}\sum_{k,l=1}^{L} W_\alpha^{(kl)}(\boldsymbol{a}_\alpha^{(k)}\boldsymbol{a}_\alpha^{(l)\top} - V[\boldsymbol{a}_\alpha^{(k)}, \boldsymbol{a}_\alpha^{(l)}]), \tag{9.29}$$

eq. (9.28) can be rewritten as the quadratic form

$$\hat{J}[\boldsymbol{u}] = (\boldsymbol{u}, \hat{\boldsymbol{M}}\boldsymbol{u}), \tag{9.30}$$

which is minimized under the constraint $\|\boldsymbol{u}\| = 1$ by the unit eigenvector $\hat{\boldsymbol{u}}$ of $\hat{\boldsymbol{M}}$ for the smallest eigenvalue. Let us call the solution $\hat{\boldsymbol{u}}$ thus obtained the *unbiased eigenvector fit*.

Assume that the true value $\bar{\boldsymbol{u}}$, which is the unit eigenvector of $\bar{\boldsymbol{M}}$ for eigenvalue 0 (see eq. (9.20)), is unique, i.e., the eigenvalue 0 is a simple root.

According to the *perturbation theorem*, the perturbation $\Delta\hat{\boldsymbol{u}} = \hat{\boldsymbol{u}} - \bar{\boldsymbol{u}}$ of $\hat{\boldsymbol{u}}$ from $\bar{\boldsymbol{u}}$ is given in the following form (see eq. (2.111)):

$$\Delta\hat{\boldsymbol{u}} = -\bar{\boldsymbol{M}}^{-}\Delta\hat{\boldsymbol{M}}\bar{\boldsymbol{u}} + O(\epsilon^2). \tag{9.31}$$

The perturbation $\Delta\hat{\boldsymbol{M}} = \hat{\boldsymbol{M}} - \bar{\boldsymbol{M}}$ of $\hat{\boldsymbol{M}}$ from $\bar{\boldsymbol{M}}$ has the following form:

$$\begin{aligned}\Delta\hat{\boldsymbol{M}} &= \frac{1}{N}\sum_{\alpha=1}^{N}\sum_{k,l=1}^{L} W_\alpha^{(kl)}(\bar{\boldsymbol{a}}_\alpha^{(k)}\Delta\boldsymbol{a}_\alpha^{(l)\top} + \Delta\boldsymbol{a}_\alpha^{(k)}\bar{\boldsymbol{a}}_\alpha^{(l)\top}) \\ &\quad + \frac{1}{N}\sum_{\alpha=1}^{N}\sum_{k,l=1}^{L} W_\alpha^{(kl)}(\Delta\boldsymbol{a}_\alpha^{(k)}\Delta\boldsymbol{a}_\alpha^{(l)\top} - \epsilon^2 V_0[\boldsymbol{a}_\alpha^{(k)}, \boldsymbol{a}_\alpha^{(l)}]).\end{aligned} \tag{9.32}$$

Suppose the estimate $\boldsymbol{u}^*$ for computing the weights $W_\alpha^{(kl)} = W_\alpha^{(kl)}(\boldsymbol{u}^*)$ is chosen to be the true value $\bar{\boldsymbol{u}}$. Ignoring terms of $O(\epsilon^2)$ and noting that $(\bar{\boldsymbol{a}}_\alpha^{(k)}, \bar{\boldsymbol{u}}) = 0$, we obtain from eq. (9.32)

$$\Delta\hat{\boldsymbol{M}}\bar{\boldsymbol{u}} = \frac{1}{N}\sum_{\alpha=1}^{N}\sum_{k,l=1}^{L} W_\alpha^{(kl)}(\Delta\boldsymbol{a}_\alpha^{(l)}, \bar{\boldsymbol{u}})\bar{\boldsymbol{a}}_\alpha^{(k)}. \tag{9.33}$$

Since $\boldsymbol{a}_\alpha^{(k)}$ is independent for each α, we obtain

$$\begin{aligned}&E[(\Delta\hat{\boldsymbol{M}}\bar{\boldsymbol{u}})(\Delta\hat{\boldsymbol{M}}\bar{\boldsymbol{u}})^\top] \\ &= \frac{1}{N^2}\sum_{\alpha,\beta=1}^{N}\sum_{k,l,m,n=1}^{L} W_\alpha^{(kl)}W_\beta^{(mn)}(\bar{\boldsymbol{u}}, E[\Delta\boldsymbol{a}_\alpha^{(l)}\Delta\boldsymbol{a}_\beta^{(n)\top}]\bar{\boldsymbol{u}})\bar{\boldsymbol{a}}_\alpha^{(k)}\bar{\boldsymbol{a}}_\beta^{(m)} \\ &= \frac{\epsilon^2}{N^2}\sum_{\alpha,\beta=1}^{N}\sum_{k,l,m,n=1}^{L} W_\alpha^{(kl)}W_\beta^{(mn)}(\bar{\boldsymbol{u}}, \delta_{\alpha\beta}V_0[\boldsymbol{a}_\alpha^{(l)}, \boldsymbol{a}_\beta^{(n)}]\bar{\boldsymbol{u}})\bar{\boldsymbol{a}}_\alpha^{(k)}\bar{\boldsymbol{a}}_\beta^{(m)} \\ &= \frac{\epsilon^2}{N^2}\sum_{\alpha=1}^{N}\sum_{k,m=1}^{L}\left(\sum_{l,n=1}^{L} W_\alpha^{(kl)}W_\alpha^{(nm)}(\bar{\boldsymbol{u}}, V_0[\boldsymbol{a}_\alpha^{(l)}, \boldsymbol{a}_\alpha^{(n)}]\bar{\boldsymbol{u}})\right)\bar{\boldsymbol{a}}_\alpha^{(k)}\bar{\boldsymbol{a}}_\alpha^{(m)} \\ &= \frac{\epsilon^2}{N^2}\sum_{\alpha=1}^{N}\sum_{k,m=1}^{L} W_\alpha^{(km)}\bar{\boldsymbol{a}}_\alpha^{(k)}\bar{\boldsymbol{a}}_\alpha^{(m)} = \frac{\epsilon^2}{N}\bar{\boldsymbol{M}},\end{aligned} \tag{9.34}$$

where the identity (9.15) is used.

From eq. (9.31), the covariance matrix $V[\hat{\boldsymbol{u}}] = E[\Delta\hat{\boldsymbol{u}}\Delta\hat{\boldsymbol{u}}^\top]$ of the unbiased eigenvector fit $\hat{\boldsymbol{u}}$ is obtained to a first approximation in the following form:

$$\begin{aligned}V[\hat{\boldsymbol{u}}] &= \bar{\boldsymbol{M}}^{-}E[(\Delta\hat{\boldsymbol{M}}\bar{\boldsymbol{u}})(\Delta\hat{\boldsymbol{M}}\bar{\boldsymbol{u}})^\top]\bar{\boldsymbol{M}}^{-} = \frac{\epsilon^2}{N}\bar{\boldsymbol{M}}^{-}\bar{\boldsymbol{M}}\bar{\boldsymbol{M}}^{-} = \frac{\epsilon^2}{N}\bar{\boldsymbol{M}}^{-} \\ &= \epsilon^2\left(\sum_{\alpha=1}^{N}\sum_{k,l=1}^{L} W_\alpha^{(kl)}\bar{\boldsymbol{a}}_\alpha^{(k)}\bar{\boldsymbol{a}}_\alpha^{(l)\top}\right)^{-}.\end{aligned} \tag{9.35}$$

The rank of this matrix is $n-1$ in general; the null space is $\{\bar{\boldsymbol{u}}\}_L$.

Comparing this with the general result in Section 7.1.5 (see eq. (7.57)), we conclude that the unbiased eigenvector fit has the same covariance matrix as the truly optimal estimate. In other words, *the unbiased least-squares approximation (9.28) does not decrease the accuracy of the solution.* In actual computation, eq. (9.35) is approximated by

$$V[\hat{\boldsymbol{u}}] = \epsilon^2 \left(\sum_{\alpha=1}^{N} \sum_{k,l=1}^{L} W_\alpha^{(kl)}(\hat{\boldsymbol{u}})(\boldsymbol{P}_{\hat{\boldsymbol{u}}}\hat{\boldsymbol{a}}_\alpha^{(k)})(\boldsymbol{P}_{\hat{\boldsymbol{u}}}\hat{\boldsymbol{a}}_\alpha^{(l)})^\top \right)^-, \tag{9.36}$$

where $\boldsymbol{P}_{\hat{\boldsymbol{u}}}$ is the n-dimensional projection matrix[2] along $\hat{\boldsymbol{u}}$. Here, $\hat{\boldsymbol{a}}_\alpha^{(k)}$ is the optimally corrected value of $\boldsymbol{a}_\alpha^{(k)}$ (see Section 5.1.2), but to a first approximation it can be approximated by the data value $\boldsymbol{a}_\alpha^{(k)}$ itself.

9.2.2 Analysis of residual

As discussed in Section 7.1.4, the residual of optimization plays an important role in *testing of hypotheses*. As in Section 7.1.4, we consider both the residual $\hat{J}[\bar{\boldsymbol{u}}]$ for the true value $\bar{\boldsymbol{u}}$ and the residual $\hat{J}[\hat{\boldsymbol{u}}]$ for the estimate $\hat{\boldsymbol{u}}$.

A. Residual for the true value

Consider the residual $\hat{J}[\bar{\boldsymbol{u}}]$ for the true value $\bar{\boldsymbol{u}}$. Since $(\bar{\boldsymbol{u}}, \bar{\boldsymbol{M}}\bar{\boldsymbol{u}}) = 0$, we see that

$$\begin{aligned}\hat{J}[\bar{\boldsymbol{u}}] &= (\bar{\boldsymbol{u}}, \hat{\boldsymbol{M}}\bar{\boldsymbol{u}}) = (\bar{\boldsymbol{u}}, (\bar{\boldsymbol{M}} + \Delta\hat{\boldsymbol{M}})\bar{\boldsymbol{u}}) = (\bar{\boldsymbol{u}}, \Delta\hat{\boldsymbol{M}}\bar{\boldsymbol{u}}) \\ &= \frac{1}{N}\sum_{\alpha=1}^{N}\sum_{k,l=1}^{L} W_\alpha^{(kl)}(\Delta\boldsymbol{a}_\alpha^{(k)}, \bar{\boldsymbol{u}})(\Delta\boldsymbol{a}_\alpha^{(l)}, \bar{\boldsymbol{u}}) \\ &\quad -\frac{\epsilon^2}{N}\sum_{\alpha=1}^{N}\sum_{k,l=1}^{L} W_\alpha^{(kl)}(\bar{\boldsymbol{u}}, V_0[\boldsymbol{a}_\alpha^{(k)}, \boldsymbol{a}_\alpha^{(l)}]\bar{\boldsymbol{u}}).\end{aligned} \tag{9.37}$$

We assume that the weights $W_\alpha^{(kl)}$ are so chosen that $W_\alpha^{(kl)} = W_\alpha^{(kl)}(\bar{\boldsymbol{u}})$. Then, the second term on the right-hand side of the above equation is $-\epsilon^2 r$ (see the identity (9.13)). If we put

$$e_\alpha^{(k)} = (\Delta\boldsymbol{a}_\alpha^{(k)}, \bar{\boldsymbol{u}}), \tag{9.38}$$

the L-vector $\boldsymbol{e}_\alpha = (e_\alpha^{(1)}, ..., e_\alpha^{(L)})^\top$ is a Gaussian random variable of mean $\boldsymbol{0}$. Its covariance matrix is

$$V[\boldsymbol{e}_\alpha] = \left(E[e_\alpha^{(k)}e_\alpha^{(l)}]\right) = \left((\bar{\boldsymbol{u}}, E[\Delta\boldsymbol{a}_\alpha^{(k)}\Delta\boldsymbol{a}_\alpha^{(l)}]\bar{\boldsymbol{u}})\right)$$

[2] The projection matrix $\mathbf{P}_{\hat{\mathbf{u}}}$ is inserted to ensure that the computed covariance matrix $V[\hat{\mathbf{u}}]$ has null space $\{\hat{\mathbf{u}}\}_L$, which is required by the normalization constraint $\|\hat{\mathbf{u}}\| = 1$.

$$= \epsilon^2 \left((\bar{\boldsymbol{u}}, V_0[\boldsymbol{a}_\alpha^{(k)}, \boldsymbol{a}_\alpha^{(l)}]\bar{\boldsymbol{u}}) \right) = \epsilon^2 \boldsymbol{W}_\alpha^-. \tag{9.39}$$

Hence, $V[\boldsymbol{e}_\alpha]^- = \boldsymbol{W}_\alpha/\epsilon^2$, so the first term on the right-hand side of eq. (9.37) is

$$\frac{1}{N}\sum_{\alpha=1}^{N}\sum_{k,l=1}^{L} W_\alpha^{(kl)} e_\alpha^{(k)} e_\alpha^{(l)} = \frac{\epsilon^2}{N}\sum_{\alpha=1}^{N}(\boldsymbol{e}_\alpha, V[\boldsymbol{e}_\alpha]^-\boldsymbol{e}_\alpha). \tag{9.40}$$

Since $V[\boldsymbol{e}_\alpha]$ $(= \epsilon^2 \boldsymbol{W}_\alpha^-)$ has rank r and the vector $\boldsymbol{e}_\alpha$ is independent for each α, the sum $\sum_{\alpha=1}^{N}(\boldsymbol{e}_\alpha, V[\boldsymbol{e}_\alpha]^-\boldsymbol{e}_\alpha)$ is a χ^2 variable with rN degrees of freedom (see eq. (3.63)). Consequently, eq. (9.37) can be written in the form

$$\hat{J}[\bar{\boldsymbol{u}}] = \frac{\epsilon^2}{N}(\chi^2_{rN} - rN), \tag{9.41}$$

where χ^2_{rN} stands for a χ^2 variable with rN degrees of freedom. The expectation and variance of the residual $\hat{J}[\bar{\boldsymbol{u}}]$ are given as follows (see eqs. (3.59)):

$$E[\hat{J}[\bar{\boldsymbol{u}}]] = 0, \qquad V[\hat{J}[\bar{\boldsymbol{u}}]] = \frac{2\epsilon^4 r}{N}. \tag{9.42}$$

It follows that if a *particular* value $\bar{\boldsymbol{u}}$ is given *independently of the data* $\{\boldsymbol{a}_\alpha^{(k)}\}$, we can construct the χ^2 test for the strong hypothesis that *the true value is* $\bar{\boldsymbol{u}}$: the strong hypothesis is rejected with significance level a% if

$$\hat{J}[\bar{\boldsymbol{u}}] > \frac{\epsilon^2}{N}(\chi^2_{rN,a} - rN). \tag{9.43}$$

B. Residual for the estimate

Consider the residual $\hat{J}[\hat{\boldsymbol{u}}]$ for the unbiased eigenvector fit $\hat{\boldsymbol{u}}$. We see that

$$\begin{aligned}\hat{J}[\hat{\boldsymbol{u}}] &= (\hat{\boldsymbol{u}}, \hat{\boldsymbol{M}}\hat{\boldsymbol{u}}) = (\bar{\boldsymbol{u}} + \Delta\hat{\boldsymbol{u}}, \hat{\boldsymbol{M}}(\bar{\boldsymbol{u}} + \Delta\hat{\boldsymbol{u}}))\\ &= (\bar{\boldsymbol{u}}, \hat{\boldsymbol{M}}\bar{\boldsymbol{u}}) + 2(\Delta\hat{\boldsymbol{u}}, \hat{\boldsymbol{M}}\bar{\boldsymbol{u}}) + (\Delta\hat{\boldsymbol{u}}, \hat{\boldsymbol{M}}\Delta\hat{\boldsymbol{u}})\\ &= \hat{J}[\bar{\boldsymbol{u}}] + 2(\Delta\hat{\boldsymbol{u}}, (\bar{\boldsymbol{M}} + \Delta\hat{\boldsymbol{M}})\bar{\boldsymbol{u}}) + (\Delta\hat{\boldsymbol{u}}, \hat{\boldsymbol{M}}\Delta\hat{\boldsymbol{u}})\\ &= \hat{J}[\bar{\boldsymbol{u}}] + 2(\Delta\hat{\boldsymbol{u}}, \Delta\hat{\boldsymbol{M}}\bar{\boldsymbol{u}}) + (\Delta\hat{\boldsymbol{u}}, \hat{\boldsymbol{M}}\Delta\hat{\boldsymbol{u}}).\end{aligned} \tag{9.44}$$

Since $\Delta\hat{\boldsymbol{u}} = O(\Delta\hat{\boldsymbol{M}}) = O(\epsilon/\sqrt{N})$ (see eqs. (9.31) and (9.32)), the last term on the right-hand side is

$$(\Delta\hat{\boldsymbol{u}}, \hat{\boldsymbol{M}}\Delta\hat{\boldsymbol{u}}) = (\Delta\hat{\boldsymbol{u}}, (\bar{\boldsymbol{M}} + \Delta\hat{\boldsymbol{M}})\Delta\hat{\boldsymbol{u}}) = (\Delta\hat{\boldsymbol{u}}, \bar{\boldsymbol{M}}\Delta\hat{\boldsymbol{u}}) + O(\frac{\epsilon^3}{N\sqrt{N}}). \tag{9.45}$$

Consider the second term on the right-hand side of eq. (9.44). According to the perturbation theorem, the smallest eigenvalue $\hat{\lambda}$ $(= \hat{J}[\hat{\boldsymbol{u}}])$ of $\hat{\boldsymbol{M}}$ has the following form (see eq. (2.108)):

$$\hat{\lambda} = (\hat{\boldsymbol{u}}, \Delta\hat{\boldsymbol{M}}\hat{\boldsymbol{u}}) + O(\Delta\hat{\boldsymbol{M}})^2 = O(\frac{\epsilon}{\sqrt{N}}). \tag{9.46}$$

Since $\hat{\boldsymbol{u}}$ is a unit vector, we have

$$(\Delta\hat{\boldsymbol{u}}, \hat{\boldsymbol{u}}) = O(\Delta\hat{\boldsymbol{u}})^2 = O(\frac{\epsilon^2}{N}). \tag{9.47}$$

Hence,

$$(\Delta\hat{\boldsymbol{u}}, \hat{\boldsymbol{M}}\hat{\boldsymbol{u}}) = \hat{\lambda}(\Delta\hat{\boldsymbol{u}}, \hat{\boldsymbol{u}}) = O(\frac{\epsilon^3}{N\sqrt{N}}). \tag{9.48}$$

On the other hand,

$$\begin{aligned}(\Delta\hat{\boldsymbol{u}}, \hat{\boldsymbol{M}}\hat{\boldsymbol{u}}) &= (\Delta\hat{\boldsymbol{u}}, (\bar{\boldsymbol{M}} + \Delta\hat{\boldsymbol{M}})(\bar{\boldsymbol{u}} + \Delta\hat{\boldsymbol{u}})) \\ &= (\Delta\hat{\boldsymbol{u}}, \Delta\hat{\boldsymbol{M}}\bar{\boldsymbol{u}}) + (\Delta\hat{\boldsymbol{u}}, \bar{\boldsymbol{M}}\Delta\hat{\boldsymbol{u}}) + O(\frac{\epsilon^3}{N\sqrt{N}}).\end{aligned} \tag{9.49}$$

Comparing eqs. (9.48) and (9.49), we conclude that

$$(\Delta\hat{\boldsymbol{u}}, \Delta\hat{\boldsymbol{M}}\bar{\boldsymbol{u}}) = -(\Delta\hat{\boldsymbol{u}}, \bar{\boldsymbol{M}}\Delta\hat{\boldsymbol{u}}) + O(\frac{\epsilon^3}{N\sqrt{N}}). \tag{9.50}$$

Substituting eqs. (9.45) and (9.50) into eq. (9.44) and using eqs. (9.37) and (9.40), we obtain

$$\begin{aligned}\hat{J}[\hat{\boldsymbol{u}}] &= \hat{J}[\bar{\boldsymbol{u}}] - (\Delta\hat{\boldsymbol{u}}, \bar{\boldsymbol{M}}\Delta\hat{\boldsymbol{u}}) + O(\frac{\epsilon^3}{N\sqrt{N}}) \\ &= \frac{\epsilon^2}{N}\left(\sum_{\alpha=1}^{N}(\boldsymbol{e}_\alpha, V[\boldsymbol{e}_\alpha]^-\boldsymbol{e}_\alpha) - (\Delta\hat{\boldsymbol{u}}, V[\hat{\boldsymbol{u}}]^-\Delta\hat{\boldsymbol{u}}) - rN\right) + O(\frac{\epsilon^3}{N\sqrt{N}}),\end{aligned} \tag{9.51}$$

where we have used the fact that $V[\hat{\boldsymbol{u}}] = \epsilon^2\bar{\boldsymbol{M}}^-/N$ (see eq. (9.35)). Since we are assuming that there exists no constraint other than $\|\hat{\boldsymbol{u}}\| = 1$, the rank of $V[\hat{\boldsymbol{u}}]$ is $n-1$ in general. Hence, the quadratic form $(\Delta\hat{\boldsymbol{u}}, V[\hat{\boldsymbol{u}}]^-\Delta\hat{\boldsymbol{u}})$ is a χ^2 variable with $n-1$ degrees of freedom (see eq. (3.61)).

From eqs. (9.31), (9.33), and (9.38), we have

$$\Delta\hat{\boldsymbol{u}} = -\frac{1}{N}\sum_{\alpha=1}^{N}\sum_{k,l=1}^{L} W_\alpha^{(kl)} e_\alpha^{(l)} \bar{\boldsymbol{M}}^- \bar{\boldsymbol{a}}_\alpha^{(k)} + O(\epsilon^2), \tag{9.52}$$

meaning that $\Delta\hat{\boldsymbol{u}}$ is obtained, to a first approximation, by a linear mapping from $\{\boldsymbol{e}_\alpha\}$. Consequently, $\hat{J}[\hat{\boldsymbol{u}}]$ can be written to a first approximation as follows (see eq. (3.64)):

$$\hat{J}[\hat{\boldsymbol{u}}] = \frac{\epsilon^2}{N}(\chi^2_{rN-n+1} - rN). \tag{9.53}$$

Its expectation and variance are

$$E[\hat{J}[\hat{\boldsymbol{u}}]] = -\frac{\epsilon^2}{N}(n-1), \qquad V[\hat{J}[\hat{\boldsymbol{u}}]] = \frac{2\epsilon^4 r}{N}\left(1 - \frac{n-1}{rN}\right). \tag{9.54}$$

From eq. (9.53), we can construct the χ^2 test for the weak hypothesis that eq. (9.1) holds for *some* value $\boldsymbol{u}$: the weak hypothesis is rejected with significance level $a\%$ if

$$\hat{J}[\hat{\boldsymbol{u}}] > \frac{\epsilon^2}{N}(\chi^2_{rN-n+1,a} - rN + n - 1). \tag{9.55}$$

Comparing eqs. (9.42) and (9.54), we observe that the expectation and variance of $\hat{J}[\hat{\boldsymbol{u}}]$ are both smaller than their respective values for $\hat{J}[\bar{\boldsymbol{u}}]$ because of the correlation etween $\hat{\boldsymbol{u}}$ and the data $\{\boldsymbol{a}_\alpha^{(k)}\}$.

9.3 Generalized Eigenvalue Fit

9.3.1 Noise level estimation

Although the unbiased eigenvalue fit seems very desirable for its unbiasedness and computational convenience, a difficulty arises if we want to compute it in a real situation: *the noise level ϵ must be estimated precisely.* If the noise is underestimated, statistical bias still remains, while if it is overestimated, statistical bias arises in the opposite direction. However, accurately predicting the noise level ϵ is very difficult in practice.

This problem does not occur for the eigenvector fit $\tilde{\boldsymbol{u}}$, since the least-squares approximation does not involve the noise level ϵ: the covariance matrices $V[\boldsymbol{a}_\alpha^{(k)}, \boldsymbol{a}_\alpha^{(k)}]$ need to be estimated only up to scale. Thus, the eigenvalue fit $\tilde{\boldsymbol{u}}$ is insensitive to the noise level ϵ, while the unbiased eigenvalue fit $\hat{\boldsymbol{u}}$ is very sensitive to it.

As pointed out in Sections 5.1.5 and 7.1.4, it is often easy to predict the form of the covariance matrix from geometric considerations, but its absolute magnitude is very difficult to predict a priori. Hence, it is desirable to solve the problem without knowing the noise level ϵ. In the following, we present such a scheme; the noise level ϵ is estimated *a posteriori*.

Note that eq. (9.29) can be written in the following form:

$$\hat{\boldsymbol{M}} = \boldsymbol{M} - \epsilon^2 \boldsymbol{N}, \tag{9.56}$$

$$\boldsymbol{N} = \frac{1}{N}\sum_{\alpha=1}^{N}\sum_{k,l=1}^{L} W_\alpha^{(kl)} V_0[\boldsymbol{a}_\alpha^{(k)}, \boldsymbol{a}_\alpha^{(l)}]. \tag{9.57}$$

Let $\boldsymbol{u}^*$ be the estimate for computing the weights: $W_\alpha^{(kl)} = W_\alpha^{(kl)}(\boldsymbol{u}^*)$. We see from the identity (9.13) that

$$(\boldsymbol{u}^*, \boldsymbol{N}\boldsymbol{u}^*) = \frac{1}{N}\sum_{\alpha=1}^{N}\sum_{k,l=1}^{L} W_\alpha^{(kl)}(\boldsymbol{u}^*, V_0[\boldsymbol{a}_\alpha^{(k)}, \boldsymbol{a}_\alpha^{(l)}]\boldsymbol{u}^*) = r. \tag{9.58}$$

Recall that we have defined the unbiased moment matrix $\hat{\boldsymbol{M}} = \boldsymbol{M} - \epsilon^2\boldsymbol{N}$ with the expectation that $\hat{\boldsymbol{M}}$ should be a better estimate of $\bar{\boldsymbol{M}}$ than $\boldsymbol{M}$. This suggests that the noise level ϵ should be estimated so that

$$\bar{\boldsymbol{M}} \approx \boldsymbol{M} - \epsilon^2\boldsymbol{N}. \tag{9.59}$$

However, exact equality may not hold. So, we introduce the following compromise. Since the true value $\bar{\boldsymbol{u}}$ satisfies $(\bar{\boldsymbol{u}}, \bar{\boldsymbol{M}}\bar{\boldsymbol{u}}) = 0$, we seek the value ϵ that satisfies

$$(\boldsymbol{u}, \hat{\boldsymbol{M}}\boldsymbol{u}) = (\boldsymbol{u}, \boldsymbol{M}\boldsymbol{u}) - \epsilon^2(\boldsymbol{u}, \boldsymbol{N}\boldsymbol{u}) = 0 \tag{9.60}$$

for some $\boldsymbol{u}$. There may exist multiple pairs $\{\boldsymbol{u}, \epsilon\}$ that satisfy eq. (9.60). From among them, we choose the one for which $\epsilon^2 = (\boldsymbol{u}, \boldsymbol{M}\boldsymbol{u})/(\boldsymbol{u}, \boldsymbol{N}\boldsymbol{u})$ is the smallest. Eq. (9.58) implies that if $\boldsymbol{u} \approx \boldsymbol{u}^*$ then $(\boldsymbol{u}, \boldsymbol{N}\boldsymbol{u}) \neq 0$. Hence, the vector $\boldsymbol{u}$ is determined by minimizing

$$I[\boldsymbol{u}] = \frac{(\boldsymbol{u}, \boldsymbol{M}\boldsymbol{u})}{(\boldsymbol{u}, \boldsymbol{N}\boldsymbol{u})}. \tag{9.61}$$

In other words, we minimize the *generalized Rayleigh quotient* (see eq. (2.103)). Since $\boldsymbol{M}$ is generally positive definite[3] in the presence of noise, eq. (9.61) is minimized under the constraint $\|\boldsymbol{u}\| = 1$ by the unit generalized eigenvector $\hat{\boldsymbol{u}}$ of the generalized eigenvalue problem

$$\boldsymbol{M}\boldsymbol{u} = c\boldsymbol{N}\boldsymbol{u} \tag{9.62}$$

for the smallest generalized eigenvalue (see eq. (2.104)). At the same time, the smallest generalized eigenvalue c gives an estimate of ϵ^2. We call the resulting solution $\hat{\boldsymbol{u}}$ the *generalized eigenvector fit*.

9.3.2 Accuracy of generalized eigenvector fit

A. Unbiasedness of generalized eigenvector fit

The bias analysis in Section 9.1.2 can be applied to $I[\boldsymbol{u}]$ by assuming that the estimate $\boldsymbol{u}^*$ for computing the weights $W_\alpha^{(kl)}$ is chosen to be the true value $\bar{\boldsymbol{u}}$. Differentiating eq. (9.61) with respect to $\boldsymbol{u}$, we obtain

$$\nabla_{\boldsymbol{u}} I[\boldsymbol{u}] = \frac{2\boldsymbol{M}\boldsymbol{u}}{(\boldsymbol{u}, \boldsymbol{N}\boldsymbol{u})} - \frac{2(\boldsymbol{u}, \boldsymbol{M}\boldsymbol{u})\boldsymbol{N}\boldsymbol{u}}{(\boldsymbol{u}, \boldsymbol{N}\boldsymbol{u})^2}. \tag{9.63}$$

[3]The moment matrix $\mathbf{M}$ is positive semi-definite by definition; it is singular if and only if noise does not exist. It follows that if the smallest eigenvalue of $\mathbf{M}$ happens to be zero, the corresponding unit eigenvector is the true value $\bar{\mathbf{u}}$, so we need not consider this case.

Since $(\bar{\boldsymbol{u}}, \boldsymbol{N}\bar{\boldsymbol{u}}) = r$ (see eq. (9.58)), we see that

$$\nabla_{\boldsymbol{u}} I[\bar{\boldsymbol{u}}] = \frac{2(\bar{\boldsymbol{M}} + \Delta\boldsymbol{M})\bar{\boldsymbol{u}}}{(\bar{\boldsymbol{u}}, \boldsymbol{N}\bar{\boldsymbol{u}})} - \frac{2(\bar{\boldsymbol{u}}, (\bar{\boldsymbol{M}} + \Delta\boldsymbol{M})\bar{\boldsymbol{u}})\boldsymbol{N}\bar{\boldsymbol{u}}}{(\bar{\boldsymbol{u}}, \boldsymbol{N}\bar{\boldsymbol{u}})^2}$$
$$= \frac{2\Delta\boldsymbol{M}\bar{\boldsymbol{u}}}{r} - \frac{2(\bar{\boldsymbol{u}}, \Delta\boldsymbol{M}\bar{\boldsymbol{u}})\boldsymbol{N}\bar{\boldsymbol{u}}}{r^2}. \tag{9.64}$$

Eqs. (9.19) and (9.57) imply $E[\Delta\boldsymbol{M}] = \epsilon^2\boldsymbol{N}$, so

$$E[\nabla_{\boldsymbol{u}} I[\bar{\boldsymbol{u}}]] = \frac{2\epsilon^2\boldsymbol{N}\bar{\boldsymbol{u}}}{r} - \frac{2\epsilon^2(\bar{\boldsymbol{u}}, \boldsymbol{N}\bar{\boldsymbol{u}})\boldsymbol{N}\bar{\boldsymbol{u}}}{r^2} = \boldsymbol{0}. \tag{9.65}$$

Consequently, $E[\boldsymbol{P}_{\bar{\boldsymbol{u}}}\nabla_{\boldsymbol{u}} I[\bar{\boldsymbol{u}}]] = \boldsymbol{0}$. Hence, the generalized eigenvector fit is a consistent estimator by the same argument as in the case of the unbiased eigenvector fit.

B. Covariance matrix of generalized eigenvector fit

The covariance matrix $V[\hat{\boldsymbol{u}}]$ of the generalized eigenvector fit $\hat{\boldsymbol{u}}$ is evaluated as follows. If $I[\boldsymbol{u}]$ takes its minimum at $\hat{\boldsymbol{u}}$, we have $\boldsymbol{P}_{\hat{\boldsymbol{u}}}\nabla_{\boldsymbol{u}} I[\hat{\boldsymbol{u}}] = \boldsymbol{0}$. Eq. (9.63) implies $(\boldsymbol{u}, \nabla_{\boldsymbol{u}} I[\boldsymbol{u}]) = 0$. Consequently, $\boldsymbol{P}_{\hat{\boldsymbol{u}}}\nabla_{\boldsymbol{u}} I[\hat{\boldsymbol{u}}] = \nabla_{\boldsymbol{u}} I[\hat{\boldsymbol{u}}]$, and the generalized eigenvector fit $\hat{\boldsymbol{u}}$ satisfies

$$\nabla_{\boldsymbol{u}} I[\hat{\boldsymbol{u}}] = \boldsymbol{0}. \tag{9.66}$$

Using eq. (9.63), we can rewrite this equation as

$$(\hat{\boldsymbol{u}}, \boldsymbol{N}\hat{\boldsymbol{u}})\boldsymbol{M}\hat{\boldsymbol{u}} = (\hat{\boldsymbol{u}}, \boldsymbol{M}\hat{\boldsymbol{u}})\boldsymbol{N}\hat{\boldsymbol{u}}. \tag{9.67}$$

Substituting $\hat{\boldsymbol{u}} = \bar{\boldsymbol{u}} + \Delta\hat{\boldsymbol{u}}$ and $\boldsymbol{M} = \bar{\boldsymbol{M}} + \Delta\boldsymbol{M}$ into this, we obtain

$$(\bar{\boldsymbol{u}}, \boldsymbol{N}\bar{\boldsymbol{u}})(\bar{\boldsymbol{M}}\Delta\hat{\boldsymbol{u}} + \Delta\boldsymbol{M}\bar{\boldsymbol{u}}) = (\bar{\boldsymbol{u}}, \Delta\boldsymbol{M}\bar{\boldsymbol{u}})\boldsymbol{N}\bar{\boldsymbol{u}} + O(\epsilon^2). \tag{9.68}$$

Since $(\bar{\boldsymbol{u}}, \boldsymbol{N}\bar{\boldsymbol{u}}) = r$, we have

$$\bar{\boldsymbol{M}}\Delta\hat{\boldsymbol{u}} = -\left(\Delta\boldsymbol{M} - \frac{(\bar{\boldsymbol{u}}, \Delta\boldsymbol{M}\bar{\boldsymbol{u}})\boldsymbol{N}}{r}\right)\bar{\boldsymbol{u}} + O(\epsilon^2). \tag{9.69}$$

Hence,

$$\Delta\hat{\boldsymbol{u}} = -\bar{\boldsymbol{M}}^{-}\Delta\hat{\boldsymbol{M}}'\bar{\boldsymbol{u}} + O(\epsilon^2), \tag{9.70}$$

where we have defined

$$\Delta\hat{\boldsymbol{M}}' = \Delta\boldsymbol{M} - \frac{(\bar{\boldsymbol{u}}, \Delta\boldsymbol{M}\bar{\boldsymbol{u}})\boldsymbol{N}}{r}. \tag{9.71}$$

Let $\Delta\hat{\boldsymbol{M}} = \hat{\boldsymbol{M}} - \bar{\boldsymbol{M}}$. Eq. (9.56) implies that

$$\Delta\hat{\boldsymbol{M}} = \boldsymbol{M} - \epsilon^2\boldsymbol{N} - \bar{\boldsymbol{M}} = \Delta\boldsymbol{M} - \epsilon^2\boldsymbol{N}. \tag{9.72}$$

Hence, eq. (9.71) can be written as

$$\Delta\hat{\boldsymbol{M}}' = \Delta\hat{\boldsymbol{M}} - \left(\frac{(\bar{\boldsymbol{u}}, \Delta\boldsymbol{M}\bar{\boldsymbol{u}})\boldsymbol{N}}{r} - \epsilon^2\right)\boldsymbol{N} = \Delta\hat{\boldsymbol{M}} - \frac{\hat{J}[\bar{\boldsymbol{u}}]}{r}, \tag{9.73}$$

where we have used the relation

$$\hat{J}[\bar{\boldsymbol{u}}] = (\bar{\boldsymbol{u}}, \Delta\hat{\boldsymbol{M}}\bar{\boldsymbol{u}}) = (\bar{\boldsymbol{u}}, (\Delta\boldsymbol{M} - \epsilon^2\boldsymbol{N})\bar{\boldsymbol{u}}) = (\bar{\boldsymbol{u}}, \Delta\boldsymbol{M}\bar{\boldsymbol{u}}) - \epsilon^2 r. \tag{9.74}$$

Since eqs. (9.42) imply $\hat{J}[\bar{\boldsymbol{u}}] \sim \epsilon^2\sqrt{2r/N}$, eq. (9.73) is written as

$$\Delta\hat{\boldsymbol{M}}' = \Delta\hat{\boldsymbol{M}} + O(\frac{\epsilon^2}{\sqrt{N}}). \tag{9.75}$$

This means that the statistical behavior of $\Delta\hat{\boldsymbol{M}}'$ is the same as that of $\Delta\hat{\boldsymbol{M}}$ if terms of $O(\epsilon^2/\sqrt{N})$ are ignored. Hence, eq. (9.34) holds for $\Delta\hat{\boldsymbol{M}}'$ as well to a first approximation, and the covariance matrix of the generalized eigenvector fit $\hat{\boldsymbol{u}}$ is also given by

$$V[\hat{\boldsymbol{u}}] = \frac{\epsilon^2}{N}\bar{\boldsymbol{M}}^- = \epsilon^2\left(\sum_{\alpha=1}^{N}\sum_{k,l=1}^{L} W_\alpha^{(kl)}\bar{\boldsymbol{a}}_\alpha^{(k)}\bar{\boldsymbol{a}}_\alpha^{(l)\top}\right)^-, \tag{9.76}$$

which can be approximated by eq. (9.36) in actual computation.

9.3.3 *Analysis of residual*

A. Residual for the true value

Consider the residual $I[\bar{\boldsymbol{u}}]$ for the true value $\bar{\boldsymbol{u}}$. Here, too, we assume that the estimate $\boldsymbol{u}^*$ for computing the weights $W_\alpha^{(kl)}$ is chosen to be the true value $\bar{\boldsymbol{u}}$. Then,

$$I[\bar{\boldsymbol{u}}] = \frac{(\bar{\boldsymbol{u}}, \boldsymbol{M}\bar{\boldsymbol{u}})}{(\bar{\boldsymbol{u}}, \boldsymbol{N}\bar{\boldsymbol{u}})} = \frac{(\bar{\boldsymbol{u}}, \hat{\boldsymbol{M}}\bar{\boldsymbol{u}}) + \epsilon^2(\bar{\boldsymbol{u}}, \boldsymbol{N}\bar{\boldsymbol{u}})}{r} = \frac{(\bar{\boldsymbol{u}}, \hat{\boldsymbol{M}}\bar{\boldsymbol{u}}) + \epsilon^2 r}{r} = \frac{\hat{J}[\bar{\boldsymbol{u}}]}{r} + \epsilon^2. \tag{9.77}$$

From eq. (9.41), we obtain

$$I[\bar{\boldsymbol{u}}] = \frac{\epsilon^2}{rN}\chi^2_{rN}. \tag{9.78}$$

Hence, the expectation and the variance of $I[\bar{\boldsymbol{u}}]$ are

$$E[I[\bar{\boldsymbol{u}}]] = \epsilon^2, \qquad V[I[\bar{\boldsymbol{u}}]] = \frac{2\epsilon^4}{rN}. \tag{9.79}$$

If the noise level ϵ is given a priori, we can construct the χ^2 test for testing the strong hypothesis: if a particular value $\bar{\boldsymbol{u}}$ is given independently of the data $\{\boldsymbol{a}_\alpha^{(k)}\}$, the strong hypothesis that *the true value is* $\bar{\boldsymbol{u}}$ is rejected with significance level $a\%$ if

$$I[\bar{\boldsymbol{u}}] > \frac{\epsilon^2}{rN}\chi^2_{rN,a}. \tag{9.80}$$

B. Residual for the estimate

Consider the residual $I[\hat{\boldsymbol{u}}]$ for the generalized eigenvector fit $\hat{\boldsymbol{u}}$. Substituting $\hat{\boldsymbol{u}} = \bar{\boldsymbol{u}} + \Delta\boldsymbol{u} = \bar{\boldsymbol{u}} + O(\epsilon)$ and noting that $(\bar{\boldsymbol{u}}, \boldsymbol{N}\bar{\boldsymbol{u}}) = r$, we obtain

$$\begin{aligned} I[\hat{\boldsymbol{u}}] &= \frac{(\hat{\boldsymbol{u}}, \boldsymbol{M}\hat{\boldsymbol{u}})}{(\hat{\boldsymbol{u}}, \boldsymbol{N}\hat{\boldsymbol{u}})} = \frac{(\hat{\boldsymbol{u}}, \hat{\boldsymbol{M}}\hat{\boldsymbol{u}}) + \epsilon^2(\hat{\boldsymbol{u}}, \boldsymbol{N}\hat{\boldsymbol{u}})}{(\bar{\boldsymbol{u}}, \boldsymbol{N}\bar{\boldsymbol{u}}) + O(\epsilon)} \\ &= \frac{(\hat{\boldsymbol{u}}, \hat{\boldsymbol{M}}\hat{\boldsymbol{u}}) + \epsilon^2((\bar{\boldsymbol{u}}, \boldsymbol{N}\bar{\boldsymbol{u}}) + O(\epsilon))}{r + O(\epsilon)} = \frac{(\hat{\boldsymbol{u}}, \hat{\boldsymbol{M}}\hat{\boldsymbol{u}})}{r} + \epsilon^2 + O(\epsilon^3). \end{aligned} \tag{9.81}$$

Eqs. (9.31), (9.70), and (9.75) imply that the statistical behavior of the generalized eigenvector fit is the same as the statistical behavior of the unbiased eigenvector fit if terms of $O(\epsilon^2/\sqrt{N})$ are ignored. Hence, the quadratic form $(\hat{\boldsymbol{u}}, \hat{\boldsymbol{M}}\hat{\boldsymbol{u}})$ in eq. (9.81) can be identified with the residual $\hat{J}[\hat{\boldsymbol{u}}]$ of the unbiased eigenvector fit under the same approximation. It follows from eq. (9.53) that to a first approximation

$$I[\hat{\boldsymbol{u}}] = \frac{\epsilon^2}{rN}\chi^2_{rN-n+1}. \tag{9.82}$$

Hence, its expectation and the variance of $I[\hat{\boldsymbol{u}}]$ are

$$E[I[\hat{\boldsymbol{u}}]] = \epsilon^2\left(1 - \frac{n-1}{rN}\right), \quad V[I[\hat{\boldsymbol{u}}]] = \frac{2\epsilon^4}{rN}\left(1 - \frac{n-1}{rN}\right). \tag{9.83}$$

Comparing eqs. (9.79) and (9.83), we observe that the expectation and variance of $I[\hat{\boldsymbol{u}}]$ are both smaller than their respective values for $I[\bar{\boldsymbol{u}}]$ due to the correlation between $\hat{\boldsymbol{u}}$ and the data $\{\boldsymbol{a}_\alpha^{(k)}\}$.

From eq. (9.82), an unbiased estimator $\hat{\epsilon}^2$ of the squared noise level ϵ^2 is obtained in the following form:

$$\hat{\epsilon}^2 = \frac{I[\hat{\boldsymbol{u}}]}{1 - (n-1)/rN}. \tag{9.84}$$

Its expectation and variance are

$$E[\hat{\epsilon}^2] = \epsilon^2, \qquad V[\hat{\epsilon}^2] = \frac{2\epsilon^4}{rN - n + 1}. \tag{9.85}$$

If the noise level ϵ is given a priori, we can construct the χ^2 test for the weak hypothesis that eq. (9.1) holds for *some* value $\boldsymbol{u}$: the weak hypothesis is rejected with significance level $a\%$ if

$$I[\hat{\boldsymbol{u}}] > \frac{\epsilon^2}{rN}\chi^2_{rN-n+1,a}. \tag{9.86}$$

In terms of the estimate $\hat{\epsilon}^2$ computed by eq. (9.84), the above χ^2 test takes the form

$$\frac{\hat{\epsilon}^2}{\epsilon^2} > \frac{\chi^2_{rN-n+1,a}}{rN-n+1}. \tag{9.87}$$

9.4 Renormalization

9.4.1 Iterations for generalized eigenvalue problem

In order to compute the generalized eigenvector fit, one must solve the generalized eigenvalue problem for the moment matrix $\boldsymbol{M}$ with respect to the matrix $\boldsymbol{N}$. If $\boldsymbol{N}$ is of full rank, this is a nonsingular generalized eigenvalue problem (see Section 2.2.4), and the solution is obtained by computing the inverse square root $\boldsymbol{N}^{-1/2}$ (see eqs. (2.93) and (2.94)). However, the domain of the data $\{\boldsymbol{a}_\alpha^{(k)}\}$ is usually constrained, i.e., the covariance matrices $V[\boldsymbol{a}_\alpha^{(k)}, \boldsymbol{a}_\alpha^{(l)}]$ are generally singular. As a result, the matrix $\boldsymbol{N}$ is generally singular, so the generalized eigenvalue problem is singular (see Section 2.2.5). It follows that one must solve a generalized eigenvalue problem of a smaller size (see eq. (2.100)). However, there exists a more efficient method if we note the following two facts:

- Not all the generalized eigenvalues and generalized eigenvectors need to be computed: only the *smallest* generalized eigenvalue and the corresponding generalized eigenvector are necessary.
- Computing the smallest generalized eigenvalue c of the problem $\boldsymbol{M}\boldsymbol{u} = c\boldsymbol{N}\boldsymbol{u}$ is equivalent to computing the value c for which the *smallest eigenvalue* of $\hat{\boldsymbol{M}} = \boldsymbol{M} - c\boldsymbol{N}$ is 0.

The second assertion is proved as follows. If c is the smallest generalized eigenvalue of the problem $\boldsymbol{M}\boldsymbol{u} = c\boldsymbol{N}\boldsymbol{u}$ and $\hat{\boldsymbol{u}}$ is the corresponding generalized eigenvector, we have

$$\hat{\boldsymbol{M}}\hat{\boldsymbol{u}} = \boldsymbol{M}\hat{\boldsymbol{u}} - c\boldsymbol{N}\hat{\boldsymbol{u}} = \boldsymbol{0}, \tag{9.88}$$

meaning that $\hat{\boldsymbol{M}}$ has eigenvalue 0 for eigenvector $\hat{\boldsymbol{u}}$. Suppose $\hat{\boldsymbol{M}}$ has another eigenvalue $-\lambda'$ (< 0) for a unit eigenvector $\boldsymbol{u}'$. Then, $\hat{\boldsymbol{M}}\boldsymbol{u}' = \boldsymbol{M}\boldsymbol{u}' - c\boldsymbol{N}\boldsymbol{u}' = -\lambda'\boldsymbol{u}'$. Hence,

$$(\boldsymbol{u}', \boldsymbol{M}\boldsymbol{u}') - c(\boldsymbol{u}', \boldsymbol{N}\boldsymbol{u}') = -\lambda'. \tag{9.89}$$

If $(\boldsymbol{u}', \boldsymbol{N}\boldsymbol{u}') = 0$, then $(\boldsymbol{u}', \boldsymbol{M}\boldsymbol{u}') = -\lambda'$, which is a contradiction because $\boldsymbol{M}$ is semi-positive definite. Hence, $(\boldsymbol{u}', \boldsymbol{N}\boldsymbol{u}') > 0$, but then

$$\frac{(\boldsymbol{u}', \boldsymbol{M}\boldsymbol{u}')}{(\boldsymbol{u}', \boldsymbol{N}\boldsymbol{u}')} = c - \frac{\lambda'}{(\boldsymbol{u}', \boldsymbol{N}\boldsymbol{u}')}, \tag{9.90}$$

which is also a contradiction, since c should be the minimum of the generalized Rayleigh quotient $(\boldsymbol{u}, \boldsymbol{M}\boldsymbol{u})/(\boldsymbol{u}, \boldsymbol{N}\boldsymbol{u})$.

Thus, we only need to compute the value c for which the smallest eigenvalue of $\hat{\boldsymbol{M}} = \boldsymbol{M} - c\boldsymbol{N}$ is 0. If $\hat{\boldsymbol{u}}$ is the corresponding unit eigenvector, we have $(\hat{\boldsymbol{u}}, \hat{\boldsymbol{M}}\hat{\boldsymbol{u}}) = 0$. Let $\boldsymbol{u}$ be the unit eigenvector of $\boldsymbol{M}$ for the smallest eigenvalue λ ($\neq 0$). Then,

$$(\boldsymbol{u}, (\boldsymbol{M} - c\boldsymbol{N})\boldsymbol{u}) = (\boldsymbol{u}, \boldsymbol{M}\boldsymbol{u}) - c(\boldsymbol{u}, \boldsymbol{N}\boldsymbol{u}) = \lambda - c(\boldsymbol{u}, \boldsymbol{N}\boldsymbol{u}). \tag{9.91}$$

It follows that if we define

$$\hat{\boldsymbol{M}} = \boldsymbol{M} - \frac{\lambda}{(\boldsymbol{u}, \boldsymbol{N}\boldsymbol{u})}\boldsymbol{N}, \tag{9.92}$$

we have $(\boldsymbol{u}, \hat{\boldsymbol{M}}\boldsymbol{u}) = 0$. However, $\boldsymbol{u}$ may not be an eigenvector of $\hat{\boldsymbol{M}}$. So, we iterate this process:

1. Let $c = 0$.

2. Let $\boldsymbol{u}$ be the unit eigenvector of the matrix

$$\hat{\boldsymbol{M}} = \boldsymbol{M} - c\boldsymbol{N} \tag{9.93}$$

   for the smallest eigenvalue, and let λ be that smallest eigenvalue.

3. If $\lambda \approx 0$, return $\boldsymbol{u}$ and c. Else, update c as follows:

$$c \leftarrow c + \frac{\lambda}{(\boldsymbol{u}, \boldsymbol{N}\boldsymbol{u})} \tag{9.94}$$

4. Go back to Step 2.

The convergence of this process can be confirmed as follows. If $\boldsymbol{u}$ is the unit eigenvector of $\hat{\boldsymbol{M}}$ for the smallest eigenvalue λ, the matrix $\hat{\boldsymbol{M}}$ is perturbed by $-\lambda\boldsymbol{N}/(\boldsymbol{u}, \boldsymbol{N}\boldsymbol{u})$ at the next iteration. According to the perturbation theorem, the corresponding eigenvalue λ is perturbed in the following form (see eq. (2.108)):

$$\lambda' = \lambda - (\boldsymbol{u}, \frac{\lambda\boldsymbol{N}}{(\boldsymbol{u}, \boldsymbol{N}\boldsymbol{u})}\boldsymbol{u}) + O(\lambda^2) = O(\lambda^2). \tag{9.95}$$

In other words, λ converges to 0 *quadratically* as in the Newton iterations, meaning that three or four iterations are usually sufficient.

9.4.2 Iterations for weight update

So far, we have assumed that the weights $W_\alpha^{(kl)}$ are approximated by using an appropriate estimate $\boldsymbol{u}^*$. Since we need not worry about the bias any longer, it can be determined by iterations: we guess the initial value $\boldsymbol{u}_0$ and compute the generalized eigenvector fit $\boldsymbol{u}_1$; using $\boldsymbol{u}_1$, we compute the generalized eigenvector fit $\boldsymbol{u}_2$ and so on. Since the generalized eigenvector fit is itself computed by iterations, the combined process requires double loops of iterations. However, the two loops can be merged in the following form:

1. Let $c = 0$ and $W_\alpha^{(kl)} = \delta_{kl}$, $\alpha = 1, ..., N$.

2. Compute the following matrices $\boldsymbol{M}$ and $\boldsymbol{N}$:

$$\boldsymbol{M} = \frac{1}{N}\sum_{\alpha=1}^{N}\sum_{k,l=1}^{L} W_\alpha^{(kl)} \boldsymbol{a}_\alpha^{(k)} \boldsymbol{a}_\alpha^{(l)\top}, \tag{9.96}$$

$$\boldsymbol{N} = \frac{1}{N}\sum_{\alpha=1}^{N}\sum_{k,l=1}^{L} W_\alpha^{(kl)} V_0[\boldsymbol{a}_\alpha^{(k)}, \boldsymbol{a}_\alpha^{(l)}]. \tag{9.97}$$

3. Compute the smallest eigenvalue λ of the matrix

$$\hat{\boldsymbol{M}} = \boldsymbol{M} - c\boldsymbol{N} \tag{9.98}$$

and the corresponding unit eigenvector $\boldsymbol{u}$.

4. If $\lambda \approx 0$, return $\boldsymbol{u}$, c, and $\hat{\boldsymbol{M}}$. Else, update c and $\boldsymbol{W}_\alpha = (W_\alpha^{(kl)})$ as follows:

$$c \leftarrow c + \frac{\lambda}{(\boldsymbol{u}, \boldsymbol{N}\boldsymbol{u})}, \qquad \boldsymbol{W}_\alpha \leftarrow \left((\boldsymbol{u}, V_0[\boldsymbol{a}_\alpha^{(k)}, \boldsymbol{a}_\alpha^{(l)}]\boldsymbol{u})\right)_r^-. \tag{9.99}$$

5. Go back to Step 2.

We call this process *renormalization.* Although the convergence is no longer quadratic, the computation is usually very efficient. Since the returned value $\boldsymbol{u}$ gives the generalized eigenvector fit $\hat{\boldsymbol{u}}$, the covariance matrix of $\hat{\boldsymbol{u}}$ is given by eq. (9.76):

$$V[\hat{\boldsymbol{u}}] = \frac{\epsilon^2}{N}(\hat{\boldsymbol{M}})_{n-1}^-. \tag{9.100}$$

Here, the rank-constrained generalized inverse is used because $\hat{\boldsymbol{M}}\hat{\boldsymbol{u}} = \boldsymbol{0}$ may not be strictly satisfied if the iterations are prematurely terminated. Since the eigenspace for the smallest eigenvalue of $\hat{\boldsymbol{M}}$ is $\{\hat{\boldsymbol{u}}\}_L$, computing $(\hat{\boldsymbol{M}})_{n-1}^-$ is equivalent to applying the projection matrix $\boldsymbol{P}_{\hat{\boldsymbol{u}}}$ as in eq. (9.76).

After renormalization, the returned constant c equals the residual $I[\hat{\boldsymbol{u}}]$ of the generalized eigenvector fit $\hat{\boldsymbol{u}}$ (see eqs. (9.61), and (9.62)). Hence, an unbiased estimator $\hat{\epsilon}^2$ of the squared noise level ϵ^2 is obtained from eq. (9.84) in the following form:

$$\hat{\epsilon}^2 = \frac{c}{1 - (n-1)/rN}. \tag{9.101}$$

Its expectation and variance are given by eqs. (9.85). It follows that the number N of necessary data for uniquely estimating $\boldsymbol{u}$ must be such that

$$N \geq \frac{n-1}{r}. \tag{9.102}$$

If $N = (n-1)/r$, the value of $\boldsymbol{u}$ is determined but the noise level ϵ cannot be estimated.

The significance of renormalization is that it produces not only an unbiased estimator $\hat{\boldsymbol{u}}$ of the parameter $\boldsymbol{u}$ by a simple numerical computation but at the same time its normalized covariance matrix $V_0[\hat{\boldsymbol{u}}]$, an unbiased estimator $\hat{\epsilon}^2$ of the squared noise level, and its variance $V[\hat{\epsilon}^2]$ as well. Let $\mu_{\max}$ be the largest eigenvalue of $V[\hat{\boldsymbol{u}}]$, and $\boldsymbol{v}_{\max}$ the corresponding unit eigenvector. The *primary deviation pair* $\{\boldsymbol{u}^+, \boldsymbol{u}^-\}$ is given as follows (see Section 4.5.3):

$$\boldsymbol{u}^+ = N[\hat{\boldsymbol{u}} + \sqrt{\mu_{\max}}\boldsymbol{v}_{\max}], \qquad \boldsymbol{u}^- = N[\hat{\boldsymbol{u}} - \sqrt{\mu_{\max}}\boldsymbol{v}_{\max}]. \tag{9.103}$$

If the value ϵ is given a priori, the hypothesis (9.1) can be tested by comparing it with the estimate $\hat{\epsilon}$ (see eq. (9.87)): the hypothesis (9.1) is rejected with significance level $a\%$ if

$$\frac{\hat{\epsilon}^2}{\epsilon^2} > \frac{\chi^2_{rN-n+1,a}}{rN-n+1}. \tag{9.104}$$

9.5 Linearization

9.5.1 Linearized algorithm

Renormalization is a numerical means to compute the generalized eigenvector fit, which is the solution of the unbiased least-squares approximation to the original optimization, but the unbiased least-squares approximation and the least-squares approximation are both designed for problems for which

1. the hypothesis is *linear*, and

2. *no constraint* is imposed other than normalization.

In many application problems, the parameter $\boldsymbol{u}$ is constrained in various ways, even though the hypothesis is linear (see Chapter 5). For example, $\boldsymbol{u}$ may be a unit vector constrained to be in a parameter space $\mathcal{U} \subset \mathcal{R}^n$. Then, the procedure for renormalization as described in the preceding section can no longer be applied. We now devise a method for overcoming this restriction by noting that the following generally holds:

- The constraint $\boldsymbol{u} \in \mathcal{U}$ is *compatible* with the hypothesis (9.1): there exists a solution $\bar{\boldsymbol{u}} \in \mathcal{U}$ that satisfies the hypothesis (9.1).

- The hypothesis (9.1) is an *overspecification*: there exists a minimum number N_0 ($\leq N$) such that equations $(\bar{\boldsymbol{a}}_\alpha^{(k)}, \boldsymbol{u}) = 0$, $k = 1, ..., L$, $\alpha = 1, ..., N_0$, can uniquely determine the solution $\bar{\boldsymbol{u}} \in \mathcal{U}$.

It follows that if the noise is small, the solution $\hat{\boldsymbol{u}}$ based on the hypothesis (9.1) *alone* is expected to be a good estimate of the true value $\bar{\boldsymbol{u}}$. This observation leads us to the following approach:

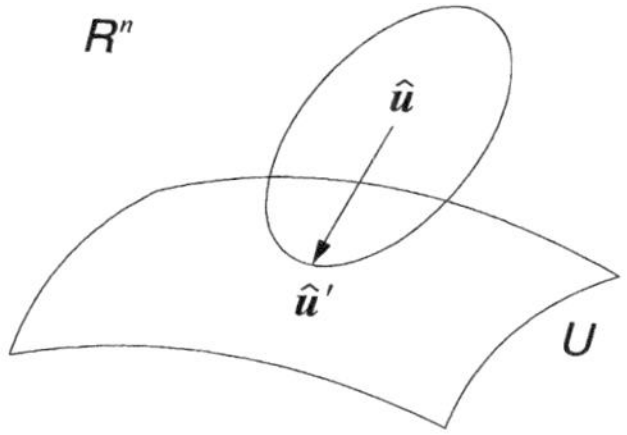

Fig. 9.2. Optimally projecting $\hat{\mathbf{u}}$ onto the parameter space $\mathcal{U} \subset \mathcal{R}^n$.

1. We apply renormalization *without considering the constraint* $\boldsymbol{u} \in \mathcal{U}$.
2. We then *optimally project* the resulting solution onto the parameter space $\mathcal{U}$ (Fig. 9.2).

Let $\hat{\boldsymbol{u}}$ be the solution obtained in Step 1, and $\hat{\boldsymbol{u}}' \in \mathcal{U}$ its projection onto $\mathcal{U}$. Suppose the manifold $\mathcal{U}$ is defined by M equations

$$F^{(m)}(\boldsymbol{u}) = 0, \qquad m = 1, ..., M, \tag{9.105}$$

together with the normalization condition $\|\boldsymbol{u}\| = 1$. Substituting $\hat{\boldsymbol{u}}' = \hat{\boldsymbol{u}} - \Delta\hat{\boldsymbol{u}}$ into eq. (9.105) and taking a linear approximation, we obtain

$$(\nabla_{\boldsymbol{u}} \hat{F}^{(m)}, \Delta\hat{\boldsymbol{u}}) = \hat{F}^{(m)}, \qquad m = 1, ..., M, \tag{9.106}$$

where $\nabla_{\boldsymbol{u}} \hat{F}^{(m)}$ and $\hat{F}^{(m)}$ are the abbreviations of $\nabla_{\boldsymbol{u}} F^{(m)}(\hat{\boldsymbol{u}})$ and $F^{(m)}(\hat{\boldsymbol{u}})$, respectively. The correction $\Delta\hat{\boldsymbol{u}}$ is determined by minimizing the Mahalanobis distance $\|\Delta\boldsymbol{u}\|_{V_0[\hat{\boldsymbol{u}}]}$ (see eq. (5.6)):

$$J = (\Delta\boldsymbol{u}, V_0[\hat{\boldsymbol{u}}]^- \Delta\boldsymbol{u}) \to \min, \qquad \Delta\hat{\boldsymbol{u}} \in \{\hat{\boldsymbol{u}}\}_L^\perp. \tag{9.107}$$

Here, we use the normalized covariance matrix $V_0[\hat{\boldsymbol{u}}]$ instead of $V[\hat{\boldsymbol{u}}]$, since the solution of the optimization is not affected if the covariance matrix is multiplied by a positive constant. Theoretically, $V_0[\hat{\boldsymbol{u}}]$ should be evaluated at $\hat{\boldsymbol{u}}'$ but we approximate it by the value at $\hat{\boldsymbol{u}}$. Note that the normalized covariance matrix $V_0[\hat{\boldsymbol{u}}]$ is given *as a by-product of renormalization* (see eq. (9.100)).

Let r be the rank of the constraint (9.105). As shown in Section 5.1.2, the first order solution of the optimization (9.107) under the constraint (9.106) is given in the following form (see eqs. (5.15) and (5.17)):

$$\Delta\hat{\boldsymbol{u}} = V_0[\hat{\boldsymbol{u}}] \sum_{m,n=1}^{M} W^{(mn)} \hat{F}^{(m)} \nabla_{\boldsymbol{u}} \hat{F}^{(n)}, \tag{9.108}$$

$$(W^{(mn)}) = \left((\nabla_{\boldsymbol{u}} \hat{F}^{(m)}, V_0[\hat{\boldsymbol{u}}] \nabla_{\boldsymbol{u}} \hat{F}^{(n)}) \right)_r^- . \tag{9.109}$$

In order to impose the normalization condition $\|\hat{\boldsymbol{u}}'\| = 1$ exactly, the actual correction of $\hat{\boldsymbol{u}}$ and its normalized covariance matrix takes the following form (see eqs. (5.25) and (5.26)):

$$\hat{\boldsymbol{u}}' = N[\hat{\boldsymbol{u}} - \Delta\hat{\boldsymbol{u}}], \qquad V_0[\hat{\boldsymbol{u}}]' = \boldsymbol{P}^{\mathcal{U}}_{\hat{\boldsymbol{u}}'} V_0[\hat{\boldsymbol{u}}] \boldsymbol{P}^{\mathcal{U}}_{\hat{\boldsymbol{u}}'}. \tag{9.110}$$

Here, $\boldsymbol{P}^{\mathcal{U}}_{\hat{\boldsymbol{u}}'}$ is the n-dimensional projection matrix onto the tangent space $T_{\hat{\boldsymbol{u}}'}(\mathcal{U})$ to the manifold $\mathcal{U}$ at $\hat{\boldsymbol{u}}'$. This correction is iterated until the constraint $\hat{\boldsymbol{u}}' \in \mathcal{U}$ is sufficiently satisfied. The normalized a posteriori covariance matrix of the final value $\hat{\boldsymbol{u}}'$ is given as follows (see eq.(5.31)):

$$V_0[\hat{\boldsymbol{u}}'] = \boldsymbol{P}^{\mathcal{U}}_{\hat{\boldsymbol{u}}'} V_0[\hat{\boldsymbol{u}}] \boldsymbol{P}^{\mathcal{U}}_{\hat{\boldsymbol{u}}'} - \sum_{m,n=1}^{M} W^{(mn)} (\boldsymbol{P}^{\mathcal{U}}_{\hat{\boldsymbol{u}}'} V_0[\hat{\boldsymbol{u}}] \nabla_{\boldsymbol{u}} \hat{F}^{(m)}) (\boldsymbol{P}^{\mathcal{U}}_{\hat{\boldsymbol{u}}'} V_0[\hat{\boldsymbol{u}}] \nabla_{\boldsymbol{u}} \hat{F}^{(n)})^\top. \tag{9.111}$$

This matrix has rank[4] $n - r - 1$. From eq. (9.111), the a posteriori covariance matrix of $\hat{\boldsymbol{u}}'$ is given in the form $V[\hat{\boldsymbol{u}}'] = \hat{\epsilon}^2 V_0[\hat{\boldsymbol{u}}']$, where $\hat{\epsilon}^2$ is an estimate of the squared noise level ϵ^2; it is given *as a by-product of renormalization* (see eq. (9.101)).

We call the above procedure *linearization*, and the resulting algorithm the *linearized algorithm*.

9.5.2 Decomposability condition

In some applications such as 3-D motion analysis, which we will study in Chapters 11 and 12, the hypothesis is nonlinear in the parameter $\boldsymbol{u}$ *but can be rearranged into the form*

$$(\bar{\boldsymbol{a}}^{(k)}_\alpha, \boldsymbol{v}) = 0, \qquad k = 1, ..., L, \tag{9.112}$$

where $\boldsymbol{v}$ is obtained from $\boldsymbol{u}$ by a nonlinear mapping $\boldsymbol{v}(\,\cdot\,)$: $\mathcal{R}^n \to \mathcal{R}^m$ in the form

$$\boldsymbol{v} = \boldsymbol{v}(\boldsymbol{u}). \tag{9.113}$$

As before, the data $\{\boldsymbol{a}^{(k)}_\alpha\}$, $k = 1, ..., L$, $\alpha = 1, ..., N$, are m-vectors and are assumed to have the form

$$\boldsymbol{a}^{(k)}_\alpha = \bar{\boldsymbol{a}}^{(k)}_\alpha + \Delta\boldsymbol{a}^{(k)}_\alpha, \tag{9.114}$$

where $\Delta\boldsymbol{a}^{(k)}_\alpha$ is a random variable of mean $\boldsymbol{0}$, independent for each α.

It appears that renormalization and linearization can be applied if $\boldsymbol{v}$ is regarded as a new variable instead of $\boldsymbol{u}$. However, we must be careful about the degree of freedom of the variable:

[4]The last "-1" in $n - r - 1$ is due to the normalization $\|\hat{\mathbf{u}}'\| = 1$.

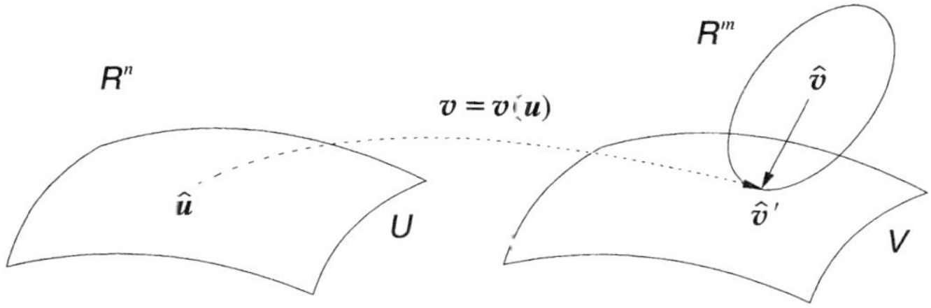

Fig. 9.3. Optimally projecting $\hat{\mathbf{v}} \in \mathcal{R}^m$ onto the manifold $\mathcal{V} \subset \mathcal{R}^m$ so that the equation $\mathbf{v}(\hat{\mathbf{u}}) = \hat{\mathbf{v}}'$ has a solution $\hat{\mathbf{u}} \in \mathcal{U}$.

1. If $m < n$, the parameter $\boldsymbol{u}$ cannot be determined even if $\boldsymbol{v}$ is obtained. In other words, the original problem is *underspecified*; the solution is indeterminate.

2. If $m = n$, eq. (9.113) determines the value $\boldsymbol{u}$ from the computed value $\boldsymbol{v}$ in general.

3. If $m > n$, eq. (9.113) is an *overspecification*; no solution may exist in general.

Let us consider the case $m > n$ more closely. If $m > n$, vector $\boldsymbol{v}(\boldsymbol{u})$ is generally constrained to be in an n'-dimensional manifold $\mathcal{V} \subset \mathcal{R}^m$ when $\boldsymbol{u}$ ranges over the parameter space $\mathcal{U} \subset \mathcal{R}^n$. Let

$$F^{(m)}(\boldsymbol{v}) = 0, \qquad m = 1, ..., m - n', \tag{9.115}$$

be the equation of the manifold $\mathcal{V}$. In order that eq. (9.113) has a solution $\boldsymbol{u}$, the optimization for $\boldsymbol{v}$ must be done under eq. (9.115). In this sense, eq. (9.115) is called the *decomposability condition.* If the decomposability condition (9.115) is imposed, renormalization can no longer be applied in the original form. As discussed in the preceding subsection, a simple way to overcome this restriction is computing an estimate $\hat{\boldsymbol{v}}$ of $\boldsymbol{v}$ without considering the decomposability condition (9.115) and then *optimally projecting* $\hat{\boldsymbol{v}}$ onto the manifold $\mathcal{V}$ (Fig. 9.3). As described earlier, this projection is done so that the Mahalanobis distance $\|\boldsymbol{v}(\boldsymbol{u}) - \boldsymbol{v}\|_{V_0[\boldsymbol{v}]}$ is minimized. The procedure is summarized as follows:

1. Compute the optimal estimate $\hat{\boldsymbol{v}}$, its normalized covariance matrix $V_0[\hat{\boldsymbol{v}}]$, and the unbiased estimator $\hat{\epsilon}^2$ of the squared noise level by applying renormalization to the hypothesis (9.112).

2. Compute the optimal correction $\Delta\hat{\boldsymbol{v}}$ so that $\hat{\boldsymbol{v}}' = \hat{\boldsymbol{v}} - \Delta\hat{\boldsymbol{v}}$ satisfies the decomposability condition (9.115); do iterations if necessary.

3. Compute the normalized a posteriori covariance matrix $V_0[\hat{\boldsymbol{v}}']$ of the resulting $\hat{\boldsymbol{v}}' \in \mathcal{V}$.

4. Solve the equation $\boldsymbol{v}(\hat{\boldsymbol{u}}) = \hat{\boldsymbol{v}}'$ for $\hat{\boldsymbol{u}} \in \mathcal{R}^n$.

5. Compute the normalized a posteriori covariance matrix $V_0[\hat{\boldsymbol{u}}]$ of the resulting solution $\hat{\boldsymbol{u}} \in \mathcal{U}$ (see eq. (3.16)); its a posteriori covariance matrix is given in the form $V[\hat{\boldsymbol{u}}] = \hat{\epsilon}^2 V_0[\hat{\boldsymbol{u}}]$.

9.6 Second Order Renormalization

9.6.1 Effective value of nonlinear data

In many problems in computer vision and robotics, the hypothesis is nonlinear *in the raw data* but can be rearranged into the form

$$(\bar{\boldsymbol{b}}_\alpha^{(k)}, \boldsymbol{u}) = 0, \qquad k = 1, ..., L, \tag{9.116}$$

where $\bar{\boldsymbol{b}}_\alpha^{(k)}$ is obtained from $\bar{\boldsymbol{a}}_\alpha$ by a nonlinear mapping $\boldsymbol{b}^{(k)}(\,\cdot\,)$: $\mathcal{R}^m \rightarrow \mathcal{R}^n$ in the form

$$\bar{\boldsymbol{b}}_\alpha^{(k)} = \boldsymbol{b}^{(k)}(\bar{\boldsymbol{a}}_\alpha). \tag{9.117}$$

In this case, if we compute

$$\boldsymbol{b}_\alpha^{(k)} = \boldsymbol{b}^{(k)}(\boldsymbol{a}_\alpha) \tag{9.118}$$

for given data $\{\boldsymbol{a}_\alpha\}$, we can apply renormalization to the converted data $\{\boldsymbol{b}_\alpha\}$. The deviation

$$\Delta\boldsymbol{b}_\alpha^{(k)} = \boldsymbol{b}_\alpha^{(k)} - E[\boldsymbol{b}_\alpha^{(k)}] \tag{9.119}$$

is a random variable of mean $\mathbf{0}$. Hence, the covariance matrix of $\boldsymbol{b}_\alpha^{(k)}$ is defined by

$$V[\boldsymbol{b}_\alpha^{(k)}, \boldsymbol{b}_\alpha^{(l)}] = E[\Delta\boldsymbol{b}_\alpha^{(k)} \Delta\boldsymbol{b}^{(l)\top}]. \tag{9.120}$$

However, such a conversion introduces a small error. As before, let the raw data be $\boldsymbol{a}_\alpha = \bar{\boldsymbol{a}}_\alpha + \Delta\boldsymbol{a}_\alpha$, the noise $\Delta\boldsymbol{a}_\alpha$ being an independent random variable of mean $\mathbf{0}$ and covariance matrix $V[\boldsymbol{a}_\alpha]$. In general, the expectation $E[\boldsymbol{b}^{(k)}(\boldsymbol{a}_\alpha)]$ is *not equal to* $\bar{\boldsymbol{b}}_\alpha^{(k)}$ (Fig. 9.4). In fact, let $b_i^{(k)}(\,\cdot\,)$, $b_{\alpha(i)}^{(k)}$, and $\bar{b}_{\alpha(i)}^{(k)}$ be the ith components of $\boldsymbol{b}^{(k)}(\,\cdot\,)$, $\boldsymbol{b}^{(k)}(\boldsymbol{a}_\alpha)$, and $\boldsymbol{b}^{(k)}(\bar{\boldsymbol{a}}_\alpha)$, respectively. We observe that

$$\begin{aligned} b_{\alpha(i)}^{(k)} &= b_i^{(k)}(\bar{\boldsymbol{a}}_\alpha + \Delta\boldsymbol{a}_\alpha) \\ &= \bar{b}_{\alpha(i)}^{(k)} + (\nabla_{\boldsymbol{a}} \bar{b}_{\alpha(i)}^{(k)}, \Delta\boldsymbol{a}_\alpha) + \frac{1}{2}(\nabla_{\boldsymbol{a}}^2 \bar{b}_{\alpha(i)}^{(k)}; \Delta\boldsymbol{a}_\alpha \Delta\boldsymbol{a}_\alpha^\top) + O(\Delta\boldsymbol{a}_\alpha)^3, \end{aligned} \tag{9.121}$$

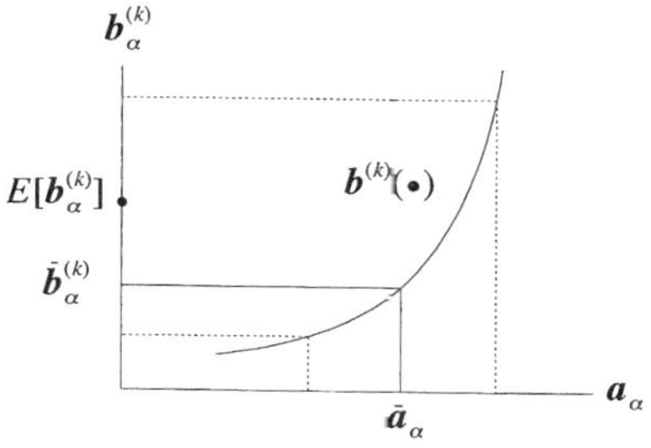

Fig. 9.4. The expectation $E[\mathbf{b}^{(k)}(\mathbf{a}_\alpha)]$ is not equal to $\bar{\mathbf{b}}_\alpha^{(k)}$.

where $\nabla_{\boldsymbol{a}} \bar{b}_{\alpha(i)}^{(k)}$ is the abbreviation of $\nabla_{\boldsymbol{a}} b_i^{(k)}(\bar{\boldsymbol{a}}_\alpha)$. The symbol $\nabla_{\boldsymbol{a}}^2 \bar{b}_{\alpha(i)}^{(k)}$ denotes the (mm)-matrix whose (pq) element is $\partial^2 b_i^{(k)}(\bar{\boldsymbol{a}}_\alpha)/\partial a_p \partial a_q$. Let ϵ be an appropriately defined noise level such that $\Delta \boldsymbol{a}_\alpha = O(\epsilon)$. Taking the expectation of eq. (9.121) and noting that the expectation of the terms of order 3 in $\Delta \boldsymbol{a}_\alpha$ vanishes, we obtain

$$E[b_i^{(k)}(\boldsymbol{a}_\alpha)] = \bar{b}_{\alpha(i)}^{(k)} + \frac{1}{2}(\nabla_{\boldsymbol{a}}^2 \bar{b}_{\alpha(i)}^{(k)}; V[\boldsymbol{a}_\alpha]) + O(\epsilon^4). \tag{9.122}$$

Thus, $E[\boldsymbol{b}^{(k)}(\boldsymbol{a}_\alpha)]$ is not equal to $\boldsymbol{b}^{(k)}(\bar{\boldsymbol{a}}_\alpha)$, although the deviation is of order $O(\epsilon^2)$.

This deviation can be canceled if we define the *effective value* $\boldsymbol{b}_\alpha^{*(k)} = (b_{\alpha(i)}^{*(k)})$ by

$$b_{\alpha(i)}^{*(k)} = b_{\alpha(i)}^{(k)} - \frac{1}{2}(\nabla_{\boldsymbol{a}}^2 b_{\alpha(i)}^{(k)}; V[\boldsymbol{a}_\alpha]), \tag{9.123}$$

where $\nabla_{\boldsymbol{a}}^2 b_{\alpha(i)}^{(k)}$ is the abbreviation of $\nabla_{\boldsymbol{a}}^2 b_i^{(k)}(\boldsymbol{a}_\alpha)$. In fact, if we note that $V[\boldsymbol{a}] = O(\epsilon^2)$ and

$$\nabla_{\boldsymbol{a}}^2 b_{\alpha(i)}^{(k)} = \nabla_{\boldsymbol{a}}^2 \bar{b}_{\alpha(i)}^{(k)} + O(\epsilon), \tag{9.124}$$

we obtain from eqs. (9.119), (9.122), and (9.123)

$$\begin{aligned} b_{\alpha(i)}^{*(k)} &= (E[b_{\alpha(i)}^{(k)}] + \Delta b_{\alpha(i)}^{(k)}) - \frac{1}{2}(\nabla_{\boldsymbol{a}}^2 b_{\alpha(i)}^{(k)}; V[\boldsymbol{a}_\alpha]) \\ &= \bar{b}_{\alpha(i)}^{(k)} + \frac{1}{2}(\nabla_{\boldsymbol{a}}^2 \bar{b}_{\alpha(i)}^{(k)} - \nabla_{\boldsymbol{a}}^2 b_{\alpha(i)}^{(k)}; V[\boldsymbol{a}_\alpha]) + O(\epsilon^4) \\ &= \bar{b}_{\alpha(i)}^{(k)} + \Delta b_{\alpha(i)} + O(\epsilon^3). \end{aligned} \tag{9.125}$$

Hence, if terms of $O(\epsilon^3)$ are ignored, $\{\boldsymbol{b}_\alpha^{*(k)}\}$ can be regarded as converted data with the interpretation that the true value $\bar{\boldsymbol{b}}_\alpha^{(k)}$ that satisfies the hypothesis (9.116) is perturbed by independent noise $\Delta \boldsymbol{b}_\alpha^{(k)}$ of mean $\mathbf{0}$ and covariance matrices $V[\boldsymbol{b}_\alpha^{(k)}, \boldsymbol{b}_\alpha^{(l)}]$, $k, l = 1, ..., L$. Since $E[O(\epsilon^3)] = O(\epsilon^4)$, we see that

$$E[\boldsymbol{b}_\alpha^{*(k)}] = \bar{\boldsymbol{b}}_\alpha^{(k)} + O(\epsilon^4). \tag{9.126}$$

Following the general theory given in Section 7.1.5, the optimal estimate $\hat{\boldsymbol{u}}$ is obtained as the solution of the optimization

$$J[\boldsymbol{u}] = \sum_{\alpha=1}^{N} \sum_{k,l=1}^{L} W_\alpha^{(kl)}(\boldsymbol{u})(\boldsymbol{b}_\alpha^{*(k)}, \boldsymbol{u})(\boldsymbol{b}_\alpha^{*(l)}, \boldsymbol{u}) \rightarrow \min, \tag{9.127}$$

$$(W_\alpha^{(kl)}(\boldsymbol{u})) = \left((\boldsymbol{u}, V[\boldsymbol{b}_\alpha^{(k)}, \boldsymbol{b}_\alpha^{(l)}]\boldsymbol{u})\right)_r^-, \tag{9.128}$$

where r is the rank of the hypothesis (9.116). Suppose $\boldsymbol{u}$ is constrained to be in an n'-dimensional parameter space $\mathcal{U} \subset \mathcal{R}^n$, and let $\hat{\boldsymbol{u}}$ be the resulting estimate. As discussed in Section 7.1.4, the residual $J[\hat{\boldsymbol{u}}]$ is a χ^2 variable with $rN - n'$ degrees of freedom, where n' is the degree of freedom of $\boldsymbol{u}$. The covariance matrix of $\hat{\boldsymbol{u}}$ is given in the following form (see eq. (7.57)):

$$V[\hat{\boldsymbol{u}}] = \left(\sum_{\alpha=1}^{N} \sum_{k,l=1}^{L} W_\alpha^{(kl)}(\hat{\boldsymbol{u}})(\boldsymbol{P}_{\hat{\boldsymbol{u}}}^{\mathcal{U}} \boldsymbol{b}_\alpha^{(k)})(\boldsymbol{P}_{\hat{\boldsymbol{u}}}^{\mathcal{U}} \boldsymbol{b}_\alpha^{(l)})^\top\right)^-. \tag{9.129}$$

Here, $\boldsymbol{P}_{\hat{\boldsymbol{u}}}^{\mathcal{U}}$ is the n-dimensional projection matrix onto the tangent space $T_{\hat{\boldsymbol{u}}}(\mathcal{U})$ to the manifold $\mathcal{U}$ at $\hat{\boldsymbol{u}}$.

9.6.2 *Second order unbiased estimation*

Suppose no constraints are imposed on the n-vector $\boldsymbol{u}$ other than normalization $\|\boldsymbol{u}\| = 1$. The least-squares approximation to eq. (9.127) is obtained if the functions $W_\alpha^{(kl)}(\boldsymbol{u})$ are replaced by constants $W_\alpha^{(kl)}$. Since multiplication of $J[\boldsymbol{u}]$ by a positive constant does not affect the solution, the least-squares approximation to eq. (9.127) can be written in the form

$$\tilde{J}[\boldsymbol{u}] = (\boldsymbol{u}, \boldsymbol{M}^*\boldsymbol{u}) \rightarrow \min, \tag{9.130}$$

where $\boldsymbol{M}^*$ is the *effective moment matrix* defined by

$$\boldsymbol{M}^* = \frac{1}{N} \sum_{\alpha=1}^{N} \sum_{k,l=1}^{L} W_\alpha^{(kl)} \boldsymbol{b}_\alpha^{*(k)} \boldsymbol{b}_\alpha^{*(l)\top}. \tag{9.131}$$

The solution under the constraint $\|\boldsymbol{u}\| = 1$ is obtained as the unit eigenvector of $\boldsymbol{M}^*$ for the smallest eigenvalue. As discussed in Section 9.1, however, the solution of the least-squares approximation is statistically biased whatever weights $W_\alpha^{(kl)}$ are used.

In order to construct an unbiased least-squares approximation, we first consider the moment matrix of the converted data $\{\boldsymbol{b}_\alpha^{(k)}\}$ defined by

$$\boldsymbol{M} = \frac{1}{N}\sum_{\alpha=1}^{N}\sum_{k,l=1}^{L} W_\alpha^{(kl)}\boldsymbol{b}_\alpha^{(k)}\boldsymbol{b}_\alpha^{(l)\top}. \tag{9.132}$$

Let $\bar{\boldsymbol{M}}$ be the unperturbed moment matrix obtained by replacing $\boldsymbol{b}_\alpha^{(k)}$ by $\bar{\boldsymbol{b}}_\alpha^{(k)}$. The hypothesis (9.116) implies that the true value $\bar{\boldsymbol{u}}$ satisfies $\bar{\boldsymbol{M}}\bar{\boldsymbol{u}} = \boldsymbol{0}$, i.e., $\bar{\boldsymbol{u}}$ is the unit eigenvector of $\bar{\boldsymbol{M}}$ for eigenvalue 0. In general, $E[\boldsymbol{M}]$ is not equal to $\bar{\boldsymbol{M}}$. Suppose we have found (nn)-matrices $\boldsymbol{N}^{(1)}$ and $\boldsymbol{N}^{(2)}$ such that the matrix

$$\hat{\boldsymbol{M}} = \boldsymbol{M} - \epsilon^2\boldsymbol{N}^{(1)} + \epsilon^4\boldsymbol{N}^{(2)} + O(\epsilon^6) \tag{9.133}$$

has expectation $\bar{\boldsymbol{M}}$, where matrices $\boldsymbol{N}^{(1)}$ and $\boldsymbol{N}^{(2)}$ do not involve the noise level ϵ. Then, the unbiased least-squares approximation is obtained in the following form:

$$\hat{J}[\boldsymbol{u}] = (\boldsymbol{u}, \hat{\boldsymbol{M}}\boldsymbol{u}) \to \min. \tag{9.134}$$

The solution under the constraint $\|\boldsymbol{u}\| = 1$ is obtained as the unit eigenvector of $\hat{\boldsymbol{M}}$ for the smallest eigenvalue.

If the noise level ϵ is not known, we let

$$\hat{\boldsymbol{M}} = \boldsymbol{M} - c\boldsymbol{N}^{(1)} + c^2\boldsymbol{N}^{(2)}, \tag{9.135}$$

and choose the constant c so that the smallest eigenvalue of $\hat{\boldsymbol{M}}$ is 0. Let $\boldsymbol{u}$ be the unit eigenvector of $\hat{\boldsymbol{M}}$ for the smallest eigenvalue λ ($\neq 0$). If c is incremented by Δc, the matrix $\hat{\boldsymbol{M}}$ changes into

$$\begin{aligned}\hat{\boldsymbol{M}}' &= \boldsymbol{M} - (c+\Delta c)\boldsymbol{N}^{(1)} + (c+\Delta c)^2\boldsymbol{N}^{(2)} \\ &= \hat{\boldsymbol{M}} - \Delta c(\boldsymbol{N}^{(1)} - 2c\boldsymbol{N}^{(2)}) + \Delta c^2\boldsymbol{N}^{(2)}.\end{aligned} \tag{9.136}$$

Since $(\boldsymbol{u}, \hat{\boldsymbol{M}}\boldsymbol{u}) = \lambda$, we see that

$$(\boldsymbol{u}, \hat{\boldsymbol{M}}'\boldsymbol{u}) = \lambda - \Delta c((\boldsymbol{u}, \boldsymbol{N}^{(1)}\boldsymbol{u}) - 2c(\boldsymbol{u}, \boldsymbol{N}^{(2)}\boldsymbol{u})) + \Delta c^2(\boldsymbol{u}, \boldsymbol{N}^{(2)}\boldsymbol{u}). \tag{9.137}$$

It follows that we can let $(\boldsymbol{u}, \hat{\boldsymbol{M}}'\boldsymbol{u}) = 0$ by choosing the increment Δc to be

$$\begin{aligned}\Delta c = \frac{1}{2(\boldsymbol{u}, \boldsymbol{N}^{(2)}\boldsymbol{u})}\Big((\boldsymbol{u}, \boldsymbol{N}^{(1)}\boldsymbol{u}) - 2c(\boldsymbol{u}, \boldsymbol{N}^{(2)}\boldsymbol{u}) \\ - \sqrt{((\boldsymbol{u}, \boldsymbol{N}^{(1)}\boldsymbol{u}) - 2c(\boldsymbol{u}, \boldsymbol{N}^{(2)}\boldsymbol{u}))^2 - 4\lambda(\boldsymbol{u}, \boldsymbol{N}^{(2)}\boldsymbol{u})}\Big).\end{aligned} \tag{9.138}$$

The covariance matrix defined by eq. (9.120) generally has the form

$$V[\boldsymbol{b}_\alpha^{(k)}, \boldsymbol{b}_\alpha^{(l)}] = \epsilon^2 V_0^{(1)}[\boldsymbol{b}_\alpha^{(k)}, \boldsymbol{b}_\alpha^{(l)}] + \epsilon^4 V_0^{(2)}[\boldsymbol{b}_\alpha^{(k)}, \boldsymbol{b}_\alpha^{(l)}] + O(\epsilon^6), \tag{9.139}$$

where the matrices $V_0^{(1)}[\boldsymbol{b}_\alpha^{(k)}, \boldsymbol{b}_\alpha^{(k)}]$ and $V_0^{(2)}[\boldsymbol{b}_\alpha^{(k)}, \boldsymbol{b}_\alpha^{(k)}]$ do not involve the noise level ϵ. By incorporating the update of the weights $W_\alpha^{(kl)}$, the renormalization procedure is described in the following form:

1. Let $c = 0$ and $W_\alpha^{(kl)} = \delta_{ij}$, $k, l = 1, ..., L$, $\alpha = 1, ..., N$.

2. Compute the matrices $\boldsymbol{M}$, $\boldsymbol{N}^{(1)}$, and $\boldsymbol{N}^{(2)}$.

3. Compute the smallest eigenvalue λ of the matrix

$$\hat{\boldsymbol{M}} = \boldsymbol{M} - c\boldsymbol{N}^{(1)} + c^2\boldsymbol{N}^{(2)}, \tag{9.140}$$

   and the corresponding unit eigenvector $\boldsymbol{u}$.

4. If $\lambda \approx 0$, return $\boldsymbol{u}$, c, and $\hat{\boldsymbol{M}}$. Else, update c and $W_\alpha^{(kl)}$ as follows:

$$D = \Big((\boldsymbol{u}, \boldsymbol{N}^{(1)}\boldsymbol{u}) - 2c(\boldsymbol{u}, \boldsymbol{N}^{(2)}\boldsymbol{u})\Big)^2 - 4\lambda(\boldsymbol{u}, \boldsymbol{N}^{(2)}\boldsymbol{u}), \tag{9.141}$$

$$c \leftarrow \begin{cases} c + \dfrac{(\boldsymbol{u}, \boldsymbol{N}^{(1)}\boldsymbol{u}) - 2c(\boldsymbol{u}, \boldsymbol{N}^{(2)}\boldsymbol{u}) - \sqrt{D}}{2(\boldsymbol{u}, \boldsymbol{N}^{(2)}\boldsymbol{u})}, & \text{if } D \geq 0, \\ c + \dfrac{\lambda}{(\boldsymbol{u}, \boldsymbol{N}^{(1)}\boldsymbol{u})}, & \text{if } D < 0, \end{cases} \tag{9.142}$$

$$W_\alpha^{(kl)} \leftarrow \Big((\boldsymbol{u}, V_0^{(1)}[\boldsymbol{b}_\alpha^{(k)}, \boldsymbol{b}_\alpha^{(l)}]\boldsymbol{u}) + c(\boldsymbol{u}, V_0^{(2)}[\boldsymbol{b}_\alpha^{(k)}, \boldsymbol{b}_\alpha^{(l)}]\boldsymbol{u})\Big)_r^-. \tag{9.143}$$

5. Go back to Step 2.

We call this process *second order renormalization*. As before, the covariance matrix of the optimal estimate $\hat{\boldsymbol{u}}$ is obtained in the following form:

$$V[\hat{\boldsymbol{u}}] = \frac{\epsilon^2}{N}(\hat{\boldsymbol{M}})_{n-1}^-. \tag{9.144}$$

The squared noise level ϵ^2 is estimated in the form

$$\hat{\epsilon}^2 = \frac{c}{1 - (n-1)/rN}. \tag{9.145}$$

Its expectation and variance are given by eqs. (9.85).

Chapter 10

Applications of Geometric Estimation

This chapter illustrates the theories in Chapters 7 and 9 by solving typical parametric fitting problems, all of which are very important in computer vision and robotics applications: fitting lines and conics to edge segments detected by an image processing operation and fitting planes to space points reconstructed by a range finder or stereo vision. In each problem, an optimal fit is computed by renormalization. At the same time, its reliability is evaluated in terms of the covariance matrix and visualized by means of the primary deviation pair. Real-data and simulation examples are also shown.

10.1 Image Line Fitting

10.1.1 Optimal line fitting

Line fitting is one of the most important processes in image understanding, because most objects in indoor environments have linear boundaries. In an image, object boundaries are usually detected by an edge operator as *edge segments*, i.e., sequences of pixels. By fitting image lines to the detected linear edge segments, we can obtain a 2-D representation of the objects. Such a 2-D representation (or "line drawing") plays a basic role in computing the 3-D interpretation of the scene. Hence, fitting an image line to pixel data is a fundamental stage for understanding the scene.

Let $\{\boldsymbol{x}_\alpha\}$, $\alpha = 1, ..., N$, be a sequence of image points to which an image line is to be fitted. Let $\{\bar{\boldsymbol{x}}_\alpha\}$ be their true positions. We write

$$\boldsymbol{x}_\alpha = \bar{\boldsymbol{x}}_\alpha + \Delta\boldsymbol{x}_\alpha, \tag{10.1}$$

and regard the noise $\Delta\boldsymbol{x}_\alpha$ as a Gaussian random variable of mean $\boldsymbol{0}$ and covariance matrix $V[\boldsymbol{x}_\alpha]$. Let $(\boldsymbol{n}, \boldsymbol{x}) = 0$ be the image line to be fitted. The problem is stated as follows.

Problem 10.1 *Estimate a unit vector* $\boldsymbol{n}$ *such that*

$$(\boldsymbol{n}, \bar{\boldsymbol{x}}_\alpha) = 0, \qquad \alpha = 1, ..., N, \tag{10.2}$$

from the data $\{\boldsymbol{x}_\alpha\}$, $\alpha = 1, ..., N$.

Eq. (10.2) is the hypothesis from which the vector $\boldsymbol{n}$ is to be estimated. The rank of this hypothesis is 1. As shown in Section 7.2.2, an optimal estimate of $\boldsymbol{n}$ can be obtained by the optimization

$$J[\boldsymbol{n}] = \sum_{\alpha=1}^{N} \frac{(\boldsymbol{n}, \boldsymbol{x}_\alpha)^2}{(\boldsymbol{n}, V[\boldsymbol{x}_\alpha]\boldsymbol{n})} \to \min \tag{10.3}$$

under the constraint $\|\boldsymbol{n}\| = 1$. The theoretical bound on the accuracy of fitting is given by the covariance matrix of the optimal estimate $\hat{\boldsymbol{n}}$ in the form

$$\bar{V}[\hat{\boldsymbol{n}}] = \left(\sum_{\alpha=1}^{N} \frac{\bar{\boldsymbol{x}}_\alpha \bar{\boldsymbol{x}}_\alpha^\top}{(\boldsymbol{n}, V[\boldsymbol{x}_\alpha]\boldsymbol{n})} \right)^-, \tag{10.4}$$

which has rank 2; its null space is $\{\boldsymbol{n}\}_L$.

Decompose the covariance matrix $V[\boldsymbol{x}_\alpha]$ into the *noise level* ϵ and the *normalized covariance matrix* $V_0[\boldsymbol{x}_\alpha]$ in the form

$$V[\boldsymbol{x}_\alpha] = \epsilon^2 V_0[\boldsymbol{x}_\alpha]. \tag{10.5}$$

If the denominator in eq. (10.3) is replaced by a constant, we obtain the least-squares approximation

$$\tilde{J}[\boldsymbol{n}] = (\boldsymbol{n}, \boldsymbol{M}\boldsymbol{n}) \to \min, \tag{10.6}$$

where the *moment matrix* $\boldsymbol{M}$ is defined by

$$\boldsymbol{M} = \frac{1}{N} \sum_{\alpha=1}^{N} W_\alpha \boldsymbol{x}_\alpha \boldsymbol{x}_\alpha^\top, \tag{10.7}$$

$$W_\alpha = \frac{1}{(\boldsymbol{n}^*, V_0[\boldsymbol{x}_\alpha]\boldsymbol{n}^*)}. \tag{10.8}$$

Here, $\boldsymbol{n}^*$ is an appropriate estimate of $\boldsymbol{n}$. The solution $\tilde{\boldsymbol{n}}$ of the optimization (10.6) is obtained as the unit eigenvector of the moment matrix $\boldsymbol{M}$ for the smallest eigenvalue.

10.1.2 Unbiased estimation and renormalization

The solution of the least-squares approximation is in general statistically biased whatever weights W_α are used. In fact, taking the expectation of the moment matrix $\boldsymbol{M}$, we see that

$$E[\boldsymbol{M}] = \frac{1}{N} \sum_{\alpha=1}^{N} W_\alpha E[(\bar{\boldsymbol{x}}_\alpha + \Delta\boldsymbol{x}_\alpha)(\bar{\boldsymbol{x}}_\alpha + \Delta\boldsymbol{x}_\alpha)^\top]$$

$$= \bar{\boldsymbol{M}} + \frac{\epsilon^2}{N} \sum_{\alpha=1}^{N} W_\alpha V_0[\boldsymbol{x}_\alpha], \tag{10.9}$$

where $\bar{\boldsymbol{M}}$ is the unperturbed moment matrix obtained by replacing $\boldsymbol{x}_\alpha$ by $\bar{\boldsymbol{x}}_\alpha$ in eq. (10.7). The hypothesis (10.2) implies that $\bar{\boldsymbol{M}}\boldsymbol{n} = \boldsymbol{0}$, i.e., $\boldsymbol{n}$ is the unit eigenvector of the unperturbed moment matrix $\bar{\boldsymbol{M}}$ for eigenvalue 0. Since $E[\boldsymbol{M}]$ is perturbed from $\bar{\boldsymbol{M}}$ by $O(\epsilon^2)$, the solution $\tilde{\boldsymbol{n}}$ of the least-squares approximation is statistically biased by $O(\epsilon^2)$ according to the perturbation theorem. However, if we define the *unbiased moment matrix*

$$\hat{\boldsymbol{M}} = \boldsymbol{M} - \epsilon^2 \boldsymbol{N}, \tag{10.10}$$

$$\boldsymbol{N} = \frac{1}{N}\sum_{\alpha=1}^{N} W_\alpha V_0[\boldsymbol{x}_\alpha], \tag{10.11}$$

we have $E[\hat{\boldsymbol{M}}] = \bar{\boldsymbol{M}}$. Hence, we obtain the unbiased least-squares approximation

$$\hat{J}[\boldsymbol{n}] = (\boldsymbol{n}, \hat{\boldsymbol{M}}\boldsymbol{n}) \to \min. \tag{10.12}$$

The solution $\hat{\boldsymbol{n}}$ is obtained as the unit eigenvector of $\hat{\boldsymbol{M}}$ for the smallest eigenvalue.

If the noise level ϵ is not known, the renormalization procedure is given as follows (see Section 9.4.2):

1. Let $c = 0$ and $W_\alpha = 1$, $\alpha = 1, ..., N$.

2. Compute the matrices $\boldsymbol{M}$ and $\boldsymbol{N}$ defined by eqs. (10.7) and (10.11), respectively.

3. Compute the smallest eigenvalue λ of the matrix

$$\hat{\boldsymbol{M}} = \boldsymbol{M} - c\boldsymbol{N}, \tag{10.13}$$

   and the corresponding unit eigenvector $\boldsymbol{n}$.

4. If $\lambda \approx 0$, return $\boldsymbol{n}$, c, and $\hat{\boldsymbol{M}}$. Else, update c and W_α as follows:

$$c \leftarrow c + \frac{\lambda}{(\boldsymbol{n}, \boldsymbol{N}\boldsymbol{n})}, \qquad W_\alpha \leftarrow \frac{1}{(\boldsymbol{n}, V_0[\boldsymbol{x}_\alpha]\boldsymbol{n})}. \tag{10.14}$$

5. Go back to Step 2.

After renormalization, the squared noise level ϵ^2 is estimated in the form

$$\hat{\epsilon}^2 = \frac{c}{1 - 2/N}. \tag{10.15}$$

Its expectation and variance are

$$E[\hat{\epsilon}^2] = \epsilon^2, \qquad V[\hat{\epsilon}^2] = \frac{2\epsilon^4}{N-2}. \tag{10.16}$$

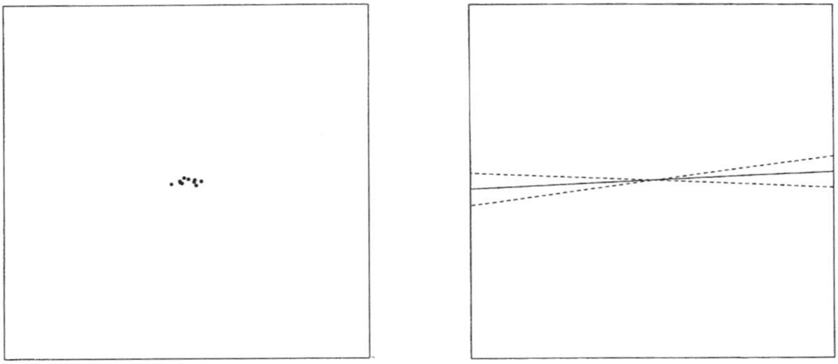

(a) (b)

Fig. 10.1. (a) Supposedly collinear image points. (b) Optimally fitted image line (solid line) and its primary deviation pair (dashed lines).

If the value ϵ is given a priori, the *collinearity test* for image points takes the following form (see eq. (7.75)):

$$\frac{\hat{\epsilon}^2}{\epsilon^2} > \frac{\chi^2_{N-2,a}}{N-2}. \tag{10.17}$$

The covariance matrix of the resulting estimate $\hat{\boldsymbol{n}}$ is obtained in the following form:

$$V[\hat{\boldsymbol{n}}] = \frac{\hat{\epsilon}^2}{N}(\hat{\boldsymbol{M}})^-_2. \tag{10.18}$$

Let $\mu_{\max}$ be the largest eigenvalue of $V[\hat{\boldsymbol{n}}]$, and $\boldsymbol{u}_{\max}$ the corresponding unit eigenvector. The primary deviation pair $\{\boldsymbol{n}^+, \boldsymbol{n}^-\}$ is given as follows (see Section 4.5.3):

$$\boldsymbol{n}^+ = N[\hat{\boldsymbol{n}} + \sqrt{\mu_{\max}}\boldsymbol{u}_{\max}], \qquad \boldsymbol{n}^- = N[\hat{\boldsymbol{n}} - \sqrt{\mu_{\max}}\boldsymbol{u}_{\max}]. \tag{10.19}$$

Example 10.1 Fig. 10.1a shows ten image points supposedly collinear and equidistant at four pixel intervals (the distance between the end points is 40 pixels). We perturbed them by adding Gaussian random noise of mean 0 and standard deviation 4 (pixels) to the x and y coordinates of each point independently. Fig. 10.1b shows the image line fitted by renormalization (solid line) and its primary deviation pair (dashed lines).

Let $(\bar{\boldsymbol{n}}, \boldsymbol{x}) = 0$ and $(\hat{\boldsymbol{n}}, \boldsymbol{x}) = 0$ be the true and the fitted image lines, respectively. Since the deviation of $\hat{\boldsymbol{n}}$ from $\bar{\boldsymbol{n}}$ is orthogonal to $\bar{\boldsymbol{n}}$ to a first approximation, the error of computation is represented by

$$\Delta\boldsymbol{n} = \boldsymbol{P}_{\bar{\boldsymbol{n}}}(\hat{\boldsymbol{n}} - \bar{\boldsymbol{n}}), \tag{10.20}$$

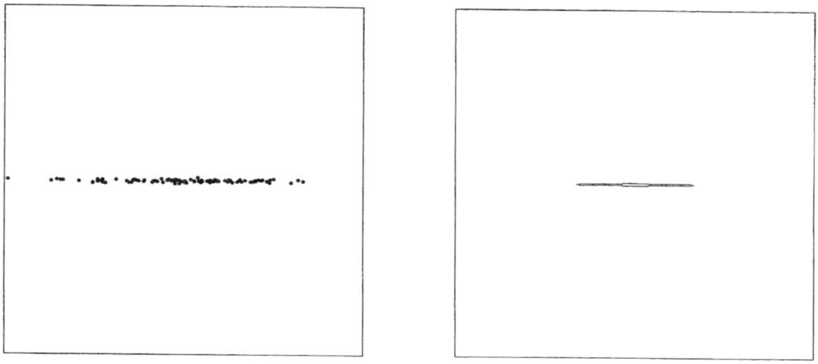

(a) (b)

Fig. 10.2. (a) Error distribution. (b) Theoretical standard confidence region.

where $\boldsymbol{P}_{\bar{\boldsymbol{n}}}$ is the projection matrix along $\bar{\boldsymbol{n}}$. In Fig. 10.2a, $\Delta\boldsymbol{n}$ is plotted in the plane perpendicular to $\bar{\boldsymbol{n}}$ for 100 trials, each time using different noise. Fig. 10.2b shows the standard confidence region computed from the theoretical expression of eq. (10.4) (see Section 4.5.3). Comparing Figs. 10.2a and 10.2b, we see that the theoretical bound on accuracy is almost attained. We also see that the error arises almost in the direction of the fitted line, confirming that eq. (7.92) is indeed a good approximation.

If the noise characteristics are symmetric with respect to the true line, the statistical bias does not appear, so the accuracy does not improve by renormalization. Recall that if the noise distribution is isotropic and homogeneous, the minimization (10.3) reduces to the least-squares optimization and the solution can be obtained analytically (see Example 7.2). Hence, we need not do iterations. However, the advantage of renormalization is that it can produce an optimal fit *irrespective of the noise distribution*. Also, it automatically estimates the noise level and gives the covariance matrix of the obtained fit.

Example 10.2 Fig. 10.3a is an edge image obtained by applying an edge operator to a portion of a real image. An image line is fitted to one edge segment by assuming that the distribution of the image noise is isotropic and identical for each pixel (see Examples 7.1 and 7.2 in Section 7.2.2); the noise level is unknown. Fig. 10.3b shows the optimally fitted image line (solid line) and its primary deviation pair (dashed lines) superimposed on the original gray-level image. This example demonstrates that by renormalization we can not only obtain an optimal fit but also visualize its reliability without knowing the noise level.

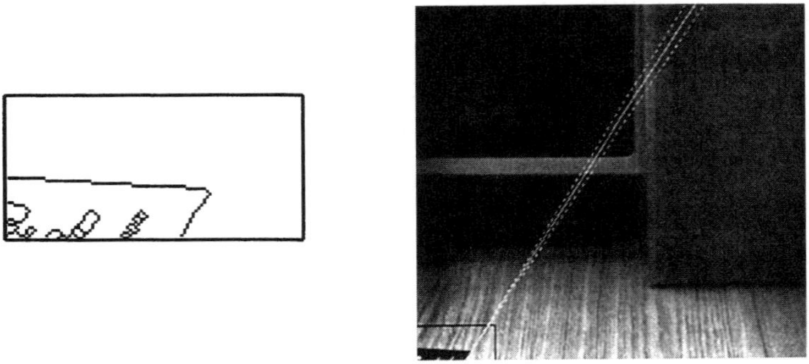

(a) (b)

Fig. 10.3. (a) Edge image. (b) Optimally fitted image line (solid line) and its primary deviation pair (dashed lines).

10.2 Conic Fitting

10.2.1 Optimal conic fitting

Conics (ellipses, parabolas, and hyperbolas) are very important image features in computer vision and robotics applications. Many industrial objects have circular and spherical shapes, and circles and spheres are projected as conics on the image plane. If a conic on the image plane is known to be a projection of a circle or an ellipse of known shape, its 3-D position can be computed analytically (but not always uniquely). In order to do such an analysis, the conic detected on the image plane must be given a mathematical representation first. Like image lines, conics can be detected as edge segments by an edge operator. Hence, the first task is to fit a conic equation to a given sequence $\{\boldsymbol{x}_\alpha\}$, $\alpha = 1, ..., N$, of image points.

In the presence of image noise, the detected image points $\{\boldsymbol{x}_\alpha\}$ are not necessarily accurate. Let $\bar{\boldsymbol{x}}_\alpha$ be the true location of the αth image point. We write

$$\boldsymbol{x}_\alpha = \bar{\boldsymbol{x}}_\alpha + \Delta\boldsymbol{x}_\alpha, \tag{10.21}$$

and regard the noise $\Delta\boldsymbol{x}_\alpha$ as a Gaussian random variable of mean $\boldsymbol{0}$ and covariance matrix $V[\boldsymbol{x}_\alpha]$. As shown in Section 4.4, a conic is represented in the form $(\boldsymbol{x}, \boldsymbol{Q}\boldsymbol{x}) = 0$ (see eq. (4.80)). Since the matrix $\boldsymbol{Q}$ can be determined only up to scale, we adopt the normalization $\|\boldsymbol{Q}\| = 1$. The problem is stated as follows:

Problem 10.2 *Estimate a* (33)*-matrix* $\boldsymbol{Q}$ *of unit norm such that*

$$(\bar{\boldsymbol{x}}_\alpha, \boldsymbol{Q}\bar{\boldsymbol{x}}_\alpha) = 0, \qquad \alpha = 1, ..., N, \tag{10.22}$$

from the data $\{\boldsymbol{x}_\alpha\}$, $\alpha = 1, ..., N$.

Eq. (10.22) is the hypothesis from which the matrix $\boldsymbol{Q}$ is to be estimated. The rank of this hypothesis is 1. Define a (33)-matrix

$$\boldsymbol{X}_\alpha = \boldsymbol{x}_\alpha \otimes \boldsymbol{x}_\alpha, \tag{10.23}$$

and let $\bar{\boldsymbol{X}}_\alpha$ be the unperturbed value of $\bar{\boldsymbol{X}}_\alpha$ obtained by replacing $\boldsymbol{x}_\alpha$ by $\bar{\boldsymbol{x}}_\alpha$. Eq. (10.22) can be written in the form

$$(\bar{\boldsymbol{X}}_\alpha; \boldsymbol{Q}) = 0. \tag{10.24}$$

Substituting eq. (10.21) into eq. (10.23), we can express $\boldsymbol{X}_\alpha$ in the form

$$\begin{aligned}\boldsymbol{X}_\alpha &= (\bar{\boldsymbol{x}}_\alpha + \Delta\boldsymbol{x}_\alpha) \otimes (\bar{\boldsymbol{x}}_\alpha + \Delta\boldsymbol{x}_\alpha) \\ &= \bar{\boldsymbol{X}}_\alpha + \Delta\boldsymbol{x}_\alpha \otimes \bar{\boldsymbol{x}}_\alpha + \bar{\boldsymbol{x}}_\alpha \otimes \Delta\boldsymbol{x}_\alpha + \Delta\boldsymbol{x}_\alpha \otimes \Delta\boldsymbol{x}_\alpha.\end{aligned} \tag{10.25}$$

Since $E[\Delta\boldsymbol{x}_\alpha] = \boldsymbol{0}$, we have

$$E[\boldsymbol{X}_\alpha] = \bar{\boldsymbol{X}}_\alpha + V[\boldsymbol{x}_\alpha]. \tag{10.26}$$

Hence, the *effective value* of $\boldsymbol{X}_\alpha$ is given as follows (see eq. (9.123)):

$$\boldsymbol{X}_\alpha^* = \boldsymbol{X}_\alpha - V[\boldsymbol{x}_\alpha]. \tag{10.27}$$

The deviation

$$\Delta\boldsymbol{X}_\alpha = \boldsymbol{X}_\alpha^* - \bar{\boldsymbol{X}}_\alpha = \Delta\boldsymbol{x}_\alpha \otimes \bar{\boldsymbol{x}}_\alpha + \bar{\boldsymbol{x}}_\alpha \otimes \Delta\boldsymbol{x}_\alpha + \Delta\boldsymbol{x}_\alpha \otimes \Delta\boldsymbol{x}_\alpha - V[\boldsymbol{x}_\alpha] \tag{10.28}$$

is a random variable of mean $\boldsymbol{O}$. Let $x_{\alpha(i)}$, $\bar{x}_{\alpha(i)}$, and $\Delta x_{\alpha(i)}$ be the ith components of $\boldsymbol{x}_\alpha$, $\bar{\boldsymbol{x}}_\alpha$, and $\Delta\boldsymbol{x}_\alpha$, respectively. The covariance tensor $\mathcal{V}[\boldsymbol{X}_\alpha]$ of $\boldsymbol{X}_\alpha$ has the following $(ijkl)$ element:

$$\begin{aligned}\mathcal{V}[\boldsymbol{X}_\alpha]_{ijkl} &= E[\Delta\boldsymbol{X}_\alpha \otimes \Delta\boldsymbol{X}_\alpha]_{ijkl} \\ &= E[\Delta x_{\alpha(j)}\Delta x_{\alpha(k)}]\bar{x}_{\alpha(i)}\bar{x}_{\alpha(l)} + E[\Delta x_{\alpha(j)}\Delta x_{\alpha(l)}]\bar{x}_{\alpha(i)}\bar{x}_{\alpha(k)} \\ &\quad + E[\Delta x_{\alpha(i)}\Delta x_{\alpha(k)}]\bar{x}_{\alpha(j)}\bar{x}_{\alpha(l)} + E[\Delta x_{\alpha(i)}\Delta x_{\alpha(l)}]\bar{x}_{\alpha(j)}\bar{x}_{\alpha(k)} \\ &\quad - E[\Delta x_{\alpha(i)}\Delta x_{\alpha(j)}]V[\boldsymbol{x}_\alpha]_{kl} - E[\Delta x_{\alpha(k)}\Delta x_{\alpha(l)}]V[\boldsymbol{x}_\alpha]_{ij} \\ &\quad + E[\Delta x_{\alpha(i)}\Delta x_{\alpha(j)}\Delta x_{\alpha(k)}\Delta x_{\alpha(l)}] + V[\boldsymbol{x}_\alpha]_{ij}V[\boldsymbol{x}_\alpha]_{kl} \\ &= V[\boldsymbol{x}_\alpha]_{ji}\bar{x}_{\alpha(i)}\bar{x}_{\alpha(k)} + V[\boldsymbol{x}_\alpha]_{jk}\bar{x}_{\alpha(i)}\bar{x}_{\alpha(l)} + V[\boldsymbol{x}_\alpha]_{il}\bar{x}_{\alpha(j)}\bar{x}_{\alpha(k)} \\ &\quad + V[\boldsymbol{x}_\alpha]_{ik}\bar{x}_{\alpha(j)}\bar{x}_{\alpha(l)} + V[\boldsymbol{x}_\alpha]_{ik}V[\boldsymbol{x}_\alpha]_{jl} + V[\boldsymbol{x}_\alpha]_{il}V[\boldsymbol{x}_\alpha]_{jk}.\end{aligned} \tag{10.29}$$

Here, we have used the following identity for Gaussian random variables (see eqs. (3.56)):

$$\begin{aligned}&E[\Delta x_{\alpha(i)}\Delta x_{\alpha(j)}\Delta x_{\alpha(k)}\Delta x_{\alpha(l)}] \\ &\qquad = V[\boldsymbol{x}_\alpha]_{ij}V[\boldsymbol{x}_\alpha]_{kl} + V[\boldsymbol{x}_\alpha]_{ik}V[\boldsymbol{x}_\alpha]_{jl} + V[\boldsymbol{x}_\alpha]_{il}V[\boldsymbol{x}_\alpha]_{jk}.\end{aligned} \tag{10.30}$$

From eq. (10.29), we obtain

$$(\boldsymbol{Q}; \mathcal{V}[\boldsymbol{X}_\alpha]\boldsymbol{Q}) = 4(\bar{\boldsymbol{x}}_\alpha, \boldsymbol{Q}V[\boldsymbol{x}_\alpha]\boldsymbol{Q}\bar{\boldsymbol{x}}_\alpha) + 2(V[\boldsymbol{x}_\alpha]\boldsymbol{Q}; \boldsymbol{Q}V[\boldsymbol{x}_\alpha]). \tag{10.31}$$

Hence, an optimal estimate of $\boldsymbol{Q}$ can be obtained as the solution of the optimization[1]

$$J[\boldsymbol{Q}] = \sum_{\alpha=1}^{N} \frac{(\boldsymbol{x}_\alpha \otimes \boldsymbol{x}_\alpha - V[\boldsymbol{x}_\alpha]; \boldsymbol{Q})^2}{4(\boldsymbol{x}_\alpha, \boldsymbol{Q}V[\boldsymbol{x}_\alpha]\boldsymbol{Q}\boldsymbol{x}_\alpha) + 2(V[\boldsymbol{x}_\alpha]\boldsymbol{Q}; \boldsymbol{Q}V[\boldsymbol{x}_\alpha])} \to \min \tag{10.32}$$

under the constraint $\|\boldsymbol{Q}\| = 1$.

Let $\hat{\boldsymbol{Q}}$ be the resulting optimal estimate. Since $\boldsymbol{Q}$ is a symmetric matrix and is normalized to $\|\boldsymbol{Q}\| = 1$, it has *five* degrees of freedom. Hence, the residual $J[\hat{\boldsymbol{Q}}]$ is a χ^2 variable with $N-5$ degrees of freedom. It follows that the number of image points must be such that

$$N \geq 5. \tag{10.33}$$

The residual is 0 when $N = 5$. For $N \geq 6$, we obtain the *conic test* for image points: the hypothesis that image points $\{\boldsymbol{x}_\alpha\}$ are on a conic is rejected with significance level $a\%$ if

$$J[\hat{\boldsymbol{Q}}] > \chi^2_{N-5,a}. \tag{10.34}$$

The theoretical bound on the accuracy of fitting is given by the covariance tensor of $\hat{\boldsymbol{Q}}$ in the form

$$\bar{\mathcal{V}}[\hat{\boldsymbol{Q}}] = \left(\sum_{\alpha=1}^{N} \frac{\mathcal{P}(\bar{\boldsymbol{x}}_\alpha \otimes \bar{\boldsymbol{x}}_\alpha) \otimes \mathcal{P}(\bar{\boldsymbol{x}}_\alpha \otimes \bar{\boldsymbol{x}}_\alpha)}{4(\bar{\boldsymbol{x}}_\alpha, \boldsymbol{Q}V[\boldsymbol{x}_\alpha]\boldsymbol{Q}\bar{\boldsymbol{x}}_\alpha) + 2(V[\boldsymbol{x}_\alpha]\boldsymbol{Q}; \boldsymbol{Q}V[\boldsymbol{x}_\alpha])}\right)^-, \tag{10.35}$$

where $\mathcal{P} = (P_{ijkl})$ is the *projection tensor* defined by

$$P_{ijkl} = \delta_{ik}\delta_{jl} - Q_{ij}Q_{kl}. \tag{10.36}$$

The covariance tensor $\bar{\mathcal{V}}[\hat{\boldsymbol{Q}}]$ has rank 5; its null space is $\{\boldsymbol{Q}\}_L$.

10.2.2 Unbiased estimation and renormalization

Decompose the covariance matrix $V[\boldsymbol{x}_\alpha]$ into the noise level ϵ and the normalized covariance matrix $V_0[\boldsymbol{x}_\alpha]$ in the form

$$V[\boldsymbol{x}_\alpha] = \epsilon^2 V_0[\boldsymbol{x}_\alpha]. \tag{10.37}$$

If the denominator in eq. (10.32) is replaced by a constant, we obtain the least-squares approximation

$$\tilde{J}[\boldsymbol{Q}] = (\boldsymbol{Q}; \mathcal{M}^*\boldsymbol{Q}) \to \min, \tag{10.38}$$

[1]In the denominator, the true values $\{\bar{\mathbf{x}}_\alpha\}$ are approximated by the data values $\{\mathbf{x}_\alpha\}$.

where $\mathcal{M}^*$ is the *effective moment tensor* defined by

$$\mathcal{M}^* = \frac{1}{N}\sum_{\alpha=1}^{N} W_\alpha(\boldsymbol{x}_\alpha \otimes \boldsymbol{x}_\alpha - V[\boldsymbol{x}_\alpha]) \otimes (\boldsymbol{x}_\alpha \otimes \boldsymbol{x}_\alpha - V[\boldsymbol{x}_\alpha]), \tag{10.39}$$

$$W_\alpha = \frac{1}{4(\boldsymbol{x}_\alpha, \boldsymbol{Q}^* V_0[\boldsymbol{x}_\alpha]\boldsymbol{Q}^*\boldsymbol{x}_\alpha) + 2\epsilon^2(V_0[\boldsymbol{x}_\alpha]\boldsymbol{Q}^*; \boldsymbol{Q}^* V_0[\boldsymbol{x}_\alpha]))}. \tag{10.40}$$

Here, $\boldsymbol{Q}^*$ is an appropriate estimate of $\boldsymbol{Q}$. The minimum is sought under the normalization constraint $\|\boldsymbol{Q}\| = 1$.

If a 6-vector $\boldsymbol{q}$ and a 66-matrix $\boldsymbol{M}^*$ are defined by casting the (33)-matrix $\boldsymbol{Q}$ and the (33)(33)-tensor $\mathcal{M}^*$ in the form

$$\boldsymbol{q} = \text{type}_6[\boldsymbol{Q}], \qquad \boldsymbol{M}^* = \text{type}_{66}[\mathcal{M}^*], \tag{10.41}$$

we have $(\boldsymbol{Q}; \mathcal{M}^*\boldsymbol{Q}) = (\boldsymbol{q}, \boldsymbol{M}^*\boldsymbol{q})$ and $\|\boldsymbol{Q}\| = \|\boldsymbol{q}\|$ (see Section 2.4.2). Hence, the solution of the minimization (10.38) is obtained as the eigenmatrix of norm 1 of tensor $\mathcal{M}^*$ (i.e., the matrix $\boldsymbol{Q}$ obtained by cast from the unit eigenvector $\boldsymbol{q}$ of the matrix $\boldsymbol{M}$) for the smallest eigenvalue. However, the solution of the least-squares approximation is statistically biased whatever weights W_α are used.

Define the *moment tensor*

$$\mathcal{M} = \frac{1}{N}\sum_{\alpha=1}^{N} W_\alpha \boldsymbol{x}_\alpha \otimes \boldsymbol{x}_\alpha \otimes \boldsymbol{x}_\alpha \otimes \boldsymbol{x}_\alpha, \tag{10.42}$$

and let $\bar{\mathcal{M}}$ be the unperturbed moment tensor obtained by replacing $\boldsymbol{x}_\alpha$ by $\bar{\boldsymbol{x}}_\alpha$. Then, the true value $\bar{\boldsymbol{Q}}$ satisfies $\bar{\mathcal{M}}\bar{\boldsymbol{Q}} = \boldsymbol{O}$, i.e., $\bar{\boldsymbol{Q}}$ is the eigenmatrix of norm 1 of $\bar{\mathcal{M}}$ for eigenvalue 0. However, $E[\mathcal{M}]$ is not equal to $\bar{\mathcal{M}}$. In fact,

$$\begin{aligned}
E[M_{ijkl}] &= \frac{1}{N}\sum_{\alpha=1}^{N} W_\alpha E[x_{\alpha(i)}x_{\alpha(j)}x_{\alpha(k)}x_{\alpha(l)}] \\
&= \frac{1}{N}\sum_{\alpha=1}^{N} W_\alpha E[(\bar{x}_{\alpha(i)} + \Delta x_{\alpha(i)})(\bar{x}_{\alpha(j)} + \Delta x_{\alpha(j)}) \\
&\qquad\qquad (\bar{x}_{\alpha(k)} + \Delta x_{\alpha(k)})(\bar{x}_{\alpha(l)} + \Delta x_{\alpha(l)})] \\
&= \frac{1}{N}\sum_{\alpha=1}^{N} W_\alpha \Big(\bar{x}_{\alpha(i)}\bar{x}_{\alpha(j)}\bar{x}_{\alpha(k)}\bar{x}_{\alpha(l)} + \bar{x}_{\alpha(i)}\bar{x}_{\alpha(j)}E[\Delta x_{\alpha(k)}\Delta x_{\alpha(l)}] \\
&\quad + \bar{x}_{\alpha(i)}\bar{x}_{\alpha(k)}E[\Delta x_{\alpha(j)}\Delta x_{\alpha(l)}] + \bar{x}_{\alpha(i)}\bar{x}_{\alpha(l)}E[\Delta x_{\alpha(j)}\Delta x_{\alpha(k)}] \\
&\quad + \bar{x}_{\alpha(j)}\bar{x}_{\alpha(k)}E[\Delta x_{\alpha(i)}\Delta x_{\alpha(l)}] + \bar{x}_{\alpha(j)}\bar{x}_{\alpha(l)}E[\Delta x_{\alpha(i)}\Delta x_{\alpha(k)}] \\
&\quad + \bar{x}_{\alpha(k)}\bar{x}_{\alpha(l)}E[\Delta x_{\alpha(i)}\Delta x_{\alpha(j)}] + E[\Delta x_{\alpha(i)}\Delta x_{\alpha(j)}\Delta x_{\alpha(k)}\Delta x_{\alpha(l)}]\Big)
\end{aligned}$$

$$= \bar{M}_{ijkl} + \frac{\epsilon^2}{N}\sum_{\alpha=1}^{N} W_\alpha \Big(\bar{x}_{\alpha(i)}\bar{x}_{\alpha(j)} V_0[\boldsymbol{x}_\alpha]_{kl} + \bar{x}_{\alpha(i)}\bar{x}_{\alpha(k)} V_0[\boldsymbol{x}_\alpha]_{jl}$$
$$+ \bar{x}_{\alpha(i)}\bar{x}_{\alpha(l)} V_0[\boldsymbol{x}_\alpha]_{jk} + \bar{x}_{\alpha(j)}\bar{x}_{\alpha(k)} V_0[\boldsymbol{x}_\alpha]_{il} + \bar{x}_{\alpha(j)}\bar{x}_{\alpha(l)} V_0[\boldsymbol{x}_\alpha]_{ik}$$
$$+ \bar{x}_{\alpha(k)}\bar{x}_{\alpha(l)} V_0[\boldsymbol{x}_\alpha]_{ij} \Big) + \frac{\epsilon^4}{N}\sum_{\alpha=1}^{N} W_\alpha \Big(V_0[\boldsymbol{x}_\alpha]_{ij} V_0[\boldsymbol{x}_\alpha]_{kl}$$
$$+ V_0[\boldsymbol{x}_\alpha]_{ik} V_0[\boldsymbol{x}_\alpha]_{jl} + V_0[\boldsymbol{x}_\alpha]_{il} V_0[\boldsymbol{x}_\alpha]_{jk} \Big). \tag{10.43}$$

Define (33)(33)-tensors $\mathcal{N}^{(1)} = (N^{(1)}_{ijkl})$ and $\mathcal{N}^{(2)} = (N^{(2)}_{ijkl})$ by

$$N^{(1)}_{ijkl} = \frac{1}{N}\sum_{\alpha=1}^{N} W_\alpha \Big(V_0[\boldsymbol{x}_\alpha]_{ij} x_{\alpha(k)} x_{\alpha(l)} + V_0[\boldsymbol{x}_\alpha]_{ik} x_{\alpha(j)} x_{\alpha(l)}$$
$$+ V_0[\boldsymbol{x}_\alpha]_{il} x_{\alpha(j)} x_{\alpha(k)} + V_0[\boldsymbol{x}_\alpha]_{jk} x_{\alpha(i)} x_{\alpha(l)}$$
$$+ V_0[\boldsymbol{x}_\alpha]_{jl} x_{\alpha(i)} x_{\alpha(j)} + V_0[\boldsymbol{x}_\alpha]_{kl} x_{\alpha(i)} x_{\alpha(j)} \Big), \tag{10.44}$$

$$N^{(2)}_{ijkl} = \frac{1}{N}\sum_{\alpha=1}^{N} W_\alpha \Big(V_0[\boldsymbol{x}_\alpha]_{ij} V_0[\boldsymbol{x}_\alpha]_{kl} + V_0[\boldsymbol{x}_\alpha]_{ik} V_0[\boldsymbol{x}_\alpha]_{jl} + V_0[\boldsymbol{x}_\alpha]_{il} V_0[\boldsymbol{x}_\alpha]_{jk} \Big). \tag{10.45}$$

Let $\bar{\mathcal{N}}^{(1)}$ be the unperturbed value of $\mathcal{N}^{(1)}$ obtained by replacing $\boldsymbol{x}_\alpha$ by $\bar{\boldsymbol{x}}_\alpha$ in eq. (10.44). Then, eq. (10.43) can be written in the following form:

$$E[\mathcal{M}] = \bar{\mathcal{M}} + \epsilon^2 \bar{\mathcal{N}}^{(1)} + \epsilon^4 \mathcal{N}^{(2)}. \tag{10.46}$$

From eqs. (10.44) and (10.45), we immediately see that

$$E[\mathcal{N}^{(1)}] = \bar{\mathcal{N}}^{(1)} + 2\epsilon^2 \mathcal{N}^{(2)}. \tag{10.47}$$

It follows that if we define the *unbiased moment tensor*

$$\hat{\mathcal{M}} = \mathcal{M} - \epsilon^2 \mathcal{N}^{(1)} + \epsilon^4 \mathcal{N}^{(2)}, \tag{10.48}$$

we have $E[\hat{\mathcal{M}}] = \bar{\mathcal{M}}$. Hence, we obtain the unbiased least-squares approximation

$$\hat{J}[\boldsymbol{Q}] = (\boldsymbol{Q}; \hat{\mathcal{M}}\boldsymbol{Q}) \to \min. \tag{10.49}$$

The solution under the constraint $\|\boldsymbol{Q}\| = 1$ is obtained as the eigenmatrix of $\hat{\mathcal{M}}$ of norm 1 for the smallest eigenvalue.

If the noise level ϵ is not known, the second order renormalization procedure is given as follows (see Section 9.6.2):

1. Let $c = 0$ and $W_\alpha = 1$, $\alpha = 1, \ldots, N$.
2. Compute the (33)(33)-tensors $\mathcal{M}$, $\mathcal{N}^{(1)}$, and $\mathcal{N}^{(2)}$ defined by eqs. (10.42), (10.44), and (10.45), respectively.

3. Compute the smallest eigenvalue λ of the (33)(33)-tensor

$$\hat{\mathcal{M}} = \mathcal{M} - c\mathcal{N}^{(1)} + c^2\mathcal{N}^{(2)}, \tag{10.50}$$

and the corresponding eigenmatrix $\boldsymbol{Q}$ of norm 1.

4. If $\lambda \approx 0$, return $\boldsymbol{Q}$, c, and $\hat{\mathcal{M}}$. Else, update c and W_α as follows:

$$D = \left((\boldsymbol{Q}; \mathcal{N}^{(1)}\boldsymbol{Q}) - 2c(\boldsymbol{Q}; \mathcal{N}^{(2)}\boldsymbol{Q})\right)^2 - 4\lambda(\boldsymbol{Q}; \mathcal{N}^{(2)}\boldsymbol{Q}), \tag{10.51}$$

$$c \leftarrow \begin{cases} c + \dfrac{(\boldsymbol{Q}; \mathcal{N}^{(1)}\boldsymbol{Q}) - 2c(\boldsymbol{Q}; \mathcal{N}^{(2)}\boldsymbol{Q}) - \sqrt{D}}{2(\boldsymbol{Q}; \mathcal{N}^{(2)}\boldsymbol{Q})}, & \text{if } D \geq 0, \\ c + \dfrac{\lambda}{(\boldsymbol{Q}; \mathcal{N}^{(1)}\boldsymbol{Q})}, & \text{if } D < 0, \end{cases} \tag{10.52}$$

$$W_\alpha \leftarrow \frac{1}{4(\boldsymbol{x}_\alpha, \boldsymbol{Q}V_0[\boldsymbol{x}_\alpha]\boldsymbol{Q}\boldsymbol{x}_\alpha) + 2c(V_0[\boldsymbol{x}_\alpha]\boldsymbol{Q}; \boldsymbol{Q}V_0[\boldsymbol{x}_\alpha])}. \tag{10.53}$$

5. Go back to Step 2.

After renormalization, the squared noise level ϵ^2 is estimated in the form

$$\hat{\epsilon}^2 = \frac{c}{1 - 5/N}. \tag{10.54}$$

Its expectation and variance are

$$E[\hat{\epsilon}^2] = \epsilon^2, \qquad V[\hat{\epsilon}^2] = \frac{2\epsilon^4}{N-5}. \tag{10.55}$$

If the value ϵ is given a priori, the conic test takes the following form (see eq. (10.34)):

$$\frac{\hat{\epsilon}^2}{\epsilon^2} > \frac{\chi^2_{N-5,a}}{N-5}. \tag{10.56}$$

The covariance tensor of the resulting estimate $\hat{\boldsymbol{Q}}$ is obtained in the following form:

$$\mathcal{V}[\hat{\boldsymbol{Q}}] = \frac{\hat{\epsilon}^2}{N}(\hat{\mathcal{M}})^-_5. \tag{10.57}$$

Let $\mu_{\max}$ be the largest eigenvalue of $\mathcal{V}[\hat{\boldsymbol{Q}}]$, and $\boldsymbol{U}_{\max}$ the corresponding eigenmatrix of norm 1. The primary deviation pair $\{\boldsymbol{Q}^+, \boldsymbol{Q}^-\}$ is given as follows (see Section 4.5.3):

$$\boldsymbol{Q}^+ = N[\hat{\boldsymbol{Q}} + \sqrt{\mu_{\max}}\boldsymbol{U}_{\max}], \qquad \boldsymbol{Q}^- = N[\hat{\boldsymbol{Q}} - \sqrt{\mu_{\max}}\boldsymbol{U}_{\max}]. \tag{10.58}$$

Example 10.3 Fig. 10.4a shows sixty equidistant image points on the ellipse

$$\frac{x^2}{50^2} + \frac{y^2}{100^2} = 1 \tag{10.59}$$

in the first quadrant in a simulated image of 512×512 pixels (lengths are measured in pixels). The interval between consecutive points is approximately 2 pixels. We added Gaussian random noise of mean 0 and standard deviation $\sigma = 0.5$ (pixel) to each of the coordinates of these points independently. Fig. 10.4b shows the conic fitted by renormalization (solid line) and its primary deviation pair (dashed lines).

Fig. 10.5a shows ten fits computed by the optimal least-squares approximation (the weights are computed from the true values), each time using different noise. The true conic is drawn in a dashed line. The existence of statistical bias is evident. Fig. 10.5b shows corresponding optimal fits computed by renormalization. We see that the statistical bias is removed.

Example 10.4 Fig. 10.6a is an edge image obtained by applying an edge operator to a portion of a real image. A conic is optimally fitted to the longest edge segment that constitutes a part of a conic by assuming that the distribution of the image noise is isotropic and identical for each pixel; the noise level is unknown. Fig. 10.6b shows the fitted conic (solid line) and its the primary deviation pair (dashed lines) superimposed on the original gray-level image. Fig. 10.7 shows similarly obtained images. We can see that the reliability of the fit rapidly decreases as the length of the conic edge segment decreases.

10.3 Space Plane Fitting by Range Sensing

10.3.1 Optimal space plane fitting

Suppose we are observing space points which are known to be on a planar surface. Let $\{\boldsymbol{r}_\alpha\}$, $\alpha = 1, ..., N$, be their locations observed in the presence of noise, and consider the problem of fitting a space plane to them. Let $\{\bar{\boldsymbol{r}}_\alpha\}$ be the true positions of $\{\boldsymbol{r}_\alpha\}$. We write

$$\boldsymbol{r}_\alpha = \bar{\boldsymbol{r}}_\alpha + \Delta\boldsymbol{r}_\alpha, \tag{10.60}$$

and regard the noise $\Delta\boldsymbol{r}_\alpha$ as a Gaussian random variable of mean $\boldsymbol{0}$ and covariance matrix $V[\boldsymbol{r}_\alpha]$. Let $\boldsymbol{\rho}_\alpha$ be the $\boldsymbol{\rho}$-representation of $\boldsymbol{r}_\alpha$, and $\bar{\boldsymbol{\rho}}_\alpha$ its true value (see eq. (4.23)). The covariance matrix of $\boldsymbol{\rho}_\alpha$ has the form

$$V[\boldsymbol{\rho}_\alpha] = V[\boldsymbol{r}_\alpha] \oplus 0 = \begin{pmatrix} V[\boldsymbol{r}_\alpha] & \\ & 0 \end{pmatrix}. \tag{10.61}$$

Let $(\boldsymbol{\nu}, \boldsymbol{\rho}) = 0$ be the $\boldsymbol{\nu}$-representation of the space plane to be fitted (see eq. (4.61)). The problem is stated as follows.

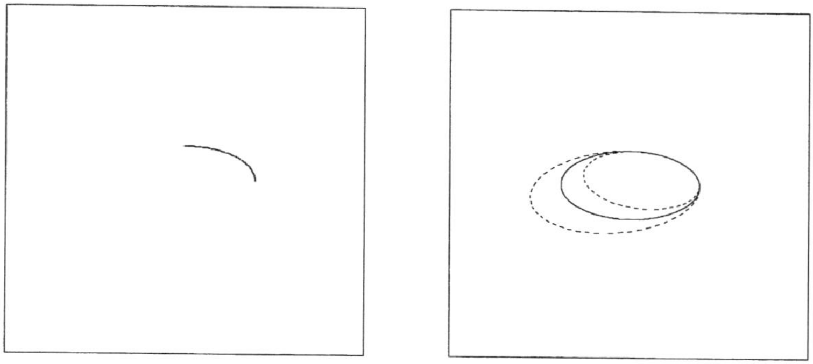

(a) (b)

Fig. 10.4. (a) Equidistant image points on an ellipse in the first quadrant. (b) Optimal fit (solid line) and its primary deviation pair (dashed lines).

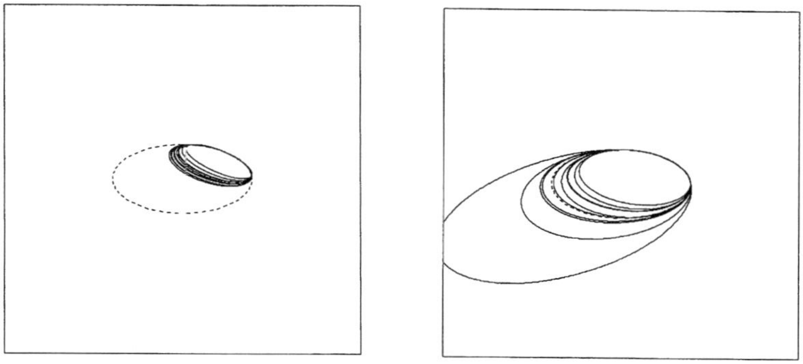

(a) (b)

Fig. 10.5. (a) Ten fits obtained by the least-squares approximation. (b) Corresponding fits obtained by renormalization.

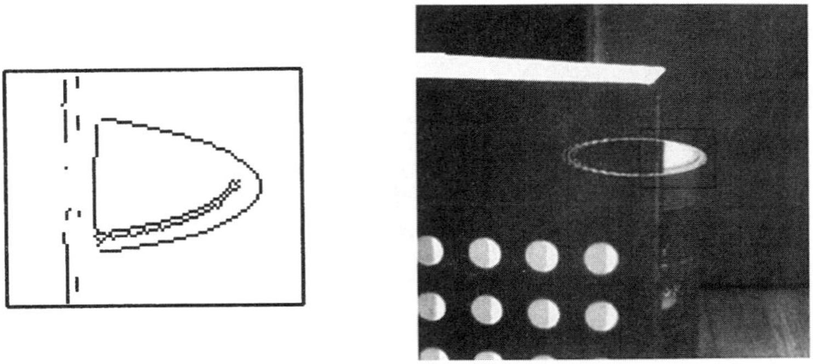

(a) (b)

Fig. 10.6. (a) Edge image. (b) Optimally fitted conic (solid line) and its primary deviation pair (dashed lines).

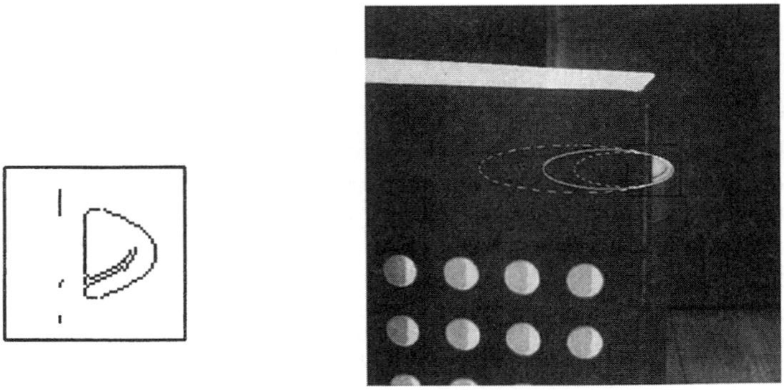

(a) (b)

Fig. 10.7. (a) Edge image. (b) Optimally fitted conic (solid line) and its primary deviation pair (dashed lines).

Problem 10.3 *Estimate a unit 4-vector* $\boldsymbol{\nu}$ *such that*

$$(\boldsymbol{\nu}, \bar{\boldsymbol{\rho}}_\alpha) = 0, \qquad \alpha = 1, ..., N, \tag{10.62}$$

from the data $\{\boldsymbol{\rho}_\alpha\}$, $\alpha = 1, ..., N$.

Eq. (10.62) is the hypothesis from which the 4-vector $\boldsymbol{\nu}$ is to be estimated: its rank is 1. As shown in Section 7.4.3, an optimal estimate of $\boldsymbol{\nu}$ can be obtained by the optimization

$$J[\boldsymbol{\nu}] = \sum_{\alpha=1}^{N} \frac{(\boldsymbol{\nu}, \boldsymbol{\rho}_\alpha)^2}{(\boldsymbol{\nu}, V[\boldsymbol{\rho}_\alpha]\boldsymbol{\nu})} \to \min \tag{10.63}$$

under the constraint $\|\boldsymbol{\nu}\| = 1$. The theoretical bound on the accuracy of fitting is given by the covariance matrix of the optimal estimate $\hat{\boldsymbol{\nu}}$ in the form

$$\bar{V}[\hat{\boldsymbol{\nu}}] = \left(\sum_{\alpha=1}^{N} \frac{\bar{\boldsymbol{\rho}}_\alpha \bar{\boldsymbol{\rho}}_\alpha^\top}{(\boldsymbol{\nu}, V[\boldsymbol{x}_\alpha]\boldsymbol{\nu})} \right)^- . \tag{10.64}$$

10.3.2 Noise model of range sensing

Range sensing means measuring the distances to objects in the scene from a fixed position without approaching them, and a device for this purpose is called a *range finder*; the data obtained by a range finder are called *range data*. Many types of range sensing are possible. A typical method is to emit a sound or radio wave and observe the phase shift between the emitted and reflected waves. In general, the error behavior of range data depends on not only the mechanical characteristics of the range finder and the accuracy of the electronic data processing involved but also the shape and position of the object to be measured. Hence, it is very difficult to give a precise error model. However, the accuracy of range data generally decreases as the distance to the object increases. Here, we assume that to a first approximation the error is proportional to the distance to the object. We also assume that the orientation in which the distance is measured can be controlled accurately.

According to this model, the error $\Delta\boldsymbol{r}_\alpha$ in eq. (10.60) occurs in the direction of $\bar{\boldsymbol{r}}$, and the covariance matrix $V[\boldsymbol{r}_\alpha]$ is modeled in the form

$$V[\boldsymbol{r}] = \epsilon^2 \bar{\boldsymbol{r}}\bar{\boldsymbol{r}}^\top, \tag{10.65}$$

where ϵ is a constant. We call ϵ the *noise level*; the standard deviation of $\|\Delta\boldsymbol{r}_\alpha\|$ is $\epsilon\|\bar{\boldsymbol{r}}_\alpha\|$. Since it is very difficult to predict the accuracy of range sensing a priori, we treat the noise level ϵ as unknown in the subsequent computation.

The covariance matrix $V[\rho_\alpha]$ is now decomposed into the noise level ϵ and the normalized covariance matrix $V_0[\rho_\alpha]$ in the form

$$V[\rho_\alpha] = \epsilon^2 V_0[\rho_\alpha], \qquad V_0[\rho_\alpha] = \begin{pmatrix} \bar{r}_\alpha \bar{r}_\alpha^\top & \\ & 0 \end{pmatrix}. \tag{10.66}$$

Let $(n, r) = d$ be the $\{n, d\}$-representation of the space plane to be fitted (see eq. (4.60)). Noting that $(n, \bar{r}_\alpha) = d$, $\alpha = 1, ..., N$, we see from eqs. (4.23) and (4.62) that

$$(\nu, V_0[\rho_\alpha]\nu) = \frac{(\bar{r}_\alpha, n)^2}{1+d^2} = \frac{d^2}{1+d^2}, \tag{10.67}$$

$$(\nu, \rho_\alpha) = \frac{\big((n, r_\alpha) - d\big)^2}{1+d^2}. \tag{10.68}$$

Substituting these into eq. (10.63) and noting that multiplication of $J[\nu]$ by a positive constant does not affect the solution, we can equivalently rephrase the optimization as finding a unit vector n and a scalar d such that

$$J[n, d] = \frac{1}{d^2}\sum_{\alpha=1}^{N}\Big((n, r_\alpha) - d\Big)^2 \to \min. \tag{10.69}$$

If the factor $1/d^2$ is ignored, this reduces to the usual east-squares optimization, minimizing the sum of the squared perpendicular distances from the data positions to the space plane (see eqs. (4.68) and (7.155)). The least-squares optimization is optimal if the noise distribution is isotropic and identical for each datum; otherwise, the solution is biased. In fact, eq. (10.69) implies that the optimal solution should have a larger value of d than the least-squares solution.

The solution of the minimization (10.69) can be obtained analytically. In fact, if we put $\tilde{n} = n/d$, the function to be minimized can be written in the following form:

$$J[\tilde{n}] = \sum_{\alpha=1}^{N}\Big((\tilde{n}, r_\alpha) - 1\Big)^2. \tag{10.70}$$

Since this is a quadratic polynomial in $\tilde{n}$, the solution is obtained by solving the following linear equation:

$$\left(\sum_{\alpha=1}^{N} r_\alpha r_\alpha^\top\right)\tilde{n} = \sum_{\alpha=1}^{N} r_\alpha. \tag{10.71}$$

The optimal estimate of the surface parameters $\{n, d\}$ is given by $\hat{n} = N[\tilde{n}]$ and $\hat{d} = 1/\|\tilde{n}\|$.

Although the solution itself can be immediately obtained, we derive the renormalization procedure in the following, because it can automatically estimate the noise level ϵ and give the covariance matrix of the obtained fit.

10.3.3 Unbiased estimation and renormalization

If the denominator in eq. (10.63) is replaced by a constant, we obtain the least-squares approximation

$$\tilde{J}[\boldsymbol{\nu}] = (\boldsymbol{\nu}, \boldsymbol{M}\boldsymbol{\nu}) \to \min, \tag{10.72}$$

where the *moment matrix* $\boldsymbol{M}$ is defined by

$$\boldsymbol{M} = \frac{1}{N}\sum_{\alpha=1}^{N} W_\alpha \boldsymbol{\rho}_\alpha \boldsymbol{\rho}_\alpha^\top, \tag{10.73}$$

$$W_\alpha = \frac{1}{(\boldsymbol{\nu}^*, V_0[\boldsymbol{\rho}_\alpha]\boldsymbol{\nu}^*)}. \tag{10.74}$$

Here, $\boldsymbol{\nu}^*$ is an appropriate estimate of $\boldsymbol{\nu}$. The solution $\tilde{\boldsymbol{\nu}}$ of the optimization (10.72) is obtained as the unit eigenvector of the moment matrix $\boldsymbol{M}$ for the smallest eigenvalue. As in the case of line fitting, the solution of the least-squares approximation is in general statistically biased whatever weights W_α are used. On the other hand, if we define the *unbiased moment matrix*

$$\hat{\boldsymbol{M}} = \boldsymbol{M} - \epsilon^2 \boldsymbol{N}, \tag{10.75}$$

$$\boldsymbol{N} = \frac{1}{N}\sum_{\alpha=1}^{N} W_\alpha V_0[\boldsymbol{\rho}_\alpha], \tag{10.76}$$

we have $E[\hat{\boldsymbol{M}}] = \bar{\boldsymbol{M}}$. Hence, we obtain the unbiased least-squares approximation

$$\hat{J}[\boldsymbol{\nu}] = (\boldsymbol{\nu}, \hat{\boldsymbol{M}}\boldsymbol{\nu}) \to \min, \tag{10.77}$$

and the solution $\hat{\boldsymbol{\nu}}$ is obtained as the unit eigenvector of $\hat{\boldsymbol{M}}$ for the smallest eigenvalue.

If the noise level ϵ is not known, we apply renormalization. In this problem, however, the normalized covariance matrix $V_0[\boldsymbol{\rho}_\alpha]$ is also unknown, since it involves the true values $\bar{\boldsymbol{r}}_\alpha$ (see eqs. (10.66)). If $\bar{\boldsymbol{r}}_\alpha$ is approximated by the data value $\boldsymbol{r}_\alpha$, the error magnitude is underestimated at those points which are detected at shorter distances than their true positions and overestimated at those points detected at longer distances. Here, we approximate $\bar{\boldsymbol{r}}_\alpha$ by the projection of $\boldsymbol{r}_\alpha$ onto the fitted space plane in the iterations. Since the projection of $\boldsymbol{r}_\alpha$ onto space plane $(\boldsymbol{n}, \boldsymbol{r}) = d$ is

$$\hat{\boldsymbol{r}}_\alpha = \frac{\hat{d}\boldsymbol{r}_\alpha}{(\hat{\boldsymbol{n}}, \boldsymbol{r}_\alpha)}, \tag{10.78}$$

the renormalization procedure is modified as follows (see eqs. (6.63) and (6.64)):

1. Compute the (44)-matrices

$$V_0^{(0)}[\rho_\alpha] = \begin{pmatrix} r_\alpha r_\alpha^\top & \\ & 0 \end{pmatrix}, \qquad \alpha = 1, ..., N, \tag{10.79}$$

   and let $c = 0$, $V_0[\rho_\alpha] = V_0^{(0)}[\rho_\alpha]$, and $W_\alpha = 1$, $\alpha = 1$, ..., N.

2. Compute the (44)-matrices M and N defined by eqs. (10.73) and (10.76), respectively.

3. Compute the smallest eigenvalue λ of the (44)-matrix

$$\hat{M} = M - cN, \tag{10.80}$$

   and the corresponding unit eigenvector ν.

4. If $\lambda \approx 0$, return ν, c, and $\hat{M}$. Else, update c, $V_0[\rho_\alpha]$, and W_α as follows, where $\kappa = (0,0,0,1)^\top$:

$$c \leftarrow c + \frac{\lambda}{(\nu, N\nu)}, \qquad V_0[\rho_\alpha] \leftarrow \frac{(\nu,\kappa)^2}{(\nu, \rho_\alpha - \kappa)^2} V_0^{(0)}[\rho_\alpha],$$

$$W_\alpha \leftarrow \frac{1}{(\nu, V_0[\rho_\alpha]\nu)}. \tag{10.81}$$

5. Go back to Step 2.

After renormalization, the squared noise level ϵ^2 is estimated in the form

$$\hat{\epsilon}^2 = \frac{c}{1 - 3/N}. \tag{10.82}$$

Its expectation and variance are

$$E[\hat{\epsilon}^2] = \epsilon^2, \qquad V[\hat{\epsilon}^2] = \frac{2\epsilon^4}{N-3}. \tag{10.83}$$

If the value ϵ is given a priori, the *coplanarity test* for space points takes the following form (see eq. (7.154)):

$$\frac{\hat{\epsilon}^2}{\epsilon^2} > \frac{\chi^2_{N-3,a}}{N-3}. \tag{10.84}$$

The covariance matrix of the resulting estimate $\hat{\nu}$ is obtained in the following form:

$$V[\hat{\nu}] = \frac{\hat{\epsilon}^2}{N}(\hat{M})_3^-. \tag{10.85}$$

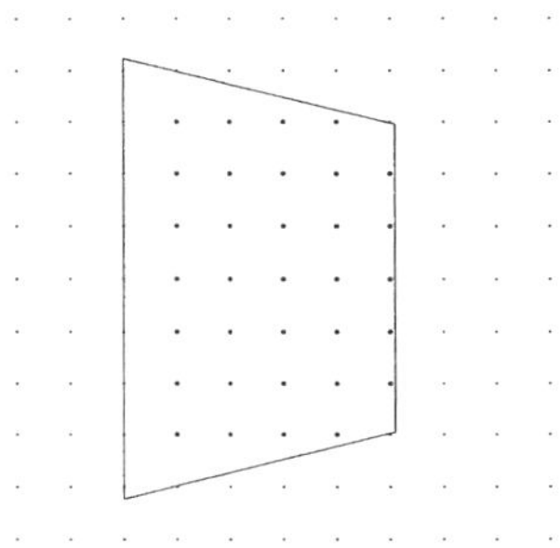

Fig. 10.8. Planar surface patch in the scene.

Let $\mu_{\max}$ be the largest eigenvalue of $V[\hat{\boldsymbol{\nu}}]$, and $\boldsymbol{\xi}_{\max}$ the corresponding unit eigenvector. The primary deviation pair $\{\boldsymbol{\nu}^{+}, \boldsymbol{\nu}^{-}\}$ is given as follows (see Section 4.5.3):

$$\boldsymbol{\nu}^{+} = N[\hat{\boldsymbol{\nu}} + \sqrt{\mu_{\max}}\boldsymbol{\xi}_{\max}], \qquad \boldsymbol{\nu}^{-} = N[\hat{\boldsymbol{\nu}} - \sqrt{\mu_{\max}}\boldsymbol{\xi}_{\max}]. \tag{10.86}$$

As we pointed out in Section 4.2.1, we must be careful about the scaling of the coordinates when we use the $\boldsymbol{\rho}$-representation $(X, Y, Z, 1)^{\top}$. Namely, we must use an appropriate unit of length for space coordinates so that X, Y, and Z are not too large or too small as compared with 1. A convenient scaling is to regard the average $r_0 = \sum_{\alpha=1}^{N} \|\boldsymbol{r}_{\alpha}\|/N$ as unit length. Once the $\{\boldsymbol{\nu}\}$-representation of the fitted space plane is obtained, its $\{\boldsymbol{n}, d\}$-representation can be obtained by computing eq. (4.63). The covariance matrix $V[\hat{\boldsymbol{n}}]$, the correlation vector $V[\hat{\boldsymbol{n}}, d]$, and the variance $V[\hat{d}]$ are computed from $V[\hat{\boldsymbol{\nu}}]$ by eqs. (4.67).

Example 10.5 Fig. 10.8 shows a planar surface patch placed in the scene and viewed from the coordinate origin, at which we assume a range finder is fixed. The dots in the figure indicate the orientations in which the distance is measured. We assume that the rays emitted outside the patch return the value ∞ and are ignored in the fitting computation. We simulated the distance measurement by adding Gaussian random noise to the exact distance to the surface in each orientation independently according to the statistical model given by eq. (10.65) for $\epsilon = 0.1$. Then, a space plane was fitted to them.

Let $\{\bar{\boldsymbol{n}}, \bar{d}\}$ and $\{\hat{\boldsymbol{n}}, \hat{d}\}$ be the true and the computed surface parameters. Since the deviation of $\hat{\boldsymbol{n}}$ from $\bar{\boldsymbol{n}}$ is orthogonal to $\bar{\boldsymbol{n}}$ to a first approximation, the error in $\hat{\boldsymbol{n}}$ can be represented by a 3-vector

$$\Delta \boldsymbol{u} = \boldsymbol{P}_{\bar{\boldsymbol{n}}}(\hat{\boldsymbol{n}} - \bar{\boldsymbol{n}}) + \frac{\hat{d} - \bar{d}}{\bar{d}}\bar{\boldsymbol{n}}, \tag{10.87}$$

where $\boldsymbol{P}_{\bar{\boldsymbol{n}}}$ is the projection matrix along $\bar{\boldsymbol{n}}$. The covariance matrix of this

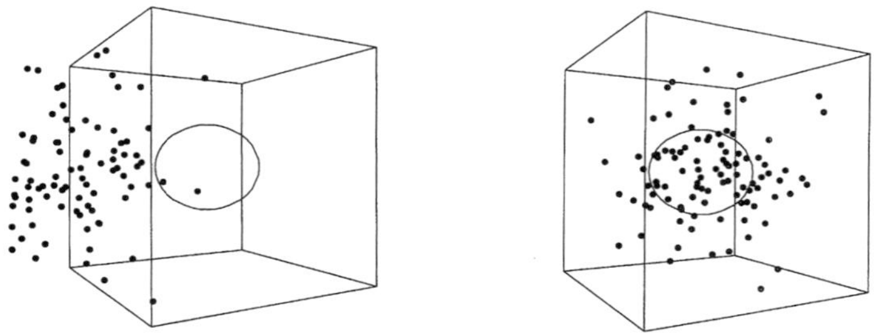

(a) (b)

Fig. 10.9. (a) Error distribution for the least-squares solution. (b) Error distribution for renormalization.

vector is given by

$$V[\Delta \boldsymbol{u}] = V[\hat{\boldsymbol{n}}] + \frac{2}{\bar{d}} S[V[\hat{\boldsymbol{n}}, \hat{d}]\bar{\boldsymbol{n}}^\top] + \frac{V[\hat{d}]}{\bar{d}^2}\bar{\boldsymbol{n}}\bar{\boldsymbol{n}}^\top. \tag{10.88}$$

In Figs. 10.9a and 10.9b, $\Delta \boldsymbol{u}$ is plotted in three dimensions for 100 trials, each time using different noise: Fig. 10.9a is for the least-squares solution given by eq. (10.69); Fig. 10.9b is for renormalization. In each figure, the ellipse indicates the standard confidence region computed from eq. (10.88) by using the theoretical expression (10.64) (see Section 4.5.3); the cubes are drawn merely for the sake of reference. We can see that statistical bias exists in the least-squares solution and the bias is removed by renormalization. We can also see from Fig. 10.9b that the theoretical bound on accuracy is almost attained. Fig. 10.10a shows a grid reconstructed by eq. (10.78) for a typical surface fit obtained by renormalization and viewed from an angle (the true position is superimposed in dashed lines); Fig. 10.10b shows its primary deviation pair.

10.4 Space Plane Fitting by Stereo Vision

10.4.1 Optimal space plane fitting

Suppose multiple feature points in the scene are observed by stereo vision, and suppose they are known to be on a planar surface. Although the 3-D structure is uniquely reconstructed by stereo vision alone, we can expect that the reliability of 3-D reconstruction can be enhanced if the knowledge that the feature points are coplanar is incorporated. This problem has practical significance, since many objects in an indoor robotic workspace, such as walls, ceilings, and floors, have planar surfaces.

An indirect but simple method is first reconstructing 3-D by stereo vision alone and computing the covariance matrices of the feature points as

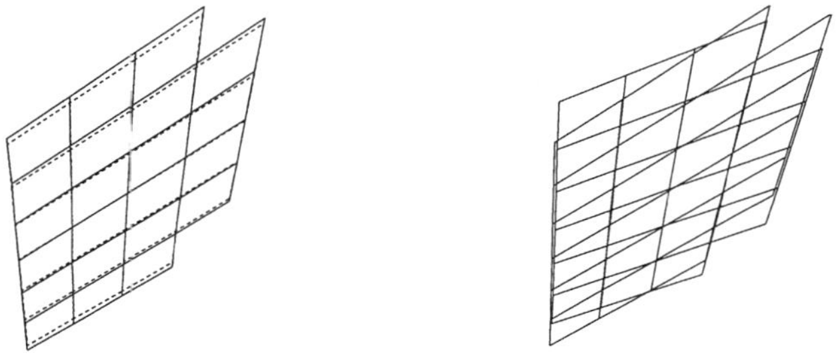

(a) (b)

Fig. 10.10. (a) An example of a fit computed by renormalization. (b) Its primary deviation pair.

described in Section 6.3. Then, a space plane is optimally fitted, as described in Section 7.4.3. Here, we consider a direct method: a space plane is optimally reconstructed directly from stereo correspondence pairs $\{\boldsymbol{x}_\alpha, \boldsymbol{x}'_\alpha\}$, $\alpha = 1, ..., N$.

As shown in Section 6.5.1, the necessary and sufficient condition for $\bar{\boldsymbol{x}}_\alpha$ and $\bar{\boldsymbol{x}}'_\alpha$ to be projections of a space point on space plane $(\boldsymbol{n}, \boldsymbol{r}) = d$ viewed by a stereo system with motion parameters $\{\boldsymbol{h}, \boldsymbol{R}\}$ is given as follows (see eqs. (6.67) and (6.71); the scale difference does not affect the result):

$$\bar{\boldsymbol{x}}'_\alpha \times \boldsymbol{A}\bar{\boldsymbol{x}}_\alpha = \boldsymbol{0}, \qquad \boldsymbol{A} = \frac{\boldsymbol{R}^\top(\boldsymbol{h}\boldsymbol{n}^\top - d\boldsymbol{I})}{\sqrt{1+d^2}}. \tag{10.89}$$

This is the hypothesis from which the surface parameters $\{\boldsymbol{n}, d\}$ are to be estimated. Define a 34-matrix $\boldsymbol{B}_\alpha$ and a 4-vector $\boldsymbol{\nu}$ by

$$\boldsymbol{B}_\alpha = \left(\boldsymbol{x}'_\alpha \times \boldsymbol{R}^\top \boldsymbol{h}\boldsymbol{x}_\alpha^\top \quad \boldsymbol{x}'_\alpha \times \boldsymbol{R}^\top \boldsymbol{x}_\alpha \right),$$

$$\boldsymbol{\nu} = \frac{1}{\sqrt{1+d^2}} \begin{pmatrix} \boldsymbol{n} \\ -d \end{pmatrix}. \tag{10.90}$$

Let $(\boldsymbol{\nu}, \boldsymbol{\rho}) = 0$ be the $\boldsymbol{\nu}$-representation of the space plane $(\boldsymbol{n}, \boldsymbol{r}) = d$. Let $\bar{\boldsymbol{B}}_\alpha$ be the unperturbed value of $\boldsymbol{B}_\alpha$ obtained by replacing $\boldsymbol{x}_\alpha$ and $\boldsymbol{x}'_\alpha$ by $\bar{\boldsymbol{x}}_\alpha$ and $\bar{\boldsymbol{x}}'_\alpha$, respectively, in the first of eqs. (10.90). Then, the first of eqs. (10.89) can be rewritten as $\bar{\boldsymbol{B}}_\alpha \boldsymbol{\nu} = \boldsymbol{0}$. Hence, the problem is stated as follows:

Problem 10.4 *Estimate a unit 4-vector* $\boldsymbol{\nu}$ *such that*

$$\bar{\boldsymbol{B}}_\alpha \boldsymbol{\nu} = \boldsymbol{0}, \qquad \alpha = 1, ..., N, \tag{10.91}$$

from the data $\{\boldsymbol{B}_\alpha\}$, $\alpha = 1, ..., N$.

If we let the kth row of the matrix $\bar{\boldsymbol{B}}_\alpha$ be $\bar{\boldsymbol{a}}_\alpha^{(k)\top}$, eq. (10.91) can be written as $(\bar{\boldsymbol{a}}_\alpha^{(k)}, \boldsymbol{\nu}) = 0$, $k = 1, 2, 3$, which has the same form as eq. (7.46). Hence, the general theory described in Section 7.1.5 can be applied.

Since the true positions $\bar{\boldsymbol{x}}_\alpha$ and $\bar{\boldsymbol{x}}'_\alpha$ satisfy the *epipolar equation* $|\bar{\boldsymbol{x}}_\alpha, \boldsymbol{h}, \boldsymbol{R}\bar{\boldsymbol{x}}'_\alpha| = 0$ (see eq. (6.4)), vectors $\bar{\boldsymbol{x}}_\alpha$, $\boldsymbol{h}$, and $\boldsymbol{R}\bar{\boldsymbol{x}}'_\alpha$ are coplanar. It follows that $\boldsymbol{R}^\top\bar{\boldsymbol{x}}_\alpha$, $\boldsymbol{R}^\top\boldsymbol{h}$, and $\bar{\boldsymbol{x}}'_\alpha$ are also coplanar. Hence, vectors $\bar{\boldsymbol{x}}'_\alpha \times \boldsymbol{R}^\top\boldsymbol{h}$ and $\bar{\boldsymbol{x}}'_\alpha \times \boldsymbol{R}^\top\bar{\boldsymbol{x}}_\alpha$ are collinear, meaning that the matrix $\bar{\boldsymbol{B}}_\alpha$ has rank 1. Consequently, only one of the three equations $(\bar{\boldsymbol{a}}_\alpha^{(k)}, \boldsymbol{\nu}) = 0$, $k = 1, 2, 3$, is linearly independent, so the rank of the hypothesis (10.91) appears to be 1. However, the epipolar equation is also an implicit hypothesis, and from the general theory given in Section 5.1.1 it can be shown that the rank[2] of the hypothesis (10.91) is 2.

The optimization (7.55) for this problem can be rewritten in the form

$$J[\boldsymbol{\nu}] = \sum_{\alpha=1}^{N}(\boldsymbol{B}_\alpha\boldsymbol{\nu}, \boldsymbol{W}_\alpha(\boldsymbol{\nu})\boldsymbol{B}_\alpha\boldsymbol{\nu}) \to \min, \tag{10.92}$$

where $\boldsymbol{W}_\alpha(\boldsymbol{\nu})$ is the (33)-matrix given by eq. (7.54). In the present problem, it can be rewritten in the following form:

$$\boldsymbol{W}_\alpha(\boldsymbol{\nu}) = \left(V[\boldsymbol{B}_\alpha\boldsymbol{\nu}]\right)_2^- . \tag{10.93}$$

Since $\boldsymbol{B}_\alpha\boldsymbol{\nu} = \boldsymbol{x}'_\alpha \times \boldsymbol{A}\boldsymbol{x}_\alpha$, we have

$$\Delta\boldsymbol{B}_\alpha\boldsymbol{\nu} = \Delta\boldsymbol{x}'_\alpha \times \boldsymbol{A}\bar{\boldsymbol{x}}_\alpha + \bar{\boldsymbol{x}}'_\alpha \times \boldsymbol{A}\Delta\boldsymbol{x}_\alpha + \Delta\boldsymbol{x}'_\alpha \times \boldsymbol{A}\Delta\boldsymbol{x}_\alpha. \tag{10.94}$$

Hence, the covariance matrix $V[\boldsymbol{B}_\alpha\boldsymbol{\nu}] = E[(\Delta\boldsymbol{B}_\alpha\boldsymbol{\nu})(\Delta\boldsymbol{B}_\alpha\boldsymbol{\nu})^\top]$ has the form

$$V[\boldsymbol{B}_\alpha\boldsymbol{\nu}] = \boldsymbol{x}'_\alpha\times\boldsymbol{A}V[\boldsymbol{x}_\alpha]\boldsymbol{A}^\top\times\boldsymbol{x}'_\alpha + (\boldsymbol{A}\boldsymbol{x}_\alpha)\times V[\boldsymbol{x}'_\alpha]\times(\boldsymbol{A}\boldsymbol{x}'_\alpha) + [V[\boldsymbol{x}'_\alpha]\times\boldsymbol{A}V[\boldsymbol{x}_\alpha]\boldsymbol{A}^\top], \tag{10.95}$$

where the symbol $[\cdot \times \cdot]$ denotes the exterior product of matrices (see eq. (2.45)).

The theoretical lower bound on the covariance matrix of the optimal estimate $\hat{\boldsymbol{\nu}}$ is given in the following form (see eq. (7.57)):

$$\bar{V}[\hat{\boldsymbol{\nu}}] = \left(\sum_{\alpha=1}^{N}\boldsymbol{P}_{\boldsymbol{\nu}}\boldsymbol{B}_\alpha^\top\boldsymbol{W}_\alpha(\boldsymbol{\nu})\boldsymbol{B}_\alpha\boldsymbol{P}_{\boldsymbol{\nu}}\right)^- . \tag{10.96}$$

Here, $\boldsymbol{P}_{\boldsymbol{\nu}}$ is the four-dimensional projection matrix along $\boldsymbol{\nu}$. The rank of $\bar{V}[\hat{\boldsymbol{\nu}}]$ is 3; its null space is $\{\boldsymbol{\nu}\}_L$. Since a space plane has three degrees of freedom, the residual

$$J[\hat{\boldsymbol{\nu}}] = \sum_{\alpha=1}^{N}(\boldsymbol{B}_\alpha\hat{\boldsymbol{\nu}}, \boldsymbol{W}_\alpha(\hat{\boldsymbol{\nu}})\boldsymbol{B}_\alpha\hat{\boldsymbol{\nu}}) \tag{10.97}$$

---

[2]This hypothesis is degenerate; see Section 7.1.1. We will give a rigorous mathematical argument about the rank of a hypothesis is Chapter 14.

is a χ^2 variable with $2N-3$ degrees of freedom. Evidently, the number N of the pairs of corresponding image points must be such that

$$N \geq 2. \tag{10.98}$$

The residual is *not* 0 when $N = 2$. For $N \geq 3$, we obtain the *coplanarity test* for feature points: the hypothesis that the observed feature points are projections of coplanar space points is rejected with significance level $a\%$ if

$$J[\hat{\boldsymbol{\nu}}] > \chi^2_{2N-3,a}. \tag{10.99}$$

Once the solution $\hat{\boldsymbol{\nu}}$ is obtained, all the feature points are optimally back projected onto the space plane $(\hat{\boldsymbol{\nu}}, \boldsymbol{\rho}) = 0$ by the procedure described in Section 6.5.1.

10.4.2 Unbiased estimation and renormalization

If the covariance matrices $V[\boldsymbol{x}_\alpha]$ and $V[\boldsymbol{x}'_\alpha]$ are decomposed into the noise level ϵ and the normalized covariance matrices $V_0[\boldsymbol{x}_\alpha]$ and $V_0[\boldsymbol{x}'_\alpha]$ in the form

$$V[\boldsymbol{x}_\alpha] = \epsilon^2 V_0[\boldsymbol{x}_\alpha], \qquad V[\boldsymbol{x}'_\alpha] = \epsilon^2 V_0[\boldsymbol{x}'_\alpha], \tag{10.100}$$

the least-squares approximation to the optimization (10.92) can be written in the form

$$\tilde{J}[\boldsymbol{\nu}] = (\boldsymbol{\nu}, \boldsymbol{M}\boldsymbol{\nu}) \to \min, \tag{10.101}$$

where $\boldsymbol{M}$ is the (44)-matrix defined by

$$\boldsymbol{M} = \begin{pmatrix} \frac{1}{N}\sum_{\alpha=1}^{N}(\boldsymbol{h}, \boldsymbol{X}_\alpha \boldsymbol{h})\boldsymbol{x}_\alpha \boldsymbol{x}_\alpha^\top & \frac{1}{N}\sum_{\alpha=1}^{N}(\boldsymbol{h}, \boldsymbol{X}_\alpha \boldsymbol{x}_\alpha)\boldsymbol{x}_\alpha \\ \frac{1}{N}\sum_{\alpha=1}^{N}(\boldsymbol{x}_\alpha, \boldsymbol{X}_\alpha \boldsymbol{h})\boldsymbol{x}_\alpha^\top & \frac{1}{N}\sum_{\alpha=1}^{N}(\boldsymbol{x}_\alpha, \boldsymbol{X}_\alpha \boldsymbol{x}_\alpha) \end{pmatrix}. \tag{10.102}$$

Here, we have defined the (44)-matrix $\boldsymbol{X}_\alpha$ by

$$\boldsymbol{X}_\alpha = \boldsymbol{R}(\boldsymbol{x}'_\alpha \times \boldsymbol{W}_\alpha \times \boldsymbol{x}'_\alpha)\boldsymbol{R}^\top. \tag{10.103}$$

The (44)-matrix $\boldsymbol{W}_\alpha$ is defined by

$$\boldsymbol{W}_\alpha = \Big(\boldsymbol{x}'_\alpha \times \boldsymbol{A}^* V_0[\boldsymbol{x}_\alpha]\boldsymbol{A}^{*\top} \times \boldsymbol{x}'_\alpha + (\boldsymbol{A}^*\boldsymbol{x}_\alpha) \times V_0[\boldsymbol{x}'_\alpha] \times (\boldsymbol{A}^*\boldsymbol{x}'_\alpha) \\ + \epsilon^2 [V_0[\boldsymbol{x}'_\alpha] \times \boldsymbol{A}^* V_0[\boldsymbol{x}_\alpha]\boldsymbol{A}^{*\top}]\Big)^-_2, \tag{10.104}$$

where $\boldsymbol{A}^*$ is the value of $\boldsymbol{A}$ obtained by replacing $\boldsymbol{n}$ and d by their appropriate estimates $\boldsymbol{n}^*$ and d^*, respectively, in the second of eqs. (10.89). The solution of the optimization (10.101) under the constraint $\|\boldsymbol{\nu}\| = 1$ is obtained as the unit eigenvector of $\boldsymbol{M}$ for the smallest eigenvalue.

It is easy to confirm that

$$E[\Delta \boldsymbol{x}'_\alpha \times \boldsymbol{W}_\alpha \times \Delta \boldsymbol{x}'_\alpha] = [\boldsymbol{W}_\alpha \times V[\boldsymbol{x}'_\alpha]]. \tag{10.105}$$

Let $\bar{\boldsymbol{X}}_\alpha$ be the unperturbed value of $\boldsymbol{X}_\alpha$ obtained by replacing $\boldsymbol{x}'_\alpha$ by $\bar{\boldsymbol{x}}'_\alpha$ in eq. (10.103). From eqs. (10.103) and (10.105), we obtain

$$E[\boldsymbol{X}_\alpha] = \bar{\boldsymbol{X}}_\alpha + \epsilon^2 \boldsymbol{Y}_\alpha, \tag{10.106}$$

where we have defined the (44)-matrix $\boldsymbol{Y}_\alpha$ by

$$\boldsymbol{Y}_\alpha = \boldsymbol{R}[\boldsymbol{W}_\alpha \times V_0[\boldsymbol{x}'_\alpha]]\boldsymbol{R}^\top. \tag{10.107}$$

Using eq. (10.106), we obtain the following identities:

$$\begin{aligned} E[(\boldsymbol{h}, \boldsymbol{X}_\alpha \boldsymbol{h})\boldsymbol{x}_\alpha \boldsymbol{x}_\alpha^\top] &= (\boldsymbol{h}, E[\boldsymbol{X}_\alpha]\boldsymbol{h})E[\boldsymbol{x}_\alpha \boldsymbol{x}_\alpha^\top] \\ &= (\boldsymbol{h}, (\bar{\boldsymbol{X}}_\alpha + \boldsymbol{Y}_\alpha)\boldsymbol{h})(\bar{\boldsymbol{x}}_\alpha \bar{\boldsymbol{x}}_\alpha^\top + \epsilon^2 V_0[\boldsymbol{x}_\alpha]) \\ &= (\boldsymbol{h}, \bar{\boldsymbol{X}}_\alpha \boldsymbol{h})\bar{\boldsymbol{x}}_\alpha \bar{\boldsymbol{x}}_\alpha^\top + \epsilon^2 (\boldsymbol{h}, \bar{\boldsymbol{X}}_\alpha \boldsymbol{h}) V_0[\boldsymbol{x}_\alpha] + \epsilon^2 (\boldsymbol{h}, \boldsymbol{Y}_\alpha \boldsymbol{h})\bar{\boldsymbol{x}}_\alpha \bar{\boldsymbol{x}}_\alpha^\top \\ &\quad + \epsilon^4 (\boldsymbol{h}, \boldsymbol{Y}_\alpha \boldsymbol{h}) V_0[\boldsymbol{x}_\alpha], \end{aligned} \tag{10.108}$$

$$\begin{aligned} E[(\boldsymbol{h}, \boldsymbol{X}_\alpha \boldsymbol{x}_\alpha)\boldsymbol{x}_\alpha] &= E[\boldsymbol{x}_\alpha \boldsymbol{x}_\alpha^\top] E[\boldsymbol{X}_\alpha^\top]\boldsymbol{h} \\ &= (\bar{\boldsymbol{x}}_\alpha \bar{\boldsymbol{x}}_\alpha^\top + \epsilon^2 V_0[\boldsymbol{x}_\alpha])(\bar{\boldsymbol{X}}_\alpha + \epsilon^2 \boldsymbol{Y}_\alpha)\boldsymbol{h} \\ &= (\bar{\boldsymbol{x}}_\alpha, \bar{\boldsymbol{X}}\boldsymbol{h})\bar{\boldsymbol{x}}_\alpha + \epsilon^2 V_0[\boldsymbol{x}_\alpha]\bar{\boldsymbol{X}}_\alpha \boldsymbol{h} + \epsilon^2 (\bar{\boldsymbol{x}}_\alpha, \boldsymbol{Y}_\alpha \boldsymbol{h})\bar{\boldsymbol{x}}_\alpha \\ &\quad + \epsilon^4 V_0[\boldsymbol{x}_\alpha]\boldsymbol{Y}_\alpha \boldsymbol{h}, \end{aligned} \tag{10.109}$$

$$\begin{aligned} E[(\boldsymbol{x}_\alpha, \boldsymbol{X}_\alpha \boldsymbol{x}_\alpha)] &= (E[\boldsymbol{x}_\alpha \boldsymbol{x}_\alpha^\top]; E[\boldsymbol{X}_\alpha]) \\ &= (\bar{\boldsymbol{x}}_\alpha \bar{\boldsymbol{x}}_\alpha + \epsilon^2 V_0[\boldsymbol{x}_\alpha]; \bar{\boldsymbol{X}}_\alpha + \epsilon^2 \boldsymbol{Y}_\alpha) \\ &= (\bar{\boldsymbol{x}}_\alpha, \bar{\boldsymbol{X}}_\alpha \bar{\boldsymbol{x}}_\alpha) + \epsilon^2 (V_0[\boldsymbol{x}_\alpha]; \bar{\boldsymbol{X}}_\alpha) + \epsilon^2 (\bar{\boldsymbol{x}}_\alpha, \boldsymbol{Y}_\alpha \boldsymbol{x}_\alpha) \\ &\quad + \epsilon^4 (V_0[\boldsymbol{x}_\alpha]; \boldsymbol{Y}_\alpha). \end{aligned} \tag{10.110}$$

Define (44)-matrices $\boldsymbol{N}^{(1)}$ and $\boldsymbol{N}^{(2)}$ by

$$\boldsymbol{N}^{(1)} = \begin{pmatrix} \frac{1}{N}\sum_{\alpha=1}^N \Big((\boldsymbol{h}, \boldsymbol{X}_\alpha \boldsymbol{h})) V_0[\boldsymbol{x}_\alpha] + (\boldsymbol{h}, \boldsymbol{Y}_\alpha \boldsymbol{h})\boldsymbol{x}_\alpha \boldsymbol{x}_\alpha^\top\Big) & \frac{1}{N}\sum_{\alpha=1}^N \Big(V_0[\boldsymbol{x}_\alpha]\boldsymbol{X}_\alpha \boldsymbol{h} + (\boldsymbol{x}_\alpha, \boldsymbol{Y}_\alpha \boldsymbol{h})\boldsymbol{x}_\alpha\Big) \\ \frac{1}{N}\sum_{\alpha=1}^N \Big((V_0[\boldsymbol{x}_\alpha]\boldsymbol{X}_\alpha \boldsymbol{h})^\top + (\boldsymbol{x}_\alpha, \boldsymbol{Y}_\alpha \boldsymbol{h})\boldsymbol{x}_\alpha^\top\Big) & \frac{1}{N}\sum_{\alpha=1}^N \Big((V_0[\boldsymbol{x}_\alpha]; \boldsymbol{X}_\alpha) + (\boldsymbol{x}_\alpha, \boldsymbol{Y}_\alpha \boldsymbol{x}_\alpha)\Big) \end{pmatrix}, \tag{10.111}$$

$$\boldsymbol{N}^{(2)} = \begin{pmatrix} \frac{1}{N}\sum_{\alpha=1}^N (\boldsymbol{h}, \boldsymbol{Y}_\alpha \boldsymbol{h}) V_0[\boldsymbol{x}_\alpha] & \frac{1}{N}\sum_{\alpha=1}^N V_0[\boldsymbol{x}_\alpha]\boldsymbol{Y}_\alpha \boldsymbol{h} \\ \frac{1}{N}\sum_{\alpha=1}^N (V_0[\boldsymbol{x}_\alpha]\boldsymbol{Y}_\alpha \boldsymbol{h})^\top & \frac{1}{N}\sum_{\alpha=1}^N (V_0[\boldsymbol{x}_\alpha]; \boldsymbol{Y}_\alpha) \end{pmatrix}. \tag{10.112}$$

Let $\bar{\boldsymbol{N}}^{(1)}$ be the unperturbed value of $\boldsymbol{N}^{(1)}$ obtained by replacing $\boldsymbol{x}_\alpha$, $\boldsymbol{x}'_\alpha$ and $\boldsymbol{X}_\alpha$ by $\bar{\boldsymbol{x}}_\alpha$, $\bar{\boldsymbol{x}}'_\alpha$, and $\bar{\boldsymbol{X}}_\alpha$, respectively, in eq. (10.111). From eqs. (10.102), (10.108)–(10.112), we see that

$$E[\boldsymbol{M}] = \bar{\boldsymbol{M}} + \epsilon^2 \bar{\boldsymbol{N}}^{(1)} + \epsilon^4 \boldsymbol{N}^{(2)}. \tag{10.113}$$

Since the true value of $\boldsymbol{\nu}$ is the unit eigenvector of $\bar{\boldsymbol{M}}$ for eigenvalue 0, eq. (10.113) implies that the least-squares solution is statistically unbiased (see Section 9.1.2).

From eqs. (10.111) and (10.112), it is easily seen that

$$E[\boldsymbol{N}^{(1)}] = \bar{\boldsymbol{N}}^{(1)} + 2\epsilon^2 \boldsymbol{N}^{(2)}. \tag{10.114}$$

It follows that if we define

$$\hat{\boldsymbol{M}} = \boldsymbol{M} - \epsilon^2 \boldsymbol{N}^{(1)} + \epsilon^4 \boldsymbol{N}^{(2)}, \tag{10.115}$$

we have $E[\hat{\boldsymbol{M}}] = \bar{\boldsymbol{M}}$. Hence, we obtain the unbiased least-squares approximation

$$\hat{J}[\boldsymbol{n}] = (\boldsymbol{\nu}, \hat{\boldsymbol{M}}\boldsymbol{\nu}) \to \min. \tag{10.116}$$

The solution $\hat{\boldsymbol{\nu}}$ is obtained as the unit eigenvector of $\hat{\boldsymbol{M}}$ for the smallest eigenvalue.

If the noise level ϵ is not known, the second order renormalization procedure is given as follows:

1. Let $c = 0$ and $\boldsymbol{W}_\alpha = \boldsymbol{I}$, $\alpha = 1, ..., N$.

2. Compute the (44)-matrices $\boldsymbol{M}$, $\boldsymbol{N}^{(1)}$, and $\boldsymbol{N}^{(2)}$ defined by eqs. (10.102), (10.111) and (10.112), respectively.

3. Compute the smallest eigenvalue λ of the (44)-matrix

$$\hat{\boldsymbol{M}} = \boldsymbol{M} - c\boldsymbol{N}^{(1)} + c^2 \boldsymbol{N}^{(2)}, \tag{10.117}$$

   and the corresponding unit eigenvector $\boldsymbol{\nu}$.

4. If $\lambda \approx 0$, return $\boldsymbol{\nu} = (\nu_1, \nu_2, \nu_3, \nu_4)^\top$, c, and $\hat{\boldsymbol{M}}$. Else, update c and $\boldsymbol{W}_\alpha$ as follows:

$$D = \Big((\boldsymbol{\nu}, \boldsymbol{N}^{(1)}\boldsymbol{\nu}) - 2c(\boldsymbol{\nu}, \boldsymbol{N}^{(2)}\boldsymbol{\nu})\Big)^2 - 4\lambda(\boldsymbol{\nu}, \boldsymbol{N}^{(2)}\boldsymbol{\nu}), \tag{10.118}$$

$$c \leftarrow \begin{cases} c + \dfrac{(\boldsymbol{\nu}, \boldsymbol{N}^{(1)}\boldsymbol{\nu}) - 2c(\boldsymbol{\nu}, \boldsymbol{N}^{(2)}\boldsymbol{\nu}) - \sqrt{D}}{2(\boldsymbol{\nu}, \boldsymbol{N}^{(2)}\boldsymbol{\nu})}, & \text{if } D \geq 0, \\ c + \dfrac{\lambda}{(\boldsymbol{\nu}, \boldsymbol{N}^{(1)}\boldsymbol{\nu})}, & \text{if } D < 0, \end{cases} \tag{10.119}$$

$$\boldsymbol{A} = \boldsymbol{R}^\top(\boldsymbol{h}(\nu_1, \nu_2, \nu_3) + \nu_4 \boldsymbol{I}), \tag{10.120}$$

$$\boldsymbol{W}_\alpha \leftarrow \Big(\boldsymbol{x}'_\alpha \times \boldsymbol{A} V_0[\boldsymbol{x}_\alpha] \boldsymbol{A}^\top \times \boldsymbol{x}'_\alpha + (\boldsymbol{A}\boldsymbol{x}_\alpha) \times V_0[\boldsymbol{x}_\alpha] \times \boldsymbol{A}\boldsymbol{x}_\alpha + c[V_0[\boldsymbol{x}'_\alpha] \times \boldsymbol{A} V_0[\boldsymbol{x}_\alpha] \boldsymbol{A}^\top]\Big)^-_2. \tag{10.121}$$

5. Go back to Step 2.

After renormalization, the squared noise level ϵ^2 is estimated in the form

$$\hat{\epsilon}^2 = \frac{c}{1 - 3/2N}. \tag{10.122}$$

Its expectation and variance are

$$E[\hat{\epsilon}^2] = \epsilon^2, \qquad V[\hat{\epsilon}^2] = \frac{2\epsilon^4}{2N - 3}. \tag{10.123}$$

If the value ϵ is given a priori, the coplanarity test takes the following form (see eq. (10.99):

$$\frac{\hat{\epsilon}^2}{\epsilon^2} > \frac{\chi^2_{2N-3,a}}{2N-3}. \tag{10.124}$$

The covariance matrix of the resulting estimate $\hat{\boldsymbol{\nu}}$ is obtained in the following form:

$$V[\hat{\boldsymbol{\nu}}] = \frac{\hat{\epsilon}^2}{N}(\hat{\boldsymbol{M}})^-_3. \tag{10.125}$$

Let $\mu_{\max}$ be the largest eigenvalue of $V[\hat{\boldsymbol{\nu}}]$, and $\boldsymbol{\xi}_{\max}$ the corresponding unit eigenvector. The primary deviation pair $\{\boldsymbol{\nu}^+, \boldsymbol{\nu}^-\}$ is given as follows (see Section 4.5.3):

$$\boldsymbol{\nu}^+ = N[\hat{\boldsymbol{\nu}} + \sqrt{\mu_{\max}}\boldsymbol{\xi}_{\max}], \qquad \boldsymbol{\nu}^- = N[\hat{\boldsymbol{\nu}} - \sqrt{\mu_{\max}}\boldsymbol{\xi}_{\max}]. \tag{10.126}$$

Once the $\boldsymbol{\nu}$-representation $(\hat{\boldsymbol{\nu}}, \boldsymbol{\rho}) = 0$ of the fitted space plane is obtained, its $\{\boldsymbol{n}, d\}$-representation $(\hat{\boldsymbol{n}}, \boldsymbol{r}) = \hat{d}$ is computed by eq. (4.63). The covariance matrix $V[\hat{\boldsymbol{n}}]$, the correlation vector $V[\hat{\boldsymbol{n}}, \hat{d}]$, and the variance $V[\hat{d}]$ of the parameters $\{\hat{\boldsymbol{n}}, \hat{d}\}$ are computed from $V[\hat{\boldsymbol{\nu}}]$ in the form of eqs. (4.67) (see Section 4.3.1).

Example 10.6 Fig. 10.11 shows simulated stereo images (512×512 pixels with focal length $f = 600$ (pixels)) of a planar grid placed in the scene. We added Gaussian noise of mean 0 and standard deviation $\sigma = 2$ (pixels) to each image coordinate of the grid points independently (so the noise level is $\epsilon = \sigma/f = 1/300$) and fitted a space plane $(\hat{\boldsymbol{n}}, \boldsymbol{r}) = \hat{d}$. As in Example 10.5, the error in the parameters $\{\hat{\boldsymbol{n}}, \hat{d}\}$ is represented by the vector $\Delta\boldsymbol{u}$ given by eq. (10.87), and its covariance matrix is given by eq. (10.88).

In Fig. 10.12, $\Delta \boldsymbol{u}$ is plotted in three-dimensions for 100 trials, each time using different noise: Fig. 10.12a is for the optimal least-squares approximation (the weights are computed from the true values); Fig. 10.12b is for renormalization. We can see that statistical bias exists in the least-squares solution and the bias is removed by renormalization. The ellipses in the figures indicate the standard confidence regions computed from eq. (10.88) by using the theoretical expression (10.96); the cubes are drawn merely for the sake of reference. We can also see from Fig. 10.12b that the theoretical bound on accuracy is almost attained.

Fig. 10.13a shows a reconstructed grid viewed from an angle. The true position is superimposed in dashed lines Fig. 10.13b shows the primary deviation pairs viewed from a different angle.

Example 10.7 Fig. 10.14 shows the real stereo images used in Example 6.6 (see Fig. 6.8), where the 3-D shape was computed without assuming any knowledge about the shape of the surface. Here, we incorporate the knowledge that the surface is planar and reconstruct it from the same feature points shown in Fig. 6.9a. Fig. 10.15 shows two views of the surface reconstructed by the procedure described in this section (solid lines) and its primary deviation pair (dashed lines). Comparing this with Fig. 6.9b, we can see that the reliability is indeed increased by the knowledge that the surface is planar.

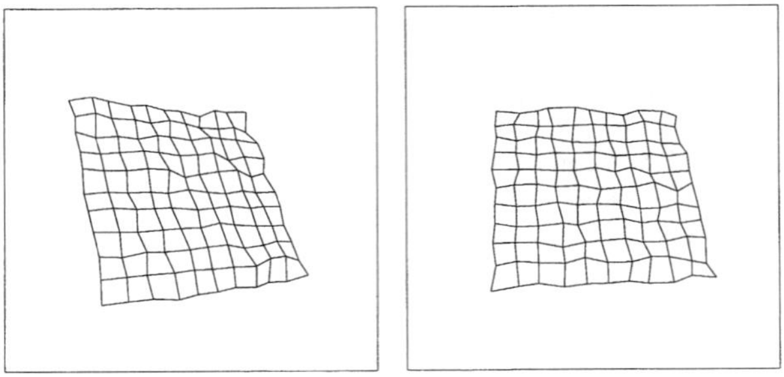

Fig. 10.11. Simulated stereo images of a planar grid.

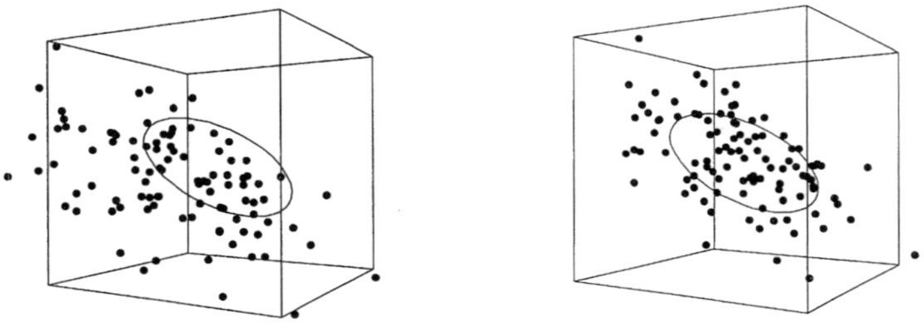

(a) (b)

Fig. 10.12. (a) Error distribution for the least-squares approximation. (b) Error distribution for renormalization.

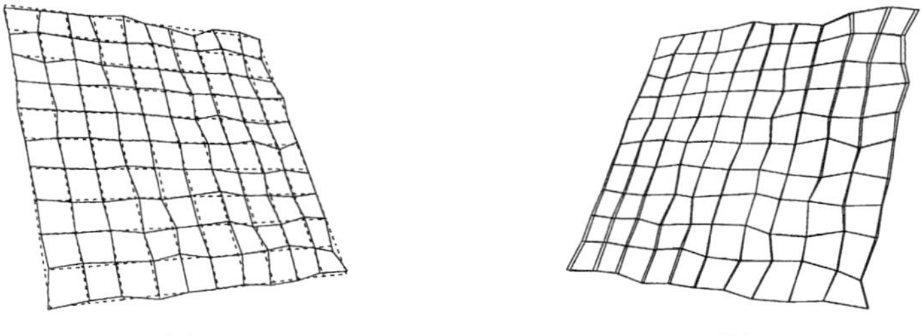

(a) (b)

Fig. 10.13. (a) Reconstructed grid (solid lines) and its true position (dashed lines). (b) Primary deviation pair of the reconstructed grid.

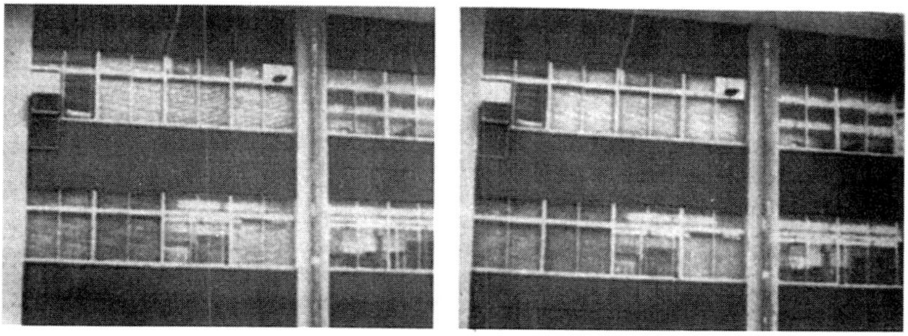

Fig. 10.14. Real stereo image (the same as Fig. 6.8).

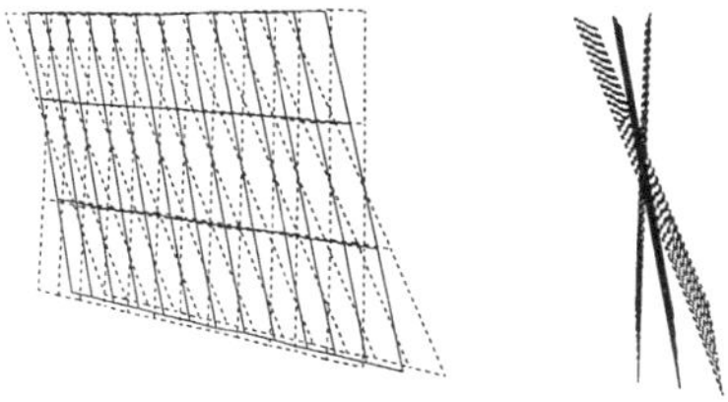

Fig. 10.15. Reconstructed surface (solid lines) and its primary deviation pair (dashed lines).

Chapter 11

3-D Motion Analysis

This chapter presents a statistically optimal algorithm for computing 3-D motion of an object from corresponding image points observed over two views. At the same time, the reliability of the computed motion parameters and the reconstructed depths are evaluated quantitatively. The analysis consists of two separate cases: the case in which the feature points are in general position in the scene, and the case in which they are known to be coplanar. The statistical properties of the theoretically optimal estimate are described first, providing a theoretical bound on the attainable accuracy. Then, the statistical properties of the solution computed by renormalization and linearization are discussed. We also discuss the *critical surface* that gives rise to ambiguity of 3-D interpretation. Finally, we formulate a statistical test for testing if the camera motion is a pure rotation or if the object is a planar surface.

11.1 General Theory

11.1.1 Camera and object motion

In Chapter 6, we saw how 3-D structure is recovered from image point correspondences observed by a stereo system with known motion parameters $\{\boldsymbol{h}, \boldsymbol{R}\}$. We now show that the motion parameters can also be computed if the correspondence is established for a sufficient number of image points. In this chapter, the stereo configuration defined in Chapter 6 is reinterpreted as a camera motion in the scene: the first and second camera positions of a stereo system are identified with the camera positions before and after the motion, respectively. However, we also use stereo terminologies interchangeably.

Let $\boldsymbol{x}_\alpha$ and $\boldsymbol{x}'_\alpha$ be corresponding image points, and $\bar{\boldsymbol{x}}_\alpha$ and $\bar{\boldsymbol{x}}'_\alpha$ their true positions. We write

$$\boldsymbol{x}_\alpha = \bar{\boldsymbol{x}}_\alpha + \Delta\boldsymbol{x}_\alpha, \qquad \boldsymbol{x}'_\alpha = \bar{\boldsymbol{x}}'_\alpha + \Delta\boldsymbol{x}'_\alpha, \tag{11.1}$$

and regard the noise terms $\Delta\boldsymbol{x}_\alpha$ and $\Delta\boldsymbol{x}'_\alpha$ as independent Gaussian random variables of mean $\boldsymbol{0}$ and covariance matrices $V[\boldsymbol{x}_\alpha]$ and $V[\boldsymbol{x}'_\alpha]$, respectively.

As discussed in Section 6.1.2, the true positions must satisfy the *epipolar equation* (6.4). Hence, the problem is stated as follows:

Problem 11.1 *Estimate the motion parameters* $\{\boldsymbol{h}, \boldsymbol{R}\}$ *that satisfy*

$$|\bar{\boldsymbol{x}}_\alpha, \boldsymbol{h}, \boldsymbol{R}\bar{\boldsymbol{x}}'_\alpha| = 0, \qquad \alpha = 1, ..., N, \tag{11.2}$$

from the data $\boldsymbol{x}_\alpha$ *and* $\boldsymbol{x}'_\alpha$, $\alpha = 1, ..., N$.

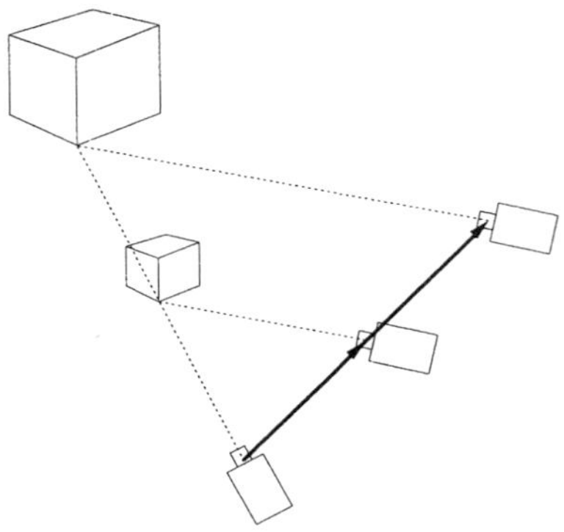

Fig. 11.1. A large camera motion relative to a large object in the distance is indistinguishable from a small camera motion relative to a small object near the camera.

Eq. (11.2) is the hypothesis from which the motion parameters $\{\boldsymbol{h}, \boldsymbol{R}\}$ are to be estimated. The rank of this hypothesis is 1. It is immediately seen that translation $\boldsymbol{h}$ can be determined only up to scale. This is intuitively interpreted as follows: as long as an image sequence is the only source of information, a large camera motion relative to large objects far away from the camera is indistinguishable from a small camera motion relative to small objects near the camera (Fig. 11.1). In order to remove this indeterminacy, we adopt the scaling $\|\boldsymbol{h}\| = 1$ if $\boldsymbol{h} \neq \boldsymbol{0}$. It follows that we must first decide whether or not $\boldsymbol{h} = \boldsymbol{0}$, i.e., whether or not the camera motion is a pure rotation. The procedure for this decision, which we call the *rotation test*, will be discussed in Section 11.7.1. Here, we assume that $\boldsymbol{h} \neq \boldsymbol{0}$ has already been confirmed.

In the following, we assume that the camera moves in a stationary scene, but the subsequent analyses can also be applied to a moving object viewed from a stationary camera (Fig. 11.2). If an object rotates around a fixed reference point $\boldsymbol{r}_G$ in the object, say its centroid, by $\boldsymbol{R}_o$ and then translates by $\boldsymbol{h}_o$, a point $\boldsymbol{r}$ in the object moves to

$$\boldsymbol{r}' = \boldsymbol{r}_G + \boldsymbol{h}_o + \boldsymbol{R}_o(\boldsymbol{r} - \boldsymbol{r}_G). \tag{11.3}$$

Comparing this with eq. (6.1), we find that the resulting image motion is the same as when the object is stationary and the camera moves with motion parameters

$$\boldsymbol{h} = \boldsymbol{r}_G - \boldsymbol{R}_o^\top(\boldsymbol{h}_o + \boldsymbol{r}_G), \qquad \boldsymbol{R} = \boldsymbol{R}_o^\top. \tag{11.4}$$

Conversely, the parameters $\{\boldsymbol{h}_o, \boldsymbol{R}_o\}$ of the object motion are computed from the parameters $\{\boldsymbol{h}, \boldsymbol{R}\}$ of the equivalent camera motion in the following form:

$$\boldsymbol{h}_o = \boldsymbol{R}^\top(\boldsymbol{r}_G - \boldsymbol{h}) - \boldsymbol{r}_G, \qquad \boldsymbol{R}_o = \boldsymbol{R}^\top. \tag{11.5}$$

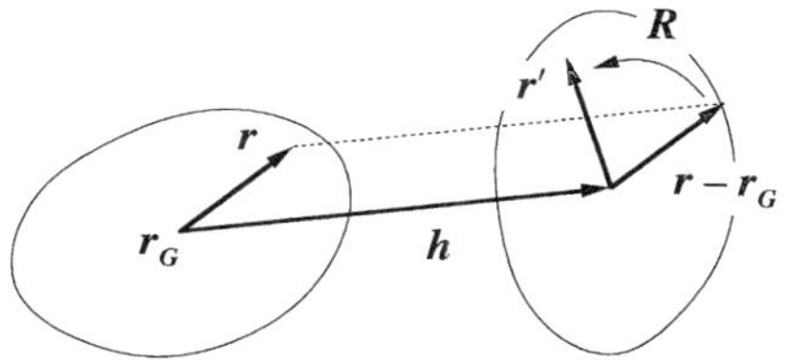

Fig. 11.2. Object motion relative to a stationary camera coordinate system.

11.1.2 Optimal estimation of motion parameters

A. Nonlinear optimization

For each α, define the 33-matrix

$$\boldsymbol{X}_\alpha = \boldsymbol{x}_\alpha \boldsymbol{x}'^{\top}_\alpha. \tag{11.6}$$

Let $\bar{\boldsymbol{X}}_\alpha$ be the unperturbed value of $\boldsymbol{X}_\alpha$ obtained by replacing $\boldsymbol{x}_\alpha$ and $\boldsymbol{x}'_\alpha$ by $\bar{\boldsymbol{x}}_\alpha$ and $\bar{\boldsymbol{x}}'_\alpha$, respectively. The epipolar equation (11.2) is expressed in the following form:

$$(\bar{\boldsymbol{X}}_\alpha; \boldsymbol{h} \times \boldsymbol{R}) = 0. \tag{11.7}$$

This equation has the same form as eq. (10.24) if $\boldsymbol{h} \times \boldsymbol{R}$ is identified with $\boldsymbol{Q}$. Hence, the subsequent analysis takes almost the same form as in Section 10.2.1, except that $\bar{\boldsymbol{X}}_\alpha$ and $\boldsymbol{h} \times \boldsymbol{R}$ are no longer symmetric matrices.

Since $\boldsymbol{x}_\alpha$ and $\boldsymbol{x}'_\alpha$ are independent, we have $E[\boldsymbol{X}_\alpha] = \bar{\boldsymbol{X}}_\alpha$. Hence, the deviation $\Delta\boldsymbol{X}_\alpha = \boldsymbol{X}_\alpha - \bar{\boldsymbol{X}}_\alpha$ is a random variable of mean $\boldsymbol{O}$. From eqs. (11.1) and (11.6), we see that

$$\Delta\boldsymbol{X}_\alpha = \Delta\boldsymbol{x}_\alpha \bar{\boldsymbol{x}}'^{\top}_\alpha + \bar{\boldsymbol{x}}_\alpha \Delta\boldsymbol{x}'^{\top}_\alpha + \Delta\boldsymbol{x}_\alpha \Delta\boldsymbol{x}'^{\top}_\alpha. \tag{11.8}$$

Let $\Delta X_{\alpha(ij)}$, $x_{\alpha(i)}$, and $x'_{\alpha(i)}$ be the (ij) element of $\Delta\boldsymbol{X}_\alpha$ and the ith components of $\boldsymbol{x}_\alpha$ and $\boldsymbol{x}'_\alpha$, respectively. The covariance tensor $\mathcal{V}[\boldsymbol{X}_\alpha]$ of $\boldsymbol{X}_\alpha$ has the following $(ijkl)$ element:

$$\begin{aligned}\mathcal{V}[\boldsymbol{X}_\alpha]_{ijkl} &= E[\Delta X_{\alpha(ij)} \Delta X_{\alpha(kl)}] \\ &= E[\Delta x_{\alpha(i)} \Delta x_{\alpha(k)}] \bar{x}'_{\alpha(j)} \bar{x}'_{\alpha(l)} + E[\Delta x'_{\alpha(j)} \Delta x'_{\alpha(l)}] \bar{x}_{\alpha(i)} \bar{x}_{\alpha(k)} \\ &\quad + E[\Delta x_{\alpha(i)} \Delta x_{\alpha(k)} \Delta x'_{\alpha(j)} \Delta x'_{\alpha(l)}] \\ &= V[\boldsymbol{x}_\alpha]_{ik} \bar{x}'_{\alpha(j)} \bar{x}'_{\alpha(l)} + V[\boldsymbol{x}'_\alpha]_{jl} \bar{x}_{\alpha(i)} \bar{x}_{\alpha(k)} + V[\boldsymbol{x}_\alpha]_{ik} V[\boldsymbol{x}'_\alpha]_{jl}. \end{aligned} \tag{11.9}$$

According to the general theory in Section 7.1.5, the optimal estimate of the motion parameters $\{\boldsymbol{h}, \boldsymbol{R}\}$ can be obtained by the following minimization (see eq. (7.55)):

$$J[\boldsymbol{h}, \boldsymbol{R}] = \sum_{\alpha=1}^{N} \frac{(\boldsymbol{X}_\alpha; \boldsymbol{h} \times \boldsymbol{R})^2}{(\boldsymbol{h} \times \boldsymbol{R}; \mathcal{V}[\boldsymbol{X}_\alpha](\boldsymbol{h} \times \boldsymbol{R}))} \to \min. \tag{11.10}$$

The minimum is sought under the constraint that $\boldsymbol{h}$ is a unit vector and $\boldsymbol{R}$ is a rotation matrix. If eq. (11.9) is substituted into eq. (11.10), the optimization can be rewritten in the following form:

$$J[\boldsymbol{h}, \boldsymbol{R}] = \sum_{\alpha=1}^{N} W_\alpha(\boldsymbol{h}, \boldsymbol{R})|\boldsymbol{x}_\alpha, \boldsymbol{h}, \boldsymbol{R}\boldsymbol{x}'_\alpha|^2 \to \min, \tag{11.11}$$

$$\begin{aligned} W_\alpha(\boldsymbol{h}, \boldsymbol{R}) = 1 \Big/ \Big(&(\boldsymbol{h} \times \boldsymbol{R}\bar{\boldsymbol{x}}'_\alpha, V[\boldsymbol{x}_\alpha](\boldsymbol{h} \times \boldsymbol{R}\bar{\boldsymbol{x}}'_\alpha)) \\ &+ (\boldsymbol{h} \times \bar{\boldsymbol{x}}_\alpha, \boldsymbol{R}V[\boldsymbol{x}'_\alpha]\boldsymbol{R}^\top(\boldsymbol{h} \times \bar{\boldsymbol{x}}_\alpha)) \\ &+ (V[\boldsymbol{x}_\alpha](\boldsymbol{h} \times \boldsymbol{R}); (\boldsymbol{h} \times \boldsymbol{R})V[\boldsymbol{x}'_\alpha])\Big). \end{aligned} \tag{11.12}$$

In actual computation, the true values $\bar{\boldsymbol{x}}_\alpha$ and $\bar{\boldsymbol{x}}'_\alpha$ are approximated by the data values $\boldsymbol{x}_\alpha$ and $\boldsymbol{x}'_\alpha$, respectively.

B. Rigidity test

Let $\{\hat{\boldsymbol{h}}, \hat{\boldsymbol{R}}\}$ be the resulting estimate of $\{\boldsymbol{h}, \boldsymbol{R}\}$. The motion parameters $\{\boldsymbol{h}, \boldsymbol{R}\}$ have *five* degrees of freedom—two for $\boldsymbol{h}$ (unit vector) and three for $\boldsymbol{R}$ (rotation matrix). Hence, the residual $J[\hat{\boldsymbol{h}}, \hat{\boldsymbol{R}}]$ is a χ^2 variable with $N - 5$ degrees of freedom (see Section 7.1.4). It follows that the number N of the pairs of corresponding image points must be such that[1]

$$N \geq 5. \tag{11.13}$$

The residual is 0 when $N = 5$. For $N \geq 6$, we obtain a *rigidity test*: the hypothesis that the camera is moving in a stationary scene (or equivalently the object in motion is rigid) is rejected with significance level $a\%$ if

$$J[\hat{\boldsymbol{h}}, \hat{\boldsymbol{R}}] > \chi^2_{N-5,a}. \tag{11.14}$$

C. Focus of expansion

Suppose the camera moves toward or away from the αth feature point P_α in the scene (Fig. 11.3a). Since $\boldsymbol{h} \times \bar{\boldsymbol{x}}_\alpha = \boldsymbol{0}$ and $\boldsymbol{h} \times \boldsymbol{R}\bar{\boldsymbol{x}}'_\alpha = \boldsymbol{0}$, the first two terms in the denominator on the right-hand side of eq. (11.12) become 0. Consequently, $W_\alpha(\boldsymbol{h}, \boldsymbol{R})$ becomes very large[2]. Geometrically, this is interpreted as follows. The expression $(\boldsymbol{h} \times \boldsymbol{R}; \mathcal{V}[\boldsymbol{X}_\alpha](\boldsymbol{h} \times \boldsymbol{R}))$ measures how likely the epipolar equation (11.7) is to be violated by noise. The epipolar equation

[1] Algorithms for computing the motion parameters from *five* pairs of points are known as *five-point algorithms*.

[2] Theoretically, $W_\alpha(\mathbf{h}, \mathbf{R})$ becomes infinite. The fact that it always remains finite is due to the approximation introduced in the computation.

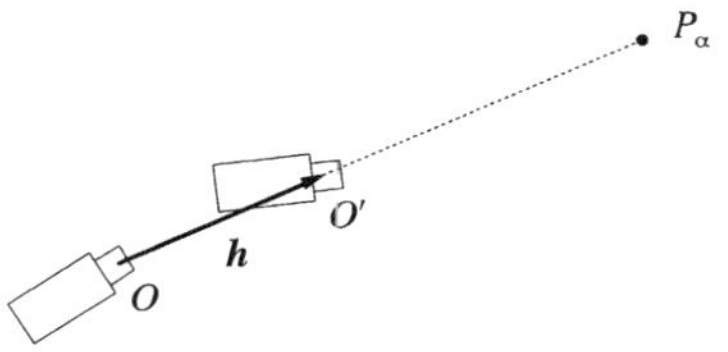

(a)

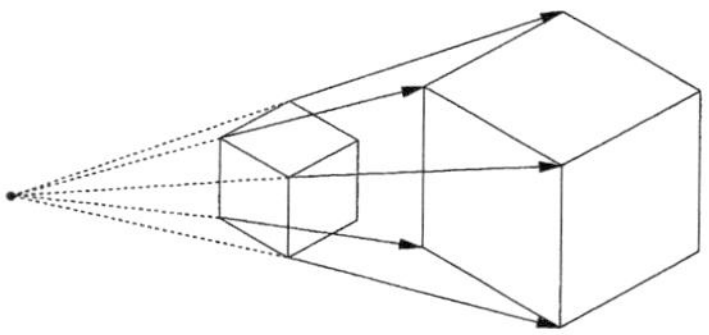
(b)

Fig. 11.3. (a) The camera moves toward or away from the αth feature point. (b) Focus of expansion.

states that the baseline and the two lines of sight that start from the viewpoints O and O' and pass through P_α are *coplanar* (see eq. (6.2)). However, if the camera moves toward or away from P_α, the two lines of sight and the baseline become *collinear*. The collinearity relation may be destroyed in the presence of noise, but the coplanarity relation still holds to a first approximation. Consequently, P_α is regarded as "robust" to noise, and thereby given a large weight. If the camera orientation does not change (i.e., $\boldsymbol{R} = \boldsymbol{I}$), the projection of P_α does not move on the image plane; all other feature points seem to move away from or toward it (Fig. 11.3b). Such an image point is known as the *focus of expansion*. Although this description does not hold if the camera rotates, we still call the projection of P_α the "focus of expansion" if O, O', and P_α are collinear.

11.1.3 Theoretical bound on accuracy

The theoretical bound on the attainable accuracy is obtained by analyzing the behavior of the function $J[\boldsymbol{h}, \boldsymbol{R}]$ near the true values $\{\bar{\boldsymbol{h}}, \bar{\boldsymbol{R}}\}$. The translation $\boldsymbol{h}$ is a unit vector, so we can write

$$\boldsymbol{h} = \bar{\boldsymbol{h}} + \Delta\boldsymbol{h}, \qquad (\bar{\boldsymbol{h}}, \Delta\boldsymbol{h}) = 0, \tag{11.15}$$

to a first approximation. Since $\boldsymbol{R}$ is a rotation matrix, a small vector $\boldsymbol{\Omega}$ exists such that

$$\boldsymbol{R} = \bar{\boldsymbol{R}} + \Delta\boldsymbol{R}, \qquad \Delta\boldsymbol{R} = \Delta\boldsymbol{\Omega} \times \bar{\boldsymbol{R}}, \tag{11.16}$$

to a first approximation (see eqs. (2.57) and (3.31)). Substituting eqs. (11.1), (11.15), and (11.16) into eq. (11.11), we obtain

$$J[\boldsymbol{h}, \boldsymbol{R}] = \sum_{\alpha=1}^{N} (W_\alpha(\bar{\boldsymbol{h}}, \bar{\boldsymbol{R}}) + O(\Delta\boldsymbol{h}, \Delta\boldsymbol{\Omega})) \Big(|\boldsymbol{x}_\alpha, \bar{\boldsymbol{h}}, \bar{\boldsymbol{R}}\boldsymbol{x}'_\alpha| + |\bar{\boldsymbol{x}}_\alpha, \Delta\boldsymbol{h}, \bar{\boldsymbol{R}}\bar{\boldsymbol{x}}'_\alpha|$$
$$+ |\bar{\boldsymbol{x}}_\alpha, \bar{\boldsymbol{h}}, \Delta\boldsymbol{R}\boldsymbol{x}'_\alpha| + O(\Delta\boldsymbol{x}_\alpha, \Delta\boldsymbol{x}'_\alpha, \Delta\boldsymbol{h}, \Delta\boldsymbol{R})^2\Big)^2$$

$$= \sum_{\alpha=1}^{N} W_\alpha(\bar{\boldsymbol{h}}, \bar{\boldsymbol{R}}) \Big(|\boldsymbol{x}_\alpha, \bar{\boldsymbol{h}}, \bar{\boldsymbol{R}}\boldsymbol{x}'_\alpha| - (\bar{\boldsymbol{x}}_\alpha \times \bar{\boldsymbol{R}}\bar{\boldsymbol{x}}'_\alpha, \Delta\boldsymbol{h}) \\ - ((\bar{\boldsymbol{x}}_\alpha, \bar{\boldsymbol{R}}\bar{\boldsymbol{x}}'_\alpha)\bar{\boldsymbol{h}} - (\bar{\boldsymbol{h}}, \bar{\boldsymbol{R}}\bar{\boldsymbol{x}}'_\alpha)\bar{\boldsymbol{x}}_\alpha, \Delta\boldsymbol{\Omega}) \Big)^2 \\ + \sum_{\alpha=1}^{N} W_\alpha(\bar{\boldsymbol{h}}, \bar{\boldsymbol{R}}) O(\Delta\boldsymbol{x}_\alpha, \Delta\boldsymbol{x}'_\alpha, \Delta\boldsymbol{h}, \Delta\boldsymbol{\Omega})^3, \qquad (11.17)$$

where $O(\cdots)^p$ denotes terms of order p or higher in $\cdots$. If we ignore the last term and define vectors

$$\bar{\boldsymbol{a}}_\alpha = \bar{\boldsymbol{x}}_\alpha \times \bar{\boldsymbol{R}}\bar{\boldsymbol{x}}'_\alpha, \qquad \bar{\boldsymbol{b}}_\alpha = (\bar{\boldsymbol{x}}_\alpha, \bar{\boldsymbol{R}}\bar{\boldsymbol{x}}'_\alpha)\bar{\boldsymbol{h}} - (\bar{\boldsymbol{h}}, \bar{\boldsymbol{R}}\bar{\boldsymbol{x}}'_\alpha)\bar{\boldsymbol{x}}_\alpha, \qquad (11.18)$$

we obtain

$$J[\boldsymbol{h}, \boldsymbol{R}] = \sum_{\alpha=1}^{N} W_\alpha(\bar{\boldsymbol{h}}, \bar{\boldsymbol{R}}) \Big(|\boldsymbol{x}_\alpha, \bar{\boldsymbol{h}}, \bar{\boldsymbol{R}}\boldsymbol{x}'_\alpha| - (\bar{\boldsymbol{a}}_\alpha, \Delta\boldsymbol{h}) - (\bar{\boldsymbol{b}}_\alpha, \Delta\boldsymbol{\Omega}) \Big)^2, \qquad (11.19)$$

where we have used the following relationship:

$$|\bar{\boldsymbol{x}}_\alpha, \bar{\boldsymbol{h}}, \Delta\boldsymbol{R}\bar{\boldsymbol{x}}'_\alpha| = (\bar{\boldsymbol{x}}_\alpha, (\bar{\boldsymbol{h}}, \bar{\boldsymbol{R}}\bar{\boldsymbol{x}}'_\alpha)\Delta\boldsymbol{\Omega} - (\bar{\boldsymbol{h}}, \Delta\boldsymbol{\Omega})\bar{\boldsymbol{R}}\bar{\boldsymbol{x}}'_\alpha). \qquad (11.20)$$

If we minimize $J[\boldsymbol{h}, \boldsymbol{R}]$ by regarding $\Delta\boldsymbol{h}$ and $\Delta\boldsymbol{\Omega}$ as variables, we need not consider the constraint $(\bar{\boldsymbol{h}}, \Delta\boldsymbol{h}) = 0$, since the epipolar equation (11.2) implies that

$$(\bar{\boldsymbol{a}}_\alpha, \bar{\boldsymbol{h}}) = |\bar{\boldsymbol{x}}_\alpha, \bar{\boldsymbol{R}}\bar{\boldsymbol{x}}_\alpha, \bar{\boldsymbol{h}}| = 0, \qquad (11.21)$$

and hence $(\bar{\boldsymbol{a}}_\alpha, \Delta\boldsymbol{h} + c\bar{\boldsymbol{h}}) = (\bar{\boldsymbol{a}}_\alpha, \Delta\boldsymbol{h})$ for an arbitrary constant c, i.e., the component of $\Delta\boldsymbol{h}$ parallel to $\bar{\boldsymbol{h}}$ has no effect on the minimization. Differentiating $J[\boldsymbol{h}, \boldsymbol{R}]$ with respect to $\Delta\boldsymbol{h}$ and $\Delta\boldsymbol{\Omega}$ and setting the result $\boldsymbol{0}$, we obtain

$$\sum_{\alpha=1}^{N} W_\alpha(\bar{\boldsymbol{h}}, \bar{\boldsymbol{R}}) \Big((\bar{\boldsymbol{a}}_\alpha, \Delta\boldsymbol{h})\bar{\boldsymbol{a}}_\alpha + (\bar{\boldsymbol{b}}_\alpha, \Delta\boldsymbol{\Omega})\bar{\boldsymbol{a}}_\alpha \Big) = \sum_{\alpha=1}^{N} W_\alpha(\bar{\boldsymbol{h}}, \bar{\boldsymbol{R}}) |\boldsymbol{x}_\alpha, \bar{\boldsymbol{h}}, \bar{\boldsymbol{R}}\boldsymbol{x}'_\alpha| \bar{\boldsymbol{a}}_\alpha, \qquad (11.22)$$

$$\sum_{\alpha=1}^{N} W_\alpha(\bar{\boldsymbol{h}}, \bar{\boldsymbol{R}}) \Big((\bar{\boldsymbol{a}}_\alpha, \Delta\boldsymbol{h})\bar{\boldsymbol{b}}_\alpha + (\bar{\boldsymbol{b}}_\alpha, \Delta\boldsymbol{\Omega})\bar{\boldsymbol{b}}_\alpha \Big) = \sum_{\alpha=1}^{N} W_\alpha(\bar{\boldsymbol{h}}, \bar{\boldsymbol{R}}) |\boldsymbol{x}_\alpha, \bar{\boldsymbol{h}}, \bar{\boldsymbol{R}}\boldsymbol{x}'_\alpha| \bar{\boldsymbol{b}}_\alpha. \qquad (11.23)$$

If we define the (66)-matrix

$$\bar{\boldsymbol{A}} = \begin{pmatrix} \sum_{\alpha=1}^{N} W_\alpha(\bar{\boldsymbol{h}}, \bar{\boldsymbol{R}}) \bar{\boldsymbol{a}}_\alpha \bar{\boldsymbol{a}}_\alpha^\top & \sum_{\alpha=1}^{N} W_\alpha(\bar{\boldsymbol{h}}, \bar{\boldsymbol{R}}) \bar{\boldsymbol{a}}_\alpha \bar{\boldsymbol{b}}_\alpha^\top \\ \sum_{\alpha=1}^{N} W_\alpha(\bar{\boldsymbol{h}}, \bar{\boldsymbol{R}}) \bar{\boldsymbol{b}}_\alpha \bar{\boldsymbol{a}}_\alpha^\top & \sum_{\alpha=1}^{N} W_\alpha(\bar{\boldsymbol{h}}, \bar{\boldsymbol{R}}) \bar{\boldsymbol{b}}_\alpha \bar{\boldsymbol{b}}_\alpha^\top \end{pmatrix}, \qquad (11.24)$$

eqs. (11.22) and (11.23) are combined into one equation in the form

$$\bar{\boldsymbol{A}} \begin{pmatrix} \Delta\boldsymbol{h} \\ \Delta\boldsymbol{\Omega} \end{pmatrix} = \sum_{\alpha=1}^{N} W_\alpha(\bar{\boldsymbol{h}}, \bar{\boldsymbol{R}}) |\boldsymbol{x}_\alpha, \bar{\boldsymbol{h}}, \bar{\boldsymbol{R}}\boldsymbol{x}'_\alpha| \begin{pmatrix} \bar{\boldsymbol{a}}_\alpha \\ \bar{\boldsymbol{b}}_\alpha \end{pmatrix}. \qquad (11.25)$$

It is easily seen from eq. (11.21) that the matrix $\bar{\boldsymbol{A}}$ has rank 5; its null space is $\{\bar{\boldsymbol{h}} \oplus \boldsymbol{0}\}_L$. The solution of eq. (11.25) is obtained in the following form:

$$\begin{pmatrix} \Delta \boldsymbol{h} \\ \Delta \boldsymbol{\Omega} \end{pmatrix} = \sum_{\alpha=1}^{N} W_\alpha(\bar{\boldsymbol{h}}, \bar{\boldsymbol{R}})|\boldsymbol{x}_\alpha, \bar{\boldsymbol{h}}, \bar{\boldsymbol{R}}\boldsymbol{x}'_\alpha|\bar{\boldsymbol{A}}^- \begin{pmatrix} \bar{\boldsymbol{a}}_\alpha \\ \bar{\boldsymbol{b}}_\alpha \end{pmatrix}. \tag{11.26}$$

The optimal estimate $\{\hat{\boldsymbol{h}}, \hat{\boldsymbol{R}}\}$ of the motion parameters $\{\boldsymbol{h}, \boldsymbol{R}\}$ is given by

$$\hat{\boldsymbol{h}} = \bar{\boldsymbol{h}} + \Delta \boldsymbol{h}, \qquad \hat{\boldsymbol{R}} = \bar{\boldsymbol{R}} + \Delta \boldsymbol{\Omega} \times \bar{\boldsymbol{R}}. \tag{11.27}$$

Their covariance matrices are defined as follows (see eq. (3.33)):

$$\begin{aligned} \bar{V}[\hat{\boldsymbol{h}}] &= E[\Delta\boldsymbol{h}\Delta\boldsymbol{h}^\top], \qquad \bar{V}[\hat{\boldsymbol{h}}, \hat{\boldsymbol{R}}] = E[\Delta\boldsymbol{h}\Delta\boldsymbol{\Omega}^\top], \\ \bar{V}[\hat{\boldsymbol{R}}, \hat{\boldsymbol{h}}] &= E[\Delta\boldsymbol{\Omega}\Delta\boldsymbol{h}^\top], \qquad \bar{V}[\hat{\boldsymbol{R}}] = E[\Delta\boldsymbol{\Omega}\Delta\boldsymbol{\Omega}^\top]. \end{aligned} \tag{11.28}$$

From eq. (11.26), we obtain

$$\begin{aligned} &\begin{pmatrix} \bar{V}[\hat{\boldsymbol{h}}] & \bar{V}[\hat{\boldsymbol{h}}, \hat{\boldsymbol{R}}] \\ \bar{V}[\hat{\boldsymbol{R}}, \hat{\boldsymbol{h}}] & \bar{V}[\hat{\boldsymbol{R}}] \end{pmatrix} = E[\begin{pmatrix} \Delta \boldsymbol{h} \\ \Delta \boldsymbol{\Omega} \end{pmatrix}\begin{pmatrix} \Delta \boldsymbol{h} \\ \Delta \boldsymbol{\Omega} \end{pmatrix}^\top] \\ &= \sum_{\alpha,\beta=1}^{N} W_\alpha(\bar{\boldsymbol{h}}, \bar{\boldsymbol{R}})\bar{W}_\beta E[|\boldsymbol{x}_\alpha, \bar{\boldsymbol{h}}, \bar{\boldsymbol{R}}\boldsymbol{x}'_\alpha|\cdot|\boldsymbol{x}_\beta, \bar{\boldsymbol{h}}, \bar{\boldsymbol{R}}\boldsymbol{x}'_\beta|]\bar{\boldsymbol{A}}^- \begin{pmatrix} \bar{\boldsymbol{a}}_\alpha \\ \bar{\boldsymbol{b}}_\alpha \end{pmatrix}\begin{pmatrix} \bar{\boldsymbol{a}}_\alpha \\ \bar{\boldsymbol{b}}_\alpha \end{pmatrix}^\top \bar{\boldsymbol{A}}^- \\ &= \bar{\boldsymbol{A}}^-\bar{\boldsymbol{A}}\bar{\boldsymbol{A}}^- = \bar{\boldsymbol{A}}^-, \end{aligned} \tag{11.29}$$

where we have used the following relationship:

$$\begin{aligned} E[|\boldsymbol{x}_\alpha, \bar{\boldsymbol{h}}, \bar{\boldsymbol{R}}\boldsymbol{x}'_\alpha| \cdot |\boldsymbol{x}_\beta, \bar{\boldsymbol{h}}, \bar{\boldsymbol{R}}\boldsymbol{x}'_\beta|] &= \delta_{\alpha\beta}E[|\boldsymbol{x}_\alpha, \bar{\boldsymbol{h}}, \bar{\boldsymbol{R}}\boldsymbol{x}'_\alpha|^2] \\ &= \delta_{\alpha\beta}E[(\boldsymbol{X}_\alpha; \bar{\boldsymbol{h}} \times \bar{\boldsymbol{R}})^2] = \delta_{\alpha\beta}E[(\Delta\boldsymbol{X}_\alpha; \bar{\boldsymbol{h}} \times \bar{\boldsymbol{R}})^2] \\ &= \delta_{\alpha\beta}(\bar{\boldsymbol{h}} \times \bar{\boldsymbol{R}}; \mathcal{V}[\boldsymbol{X}_\alpha](\bar{\boldsymbol{h}} \times \bar{\boldsymbol{R}})) = \frac{\delta_{\alpha\beta}}{W_\alpha(\bar{\boldsymbol{h}}, \bar{\boldsymbol{R}})}. \end{aligned} \tag{11.30}$$

The covariance matrices $\bar{V}[\hat{\boldsymbol{h}}]$, $\bar{V}[\hat{\boldsymbol{h}}, \hat{\boldsymbol{R}}]$ $(= \bar{V}[\hat{\boldsymbol{R}}, \hat{\boldsymbol{h}}])$, and $\bar{V}[\hat{\boldsymbol{R}}]$ thus obtained give a theoretical bound on the attainable accuracy of estimating $\{\boldsymbol{h}, \boldsymbol{R}\}$. In actual computation, eq. (11.29) is approximated by

$$\begin{aligned} &\begin{pmatrix} V[\hat{\boldsymbol{h}}] & V[\hat{\boldsymbol{h}}, \hat{\boldsymbol{R}}] \\ V[\hat{\boldsymbol{R}}, \hat{\boldsymbol{h}}] & V[\hat{\boldsymbol{R}}] \end{pmatrix} \\ &= \begin{pmatrix} \sum_{\alpha=1}^N W_\alpha(\hat{\boldsymbol{h}}, \hat{\boldsymbol{R}})(\boldsymbol{P}_{\hat{\boldsymbol{h}}}\hat{\boldsymbol{a}}_\alpha)(\boldsymbol{P}_{\hat{\boldsymbol{h}}}\hat{\boldsymbol{a}}_\alpha)^\top & \sum_{\alpha=1}^N W_\alpha(\hat{\boldsymbol{h}}, \hat{\boldsymbol{R}})(\boldsymbol{P}_{\hat{\boldsymbol{h}}}\hat{\boldsymbol{a}}_\alpha)\hat{\boldsymbol{b}}_\alpha^\top \\ \sum_{\alpha=1}^N W_\alpha(\hat{\boldsymbol{h}}, \hat{\boldsymbol{R}})\hat{\boldsymbol{b}}_\alpha(\boldsymbol{P}_{\hat{\boldsymbol{h}}}\hat{\boldsymbol{a}}_\alpha)^\top & \sum_{\alpha=1}^N W_\alpha(\hat{\boldsymbol{h}}, \hat{\boldsymbol{R}})\hat{\boldsymbol{b}}_\alpha\hat{\boldsymbol{b}}_\alpha^\top \end{pmatrix}^-, \end{aligned} \tag{11.31}$$

where $\{\hat{\boldsymbol{h}}, \hat{\boldsymbol{R}}\}$ are the estimated motion parameters and

$$\hat{\boldsymbol{a}}_\alpha = \hat{\boldsymbol{x}}_\alpha \times \hat{\boldsymbol{R}}\hat{\boldsymbol{x}}'_\alpha, \qquad \hat{\boldsymbol{b}}_\alpha = (\hat{\boldsymbol{x}}_\alpha, \hat{\boldsymbol{R}}\hat{\boldsymbol{x}}'_\alpha)\hat{\boldsymbol{h}} - (\hat{\boldsymbol{h}}, \hat{\boldsymbol{R}}\hat{\boldsymbol{x}}'_\alpha)\hat{\boldsymbol{x}}_\alpha. \tag{11.32}$$

Here, $\hat{\boldsymbol{x}}_\alpha$ and $\hat{\boldsymbol{x}}'_\alpha$ are, respectively, the optimally corrected positions of $\boldsymbol{x}_\alpha$ and $\boldsymbol{x}'_\alpha$ (see eqs. (6.20)).

11.2 Linearization and Renormalization

11.2.1 Linearization

The optimization (11.11) is nonlinear, requiring numerical search. However, the hypothesis (11.2) (or equivalently (11.7)) is linear in the *essential matrix* $\boldsymbol{G} = \boldsymbol{h} \times \boldsymbol{R}$ (see eq. (6.7)), and the epipolar equation (11.2) is expressed in the form $(\bar{\boldsymbol{x}}_\alpha, \boldsymbol{G}\bar{\boldsymbol{x}}'_\alpha) = 0$ (see eq. (6.8)). Hence, the linearization technique described in Section 9.5.2 can be applied. To be specific, Problem 11.1 is decomposed into the following two subproblems:

Problem 11.2 *Estimate a matrix* $\boldsymbol{G}$ *such that*

$$(\bar{\boldsymbol{x}}_\alpha, \boldsymbol{G}\bar{\boldsymbol{x}}'_\alpha) = 0, \qquad \alpha = 1, .., N, \tag{11.33}$$

from the data $\boldsymbol{x}_\alpha$ *and* $\boldsymbol{x}_\alpha$, $\alpha = 1, \ldots, N$.

Problem 11.3 *Decompose the matrix* $\boldsymbol{G}$ *into motion parameters* $\{\boldsymbol{h}, \boldsymbol{R}\}$ *in such a way that*

$$\boldsymbol{G} = \boldsymbol{h} \times \boldsymbol{R}. \tag{11.34}$$

Consider Problem 11.2 first. Eq. (11.33) is the hypothesis from which the essential matrix $\boldsymbol{G}$ is to be estimated. The rank of this hypothesis is 1. Since the scale of $\boldsymbol{G}$ is indeterminate, we normalize it so that the resulting translation $\boldsymbol{h}$ is a unit vector. Note the following relationship (see eqs. (2.37) and (2.142)):

$$\begin{aligned}\|\boldsymbol{G}\|^2 &= \mathrm{tr}\left((\boldsymbol{h} \times \boldsymbol{R})(\boldsymbol{h} \times \boldsymbol{R})^\top\right) = \mathrm{tr}\left((\boldsymbol{h} \times \boldsymbol{I})\boldsymbol{R}\boldsymbol{R}^\top(\boldsymbol{h} \times \boldsymbol{I})^\top\right) \\ &= \mathrm{tr}(\|\boldsymbol{h}\|^2\boldsymbol{I} - \boldsymbol{h}\boldsymbol{h}^\top) = 2\|\boldsymbol{h}\|^2. \end{aligned} \tag{11.35}$$

Hence, normalizing $\boldsymbol{h}$ to $\|\boldsymbol{h}\| = 1$ is equivalent to normalizing $\boldsymbol{G}$ to $\|\boldsymbol{G}\| = \sqrt{2}$. If the essential matrix $\boldsymbol{G} = \boldsymbol{h} \times \boldsymbol{R}$ is regarded as a variable in eqs. (11.11) and (11.12), it can be optimally estimated by the following optimization:

$$\begin{aligned} J[\boldsymbol{G}] &= \sum_{\alpha=1}^{N} \frac{(\boldsymbol{x}_\alpha, \boldsymbol{G}\boldsymbol{x}'_\alpha)^2}{(\boldsymbol{x}'_\alpha, \boldsymbol{G}^\top V[\boldsymbol{x}_\alpha]\boldsymbol{G}\boldsymbol{x}'_\alpha) + (\boldsymbol{x}_\alpha, \boldsymbol{G}V[\boldsymbol{x}'_\alpha]\boldsymbol{G}^\top\boldsymbol{x}_\alpha) + (V[\boldsymbol{x}_\alpha]\boldsymbol{G}; \boldsymbol{G}V[\boldsymbol{x}'_\alpha])} \\ &\to \min. \end{aligned} \tag{11.36}$$

Let $\hat{\boldsymbol{G}}$ be the resulting estimate. Since $\boldsymbol{G}$ has eight degrees of freedom (the nine elements are normalized), the residual $J[\hat{\boldsymbol{G}}]$ is a χ^2 variable with $N - 8$ degrees of freedom. It follows that the number N of the pairs of corresponding image points must be such that

$$N \geq 8. \tag{11.37}$$

The residual is 0 when $N = 8$. In other words, the linearization technique requires *three extra pairs of image points*[3] as compared with direct optimization (11.10) (see eq. (11.13)).

If the covariance matrices $V[\boldsymbol{x}_\alpha]$ and $V[\boldsymbol{x}'_\alpha]$ are decomposed into the *noise level* ϵ and the *normalized covariance matrices* $V_0[\boldsymbol{x}_\alpha]$ and $V_0[\boldsymbol{x}'_\alpha]$ in the form

$$V[\boldsymbol{x}_\alpha] = \epsilon^2 V_0[\boldsymbol{x}_\alpha], \qquad V[\boldsymbol{x}'_\alpha] = \epsilon^2 V_0[\boldsymbol{x}'_\alpha], \tag{11.38}$$

the least-squares approximation to (11.36) has the form

$$\tilde{J}[\boldsymbol{G}] = (\boldsymbol{G}; \mathcal{M}\boldsymbol{G}) \to \min, \tag{11.39}$$

where the *moment tensor* $\mathcal{M}$ is defined by

$$\mathcal{M} = \frac{1}{N}\sum_{\alpha=1}^{N} W_\alpha \boldsymbol{x}_\alpha \otimes \boldsymbol{x}'_\alpha \otimes \boldsymbol{x}_\alpha \otimes \boldsymbol{x}'_\alpha, \tag{11.40}$$

$$W_\alpha = 1 \Big/ \Big((\boldsymbol{x}'_\alpha, \boldsymbol{G}^{*\top} V_0[\boldsymbol{x}_\alpha] \boldsymbol{G}^* \boldsymbol{x}'_\alpha) + (\boldsymbol{x}_\alpha, \boldsymbol{G}^* V_0[\boldsymbol{x}'_\alpha] \boldsymbol{G}^{*\top} \boldsymbol{x}_\alpha) + \epsilon^2 (V_0[\boldsymbol{x}_\alpha]\boldsymbol{G}^*; \boldsymbol{G}^* V_0[\boldsymbol{x}'_\alpha]) \Big). \tag{11.41}$$

Here, $\boldsymbol{G}^*$ is an appropriate estimate of $\boldsymbol{G}$. The minimum is sought under the normalization $\|\boldsymbol{G}\| = \sqrt{2}$.

If a 9-vector $\boldsymbol{g}$ and a 99-matrix $\boldsymbol{M}$ are defined by casting the 33-matrix $\boldsymbol{G}$ and the 3333-tensor $\mathcal{M}$, respectively, in the form[4]

$$\boldsymbol{g} = \text{type}_9[\boldsymbol{G}], \qquad \boldsymbol{M} = \text{type}_{99}[\mathcal{M}], \tag{11.42}$$

we have $(\boldsymbol{G}; \mathcal{M}\boldsymbol{G}) = (\boldsymbol{g}, \boldsymbol{M}\boldsymbol{g})$ and $\|\boldsymbol{G}\| = \|\boldsymbol{g}\|$ (see Section 2.4.2). Hence, the solution of the optimization (11.39) under the constraint $\|\boldsymbol{G}\| = \sqrt{2}$ is obtained as the eigenmatrix of norm $\sqrt{2}$ of tensor $\mathcal{M}$ (i.e., the matrix $\boldsymbol{G}$ obtained by cast from the eigenvector $\boldsymbol{g}$ of norm $\sqrt{2}$ of the matrix $\boldsymbol{M}$) for the smallest eigenvalue.

Let $\bar{\mathcal{M}}$ be the unperturbed moment tensor obtained by replacing $\boldsymbol{x}_\alpha$ and $\boldsymbol{x}'_\alpha$ by $\bar{\boldsymbol{x}}_\alpha$ and $\bar{\boldsymbol{x}}'_\alpha$, respectively, in eq. (11.40). Eq. (11.33) implies that the true value $\bar{\boldsymbol{G}}$ satisfies $\bar{\mathcal{M}}\bar{\boldsymbol{G}} = \boldsymbol{O}$, i.e., $\bar{\boldsymbol{G}}$ is the eigenmatrix $\bar{\mathcal{M}}$ for eigenvalue 0. However, $E[\mathcal{M}]$ is generally not equal to $\bar{\mathcal{M}}$. In fact,

$$\begin{aligned} E[x_{\alpha(i)}x_{\alpha(k)}] &= E[(\bar{x}_{\alpha(i)} + \Delta x_{\alpha(i)})(\bar{x}_{\alpha(k)} + \Delta x_{\alpha(k)})] \\ &= \bar{x}_{\alpha(i)}\bar{x}_{\alpha(k)} + E[\Delta x_{\alpha(i)}\Delta x_{\alpha(k)}] = \bar{x}_{\alpha(i)}\bar{x}_{\alpha(k)} + \epsilon^2 V_0[\boldsymbol{x}_\alpha]_{ij}, \end{aligned} \tag{11.43}$$

---

[3] Algorithms for computing the motion parameters from *eight* pairs of points by using the linearization technique are known as *eight-point algorithms*.

[4] Here, $\mathbf{G}$ is a 33-matrix and $\mathcal{M}$ is a 3333-tensor, so they are respectively cast into a 9-vector and a 99-matrix, whereas $\mathbf{Q}$ in Section 10.2.2 is a (33)-matrix and $\mathcal{M}$ is a (33)(33)-tensor, so they are respectively cast into a 6-vector and a 66-matrix; see eq. (10.41).

and similarly

$$E[x'_{\alpha(j)}x'_{\alpha(l)}] = \bar{x}'_{\alpha(j)}\bar{x}'_{\alpha(l)} + \epsilon^2 V_0[\boldsymbol{x}'_\alpha]_{jl}. \tag{11.44}$$

It follows that

$$\begin{aligned} E[M_{ijkl}] &= \frac{1}{N}\sum_{\alpha=1}^{N} W_\alpha E[x_{\alpha(i)}x_{\alpha(k)}]E[x'_{\alpha(j)}x_{\alpha(l)}] \\ &= \bar{M}_{ijkl} + \frac{\epsilon^2}{N}\sum_{\alpha=1}^{N} W_\alpha \left(V_0[\boldsymbol{x}_\alpha]_{ik}\bar{x}'_j\bar{x}'_l + V_0[\boldsymbol{x}'_\alpha]_{jl}\bar{x}_i\bar{x}_k\right) \\ &\quad + \frac{\epsilon^4}{N}\sum_{\alpha=1}^{N} W_\alpha V_0[\boldsymbol{x}_\alpha]_{ik}V_0[\boldsymbol{x}'_\alpha]_{jl}. \end{aligned} \tag{11.45}$$

Hence, the solution of the least-squares approximation is statistically biased whatever weights W_α are used.

11.2.2 Unbiased estimation and renormalization

Constructing the unbiased least-squares approximation requires defining an unbiased moment tensor $\hat{\mathcal{M}}$ such that $E[\hat{\mathcal{M}}] = \bar{\mathcal{M}}$. Define 3333-tensors $\mathcal{N}^{(1)} = (N^{(1)}_{ijkl})$ and $\mathcal{N}^{(2)} = (N^{(2)}_{ijkl})$ by

$$N^{(1)}_{ijkl} = \frac{1}{N}\sum_{\alpha=1}^{N} W_\alpha \left(V_0[\boldsymbol{x}_\alpha]_{ik}x'_{\alpha(j)}x'_{\alpha(l)} + V_0[\boldsymbol{x}'_\alpha]_{jl}x_{\alpha(i)}x_{\alpha(k)}\right), \tag{11.46}$$

$$N^{(2)}_{ijkl} = \frac{1}{N}\sum_{\alpha=1}^{N} W_\alpha V_0[\boldsymbol{x}_\alpha]_{ik}V_0[\boldsymbol{x}'_\alpha]_{jl}. \tag{11.47}$$

Let $\bar{\mathcal{N}}^{(1)}$ be the unperturbed value of $\mathcal{N}^{(1)}$ obtained by replacing $\boldsymbol{x}_\alpha$ and $\boldsymbol{x}'_\alpha$ by $\bar{\boldsymbol{x}}_\alpha$ and $\bar{\boldsymbol{x}}'_\alpha$, respectively, in eq. (11.46). Then, eq. (11.45) can be written in the following form:

$$E[\mathcal{M}] = \bar{\mathcal{M}} + \epsilon^2\bar{\mathcal{N}}^{(1)} + \epsilon^4\mathcal{N}^{(2)}. \tag{11.48}$$

It is immediately seen from eqs. (11.46) and (11.47) that

$$E[\mathcal{N}^{(1)}] = \bar{\mathcal{N}}^{(1)} + 2\epsilon^2\mathcal{N}^{(2)}. \tag{11.49}$$

It follows that if we define

$$\hat{\mathcal{M}} = \mathcal{M} - \epsilon^2\mathcal{N}^{(1)} + \epsilon^4\mathcal{N}^{(2)}, \tag{11.50}$$

we have $E[\hat{\mathcal{M}}] = \bar{\mathcal{M}}$. Hence, we obtain the unbiased least-squares approximation

$$\hat{J}[\boldsymbol{G}] = (\boldsymbol{G}; \hat{\mathcal{M}}\boldsymbol{G}) \to \min. \tag{11.51}$$

The solution under the constraint $\|\boldsymbol{G}\| = \sqrt{2}$ is obtained as the eigenmatrix of norm $\sqrt{2}$ of $\hat{\mathcal{M}}$ for the smallest eigenvalue.

If the noise level ϵ is not known, the second order renormalization procedure is given as follows (see Section 9.6.2):

1. Let $c = 0$ and $W_\alpha = 1$, $\alpha = 1, ..., N$.

2. Compute the 3333-tensors $\mathcal{M}$, $\mathcal{N}^{(1)}$, and $\mathcal{N}^{(1)}$ defined by eqs. (11.40), (11.46), and (11.47), respectively.

3. Compute the smallest eigenvalue λ of the 3333-tensor

$$\hat{\mathcal{M}} = \mathcal{M} - c\mathcal{N}^{(1)} + c^2\mathcal{N}^{(2)}, \tag{11.52}$$

and the corresponding eigenmatrix $\boldsymbol{G}$ of norm $\sqrt{2}$.

4. If $\lambda \approx 0$, return $\boldsymbol{G}$, c, and $\hat{\mathcal{M}}$. Else, update the constant c and the weights W_α as follows:

$$D = \Big((\boldsymbol{G}; \mathcal{N}^{(1)}\boldsymbol{G}) - 2c(\boldsymbol{G}; \mathcal{N}^{(2)}\boldsymbol{G})\Big)^2 - 8\lambda(\boldsymbol{G}; \mathcal{N}^{(2)}\boldsymbol{G}), \tag{11.53}$$

$$c \leftarrow \begin{cases} c + \dfrac{(\boldsymbol{G}; \mathcal{N}^{(1)}\boldsymbol{G}) - 2c(\boldsymbol{G}; \mathcal{N}^{(2)}\boldsymbol{G}) - \sqrt{D}}{2(\boldsymbol{G}; \mathcal{N}^{(2)}\boldsymbol{G})}, & \text{If } D \geq 0, \\ c + \dfrac{2\lambda}{(\boldsymbol{G}; \mathcal{N}^{(1)}\boldsymbol{G})}, & \text{If } D < 0, \end{cases} \tag{11.54}$$

$$W_\alpha \leftarrow 1 \Big/ \Big((\boldsymbol{x}'_\alpha, \boldsymbol{G}^\top V_0[\boldsymbol{x}_\alpha]\boldsymbol{G}\boldsymbol{x}'_\alpha) + (\boldsymbol{x}_\alpha, \boldsymbol{G}V_0[\boldsymbol{x}'_\alpha]\boldsymbol{G}^\top \boldsymbol{x}_\alpha) + c(V_0[\boldsymbol{x}_\alpha]\boldsymbol{G}; \boldsymbol{G}V_0[\boldsymbol{x}'_\alpha]) \Big). \tag{11.55}$$

5. Go back to Step 2.

After renormalization, the squared noise level ϵ^2 is estimated in the form

$$\hat{\epsilon}^2 = \frac{c}{1 - 8/N}. \tag{11.56}$$

Its expectation and variance are

$$E[\hat{\epsilon}^2] = \epsilon^2, \qquad V[\hat{\epsilon}^2] = \frac{2\epsilon^4}{N - 8}. \tag{11.57}$$

The covariance tensor $\mathcal{V}[\boldsymbol{G}]$ of the resulting estimate $\boldsymbol{G}$ is obtained in the following form:

$$\mathcal{V}[\boldsymbol{G}] = \frac{\hat{\epsilon}^2}{N}(\hat{\mathcal{M}})_8^-. \tag{11.58}$$

11.3 Optimal Correction and Decomposition

11.3.1 *Correction of the essential matrix*

A. Decomposability condition

We now consider Problem 11.3. Here, the crucial fact is that *not every matrix* $\boldsymbol{G}$ *can be decomposed into the form* $\boldsymbol{G} = \boldsymbol{h} \times \boldsymbol{R}$. A matrix $\boldsymbol{G}$ is said to be *decomposable* if there exist a unit vector $\boldsymbol{h}$ and a rotation matrix $\boldsymbol{R}$ such that $\boldsymbol{G} = \boldsymbol{h} \times \boldsymbol{R}$. It can be proved that a matrix $\boldsymbol{G}$ is decomposable if and only if its *singular values* are 1, 1, and 0 (see Section 2.3.1), which can equivalently be written as

$$\det \boldsymbol{G} = 0, \qquad \|\boldsymbol{G}\| = \sqrt{2}, \qquad \|\boldsymbol{G}\boldsymbol{G}^\top\| = \sqrt{2}. \tag{11.59}$$

We call these equations the *decomposability condition.*

B. Correction for decomposability

The optimal estimate of $\boldsymbol{G}$ computed by renormalization may not be exactly decomposable, so it must be optimally corrected into $\hat{\boldsymbol{G}}$ that satisfies eqs. (11.59) (see Section 9.5.2). Since the normalization $\|\boldsymbol{G}\| = \sqrt{2}$ is imposed by renormalization, we only need to impose

$$\det \hat{\boldsymbol{G}} = 0, \qquad \|\hat{\boldsymbol{G}}\hat{\boldsymbol{G}}^\top\|^2 = 2. \tag{11.60}$$

The rank of this constraint is 2.

Let $\hat{\boldsymbol{G}} = \boldsymbol{G} - \Delta\boldsymbol{G}$. In linear approximation, the first constraint $\det(\boldsymbol{G} - \Delta\boldsymbol{G}) = 0$ can be written as $\det \boldsymbol{G} - \mathrm{tr}(\boldsymbol{G}^\dagger \Delta\boldsymbol{G}) = 0$ (see eq. (2.17)), which can be rewritten in the form

$$(\boldsymbol{G}^{\dagger\top}; \Delta\boldsymbol{G}) = \det \boldsymbol{G}, \tag{11.61}$$

where $\boldsymbol{G}^\dagger$ is the *cofactor matrix* of $\boldsymbol{G}$ (see eq. (2.18)). In linear approximation, the second of eqs. (11.60) can be written as

$$(\boldsymbol{G}\boldsymbol{G}^\top\boldsymbol{G}; \Delta\boldsymbol{G}) = \frac{1}{4}\|\boldsymbol{G}\boldsymbol{G}^\top\|^2 - \frac{1}{2}. \tag{11.62}$$

In order not to violate the normalization $\|\boldsymbol{G}\| = \sqrt{2}$, the increment $\Delta\boldsymbol{G}$ must be such that $\Delta\boldsymbol{G} \in \{\boldsymbol{G}\}_L^\perp$ to a first approximation. The increment $\Delta\boldsymbol{G}$ is optimally determined by minimizing the Mahalanobis distance $\|\Delta\boldsymbol{G}\|_{\mathcal{V}[\boldsymbol{G}]}$ (see Section 9.5.2), i.e.,

$$(\Delta\boldsymbol{G}; \mathcal{V}[\boldsymbol{G}]^- \Delta\boldsymbol{G}) \to \min, \qquad \Delta\boldsymbol{G} \in \{\boldsymbol{G}\}_L^\perp, \tag{11.63}$$

under the linearized constraints (11.61) and (11.62). This optimization requires the covariance tensor $\mathcal{V}[\boldsymbol{G}]$, which is given by eq. (11.58) as a by-product of renormalization. Since multiplication of the covariance tensor $\mathcal{V}[\boldsymbol{G}]$ by a positive constant does not affect the solution, it can be replaced by the *normalized covariance tensor* $\mathcal{V}_0[\boldsymbol{G}]$ defined by setting $\hat{\epsilon} = 1$ in eq. (11.58). The first order solution is given as follows (see eqs. (5.12), (5.14) and (5.15)):

$$\Delta \boldsymbol{G} = \lambda_1 \mathcal{V}_0[\boldsymbol{G}]\boldsymbol{G}^{\dagger\top} + \lambda_2 \mathcal{V}_0[\boldsymbol{G}](\boldsymbol{G}\boldsymbol{G}^\top \boldsymbol{G}), \tag{11.64}$$

$$\begin{pmatrix} \lambda_1 \\ \lambda_2 \end{pmatrix} = \boldsymbol{W} \begin{pmatrix} \det \boldsymbol{G} \\ (\|\boldsymbol{G}\boldsymbol{G}^\top\|^2 - 2)/4 \end{pmatrix}, \tag{11.65}$$

$$\boldsymbol{W} = \begin{pmatrix} (\boldsymbol{G}^{\dagger\top}; \mathcal{V}_0[\boldsymbol{G}]\boldsymbol{G}^{\dagger\top}) & (\boldsymbol{G}^{\dagger\top}; \mathcal{V}_0[\boldsymbol{G}](\boldsymbol{G}\boldsymbol{G}^\top \boldsymbol{G})) \\ (\boldsymbol{G}\boldsymbol{G}^\top \boldsymbol{G}; \mathcal{V}_0[\boldsymbol{G}]\boldsymbol{G}^{\dagger\top}) & (\boldsymbol{G}\boldsymbol{G}^\top \boldsymbol{G}; \mathcal{V}_0[\boldsymbol{G}](\boldsymbol{G}\boldsymbol{G}^\top \boldsymbol{G})) \end{pmatrix}^{-1}. \tag{11.66}$$

Since the constraint $\Delta \boldsymbol{G} \in \{\boldsymbol{G}\}_L^\perp$ ensures $\|\boldsymbol{G}\| = \sqrt{2}$ only to a first approximation, the actual correction takes the form

$$\hat{\boldsymbol{G}} = \sqrt{2} N[\boldsymbol{G} - \Delta \boldsymbol{G}], \tag{11.67}$$

where $N[\,\cdot\,]$ is the normalization operator (see eq. (5.25)).

This correction is iterated until the decomposability condition (11.60) is sufficiently satisfied. In this process, the normalized covariance tensor $\mathcal{V}_0[\boldsymbol{G}]$ also needs to be updated, since its null space should change as $\boldsymbol{G}$ changes (see eq. (5.26)). So, it is projected in the following form:

$$\hat{\mathcal{V}}_0[\boldsymbol{G}]_{ijkl} = \sum_{m,n,p,q=1}^{3} \hat{P}_{ijmn} \hat{P}_{klpq} \mathcal{V}_0[\boldsymbol{G}]_{mnpq}. \tag{11.68}$$

Here, $\mathcal{V}_0[\boldsymbol{G}]_{ijkl}$ and $\hat{\mathcal{V}}_0[\boldsymbol{G}]_{ijkl}$ are the $(ijkl)$ elements of $\mathcal{V}_0[\boldsymbol{G}]$ and $\hat{\mathcal{V}}_0[\boldsymbol{G}]$, respectively, and $\mathcal{P}_{\hat{\boldsymbol{g}}} = (\hat{P}_{ijkl})$ is the orthogonal projection tensor onto $\{\hat{\boldsymbol{G}}\}_L^\perp$ defined by

$$\hat{P}_{ijkl} = \delta_{ik}\delta_{jl} - \frac{1}{2}\hat{G}_{ij}\hat{G}_{kl}. \tag{11.69}$$

C. Singularity of the decomposability condition

Since $\boldsymbol{G}$ has nine elements, the three equations (11.59) should constrain $\hat{\boldsymbol{G}}$ to be in a *six*-dimensional manifold in the nine-dimensional parameter space. However, a decomposable matrix $\hat{\boldsymbol{G}}$ is specified by a unit vector $\boldsymbol{h}$ and a rotation matrix $\boldsymbol{R}$, so $\hat{\boldsymbol{G}}$ has *five* degrees of freedom. Where does this discrepancy come from?

This anomaly originates from the fact that the constraint given by eqs. (11.59) is *singular* in the sense discussed in Section 5.1.1. In general, the three equations $\det \boldsymbol{G} = c_1$, $\|\boldsymbol{G}\| = c_2$, and $\|\boldsymbol{G}\boldsymbol{G}^\top\| = c_3$ constrain $\boldsymbol{G}$ to be in a *six*-dimensional manifold. However, the particular values $c_1 = 0$, $c_2 = \sqrt{2}$, and $c_3 = \sqrt{2}$ happen to be *critical values*, at which the six-dimensional

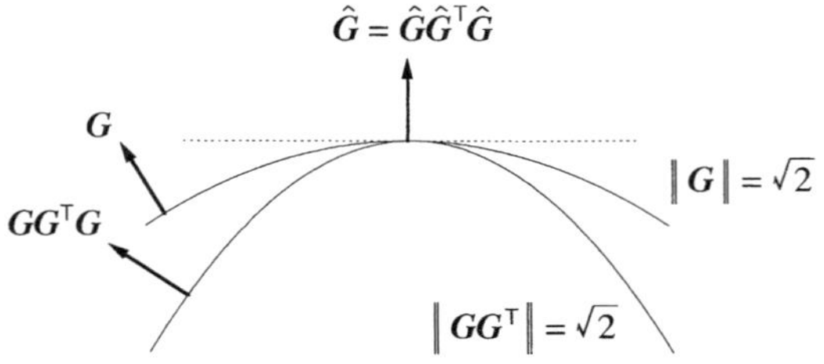

Fig. 11.4. The manifold defined by $\|\mathbf{G}\| = \sqrt{2}$ and the manifold defined by $\|\mathbf{G}\mathbf{G}^\top\| = \sqrt{2}$ do not meet transversally.

manifold *degenerates* into *five* dimensions. To be specific, the manifold defined by $\|\boldsymbol{G}\| = \sqrt{2}$ and the manifold defined by $\|\boldsymbol{G}\boldsymbol{G}^\top\| = \sqrt{2}$ *do not meet transversally* in $\mathcal{R}^9$ (see Section 3.2.1).

As an analogy, imagine two (two-dimensional) spheres in three dimensions. If they meet, their intersection is generally a (one-dimensional) circle. However, as the two spheres move away from each other, they eventually reach a critical state, at which their intersection is a (zero-dimensional) point. If they move further apart, there no longer exists an intersection[5].

The critical nature of eqs. (11.59) is easily detected from eq. (11.62). In fact, in the limit in which $\hat{\boldsymbol{G}}$ is decomposable, the "surface normal" $\hat{\boldsymbol{G}}\hat{\boldsymbol{G}}^\top\hat{\boldsymbol{G}}$ coincides with $\hat{\boldsymbol{G}}$ (see the next subsection). Hence, the linear subspace defined by eqs. (11.62) coincides with $\{\hat{\boldsymbol{G}}\}_L^\perp$ (Fig. 11.4), just as two spheres in contact have a common normal at the contact point, sharing a common tangent plane. As a result of this singularity, the convergence of the iterations in the form of (11.67) is no longer quadratic (see Section 5.1.3).

11.3.2 Decomposition into motion parameters

If $\boldsymbol{G}$ is decomposable into motion parameters $\{\boldsymbol{h}, \boldsymbol{R}\}$, the decomposition is done as follows. First, note the following identities (see eqs. (2.36) and (2.48)):

$$\boldsymbol{G}^\top\boldsymbol{h} = ((\boldsymbol{h}\times\boldsymbol{I})\boldsymbol{R})^\top\boldsymbol{h} = -\boldsymbol{R}^\top\boldsymbol{h}\times\boldsymbol{h} = \boldsymbol{0}, \tag{11.70}$$

$$\boldsymbol{G}\boldsymbol{G}^\top = (\boldsymbol{h}\times\boldsymbol{R})(\boldsymbol{h}\times\boldsymbol{R})^\top = (\boldsymbol{h}\times\boldsymbol{I})\boldsymbol{R}\boldsymbol{R}^\top(\boldsymbol{h}\times\boldsymbol{I})^\top(\boldsymbol{h}\times\boldsymbol{I})(\boldsymbol{h}\times\boldsymbol{I})^\top = \boldsymbol{P}_{\boldsymbol{h}}. \tag{11.71}$$

Here, $\boldsymbol{P}_{\boldsymbol{h}}$ is the projection matrix along $\boldsymbol{h}$. Eq. (11.70) implies that $\boldsymbol{h}$ is computed up to sign as the unit eigenvector of the symmetric matrix $\boldsymbol{G}\boldsymbol{G}^\top$ for eigenvalue 0, or the smallest eigenvalue in practice.

[5] Such a critical phenomenon is characteristic of a *real* space and disappears in a *complex* space. For example, disjoint spheres in a real space have *imaginary intersections* in a complex space.

Let Z_α and Z'_α be the depths of the αth feature point P_α before and after the camera motion, respectively (see eq. (6.35)). From

$$Z_\alpha \bar{\boldsymbol{x}}_\alpha = \boldsymbol{h} + Z'_\alpha \boldsymbol{R}\bar{\boldsymbol{x}}'_\alpha, \tag{11.72}$$

we obtain

$$Z_\alpha \boldsymbol{h} \times \bar{\boldsymbol{x}}_\alpha = Z'_\alpha \boldsymbol{h} \times \boldsymbol{R}\bar{\boldsymbol{x}}'_\alpha. \tag{11.73}$$

Since $Z_\alpha > 0$ and $Z'_\alpha > 0$, we have

$$|\boldsymbol{h}, \bar{\boldsymbol{x}}_\alpha, \boldsymbol{G}\bar{\boldsymbol{x}}'_\alpha| = (\boldsymbol{h} \times \bar{\boldsymbol{x}}_\alpha, \boldsymbol{h} \times \boldsymbol{R}\bar{\boldsymbol{x}}'_\alpha) = \frac{Z'_\alpha}{Z_\alpha}\|\boldsymbol{h} \times \boldsymbol{R}\bar{\boldsymbol{x}}'_\alpha\|^2 > 0, \tag{11.74}$$

unless $\boldsymbol{h} \times \bar{\boldsymbol{x}}_\alpha = \boldsymbol{h} \times \boldsymbol{R}\bar{\boldsymbol{x}}'_\alpha = \boldsymbol{0}$, in which case P_α is the focus of expansion (see 11.3). It follows that the sign of $\boldsymbol{h}$ is determined by the following criterion:

$$\sum_{\alpha=1}^{N} \mathrm{sgn}(|\boldsymbol{h}, \boldsymbol{x}_\alpha, \boldsymbol{G}\boldsymbol{x}'_\alpha|) > 0. \tag{11.75}$$

On the other hand, if we define the matrix

$$\boldsymbol{K} = -\boldsymbol{h} \times \boldsymbol{G}, \tag{11.76}$$

the following identity is obtained (see eq. (2.48)):

$$\boldsymbol{K} = -\boldsymbol{h} \times (\boldsymbol{h} \times \boldsymbol{R}) = -(\boldsymbol{h} \times \boldsymbol{I})(\boldsymbol{h} \times \boldsymbol{I})\boldsymbol{R} = \boldsymbol{P}_{\boldsymbol{h}}\boldsymbol{R}. \tag{11.77}$$

If we define unit vectors $\boldsymbol{v}_1$ and $\boldsymbol{v}_2$ such that $\boldsymbol{V}_{\boldsymbol{h}} = (\boldsymbol{v}_1, \boldsymbol{v}_2, \boldsymbol{h})$ is an orthogonal matrix, we obtain the identity

$$\boldsymbol{P}_{\boldsymbol{h}} = \boldsymbol{V}_{\boldsymbol{h}} \mathrm{diag}(1,1,0) \boldsymbol{V}_{\boldsymbol{h}}^\top, \tag{11.78}$$

which is the spectral decomposition of $\boldsymbol{P}_{\boldsymbol{h}}$ (see eq. (2.62)). Hence, eq. (11.77) can be written as

$$\boldsymbol{K} = \boldsymbol{V}_{\boldsymbol{h}} \mathrm{diag}(1,1,0) \boldsymbol{V}_{\boldsymbol{h}}^\top \boldsymbol{R} = (\boldsymbol{v}_1, \boldsymbol{v}_2, \boldsymbol{h})\boldsymbol{\Lambda}(\boldsymbol{R}^\top\boldsymbol{v}_1, \boldsymbol{R}^\top\boldsymbol{v}_2, \boldsymbol{R}^\top\boldsymbol{v})^\top. \tag{11.79}$$

Let

$$\boldsymbol{K} = \boldsymbol{V}\boldsymbol{\Lambda}\boldsymbol{U}^\top \tag{11.80}$$

be the singular value decomposition of $\boldsymbol{K}$ (see eq. (2.114)). Comparing eqs. (11.79) and (11.80), we conclude that

$$\boldsymbol{V} = (\varepsilon\boldsymbol{v}_1, \varepsilon\boldsymbol{v}_2, \varepsilon'\boldsymbol{h}), \qquad \boldsymbol{U} = (\varepsilon\boldsymbol{R}^\top\boldsymbol{v}_1, \varepsilon\boldsymbol{R}^\top\boldsymbol{v}_2, \varepsilon''\boldsymbol{R}\boldsymbol{h}), \tag{11.81}$$

where ε, ε', and ε'' are 1 or -1 independently. If $\varepsilon'\varepsilon'' = 1$, we have the following relation (see eqs. (2.6)):

$$\boldsymbol{V}\boldsymbol{U}^\top = \boldsymbol{v}_1\boldsymbol{v}_1^\top\boldsymbol{R} + \boldsymbol{v}_2\boldsymbol{v}_2^\top\boldsymbol{R} + \boldsymbol{h}\boldsymbol{h}^\top\boldsymbol{R} = (\boldsymbol{v}_1\boldsymbol{v}_1^\top + \boldsymbol{v}_2\boldsymbol{v}_2^\top + \boldsymbol{h}\boldsymbol{h}^\top)\boldsymbol{R} = \boldsymbol{R}. \tag{11.82}$$

If $\varepsilon'\varepsilon'' = -1$, we have

$$\boldsymbol{V}\mathrm{diag}(1,1,-1)\boldsymbol{U}^\top = \boldsymbol{v}_1\boldsymbol{v}_1^\top\boldsymbol{R} + \boldsymbol{v}_2\boldsymbol{v}_2^\top\boldsymbol{R} + \boldsymbol{h}\boldsymbol{h}^\top\boldsymbol{R} = \boldsymbol{R}. \tag{11.83}$$

In either case,

$$\boldsymbol{R} = \boldsymbol{V}\mathrm{diag}(1,1,\det(\boldsymbol{V}\boldsymbol{U}^\top))\boldsymbol{U}^\top. \tag{11.84}$$

In summary, the motion parameters $\{\boldsymbol{h}, \boldsymbol{R}\}$ are computed from the essential matrix $\boldsymbol{G}$ by the following procedure:

1. Let $\boldsymbol{h}$ be the unit eigenvector of matrix $\boldsymbol{G}\boldsymbol{G}^\top$ for the smallest eigenvalue.

2. Adjust the sign of $\boldsymbol{h}$ so that $\sum_{\alpha=1}^N |\boldsymbol{h}, \boldsymbol{x}_\alpha, \boldsymbol{G}\boldsymbol{x}'_\alpha| > 0$.

3. Compute the singular value decomposition of the matrix $\boldsymbol{K} = -\boldsymbol{h}\times\boldsymbol{G}$ in the form $\boldsymbol{K} = \boldsymbol{V}\boldsymbol{\Lambda}\boldsymbol{U}^\top$.

4. Compute $\boldsymbol{R} = \boldsymbol{V}\mathrm{diag}(1,1,\det(\boldsymbol{V}\boldsymbol{U}^\top))\boldsymbol{U}^\top$.

Note that Steps 3 and 4 can also be viewed as the following least-squares optimization (see Section 2.3.5):

$$\|\boldsymbol{R} - \boldsymbol{K}\| \to \min. \tag{11.85}$$

On the other hand, there exists an alternative method, which is more efficient. Let $\boldsymbol{G} = (\boldsymbol{g}_1, \boldsymbol{g}_2, \boldsymbol{g}_3)$ and $\boldsymbol{R} = (\boldsymbol{r}_1, \boldsymbol{r}_2, \boldsymbol{r}_3)$. Since $\boldsymbol{G} = \boldsymbol{h}\times\boldsymbol{R}$, we have $\boldsymbol{g}_i = \boldsymbol{h}\times\boldsymbol{r}_i$, i = 1, 2, 3, from which the following identity is obtained:

$$\boldsymbol{r}_i = \boldsymbol{g}_i\times\boldsymbol{h} + \boldsymbol{g}_{i+1}\times\boldsymbol{g}_{i+2}. \tag{11.86}$$

Here, the indices are computed modulo 3 (e.g., $\boldsymbol{g}_4$ means $\boldsymbol{g}_1$). This identity is easily confirmed by substituting $\boldsymbol{g}_i = \boldsymbol{h}\times\boldsymbol{r}_i$ and using the relationship $\boldsymbol{r}_{i+1}\times\boldsymbol{r}_{i+2} = \boldsymbol{r}_i$. The reason why eq. (11.86) has not been used is that eq. (11.86) holds *if and only if* $\boldsymbol{G}$ *is decomposable*. Hence, the computation is *fragile* in the sense that if $\boldsymbol{G}$ is not decomposable, the computed matrix $\boldsymbol{R}$ is not necessarily a rotation matrix. In contrast, the method of singular value decomposition produces an exact rotation matrix $\boldsymbol{R}$ even if $\boldsymbol{G}$ is not decomposable, say, as a result of a premature termination of the correction iterations described in the preceding subsection. In this sense, the computation is *robust*. The same applies to the computation of $\boldsymbol{h}$: many other methods exist for computing $\boldsymbol{h}$ if $\boldsymbol{G}$ is decomposable, and some are more efficient, but the above method is robust, always producing a unit vector $\boldsymbol{h}$ very close to the true solution even if $\boldsymbol{G}$ is not decomposable.

11.4 Reliability of 3-D Reconstruction

11.4.1 Depth and its variance

Let $\{\hat{\boldsymbol{h}}, \hat{\boldsymbol{R}}\}$ be the motion parameters computed by the procedure described so far. From them, the 3-D positions of the feature points are reconstructed just as in the case of stereo vision. Namely, each corresponding pair $\{\boldsymbol{x}_\alpha, \boldsymbol{x}'_\alpha\}$ is optimally corrected into $\{\hat{\boldsymbol{x}}_\alpha, \hat{\boldsymbol{x}}'_\alpha\}$ so that the epipolar equation $(\hat{\boldsymbol{x}}_\alpha, \hat{\boldsymbol{G}}\hat{\boldsymbol{x}}'_\alpha) = 0$ is satisfied, where $\hat{\boldsymbol{G}} = \hat{\boldsymbol{h}} \times \hat{\boldsymbol{R}}$. The correction is given as follows (see eqs. (6.20)):

$$\Delta \boldsymbol{x}_\alpha = \frac{(\boldsymbol{x}_\alpha, \hat{\boldsymbol{G}}\boldsymbol{x}'_\alpha) V_0[\boldsymbol{x}_\alpha] \hat{\boldsymbol{G}}\boldsymbol{x}'_\alpha}{(\boldsymbol{x}'_\alpha, \hat{\boldsymbol{G}}^\top V_0[\boldsymbol{x}_\alpha]\hat{\boldsymbol{G}}\boldsymbol{x}'_\alpha) + (\boldsymbol{x}_\alpha, \hat{\boldsymbol{G}} V_0[\boldsymbol{x}'_\alpha]\hat{\boldsymbol{G}}^\top \boldsymbol{x}_\alpha)},$$

$$\Delta \boldsymbol{x}'_\alpha = \frac{(\boldsymbol{x}_\alpha, \hat{\boldsymbol{G}}\boldsymbol{x}'_\alpha) V_0[\boldsymbol{x}'_\alpha] \hat{\boldsymbol{G}}^\top \boldsymbol{x}_\alpha}{(\boldsymbol{x}'_\alpha, \hat{\boldsymbol{G}}^\top V_0[\boldsymbol{x}_\alpha]\hat{\boldsymbol{G}}\boldsymbol{x}'_\alpha) + (\boldsymbol{x}_\alpha, \hat{\boldsymbol{G}} V_0[\boldsymbol{x}'_\alpha]\hat{\boldsymbol{G}}^\top \boldsymbol{x}_\alpha)}. \tag{11.87}$$

As discussed in Chapter 6, this correction has the following geometric interpretations (they are equivalent to each other):

- Image points $\boldsymbol{x}_\alpha$ and $\boldsymbol{x}'_\alpha$ are moved to $\hat{\boldsymbol{x}}_\alpha$ and $\hat{\boldsymbol{x}}'_\alpha$, respectively, so that $\hat{\boldsymbol{x}}_\alpha$ is *on the epipolar of* $\hat{\boldsymbol{x}}'_\alpha$ in the first image and $\hat{\boldsymbol{x}}'_\alpha$ is *on the epipolar of* $\hat{\boldsymbol{x}}_\alpha$ in the second image (see Section 6.1.2).
- Image points $\boldsymbol{x}_\alpha$ and $\boldsymbol{x}'_\alpha$ are moved to $\hat{\boldsymbol{x}}_\alpha$ and $\hat{\boldsymbol{x}}'_\alpha$, respectively, so that *the lines of sight of* $\hat{\boldsymbol{x}}_\alpha$ *and* $\hat{\boldsymbol{x}}'_\alpha$ *meet in the scene* (see Fig. 6.4).
- The direct sum $\boldsymbol{x}_\alpha \oplus \boldsymbol{x}'_\alpha$ is identified with a point in the four-dimensional space $\mathcal{R}^4 = \mathcal{R}^2 \oplus \mathcal{R}^2$ and moved to $\hat{\boldsymbol{x}}_\alpha \oplus \hat{\boldsymbol{x}}'_\alpha$ so that $\hat{\boldsymbol{x}}_\alpha \oplus \hat{\boldsymbol{x}}'_\alpha$ is on the three-dimensional manifold defined by $(\boldsymbol{x}, \hat{\boldsymbol{G}}\boldsymbol{x}') = 0$ (see Fig. 6.10).

According to the general theory in Section 7.1.2, we obtain the following relationship for the residual (see eq. (11.11)):

$$J_0[\hat{\boldsymbol{h}}, \hat{\boldsymbol{R}}] = \sum_{\alpha=1}^{N} \Big((\boldsymbol{x}_\alpha - \hat{\boldsymbol{x}}_\alpha, V_0[\boldsymbol{x}_\alpha]^- (\boldsymbol{x}_\alpha - \hat{\boldsymbol{x}}_\alpha)) + (\boldsymbol{x}'_\alpha - \hat{\boldsymbol{x}}'_\alpha, V_0[\boldsymbol{x}'_\alpha]^- (\boldsymbol{x}'_\alpha - \hat{\boldsymbol{x}}'_\alpha)) \Big)$$
$$= \sum_{\alpha=1}^{N} \frac{(\boldsymbol{x}_\alpha, \hat{\boldsymbol{G}}\boldsymbol{x}'_\alpha)^2}{(\hat{\boldsymbol{x}}'_\alpha, \hat{\boldsymbol{G}}^\top V_0[\boldsymbol{x}_\alpha]\hat{\boldsymbol{G}}\hat{\boldsymbol{x}}'_\alpha) + (\hat{\boldsymbol{x}}_\alpha, \hat{\boldsymbol{G}} V_0[\boldsymbol{x}'_\alpha]\hat{\boldsymbol{G}}^\top \hat{\boldsymbol{x}}_\alpha)}. \tag{11.88}$$

Since $J_0[\hat{\boldsymbol{h}}, \hat{\boldsymbol{R}}]/\epsilon^2$ is a χ^2 variable with $N-5$ degrees of freedom, an unbiased estimator of the squared noise level ϵ^2 is obtained in the form

$$\hat{\epsilon}^2 = \frac{J_0[\hat{\boldsymbol{h}}, \hat{\boldsymbol{R}}]}{N-5}. \tag{11.89}$$

Its expectation and variance are

$$E[\hat{\epsilon}^2] = \epsilon^2, \qquad V[\hat{\epsilon}^2] = \frac{2\epsilon^4}{N-5}. \tag{11.90}$$

The rigidity test (11.14) can be rewritten as

$$\frac{\hat{\epsilon}^2}{\epsilon^2} > \frac{\chi^2_{N-5,a}}{N-5}. \tag{11.91}$$

Using the corrected values $\hat{\boldsymbol{x}}_\alpha$ and $\hat{\boldsymbol{x}}'_\alpha$, we can compute the depths as follows (see eqs. (6.38)):

$$\hat{Z}_\alpha = \frac{(\hat{\boldsymbol{h}} \times \hat{\boldsymbol{R}}\hat{\boldsymbol{x}}'_\alpha, \hat{\boldsymbol{x}}_\alpha \times \hat{\boldsymbol{R}}\hat{\boldsymbol{x}}'_\alpha)}{\|\hat{\boldsymbol{x}}_\alpha \times \hat{\boldsymbol{R}}\hat{\boldsymbol{x}}'_\alpha\|^2},$$

$$\hat{Z}'_\alpha = \frac{(\hat{\boldsymbol{h}} \times \hat{\boldsymbol{x}}_\alpha, \hat{\boldsymbol{x}}_\alpha \times \hat{\boldsymbol{R}}\hat{\boldsymbol{x}}'_\alpha)}{\|\hat{\boldsymbol{x}}_\alpha \times \hat{\boldsymbol{R}}\hat{\boldsymbol{x}}'_\alpha\|^2}. \tag{11.92}$$

The corresponding space point is reconstructed to be

$$\hat{\boldsymbol{r}}_\alpha = \hat{Z}_\alpha \hat{\boldsymbol{x}}_\alpha. \tag{11.93}$$

Up to now, there exist *two* solutions, since renormalization computes the essential matrix $\boldsymbol{G}$ *up to sign*. It is easily seen that if one solution is $\{\boldsymbol{h}, \boldsymbol{R}\}$, Z_α, and Z'_α, the other solution[6] is $\{-\boldsymbol{h}, \boldsymbol{R}\}$, $-Z_\alpha$, and $-Z'_\alpha$. Hence, the correct solution is chosen by imposing the constraint

$$\sum_{\alpha=1}^{N}(\mathrm{sgn}(\hat{Z}_\alpha) + \mathrm{sgn}(\hat{Z}'_\alpha)) > 0. \tag{11.94}$$

It appears that we could alternatively impose $\sum_{\alpha=1}^{N}(\hat{Z}_\alpha + \hat{Z}'_\alpha) > 0$. However, this would be dangerous because there exists a possibility that the depth of a feature point located very far away in front of the camera ($Z_\alpha \approx \infty$) is estimated to be very far *behind* the camera ($\hat{Z}_\alpha \approx -\infty$) in the presence of image noise; even one such anomaly could reverse the sign of $\sum_{\alpha=1}^{N}(\hat{Z}_\alpha + \hat{Z}'_\alpha)$.

The reliability of the reconstructed space points is affected by the following two sources of errors:

1. errors in the observed image points $\boldsymbol{x}_\alpha$ and $\boldsymbol{x}'_\alpha$, $\alpha = 1, ..., N$;

2. errors in the computed motion parameters $\{\hat{\boldsymbol{h}}, \hat{\boldsymbol{R}}\}$.

Strictly speaking, errors in $\{\hat{\boldsymbol{h}}, \hat{\boldsymbol{R}}\}$ are correlated with errors in $\boldsymbol{x}_\alpha$ and $\boldsymbol{x}'_\alpha$, since the motion parameters $\{\hat{\boldsymbol{h}}, \hat{\boldsymbol{R}}\}$ are computed from $\boldsymbol{x}_\alpha$ and $\boldsymbol{x}'_\alpha$, $\alpha = 1$,

[6] Eq. (11.74) only guarantees that Z_α and Z'_α have the same sign.

..., N. However, if we focus on a particular feature point, it can be regarded as approximately independent of errors in $\{\hat{\boldsymbol{h}}, \hat{\boldsymbol{R}}\}$ if the number of feature points is large. Hence, the effect of image noise at individual image points and the effect of errors in $\{\hat{\boldsymbol{h}}, \hat{\boldsymbol{R}}\}$ can be treated separately. In other words, we can assume that the motion parameters are accurate when analyzing the effect of image noise, while we can assume that image noise does not exist when analyzing the effect of errors in the motion parameters.

Consider the effect of image noise, assuming that the motion parameters $\{\hat{\boldsymbol{h}}, \hat{\boldsymbol{R}}\}$ are accurate. The covariance matrices of the corrected positions $\hat{\boldsymbol{x}}_\alpha$ and $\hat{\boldsymbol{x}}'_\alpha$ are given by eqs. (6.21), i.e.,

$$V_i[\hat{\boldsymbol{x}}_\alpha] = \hat{\epsilon}^2 \left(V_0[\boldsymbol{x}_\alpha] - \frac{(V_0[\boldsymbol{x}_\alpha]\hat{\boldsymbol{G}}\hat{\boldsymbol{x}}'_\alpha)(V_0[\boldsymbol{x}_\alpha]\hat{\boldsymbol{G}}\hat{\boldsymbol{x}}'_\alpha)^\top}{(\hat{\boldsymbol{x}}'_\alpha, \hat{\boldsymbol{G}}^\top V_0[\boldsymbol{x}_\alpha]\hat{\boldsymbol{G}}\hat{\boldsymbol{x}}'_\alpha) + (\hat{\boldsymbol{x}}_\alpha, \hat{\boldsymbol{G}} V_0[\boldsymbol{x}'_\alpha]\hat{\boldsymbol{G}}^\top \hat{\boldsymbol{x}}_\alpha)} \right),$$

$$V_i[\hat{\boldsymbol{x}}'_\alpha] = \hat{\epsilon}^2 \left(V_0[\boldsymbol{x}'_\alpha] - \frac{(V_0[\boldsymbol{x}'_\alpha]\hat{\boldsymbol{G}}^\top \hat{\boldsymbol{x}}_\alpha)(V_0[\boldsymbol{x}'_\alpha]\hat{\boldsymbol{G}}^\top \hat{\boldsymbol{x}}_\alpha)^\top}{(\hat{\boldsymbol{x}}'_\alpha, \hat{\boldsymbol{G}}^\top V_0[\boldsymbol{x}_\alpha]\hat{\boldsymbol{G}}\hat{\boldsymbol{x}}'_\alpha) + (\hat{\boldsymbol{x}}_\alpha, \hat{\boldsymbol{G}} V_0[\boldsymbol{x}'_\alpha]\hat{\boldsymbol{G}}^\top \hat{\boldsymbol{x}}_\alpha)} \right),$$

$$V_i[\hat{\boldsymbol{x}}_\alpha, \hat{\boldsymbol{x}}'_\alpha] = -\frac{\hat{\epsilon}^2 (V_0[\boldsymbol{x}_\alpha]\hat{\boldsymbol{G}}\hat{\boldsymbol{x}}'_\alpha)(V_0[\boldsymbol{x}'_\alpha]\hat{\boldsymbol{G}}^\top \hat{\boldsymbol{x}}_\alpha)^\top}{(\hat{\boldsymbol{x}}'_\alpha, \hat{\boldsymbol{G}}^\top V_0[\boldsymbol{x}_\alpha]\hat{\boldsymbol{G}}\hat{\boldsymbol{x}}'_\alpha) + (\hat{\boldsymbol{x}}_\alpha, \hat{\boldsymbol{G}} V_0[\boldsymbol{x}'_\alpha]\hat{\boldsymbol{G}}^\top \hat{\boldsymbol{x}}_\alpha)} = V_i[\hat{\boldsymbol{x}}'_\alpha, \hat{\boldsymbol{x}}_\alpha]^\top, \tag{11.95}$$

where $\hat{\epsilon}^2$ is the estimate of ϵ^2 given by eq. (11.89). The subscript i indicates that we are considering image noise only. The variance $V_i[\hat{Z}_\alpha]$ and the covariance vector $V_i[\hat{\boldsymbol{x}}_\alpha, \hat{Z}_\alpha]$ are computed by eq. (6.48), i.e.,

$$V_i[\hat{Z}_\alpha] = \frac{1}{\|\hat{\boldsymbol{x}}_\alpha \times \hat{\boldsymbol{R}}\hat{\boldsymbol{x}}'_\alpha\|^2} \Big(\hat{Z}_\alpha^2 (\hat{\boldsymbol{m}}_\alpha, V_i[\hat{\boldsymbol{x}}_\alpha]\hat{\boldsymbol{m}}_\alpha)$$
$$- 2\hat{Z}_\alpha \hat{Z}'_\alpha (\hat{\boldsymbol{m}}_\alpha, V_i[\hat{\boldsymbol{x}}_\alpha, \hat{\boldsymbol{x}}'_\alpha]\hat{\boldsymbol{R}}^\top \hat{\boldsymbol{m}}_\alpha) + \hat{Z}'^2_\alpha (\hat{\boldsymbol{m}}_\alpha, \hat{\boldsymbol{R}} V_i[\hat{\boldsymbol{x}}'_\alpha]\hat{\boldsymbol{R}}^\top \hat{\boldsymbol{m}}_\alpha) \Big),$$

$$V_i[\hat{\boldsymbol{x}}_\alpha, \hat{Z}_\alpha] = -\frac{(\hat{Z}_\alpha V_i[\hat{\boldsymbol{x}}_\alpha] - \hat{Z}'_\alpha V_i[\hat{\boldsymbol{x}}_\alpha, \hat{\boldsymbol{x}}'_\alpha]\hat{\boldsymbol{R}}^\top)\hat{\boldsymbol{m}}_\alpha}{(\hat{\boldsymbol{m}}_\alpha, \hat{\boldsymbol{x}}_\alpha)}, \tag{11.96}$$

where we put

$$\hat{\boldsymbol{m}}_\alpha = N[\hat{\boldsymbol{h}} \times \hat{\boldsymbol{x}}_\alpha] \times \hat{\boldsymbol{R}}\hat{\boldsymbol{x}}'_\alpha. \tag{11.97}$$

The covariance matrix $V_i[\hat{\boldsymbol{r}}_\alpha]$ of the reconstructed space point $\hat{\boldsymbol{r}}_\alpha$ is given by eq. (6.50), i.e.,

$$V_i[\boldsymbol{r}] = \hat{Z}_\alpha^2 V_i[\hat{\boldsymbol{x}}_\alpha] + 2\hat{Z}_\alpha S[V_i[\hat{\boldsymbol{x}}_\alpha, \hat{Z}_\alpha]\hat{\boldsymbol{x}}^\top] + V_i[\hat{Z}_\alpha]\hat{\boldsymbol{x}}\hat{\boldsymbol{x}}^\top, \tag{11.98}$$

where $S[\,\cdot\,]$ is the symmetrization operator (see eqs. (2.205)).

11.4.2 Effect of errors in the motion parameters

We now consider the effect of errors in the motion parameters $\{\hat{\boldsymbol{h}}, \hat{\boldsymbol{R}}\}$, assuming that $\boldsymbol{x}_\alpha$ and $\boldsymbol{x}'_\alpha$, $\alpha = 1, ..., N$, are all accurate. This analysis was already done in Section 6.7, but there the translation $\hat{\boldsymbol{h}}$ and the rotation $\hat{\boldsymbol{R}}$ were assumed independent. If correlations exist between them, the computation of the covariance matrix $V_m[\hat{\boldsymbol{r}}_\alpha]$ of the reconstructed space point $\hat{\boldsymbol{r}}_\alpha$ must undergo the following stages of computation (the subscript m indicates that we are considering errors in the motion parameters only).

A. Epipolar equation

Since image noise is assumed not to exist, we have $\hat{\boldsymbol{x}}_\alpha = \bar{\boldsymbol{x}}_\alpha$ and $\hat{\boldsymbol{x}}'_\alpha = \bar{\boldsymbol{x}}'_\alpha$. Hence, the epipolar equation $(\hat{\boldsymbol{x}}_\alpha, \bar{\boldsymbol{h}} \times \bar{\boldsymbol{R}}\hat{\boldsymbol{x}}'_\alpha) = 0$ is satisfied. However, if the perturbed motion parameters $\hat{\boldsymbol{h}} = \bar{\boldsymbol{h}} + \Delta\boldsymbol{h}$ and $\hat{\boldsymbol{R}} = \bar{\boldsymbol{R}} + \Delta\boldsymbol{R}$ are substituted for their true values $\{\bar{\boldsymbol{h}}, \bar{\boldsymbol{R}}\}$, the quantity

$$\hat{e}_\alpha = (\hat{\boldsymbol{x}}_\alpha, \hat{\boldsymbol{h}} \times \hat{\boldsymbol{R}}\hat{\boldsymbol{x}}'_\alpha) \tag{11.99}$$

is no longer zero. Writing $\Delta\boldsymbol{R} = \Delta\boldsymbol{\Omega} \times \bar{\boldsymbol{R}}$, we obtain to a first approximation

$$\hat{e}_\alpha = (\hat{\boldsymbol{x}}_\alpha, \Delta\boldsymbol{h} \times \bar{\boldsymbol{R}}\hat{\boldsymbol{x}}'_\alpha + \bar{\boldsymbol{h}} \times \Delta\boldsymbol{R}\hat{\boldsymbol{x}}'_\alpha) = -(\hat{\boldsymbol{a}}_\alpha, \Delta\boldsymbol{h}) - (\hat{\boldsymbol{b}}_\alpha, \Delta\boldsymbol{\Omega}), \tag{11.100}$$

where[7]

$$\hat{\boldsymbol{a}}_\alpha = \hat{\boldsymbol{x}}_\alpha \times \hat{\boldsymbol{R}}\hat{\boldsymbol{x}}'_\alpha, \qquad \hat{\boldsymbol{b}}_\alpha = (\hat{\boldsymbol{x}}_\alpha, \hat{\boldsymbol{R}}\hat{\boldsymbol{x}}'_\alpha)\hat{\boldsymbol{h}} - (\hat{\boldsymbol{h}}, \hat{\boldsymbol{R}}\hat{\boldsymbol{x}}'_\alpha)\hat{\boldsymbol{x}}_\alpha. \tag{11.101}$$

Hence,

$$V_m[\hat{e}_\alpha] = (\hat{\boldsymbol{a}}_\alpha, V[\hat{\boldsymbol{h}}]\hat{\boldsymbol{a}}_\alpha) + 2(\hat{\boldsymbol{a}}_\alpha, V[\hat{\boldsymbol{h}}, \hat{\boldsymbol{R}}]\hat{\boldsymbol{b}}_\alpha) + (\hat{\boldsymbol{b}}_\alpha, V[\hat{\boldsymbol{R}}]\hat{\boldsymbol{b}}_\alpha),$$

$$V_m[\hat{\boldsymbol{h}}, \hat{e}_\alpha] = -V[\hat{\boldsymbol{h}}]\hat{\boldsymbol{a}}_\alpha - V[\hat{\boldsymbol{h}}, \hat{\boldsymbol{R}}]\hat{\boldsymbol{b}}_\alpha,$$

$$V_m[\hat{\boldsymbol{R}}, \hat{e}_\alpha] = -V[\hat{\boldsymbol{R}}, \hat{\boldsymbol{h}}]\hat{\boldsymbol{a}}_\alpha - V[\hat{\boldsymbol{R}}]\hat{\boldsymbol{b}}_\alpha, \tag{11.102}$$

where the covariance matrices $V[\hat{\boldsymbol{h}}]$, $V[\hat{\boldsymbol{h}}, \hat{\boldsymbol{R}}]$, and $V[\hat{\boldsymbol{R}}]$ are computed by eq. (11.31). In computing eq. (11.31), we need the value ϵ^2 because $W_\alpha(\hat{\boldsymbol{h}}, \hat{\boldsymbol{R}})$ involves the covariance matrices $V[\boldsymbol{x}_\alpha] = \epsilon^2 V_0[\boldsymbol{x}_\alpha]$ and $V[\boldsymbol{x}'_\alpha] = \epsilon^2 V_0[\boldsymbol{x}'_\alpha]$ (see eq. (11.12)). It is estimated by eq. (11.89).

B. Correction of image points

Since image points $\hat{\boldsymbol{x}}_\alpha$ and $\hat{\boldsymbol{x}}'_\alpha$ do not satisfy the epipolar equation for the perturbed motion parameters, they are respectively corrected into $\hat{\boldsymbol{x}}_\alpha - \Delta\hat{\boldsymbol{x}}_\alpha$ and

[7] Strictly speaking, $\{\hat{\mathbf{h}}, \hat{\mathbf{R}}\}$ in eqs. (11.101) should be $\{\bar{\mathbf{h}}, \bar{\mathbf{R}}\}$, but the use of $\{\hat{\mathbf{h}}, \hat{\mathbf{R}}\}$ introduces only a second order difference.

$\hat{\boldsymbol{x}}'_\alpha - \Delta\hat{\boldsymbol{x}}'_\alpha$ according to eqs. (11.87). To a first approximation, the correction has the following form:

$$\Delta\hat{\boldsymbol{x}}_\alpha = \frac{\hat{e}_\alpha V_0[\boldsymbol{x}_\alpha]\hat{\boldsymbol{G}}\hat{\boldsymbol{x}}'_\alpha}{(\hat{\boldsymbol{x}}'_\alpha, \hat{\boldsymbol{G}}^\top V_0[\boldsymbol{x}_\alpha]\hat{\boldsymbol{G}}\hat{\boldsymbol{x}}'_\alpha) + (\hat{\boldsymbol{x}}_\alpha, \hat{\boldsymbol{G}}V_0[\boldsymbol{x}'_\alpha]\hat{\boldsymbol{G}}^\top \hat{\boldsymbol{x}}_\alpha)},$$

$$\Delta\hat{\boldsymbol{x}}'_\alpha = \frac{\hat{e}_\alpha V_0[\boldsymbol{x}'_\alpha]\hat{\boldsymbol{G}}^\top \hat{\boldsymbol{x}}_\alpha}{(\hat{\boldsymbol{x}}'_\alpha, \hat{\boldsymbol{G}}^\top V_0[\boldsymbol{x}_\alpha]\hat{\boldsymbol{G}}\hat{\boldsymbol{x}}'_\alpha) + (\hat{\boldsymbol{x}}_\alpha, \hat{\boldsymbol{G}}V_0[\boldsymbol{x}'_\alpha]\hat{\boldsymbol{G}}^\top \hat{\boldsymbol{x}}_\alpha)}. \tag{11.103}$$

Hence, we obtain

$$V_m[\hat{\boldsymbol{x}}_\alpha] = \frac{V_m[\hat{e}_\alpha](V_0[\boldsymbol{x}_\alpha]\hat{\boldsymbol{G}}\hat{\boldsymbol{x}}'_\alpha)(V_0[\boldsymbol{x}_\alpha]\hat{\boldsymbol{G}}\hat{\boldsymbol{x}}'_\alpha)^\top}{\left((\hat{\boldsymbol{x}}'_\alpha, \hat{\boldsymbol{G}}^\top V_0[\boldsymbol{x}_\alpha]\hat{\boldsymbol{G}}\hat{\boldsymbol{x}}'_\alpha) + (\hat{\boldsymbol{x}}_\alpha, \hat{\boldsymbol{G}}V_0[\boldsymbol{x}'_\alpha]\hat{\boldsymbol{G}}^\top \hat{\boldsymbol{x}}_\alpha)\right)^2},$$

$$V_m[\hat{\boldsymbol{x}}'_\alpha] = \frac{V_m[\hat{e}_\alpha](V_0[\boldsymbol{x}'_\alpha]\hat{\boldsymbol{G}}^\top \hat{\boldsymbol{x}}_\alpha)(V_0[\boldsymbol{x}'_\alpha]\hat{\boldsymbol{G}}^\top \hat{\boldsymbol{x}}_\alpha)^\top}{\left((\hat{\boldsymbol{x}}'_\alpha, \hat{\boldsymbol{G}}^\top V_0[\boldsymbol{x}_\alpha]\hat{\boldsymbol{G}}\hat{\boldsymbol{x}}'_\alpha) + (\hat{\boldsymbol{x}}_\alpha, \hat{\boldsymbol{G}}V_0[\boldsymbol{x}'_\alpha]\hat{\boldsymbol{G}}^\top \hat{\boldsymbol{x}}_\alpha)\right)^2},$$

$$V_m[\hat{\boldsymbol{x}}_\alpha, \hat{\boldsymbol{x}}'_\alpha] = \frac{V_m[\hat{e}_\alpha](V_0[\boldsymbol{x}_\alpha]\hat{\boldsymbol{G}}\hat{\boldsymbol{x}}'_\alpha)(V_0[\boldsymbol{x}'_\alpha]\hat{\boldsymbol{G}}^\top \hat{\boldsymbol{x}}_\alpha)^\top}{\left((\hat{\boldsymbol{x}}'_\alpha, \hat{\boldsymbol{G}}^\top V_0[\boldsymbol{x}_\alpha]\hat{\boldsymbol{G}}\hat{\boldsymbol{x}}'_\alpha) + (\hat{\boldsymbol{x}}_\alpha, \hat{\boldsymbol{G}}V_0[\boldsymbol{x}'_\alpha]\hat{\boldsymbol{G}}^\top \hat{\boldsymbol{x}}_\alpha)\right)^2},$$

$$V_m[\hat{\boldsymbol{h}}, \hat{\boldsymbol{x}}_\alpha] = -\frac{V_m[\hat{\boldsymbol{h}}, \hat{e}_\alpha](V_0[\boldsymbol{x}_\alpha]\hat{\boldsymbol{G}}\hat{\boldsymbol{x}}'_\alpha)^\top}{(\hat{\boldsymbol{x}}'_\alpha, \hat{\boldsymbol{G}}^\top V_0[\boldsymbol{x}_\alpha]\hat{\boldsymbol{G}}\hat{\boldsymbol{x}}'_\alpha) + (\hat{\boldsymbol{x}}_\alpha, \hat{\boldsymbol{G}}V_0[\boldsymbol{x}'_\alpha]\hat{\boldsymbol{G}}^\top \hat{\boldsymbol{x}}_\alpha)},$$

$$V_m[\hat{\boldsymbol{h}}, \hat{\boldsymbol{x}}'_\alpha] = -\frac{V_m[\hat{\boldsymbol{h}}, \hat{e}_\alpha](V_0[\boldsymbol{x}'_\alpha]\hat{\boldsymbol{G}}^\top \hat{\boldsymbol{x}}_\alpha)^\top}{(\hat{\boldsymbol{x}}'_\alpha, \hat{\boldsymbol{G}}^\top V_0[\boldsymbol{x}_\alpha]\hat{\boldsymbol{G}}\hat{\boldsymbol{x}}'_\alpha) + (\hat{\boldsymbol{x}}_\alpha, \hat{\boldsymbol{G}}V_0[\boldsymbol{x}'_\alpha]\hat{\boldsymbol{G}}^\top \hat{\boldsymbol{x}}_\alpha)},$$

$$V_m[\hat{\boldsymbol{R}}, \hat{\boldsymbol{x}}_\alpha] = -\frac{V_0[\hat{\boldsymbol{R}}, \hat{e}_\alpha](V_0[\boldsymbol{x}_\alpha]\hat{\boldsymbol{G}}\hat{\boldsymbol{x}}'_\alpha)^\top}{(\hat{\boldsymbol{x}}'_\alpha, \hat{\boldsymbol{G}}^\top V_0[\boldsymbol{x}_\alpha]\hat{\boldsymbol{G}}\hat{\boldsymbol{x}}'_\alpha) + (\hat{\boldsymbol{x}}_\alpha, \hat{\boldsymbol{G}}V_0[\boldsymbol{x}'_\alpha]\hat{\boldsymbol{G}}^\top \hat{\boldsymbol{x}}_\alpha)},$$

$$V_m[\hat{\boldsymbol{R}}, \hat{\boldsymbol{x}}'_\alpha] = -\frac{V_m[\hat{\boldsymbol{h}}, \hat{e}_\alpha](V_0[\boldsymbol{x}'_\alpha]\hat{\boldsymbol{G}}^\top \hat{\boldsymbol{x}}_\alpha)^\top}{(\hat{\boldsymbol{x}}'_\alpha, \hat{\boldsymbol{G}}^\top V_0[\boldsymbol{x}_\alpha]\hat{\boldsymbol{G}}\hat{\boldsymbol{x}}'_\alpha) + (\hat{\boldsymbol{x}}_\alpha, \hat{\boldsymbol{G}}V_0[\boldsymbol{x}'_\alpha]\hat{\boldsymbol{G}}^\top \hat{\boldsymbol{x}}_\alpha)}. \tag{11.104}$$

C. Depth variance

Recall that the depths $\hat{Z}_\alpha$ and $\hat{Z}'_\alpha$ have been determined from following relationship (see eq. (6.35)):

$$\hat{Z}_\alpha \hat{\boldsymbol{x}}_\alpha = \hat{\boldsymbol{h}} + \hat{Z}'_\alpha \hat{\boldsymbol{R}}\hat{\boldsymbol{x}}'_\alpha. \tag{11.105}$$

If $\hat{\boldsymbol{h}}$ and $\hat{\boldsymbol{R}}$ are perturbed by $\Delta\boldsymbol{h}$ and $\Delta\boldsymbol{\Omega} \times \hat{\boldsymbol{R}}$, respectively, the depths $\hat{Z}_\alpha$ and $\hat{Z}'_\alpha$ are accordingly perturbed by ΔZ_α and $\Delta Z'_\alpha$, respectively. To a first approximation, we have

$$\Delta Z_\alpha \hat{\boldsymbol{x}}_\alpha = \Delta Z'_\alpha \hat{\boldsymbol{R}}\hat{\boldsymbol{x}}'_\alpha + \Delta\boldsymbol{h} - \hat{Z}'_\alpha(\hat{\boldsymbol{R}}\hat{\boldsymbol{x}}'_\alpha) \times \Delta\boldsymbol{\Omega} - \hat{Z}_\alpha \Delta\hat{\boldsymbol{x}}_\alpha + \hat{Z}'_\alpha \hat{\boldsymbol{R}}\Delta\hat{\boldsymbol{x}}'_\alpha. \tag{11.106}$$

Taking the vector product with $\hat{\boldsymbol{R}}\hat{\boldsymbol{x}}'_\alpha$ on both sides, we obtain

$$\Delta Z_\alpha(\hat{\boldsymbol{R}}\hat{\boldsymbol{x}}'_\alpha) \times \hat{\boldsymbol{x}}_\alpha = (\hat{\boldsymbol{R}}\hat{\boldsymbol{x}}'_\alpha) \times \Delta\boldsymbol{h} - \hat{Z}'_\alpha(\hat{\boldsymbol{R}}\hat{\boldsymbol{x}}'_\alpha) \times ((\hat{\boldsymbol{R}}\hat{\boldsymbol{x}}'_\alpha) \times \Delta\boldsymbol{\Omega})$$
$$-\hat{Z}_\alpha(\hat{\boldsymbol{R}}\hat{\boldsymbol{x}}'_\alpha) \times \Delta\hat{\boldsymbol{x}}_\alpha + \hat{Z}'_\alpha(\hat{\boldsymbol{R}}\hat{\boldsymbol{x}}'_\alpha) \times \hat{\boldsymbol{R}}\Delta\hat{\boldsymbol{x}}'_\alpha. \tag{11.107}$$

Taking the inner product with

$$\hat{\boldsymbol{n}}_\alpha = N[\hat{\boldsymbol{h}} \times \hat{\boldsymbol{x}}_\alpha] \tag{11.108}$$

on both sides of eq. (11.107), we obtain

$$\Delta Z_\alpha(\hat{\boldsymbol{n}}_\alpha, (\hat{\boldsymbol{R}}\hat{\boldsymbol{x}}'_\alpha) \times \hat{\boldsymbol{x}}_\alpha) = (\hat{\boldsymbol{n}}_\alpha, (\hat{\boldsymbol{R}}\hat{\boldsymbol{x}}'_\alpha) \times \Delta\boldsymbol{h}) + \hat{Z}'_\alpha\|\hat{\boldsymbol{x}}'_\alpha\|^2(\hat{\boldsymbol{n}}_\alpha, \Delta\boldsymbol{\Omega})$$
$$-\hat{Z}_\alpha(\hat{\boldsymbol{n}}_\alpha, (\hat{\boldsymbol{R}}\hat{\boldsymbol{x}}'_\alpha) \times \Delta\hat{\boldsymbol{x}}_\alpha) + \hat{Z}'_\alpha(\hat{\boldsymbol{n}}_\alpha, (\hat{\boldsymbol{R}}\hat{\boldsymbol{x}}'_\alpha) \times \hat{\boldsymbol{R}}\Delta\hat{\boldsymbol{x}}'_\alpha). \tag{11.109}$$

If we define

$$\hat{\boldsymbol{m}}_\alpha = \hat{\boldsymbol{n}}_\alpha \times \hat{\boldsymbol{R}}\hat{\boldsymbol{x}}'_\alpha = N[\hat{\boldsymbol{x}} \times \hat{\boldsymbol{h}}] \times \hat{\boldsymbol{R}}\hat{\boldsymbol{x}}'_\alpha, \tag{11.110}$$

we obtain

$$\Delta Z_\alpha = \frac{(\hat{\boldsymbol{m}}_\alpha, \Delta\boldsymbol{h}) + \hat{Z}'_\alpha\|\hat{\boldsymbol{x}}'_\alpha\|^2(\hat{\boldsymbol{n}}_\alpha, \Delta\boldsymbol{\Omega}) - \hat{Z}_\alpha(\hat{\boldsymbol{m}}_\alpha, \Delta\hat{\boldsymbol{x}}_\alpha) + \hat{Z}'_\alpha(\hat{\boldsymbol{R}}^\top\hat{\boldsymbol{m}}_\alpha, \Delta\hat{\boldsymbol{x}}'_\alpha)}{(\hat{\boldsymbol{m}}_\alpha, \hat{\boldsymbol{x}}_\alpha)}. \tag{11.111}$$

Noting eq. (6.47), we obtain the variance $V_m[\hat{Z}_\alpha]$ and the covariance vector $V_m[\hat{\boldsymbol{x}}_\alpha, \hat{Z}_\alpha]$ in the following form:

$$V_m[\hat{Z}_\alpha] = \frac{1}{\|\hat{\boldsymbol{x}}_\alpha \times \hat{\boldsymbol{R}}\hat{\boldsymbol{x}}'_\alpha\|^2}\Big((\hat{\boldsymbol{m}}_\alpha, V[\hat{\boldsymbol{h}}]\hat{\boldsymbol{m}}_\alpha) + \hat{Z}'^2_\alpha\|\hat{\boldsymbol{x}}'_\alpha\|^4(\hat{\boldsymbol{n}}_\alpha, V[\hat{\boldsymbol{R}}]\hat{\boldsymbol{n}}_\alpha)$$
$$+\hat{Z}^2_\alpha(\hat{\boldsymbol{m}}_\alpha, V_m[\hat{\boldsymbol{x}}_\alpha]\hat{\boldsymbol{m}}_\alpha) + \hat{Z}'^2_\alpha(\hat{\boldsymbol{m}}_\alpha, \hat{\boldsymbol{R}}V_m[\hat{\boldsymbol{x}}'_\alpha]\hat{\boldsymbol{R}}^\top\hat{\boldsymbol{m}}_\alpha)$$
$$+2\hat{Z}'_\alpha\|\hat{\boldsymbol{x}}'_\alpha\|^2(\hat{\boldsymbol{m}}_\alpha, V[\hat{\boldsymbol{h}}, \hat{\boldsymbol{R}}]\hat{\boldsymbol{n}}_\alpha) - 2\hat{Z}_\alpha(\hat{\boldsymbol{m}}_\alpha, V_m[\hat{\boldsymbol{h}}, \hat{\boldsymbol{x}}_\alpha]\hat{\boldsymbol{m}}_\alpha)$$
$$+2\hat{Z}'_\alpha(\hat{\boldsymbol{m}}_\alpha, V_m[\hat{\boldsymbol{h}}, \hat{\boldsymbol{x}}'_\alpha]\hat{\boldsymbol{R}}^\top\hat{\boldsymbol{m}}_\alpha) - 2\hat{Z}_\alpha\hat{Z}'_\alpha\|\hat{\boldsymbol{x}}'_\alpha\|^2(\hat{\boldsymbol{n}}_\alpha, V_m[\hat{\boldsymbol{R}}, \hat{\boldsymbol{x}}_\alpha]\hat{\boldsymbol{m}}_\alpha)$$
$$+2\hat{Z}'^2_\alpha\|\hat{\boldsymbol{x}}'_\alpha\|^2(\hat{\boldsymbol{n}}_\alpha, V_m[\hat{\boldsymbol{R}}, \hat{\boldsymbol{x}}'_\alpha]\hat{\boldsymbol{R}}^\top\hat{\boldsymbol{m}}_\alpha)$$
$$-\,2\hat{Z}_\alpha\hat{Z}'_\alpha(\hat{\boldsymbol{m}}_\alpha, V_m[\hat{\boldsymbol{x}}_\alpha, \hat{\boldsymbol{x}}'_\alpha]\hat{\boldsymbol{R}}^\top\hat{\boldsymbol{m}}_\alpha)\Big),$$

$$V_m[\hat{\boldsymbol{x}}_\alpha, \hat{Z}_\alpha] = \frac{1}{(\hat{\boldsymbol{m}}_\alpha, \hat{\boldsymbol{x}}_\alpha)}\Big(V_m[\hat{\boldsymbol{x}}_\alpha, \hat{\boldsymbol{h}}]\hat{\boldsymbol{m}}_\alpha + \hat{Z}'_\alpha\|\hat{\boldsymbol{x}}'_\alpha\|^2V_m[\hat{\boldsymbol{x}}_\alpha, \hat{\boldsymbol{R}}]\hat{\boldsymbol{n}}_\alpha$$
$$-\,\hat{Z}_\alpha V_m[\hat{\boldsymbol{x}}_\alpha]\hat{\boldsymbol{m}}_\alpha + \hat{Z}'_\alpha V_m[\hat{\boldsymbol{x}}_\alpha, \hat{\boldsymbol{x}}']\hat{\boldsymbol{R}}^\top\hat{\boldsymbol{m}}_\alpha\Big). \tag{11.112}$$

Finally, the covariance matrix $V_m[\hat{\boldsymbol{r}}_\alpha]$ of the reconstructed space point $\hat{\boldsymbol{r}}_\alpha = \hat{Z}_\alpha\hat{\boldsymbol{x}}_\alpha$ is given by

$$V_m[\hat{\boldsymbol{r}}_\alpha] = \hat{Z}^2_\alpha V_m[\hat{\boldsymbol{x}}_\alpha] + 2\hat{Z}_\alpha S[V_m[\hat{\boldsymbol{x}}_\alpha, \hat{Z}_\alpha]\hat{\boldsymbol{x}}_\alpha^\top] + V_m[\hat{Z}_\alpha]\hat{\boldsymbol{x}}_\alpha\hat{\boldsymbol{x}}_\alpha^\top. \tag{11.113}$$

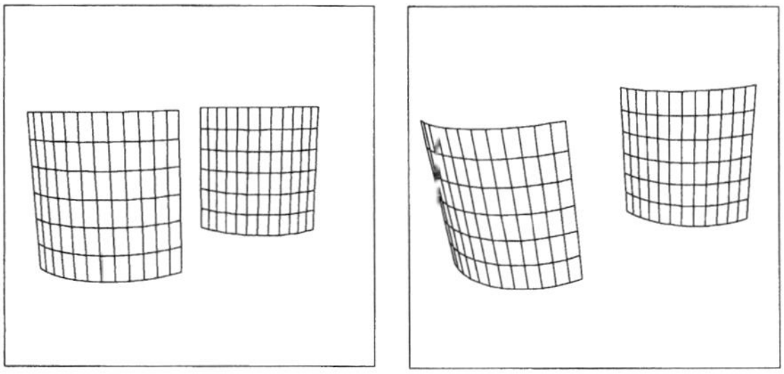

Fig. 11.5. Simulated images of two objects.

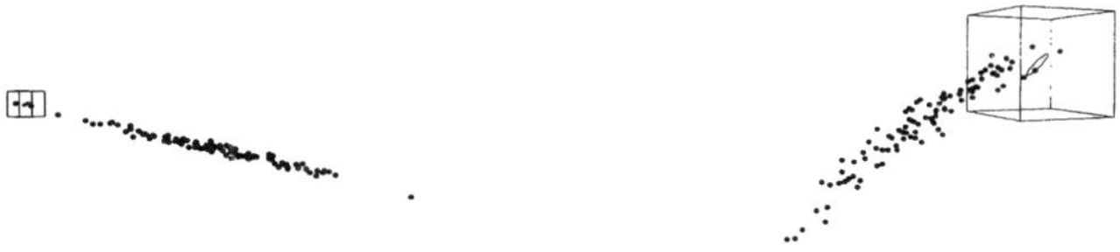

(a) (b)

Fig. 11.6. Least-squares approximation with optimal weights: (a) errors in translation; (b) errors in rotation.

Example 11.1 Fig. 11.5 shows simulated images (512×512 pixels with focal length $f = 600$ (pixels)) of two cylindrical grids in the scene viewed from different angles. We added Gaussian noise of mean 0 and standard deviation $\sigma = 1$ (pixels) to the x- and y-coordinates of each grid point independently, so the noise level is $\epsilon = \sigma/f = 1/600$ (this value is not used in the reconstruction computation). Let $\{\bar{\boldsymbol{h}}, \bar{\boldsymbol{R}}\}$ and $\{\hat{\boldsymbol{h}}, \hat{\boldsymbol{R}}\}$ be the true and the computed motion parameters, respectively. Since $\hat{\boldsymbol{h}}$ is a unit vector, the error in translation is measured by

$$\Delta \boldsymbol{h} = \boldsymbol{P}_{N[\bar{\boldsymbol{h}}]}(\hat{\boldsymbol{h}} - N[\bar{\boldsymbol{h}}]), \tag{11.114}$$

where $\boldsymbol{P}_{N[\bar{\boldsymbol{h}}]}$ is the projection matrix along $N[\bar{\boldsymbol{h}}]$. The error in rotation is measured by

$$\Delta \boldsymbol{\Omega} = \Delta\Omega \tilde{\boldsymbol{l}}, \tag{11.115}$$

where $\Delta\Omega$ and $\tilde{\boldsymbol{l}}$ are the angle and axis (unit vector) of the relative rotation $\hat{\boldsymbol{R}}\bar{\boldsymbol{R}}^{-1}$ $(= \hat{\boldsymbol{R}}\bar{\boldsymbol{R}}^{\top})$.

In Figs. 11.6–11.8, $\Delta \boldsymbol{h}$ and $\Delta \boldsymbol{\Omega}$ are plotted in three-dimensions for 100 trials, each time using different noise. The ellipses in these figures indicate the standard confidence regions[8] defined by $\bar{V}[\hat{\boldsymbol{h}}]$ and $\bar{V}[\hat{\boldsymbol{R}}]$ (see eq. (11.29)); the cubes are drawn merely for the sake of reference. Fig. 11.6 is for the

[8]Since $\bar{V}[\hat{\mathbf{h}}]$ has rank 2 and $\bar{V}[\hat{\mathbf{R}}]$ has rank 3, the standard confidence region is a space conic for translation and an ellipsoid for rotation; see Section 4.5.3.

least-squares approximation with optimal weights (computed from the true motion parameters $\{\bar{\boldsymbol{h}}, \bar{\boldsymbol{R}}\}$); Fig. 11.7 is for renormalization; Fig. 11.8 is for renormalization followed by optimal correction of the essential matrix. The corresponding standard confidence regions and reference cubes have the same absolute sizes. From Fig. 11.8, we can see that the theoretical bound on the accuracy of the motion parameters given by eq. (11.29) is almost attained if renormalization and optimal correction are combined.

Fig. 11.9 shows a reconstructed shape viewed from a different angle. The true shape is superimposed in dashed lines. Fig. 11.10a shows the standard confidence regions computed from the total covariance matrix $V_t[\hat{\boldsymbol{r}}_\alpha] = V_i[\hat{\boldsymbol{r}}_\alpha] + V_m[\hat{\boldsymbol{r}}_\alpha]$ around the reconstructed grid points, where $V_i[\hat{\boldsymbol{r}}_\alpha]$ is the covariance matrix of $\hat{\boldsymbol{r}}_\alpha$ due to image noise (see eq. (11.98)) and $V_m[\hat{\boldsymbol{r}}_\alpha]$ is the covariance matrix of $\hat{\boldsymbol{r}}_\alpha$ due to errors in the computed motion parameters (see eq. (11.113)). Fig. 11.10b shows the shapes that envelop the primary deviation pairs of the grid points.

11.5 Critical Surfaces

11.5.1 *Weak critical surfaces*

The linearization technique introduced in Section 11.2.1 is based on the fact that the essential matrix $\boldsymbol{G}$ is an eigenmatrix of the unperturbed moment tensor $\bar{\mathcal{M}}$ for eigenvalue 0, which we have implicitly assumed to be a *simple* root. If it is a multiple root, infinitely many eigenmatrices exist. Such ambiguity occurs if and only if the epipolar equation (11.33) is satisfied by a false essential matrix $\tilde{\boldsymbol{G}}$ ($\neq \boldsymbol{G}$). In other words, ambiguity occurs if and only if space points $\boldsymbol{r}_\alpha = Z_\alpha \bar{\boldsymbol{x}}_\alpha$ and $\boldsymbol{r}'_\alpha = Z'_\alpha \bar{\boldsymbol{x}}'_\alpha$ satisfy

$$(\boldsymbol{r}_\alpha, \tilde{\boldsymbol{G}}\boldsymbol{r}'_\alpha) = 0, \qquad \alpha = 1, ..., N, \tag{11.116}$$

for $\tilde{\boldsymbol{G}} \neq \boldsymbol{G}$. If so, the linearization technique in Section 11.2.1 does not work, since any linear combination of $\tilde{\boldsymbol{G}}$ and $\boldsymbol{G}$ is an eigenmatrix of $\bar{\mathcal{M}}$ for eigenvalue 0.

Eliminating $\boldsymbol{r}'_\alpha$ by using the relation $\boldsymbol{r}_\alpha = \boldsymbol{h} + \boldsymbol{R}\boldsymbol{r}'_\alpha$ (see eq. (6.35)), we can rewrite eq. (11.116) in the form

$$(\boldsymbol{r}_\alpha, \boldsymbol{G}'(\boldsymbol{r}_\alpha - \boldsymbol{h})) = 0, \qquad \alpha = 1, ..., N, \tag{11.117}$$

where

$$\boldsymbol{G}' = \tilde{\boldsymbol{G}}\boldsymbol{R}^\top. \tag{11.118}$$

Eq. (11.117) implies that the 3-D interpretation is ambiguous if and only if all feature points are on a surface whose equation has the form

$$(\boldsymbol{r}, \boldsymbol{G}'(\boldsymbol{r} - \boldsymbol{h})) = 0 \tag{11.119}$$

(a) (b)

Fig. 11.7. Renormalization: (a) errors in translation; (b) errors in rotation.

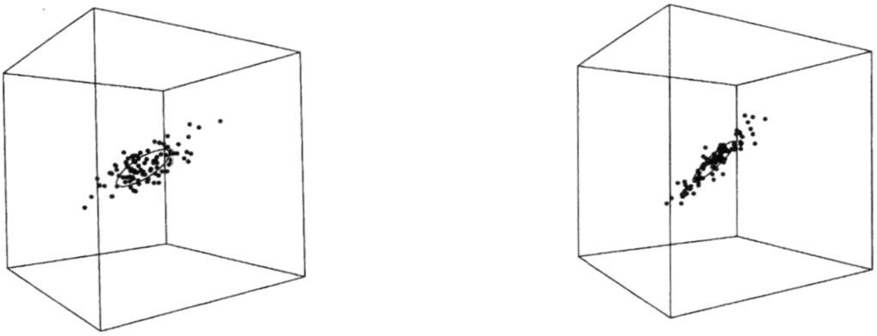

(a) (b)

Fig. 11.8. Renormalization followed by optimal correction: (a) errors in translation; (b) errors in rotation.

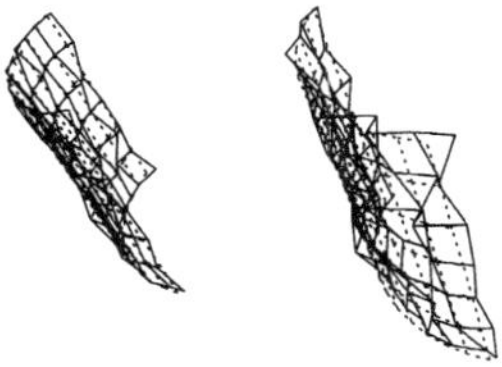

Fig. 11.9. Reconstructed shape. The true shape is superimposed in dashed lines.

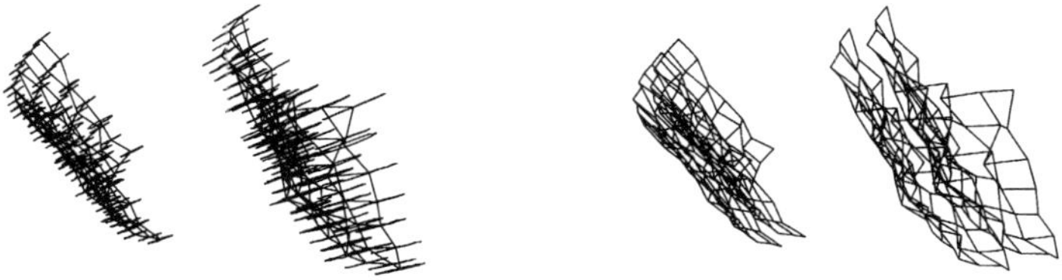

(a) (b)

Fig. 11.10. (a) Reconstructed shape and the standard confidence regions of grid points. (b) Two shapes corresponding to the primary deviation pairs of grid points.

for $\boldsymbol{G}' \neq \boldsymbol{G}\boldsymbol{R}^\top$. Since this equation is identically satisfied by $\boldsymbol{r} = \boldsymbol{0}$ and $\boldsymbol{r} = \boldsymbol{h}$, the origins O and O' of the two camera coordinate systems are both on this surface. If $S[\boldsymbol{G}']$ is nonsingular, eq. (11.119) defines a centered quadric of the following form (see eq. (4.98)):

$$(\boldsymbol{r} - \boldsymbol{r}_C, \boldsymbol{S}(\boldsymbol{r} - \boldsymbol{r}_C)) = 1. \tag{11.120}$$

The center $\boldsymbol{r}_C$ and the matrix $\boldsymbol{S}$ are given by

$$\boldsymbol{r}_C = \frac{1}{2}S[\boldsymbol{G}']^{-1}\boldsymbol{G}'\boldsymbol{h}, \qquad \boldsymbol{S} = \frac{4S[\boldsymbol{G}']}{(\boldsymbol{G}'\boldsymbol{h}, S[\boldsymbol{G}']^{-1}\boldsymbol{G}'\boldsymbol{h})}. \tag{11.121}$$

If a surface in the scene has the form of eq. (11.119) for some 33-matrix $\boldsymbol{G}'$, we call it a *weak critical surface*.

11.5.2 Strong critical surfaces

If the observed feature points are all on a weak critical surface, the epipolar equation (11.33) is satisfied by infinitely many false essential matrices. However, the true essential matrix $\boldsymbol{G}$ must be decomposable, so the correct essential matrix $\boldsymbol{G}$ can be chosen *if the false essential matrices are not decomposable.* In fact, the correct motion parameters $\{\boldsymbol{h}, \boldsymbol{R}\}$ are obtained by the direct optimization (11.11). On the other hand, if the epipolar equation (11.33) is satisfied by a false essential matrix *that is decomposable*, the problem is inherently ambiguous. This occurs if and only if all the feature points are on a surface whose equation has the form of eq. (11.119) for which the matrix $\tilde{\boldsymbol{G}}$ in eq. (11.118) can be decomposed into false motion parameters $\{\tilde{\boldsymbol{h}}, \tilde{\boldsymbol{R}}\}$. This condition is equivalent to saying that there exist a vector $\tilde{\boldsymbol{h}}$ ($\neq \boldsymbol{h}$) and a rotation matrix $\tilde{\boldsymbol{R}}$ ($\neq \boldsymbol{R}$) such that

$$\boldsymbol{G}' = \tilde{\boldsymbol{h}} \times \tilde{\boldsymbol{R}}\boldsymbol{R}^\top. \tag{11.122}$$

Let us call a surface that satisfies this condition a *strong critical surface.* By definition, a strong critical surface is also a weak critical surface and hence a quadric that passes through the origins O and O' of the two camera coordinate systems. However, the converse does not necessarily hold. In fact, eq. (11.122) implies

$$\boldsymbol{G}'^\top\tilde{\boldsymbol{h}} = \boldsymbol{0}, \tag{11.123}$$

and hence eq. (11.119) is identically satisfied by $\boldsymbol{r} = c\tilde{\boldsymbol{h}}$ for an arbitrary constant c. In other words, the space line $\boldsymbol{r} \times \tilde{\boldsymbol{h}} = \boldsymbol{0}$ is entirely contained in that surface. This implies that a strong critical surface is a *hyperboloid of one sheet* or its degeneracy (Fig 11.11a).

A critical surface is not a mere theoretical construct; it inevitably appears in practical applications. Indeed, *all planar surfaces are strong critical surfaces* and hence weak critical surfaces as well. This may sound contradictory,

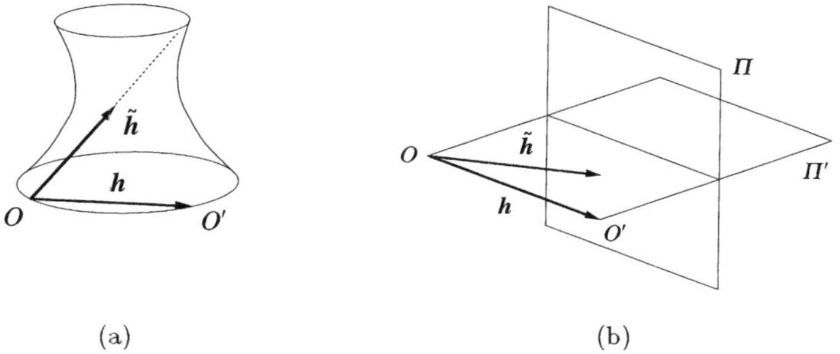

Fig. 11.11. (a) Hyperboloid of one sheet. (b) Two planar surfaces as a degenerate hyperboloid.

since a strong critical surface must be a hyperboloid passing through O and O'. By a careful analysis, however, it can be shown that an arbitrary space plane Π that does not pass though O or O' can be coupled with another space plane Π' that passes through O and O' and intersects with Π *perpendicularly* in such a way that Π and Π' constitute a *degenerate hyperboloid* (Fig. 11.11b). Consequently, the 3-D interpretation is inherently ambiguous if all feature points are on a planar surface.

In the following, we construct an algorithm for computing all possible interpretations of a planar surface scene. In order to invoke this algorithm, however, we must first test if the feature points we are observing are on a planar surface. This *planarity test* is derived from a statistical analysis of the image noise.

11.6 3-D Reconstruction from Planar Surface Motion

11.6.1 Optimal solution and planarity test

As shown in Sections 6.5.1 and 10.4.1, the necessary and sufficient condition for $\bar{\boldsymbol{x}}_\alpha$ and $\bar{\boldsymbol{x}}'_\alpha$ to be projections of a space point on space plane $(\boldsymbol{n}, \boldsymbol{r}) = d$ viewed by two cameras with motion parameters $\{\boldsymbol{h}, \boldsymbol{R}\}$ is $\bar{\boldsymbol{x}}'_\alpha \times \boldsymbol{A}\bar{\boldsymbol{x}}_\alpha = \boldsymbol{0}$, where $\boldsymbol{A} = \boldsymbol{R}^\top(\boldsymbol{h}\boldsymbol{n}^\top - d\boldsymbol{I})$ (see eqs. (6.67) and (6.71)). Hence, the 3-D analysis can be decomposed into the following two subproblems:

Problem 11.4 *Estimate a matrix $\boldsymbol{A}$ such that*

$$\bar{\boldsymbol{x}}'_\alpha \times \boldsymbol{A}\bar{\boldsymbol{x}}_\alpha = \boldsymbol{0}, \qquad \alpha = 1, ..., N, \tag{11.124}$$

from the data $\boldsymbol{x}_\alpha$ and $\boldsymbol{x}'_\alpha$, $\alpha = 1, ..., N$.

Problem 11.5 *Decompose the matrix $\boldsymbol{A}$ into surface and motion parameters $\{\boldsymbol{n}, d\}$ and $\{\boldsymbol{h}, \boldsymbol{R}\}$ in such a way that*

$$\boldsymbol{A} = \boldsymbol{R}^\top(\boldsymbol{h}\boldsymbol{n}^\top - d\boldsymbol{I}). \tag{11.125}$$

Consider Problem 11.4 first. Eq. (11.124) is the hypothesis from which $\{\boldsymbol{h}, \boldsymbol{R}\}$ and $\{\boldsymbol{n}, d\}$ are to be estimated. The rank[9] of this hypothesis is 2. Since eq. (11.124) is linear in $\boldsymbol{A}$, the scale of the matrix $\boldsymbol{A}$ is indeterminate. This is a consequence of the fact that a small camera motion near the surface is indistinguishable from a large camera motion far away from the surface, just as in the case of non-coplanar feature points (see Fig. 11.1). In order to remove this indeterminacy, we adopt the normalization $\|\boldsymbol{A}\| = 1$.

For each α, define the 33-matrix

$$\boldsymbol{X}_\alpha^{(k)} = \boldsymbol{e}^{(k)} \times \boldsymbol{x}'_\alpha \boldsymbol{x}_\alpha^\top, \tag{11.126}$$

where $\boldsymbol{e}^{(1)} = (1,0,0)^\top$, $\boldsymbol{e}^{(2)} = (0,1,0,)^\top$, and $\boldsymbol{e}^{(3)} = (0,0,1)^\top$. Let $\bar{\boldsymbol{X}}_\alpha^{(k)}$ be the unperturbed value of $\boldsymbol{X}_\alpha^{(k)}$ obtained by replacing $\boldsymbol{x}_\alpha$ and $\boldsymbol{x}'_\alpha$ by $\bar{\boldsymbol{x}}_\alpha$ and $\bar{\boldsymbol{x}}'_\alpha$, respectively. Eq. (11.124) can be written in the following form:

$$(\bar{\boldsymbol{X}}_\alpha^{(k)}; \boldsymbol{A}) = 0, \qquad k = 1,2,3. \tag{11.127}$$

Let $X_{\alpha(ij)}^{(k)}$, $x_{\alpha(i)}$, and $x'_{\alpha(i)}$ be the (ij) element of $\boldsymbol{X}_\alpha^{(k)}$ and the ith components of $\boldsymbol{x}_\alpha$ and $\boldsymbol{x}'_\alpha$, respectively. In elements, eq. (11.126) can be written as

$$X_{\alpha(ij)}^{(k)} = \sum_{l=1}^{3} \epsilon_{ikl} x'_{\alpha(l)} x_{\alpha(j)}, \tag{11.128}$$

where ϵ_{ijk} is the Eddington epsilon (see Section 2.1.3). Since $\boldsymbol{x}_\alpha$ and $\boldsymbol{x}'_\alpha$ are assumed independent, we have $E[\boldsymbol{X}_\alpha^{(k)}] = \bar{\boldsymbol{X}}_\alpha^{(k)}$. The deviation $\Delta\boldsymbol{X}_\alpha^{(k)} = \boldsymbol{X}_\alpha^{(k)} - \bar{\boldsymbol{X}}_\alpha^{(k)}$ has the following (ij) element:

$$\Delta X_{\alpha(ij)}^{(k)} = \sum_{l=1}^{3} \epsilon_{ikl} (\Delta x'_{\alpha(l)} \bar{x}_{\alpha(j)} + \bar{x}'_{\alpha(l)} \Delta x_{\alpha(j)} + \Delta x'_{\alpha(l)} \Delta x_{\alpha(j)}). \tag{11.129}$$

Let $\mathcal{V}[\boldsymbol{X}_\alpha^{(m)}, \boldsymbol{X}_\alpha^{(n)}] = E[\Delta\boldsymbol{X}_\alpha^{(m)} \otimes \Delta\boldsymbol{X}_\alpha^{(n)}]$ be the covariance tensor of $\boldsymbol{X}_\alpha^{(m)}$ and $\boldsymbol{X}_\alpha^{(n)}$. Since $\Delta\boldsymbol{x}_\alpha$ and $\Delta\boldsymbol{x}'_\alpha$ are independent random variables of mean $\boldsymbol{0}$, the covariance tensor $\mathcal{V}[\boldsymbol{X}_\alpha^{(m)}, \boldsymbol{X}_\alpha^{(n)}]$ has the following $(ijkl)$ element:

$$\begin{aligned}\mathcal{V}[\boldsymbol{X}_\alpha^{(m)}, \boldsymbol{X}_\alpha^{(n)}]_{ijkl} &= \sum_{p,q=1}^{3} \epsilon_{imp}\epsilon_{knq} \Big(E[\Delta x_{\alpha(j)} \Delta x_{\alpha(l)}] \bar{x}'_{\alpha(p)} \bar{x}'_{\alpha(q)} \\ &\quad + E[\Delta x'_{\alpha(p)} \Delta x'_{\alpha(q)}] \bar{x}_{\alpha(j)} \bar{x}_{\alpha(l)} + E[\Delta x_{\alpha(j)} \Delta x_{\alpha(l)}] E[\Delta x'_{\alpha(p)} \Delta x'_{\alpha(q)}] \Big) \\ &= \sum_{p,q=1}^{3} \epsilon_{imp}\epsilon_{knq} \Big(V[\boldsymbol{x}_\alpha]_{jl} \bar{x}'_{\alpha(p)} \bar{x}'_{\alpha(q)} + V[\boldsymbol{x}'_\alpha]_{pq} \bar{x}_{\alpha(j)} \bar{x}_{\alpha(l)} + V[\boldsymbol{x}_\alpha]_{jl} V[\boldsymbol{x}'_\alpha]_{pq} \Big).\end{aligned} \tag{11.130}$$

[9]This hypothesis is degenerate; see Section 6.5.1.

According to the general theory in Section 7.1.5, the matrix $\boldsymbol{A}$ can be optimally estimated by the following optimization (see eq. (7.55)):

$$J[\boldsymbol{A}] = \sum_{\alpha=1}^{N} \sum_{k,l=1}^{3} W_{\alpha}^{(kl)}(\boldsymbol{A})(\boldsymbol{X}_{\alpha}^{(k)}; \boldsymbol{A})(\boldsymbol{X}_{\alpha}^{(l)}; \boldsymbol{A}) \to \min. \tag{11.131}$$

Here, the (33)-matrix $\boldsymbol{W}_{\alpha}(\boldsymbol{A}) = (W_{\alpha}^{(kl)}(\boldsymbol{A}))$ is defined by $\boldsymbol{W}_{\alpha}(\boldsymbol{A}) = \left((\boldsymbol{A}; \mathcal{V}[\boldsymbol{X}_{\alpha}^{(k)}, \boldsymbol{X}_{\alpha}^{(l)}]\boldsymbol{A})\right)_2^-$, which can be expressed in the form

$$\boldsymbol{W}_{\alpha}(\boldsymbol{A}) = \Big(\boldsymbol{x}'_{\alpha} \times \boldsymbol{A}V[\boldsymbol{x}_{\alpha}]\boldsymbol{A}^{\top} \times \boldsymbol{x}'_{\alpha} + (\boldsymbol{A}\boldsymbol{x}_{\alpha}) \times V[\boldsymbol{x}'_{\alpha}] \times (\boldsymbol{A}\boldsymbol{x}_{\alpha}) \\ + [\boldsymbol{A}V[\boldsymbol{x}_{\alpha}]\boldsymbol{A}^{\top} \times V[\boldsymbol{x}'_{\alpha}]]\Big)_2^-, \tag{11.132}$$

where the symbol $[\cdot \times \cdot]$ denotes the exterior product of matrices (see eq. (2.45)). The minimum of $J[\boldsymbol{A}]$ is sought under the constraint $\|\boldsymbol{A}\| = 1$.

Let $\hat{\boldsymbol{A}}$ be the resulting estimate. Since $\boldsymbol{A}$ has eight degrees of freedom (its nine elements are normalized), the residual $J[\hat{\boldsymbol{A}}]$ is a χ^2 variable with $2(N-4)$ degrees of freedom. It follows that the number N of the pairs of corresponding image points must be such that

$$N \geq 4. \tag{11.133}$$

The residual is 0 when $N = 4$. For $N \geq 5$, we obtain a *planarity test*: the hypothesis that the feature points are collinear is rejected with significance level $a\%$ if

$$J[\hat{\boldsymbol{A}}] > \chi^2_{2(N-4),a}. \tag{11.134}$$

Note that although the general motion algorithm breaks down if all feature points are coplanar, the planar surface algorithm can be applied to general motion as well, always yielding an (incorrect) matrix $\boldsymbol{A}$. It follows that if the feature points are nearly coplanar, it is safer to apply the planar surface algorithm than the general motion algorithm; it should be switched to the general motion algorithm *only when assuming planarity is not compatible with the image data to a convincing degree*. In practice, however, the opposite approach has often been adopted: the general motion algorithm is used first by assuming non-planarity and switched to the planar surface algorithm *only when assuming non-planarity causes a computational breakdown*. To be specific, it has been customary to abandon the general motion algorithm when the second smallest eigenvalue of the moment tensor $\mathcal{M}$ given by eq. (11.40) is sufficiently close to its smallest eigenvalue, and the decision criterion has been set arbitrarily. In contrast, the use of the planarity test as described above has a solid statistical foundation. In Chapter 13, we present an alternative criterion which does not require the significance level.

11.6.2 Unbiased estimation and renormalization

The least-squares approximation to the optimization (11.131) based on the normalized covariance matrices $V_0[\boldsymbol{x}_\alpha]$ and $V_0[\boldsymbol{x}'_\alpha]$ can be written in the following form:

$$\tilde{J}[\boldsymbol{A}] = (\boldsymbol{A}; \mathcal{M}\boldsymbol{A}) \to \min. \tag{11.135}$$

The moment tensor $\mathcal{M}$ is defined by

$$\mathcal{M} = \frac{1}{N}\sum_{\alpha=1}^{N}\sum_{k,l=1}^{3} W_\alpha^{(kl)} \boldsymbol{X}_\alpha^{(k)} \otimes \boldsymbol{X}_\alpha^{(l)}, \tag{11.136}$$

$$\boldsymbol{W}_\alpha = \Big(\boldsymbol{x}'_\alpha \times \boldsymbol{A}^* V_0[\boldsymbol{x}_\alpha]\boldsymbol{A}^{*\top} \times \boldsymbol{x}'_\alpha + (\boldsymbol{A}^*\boldsymbol{x}_\alpha) \times V_0[\boldsymbol{x}'_\alpha] \times (\boldsymbol{A}^*\boldsymbol{x}_\alpha)$$

$$+ \epsilon^2[\boldsymbol{A}^* V_0[\boldsymbol{x}_\alpha]\boldsymbol{A}^{*\top} \times V_0[\boldsymbol{x}'_\alpha]]\Big)_2^-, \tag{11.137}$$

where $\boldsymbol{A}^*$ is an appropriate estimate of $\boldsymbol{A}$. The solution of (11.135) under the constraint $\|\boldsymbol{A}\| = 1$ is obtained as the eigenmatrix of norm 1 of tensor $\mathcal{M}$ for the smallest eigenvalue. The eigenvalues and eigenmatrices of tensor $\mathcal{M}$ are computed by casting $\boldsymbol{A}$ and $\mathcal{M}$ into a 9-vector and a 99-matrix, respectively, as in the case of the essential matrix $\boldsymbol{G}$ (see eqs. (11.42)).

The hypothesis (11.127) implies that the true value $\bar{\boldsymbol{A}}$ satisfies $\bar{\mathcal{M}}\bar{\boldsymbol{A}} = \boldsymbol{O}$, i.e., $\bar{\boldsymbol{A}}$ is the eigenmatrix of $\bar{\mathcal{M}}$ for eigenvalue 0. However, $E[\mathcal{M}]$ is generally not equal to $\bar{\mathcal{M}}$. In fact, substituting $\boldsymbol{X}_\alpha^{(k)} = \bar{\boldsymbol{X}}_\alpha^{(k)} + \Delta\boldsymbol{X}_\alpha^{(k)}$ into eq. (11.136) and noting that $E[\Delta\boldsymbol{X}_\alpha^{(k)}] = \boldsymbol{O}$, we see that

$$E[\mathcal{M}] = \frac{1}{N}\sum_{\alpha=1}^{N}\sum_{k,l=1}^{3} W_\alpha^{(kl)} E[\bar{\boldsymbol{X}}_\alpha^{(k)} \otimes \bar{\boldsymbol{X}}_\alpha^{(l)}]$$

$$+\frac{1}{N}\sum_{\alpha=1}^{N}\sum_{k,l=1}^{3} W_\alpha^{(kl)} E[\Delta\boldsymbol{X}_\alpha^{(k)} \otimes \Delta\boldsymbol{X}_\alpha^{(l)}]$$

$$= \bar{\mathcal{M}} + \frac{1}{N}\sum_{\alpha=1}^{N}\sum_{k,l=1}^{3} W_\alpha^{(kl)} \mathcal{V}[\boldsymbol{X}_\alpha^{(k)}, \boldsymbol{X}_\alpha^{(l)}]. \tag{11.138}$$

Hence, the solution of the least-squares approximation is statistically biased whatever weights W_α are used.

Define 3333-tensors $\mathcal{N}^{(1)} = (N_{ijkl}^{(1)})$ and $\mathcal{N}^{(2)} = (N_{ijkl}^{(2)})$ by

$$N_{ijkl}^{(1)} = \frac{1}{N}\sum_{\alpha=1}^{N}\sum_{m,n,p,q=1}^{3} \epsilon_{imp}\epsilon_{knq} W_\alpha^{(mn)} \Big(V_0[\boldsymbol{x}_\alpha]_{jl} x'_{\alpha(p)} x'_{\alpha(q)}$$

$$+ V_0[\boldsymbol{x}'_\alpha]_{pq} x_{\alpha(j)} x_{\alpha(l)}\Big), \tag{11.139}$$

$$N^{(2)}_{ijkl} = \frac{1}{N}\sum_{\alpha=1}^{N}\sum_{m,n,p,q=1}^{3} \epsilon_{imp}\epsilon_{knq} W^{(mn)}_{\alpha} V_0[\boldsymbol{x}_\alpha]_{jl} V_0[\boldsymbol{x}'_\alpha]_{pq}. \tag{11.140}$$

Let $\bar{\mathcal{N}}^{(1)}$ be the unperturbed value of $\mathcal{N}^{(1)}$ obtained by replacing $\boldsymbol{x}_\alpha$ and $\boldsymbol{x}'_\alpha$ by $\bar{\boldsymbol{x}}_\alpha$ and $\bar{\boldsymbol{x}}'_\alpha$, respectively, in eq. (11.139). Eq. (11.138) can be written in the following form:

$$E[\mathcal{M}] = \bar{\mathcal{M}} + \epsilon^2 \bar{\mathcal{N}}^{(1)} + \epsilon^4 \mathcal{N}^{(2)}. \tag{11.141}$$

It is immediately seen from eqs. (11.139) and (11.140) that

$$E[\mathcal{N}^{(1)}] = \bar{\mathcal{N}}^{(1)} + 2\epsilon^2 \mathcal{N}^{(2)}. \tag{11.142}$$

It follows that if we define

$$\hat{\mathcal{M}} = \mathcal{M} - \epsilon^2 \mathcal{N}^{(1)} + \epsilon^4 \mathcal{N}^{(2)}, \tag{11.143}$$

we have $E[\hat{\mathcal{M}}] = \bar{\mathcal{M}}$. Hence, we obtain the unbiased least-squares approximation

$$\hat{J}[\boldsymbol{A}] = (\boldsymbol{A}; \hat{\mathcal{M}}\boldsymbol{A}) \to \min. \tag{11.144}$$

The solution under the constraint $\|\boldsymbol{A}\| = 1$ is obtained as the eigenmatrix of unit norm of $\hat{\mathcal{M}}$ for the smallest eigenvalue.

If the noise level ϵ is not known, the second order renormalization procedure is given as follows (see Section 9.6.2):

1. Let $c = 0$ and $\boldsymbol{W}_\alpha = \boldsymbol{I}$, $\alpha = 1, ..., N$.

2. Compute the 3333-tensors $\mathcal{M}$, $\mathcal{N}^{(1)}$, and $\mathcal{N}^{(2)}$ defined by eqs. (11.136), (11.139), and (11.140), respectively.

3. Compute the smallest eigenvalue λ of the 3333-tensor

$$\hat{\mathcal{M}} = \mathcal{M} - c\mathcal{N}^{(1)} + c^2 \mathcal{N}^{(2)}, \tag{11.145}$$

and the corresponding eigenmatrix $\boldsymbol{A}$ of unit norm.

4. If $\lambda \approx 0$, return $\boldsymbol{A}$, c, and $\hat{\mathcal{M}}$. Else, update c and $\boldsymbol{W}_\alpha$ as follows:

$$D = \Big((\boldsymbol{A}; \mathcal{N}^{(1)}\boldsymbol{A}) - 2c(\boldsymbol{A}; \mathcal{N}^{(2)}\boldsymbol{A})\Big)^2 - 4\lambda(\boldsymbol{A}; \mathcal{N}^{(2)}\boldsymbol{A}), \tag{11.146}$$

$$c \leftarrow \begin{cases} c + \dfrac{(\boldsymbol{A}; \mathcal{N}^{(1)}\boldsymbol{A}) - 2c(\boldsymbol{A}; \mathcal{N}^{(2)}\boldsymbol{A}) - \sqrt{D}}{2(\boldsymbol{A}; \mathcal{N}^{(2)}\boldsymbol{A})}, & \text{If } D \geq 0, \\ c + \dfrac{\lambda}{(\boldsymbol{A}; \mathcal{N}^{(1)}\boldsymbol{A})}, & \text{If } D < 0, \end{cases} \tag{11.147}$$

$$\boldsymbol{W}_\alpha \leftarrow \Big(\boldsymbol{x}'_\alpha \times \boldsymbol{A}V_0[\boldsymbol{x}_\alpha]\boldsymbol{A}^\top \times \boldsymbol{x}'_\alpha + (\boldsymbol{A}\boldsymbol{x}_\alpha) \times V_0[\boldsymbol{x}'_\alpha] \times (\boldsymbol{A}\boldsymbol{x}_\alpha) + c[\boldsymbol{A}V_0[\boldsymbol{x}_\alpha]\boldsymbol{A}^\top \times V_0[\boldsymbol{x}'_\alpha]]\Big)^-_2. \tag{11.148}$$

5. Go back to Step 2.

After renormalization, the squared noise level ϵ^2 is estimated in the form

$$\hat{\epsilon}^2 = \frac{c}{1 - 4/N}. \tag{11.149}$$

Its expectation and variance are given by

$$E[\hat{\epsilon}^2] = \epsilon^2, \qquad V[\hat{\epsilon}^2] = \frac{\epsilon^4}{N - 4}. \tag{11.150}$$

The covariance tensor $\mathcal{V}[\hat{\boldsymbol{A}}]$ of the resulting estimate $\hat{\boldsymbol{A}}$ is obtained in the following form:

$$\mathcal{V}[\hat{\boldsymbol{A}}] = \frac{\hat{\epsilon}^2}{N}(\hat{\mathcal{M}})_8^-. \tag{11.151}$$

In terms of the estimate $\hat{\epsilon}^2$ computed by eq. (11.149), the planarity test given by eq. (11.134) can be rewritten in the form

$$\frac{\hat{\epsilon}^2}{\epsilon^2} > \frac{\chi^2_{2(N-4),a}}{2(N-4)}. \tag{11.152}$$

We can interpret this as follows: we compare the estimate $\hat{\epsilon}$ computed under the hypothesis that the feature points are all coplanar with the noise level ϵ expected from the accuracy of the image processing operations for locating feature points.

11.6.3 Computation of surface and motion parameters

A. Decomposition of the matrix of the image transformation

We now solve Problem 11.5. The important fact is that unlike the decomposition of the essential matrix $\boldsymbol{G}$, *no decomposability condition needs to be imposed.* In other words, there always exist $\{\boldsymbol{n}, d\}$ and $\{\boldsymbol{h}, \boldsymbol{R}\}$ that satisfy eq. (11.125), although the decomposition may not be unique. This is easily understood by counting the degrees of freedom. The matrix $\boldsymbol{A}$ has eight degrees of freedom, since its elements are normalized. The surface parameters $\{\boldsymbol{n}, d\}$ have three degrees of freedom; the motion parameters $\{\boldsymbol{h}, \boldsymbol{R}\}$ have five degrees of freedom. Thus, the matrix $\boldsymbol{A}$ has the same degrees of freedom as the surface and motion parameters.

In order to solve Problem 11.5, the following two types of camera motion must be distinguished (Fig. 11.12):

Case 1 The camera moves on one side of the planar surface; the two images are views from the same side of the surface.

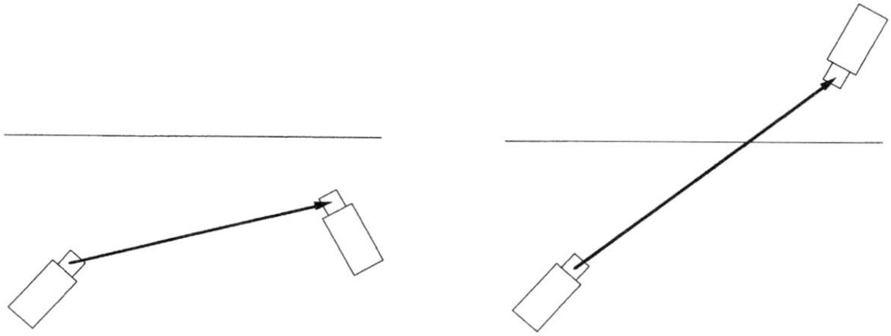

(a) (b)

Fig. 11.12. (a) Case 1: the camera moves on one side of the planar surface. (b) Case 2: the camera penetrates through the planar surface.

Case 2 The camera penetrates through the planar surface; the second image is a view from the opposite side of the surface.

Although the normalization $\|\boldsymbol{A}\| = 1$ is imposed in the renormalization computation, we rescale $\boldsymbol{A}$ so that $\det \boldsymbol{A} = 1$ Then, the surface and motion parameters $\{\boldsymbol{n}, d\}$ and $\{\boldsymbol{h}, \boldsymbol{R}\}$ are computed by the following procedure:

1. Let $\lambda_1 \geq \lambda_2 \geq \lambda_3$ (> 0) be the eigenvalues of matrix $\boldsymbol{A}^\top \boldsymbol{A}$, and $\{\boldsymbol{u}_1, \boldsymbol{u}_2, \boldsymbol{u}_3\}$ the orthonormal set of the corresponding eigenvectors[10]. Let $\sigma_i = \sqrt{\lambda_i}$ (> 0), $i = 1, 2, 3$.

2. The surface parameters $\{\boldsymbol{n}, d\}$ in Case 1 are given by

$$\boldsymbol{n} = N[\sqrt{\sigma_1^2 - \sigma_2^2}\boldsymbol{u}_1 \pm \sqrt{\sigma_2^2 - \sigma_3^2}\boldsymbol{u}_3], \qquad d = \frac{\sigma_2}{\sigma_1 - \sigma_3}. \tag{11.153}$$

   In Case 2, the distance d is replaced by

$$d = \frac{\sigma_2}{\sigma_1 + \sigma_3}. \tag{11.154}$$

3. The motion parameters $\{\boldsymbol{h}, \boldsymbol{R}\}$ in Case 1 are given by

$$\boldsymbol{h} = N[-\sigma_3\sqrt{\sigma_1^2 - \sigma_2^2}\boldsymbol{u}_1 \pm \sigma_1\sqrt{\sigma_2^2 - \sigma_3^2}\boldsymbol{u}_3], \tag{11.155}$$

$$\boldsymbol{R} = \frac{1}{\sigma_2}\left(\boldsymbol{I} + \sigma_2^3 \boldsymbol{p}\boldsymbol{h}^\top\right)\boldsymbol{A}^\top, \tag{11.156}$$

[10]In other words, $\{\mathbf{u}_i\}$ is the *left orthonormal system* of $\mathbf{A}$ for *singular values* $\{\sqrt{\lambda_i}\}$; see Section 2.3.1.

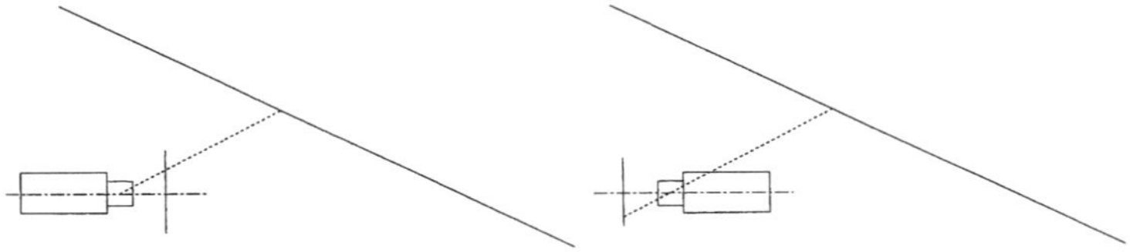

Fig. 11.13. A surface behind the camera can be also observed.

and in Case 2 by

$$h = N[\sigma_3\sqrt{\sigma_1^2 - \sigma_2^2}u_1 \pm \sigma_1\sqrt{\sigma_2^2 - \sigma_3^2}u_3], \tag{11.157}$$

$$R = \frac{1}{\sigma_2}\left(-I + \sigma_2^3 p h^\top\right) A^\top, \tag{11.158}$$

where the double signs $\pm$ correspond to that in eqs. (11.153).

4. For each solution, another solution is obtained by changing the signs of n and v simultaneously.

B. Geometry of ambiguity

The above procedure yields *eight* solutions in general. This ambiguity is partly explained by noting the following two facts:

- According to our camera imaging model, a surface *behind* the camera can be observed as well as a surface *in front* of the camera[11] (Fig. 11.13).
- The surface and motion parameters are computed from *the matrix A of the image transformation*, not *individual feature points*.

Suppose one solution $\{n, d\}$ and $\{h, R\}$ is obtained.

Type 1 If we

1. move the first camera to the other side of the planar surface, and
2. reverse the translation (Fig. 11.14),

the new surface and motion parameters are $\{-n, d\}$ and $\{-h, R\}$, respectively. We can see from eq. (11.125) that the matrix A of the image transformation does not change.

[11]This is not a mere mathematical artifact due to the abstract perspective projection model; this also occurs to a real camera. For example, if the optical axis is oriented upward but not vertical, the horizon can be seen if the field of view is sufficiently large.

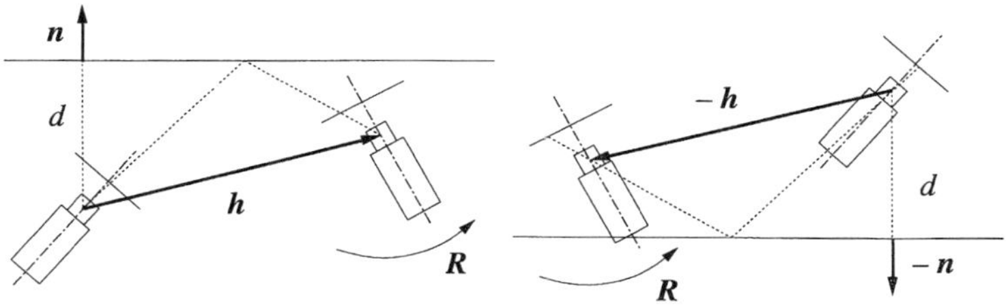

Fig. 11.14. The first camera is moved to the other side of the planar surface, and the camera translation is reversed.

Type 2 If we

1. reverse the three coordinate axis orientations of the second camera, and
2. move it to the mirror image position with respect to the planar surface (Fig. 11.15),

the new motion parameters[12] $\{\boldsymbol{h}', \boldsymbol{R}'\}$ are given by

$$\boldsymbol{h}' = \boldsymbol{J_n}\boldsymbol{h} + 2d\boldsymbol{n}, \qquad \boldsymbol{R}' = -\boldsymbol{R}\boldsymbol{J_n}, \tag{11.159}$$

where $\boldsymbol{J_n}$ is the linear mapping of the reflection with respect to the plane with surface unit normal $\boldsymbol{n}$ (Fig. 11.16): it has the expression

$$\boldsymbol{J_n} = \boldsymbol{I} - 2\boldsymbol{n}\boldsymbol{n}^\top. \tag{11.160}$$

The matrix of the resulting image transformation is

$$\begin{aligned}\boldsymbol{A}' &= \boldsymbol{R}'^\top(\boldsymbol{h}'\boldsymbol{n}^\top - d\boldsymbol{I}) = -\boldsymbol{R}^\top\boldsymbol{J}_{\boldsymbol{n}}^\top\Big((\boldsymbol{J_n}\boldsymbol{h}+2d\boldsymbol{n})\boldsymbol{n}^\top - d\boldsymbol{I})\Big) \\ &= -\boldsymbol{R}^\top\boldsymbol{J_n}\Big(\boldsymbol{J_n}\boldsymbol{n}\boldsymbol{n}^\top - d(\boldsymbol{I}-2\boldsymbol{n}\boldsymbol{n}^\top)\Big) = -\boldsymbol{R}^\top\boldsymbol{J_n}(\boldsymbol{J_n}\boldsymbol{n}\boldsymbol{n}^\top - d\boldsymbol{J_n}) \\ &= -\boldsymbol{R}^\top(\boldsymbol{n}\boldsymbol{n}^\top - d\boldsymbol{I}) = -\boldsymbol{A},\end{aligned} \tag{11.161}$$

where we have used the identities $\boldsymbol{J}_{\boldsymbol{n}}^\top = \boldsymbol{J_n}$ and $\boldsymbol{J}_{\boldsymbol{n}}^2 = \boldsymbol{I}$. Thus, the two motions have the same matrix of image transformation up to sign[13].

C. Resolution of ambiguity

Although eight solutions exist in general, we can reduce the number of solutions if we can tell whether the motion belongs to Case 1 or Case 2. In fact,

[12] The camera orientation obtained after two reflections is described by a (proper) orthogonal matrix.

[13] Note that multiplication of the matrix **A** by a nonzero (positive or negative) constant does not affect the 3-D interpretation.

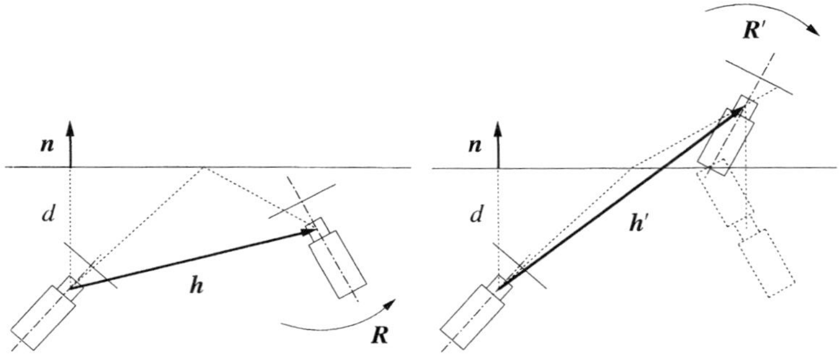

Fig. 11.15. The second camera is reversed and moved to the mirror image position with respect to the planar surface.

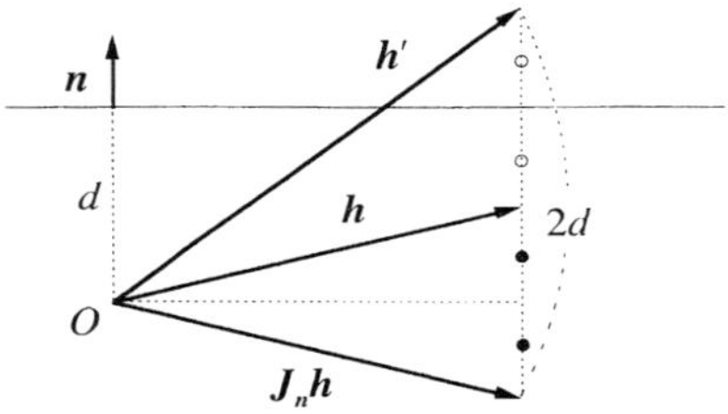

Fig. 11.16. Vectors **h** and **h**′ are the mirror images of each other with respect to the plane $(\mathbf{n}, \mathbf{r}) = d$.

if we impose the condition that all the feature points have positive depths before and after the camera motion, the number of solution reduces to at most two—in most cases one. This is because spurious interpretations are in most cases such that the feature points are all behind the camera or some of them are behind the camera. In the latter case, the *vanishing line* separates some feature points from the rest, which is impossible because only one side of it is visible (Fig. 11.17; see Section 4.6.2).

The true position $\bar{\boldsymbol{r}}_\alpha$ of the αth feature point satisfies the following equation (see eq. (11.72)):

$$\bar{\boldsymbol{r}}_\alpha = \bar{Z}_\alpha \bar{\boldsymbol{x}}_\alpha = \boldsymbol{h} + \bar{Z}'_\alpha \boldsymbol{R}\bar{\boldsymbol{x}}'_\alpha. \tag{11.162}$$

If this point is on space plane $(\boldsymbol{n}, \boldsymbol{r}) = d$, the depths $\bar{Z}_\alpha$ and $\bar{Z}'_\alpha$ can be determined by substituting eq. (11.162) into $(\boldsymbol{n}, \bar{\boldsymbol{r}}_\alpha) = d$: we obtain

$$\bar{Z}_\alpha = \frac{d}{(\boldsymbol{n}, \bar{\boldsymbol{x}}'_\alpha)}, \qquad \bar{Z}'_\alpha = \frac{d - (\boldsymbol{n}, \boldsymbol{h})}{(\boldsymbol{n}, \boldsymbol{R}\bar{\boldsymbol{x}}'_\alpha)}. \tag{11.163}$$

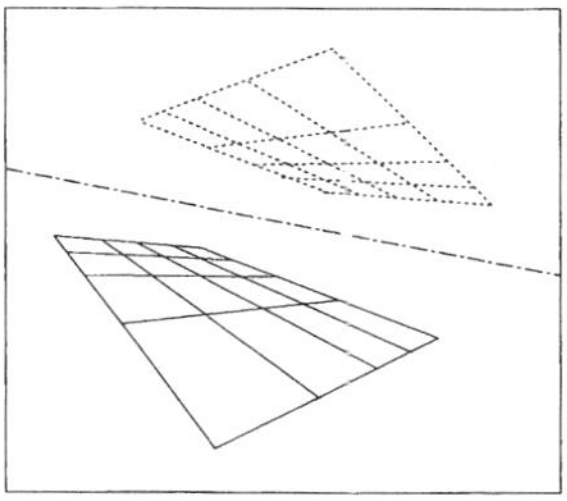

Fig. 11.17. The vanishing line cannot pass through the projected feature points; only one side of it is visible.

Hence, the condition to be imposed is

$$\frac{d}{(\boldsymbol{n}, \boldsymbol{x}'_\alpha)} > 0, \qquad \frac{d - (\boldsymbol{n}, \boldsymbol{h})}{(\boldsymbol{n}, \boldsymbol{R}\boldsymbol{x}'_\alpha)} > 0. \tag{11.164}$$

If the correct surface and motion parameters are chosen, eqs. (11.162) and (11.163) determine the 3-D position of each feature point. However, if the data values $\boldsymbol{x}_\alpha$ and $\boldsymbol{x}'_\alpha$ are substituted for $\bar{\boldsymbol{x}}_\alpha$ and $\bar{\boldsymbol{x}}'_\alpha$ in them, eq. (11.162) does not hold exactly. Geometrically, this is equivalent to saying that the two lines of sight defined by $\boldsymbol{x}_\alpha$ and $\boldsymbol{x}'_\alpha$ may not meet; if they do, the intersection may not be on the space plane $(\boldsymbol{n}, \boldsymbol{r}) = d$. As described in Section 6.5.1, this problem can be resolved by the *optimal back projection*: the corresponding image points $\boldsymbol{x}_\alpha$ and $\boldsymbol{x}'_\alpha$ are optimally corrected into $\hat{\boldsymbol{x}}_\alpha$ and $\hat{\boldsymbol{x}}'_\alpha$ so that their lines of sight intersect exactly on the space plane $(\boldsymbol{n}, \boldsymbol{r}) = d$ (see Fig. 6.12). The correction takes the following form (see eqs. (6.74) and (6.75)):

$$\hat{\boldsymbol{x}}_\alpha = \boldsymbol{x}_\alpha - (V_0[\boldsymbol{x}_\alpha]\boldsymbol{A}^\top \times \boldsymbol{x}'_\alpha)\boldsymbol{W}_\alpha(\boldsymbol{x}'_\alpha \times \boldsymbol{A}\boldsymbol{x}_\alpha),$$

$$\hat{\boldsymbol{x}}'_\alpha = \boldsymbol{x}'_\alpha + (V_0[\boldsymbol{x}'_\alpha] \times (\boldsymbol{A}\boldsymbol{x}_\alpha))\boldsymbol{W}_\alpha(\boldsymbol{x}'_\alpha \times \boldsymbol{A}\boldsymbol{x}_\alpha). \tag{11.165}$$

$$\boldsymbol{W}_\alpha = \left(\boldsymbol{x}'_\alpha \times \boldsymbol{A}V_0[\boldsymbol{x}_\alpha]\boldsymbol{A}^\top \times \boldsymbol{x}'_\alpha + (\boldsymbol{A}\boldsymbol{x}_\alpha) \times V_0[\boldsymbol{x}'_\alpha] \times (\boldsymbol{A}\boldsymbol{x}_\alpha)\right)^-_2. \tag{11.166}$$

This correction is iterated until eq. (11.124) is sufficiently satisfied. Then, the 3-D position of the αth feature point is given by

$$\boldsymbol{r}_\alpha = \frac{d\hat{\boldsymbol{x}}_\alpha}{(\boldsymbol{n}, \hat{\boldsymbol{x}}_\alpha)}. \tag{11.167}$$

Example 11.2 Fig. 11.18 shows two simulated images (512×512 pixels) of a planar grid in the scene viewed from a moving camera. The focal length is assumed to be $f = 600$ (pixels). We added Gaussian noise of mean 0 and standard deviation $\sigma = 5$ (pixels) to the x and y coordinates of each grid point

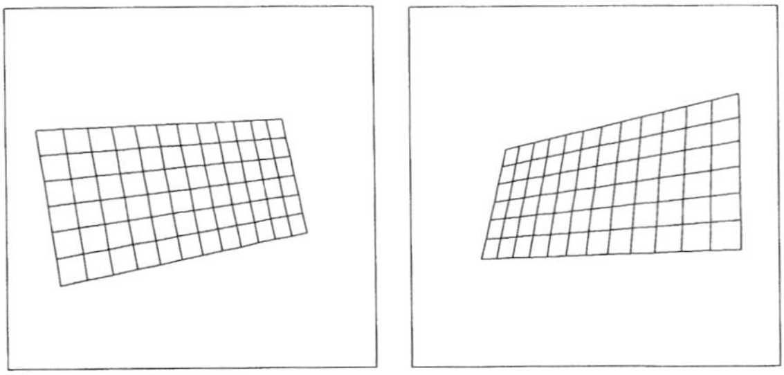

Fig. 11.18. Motion images of a planar grid.

independently, so the noise level is $\epsilon = \sigma/f = 1/120$ (this value is not used in the reconstruction computation). We assume that the motion belongs to Case 1. We also assume that the correct solution can be distinguished from the spurious solution. Let $\{\boldsymbol{n}, d\}$ and $\{\boldsymbol{h}, \boldsymbol{R}\}$ be the computed surface and motion parameters. Their deviation from the true values $\{\bar{\boldsymbol{n}}, \bar{d}\}$ and $\{\bar{\boldsymbol{h}}, \bar{\boldsymbol{R}}\}$ is measured as follows (see eqs. (10.87), (11.114), and (11.115)):

- The error in the surface parameters is represented by the 3-vector

$$\Delta \boldsymbol{u} = \boldsymbol{P}_{\bar{\boldsymbol{n}}}(\boldsymbol{n} - \bar{\boldsymbol{n}}) + \frac{\|\bar{\boldsymbol{h}}\|d - \bar{d}}{\bar{d}}\bar{\boldsymbol{n}}. \tag{11.168}$$

- The error in translation is represented by the 3-vector

$$\Delta \boldsymbol{h} = \boldsymbol{P}_{N[\bar{\boldsymbol{h}}]}(\boldsymbol{h} - N[\bar{\boldsymbol{h}}]). \tag{11.169}$$

- The error in rotation is represented by the 3-vector

$$\Delta \boldsymbol{\Omega} = \Delta\tilde{\Omega}\tilde{\boldsymbol{l}}, \tag{11.170}$$

  where $\tilde{\Omega}$ and $\tilde{\boldsymbol{l}}$ are, respectively, the axis and angle of the relative rotation $\tilde{\boldsymbol{R}} = \boldsymbol{R}\bar{\boldsymbol{R}}^{-1}$ $(= \boldsymbol{R}\bar{\boldsymbol{R}}^{\top})$.

In Figs. 11.19 and 11.20, $\Delta\boldsymbol{u}$, $\Delta\boldsymbol{h}$, and $\Delta\boldsymbol{\Omega}$ are plotted in three dimensions for 100 trials, each time using different noise. Fig. 11.19 is for the least-squares approximation (the weights $W_\alpha^{(kl)}$ are computed from the true surface and motion parameters $\{\bar{\boldsymbol{n}}, \bar{d}\}$ and $\{\bar{\boldsymbol{h}}, \bar{\boldsymbol{R}}\}$); Fig. 11.20 is for renormalization. Comparing Figs. 11.19 and 11.20, we can see that the least-squares solution has statistical bias, which is removed by renormalization.

Fig. 11.21a shows one example of a grid reconstructed by directly applying eq. (11.167) to $\boldsymbol{x}_\alpha$; Fig. 11.21b shows the corresponding surface reconstructed by the optimal correction of the feature points. In both figures, the true position is superimposed in dashed lines. We can see that the correction enhances the accuracy of 3-D reconstruction.

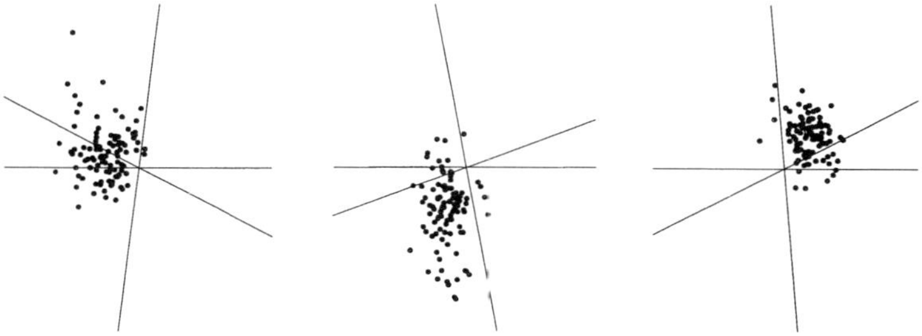

(a) (b) (c)

Fig. 11.19. Least-squares approximation. (a) Errors in the surface parameters. (b) Errors in translation. (c) Errors in rotation.

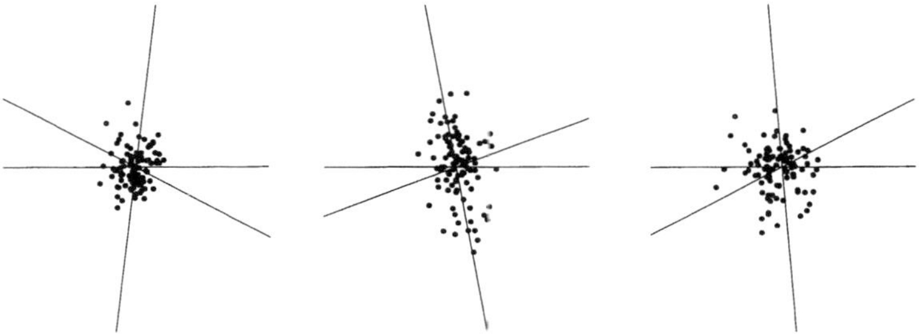

(a) (b) (c)

Fig. 11.20. Renormalization. (a) Errors in the surface parameters. (b) Errors in translation. (c) Errors in rotation.

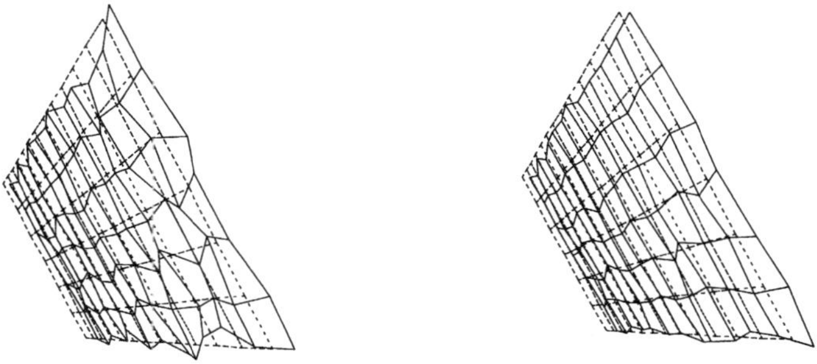

(a) (b)

Fig. 11.21. (a) Simple back projection. (b) Optimal back projection.

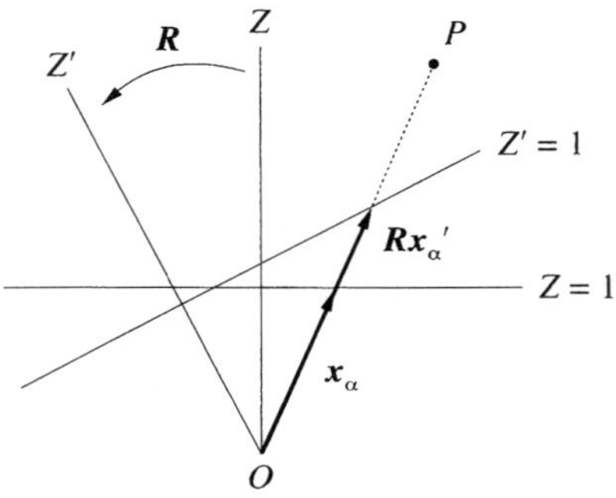

Fig. 11.22. Camera rotation with no translation.

11.7 Camera Rotation and Information

11.7.1 *Rotation test*

A. Rotation estimation

So far we have assumed that $\boldsymbol{h} \neq \boldsymbol{0}$. If $\boldsymbol{h} = \boldsymbol{0}$, the camera motion is a pure rotation around the center of the lens, and the incoming rays of light are the same before and after the camera rotation (Fig. 11.22). Consequently, no 3-D information can be obtained. In fact, if no image noise exists, eq. (11.72) implies that the corresponding image points $\bar{\boldsymbol{x}}_\alpha$ and $\bar{\boldsymbol{x}}'_\alpha$ are related by

$$Z_\alpha \bar{\boldsymbol{x}}_\alpha = Z'_\alpha \boldsymbol{R} \bar{\boldsymbol{x}}'_\alpha. \tag{11.171}$$

Thus, the depths Z_α and Z'_α are completely indeterminate.

It follows that in order to do a 3-D motion analysis, we need to test whether $\boldsymbol{h} = \boldsymbol{0}$ or not in advance. This *rotation test* can be done in the form of a χ^2 test by first hypothesizing $\boldsymbol{h} = \boldsymbol{0}$ and then testing if the discrepancy of the observation from the hypothesis is small enough to be accounted for by the statistical behavior of the image noise.

Eq. (11.171) is equivalent to

$$\bar{\boldsymbol{x}}_\alpha \propto \boldsymbol{R} \bar{\boldsymbol{x}}'_\alpha. \tag{11.172}$$

In the presence of noise, the observed image points $\boldsymbol{x}_\alpha$ and $\boldsymbol{x}'_\alpha$, $\alpha = 1, ..., N$, may not exactly satisfy this condition. So, we consider the following problem:

Problem 11.6 *Estimate a rotation matrix* $\boldsymbol{R}$ *such that*

$$\bar{\boldsymbol{x}}'_\alpha \times \boldsymbol{R} \bar{\boldsymbol{x}}_\alpha = \boldsymbol{0}, \qquad \alpha = 1, ..., N, \tag{11.173}$$

from the data $\boldsymbol{x}_\alpha$ *and* $\boldsymbol{x}'_\alpha$, $\alpha = 1, ..., N$.

Evidently, eq. (11.172) is equivalent to eq. (11.173). This problem is formally identical with Problem 11.4, so the matrix $\boldsymbol{R}$ can be optimally estimated by the following optimization (see eq. (11.131)):

$$J[\boldsymbol{R}] = \sum_{\alpha=1}^{N} \sum_{k,l=1}^{3} W_{\alpha}^{(kl)}(\boldsymbol{R})(\boldsymbol{X}_{\alpha}^{(k)}; \boldsymbol{R})(\boldsymbol{X}_{\alpha}^{(l)}; \boldsymbol{R}) \to \min. \tag{11.174}$$

$$\boldsymbol{W}_{\alpha}(\boldsymbol{R}) = \Big(\boldsymbol{x}_{\alpha} \times \boldsymbol{R}V[\boldsymbol{x}'_{\alpha}]\boldsymbol{R}^{\top} \times \boldsymbol{x}_{\alpha} + (\boldsymbol{R}\boldsymbol{x}'_{\alpha}) \times V[\boldsymbol{x}_{\alpha}] \times (\boldsymbol{R}\boldsymbol{x}'_{\alpha}) + [\boldsymbol{R}V[\boldsymbol{x}'_{\alpha}]\boldsymbol{R}^{\top} \times V[\boldsymbol{x}_{\alpha}]]\Big)_2^{-}, \tag{11.175}$$

The only difference from eq. (11.131) is that the minimum is sought under the constraint that $\boldsymbol{R}$ *is a rotation matrix.*

B. χ^2 test

Let $\hat{\boldsymbol{R}}$ be the resulting estimate. Since $\boldsymbol{R}$ has three degrees of freedom, the residual $J[\hat{\boldsymbol{R}}]$ is a χ^2 variable with $2N-3$ degrees of freedom. It follows that the number N of the pairs of corresponding image points must be such that

$$N \geq 2. \tag{11.176}$$

However, the residual is *not* 0 when $N = 2$. The rotation test can be done in the form of the standard χ^2 test: the hypothesis that the camera motion is a pure rotation is rejected with significance level $a\%$ if

$$J[\hat{\boldsymbol{R}}] > \chi^2_{2N-3,a}. \tag{11.177}$$

Decompose the covariance matrices $V[\boldsymbol{x}_{\alpha}]$ and $V[\boldsymbol{x}'_{\alpha}]$ into the noise level ϵ and the normalized covariance matrices $V_0[\boldsymbol{x}_{\alpha}]$ and $V_0[\boldsymbol{x}'_{\alpha}]$ as shown in eqs. (11.38). Let $J_0[\hat{\boldsymbol{R}}]$ be the normalized residual obtained by replacing $V[\boldsymbol{x}_{\alpha}]$ and $V[\boldsymbol{x}'_{\alpha}]$ by $V_0[\boldsymbol{x}_{\alpha}]$ and $V_0[\boldsymbol{x}'_{\alpha}]$, respectively, in the expression for $J[\hat{\boldsymbol{R}}]$. An unbiased estimator of ϵ^2 under the hypothesis $\boldsymbol{h} = \boldsymbol{0}$ is obtained in the form

$$\hat{\epsilon}^2 = \frac{J_0[\hat{\boldsymbol{R}}]}{2N-3}. \tag{11.178}$$

Its expectation and variance under the hypothesis are

$$E[\hat{\epsilon}^2] = \epsilon^2, \qquad V[\hat{\epsilon}^2] = \frac{2\epsilon^4}{2N-3}. \tag{11.179}$$

In terms of the estimate $\hat{\epsilon}^2$, the rotation test given by eq. (11.177) takes the following form:

$$\frac{\hat{\epsilon}^2}{\epsilon^2} > \frac{\chi^2_{2N-3,a}}{2N-3}. \tag{11.180}$$

The interpretation is the same as in the case of the planarity test (see eq. (11.152)): we compare the estimate $\hat{\epsilon}$ computed under the hypothesis $\boldsymbol{h} = \boldsymbol{0}$ with the noise level ϵ expected from the accuracy of the image processing operations for locating feature points.

C. Estimation of the residual

Since the optimization (11.174) is nonlinear, we need numerical search in the parameter space for $\boldsymbol{R}$, which is computationally costly. However, an approximately optimal solution can be computed easily. Normalizing both sides of eq. (11.171) into unit vectors and noting that $N[\boldsymbol{R}\boldsymbol{x}'_\alpha] = \boldsymbol{R}N[\boldsymbol{x}'_\alpha]$, we obtain

$$N[\bar{\boldsymbol{x}}_\alpha] = \boldsymbol{R}N[\boldsymbol{x}'_\alpha]. \tag{11.181}$$

If vectors $\boldsymbol{x}_\alpha$ and $\boldsymbol{x}'_\alpha$, $\alpha = 1, ..., N$, are normalized into

$$\boldsymbol{m}_\alpha = N[\boldsymbol{x}_\alpha], \qquad \boldsymbol{m}'_\alpha = N[\boldsymbol{x}'_\alpha], \tag{11.182}$$

the rotation $\boldsymbol{R}$ can be estimated by the least-squares optimization

$$J[\boldsymbol{R}] = \frac{1}{N}\sum_{\alpha=1}^{N} W_\alpha \|\boldsymbol{m}_\alpha - \boldsymbol{R}\boldsymbol{m}'_\alpha\|^2 \rightarrow \min, \tag{11.183}$$

where W_α are arbitrary positive weights. If we define the *correlation matrix*

$$\boldsymbol{C} = \sum_{\alpha=1}^{N} W_\alpha \boldsymbol{m}_\alpha \boldsymbol{m}'^{\top}_\alpha, \tag{11.184}$$

the optimization (11.183) is equivalent to the following maximization (see eqs. (2.157) and (2.158)):

$$(\boldsymbol{C}; \boldsymbol{R}) \rightarrow \max. \tag{11.185}$$

Let

$$\boldsymbol{C} = \boldsymbol{V}\boldsymbol{\Lambda}\boldsymbol{U}^\top \tag{11.186}$$

be the singular value decomposition of $\boldsymbol{C}$ (see eq. (2.114)). The solution of the optimization (11.185) is given as follows (see eq. (2.160)):

$$\hat{\boldsymbol{R}} = \boldsymbol{V}\mathrm{diag}(1, 1, \det(\boldsymbol{V}\boldsymbol{U}^\top))\boldsymbol{U}^\top. \tag{11.187}$$

This may not be the truly optimal solution of (11.174) but can be used to compute a good approximation to the *residual*, because to a first approximation the function $J_0[\boldsymbol{R}]$ is *stationary* in the neighborhood of $\hat{\boldsymbol{R}}$.

11.7.2 Information in motion images

The rotation test described in the preceding section can also be given the following intuitive interpretation. If $\|\boldsymbol{h}\| = 0$, no 3-D information can be obtained, and the residual $J_0[\hat{\boldsymbol{R}}]$ is very small (exactly 0 if noise does not exist). As in the case of stereo vision, the reliability of 3-D reconstruction generally increases as $\|\boldsymbol{h}\|$ increases. If $\|\boldsymbol{h}\|$ increases, the residual $J_0[\hat{\boldsymbol{R}}]$ also increases. This implies that we can view $J_0[\hat{\boldsymbol{R}}]$ as the *information* in the two images. From this viewpoint, we observe the following:

- The information is minimum when $\boldsymbol{h} = \boldsymbol{0}$; it generally increases as $\|\boldsymbol{h}\|$ increases.
- If $\boldsymbol{h}$ is the same, the information increases as the variation of the depth becomes larger[14].
- The information is unchanged if an arbitrary rotation $\boldsymbol{R}$ is added to the camera motion[15].
- The information increases as the number N of the feature points increases.

In general, the accuracy of 3-D interpretation decreases as the image noise increases. However, the above observation implies that the negative effect of image noise cannot be measured simply by its magnitude: *it must be weighed against the information in the two images*. In other words, 3-D interpretation can be disrupted by image noise of a very small magnitude if the information is very small, while 3-D interpretation is robust to image noise of a large magnitude if the information is sufficiently large. This viewpoint is further extended in a more general framework in Chapter 13.

Example 11.3 Fig. 11.23 shows two images of a scene with small depth variance viewed from a translating camera. It is widely known that 3-D interpretation of this type of motion images is very difficult in the presence of noise, because similar images would be observed by a camera rotating around the center of the lens (Fig. 11.24), in which case 3-D interpretation is theoretically impossible. For images such as shown in Fig. 11.23, accurate 3-D interpretation is possible only if the image noise is very small. In other words, *the image noise must be small as compared with the information in the images*. Thus, the image noise magnitude is a very deceptive measure of the reliability of the 3-D interpretation.

---

[14] Humans in motion can perceive 3-D structures from the retinal image motion caused by the difference in depth. Psychologists call this phenomenon *motion parallax*.

[15] We assume that all visible feature points are still visible in the image frame after the rotation.

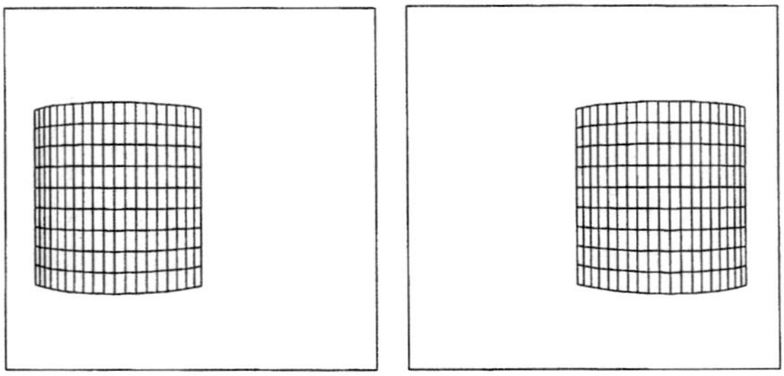

Fig. 11.23. Motion images of a scene viewed from a translating camera.

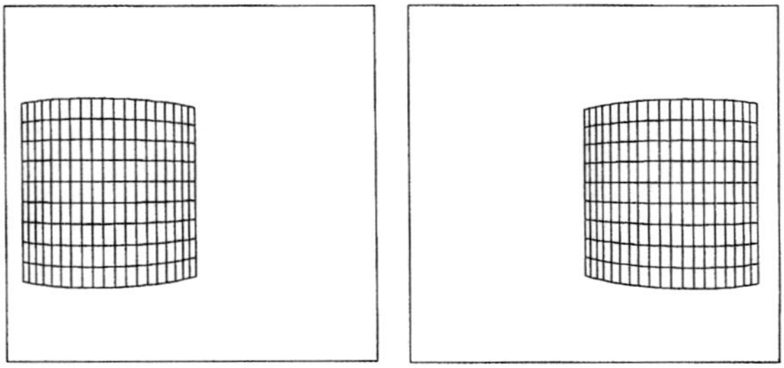

Fig. 11.24. Motion images of a scene viewed from a rotating camera.

Chapter 12

3-D Interpretation of Optical Flow

In this chapter, we study 3-D interpretation of small (theoretically infinitesimal) image motion called *optical flow*. We begin with the problem of how to detect optical flow from gray-level images and how to evaluate its reliability. We then turn to 3-D reconstruction from optical flow. The entire analysis is based on the *epipolar equation*, which is obtained from the epipolar equation for finite motion by taking the limit of infinitesimal motion. The subsequent analysis almost parallels the finite motion analysis given in Chapter 11: we derive a theoretical bound on the attainable accuracy of the motion parameters and present a numerical scheme for 3-D reconstruction by using renormalization and linearization. The critical surface of optical flow, the rotation test, and the planarity test are also discussed in the same way as in the case of finite motion.

12.1 Optical Flow Detection

12.1.1 Gradient equation

If a smoothly moving camera takes a sequence of images of a stationary object (or equivalently a stationary camera takes a sequence of images of a smoothly moving object), the difference between consecutive image frames is very small and defines interframe displacements of image points called *optical flow*. As compared with feature matching, which is necessary for 3-D analysis of finite motion, optical flow has the advantage that the flow can be detected densely (usually at each pixel) by a homogeneous image operation. Hence, it is widely expected that optical flow can be used as an important source of information for many practical purposes such as real time 3-D interpretation of the environment for robot navigation.

The basic assumption for detecting optical flow is that *corresponding image points have the same gray level*. If the image motion is idealized as continuous, this assumption takes the following form:

$$I_x u + I_y v + I_t = 0. \qquad (12.1)$$

Here, I and (u, v) are, respectively, the image intensity and the optical flow in position (x, y) at time t. The subscripts x, y, and t denote partial derivatives $\partial/\partial x$, $\partial/\partial y$, and $\partial/\partial t$, respectively. A differential equation of the form of eq. (12.1) is generally known as the *conservation equation*, stating that quantity I does not change in the course of motion. Let us call I_x, I_y, and

I_t the *(spatio-temporal) gradient values*, and eq. (12.1) the *(spatio-temporal) gradient equation.*

The gradient equation (12.1) alone is insufficient to determine the two unknowns u and v. This indeterminacy can be resolved *if the flow is assumed to be constant over a small region in the image*. Suppose the flow is constant in an image region that contains N pixels, to which we refer by serial numbers $\alpha = 1, ..., N$. We write the gradient values at the αth pixel as $I_{x\alpha}$, $I_{y\alpha}$, and $I_{t\alpha}$ They can be computed by applying a digital filter in the (spatio-temporal) neighborhood of the pixel in question; different computational schemes are obtained depending on what kind of filter is used. Whatever filter is used, however, the computed value is an approximation to the true value. Hence, optical flow detection can be viewed as the following parametric fitting:

Problem 12.1 *Estimate the flow components u and v from observed gradient values $\{I_{x\alpha}, I_{y\alpha}, I_{t\alpha}\}$ in such a way that their (unknown) true values $\{\bar{I}_{x\alpha}, \bar{I}_{y\alpha}, \bar{I}_{t\alpha}\}$ satisfy*

$$\bar{I}_{x\alpha}u + \bar{I}_{y\alpha}v + \bar{I}_{t\alpha} = 0, \qquad \alpha = 1, ..., N. \tag{12.2}$$

Eq. (12.2) is the hypothesis from which the optical flow is to be estimated; the rank of this hypothesis is 1. Problem 12.1 is formally equivalent to the problem of fitting a space plane $uX+vY+Z=0$ to space points $(I_{x\alpha}, I_{y\alpha}, I_{t\alpha})$, $\alpha = 1, ..., N$, in the XYZ space. We write the observed gradient values as

$$I_{x\alpha} = \bar{I}_{x\alpha} + \Delta I_{x\alpha}, \quad I_{y\alpha} = \bar{I}_{y\alpha} + \Delta I_{y\alpha}, \quad I_{t\alpha} = \bar{I}_{t\alpha} + \Delta I_{t\alpha}, \tag{12.3}$$

and regard the noise terms $\Delta I_{x\alpha}$, $\Delta I_{y\alpha}$, and $\Delta I_{t\alpha}$ as Gaussian random variables of means 0, independent for each α. Let their variances and covariances be

$$E[\Delta I_{x\alpha}^2] = \sigma_{x\alpha}^2, \quad E[\Delta I_{y\alpha}^2] = \sigma_{y\alpha}^2, \quad E[\Delta I_{t\alpha}^2] = \sigma_{t\alpha}^2,$$
$$E[\Delta I_{x\alpha}\Delta I_{y\alpha}] = \gamma_{xy\alpha}, \quad E[\Delta I_{x\alpha}\Delta I_{t\alpha}] = \gamma_{xt\alpha}, \quad E[\Delta I_{y\alpha}\Delta I_{t\alpha}] = \gamma_{yt\alpha}. \tag{12.4}$$

If we write

$$\boldsymbol{a}_\alpha = \begin{pmatrix} I_{x\alpha} \\ I_{y\alpha} \\ I_{t\alpha} \end{pmatrix}, \qquad \boldsymbol{u} = \begin{pmatrix} u \\ v \\ 1 \end{pmatrix}, \tag{12.5}$$

the gradient equation (12.2) can be written in the form

$$(\bar{\boldsymbol{a}}_\alpha, \boldsymbol{u}) = 0, \tag{12.6}$$

where $\bar{\boldsymbol{a}}_\alpha$ denotes the true value of $\boldsymbol{a}_\alpha$. Vector $\boldsymbol{a}_\alpha$ can be interpreted as the *space-time gradient* of the image intensity I at the αth pixel in the xyt space time. It is normal to the surface on which the intensity I has a constant value (Fig. 12.1). Vector $\boldsymbol{u}$ can be interpreted as the *space-time velocity*. It is tangent to the *world line* in the xyt space time. It follows that the gradient

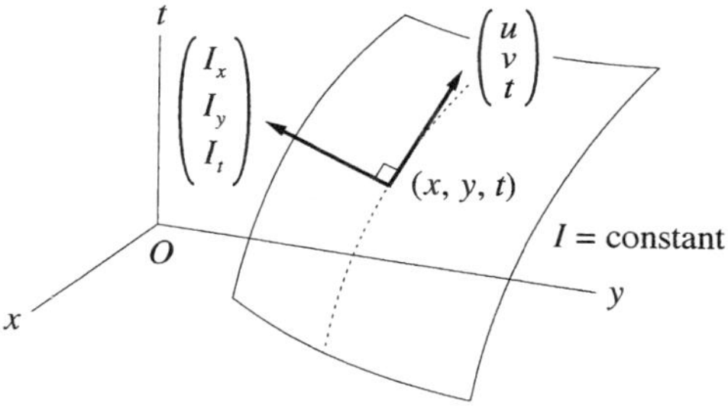

Fig. 12.1. The image intensity I is constant along the world line in the xyt space time.

equation (12.6) (or equivalently eq. (12.2)) states that in the absence of noise *the image intensity I is constant along the world line in the xyt space time.*

The covariance matrix of vector $\boldsymbol{a}_\alpha$ is given by

$$V[\boldsymbol{a}_\alpha] = \begin{pmatrix} \sigma_{x\alpha}^2 & \gamma_{xy\alpha} & \gamma_{xt\alpha} \\ \gamma_{xy\alpha} & \sigma_{y\alpha}^2 & \gamma_{yt\alpha} \\ \gamma_{xt\alpha} & \gamma_{yt\alpha} & \sigma_{t\alpha}^2 \end{pmatrix}. \tag{12.7}$$

As discussed in Section 7.1.1, the optimal estimate of $\boldsymbol{u}$ can be computed by the optimization

$$J[\boldsymbol{u}] = \sum_{\alpha=1}^{N} \frac{(\boldsymbol{a}_\alpha, \boldsymbol{u})^2}{(\boldsymbol{u}, V[\boldsymbol{a}_\alpha]\boldsymbol{u})} \to \min \tag{12.8}$$

under the constraint $(\boldsymbol{k}, \boldsymbol{u}) = 1$. Let $\hat{\boldsymbol{u}}$ be the resulting estimate. Theoretically, the residual $J[\hat{\boldsymbol{u}}]$ is a χ^2 variable of $N-2$ degrees of freedom (so, we need at least two pixels), and the hypothesis of the constancy of gray levels in the course of motion can be tested by the standard χ^2 test, provided the statistical model given by eqs. (12.3) and (12.4) is strictly true, which is very difficult to ascertain for real images.

12.1.2 Reliability of optical flow

According to the general theory in Section 7.1.3, the theoretical covariance matrix of the optimal estimate $\hat{\boldsymbol{u}}$ computed by the optimization (12.8) is given by

$$\bar{V}[\hat{\boldsymbol{u}}] = \left(\sum_{\alpha=1}^{N} \frac{(\boldsymbol{P}_{\boldsymbol{k}}\bar{\boldsymbol{a}}_\alpha)(\boldsymbol{P}_{\boldsymbol{k}}\bar{\boldsymbol{a}}_\alpha)^\top}{(\bar{\boldsymbol{u}}, V[\boldsymbol{a}_\alpha]\bar{\boldsymbol{u}})} \right)^-, \tag{12.9}$$

where $\bar{\boldsymbol{u}}$ is the true value of $\boldsymbol{u}$. This covariance matrix gives a theoretical bound on the attainable accuracy of optical flow detection. In practice, the

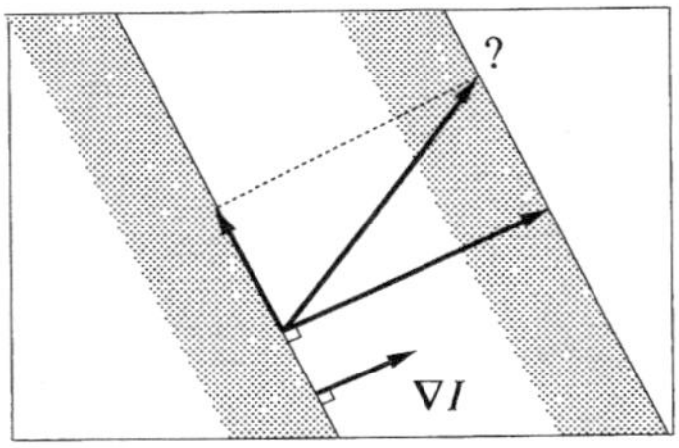

Fig. 12.2. The aperture problem: the tangential component is ambiguous when a gray-level image is moving.

above expression is approximated by

$$V[\hat{\boldsymbol{u}}] = \left(\sum_{\alpha=1}^{N} \frac{(\boldsymbol{P_k a_\alpha})(\boldsymbol{P_k a_\alpha})^\top}{(\hat{\boldsymbol{u}}, V[\boldsymbol{a}_\alpha]\hat{\boldsymbol{u}})} \right)^{-}. \tag{12.10}$$

Let $(\hat{u}, \hat{v})$ be the estimated flow, and write $\nabla I_\alpha = (I_{x\alpha}, I_{y\alpha})^\top$. Eq. (12.10) can be written as

$$V[\hat{\boldsymbol{u}}] = \begin{pmatrix} V[\hat{u}, \hat{v}] & 0 \\ 0 & 0 \end{pmatrix}, \tag{12.11}$$

where $V[\hat{u}, \hat{v}]$ is a (22)-matrix given by

$$\begin{aligned} V[\hat{u}, \hat{v}] &= \left(\sum_{\alpha=1}^{N} W_\alpha (\nabla I_\alpha)(\nabla I_\alpha)^\top \right)^{-1} \\ &= \begin{pmatrix} \sum_{\alpha=1}^{N} W_\alpha I_{x\alpha}^2 & \sum_{\alpha=1}^{N} W_\alpha I_{x\alpha} I_{y\alpha} \\ \sum_{\alpha=1}^{N} W_\alpha I_{y\alpha} I_{x\alpha} & \sum_{\alpha=1}^{N} W_\alpha I_{y\alpha}^2 \end{pmatrix}^{-1}, \end{aligned} \tag{12.12}$$

where

$$W_\alpha = \frac{1}{\sigma_{x\alpha}^2 \hat{u}^2 + 2\gamma_{xy\alpha}\hat{u}\hat{v} + \sigma_{y\alpha}^2 \hat{v}^2 + 2(\gamma_{xt\alpha}\hat{u} + \gamma_{yt\alpha}\hat{v}) + \sigma_{t\alpha}^2}. \tag{12.13}$$

The inverse matrix in eq. (12.12) may not always exist. It is easily seen that the inverse exists if and only if the rank of $\{\nabla I_\alpha\}$ is two, i.e., the intensity gradient vectors $\{\nabla I_\alpha\}$ are not all parallel. If $\{\nabla I_\alpha\}$ are all parallel, the isointensity contour of the gray level consists of parallel lines, so the motion along the isointensity lines (called *tangential flow*) is indiscernible; only the motion in the direction of the intensity gradient (called the *normal flow*) is visible (Fig. 12.2). For such a flow, the covariance matrix diverges to infinity. This anomaly, known as the *aperture problem*, can be understood easily if we recall the space-time description: since the constraint on the space-time velocity $\boldsymbol{u}$ is its orthogonality to the space-time gradient $\boldsymbol{a}_\alpha$, we need gradient values that have different orientations at least at two pixels (Fig. 12.3).

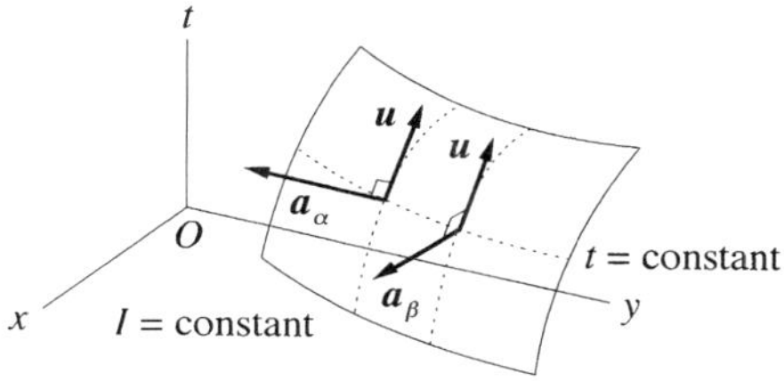

Fig. 12.3. The space-time velocity **u** can be determined if and only if the space component of the space-time gradient has different orientations at least at two pixels.

The aperture problem always occurs in a region where the image intensity is constant or changes linearly. Hence, the accuracy of the detected flow is generally low. Since the interframe camera motion relative to the object is very small, 3-D interpretation based on optical flow is sensitively affected by image noise. In the following sections, we cope with this difficulty by deriving a theoretically optimal technique for maximizing the accuracy of the 3-D reconstruction.

Example 12.1 If the noise in the gradient values is isotropic and homogeneous in space and if the time components are independent of the space components, the noise is characterized by two constants σ_s and σ_t:

$$\sigma_{x\alpha}^2 = \sigma_{y\alpha}^2 = \sigma_s^2, \quad \sigma_{t\alpha}^2 = \sigma_t^2, \quad \gamma_{xy\alpha} = \gamma_{xt\alpha} = \gamma_{yt\alpha} = 0. \tag{12.14}$$

Then, the minimization (12.8) reduces to

$$J_0[u, v] = \sum_{\alpha=1}^{N} \frac{(I_{x\alpha}u + I_{y\alpha}v + I_{t\alpha})^2}{(u^2 + v^2 + \gamma^2)} \to \min, \tag{12.15}$$

where $\gamma = \sigma_t/\sigma_s$. Eq. (12.12) reduces to

$$V[\hat{u}, \hat{v}] = \sigma^2(\hat{u}^2 + \hat{v}^2 + \gamma^2) \begin{pmatrix} \sum_{\alpha=1}^N I_{x\alpha}^2 & \sum_{\alpha=1}^N I_{x\alpha}I_{y\alpha} \\ \sum_{\alpha=1}^N I_{y\alpha}I_{x\alpha} & \sum_{\alpha=1}^N I_{y\alpha}^2 \end{pmatrix}^{-1}. \tag{12.16}$$

An unbiased estimator of σ_s^2 is obtained in the form

$$\hat{\sigma}_s^2 = \frac{J_0[\hat{u}, \hat{v}]}{N-2}. \tag{12.17}$$

Its expectation and variance are given by

$$E[\hat{\sigma}_s^2] = \sigma_s^2, \qquad V[\hat{\sigma}_s^2] = \frac{2\sigma_s^4}{N-2}. \tag{12.18}$$

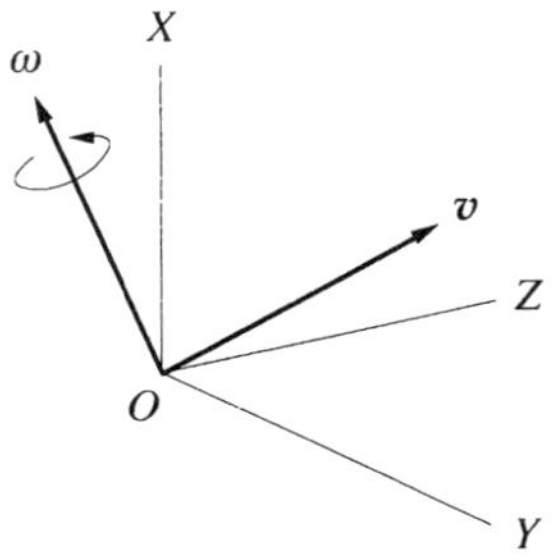

Fig. 12.4. Instantaneous motion parameters.

12.2 Theoretical Basis of 3-D Interpretation

12.2.1 Optical flow equation

Suppose the camera is smoothly moving in a stationary scene. An instantaneous camera motion is a composition of an instantaneous translation of the center of the lens and an instantaneous rotation around it. Let $\boldsymbol{l}$ (unit vector) be the instantaneous axis of rotation, and ω the angular velocity around it. The instantaneous camera motion is specified by the *translation velocity* $\boldsymbol{v}$ and the *rotation velocity* $\boldsymbol{\omega} = \omega \boldsymbol{l}$ (see Section 2.1.5); we call $\{\boldsymbol{v}, \boldsymbol{\omega}\}$ the (instantaneous) *motion parameters* (Fig. 12.4).

Viewed from the camera in motion, a space point $\boldsymbol{r}$ is rotating around the center of the lens with rotation velocity $-\boldsymbol{\omega}$ and translating with translation velocity $-\boldsymbol{v}$. Hence, the velocity of $\boldsymbol{r}$ relative to the camera has the following form (see eq. (2.59)):

$$\dot{\boldsymbol{r}} = -\boldsymbol{v} - \boldsymbol{\omega} \times \boldsymbol{r}. \tag{12.19}$$

Throughout this chapter, we use dots to denote time derivatives.

As in the case of finite motion, an infinitesimal object motion relative to a stationary camera is equivalently treated as an infinitesimal motion of the camera relative to the object. Suppose an object is rotating around a reference point $\boldsymbol{r}_G$ arbitrarily fixed in the object, say its centroid, with rotation velocity $\boldsymbol{\omega}_o$ and translating with translation velocity $\boldsymbol{v}_o$. The velocity of an arbitrary point $\boldsymbol{r}$ in the object is

$$\dot{\boldsymbol{r}} = \boldsymbol{v}_0 + \boldsymbol{\omega}_0 \times (\boldsymbol{r} - \boldsymbol{r}_G). \tag{12.20}$$

Comparing this with eq. (12.19), we find that the motion parameters of the equivalent camera motion relative to the object are given as follows (see eqs. (11.4)):

$$\boldsymbol{v} = -\boldsymbol{v}_o + \boldsymbol{\omega}_o \times \boldsymbol{r}_G, \qquad \boldsymbol{\omega} = -\boldsymbol{\omega}_o. \tag{12.21}$$

Conversely, the parameters $\{\boldsymbol{v}_o, \boldsymbol{\omega}_o\}$ of the object motion are computed from the parameters $\{\boldsymbol{v}, \boldsymbol{\omega}\}$ of the equivalent camera motion in the following form

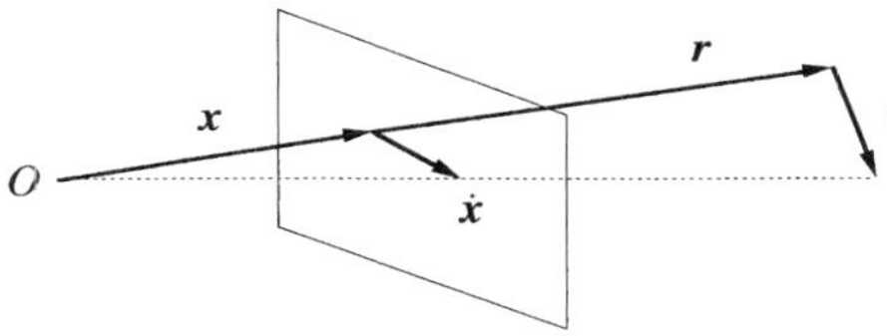

Fig. 12.5. Optical flow.

(see eqs. (11.5)):

$$\boldsymbol{v}_o = -\boldsymbol{v} - \boldsymbol{\omega} \times \boldsymbol{r}_G, \qquad \boldsymbol{\omega}_o = -\boldsymbol{\omega}. \tag{12.22}$$

Consider an image point $\boldsymbol{x}$. The corresponding space point $\boldsymbol{r}$ can be written in the following form:

$$\boldsymbol{x} = \frac{\boldsymbol{r}}{Z(\boldsymbol{x})}. \tag{12.23}$$

Since $(\boldsymbol{k}, \boldsymbol{x}) = 1$ for $\boldsymbol{k} = (0, 0, 1)^\top$, the depth $Z(\boldsymbol{x})$ is given by

$$Z(\boldsymbol{x}) = (\boldsymbol{k}, \boldsymbol{r}). \tag{12.24}$$

Suppose the space point $\boldsymbol{r}$ has velocity $\dot{\boldsymbol{r}}$ relative to the camera (Fig. 12.5). Differentiating eq. (12.23) with respect to time, we obtain

$$\dot{\boldsymbol{x}} = \frac{\dot{\boldsymbol{r}}}{Z(\boldsymbol{x})} - \frac{\dot{Z}(\boldsymbol{x})\boldsymbol{r}}{Z(\boldsymbol{x})^2} = \frac{\dot{\boldsymbol{r}}}{Z(\boldsymbol{x})} - \frac{(\boldsymbol{k}, \dot{\boldsymbol{r}})\boldsymbol{r}}{Z(\boldsymbol{x})^2} = \frac{\boldsymbol{Q}_{\boldsymbol{x}}\dot{\boldsymbol{r}}}{Z(\boldsymbol{x})}, \tag{12.25}$$

where we have defined

$$\boldsymbol{Q}_{\boldsymbol{x}} = \boldsymbol{I} - \boldsymbol{x}\boldsymbol{k}^\top. \tag{12.26}$$

Note the following identities:

$$\boldsymbol{Q}_{\boldsymbol{x}}\boldsymbol{x} = \boldsymbol{0}, \qquad \boldsymbol{Q}_{\boldsymbol{x}}\dot{\boldsymbol{x}} = \dot{\boldsymbol{x}},$$

$$\boldsymbol{Q}_{\boldsymbol{x}}^2 = \boldsymbol{Q}_{\boldsymbol{x}}, \qquad \boldsymbol{P}_{\boldsymbol{k}}\boldsymbol{Q}_{\boldsymbol{x}} = \boldsymbol{Q}_{\boldsymbol{x}}, \qquad \boldsymbol{Q}_{\boldsymbol{x}}\boldsymbol{P}_{\boldsymbol{k}} = \boldsymbol{P}_{\boldsymbol{k}}. \tag{12.27}$$

Substituting eq. (12.19) into eq. (12.25), we obtain the following expression for the image velocity $\dot{\boldsymbol{x}}$:

$$\dot{\boldsymbol{x}} = -\boldsymbol{Q}_{\boldsymbol{x}}\left(\frac{\boldsymbol{v}}{Z(\boldsymbol{x})} + \boldsymbol{\omega} \times \boldsymbol{x}\right). \tag{12.28}$$

We call this the *optical flow equation*. In the following, we call $\dot{\boldsymbol{x}}$ simply the *optical flow* at $\boldsymbol{x}$.

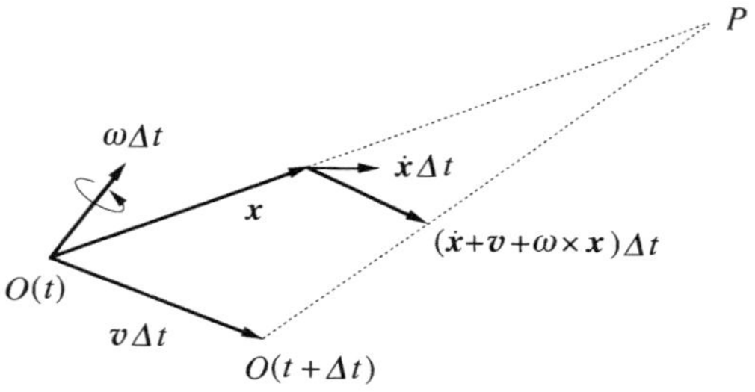

Fig. 12.6. Geometric interpretation of the epipolar equation.

12.2.2 Epipolar equation for optical flow

Consider an infinitesimal image motion. After Δt seconds, image point $\boldsymbol{x}$ moves to

$$\boldsymbol{x}' = \boldsymbol{x} + \dot{\boldsymbol{x}}\Delta t + O(\Delta t^2). \tag{12.29}$$

The motion parameters $\{\boldsymbol{h},\ \boldsymbol{R}\}$ are expressed in the following form (see eq. (2.58)):

$$\boldsymbol{h} = \boldsymbol{v}\Delta t + O(\Delta t^2), \qquad \boldsymbol{R} = \boldsymbol{I} + \boldsymbol{\omega} \times \boldsymbol{I}\Delta t + O(\Delta t^2). \tag{12.30}$$

From eqs. (12.29) and (12.30), we obtain

$$|\boldsymbol{x}, \boldsymbol{h}, \boldsymbol{R}\boldsymbol{x}'| = -e(\boldsymbol{x})\Delta t^2 + O(\Delta t)^3, \tag{12.31}$$

where

$$e(\boldsymbol{x}) = |\boldsymbol{x}, \dot{\boldsymbol{x}}, \boldsymbol{v}| + (\boldsymbol{v} \times \boldsymbol{x}, \boldsymbol{\omega} \times \boldsymbol{x}). \tag{12.32}$$

Thus, the epipolar equation $|\boldsymbol{x}, \boldsymbol{h}, \boldsymbol{R}\boldsymbol{x}'| = 0$ reduces to

$$e(\boldsymbol{x}) = 0. \tag{12.33}$$

The epipolar equation $|\boldsymbol{x}, \boldsymbol{h}, \boldsymbol{R}\boldsymbol{x}'| = 0$ states that the camera translation $\boldsymbol{h}$ and the two vectors $\boldsymbol{x}$ and $\boldsymbol{R}\boldsymbol{x}'$, which define the lines of sight of a feature point before and after the camera motion, must be coplanar (see Fig. 6.1). Since eq. (12.33) is obtained as the limit of infinitesimal camera motion, it should describe the condition that the camera translation velocity $\boldsymbol{v}$, the vector $\boldsymbol{x}$, which defines the line of sight of a feature point, and the velocity of $\boldsymbol{x}$ must be coplanar (Fig. 12.6). This is easily confirmed if we note that the velocity of vector $\boldsymbol{x}$ relative to a stationary scene is $\dot{\boldsymbol{x}} + \boldsymbol{v} + \boldsymbol{\omega} \times \boldsymbol{x}$; the vector $\boldsymbol{x}$ is moving with velocity $\dot{\boldsymbol{x}}$ relative to the camera, which is translating with translation velocity $\boldsymbol{v}$ and rotating with rotation velocity $\boldsymbol{\omega}$. It is immediately seen that eq. (12.32) can be rewritten in the form

$$e(\boldsymbol{x}) = |\boldsymbol{x}, \dot{\boldsymbol{x}} + \boldsymbol{v} + \boldsymbol{\omega} \times \boldsymbol{x}, \boldsymbol{v}|, \tag{12.34}$$

stating that three vectors $\boldsymbol{x}$, $\dot{\boldsymbol{x}} + \boldsymbol{v} + \boldsymbol{\omega} \times \boldsymbol{x}$, and $\boldsymbol{v}$ are coplanar. We call eq. (12.33) the *epipolar equation* for optical flow.

The fact that eq. (12.33) is the necessary and sufficient condition for 3-D reconstruction is alternatively confirmed as follows. First, note that the optical flow equation (12.28) can be rewritten in the following form:

$$\dot{\boldsymbol{x}} + \boldsymbol{Q}_{\boldsymbol{x}}(\boldsymbol{\omega} \times \boldsymbol{x}) = -\frac{\boldsymbol{Q}_{\boldsymbol{x}}\boldsymbol{v}}{Z(\boldsymbol{x})}. \tag{12.35}$$

It follows that for given vectors $\dot{\boldsymbol{x}}$, $\boldsymbol{\omega}$, and $\boldsymbol{v}$, the depth $Z(\boldsymbol{x})$ that satisfies eq. (12.35) exists if and only if vectors $\dot{\boldsymbol{x}} + \boldsymbol{Q}_{\boldsymbol{x}}(\boldsymbol{\omega} \times \boldsymbol{x})$ and $\boldsymbol{Q}_{\boldsymbol{x}}\boldsymbol{v}$ are parallel, i.e.,

$$\left(\dot{\boldsymbol{x}} + \boldsymbol{Q}_{\boldsymbol{x}}(\boldsymbol{\omega} \times \boldsymbol{x})\right) \times \boldsymbol{Q}_{\boldsymbol{x}}\boldsymbol{v} = \boldsymbol{0}. \tag{12.36}$$

By definition, $\dot{\boldsymbol{x}} \in \{\boldsymbol{k}\}_L^{\perp}$. The defining equation (12.26) of matrix $\boldsymbol{Q}_{\boldsymbol{x}}$ implies that $\boldsymbol{Q}_{\boldsymbol{x}}\boldsymbol{a} \in \{\boldsymbol{k}\}_L^{\perp}$ for an arbitrary vector $\boldsymbol{a}$. Hence, the left-hand side of eq. (12.36) has the form $c\boldsymbol{k}$ for some constant c. Since $(\boldsymbol{x}, \boldsymbol{k}) = 1$, the constant c can be determined by computing the inner product of the left-hand side of eq. (12.36) with $\boldsymbol{x}$: we obtain

$$\begin{aligned} c &= (\boldsymbol{x}, \left(\dot{\boldsymbol{x}} + \boldsymbol{Q}_{\boldsymbol{x}}(\boldsymbol{\omega} \times \boldsymbol{x})\right) \times \boldsymbol{Q}_{\boldsymbol{x}}\boldsymbol{v}) = |\boldsymbol{x}, \dot{\boldsymbol{x}}, \boldsymbol{Q}_{\boldsymbol{x}}\boldsymbol{v}| + |\boldsymbol{x}, \boldsymbol{Q}_{\boldsymbol{x}}(\boldsymbol{\omega} \times \boldsymbol{x}), \boldsymbol{Q}_{\boldsymbol{x}}\boldsymbol{v}| \\ &= |\boldsymbol{x}, \dot{\boldsymbol{x}}, \boldsymbol{v} - (\boldsymbol{k}, \boldsymbol{v})\boldsymbol{x}| + |\boldsymbol{x}, \boldsymbol{\omega} \times \boldsymbol{x} - (\boldsymbol{k}, \boldsymbol{\omega} \times \boldsymbol{x})\boldsymbol{x}, \boldsymbol{v} - (\boldsymbol{k}, \boldsymbol{v})\boldsymbol{x}| \\ &= |\boldsymbol{x}, \dot{\boldsymbol{x}}, \boldsymbol{v}| + (\boldsymbol{v} \times \boldsymbol{x}, \boldsymbol{\omega} \times \boldsymbol{x}) = e(\boldsymbol{x}). \end{aligned} \tag{12.37}$$

It follows that eq. (12.36) can be equivalently written as $e(\boldsymbol{x})\boldsymbol{k} = \boldsymbol{0}$, meaning that the depth $Z(\boldsymbol{x})$ exists if and only if $e(\boldsymbol{x}) = 0$.

12.2.3 3-D analysis from optical flow

In the above analysis, we have assumed that no image noise exists. In the presence of noise, the epipolar equation $e(\boldsymbol{x}) = 0$ does not necessarily hold. Let $\bar{\dot{\boldsymbol{x}}}$ be the flow that should be observed in the absence of noise. We write

$$\dot{\boldsymbol{x}} = \bar{\dot{\boldsymbol{x}}} + \Delta\dot{\boldsymbol{x}}, \tag{12.38}$$

and regard the noise term $\Delta\dot{\boldsymbol{x}}$ as a random variable of mean $\boldsymbol{0}$ and covariance matrix $V[\dot{\boldsymbol{x}}]$, independent at each pixel. If the optical flow is detected by the optimization (12.15), the covariance matrix $V[\dot{\boldsymbol{x}}]$ can be identified with the matrix $V[\hat{\boldsymbol{u}}]$ given by eq. (12.11). As in the case of the finite motion analysis, the problem of computing 3-D structures from optical flow is stated as follows:

Problem 12.2 *Estimate the motion parameters* $\{\boldsymbol{v}, \boldsymbol{\omega}\}$ *that satisfy*

$$|\boldsymbol{x}, \bar{\dot{\boldsymbol{x}}}, \boldsymbol{v}| + (\boldsymbol{v} \times \boldsymbol{x}, \boldsymbol{\omega} \times \boldsymbol{x}) = 0 \tag{12.39}$$

from the observed flow $\dot{\boldsymbol{x}}$.

Eq. (12.39) is the hypothesis from which the motion parameters $\{\boldsymbol{v}, \boldsymbol{\omega}\}$ are to be estimated. The rank of this hypothesis is 1. Since eq. (12.39) is homogeneous in $\boldsymbol{v}$, the absolute magnitude of the translation velocity $\boldsymbol{v}$ is indeterminate. As in the case of finite motion, this is due to the fact that as long as the image motion is the only source of information, a small camera motion near a small object is indistinguishable from a large camera motion far away from a large object (see Fig. 11.1). The decision criterion (the *rotation test*) to test whether or not $\boldsymbol{v} = \boldsymbol{0}$, i.e., whether or not the camera motion is a pure rotation, will be discussed in Section 12.9.2. In the following, we normalize $\boldsymbol{v}$ into $\|\boldsymbol{v}\| = 1$, assuming that $\boldsymbol{v} \neq \boldsymbol{0}$ has already been confirmed.

12.3 Optimal Estimation of Motion Parameters

12.3.1 Optimal estimation

A. Flow matrix

Eq. (12.32) can be rewritten in the following form:

$$e(\boldsymbol{x}) = (\dot{\boldsymbol{x}}, \boldsymbol{v} \times \boldsymbol{x}) + \left(\left(\boldsymbol{x}, (\boldsymbol{v}, \boldsymbol{\omega})\boldsymbol{I} - S[\boldsymbol{v}\boldsymbol{\omega}^\top]\right) \boldsymbol{x}\right)$$
$$= (\dot{\boldsymbol{x}}\boldsymbol{x}^\top; \boldsymbol{v} \times \boldsymbol{I}) + (\boldsymbol{x}\boldsymbol{x}^\top; (\boldsymbol{v}, \boldsymbol{\omega})\boldsymbol{I} - S[\boldsymbol{v}\boldsymbol{\omega}^\top]). \quad (12.40)$$

The symbol $S[\,\cdot\,]$ denotes the symmetrization operator (see eqs. (2.205)). Define 33-matrices $\boldsymbol{X}$ and $\boldsymbol{F}$ by

$$\boldsymbol{X} = \boldsymbol{x}\boldsymbol{x}^\top + A[\dot{\boldsymbol{x}}\boldsymbol{x}^\top], \quad (12.41)$$

$$\boldsymbol{F} = (\boldsymbol{v}, \boldsymbol{\omega})\boldsymbol{I} - S[\boldsymbol{v}\boldsymbol{\omega}^\top] + \boldsymbol{v} \times \boldsymbol{I}, \quad (12.42)$$

where the symbol $A[\,\cdot\,]$ denotes the antisymmetrization operator (see eqs. (2.205)). We call $\boldsymbol{F}$ the *flow matrix*. Let $\bar{\boldsymbol{X}}$ be the unperturbed value of $\boldsymbol{X}$ obtained by replacing $\dot{\boldsymbol{x}}$ by $\bar{\dot{\boldsymbol{x}}}$ in eq. (12.41). It is easily seen that eq. (12.39) can be written in the form

$$(\bar{\boldsymbol{X}}; \boldsymbol{F}) = 0. \quad (12.43)$$

This equation has the same form as eq. (11.7). Hence, the subsequent analysis takes essentially the same form as in the case of finite motion analysis.

Eq. (12.41) is linear in $\dot{\boldsymbol{x}}$, so $E[\boldsymbol{X}] = \bar{\boldsymbol{X}}$ and the deviation $\Delta\boldsymbol{X} = \boldsymbol{X} - \bar{\boldsymbol{X}}$ is a random variable of mean $\boldsymbol{O}$. From eq. (12.41), we see that

$$\Delta\boldsymbol{X} = A[\Delta\dot{\boldsymbol{x}}\boldsymbol{x}^\top] = \frac{1}{2}(\Delta\dot{\boldsymbol{x}}\boldsymbol{x}^\top - \boldsymbol{x}\Delta\dot{\boldsymbol{x}}^\top). \quad (12.44)$$

Let $\mathcal{V}[\boldsymbol{X}]$ be the covariance tensor of $\boldsymbol{X}$. It has the following $(ijkl)$ element:

$$\mathcal{V}[\boldsymbol{X}]_{ijkl} = E[\Delta X_{ij}\Delta X_{kl}] = \frac{1}{4}E[(\Delta\dot{x}_i x_j - x_i\Delta\dot{x}_j)(\Delta\dot{x}_k x_l - x_k\Delta\dot{x}_l)]$$
$$= \frac{1}{4}(V[\dot{\boldsymbol{x}}]_{ik}x_j x_l - V[\dot{\boldsymbol{x}}]_{il}x_j x_k - V[\dot{\boldsymbol{x}}]_{jk}x_i x_l + V[\dot{\boldsymbol{x}}]_{jl}x_i x_k). \quad (12.45)$$

Since $\Delta \boldsymbol{X}$ is a [33]-matrix, the covariance tensor $\mathcal{V}[\boldsymbol{X}]$ is a [33][33]-tensor. According to the general theory in Section 7.1.5, an optimal estimate of the motion parameters $\{\boldsymbol{v}, \boldsymbol{\omega}\}$ can be obtained by the following optimization (see eq. (7.55)):

$$J[\boldsymbol{v}, \boldsymbol{\omega}] = \int_S \frac{(\boldsymbol{X}; \boldsymbol{F})}{(\boldsymbol{F}; \mathcal{V}[\boldsymbol{X}]\boldsymbol{F})} dxdy \to \min. \tag{12.46}$$

The minimum is sought under the constraint that $\|\boldsymbol{v}\| = 1$. Here, the integral $\int_S dxdy$ is a symbolic notation for summation over all the pixels in the region S where the optical flow is defined.

From eq. (12.42), we see that

$$A[\boldsymbol{F}] = \boldsymbol{v} \times \boldsymbol{I}. \tag{12.47}$$

Since $\mathcal{V}[\boldsymbol{X}]$ is a [33][33]-tensor, all (33)-matrices belong to its null space (see Section 2.4.3). Hence, eq. (12.45) implies that

$$\begin{aligned}(\boldsymbol{F}; \mathcal{V}[\boldsymbol{X}]\boldsymbol{F}) &= (S[\boldsymbol{F}] + A[\boldsymbol{F}]; \mathcal{V}[\boldsymbol{X}](S[\boldsymbol{F}] + A[\boldsymbol{F}])) = (A[\boldsymbol{F}]; \mathcal{V}[\boldsymbol{X}]A[\boldsymbol{F}]) \\ &= (\boldsymbol{x}, A[\boldsymbol{F}]^\top V[\dot{\boldsymbol{x}}]A[\boldsymbol{F}]\boldsymbol{x}) = (\boldsymbol{v}, (\boldsymbol{x} \times V[\dot{\boldsymbol{x}}] \times \boldsymbol{x})\boldsymbol{v}). \end{aligned} \tag{12.48}$$

Consequently, the optimization (12.46) can be rewritten in the following form:

$$J[\boldsymbol{v}, \boldsymbol{\omega}] = \int_S W(\boldsymbol{x}, \boldsymbol{v}) \Big(|\boldsymbol{x}, \dot{\boldsymbol{x}}, \boldsymbol{v}| + (\boldsymbol{v} \times \boldsymbol{x}, \boldsymbol{\omega} \times \boldsymbol{x})\Big)^2 dxdy \to \min, \tag{12.49}$$

$$W(\boldsymbol{x}, \boldsymbol{v}) = \frac{1}{(\boldsymbol{v}, (\boldsymbol{x} \times V[\dot{\boldsymbol{x}}] \times \boldsymbol{x})\boldsymbol{v})}. \tag{12.50}$$

Example 12.2 If each flow component is perturbed independently by Gaussian noise of mean 0 and variance ϵ^2, the covariance matrix of $\dot{\boldsymbol{x}}$ is $V[\dot{\boldsymbol{x}}] = \epsilon^2 \boldsymbol{P_k}$. If we define a (33)-matrix $\boldsymbol{S_x}$ by

$$\boldsymbol{S_x} = \boldsymbol{x} \times \boldsymbol{P_k} \times \boldsymbol{x} = \begin{pmatrix} 1 & & -x \\ & 1 & -y \\ -x & -y & x^2 + y^2 \end{pmatrix}, \tag{12.51}$$

eq. (12.50) can be written in the following form:

$$W(\boldsymbol{x}, \boldsymbol{v}) = \frac{1}{\epsilon^2 (\boldsymbol{v}, \boldsymbol{S_x} \boldsymbol{v})}. \tag{12.52}$$

B. Rigidity test

Let $\{\hat{\boldsymbol{v}}, \hat{\boldsymbol{\omega}}\}$ be the optimal estimate of $\{\boldsymbol{v}, \boldsymbol{\omega}\}$ determined by the optimization (12.49). The motion parameters $\{\boldsymbol{v}, \boldsymbol{\omega}\}$ have *five* degrees of freedom—two for $\boldsymbol{v}$ (unit vector) and three for $\boldsymbol{\omega}$. Hence, the residual $J[\hat{\boldsymbol{v}}, \hat{\boldsymbol{\omega}}]$ is a χ^2 variable with $S - 5$ degrees of freedom, where S is the number of pixels at which

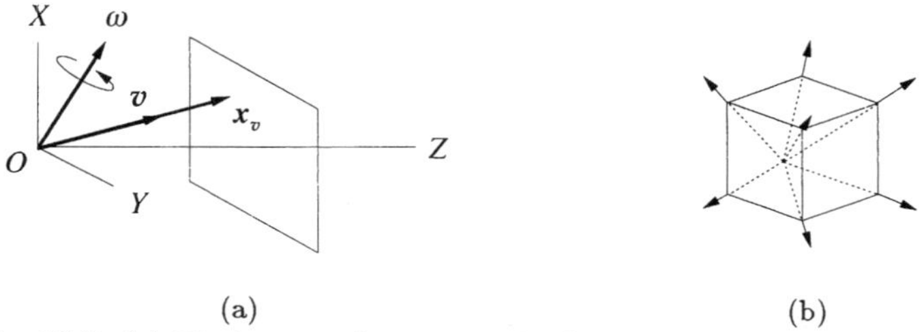

Fig. 12.7. (a) The image point seen in the direction of the camera translation. (b) Focus of expansion.

the optical flow is defined. It follows that optical flow must be observed at least at *five* pixels. This is in agreement with the case of finite motion, but this is merely a theoretical result. In an actual application, the flow must be observed at a large number of pixels, since it is much more vulnerable to image noise than the point correspondence for finite motion. We can also obtain, at least in principle, a *rigidity test*: the hypothesis that the camera is moving in a stationary scene (or equivalently the object in motion is rigid) is rejected with significance level a% if

$$J[\hat{\boldsymbol{v}}, \hat{\boldsymbol{\omega}}] > \chi^2_{S-5,a}. \tag{12.53}$$

C. Focus of expansion

Let $\boldsymbol{x}_{\boldsymbol{v}}$ be the image point whose line of sight is in the direction of the translation velocity $\boldsymbol{v}$ (Fig. 12.7a). Since $\boldsymbol{x}_{\boldsymbol{v}} \propto \boldsymbol{v}$, eq. (12.50) implies that $W(\boldsymbol{x}_{\boldsymbol{v}}, \boldsymbol{v}) = \infty$. It is immediately seen from eq. (12.32) that $e(\boldsymbol{x}_{\boldsymbol{v}}) = 0$ holds irrespective of the values of $\dot{\boldsymbol{x}}$ and $\boldsymbol{\omega}$. Hence, $\boldsymbol{x}_{\boldsymbol{v}}$ is a *singularity* of the optimization (12.49). If the camera is not rotating (i.e., $\boldsymbol{\omega} = \boldsymbol{0}$), the optical flow seems to diverge from or converge to $\boldsymbol{x}_{\boldsymbol{v}}$. Such a point is known as the *focus of expansion* (Fig. 12.7b). Although this description does not hold if the camera is rotating, we still call $\boldsymbol{x}_{\boldsymbol{v}}$ the *focus of expansion*. The underlying geometry is the same as in the case of the finite motion analysis in Section 11.1.2 (see Fig. 11.3).

In an actual application, a measure must be taken to avoid this singularity. For example, each time a candidate value of $\boldsymbol{v}$ is computed, a neighborhood of the corresponding focus of expansion is excluded or the function $W(\boldsymbol{x}, \boldsymbol{v})$ is replaced by a constant in that neighborhood in the next optimization step.

12.3.2 Theoretical bound on accuracy

The theoretical bound on the attainable accuracy is obtained by analyzing the behavior of the function $J[\boldsymbol{v}, \boldsymbol{\omega}]$ near the true values $\{\bar{\boldsymbol{v}}, \bar{\boldsymbol{\omega}}\}$. Substituting

$\boldsymbol{v} = \bar{\boldsymbol{v}} + \Delta\boldsymbol{v}$, $\boldsymbol{\omega} = \bar{\boldsymbol{\omega}} + \Delta\boldsymbol{\omega}$, and eq. (12.38) into eq. (12.49), we obtain

$$J[\boldsymbol{v}, \boldsymbol{\omega}] = \int_S (W(\boldsymbol{x}, \bar{\boldsymbol{v}}) + O(\Delta\boldsymbol{v}))\Big(|\boldsymbol{x}, \dot{\boldsymbol{x}}, \bar{\boldsymbol{v}}| + |\boldsymbol{x}, \bar{\dot{\boldsymbol{x}}}, \Delta\boldsymbol{v}| + (\bar{\boldsymbol{v}} \times \boldsymbol{x}, \bar{\boldsymbol{\omega}} \times \boldsymbol{x})$$
$$+ (\Delta\boldsymbol{v} \times \boldsymbol{x}, \boldsymbol{\omega} \times \boldsymbol{x}) + (\boldsymbol{v} \times \boldsymbol{x}, \Delta\boldsymbol{\omega} \times \boldsymbol{x}) + O(\Delta\dot{\boldsymbol{x}}, \Delta\boldsymbol{v}, \Delta\boldsymbol{\omega})^2\Big)^2 dxdy$$
$$= \int_S W(\boldsymbol{x}, \bar{\boldsymbol{v}})\Big(\bar{e}(\boldsymbol{x}) + (\bar{\boldsymbol{a}}(\boldsymbol{x}), \Delta\boldsymbol{v}) + (\bar{\boldsymbol{b}}(\boldsymbol{x}), \Delta\boldsymbol{\omega})\Big)^2 dxdy$$
$$+ \int_S W(\boldsymbol{x}, \bar{\boldsymbol{v}}) O(\Delta\dot{\boldsymbol{x}}, \Delta\boldsymbol{v}, \Delta\boldsymbol{\omega})^3 dxdy, \tag{12.54}$$

where we have defined

$$\bar{e}(\boldsymbol{x}) = |\boldsymbol{x}, \dot{\boldsymbol{x}}, \bar{\boldsymbol{v}}| + (\bar{\boldsymbol{v}} \times \boldsymbol{x}, \bar{\boldsymbol{\omega}} \times \boldsymbol{x}), \tag{12.55}$$

$$\bar{\boldsymbol{a}}(\boldsymbol{x}) = \boldsymbol{x} \times \bar{\dot{\boldsymbol{x}}} + \|\boldsymbol{x}\|^2 \bar{\boldsymbol{\omega}} - (\boldsymbol{x}, \bar{\boldsymbol{\omega}})\boldsymbol{x},$$
$$\bar{\boldsymbol{b}}(\boldsymbol{x}) = \|\boldsymbol{x}\|^2 \bar{\boldsymbol{v}} - (\boldsymbol{x}, \bar{\boldsymbol{v}})\boldsymbol{x}. \tag{12.56}$$

Ignoring $\int_S W(\boldsymbol{x}, \bar{\boldsymbol{v}}) O(\Delta\boldsymbol{x}, \Delta\boldsymbol{v}, \Delta\boldsymbol{\omega})^3 dxdy$ in eq. (12.54), we obtain

$$J[\boldsymbol{v}, \boldsymbol{\omega}] = \int_S W(\boldsymbol{x}, \bar{\boldsymbol{v}})\Big(\bar{e}(\boldsymbol{x}) + (\bar{\boldsymbol{a}}(\boldsymbol{x}), \Delta\boldsymbol{v}) + (\bar{\boldsymbol{b}}(\boldsymbol{x}), \Delta\boldsymbol{\omega})\Big)^2 dxdy. \tag{12.57}$$

Since $\boldsymbol{v}$ is normalized into a unit vector, the deviation $\Delta\boldsymbol{v}$ should be such that $(\bar{\boldsymbol{v}}, \Delta\boldsymbol{v}) = 0$ to a first approximation. However, this constraint need not be imposed, because the epipolar equation implies that

$$(\bar{\boldsymbol{a}}(\boldsymbol{x}), \bar{\boldsymbol{v}}) = (\boldsymbol{x} \times \bar{\dot{\boldsymbol{x}}}, \bar{\boldsymbol{v}}) + \|\boldsymbol{x}\|^2 (\bar{\boldsymbol{v}}, \bar{\boldsymbol{\omega}}) - (\boldsymbol{x}, \bar{\boldsymbol{v}})(\boldsymbol{x}, \bar{\boldsymbol{\omega}})$$
$$= |\boldsymbol{x}, \bar{\dot{\boldsymbol{x}}}, \bar{\boldsymbol{v}}| + (\bar{\boldsymbol{v}} \times \boldsymbol{x}, \bar{\boldsymbol{\omega}} \times \boldsymbol{x}) = 0. \tag{12.58}$$

Hence, $(\bar{\boldsymbol{a}}(\boldsymbol{x}), \Delta\boldsymbol{v} + c\bar{\boldsymbol{v}}) = (\bar{\boldsymbol{a}}(\boldsymbol{x}), \Delta\boldsymbol{v})$ for an arbitrary constant c. In other words, the component of $\Delta\boldsymbol{v}$ parallel to $\bar{\boldsymbol{v}}$ has no effect on the minimization. Differentiating $J[\boldsymbol{v}, \boldsymbol{\omega}]$ with respect to $\Delta\boldsymbol{v}$ and $\Delta\boldsymbol{\omega}$ and setting the result $\boldsymbol{0}$, we obtain

$$\int_S W(\boldsymbol{x}, \bar{\boldsymbol{v}})\Big((\bar{\boldsymbol{a}}(\boldsymbol{x}), \Delta\boldsymbol{v})\bar{\boldsymbol{a}}(\boldsymbol{x}) + (\bar{\boldsymbol{b}}(\boldsymbol{x}), \Delta\boldsymbol{\omega})\bar{\boldsymbol{a}}(\boldsymbol{x})\Big) dxdy$$
$$= -\int_S W(\boldsymbol{x}, \bar{\boldsymbol{v}}) \bar{e}(\boldsymbol{x}) \bar{\boldsymbol{a}}(\boldsymbol{x}) dxdy, \tag{12.59}$$

$$\int_S W(\boldsymbol{x}, \bar{\boldsymbol{v}})\Big((\bar{\boldsymbol{a}}(\boldsymbol{x}), \Delta\boldsymbol{v})\bar{\boldsymbol{b}}(\boldsymbol{x}) + (\bar{\boldsymbol{b}}(\boldsymbol{x}), \Delta\boldsymbol{\omega})\bar{\boldsymbol{b}}(\boldsymbol{x})\Big) dxdy$$
$$= -\int_S W(\boldsymbol{x}, \bar{\boldsymbol{v}}) \bar{e}(\boldsymbol{x}) \bar{\boldsymbol{b}}(\boldsymbol{x}) dxdy. \tag{12.60}$$

If we define (66)-matrix

$$\bar{\boldsymbol{A}} = \begin{pmatrix} \int_S W(\boldsymbol{x}, \bar{\boldsymbol{v}})\bar{\boldsymbol{a}}(\boldsymbol{x})\bar{\boldsymbol{a}}(\boldsymbol{x})^\top dxdy & \int_S W(\boldsymbol{x}, \bar{\boldsymbol{v}})\bar{\boldsymbol{a}}(\boldsymbol{x})\bar{\boldsymbol{b}}(\boldsymbol{x})^\top dxdy \\ \int_S W(\boldsymbol{x}, \bar{\boldsymbol{v}})\bar{\boldsymbol{b}}(\boldsymbol{x})\bar{\boldsymbol{a}}(\boldsymbol{x})^\top dxdy & \int_S W(\boldsymbol{x}, \bar{\boldsymbol{v}})\bar{\boldsymbol{b}}(\boldsymbol{x})\bar{\boldsymbol{b}}(\boldsymbol{x})^\top dxdy \end{pmatrix}, \tag{12.61}$$

eqs. (12.59) and (12.60) are combined into one equation in the form

$$\bar{\boldsymbol{A}} \begin{pmatrix} \Delta \boldsymbol{v} \\ \Delta \boldsymbol{\omega} \end{pmatrix} = -\int_S W(\boldsymbol{x}, \bar{\boldsymbol{v}})\bar{e}(\boldsymbol{x}) \begin{pmatrix} \bar{\boldsymbol{a}}(\boldsymbol{x}) \\ \bar{\boldsymbol{b}}(\boldsymbol{x}) \end{pmatrix} dxdy. \tag{12.62}$$

It is easily seen from eq. (12.58) that the matrix $\bar{\boldsymbol{A}}$ has rank 5; its null space is $\{\bar{\boldsymbol{v}} \oplus \boldsymbol{0}\}_L$. The solution of eq. (12.62) is obtained in the following form:

$$\begin{pmatrix} \Delta \hat{\boldsymbol{v}} \\ \Delta \hat{\boldsymbol{\omega}} \end{pmatrix} = -\int_S W(\boldsymbol{x}, \bar{\boldsymbol{v}})\bar{e}(\boldsymbol{x})\bar{\boldsymbol{A}}^- \begin{pmatrix} \bar{\boldsymbol{a}}(\boldsymbol{x}) \\ \bar{\boldsymbol{b}}(\boldsymbol{x}) \end{pmatrix} dxdy. \tag{12.63}$$

The optimal estimate $\{\hat{\boldsymbol{v}}, \hat{\boldsymbol{\omega}}\}$ of the motion parameters $\{\boldsymbol{v}, \boldsymbol{\omega}\}$ is given by

$$\hat{\boldsymbol{v}} = \bar{\boldsymbol{v}} + \Delta\hat{\boldsymbol{v}}, \qquad \hat{\boldsymbol{\omega}} = \bar{\boldsymbol{\omega}} + \Delta\hat{\boldsymbol{\omega}}. \tag{12.64}$$

Their covariance matrices are obtained from eq. (12.63) in the following form:

$$\begin{aligned} \begin{pmatrix} \bar{V}[\hat{\boldsymbol{v}}] & \bar{V}[\hat{\boldsymbol{v}}, \hat{\boldsymbol{\omega}}] \\ \bar{V}[\hat{\boldsymbol{\omega}}, \hat{\boldsymbol{v}}] & \bar{V}[\hat{\boldsymbol{\omega}}] \end{pmatrix} &= E\left[\begin{pmatrix} \Delta \hat{\boldsymbol{v}} \\ \Delta \hat{\boldsymbol{\omega}} \end{pmatrix}\begin{pmatrix} \Delta \hat{\boldsymbol{v}} \\ \Delta \hat{\boldsymbol{\omega}} \end{pmatrix}^\top\right] \\ &= \int_S\int_S W(\boldsymbol{x}, \bar{\boldsymbol{v}})W(\boldsymbol{x}', \bar{\boldsymbol{v}})E[\bar{e}(\boldsymbol{x})\bar{e}(\boldsymbol{x}')]\bar{\boldsymbol{A}}^- \\ &\quad \begin{pmatrix} \bar{\boldsymbol{a}}(\boldsymbol{x}) \\ \bar{\boldsymbol{b}}(\boldsymbol{x}) \end{pmatrix}\begin{pmatrix} \bar{\boldsymbol{a}}(\boldsymbol{x}') \\ \bar{\boldsymbol{b}}(\boldsymbol{x}') \end{pmatrix}^\top \bar{\boldsymbol{A}}^- dxdydx'dy' \\ &= \bar{\boldsymbol{A}}^-\bar{\boldsymbol{A}}\bar{\boldsymbol{A}}^- = \bar{\boldsymbol{A}}^-. \end{aligned} \tag{12.65}$$

Here, we have used the relationship

$$\begin{aligned} E[\bar{e}(\boldsymbol{x})\bar{e}(\boldsymbol{x}')] &= \delta(\boldsymbol{x} - \boldsymbol{x}')E[\bar{e}(\boldsymbol{x})^2] = \delta(\boldsymbol{x} - \boldsymbol{x}')E[(\boldsymbol{X}; \bar{\boldsymbol{F}})^2] \\ &= \delta(\boldsymbol{x} - \boldsymbol{x}')E[(\Delta\boldsymbol{X}; \bar{\boldsymbol{F}})^2] = \delta(\boldsymbol{x} - \boldsymbol{x}')(\bar{\boldsymbol{F}}; \mathcal{V}[\boldsymbol{X}]\bar{\boldsymbol{F}}) = \frac{\delta(\boldsymbol{x} - \boldsymbol{x}')}{W(\boldsymbol{x}, \bar{\boldsymbol{v}})}, \end{aligned} \tag{12.66}$$

where $\bar{\boldsymbol{F}}$ is the unperturbed flow matrix $\boldsymbol{F}$ obtained by replacing $\boldsymbol{v}$ and $\boldsymbol{\omega}$ by $\bar{\boldsymbol{v}}$ and $\bar{\boldsymbol{\omega}}$, respectively, in eq. (12.42). The function $\delta(\boldsymbol{x})$ takes value 1 if $\boldsymbol{x} = \boldsymbol{0}$ and 0 otherwise.

The covariance matrices $\bar{V}[\hat{\boldsymbol{v}}]$, $\bar{V}[\hat{\boldsymbol{v}}, \hat{\boldsymbol{\omega}}]$ $(= \bar{V}[\hat{\boldsymbol{\omega}}, \hat{\boldsymbol{v}}]^\top)$, and $\bar{V}[\hat{\boldsymbol{\omega}}]$ thus obtained give a theoretical bound on the attainable accuracy of estimating $\{\boldsymbol{v}, \boldsymbol{\omega}\}$. In actual computation, eq. (12.65) is approximated by

$$\begin{pmatrix} V[\hat{\boldsymbol{v}}] & V[\hat{\boldsymbol{v}}, \hat{\boldsymbol{\omega}}] \\ V[\hat{\boldsymbol{\omega}}, \hat{\boldsymbol{v}}] & V[\hat{\boldsymbol{\omega}}] \end{pmatrix} =$$

$$\begin{pmatrix} \int_S W(\boldsymbol{x}, \hat{\boldsymbol{v}})(\boldsymbol{P}_{\hat{\boldsymbol{v}}}\hat{\boldsymbol{a}}(\boldsymbol{x}))(\boldsymbol{P}_{\hat{\boldsymbol{v}}}\hat{\boldsymbol{a}}(\boldsymbol{x}))^\top dxdy & \int_S W(\boldsymbol{x}, \hat{\boldsymbol{v}})(\boldsymbol{P}_{\hat{\boldsymbol{v}}}\hat{\boldsymbol{a}}(\boldsymbol{x}))\hat{\boldsymbol{b}}(\boldsymbol{x})^\top dxdy \\ \int_S W(\boldsymbol{x}, \hat{\boldsymbol{v}})\hat{\boldsymbol{b}}(\boldsymbol{x})(\boldsymbol{P}_{\hat{\boldsymbol{v}}}\hat{\boldsymbol{a}}(\boldsymbol{x}))^\top dxdy & \int_S W(\boldsymbol{x}, \hat{\boldsymbol{v}})\hat{\boldsymbol{b}}(\boldsymbol{x})\hat{\boldsymbol{b}}(\boldsymbol{x})^\top dxdy \end{pmatrix}^-, \tag{12.67}$$

where $\{\hat{\boldsymbol{v}}, \hat{\boldsymbol{\omega}}\}$ are the estimated motion parameters and

$$\hat{\boldsymbol{a}}(\boldsymbol{x}) = \boldsymbol{x} \times \hat{\dot{\boldsymbol{x}}} + \|\boldsymbol{x}\|^2\hat{\boldsymbol{\omega}} - (\boldsymbol{x}, \hat{\boldsymbol{\omega}})\boldsymbol{x},$$

$$\hat{\boldsymbol{b}}(\boldsymbol{x}) = \|\boldsymbol{x}\|^2\hat{\boldsymbol{v}} - (\boldsymbol{x}, \hat{\boldsymbol{v}})\boldsymbol{x}. \tag{12.68}$$

Here, $\hat{\dot{\boldsymbol{x}}}$ is the optimally corrected flow (this correction will be discussed later).

12.4 Linearization and Renormalization

12.4.1 Linearization

The optimization (12.49) is nonlinear, requiring numerical search. However, the hypothesis (12.43) is linear in the flow matrix $\boldsymbol{F}$. Hence, the linearization technique described in Section 9.5.2 can be applied. To be specific, Problem 12.2 is decomposed into the following two subproblems:

Problem 12.3 *Estimate a matrix $\boldsymbol{F}$ such that*

$$(\bar{\boldsymbol{X}}; \boldsymbol{F}) = 0 \tag{12.69}$$

from the observed flow $\dot{\boldsymbol{x}}$.

Problem 12.4 *Decompose the matrix $\boldsymbol{F}$ into motion parameters $\{\boldsymbol{v}, \boldsymbol{\omega}\}$ in such a way that*

$$\boldsymbol{F} = (\boldsymbol{v}, \boldsymbol{\omega})\boldsymbol{I} - S[\boldsymbol{v}\boldsymbol{\omega}^\top] + \boldsymbol{v} \times \boldsymbol{I}. \tag{12.70}$$

Consider Problem 12.3 first. Eq. (12.69) is the hypothesis from which the flow matrix $\boldsymbol{F}$ is to be estimated; the rank of this hypothesis is 1. Eq. (12.47) implies that the translation velocity $\boldsymbol{v}$ is obtained by the following cast (see eqs. (2.182)):

$$\boldsymbol{v} = \text{type}_3[A[\boldsymbol{F}]]. \tag{12.71}$$

It follows that the normalization $\|\boldsymbol{v}\| = 1$ is equivalent to the following normalization (see eqs. (2.183)):

$$\|A[\boldsymbol{F}]\| = \sqrt{2}. \tag{12.72}$$

If we regard $\boldsymbol{F}$ as a variable in the optimization (12.46) and use eq. (12.45), we see that the flow matrix $\boldsymbol{F}$ can be optimally estimated by

$$J[\boldsymbol{F}] = \int_S \frac{(\boldsymbol{X}; \boldsymbol{F})}{(\boldsymbol{x}, A[\boldsymbol{F}]^\top V[\dot{\boldsymbol{x}}]A[\boldsymbol{F}]\boldsymbol{x})} dxdy \to \min \tag{12.73}$$

under the constraint $\|A[\boldsymbol{F}]\| = \sqrt{2}$. Let $\hat{\boldsymbol{F}}$ be the resulting estimate of $\boldsymbol{F}$. Since the flow matrix $\boldsymbol{F}$ has eight degrees of freedom (its nine elements are constrained by the normalization (12.72)), the residual $J[\hat{\boldsymbol{F}}]$ is a χ^2 variable with $S - 8$ degrees of freedom. It follows that the flow matrix can be determined in principle if optical flow is observed at least at *eight* pixels. This is in agreement with the finite motion analysis, but in an actual application a large number of data are necessary for robust computation, as pointed out earlier.

If the covariance matrix $V[\dot{\boldsymbol{x}}]$ is decomposed into the noise level ϵ and the normalized covariance matrix $V_0[\dot{\boldsymbol{x}}]$ in the form

$$V[\dot{\boldsymbol{x}}] = \epsilon^2 V_0[\dot{\boldsymbol{x}}], \tag{12.74}$$

the covariance tensor $\mathcal{V}[\boldsymbol{X}]$ is accordingly decomposed in the form

$$\mathcal{V}[\boldsymbol{X}] = \epsilon^2 \mathcal{V}_0[\boldsymbol{X}]. \tag{12.75}$$

The least-squares approximation to the optimization (12.73) has the following form:

$$\tilde{J}[\boldsymbol{F}] = (\boldsymbol{F}; \mathcal{M}\boldsymbol{F}) \to \min. \tag{12.76}$$

Here, the *moment tensor* $\mathcal{M}$ is defined by

$$\mathcal{M} = \frac{1}{S}\int_S W(\boldsymbol{x})\boldsymbol{X} \otimes \boldsymbol{X}\, dxdy, \tag{12.77}$$

$$W(\boldsymbol{x}) = \frac{1}{(\boldsymbol{x}, A[\boldsymbol{F}^*]^\top V_0[\dot{\boldsymbol{x}}]A[\boldsymbol{F}^*]\boldsymbol{x})}, \tag{12.78}$$

where $\boldsymbol{F}^*$ is an appropriate estimate of $\boldsymbol{F}$.

If a 9-vector $\boldsymbol{f}$ and a 99-matrix $\boldsymbol{M}$ are respectively defined by casting the 33-matrix $\boldsymbol{F}$ and the 3333-tensor $\mathcal{M}$ in the form

$$\boldsymbol{f} = \text{type}_9[\boldsymbol{F}], \qquad \boldsymbol{M} = \text{type}_{99}[\mathcal{M}], \tag{12.79}$$

we have $(\boldsymbol{f}; \mathcal{M}\boldsymbol{f}) = (\boldsymbol{F}, \boldsymbol{M}\boldsymbol{F})$ (see Section 2.4.2). Hence, the solution of the optimization (12.76) is obtained as the eigenmatrix of tensor $\mathcal{M}$ (i.e., the matrix $\boldsymbol{F}$ obtained by cast from the eigenvector $\boldsymbol{f}$ of the matrix $\boldsymbol{M}$) for the smallest eigenvalue. The scale of the matrix $\boldsymbol{F}$ is adjusted so that $\|A[\boldsymbol{F}]\| = \sqrt{2}$.

12.4.2 Unbiased estimation and renormalization

Let $\bar{\mathcal{M}}$ be the unperturbed moment tensor obtained by replacing $\boldsymbol{X}$ by $\bar{\boldsymbol{X}}$ in eq. (12.77). Eq. (12.69) implies that the true flow matrix $\bar{\boldsymbol{F}}$ satisfies $\bar{\mathcal{M}}\bar{\boldsymbol{F}} = \boldsymbol{O}$, i.e., $\bar{\boldsymbol{F}}$ is the eigenmatrix $\bar{\mathcal{M}}$ for eigenvalue 0. However, the expectation $E[\mathcal{M}]$ of $\mathcal{M}$ is generally not equal to $\bar{\mathcal{M}}$. In fact,

$$E[\mathcal{M}] = \frac{1}{S}\int_S W(\boldsymbol{x})E[(\bar{\boldsymbol{X}} + \Delta\boldsymbol{X}) \otimes (\bar{\boldsymbol{X}} + \Delta\boldsymbol{X})]dxdy$$

$$= \frac{1}{S}\int_S W(\boldsymbol{x})\left(\bar{\boldsymbol{X}} \otimes \bar{\boldsymbol{X}} + E[\Delta \boldsymbol{X} \otimes \Delta \boldsymbol{X}]\right) dxdy$$
$$= \bar{\mathcal{M}} + \frac{\epsilon^2}{S}\int_S W(\boldsymbol{x})\mathcal{V}_0[\boldsymbol{X}]dxdy. \tag{12.80}$$

Hence, the solution of the least-squares approximation is statistically biased whatever estimate $\boldsymbol{F}^*$ is chosen.

Define a [33][33]-tensor $\mathcal{N}$ by

$$\mathcal{N} = \frac{1}{S}\int_S W(\boldsymbol{x})\mathcal{V}_0[\boldsymbol{X}]dxdy. \tag{12.81}$$

Eq. (12.80) implies that if we define

$$\hat{\mathcal{M}} = \mathcal{M} - \epsilon^2\mathcal{N}, \tag{12.82}$$

we have $E[\hat{\mathcal{M}}] = \bar{\mathcal{M}}$. Hence, we obtain the unbiased least-squares approximation

$$\hat{J}[\boldsymbol{F}] = (\boldsymbol{F}; \hat{\mathcal{M}}\boldsymbol{F}) \to \min. \tag{12.83}$$

The solution is obtained as the eigenmatrix of tensor $\hat{\mathcal{M}}$ for the smallest eigenvalue. The scale of the matrix $\boldsymbol{F}$ is adjusted so that $\|A[\boldsymbol{F}]\| = \sqrt{2}$.

If the noise level ϵ is not known, the renormalization procedure is given as follows[1] (see Section 9.4.2):

1. Let $c = 0$ and $W(\boldsymbol{x}) = 1$.

2. Compute the 3333-tensor $\mathcal{M}$ and the [33][33]-tensor $\mathcal{N}$ defined by eqs. (12.77) and (12.81), respectively.

3. Compute the smallest eigenvalue λ of the [33][33]-tensor

$$\hat{\mathcal{M}} = \mathcal{M} - c\mathcal{N}, \tag{12.84}$$

   and the corresponding eigenmatrix $\boldsymbol{F}$ scaled so that $\|A[\boldsymbol{F}]\| = \sqrt{2}$.

4. If $\lambda \approx 0$, return $\boldsymbol{F}$, c, and $\hat{\mathcal{M}}$. Else, update the constant c and the function $W(\boldsymbol{x})$ as follows:

$$c \leftarrow c + \frac{\lambda\|\boldsymbol{F}\|^2}{(\boldsymbol{F}; \mathcal{N}\boldsymbol{F})}, \qquad W(\boldsymbol{x}) \leftarrow \frac{1}{(\boldsymbol{x}, A[\boldsymbol{F}]^\top V_0[\dot{\boldsymbol{x}}]A[\boldsymbol{F}]\boldsymbol{x})}. \tag{12.85}$$

5. Go back to Step 2.

[1] In Section 9.4, renormalization was introduced by assuming that no constraint other than normalization existed. It is easy to confirm that the same procedure can be applied if the normalization $\|A[\mathbf{F}]\| = \sqrt{2}$ is imposed instead of $\|\mathbf{F}\| = 1$.

After renormalization, the squared noise level ϵ^2 is estimated in the form

$$\hat{\epsilon}^2 = \frac{c}{1 - 8/S}. \tag{12.86}$$

Its expectation and variance are given by

$$E[\hat{\epsilon}^2] = \epsilon^2, \qquad V[\hat{\epsilon}^2] = \frac{2\epsilon^4}{S-8}. \tag{12.87}$$

Since the number S of the pixels at which the optical flow is defined is usually very large[2], the variance $V[\hat{\epsilon}^2]$ is very small. Hence, $\hat{\epsilon}^2$ is expected to be a very good estimate of ϵ^2.

The constraint (12.72) is linearized as follows:

$$(A[\boldsymbol{F}]; A[\Delta\boldsymbol{F}]) = (A[\boldsymbol{F}]; \Delta\boldsymbol{F}) = 0. \tag{12.88}$$

Hence, the null space of $\boldsymbol{F}$ is $\{A[\boldsymbol{F}]\}_L$. The projection tensor $\mathcal{P} = (P_{ijkl})$ onto $\{A[\boldsymbol{F}]\}_L^\perp$ is given by

$$P_{ijkl} = \delta_{ik}\delta_{jl} - \frac{1}{2}A[F_{ij}]A[F_{kl}], \tag{12.89}$$

where $A[F_{ij}] = (F_{ij} - F_{ji})/2$ (= the (ij) element of $A[\boldsymbol{F}]$). Let $\hat{\mathcal{M}} = (\hat{M}_{ijkl})$ be the moment tensor obtained after renormalization. Define a [33][33]-tensor $\hat{\mathcal{M}}' = (\hat{M}'_{ijkl})$ by

$$\hat{M}'_{ijkl} = \sum_{m,n,p,q=1}^{3} P_{ijmn}P_{klpq}\hat{M}_{mnpq}. \tag{12.90}$$

The covariance tensor $\mathcal{V}[\boldsymbol{F}]$ of the resulting estimate $\boldsymbol{F}$ is obtained in the following form (see eq. (7.30)):

$$\mathcal{V}[\boldsymbol{F}] = \frac{\hat{\epsilon}^2}{S}(\hat{\mathcal{M}}')_8^-. \tag{12.91}$$

12.5 Optimal 3-D Reconstruction

12.5.1 *Optimal correction and decomposition*

A. Decomposability condition

We now consider Problem 12.4. Namely, we decompose the flow matrix $\boldsymbol{F}$ into motion parameters $\{\boldsymbol{v}, \boldsymbol{\omega}\}$ such that eq. (12.70) holds. As in the case of the essential matrix $\boldsymbol{G}$ for finite motion (see Section 11.3.1), *not every matrix*

[2]For example, if optical flow is defined at every pixel of a 512×512 image frame, we have $S = 512 \times 512 \approx 3 \times 10^5$.

$\boldsymbol{F}$ can be decomposed in the form of eq. (12.70): a (33)-matrix $\boldsymbol{F}$ has the form of eq. (12.70) if and only if

$$\boldsymbol{K} = \frac{1}{2}\mathrm{tr}\boldsymbol{K}(\boldsymbol{I} - \boldsymbol{v}\boldsymbol{v}^\top) + 2S[\boldsymbol{K}\boldsymbol{v}\boldsymbol{v}^\top], \qquad \|\boldsymbol{v}\| = 1, \tag{12.92}$$

where

$$\boldsymbol{K} = S[\boldsymbol{F}], \qquad \boldsymbol{v} = \mathrm{type}_3[A[\boldsymbol{F}]]. \tag{12.93}$$

In fact, if eq. (12.70) holds, it is easy to confirm eqs. (12.92) by direct substitution. Conversely, if eqs. (12.92) hold, it is immediately observed that

$$(\boldsymbol{v}, \boldsymbol{K}\boldsymbol{v}) = 0. \tag{12.94}$$

If we let

$$\boldsymbol{\omega} = \frac{1}{2}(\mathrm{tr}\boldsymbol{K})\boldsymbol{v} - 2\boldsymbol{K}\boldsymbol{v}, \tag{12.95}$$

eq. (12.70) can be easily confirmed by direct substitution. We call eqs. (12.92) and (12.93) the *decomposability condition* for optical flow. We say that a flow matrix $\boldsymbol{F}$ is *decomposable* if it satisfies the decomposability condition.

B. Correction of the flow matrix

The flow matrix $\boldsymbol{F}$ computed by renormalization may not satisfy the decomposability condition (12.92) exactly. Hence, it must be corrected into $\hat{\boldsymbol{F}}$ that satisfies

$$\hat{\boldsymbol{K}} = \frac{1}{2}\mathrm{tr}\hat{\boldsymbol{K}}(\boldsymbol{I} - \hat{\boldsymbol{v}}\hat{\boldsymbol{v}}^\top) + 2S[\hat{\boldsymbol{K}}\hat{\boldsymbol{v}}\hat{\boldsymbol{v}}^\top], \tag{12.96}$$

where

$$\hat{\boldsymbol{K}} = S[\hat{\boldsymbol{F}}], \qquad \hat{\boldsymbol{v}} = \mathrm{type}_3[A[\hat{\boldsymbol{F}}]]. \tag{12.97}$$

Constrained by the normalization $\|A[\boldsymbol{F}]\| = \sqrt{2}$, the flow matrix $\boldsymbol{F}$ has eight degrees of freedom, while the motion parameters $\{\boldsymbol{v}, \boldsymbol{\omega}\}$ have five degrees of freedom. Hence, only three of the six[3] component equations of (12.96) are independent. In other words, the rank of the constraint given by eqs. (12.96) and (12.97) is 3.

Substituting $\hat{\boldsymbol{F}} = \boldsymbol{F} - \Delta\boldsymbol{F}$ into eq. (12.96) and taking a linear approximation, we obtain

$$\boldsymbol{K} - \frac{1}{2}\mathrm{tr}\boldsymbol{K}(\boldsymbol{I} - \boldsymbol{v}\boldsymbol{v}^\top) - 2S[\boldsymbol{K}\boldsymbol{v}\boldsymbol{v}^\top] - \mathrm{tr}\boldsymbol{K}S[\Delta\boldsymbol{v}\boldsymbol{v}^\top] + 2S[\boldsymbol{K}\Delta\boldsymbol{v}\boldsymbol{v}^\top$$
$$+\boldsymbol{K}\boldsymbol{v}\Delta\boldsymbol{v}^\top] - \Delta\boldsymbol{K} + \frac{1}{2}\mathrm{tr}\Delta\boldsymbol{K}(\boldsymbol{I} - \boldsymbol{v}\boldsymbol{v}^\top) + 2S[\Delta\boldsymbol{K}\boldsymbol{v}\boldsymbol{v}^\top] = \boldsymbol{O}, \tag{12.98}$$

[3]Since eq. (12.96) is an equality between (33)-matrices, it gives six element-wise equalities.

where $\Delta \boldsymbol{K} = S[\Delta \boldsymbol{F}]$ and $\Delta \boldsymbol{v} = \text{type}_3[A[\Delta \boldsymbol{F}]]$. If a (33)3-tensor $\mathcal{A} = (A_{ijk})$, a (33)(33)-tensor $\mathcal{B} = (B_{ijkl})$, and a (33)-matrix $\boldsymbol{D}$ are defined by

$$A_{ijk} = \frac{1}{2}\text{tr}\boldsymbol{K}(\delta_{ik}v_j + \delta_{jk}v_i) - K_{ik}v_j - K_{jk}v_i - \delta_{jk}\sum_{l=1}^{3} K_{il}v_l - \delta_{ik}\sum_{l=1}^{3} K_{jl}v_l, \tag{12.99}$$

$$B_{ijkl} = \frac{1}{2}(\delta_{ik}\delta_{jl}+\delta_{il}\delta_{jk}-\delta_{ij}\delta_{kl}+\delta_{kl}v_iv_j-\delta_{ik}v_lv_j-\delta_{jl}v_kv_i-\delta_{il}v_kv_j-\delta_{jk}v_lv_i), \tag{12.100}$$

$$\boldsymbol{D} = \boldsymbol{K} - \frac{1}{2}\text{tr}\boldsymbol{K}(\boldsymbol{I} - \boldsymbol{v}\boldsymbol{v}^\top) - 2S[\boldsymbol{K}\boldsymbol{v}\boldsymbol{v}^\top], \tag{12.101}$$

eq. (12.98) can be written in the following form:

$$\mathcal{A}\Delta\boldsymbol{v} + \mathcal{B}\Delta\boldsymbol{K} = \boldsymbol{D}. \tag{12.102}$$

If a (33)33-tensor $\mathcal{C} = (C_{ijkl})$ is defined by

$$\mathcal{C} = \frac{1}{2}\text{type}_{33[33]}[\mathcal{A}] + \mathcal{B}, \tag{12.103}$$

or in elements

$$C_{ijkl} = -\frac{1}{2}\sum_{m=1}^{3}\epsilon_{klm}A_{ijm} + B_{ijkl}, \tag{12.104}$$

eq. (12.102) can be rewritten in the following form :

$$\mathcal{C}\Delta\boldsymbol{F} = \boldsymbol{D}. \tag{12.105}$$

Hence, the optimal correction $\Delta\boldsymbol{F}$ is determined by minimizing the Mahalanobis distance $\|\Delta\boldsymbol{F}\|_{\mathcal{V}[\boldsymbol{F}]}$ (see Section 9.5.2), i.e.,

$$(\Delta\boldsymbol{F}, \mathcal{V}[\boldsymbol{F}]^-\Delta\boldsymbol{F}) \to \min, \qquad \Delta\boldsymbol{F} \in \{A[\boldsymbol{F}]\}_L^\perp, \tag{12.106}$$

under the linearized constraint (12.105). The covariance tensor $\mathcal{V}[\boldsymbol{F}]$ of the flow matrix $\boldsymbol{F}$ is given as a by-product of the renormalization procedure (see eq. (12.91)). Since multiplication of the covariance tensor $\mathcal{V}[\boldsymbol{F}]$ by a positive constant does not affect the solution, it can be replaced by the *normalized covariance tensor* $\mathcal{V}_0[\boldsymbol{F}]$ defined by setting $\hat{\epsilon} = 1$ in eq. (12.91).

The first order solution is obtained as follows:

$$\Delta F_{ij} = \sum_{k,l,m,n,p,q=1}^{3} W_{mnpq}C_{mnkl}\mathcal{V}_0[\boldsymbol{F}]_{ijkl}D_{pq}. \tag{12.107}$$

Here, $\mathcal{W} = (W_{ijkl})$ is a (33)(33)-tensor defined by

$$\mathcal{W} = (\mathcal{V})_3^-, \tag{12.108}$$

where $\mathcal{V} = (V_{ijkl})$ is a (33)(33)-tensor given by

$$V_{ijkl} = \sum_{m,n,p,q=1}^{3} C_{ijmn} C_{klpq} \mathcal{V}_0[\boldsymbol{F}]_{mnpq}. \tag{12.109}$$

Since the constraint $\Delta \boldsymbol{F} \in \{A[\boldsymbol{F}]\}_L^\perp$ ensures the constraint $\|A[\boldsymbol{F}]\| = \sqrt{2}$ only to a first approximation, the actual correction takes the following form:

$$\hat{\boldsymbol{F}} = \frac{\sqrt{2}(\boldsymbol{F} - \Delta \boldsymbol{F})}{\|A[\boldsymbol{F} - \Delta \boldsymbol{F}]\|}. \tag{12.110}$$

This correction is iterated until the matrix $\boldsymbol{D}$ becomes sufficiently close to $\boldsymbol{O}$. In this process, the normalized covariance tensor $\mathcal{V}_0[\boldsymbol{F}]$ needs to be updated, since its null space should change as $\boldsymbol{F}$ changes (see eq. (5.26)). Let $\hat{\mathcal{P}} = (\hat{P}_{ijkl})$ be the orthogonal projection tensor onto $\{A[\hat{\boldsymbol{F}}]\}_L^\perp$ (see eq. (12.89)). We let

$$\hat{\mathcal{V}}_0[\boldsymbol{F}]_{ijkl} = \hat{P}_{ijmn} \hat{P}_{klpq} \mathcal{V}_0[\boldsymbol{F}]_{mnpq}, \tag{12.111}$$

where $\mathcal{V}_0[\boldsymbol{F}]_{ijkl}$ and $\hat{\mathcal{V}}_0[\boldsymbol{F}]_{ijkl}$ are the $(ijkl)$ elements of $\mathcal{V}_0[\boldsymbol{F}]$ and $\hat{\mathcal{V}}_0[\boldsymbol{F}]$, respectively.

C. Decomposition into motion parameters

If the corrected flow matrix $\hat{\boldsymbol{F}}$ satisfies the decomposability condition (12.92), the decomposition into motion parameters $\{\boldsymbol{v}, \boldsymbol{\omega}\}$ can be done in principle by the second of eqs. (12.93) and eq. (12.95). However, these hold only when $\hat{\boldsymbol{F}}$ is decomposable. As in the case of finite motion, we should use a robust expression that yields a good approximation even if $\hat{\boldsymbol{F}}$ does not satisfy eq. (12.92), say, as a result of a premature termination of the correction iterations described above. We compute the translation velocity $\boldsymbol{v}$ by the second of eqs. (12.97) and determine the rotation velocity $\boldsymbol{\omega}$ by the least-squares optimization

$$\|S[\hat{\boldsymbol{F}}] - (\boldsymbol{v}, \boldsymbol{\omega})\boldsymbol{I} + S[\boldsymbol{v}\boldsymbol{\omega}^\top]\|^2 \to \min. \tag{12.112}$$

Then, we obtain

$$\boldsymbol{v} = \mathrm{type}_3[A[\hat{\boldsymbol{F}}]], \qquad \boldsymbol{\omega} = \frac{1}{2}\left(\mathrm{tr}\hat{\boldsymbol{F}} + 3(\boldsymbol{v}, \hat{\boldsymbol{F}}\boldsymbol{v})\right)\boldsymbol{v} - 2S[\hat{\boldsymbol{F}}]\boldsymbol{v}. \tag{12.113}$$

The second equation is equivalent to eq. (12.95) if $\hat{\boldsymbol{F}}$ is decomposable.

12.5.2 *Optimal correction of optical flow*

A. Epipolar of optical flow

Since the epipolar equation $e(\boldsymbol{x}) = 0$ is linear in $\dot{\boldsymbol{x}}$, it defines a plane in a three-dimensional space if $\dot{\boldsymbol{x}}$ is regarded as a variable. The constraint $(\boldsymbol{k}, \dot{\boldsymbol{x}})$

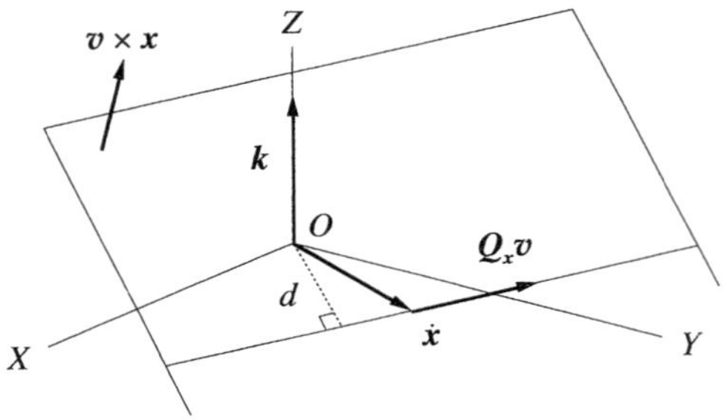

Fig. 12.8. Epipolar of optical flow.

= 0 also defines a plane. Hence, the set of all vectors $\dot{\boldsymbol{x}}$ that satisfy the epipolar equation for fixed $\boldsymbol{x}$, $\boldsymbol{v}$, and $\boldsymbol{\omega}$ is a line: it is the intersection of the plane defined by $(\boldsymbol{k}, \dot{\boldsymbol{x}}) = 0$ with the plane defined by the epipolar equation $e(\boldsymbol{x}) = 0$ (Fig. 12.8). We call this line the *epipolar* of the optical flow $\dot{\boldsymbol{x}}$ at $\boldsymbol{x}$. From eqs. (12.36) and (12.37), we see that the equation of the epipolar is

$$\dot{\boldsymbol{x}} \times \boldsymbol{Q}_{\boldsymbol{x}}\boldsymbol{v} = -(\boldsymbol{v} \times \boldsymbol{x}, \boldsymbol{\omega} \times \boldsymbol{x})\boldsymbol{k}. \tag{12.114}$$

The orientation of this line is $\boldsymbol{Q}_{\boldsymbol{x}}\boldsymbol{v}$, and its distance d from the origin O is

$$d = \frac{|(\boldsymbol{v} \times \boldsymbol{x}, \boldsymbol{\omega} \times \boldsymbol{x})|}{\|\boldsymbol{Q}_{\boldsymbol{x}}\boldsymbol{v}\|}. \tag{12.115}$$

B. Correction of optical flow

Let $\{\hat{\boldsymbol{v}}, \hat{\boldsymbol{\omega}}\}$ be the motion parameters computed from optical flow. Although they are estimated so that the flow satisfies the epipolar equation $e(\boldsymbol{x}) = 0$ *on the average*, individual flow components may not necessary satisfy it. In order that the depth can be defined at each point, the observed flow $\dot{\boldsymbol{x}}$ must be corrected into the value $\hat{\dot{\boldsymbol{x}}}$ that is compatible with the epipolar equation for the computed motion parameters $\{\hat{\boldsymbol{v}}, \hat{\boldsymbol{\omega}}\}$. This correction has the following geometric interpretations (they are equivalent to each other):

- The vector $\dot{\boldsymbol{x}}$ is moved to $\hat{\dot{\boldsymbol{x}}}$ so that $\hat{\dot{\boldsymbol{x}}}$ is on the epipolar at $\boldsymbol{x}$ (see Fig. 12.9).
- The vector $\dot{\boldsymbol{x}}$ is moved to $\hat{\dot{\boldsymbol{x}}}$ so that *the lines of sight of* $\boldsymbol{x}(t)$ *and* $\boldsymbol{x}(t+\Delta t)$ *meet at a point in the scene* to a first approximation (see Fig. 12.6).

If we let

$$\hat{e}(\boldsymbol{x}) = |\boldsymbol{x}, \dot{\boldsymbol{x}}, \hat{\boldsymbol{v}}| + (\hat{\boldsymbol{v}} \times \boldsymbol{x}, \hat{\boldsymbol{\omega}} \times \boldsymbol{x}), \tag{12.116}$$

this is not zero in general. It is easy to see that the right-hand side vanishes if $\dot{\boldsymbol{x}}$ is replaced by $\hat{\dot{\boldsymbol{x}}} = \dot{\boldsymbol{x}} - \Delta\dot{\boldsymbol{x}}$ in such a way that

$$(\hat{\boldsymbol{v}} \times \boldsymbol{x}, \Delta\dot{\boldsymbol{x}}) = \hat{e}(\boldsymbol{x}). \tag{12.117}$$

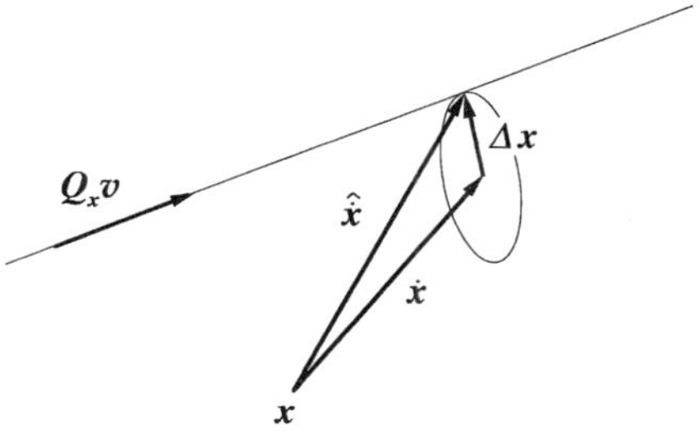

Fig. 12.9. Optical correction of optical flow.

The rank of this constraint is 1. An optimal value of $\Delta\dot{\boldsymbol{x}}$ is determined by minimizing the Mahalanobis distance $\|\Delta\dot{\boldsymbol{x}}\|_{V_0[\dot{\boldsymbol{x}}]}$, i.e.,

$$(\Delta\dot{\boldsymbol{x}}, V_0[\dot{\boldsymbol{x}}]^-\Delta\dot{\boldsymbol{x}}) \to \min, \qquad \Delta\dot{\boldsymbol{x}} \in \{\boldsymbol{k}\}_L^\perp, \tag{12.118}$$

under the constraint (12.117). The solution is given by

$$\Delta\dot{\boldsymbol{x}} = \frac{\hat{e}(\boldsymbol{x})V_0[\dot{\boldsymbol{x}}](\hat{\boldsymbol{v}} \times \boldsymbol{x})}{(\hat{\boldsymbol{v}} \times \boldsymbol{x}, V_0[\dot{\boldsymbol{x}}](\hat{\boldsymbol{v}} \times \boldsymbol{x}))}. \tag{12.119}$$

It follows that the correction has the form

$$\hat{\dot{\boldsymbol{x}}} = \dot{\boldsymbol{x}} - \frac{\hat{e}(\boldsymbol{x})V_0[\dot{\boldsymbol{x}}](\hat{\boldsymbol{v}} \times \boldsymbol{x})}{(\hat{\boldsymbol{v}}, (\boldsymbol{x} \times V_0[\dot{\boldsymbol{x}}] \times \boldsymbol{x})\hat{\boldsymbol{v}})}. \tag{12.120}$$

Since the epipolar equation $e(\boldsymbol{x}) = 0$ is linear in $\dot{\boldsymbol{x}}$, no approximation has been introduced into the constraint (12.117). Hence, the above correction need not be iterated.

The residual $(\Delta\dot{\boldsymbol{x}}, V_0[\dot{\boldsymbol{x}}]^-\Delta\dot{\boldsymbol{x}})$ of the optimization (12.118) is given by

$$\hat{\epsilon}(\boldsymbol{x})^2 = \frac{\hat{e}(\boldsymbol{x})^2}{(\hat{\boldsymbol{v}}, (\boldsymbol{x} \times V_0[\dot{\boldsymbol{x}}] \times \boldsymbol{x})\hat{\boldsymbol{v}})}. \tag{12.121}$$

From eqs. (12.49) and (12.50), we can see that

$$J_0[\hat{\boldsymbol{v}}, \hat{\boldsymbol{\omega}}] = \int_S \hat{\epsilon}(\boldsymbol{x})^2 dxdy \tag{12.122}$$

equals ϵ^2 times the residual of the optimization (12.49). Hence, $J_0[\hat{\boldsymbol{v}}, \hat{\boldsymbol{\omega}}]/\epsilon^2$ is a χ^2 variable with $S - 5$ degrees of freedom. An unbiased estimator of the squared noise level ϵ^2 is obtained in the form

$$\hat{\epsilon}^2 = \frac{J_0[\hat{\boldsymbol{v}}, \hat{\boldsymbol{\omega}}]}{S - 5}. \tag{12.123}$$

Its expectation and variance are

$$E[\hat{\epsilon}^2] = \epsilon^2, \qquad V[\hat{\epsilon}^2] = \frac{2\epsilon^4}{S-5}. \tag{12.124}$$

The rigidity test (12.53) can be rewritten as

$$\frac{\hat{\epsilon}^2}{\epsilon^2} > \frac{\chi^2_{S-5,a}}{S-5}. \tag{12.125}$$

12.5.3 *Computation of depth*

A. Depth equation

From eq. (12.119), we see that

$$\begin{aligned}(\boldsymbol{Q_x}\hat{\boldsymbol{v}}, V_0[\dot{\boldsymbol{x}}]^- \Delta\dot{\boldsymbol{x}}) &= \frac{\hat{e}(\boldsymbol{x})(\boldsymbol{Q_x}\hat{\boldsymbol{v}}, V_0[\dot{\boldsymbol{x}}]^- V_0[\dot{\boldsymbol{x}}](\hat{\boldsymbol{v}} \times \boldsymbol{x}))}{(\hat{\boldsymbol{v}} \times \boldsymbol{x}, V_0[\dot{\boldsymbol{x}}](\hat{\boldsymbol{v}} \times \boldsymbol{x}))} \\ &= \frac{\hat{e}(\boldsymbol{x})(\boldsymbol{Q_x}\hat{\boldsymbol{v}}, \boldsymbol{P_k}(\hat{\boldsymbol{v}} \times \boldsymbol{x}))}{(\hat{\boldsymbol{v}} \times \boldsymbol{x}, V_0[\dot{\boldsymbol{x}}](\hat{\boldsymbol{v}} \times \boldsymbol{x}))} = \frac{\hat{e}(\boldsymbol{x})(\boldsymbol{P_k}\boldsymbol{Q_x}\hat{\boldsymbol{v}}, \hat{\boldsymbol{v}} \times \boldsymbol{x})}{(\hat{\boldsymbol{v}} \times \boldsymbol{x}, V_0[\dot{\boldsymbol{x}}](\hat{\boldsymbol{v}} \times \boldsymbol{x}))} \\ &= \frac{\hat{e}(\boldsymbol{x})(\boldsymbol{Q_x}\hat{\boldsymbol{v}}, \hat{\boldsymbol{v}} \times \boldsymbol{x})}{(\hat{\boldsymbol{v}} \times \boldsymbol{x}, V_0[\dot{\boldsymbol{x}}](\hat{\boldsymbol{v}} \times \boldsymbol{x}))} = \frac{\hat{e}(\boldsymbol{x})(\hat{\boldsymbol{v}} - (\hat{\boldsymbol{v}}, \boldsymbol{k})\boldsymbol{x}, \hat{\boldsymbol{v}} \times \boldsymbol{x})}{(\hat{\boldsymbol{v}} \times \boldsymbol{x}, V_0[\dot{\boldsymbol{x}}](\hat{\boldsymbol{v}} \times \boldsymbol{x}))} = 0,\end{aligned} \tag{12.126}$$

where we have used eqs. (12.26) and (12.27). If the corrected flow $\hat{\dot{\boldsymbol{x}}} = \dot{\boldsymbol{x}} - \Delta\dot{\boldsymbol{x}}$ satisfies the epipolar equation for the computed motion parameters $\{\hat{\boldsymbol{v}}, \hat{\boldsymbol{\omega}}\}$, there exists a depth $\hat{Z}(\boldsymbol{x})$ such that

$$\dot{\boldsymbol{x}} - \Delta\dot{\boldsymbol{x}} = -\boldsymbol{Q_x}\left(\frac{\hat{\boldsymbol{v}}}{\hat{Z}(\boldsymbol{x})} + \hat{\boldsymbol{\omega}} \times \boldsymbol{x}\right). \tag{12.127}$$

Noting the second of eqs. (12.27), we can rewrite the above equation in the following form:

$$\Delta\dot{\boldsymbol{x}} = \boldsymbol{Q_x}\left(\dot{\boldsymbol{x}} + \hat{\boldsymbol{\omega}} \times \boldsymbol{x} + \frac{\hat{\boldsymbol{v}}}{\hat{Z}(\boldsymbol{x})}\right). \tag{12.128}$$

Eq. (12.126) implies

$$(\boldsymbol{Q_x}\hat{\boldsymbol{v}}, V_0[\dot{\boldsymbol{x}}]^- \boldsymbol{Q_x}\left(\dot{\boldsymbol{x}} + \hat{\boldsymbol{\omega}} \times \boldsymbol{x} + \frac{\hat{\boldsymbol{v}}}{\hat{Z}(\boldsymbol{x})}\right)) = 0, \tag{12.129}$$

from which the depth $\hat{Z}(\boldsymbol{x})$ is given in the following form:

$$\hat{Z}(\boldsymbol{x}) = -\frac{(\hat{\boldsymbol{v}}, \boldsymbol{Q_x}^\top V_0[\dot{\boldsymbol{x}}]^- \boldsymbol{Q_x}\hat{\boldsymbol{v}})}{(\hat{\boldsymbol{v}}, \boldsymbol{Q_x}^\top V_0[\dot{\boldsymbol{x}}]^- \boldsymbol{Q_x}(\dot{\boldsymbol{x}} + \hat{\boldsymbol{\omega}} \times \boldsymbol{x}))}. \tag{12.130}$$

We call this the *depth equation*. We observe the following:

- Eq. (12.28) implies that $\boldsymbol{Q}_{\boldsymbol{x}}(\dot{\boldsymbol{x}} + \hat{\boldsymbol{\omega}} \times \boldsymbol{x}) = \boldsymbol{0}$ if $\hat{\boldsymbol{v}} = \boldsymbol{0}$. Hence, the depth $\hat{Z}(\boldsymbol{x})$ is indeterminate if $\hat{\boldsymbol{v}} = \boldsymbol{0}$.

- Let $\boldsymbol{x}_{\hat{\boldsymbol{v}}}$ ($\propto \hat{\boldsymbol{v}}$) be the *focus of expansion* for the computed motion parameters $\{\hat{\boldsymbol{v}}, \hat{\boldsymbol{\omega}}\}$ (see Section 12.3.1). Since $\boldsymbol{Q}_{\boldsymbol{x}_{\hat{v}}}\hat{\boldsymbol{v}} \propto \boldsymbol{Q}_{\boldsymbol{x}_{\hat{v}}}\boldsymbol{x}_{\hat{\boldsymbol{v}}} = \boldsymbol{0}$ (see the first of eqs. (12.27)), the depth at the focus of expansion $\boldsymbol{x}_{\hat{\boldsymbol{v}}}$ is indeterminate.

Up to now, there exist *two* solutions, since renormalization computes the flow matrix $\boldsymbol{F}$ only up to sign. It is easily seen that if one solution is $\{\boldsymbol{v}, \boldsymbol{\omega}\}$ and $Z(\boldsymbol{x})$, the other solution is $\{-\boldsymbol{v}, \boldsymbol{\omega}\}$ and $-Z(\boldsymbol{x})$. Hence, the correct solution is chosen by imposing the constraint

$$\int_S \mathrm{sgn}(\hat{Z}(\boldsymbol{x}))dxdy > 0. \tag{12.131}$$

As in the case of finite motion (see eq. (11.94)), the use of a seemingly equivalent condition $\int_S \hat{Z}(\boldsymbol{x})dxdy > 0$ is dangerous because a large depth ($Z(\boldsymbol{x}) \approx \infty$) can be estimated to be very far *behind* the camera ($\hat{Z}(\boldsymbol{x}) \approx -\infty$) in the presence of image noise.

B. Reinterpretation of optimal estimation

Eq. (12.128) implies that

$$(\Delta\dot{\boldsymbol{x}}, V_0[x]^-\Delta\dot{\boldsymbol{x}}) = (\dot{\boldsymbol{x}} + \hat{\boldsymbol{\omega}} \times \boldsymbol{x} + \frac{\hat{\boldsymbol{v}}}{\hat{Z}(\boldsymbol{x})}, \boldsymbol{Q}_{\boldsymbol{x}}^\top V_0[\dot{x}]^- \boldsymbol{Q}_{\boldsymbol{x}}\left(\dot{\boldsymbol{x}} + \hat{\boldsymbol{\omega}} \times \boldsymbol{x} + \frac{\hat{\boldsymbol{v}}}{\hat{Z}(\boldsymbol{x})}\right)). \tag{12.132}$$

If eq. (12.130) is substituted into the right-hand side of this equation, it should be equal to the right-hand side of eq. (12.121). On the other hand, it is easily seen that the depth $\hat{Z}(\boldsymbol{x})$ given by eq. (12.130) *minimizes* the right-hand side of eq. (12.132) viewed as a function of $\hat{Z}(\boldsymbol{x})$. Thus, we conclude that

$$\frac{\hat{e}(\boldsymbol{x})^2}{(\hat{\boldsymbol{v}}, (\boldsymbol{x} \times V_0[\dot{x}] \times \boldsymbol{x})\hat{\boldsymbol{v}})} = \min_{Z(\boldsymbol{x})}(\dot{\boldsymbol{x}} + \frac{\hat{\boldsymbol{v}}}{Z(\boldsymbol{x})} + \hat{\boldsymbol{\omega}} \times \boldsymbol{x}, \boldsymbol{Q}_{\boldsymbol{x}}^\top V_0[\dot{x}]^- \boldsymbol{Q}_{\boldsymbol{x}}\left(\dot{\boldsymbol{x}} + \frac{\hat{\boldsymbol{v}}}{Z(\boldsymbol{x})} + \hat{\boldsymbol{\omega}} \times \boldsymbol{x}\right)), \tag{12.133}$$

and the minimum is attained by the depth $\hat{Z}(\boldsymbol{x})$ given by eq. (12.130). It follows that the optimization (12.122) is equivalent to

$$J_0[\boldsymbol{v}, \boldsymbol{\omega}] = \int_S \min_{Z(\boldsymbol{x})} \left\| \dot{\boldsymbol{x}} + \boldsymbol{Q}_{\boldsymbol{x}}\left(\frac{\boldsymbol{v}}{Z(\boldsymbol{x})} + \boldsymbol{\omega} \times \boldsymbol{x}\right)\right\|^2_{V_0[\dot{\boldsymbol{x}}]} dxdy \to \min. \tag{12.134}$$

Example 12.3 If each flow component is perturbed independently by Gaussian noise of mean 0 and variance ϵ^2, the normalized covariance matrix of $\dot{\boldsymbol{x}}$ is $V_0[\dot{\boldsymbol{x}}] = \boldsymbol{P_k}$. From the fourth of eqs. (12.27), we see that

$$\boldsymbol{Q}_{\boldsymbol{x}}^\top \boldsymbol{P_k} \boldsymbol{Q}_{\boldsymbol{x}} = \boldsymbol{Q}_{\boldsymbol{x}}^\top \boldsymbol{Q}_{\boldsymbol{x}} = \boldsymbol{S}_{\boldsymbol{x}}, \tag{12.135}$$

where the matrix $\boldsymbol{S}_{\boldsymbol{x}}$ is defined by eq. (12.51). Hence, eq. (12.130) can be written as

$$\hat{Z}(\boldsymbol{x}) = -\frac{(\hat{\boldsymbol{v}}, \boldsymbol{S}_{\boldsymbol{x}}\hat{\boldsymbol{v}})}{(\hat{\boldsymbol{v}}, \boldsymbol{S}_{\boldsymbol{x}}(\dot{\boldsymbol{x}} + \hat{\boldsymbol{\omega}} \times \boldsymbol{x}))}. \tag{12.136}$$

Eq. (12.133) implies the following identity:

$$\min_{Z(\boldsymbol{x})} \left\| \dot{\boldsymbol{x}} + \boldsymbol{Q}_{\boldsymbol{x}} \left(\frac{\hat{\boldsymbol{v}}}{Z(\boldsymbol{x})} + \hat{\boldsymbol{\omega}} \times \boldsymbol{x} \right) \right\|^2 = \frac{\hat{e}(\boldsymbol{x})^2}{(\hat{\boldsymbol{v}}, \boldsymbol{S}_{\boldsymbol{x}}\hat{\boldsymbol{v}})}. \tag{12.137}$$

The minimum is attained by the depth $\hat{Z}(\boldsymbol{x})$ given by eq. (12.136). It follows that the optimization (12.134) with respect to the Mahalanobis distance reduces to the following optimization with respect to the Euclidean distance:

$$\begin{aligned} J_0[\boldsymbol{v}, \boldsymbol{\omega}] &= \int_S \min_{Z(\boldsymbol{x})} \left\| \dot{\boldsymbol{x}} + \boldsymbol{Q}_{\boldsymbol{x}} \left(\frac{\boldsymbol{v}}{Z(\boldsymbol{x})} + \boldsymbol{\omega} \times \boldsymbol{x} \right) \right\|^2 dxdy \\ &= \int_S \frac{e(\boldsymbol{x})^2 dxdy}{(\boldsymbol{v}, \boldsymbol{S}_{\boldsymbol{x}}\boldsymbol{v})} \to \min. \end{aligned} \tag{12.138}$$

12.6 Reliability of 3-D Reconstruction

12.6.1 Effect of image noise

Errors in the 3-D shape reconstructed by the depth equation (12.130) originate from the following two sources:

1. errors in the observed flow $\dot{\boldsymbol{x}}$;

2. errors in the computed motion parameters $\{\hat{\boldsymbol{v}}, \hat{\boldsymbol{\omega}}\}$.

Strictly speaking, these two sources of error are correlated, since $\{\hat{\boldsymbol{v}}, \hat{\boldsymbol{\omega}}\}$ are computed from the flow $\dot{\boldsymbol{x}}$. However, since $\{\hat{\boldsymbol{v}}, \hat{\boldsymbol{\omega}}\}$ are estimated by optimization over all the pixels at which the flow is defined, the correlation between $\{\hat{\boldsymbol{v}}, \hat{\boldsymbol{\omega}}\}$ and $\dot{\boldsymbol{x}}$ *at a particular pixel* is expected to be very small. Hence, as in the case of finite motion, errors in 3-D reconstruction can be treated to a first approximation as the sum of the errors from these two sources.

First, consider the effect of image noise, assuming that the motion parameters $\{\hat{\boldsymbol{v}}, \hat{\boldsymbol{\omega}}\}$ are accurate. If the observed flow $\dot{\boldsymbol{x}}$ is corrected into $\hat{\dot{\boldsymbol{x}}}$

by eq. (12.120), its a posteriori covariance matrix is given as follows (see eq. (5.46)):

$$V_i[\hat{\dot{\boldsymbol{x}}}] = \hat{\epsilon}^2 \left(V_0[\dot{\boldsymbol{x}}] - \frac{(V_0[\dot{\boldsymbol{x}}](\hat{\boldsymbol{v}} \times \boldsymbol{x}))(V_0[\dot{\boldsymbol{x}}](\hat{\boldsymbol{v}} \times \boldsymbol{x}))^\top}{(\hat{\boldsymbol{v}} \times \boldsymbol{x}, V_0[\dot{\boldsymbol{x}}](\hat{\boldsymbol{v}} \times \boldsymbol{x}))} \right). \tag{12.139}$$

Here, $\hat{\epsilon}^2$ is the value estimated by eq. (12.123). The subscript i indicates that we are considering the effect of image noise only. It is easily seen that the above matrix has rank 1; its range and null space are $\{(\boldsymbol{x} \times \hat{\boldsymbol{v}}) \times \boldsymbol{k}\}_L = \{\boldsymbol{Q}_{\boldsymbol{x}}\hat{\boldsymbol{v}}\}_L$ and $\{\boldsymbol{x} \times \hat{\boldsymbol{v}}, \boldsymbol{k}\}_L$, respectively (see eq. (12.26)). This is a consequence of the fact that errors in $\hat{\dot{\boldsymbol{x}}}$ are constrained to be along the epipolar at $\boldsymbol{x}$ (see Fig. 12.9). It follows that the covariance matrix $V_i[\hat{\dot{\boldsymbol{x}}}]$ has the following form:

$$V_i[\hat{\dot{\boldsymbol{x}}}] = \text{constant} \times (\boldsymbol{Q}_{\boldsymbol{x}}\hat{\boldsymbol{v}})(\boldsymbol{Q}_{\boldsymbol{x}}\hat{\boldsymbol{v}})^\top. \tag{12.140}$$

Recall that the depth $\hat{Z}(\boldsymbol{x})$ has been computed from the following relationship (see eq. (12.127)):

$$\hat{\dot{\boldsymbol{x}}} = -\boldsymbol{Q}_{\boldsymbol{x}} \left(\frac{\hat{\boldsymbol{v}}}{\hat{Z}(\boldsymbol{x})} + \hat{\boldsymbol{\omega}} \times \boldsymbol{x} \right). \tag{12.141}$$

Since the corrected flow $\hat{\dot{\boldsymbol{x}}}$ is computed from the observed flow, it is a random variable. If it is perturbed by $\Delta\hat{\dot{\boldsymbol{x}}}$, the depth $\hat{Z}(\boldsymbol{x})$ is accordingly perturbed by $\Delta Z(\boldsymbol{x})$. To a first approximation, we have

$$\Delta\hat{\dot{\boldsymbol{x}}} = \frac{\Delta Z(\boldsymbol{x})}{\hat{Z}(\boldsymbol{x})^2}\boldsymbol{Q}_{\boldsymbol{x}}\hat{\boldsymbol{v}}. \tag{12.142}$$

Hence,

$$\mathrm{tr}V_i[\hat{\dot{\boldsymbol{x}}}] = E[\|\Delta\hat{\dot{\boldsymbol{x}}}\|^2] = \frac{E[\Delta Z(\boldsymbol{x})^2]}{\hat{Z}(\boldsymbol{x})^4}(\hat{\boldsymbol{v}}, \boldsymbol{Q}_{\boldsymbol{x}}^\top\boldsymbol{Q}_{\boldsymbol{x}}\hat{\boldsymbol{v}}) = \frac{V_i[\hat{Z}(\boldsymbol{x})]}{\hat{Z}(\boldsymbol{x})^4}(\hat{\boldsymbol{v}}, \boldsymbol{S}_{\boldsymbol{x}}\hat{\boldsymbol{v}}), \tag{12.143}$$

where the matrix $\boldsymbol{S}_{\boldsymbol{x}}$ is defined by eq. (12.51) (see eqs. (12.135)). Consequently, the variance $V_i[\hat{Z}(\boldsymbol{x})]$ of the computed depth $\hat{Z}(\boldsymbol{x})$ is given by

$$V_i[\hat{Z}(\boldsymbol{x})] = \frac{\hat{Z}(\boldsymbol{x})^4 \mathrm{tr}V_i[\hat{\dot{\boldsymbol{x}}}]}{(\hat{\boldsymbol{v}}, \boldsymbol{S}_{\boldsymbol{x}}\hat{\boldsymbol{v}})}. \tag{12.144}$$

12.6.2 Effect of errors in the motion parameters

Consider the effect of errors in the motion parameters $\{\hat{\boldsymbol{v}}, \hat{\boldsymbol{\omega}}\}$, assuming that the flow $\dot{\boldsymbol{x}}$ is accurate. Let $\hat{\boldsymbol{v}} = \bar{\boldsymbol{v}} + \Delta\boldsymbol{v}$ and $\hat{\boldsymbol{\omega}} = \bar{\boldsymbol{\omega}} = \bar{\boldsymbol{\omega}} + \Delta\boldsymbol{\omega}$. Since the epipolar equation $e(\boldsymbol{x}) = 0$ is satisfied by the true values $\{\bar{\boldsymbol{v}}, \bar{\boldsymbol{\omega}}\}$, we have to a first approximation

$$\begin{aligned}
\hat{e}(\boldsymbol{x}) &= |\boldsymbol{x}, \dot{\boldsymbol{x}}, \bar{\boldsymbol{v}} + \Delta\boldsymbol{v}| + ((\bar{\boldsymbol{v}} + \Delta\boldsymbol{v}) \times \boldsymbol{x}, (\bar{\boldsymbol{\omega}} + \Delta\boldsymbol{\omega}) \times \boldsymbol{x}) \\
&= |\boldsymbol{x}, \dot{\boldsymbol{x}}, \Delta\boldsymbol{v}| + (\Delta\boldsymbol{v} \times \boldsymbol{x}, \bar{\boldsymbol{\omega}} \times \boldsymbol{x}) + (\bar{\boldsymbol{v}} \times \boldsymbol{x}, \Delta\boldsymbol{\omega} \times \boldsymbol{x}) \\
&= (\hat{\boldsymbol{a}}(\boldsymbol{x}), \Delta\boldsymbol{v}) + (\hat{\boldsymbol{b}}(\boldsymbol{x}), \Delta\boldsymbol{\omega}),
\end{aligned} \tag{12.145}$$

where[4]

$$\hat{\boldsymbol{a}}(\boldsymbol{x}) = \boldsymbol{x} \times \dot{\boldsymbol{x}} + \|\boldsymbol{x}\|^2 \hat{\boldsymbol{\omega}} - (\boldsymbol{x}, \hat{\boldsymbol{\omega}})\boldsymbol{x},$$

$$\hat{\boldsymbol{b}}(\boldsymbol{x}) = \|\boldsymbol{x}\|^2 \hat{\boldsymbol{v}} - (\boldsymbol{x}, \hat{\boldsymbol{v}})\boldsymbol{x}. \tag{12.146}$$

From eq. (12.145), we obtain the following relationships (the subscript m indicates that we are considering errors in the motion parameters only):

$$V_m[\hat{e}(\boldsymbol{x})] = (\hat{\boldsymbol{a}}(\boldsymbol{x}), V[\hat{\boldsymbol{v}}]\hat{\boldsymbol{a}}(\boldsymbol{x})) + 2(\hat{\boldsymbol{a}}(\boldsymbol{x}), V[\hat{\boldsymbol{v}}, \hat{\boldsymbol{\omega}}]\hat{\boldsymbol{b}}(\boldsymbol{x})) + (\hat{\boldsymbol{b}}(\boldsymbol{x}), V[\hat{\boldsymbol{\omega}}]\hat{\boldsymbol{b}}(\boldsymbol{x})),$$

$$V_m[\hat{\boldsymbol{v}}, \hat{e}(\boldsymbol{x})] = E[\hat{e}(\boldsymbol{x})\Delta \boldsymbol{v}] = V[\hat{\boldsymbol{v}}]\hat{\boldsymbol{a}}(\boldsymbol{x}) + V[\hat{\boldsymbol{v}}, \hat{\boldsymbol{\omega}}]\hat{\boldsymbol{b}}(\boldsymbol{x}),$$

$$V_m[\hat{\boldsymbol{\omega}}, \hat{e}(\boldsymbol{x})] = E[\hat{e}(\boldsymbol{x})\Delta \boldsymbol{\omega}] = V[\hat{\boldsymbol{\omega}}, \hat{\boldsymbol{v}}]\hat{\boldsymbol{a}}(\boldsymbol{x}) + V[\hat{\boldsymbol{\omega}}]\hat{\boldsymbol{b}}(\boldsymbol{x}). \tag{12.147}$$

The covariance matrices $V[\hat{\boldsymbol{v}}]$, $V[\hat{\boldsymbol{v}}, \hat{\boldsymbol{\omega}}]$, and $V[\hat{\boldsymbol{\omega}}]$ are computed by eq. (12.67). In computing eq. (12.67), we need the value of ϵ^2 because $W(\boldsymbol{x}, \hat{\boldsymbol{v}})$ involves the covariance matrix $V[\dot{\boldsymbol{x}}] = \epsilon^2 V_0[\dot{\boldsymbol{x}}]$ (see eq. (12.50)). It is estimated by eq. (12.123).

If the motion parameters are not accurate, the flow $\dot{\boldsymbol{x}}$ is corrected by eq. (12.120) even though $\dot{\boldsymbol{x}}$ itself is accurate. From eq. (12.119), the a posteriori covariance matrix $V_m[\hat{\dot{\boldsymbol{x}}}]$ is obtained to a first approximation in the following form:

$$V_m[\hat{\dot{\boldsymbol{x}}}] = \frac{V_m[\hat{e}(\boldsymbol{x})](V_0[\dot{\boldsymbol{x}}](\hat{\boldsymbol{v}} \times \boldsymbol{x}))(V_0[\dot{\boldsymbol{x}}](\hat{\boldsymbol{v}} \times \boldsymbol{x}))^\top}{(\hat{\boldsymbol{v}} \times \boldsymbol{x}, V_0[\dot{\boldsymbol{x}}](\hat{\boldsymbol{v}} \times \boldsymbol{x}))^2}. \tag{12.148}$$

We also have

$$V_m[\hat{\boldsymbol{v}}, \hat{\dot{\boldsymbol{x}}}] = \frac{V_m[\hat{\boldsymbol{v}}, \hat{e}(\boldsymbol{x})](V_0[\dot{\boldsymbol{x}}](\hat{\boldsymbol{v}} \times \boldsymbol{x}))^\top}{(\hat{\boldsymbol{v}} \times \boldsymbol{x}, V_0[\dot{\boldsymbol{x}}](\hat{\boldsymbol{v}} \times \boldsymbol{x}))},$$

$$V_m[\hat{\boldsymbol{\omega}}, \hat{\dot{\boldsymbol{x}}}] = \frac{V_m[\hat{\boldsymbol{\omega}}, \hat{e}(\boldsymbol{x})](V_0[\dot{\boldsymbol{x}}](\hat{\boldsymbol{v}} \times \boldsymbol{x}))^\top}{(\hat{\boldsymbol{v}} \times \boldsymbol{x}, V_0[\dot{\boldsymbol{x}}](\hat{\boldsymbol{v}} \times \boldsymbol{x}))}. \tag{12.149}$$

From eq. (12.141), we see that the perturbations $\Delta \boldsymbol{v}$, $\Delta \boldsymbol{\omega}$, and $\Delta \hat{\dot{\boldsymbol{x}}}$ are related to the perturbation $\Delta Z(\boldsymbol{x})$ to a first approximation in the form

$$\Delta \hat{\dot{\boldsymbol{x}}} = -\boldsymbol{Q}_{\boldsymbol{x}} \left(\frac{\Delta \boldsymbol{v}}{\hat{Z}(\boldsymbol{x})} - \frac{\Delta Z(\boldsymbol{x}) \hat{\boldsymbol{v}}}{\hat{Z}(\boldsymbol{x})^2} + \Delta \boldsymbol{\omega} \times \boldsymbol{x} \right), \tag{12.150}$$

from which we have

$$\frac{\Delta Z(\boldsymbol{x})}{\hat{Z}(\boldsymbol{x})^2} \boldsymbol{Q}_{\boldsymbol{x}} \hat{\boldsymbol{v}} = \boldsymbol{Q}_{\boldsymbol{x}} \left(\Delta \hat{\dot{\boldsymbol{x}}} + \frac{\Delta \boldsymbol{v}}{\hat{Z}(\boldsymbol{x})} + \Delta \boldsymbol{\omega} \times \boldsymbol{x} \right). \tag{12.151}$$

---

[4]Strictly speaking, the motion parameters $\{\hat{\mathrm{v}}, \hat{\omega}\}$ in eqs. (12.146) should be $\{\bar{\mathrm{v}}, \bar{\omega}\}$, but this approximation introduces only a second order difference.

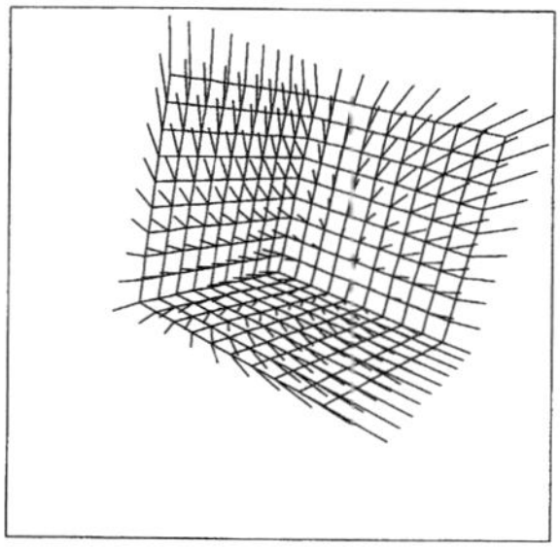

Fig. 12.10. Simulated optical flow.

Taking the expectation of the squared norm on both sides, we obtain

$$\frac{V_m[\hat{Z}(\boldsymbol{x})]\|\boldsymbol{Q_x}\hat{\boldsymbol{v}}\|^2}{\hat{Z}(\boldsymbol{x})^4} = E[\left\|\boldsymbol{Q_x}\left(\Delta\dot{\hat{\boldsymbol{x}}} + \frac{\Delta\boldsymbol{v}}{\hat{Z}(\boldsymbol{x})} + \Delta\boldsymbol{\omega}\times\boldsymbol{x}\right)\right\|^2], \tag{12.152}$$

which can be rewritten as

$$\frac{V_m[\hat{Z}(\boldsymbol{x})](\hat{\boldsymbol{v}}, \boldsymbol{Q}_{\boldsymbol{x}}^\top\boldsymbol{Q_x}\hat{\boldsymbol{v}})}{\hat{Z}(\boldsymbol{x})^4}$$
$$= (E[\left(\Delta\dot{\hat{\boldsymbol{x}}} + \frac{\Delta\boldsymbol{v}}{\hat{Z}(\boldsymbol{x})} + \Delta\boldsymbol{\omega}\times\boldsymbol{x}\right)\left(\Delta\dot{\hat{\boldsymbol{x}}} + \frac{\Delta\boldsymbol{v}}{\hat{Z}(\boldsymbol{x})} + \Delta\boldsymbol{\omega}\times\boldsymbol{x}\right)^\top]; \boldsymbol{Q}_{\boldsymbol{x}}^\top\boldsymbol{Q_x}). \tag{12.153}$$

From this, the variance $V_m[\hat{Z}(\boldsymbol{x})]$ of the depth $\hat{Z}(\boldsymbol{x})$ is obtained in the following form:

$$V_m[\hat{Z}(\boldsymbol{x})] = \frac{\hat{Z}(\boldsymbol{x})^4}{(\hat{\boldsymbol{v}}, \boldsymbol{S_x}\hat{\boldsymbol{v}})}\Bigg((V[\dot{\hat{\boldsymbol{x}}}]; \boldsymbol{S_x}) + \frac{(V[\hat{\boldsymbol{v}}]; \boldsymbol{S_x})}{\hat{Z}(\boldsymbol{x})} + (\boldsymbol{x}\times V[\hat{\boldsymbol{\omega}}]\times\boldsymbol{x}; \boldsymbol{S_x})$$
$$+\frac{2(V_m[\hat{\boldsymbol{v}}, \dot{\hat{\boldsymbol{x}}}]; \boldsymbol{S_x})}{\hat{Z}(\boldsymbol{x})} - 2(\boldsymbol{x}\times V_m[\hat{\boldsymbol{\omega}}, \dot{\hat{\boldsymbol{x}}}]; \boldsymbol{S_x}) - \frac{2(\boldsymbol{x}\times V[\hat{\boldsymbol{\omega}}, \hat{\boldsymbol{v}}]; \boldsymbol{S_x})}{\hat{Z}(\boldsymbol{x})}\Bigg). \tag{12.154}$$

Example 12.4 Fig. 12.10 shows a simulated image (512×512 pixels with focal length $f = 600$ (pixels)) of three planar grids in the scene viewed from a moving camera. We synthesized optical flow at the grid points according to eq. (12.28) and added Gaussian noise of mean 0 and standard deviation $\sigma = 1$ (pixels) to the x and y flow components at each point independently. Hence, the noise level is $\epsilon = \sigma/f = 1/600$ (this value is not used in the reconstruction computation). Let $\{\bar{\boldsymbol{v}}, \bar{\boldsymbol{\omega}}\}$ and $\{\hat{\boldsymbol{v}}, \hat{\boldsymbol{\omega}}\}$ be the true and the computed motion

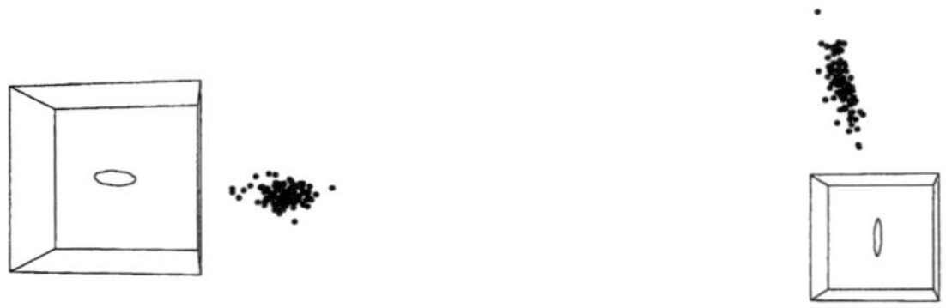

(a) (b)

Fig. 12.11. Least-squares approximation with optimal weights: (a) errors in translation; (b) errors in rotation.

(a) (b)

Fig. 12.12. Renormalization: (a) errors in translation; (b) errors in rotation.

parameters, respectively. Since $\boldsymbol{v}$ is a unit vector, the error in translation is measured by

$$\Delta \boldsymbol{v} = \boldsymbol{P}_{N[\bar{\boldsymbol{v}}]}(\hat{\boldsymbol{v}} - N[\bar{\boldsymbol{v}}]). \qquad (12.155)$$

The error in rotation is measured by

$$\Delta \boldsymbol{\omega} = \hat{\boldsymbol{\omega}} - \bar{\boldsymbol{\omega}}. \qquad (12.156)$$

In Figs. 12.11–12.13, vectors $\Delta \boldsymbol{v}$ and $\Delta \boldsymbol{\omega}$ are plotted in three dimensions for 100 trials, each time using different noise. The ellipses in these figures indicate the standard confidence regions[5] defined by $\bar{V}[\hat{\boldsymbol{v}}]$ and $\bar{V}[\hat{\boldsymbol{\omega}}]$ (see eq. (12.65)); the cubes are drawn merely for the sake of reference. Fig. 12.11 is for the least-squares approximation with optimal weights (computed from the true motion parameters $\{\bar{\boldsymbol{v}}, \bar{\boldsymbol{\omega}}\}$); Fig. 12.12 is for renormalization; Fig. 12.13 is for renormalization followed by optimal correction of the flow matrix. The corresponding standard confidence regions and reference cubes have the same absolute sizes. From Fig. 12.13, we can see that the accuracy almost attains the theoretical bound given by eq. (12.65) if renormalization and the optimal correction are combined.

Fig. 12.14 shows a reconstructed shape viewed from a different angle. The true shape is superimposed in dashed lines. Fig. 12.15a shows the standard confidence regions, which degenerate into line segments. They are computed from the total variance $V_t[\hat{Z}(\boldsymbol{x})] = V_i[\hat{Z}(\boldsymbol{x})] + V_m[\hat{Z}(\boldsymbol{x})]$ around the reconstructed grid points: $V_i[\hat{Z}(\boldsymbol{x})]$ is the variance due to the image noise (see

[5]Since $\bar{V}[\hat{\mathbf{v}}]$ has rank 2 and $\bar{V}[\hat{\omega}]$ has rank 3, the standard confidence region is a space conic for the translation velocity and an ellipsoid for the rotation velocity; see Section 4.5.3.

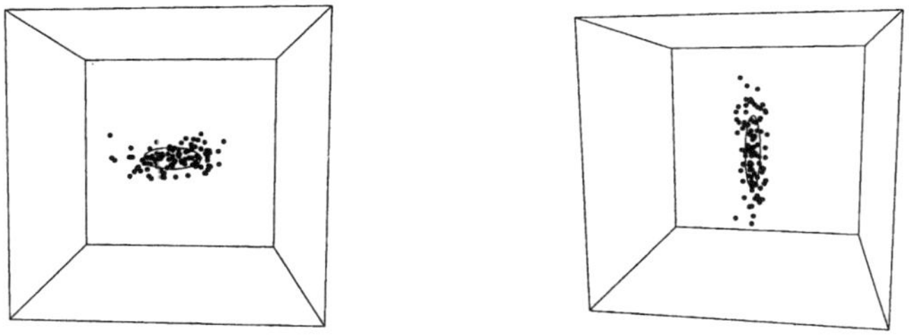

(a) (b)

Fig. 12.13. Renormalization followed by optimal correction: (a) errors in translation; (b) errors in rotation.

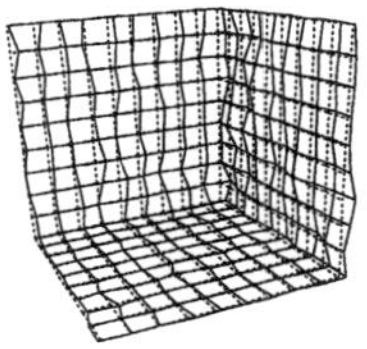

Fig. 12.14. Reconstructed shape. The true shape is superimposed in dashed lines.

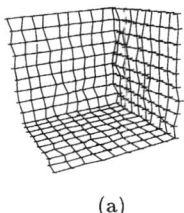

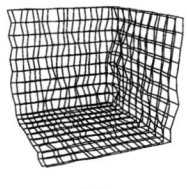

(a) (b)

Fig. 12.15. (a) Reconstructed shape and the standard confidence regions of grid points. (b) Two shapes corresponding to the primary deviation pairs of grid points.

eq. (12.144)); $V_m[\hat{Z}(\boldsymbol{x})]$ is the variance due to errors in the computed motion parameters (see eq. (12.154)). Fig. 12.15b shows the shapes that envelop the primary deviation pairs of the grid points.

12.7 Critical Surfaces for Optical flow

12.7.1 Weak critical surfaces

As in the case of finite motion, the linearization technique described in Section 12.4.1 is based on the fact that the flow matrix $\boldsymbol{F}$ is an eigenmatrix of the unperturbed moment tensor $\bar{\mathcal{M}}$ for eigenvalue 0, which we have implicitly assumed to be a *simple* root. If it is a multiple root, infinitely many eigenmatrices exist. This occurs if and only if the epipolar equation (12.69) is satisfied by a false flow matrix $\tilde{\boldsymbol{F}}$ ($\neq \boldsymbol{F}$). In other words, ambiguity occurs if and only if the object surface $\boldsymbol{r} = Z(\bar{\boldsymbol{x}})\bar{\boldsymbol{x}}$ is such that

$$(\bar{\boldsymbol{X}}; \tilde{\boldsymbol{F}}) = 0 \tag{12.157}$$

for $\tilde{\boldsymbol{F}} \neq \boldsymbol{F}$. For such a surface, the linearization technique in Section 12.4.1 does not work, since any linear combination of $\tilde{\boldsymbol{F}}$ and $\boldsymbol{F}$ is an eigenmatrix of $\bar{\mathcal{M}}$ for eigenvalue 0.

If we put $\tilde{\boldsymbol{v}} = \mathrm{type}_3[A[\tilde{\boldsymbol{F}}]]$, we can write $A[\tilde{\boldsymbol{F}}] = \tilde{\boldsymbol{v}} \times \boldsymbol{I}$. It follows that

$$\begin{aligned}(\bar{\boldsymbol{X}}; \tilde{\boldsymbol{F}}) &= (\boldsymbol{x}\boldsymbol{x}^\top + A[\dot{\bar{\boldsymbol{x}}}\boldsymbol{x}^\top]; S[\tilde{\boldsymbol{F}}] + A[\tilde{\boldsymbol{F}}]) = (\boldsymbol{x}\boldsymbol{x}^\top; S[\tilde{\boldsymbol{F}}]) + (\dot{\bar{\boldsymbol{x}}}\boldsymbol{x}^\top; \tilde{\boldsymbol{v}} \times \boldsymbol{I}) \\ &= (\boldsymbol{x}, S[\tilde{\boldsymbol{F}}]\boldsymbol{x}) + (\dot{\bar{\boldsymbol{x}}}, \tilde{\boldsymbol{v}} \times \boldsymbol{x}).\end{aligned} \tag{12.158}$$

Substituting eq. (12.28) into this and replacing $\boldsymbol{x}$ by $\boldsymbol{r}/Z(\boldsymbol{x})$, we obtain

$$\begin{aligned}(\bar{\boldsymbol{X}}; \tilde{\boldsymbol{F}}) &= (\boldsymbol{x}, S[\tilde{\boldsymbol{F}}]\boldsymbol{x}) - (\boldsymbol{Q}_{\boldsymbol{x}}\left(\frac{\boldsymbol{v}}{Z(\boldsymbol{x})} + \boldsymbol{\omega} \times \boldsymbol{x}\right), \tilde{\boldsymbol{v}} \times \boldsymbol{x}) \\ &= \frac{(\boldsymbol{r}, S[\tilde{\boldsymbol{F}}]\boldsymbol{r})}{Z(\boldsymbol{x})^2} - (\frac{\boldsymbol{v}}{Z(\boldsymbol{x})} + \frac{\boldsymbol{\omega} \times \boldsymbol{r}}{Z(\boldsymbol{x})}, \frac{\boldsymbol{Q}_{\boldsymbol{x}}^\top(\tilde{\boldsymbol{v}} \times \boldsymbol{r})}{Z(\boldsymbol{x})}) \\ &= \frac{(\boldsymbol{r}, S[\tilde{\boldsymbol{F}}]\boldsymbol{r}) - (\boldsymbol{v}, \tilde{\boldsymbol{v}} \times \boldsymbol{r}) - (\boldsymbol{r}, \left((\boldsymbol{\omega}, \tilde{\boldsymbol{v}})\boldsymbol{I} - S[\boldsymbol{\omega}\tilde{\boldsymbol{v}}^\top]\right)\boldsymbol{r})}{Z(\boldsymbol{x})^2},\end{aligned} \tag{12.159}$$

where we have used the identity $\boldsymbol{Q}_{\boldsymbol{x}}^\top(\tilde{\boldsymbol{v}} \times \boldsymbol{x}) = \tilde{\boldsymbol{v}} \times \boldsymbol{x}$. Hence, eq. (12.157) holds if and only if the object surface has the form

$$(\boldsymbol{r}, \boldsymbol{K}'\boldsymbol{r}) = (\boldsymbol{v} \times \tilde{\boldsymbol{v}}, \boldsymbol{r}), \tag{12.160}$$

where

$$\boldsymbol{K}' = S[\tilde{\boldsymbol{F}}] - (\boldsymbol{\omega}, \tilde{\boldsymbol{v}})\boldsymbol{I} + S[\boldsymbol{\omega}\tilde{\boldsymbol{v}}^\top]. \tag{12.161}$$

Since eq. (12.160) is identically satisfied by $\boldsymbol{r} = \boldsymbol{0}$, the origin O of the camera coordinate system is on this surface. If $\boldsymbol{K}'$ is nonsingular, eq. (12.160) defines a centered quadric of the following form (see eq. (4.98)):

$$(\boldsymbol{r} - \boldsymbol{r}_C, \boldsymbol{S}(\boldsymbol{r} - \boldsymbol{r}_C)) = 1. \tag{12.162}$$

The center $\boldsymbol{r}_C$ and the matrix $\boldsymbol{S}$ are respectively given by

$$\boldsymbol{r}_C = \frac{1}{2}\boldsymbol{K}'^{-1}(\boldsymbol{v} \times \tilde{\boldsymbol{v}}), \qquad \boldsymbol{S} = \frac{4\boldsymbol{K}'}{(\boldsymbol{v} \times \tilde{\boldsymbol{v}}, \boldsymbol{K}'^{-1}(\boldsymbol{v} \times \tilde{\boldsymbol{v}}))}. \tag{12.163}$$

We call a surface whose equation has the form of eq. (12.160) for some (33)-matrix $\boldsymbol{K}'$ a *weak critical surface*.

12.7.2 Strong critical surfaces

If the object surface is included in a weak critical surface, the epipolar equation (12.69) is satisfied by infinitely many false flow matrices. However, since the true flow matrix $\boldsymbol{F}$ must be decomposable, the correct flow matrix $\boldsymbol{F}$ can be chosen *if the false flow matrices are not decomposable*. In fact, the correct motion parameters $\{\boldsymbol{v}, \boldsymbol{\omega}\}$ can be obtained by the direct optimization (12.49). However, if the epipolar equation (12.69) is satisfied by a false flow matrix *that is decomposable*, the problem is inherently ambiguous. This occurs if and only if the object surface has the form of eq. (12.160) for which the matrix $\tilde{\boldsymbol{F}}$ in eq. (12.161) can be decomposed into false motion parameters $\{\tilde{\boldsymbol{v}}, \tilde{\boldsymbol{\omega}}\}$. This condition is equivalent to saying that there exist vectors $\tilde{\boldsymbol{v}}$ ($\neq \boldsymbol{v}$) and $\tilde{\boldsymbol{\omega}}$ ($\neq \boldsymbol{\omega}$) such that

$$\boldsymbol{K}' = (\tilde{\boldsymbol{v}}, \tilde{\boldsymbol{\omega}} - \boldsymbol{\omega})\boldsymbol{I} - S[\tilde{\boldsymbol{v}}(\tilde{\boldsymbol{\omega}} - \boldsymbol{\omega})^\top]. \tag{12.164}$$

Let us call a surface that satisfies this condition a *strong critical surface*. By definition, a strong critical surface is also a weak critical surface and hence a quadric that passes through the origin O of the camera coordinate system. However, the converse does not necessarily hold. In fact, eq. (12.164) implies

$$(\tilde{\boldsymbol{v}}, \boldsymbol{K}'\tilde{\boldsymbol{v}}) = 0, \tag{12.165}$$

and hence eq. (12.160) is identically satisfied by $\boldsymbol{r} = c\tilde{\boldsymbol{v}}$ for an arbitrary constant c. In other words, the space line $\boldsymbol{r} \times \tilde{\boldsymbol{v}} = \boldsymbol{0}$ is entirely contained in that surface. This implies that a strong critical surface is a *hyperboloid of one sheet* or its degeneracy (Fig. 12.16a).

As in the case of finite motion, *all planar surfaces are strong critical surfaces* and hence weak critical surfaces as well: an arbitrary space plane Π that does not pass through the origin O can be coupled with another space plane Π' that passes through O and intersects with Π *perpendicularly* in such a way that Π and Π' constitute a *degenerate hyperboloid* (Fig. 12.16b). Consequently, the 3-D interpretation of a planar surface is inherently ambiguous. In

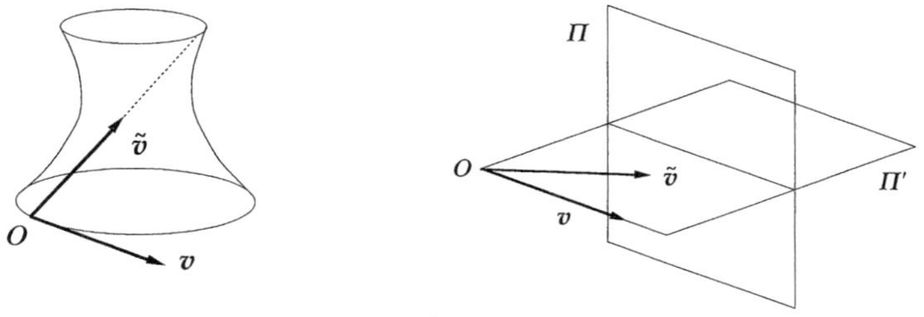

(a) (b)

Fig. 12.16. (a) Hyperboloid of one sheet. (b) Two planar surfaces as a degenerate hyperboloid.

the following, we construct an algorithm for computing all possible interpretations of a planar surface scene. In order to invoke this algorithm, however, we must first test if the object surface we are observing is a planar surface. This *planarity test* is derived from a statistical analysis of the image noise.

12.8 Analysis of Planar Surface Optical Flow

12.8.1 Optical flow equation for a planar surface

If space plane $(\boldsymbol{n}, \boldsymbol{r}) = d$ is viewed from a camera moving with motion parameters $\{\boldsymbol{h},\ \boldsymbol{R}\}$, the projected image motion has the following form (see eqs. (6.66) and (6.67)):

$$\boldsymbol{x}' = \frac{\boldsymbol{A}\boldsymbol{x}}{(\boldsymbol{k}, \boldsymbol{A}\boldsymbol{x})}, \qquad \boldsymbol{A} = \boldsymbol{R}^\top(\boldsymbol{h}\boldsymbol{n}^\top - d\boldsymbol{I}). \tag{12.166}$$

If the camera is stationary, we have $\boldsymbol{h} = \boldsymbol{0}$ and $\boldsymbol{R} = \boldsymbol{I}$, so $\boldsymbol{A} = -d\boldsymbol{I}$. It follows that for an infinitesimal camera motion the matrix $\boldsymbol{A}$ has the form

$$\boldsymbol{A} = -d(\boldsymbol{I} + \boldsymbol{W}\Delta t) + O(\Delta t^2). \tag{12.167}$$

Substitution of this into the first of eqs. (12.166) yields

$$\begin{aligned} \boldsymbol{x}' &= \frac{\boldsymbol{x} + \boldsymbol{W}\boldsymbol{x}\Delta t + O(\Delta t^2)}{(\boldsymbol{k}, \boldsymbol{x}) + (\boldsymbol{k}, \boldsymbol{W}\boldsymbol{x})\Delta t + O(\Delta t^2)} \\ &= \boldsymbol{x} + \Big(\boldsymbol{W}\boldsymbol{x} - (\boldsymbol{k}, \boldsymbol{W}\boldsymbol{x})\boldsymbol{x}\Big)\Delta t + O(\Delta t^2). \end{aligned} \tag{12.168}$$

Hence, the optical flow $\dot{\boldsymbol{x}} = \lim_{\Delta t \to 0}(\boldsymbol{x}' - \boldsymbol{x})/\Delta t$ has the form

$$\dot{\boldsymbol{x}} = \boldsymbol{W}\boldsymbol{x} - (\boldsymbol{k}, \boldsymbol{W}\boldsymbol{x})\boldsymbol{x}. \tag{12.169}$$

We call $\boldsymbol{W}$ the *flow matrix* of planar surface motion.

Eq. (12.169) implies that the flow $\dot{\boldsymbol{x}}$ remains the same if the flow matrix $\boldsymbol{W}$ is replaced $\boldsymbol{W} + c\boldsymbol{I}$ for an arbitrary constant c. This indeterminacy originates from the fact that the image motion given by the first of eqs. (12.166) is unchanged if the matrix $\boldsymbol{A}$ is replaced by $c\boldsymbol{A}$ for an arbitrary constant c. In order to remove this indeterminacy, we hereafter impose the following normalization:

$$\mathrm{tr}\boldsymbol{W} = 0. \tag{12.170}$$

If the camera motion is infinitesimal, the motion parameters $\{\boldsymbol{h}, \boldsymbol{R}\}$ have the following form (see eqs. (12.30)):

$$\boldsymbol{R} = \boldsymbol{I} + \boldsymbol{\omega} \times \boldsymbol{I}\Delta t + O(\Delta t^2), \qquad \boldsymbol{h} = \boldsymbol{v}\Delta t + O(\Delta t^2). \tag{12.171}$$

Substituting these into the second of eqs. (12.166), we obtain

$$\begin{aligned} \boldsymbol{A} &= \left(\boldsymbol{I} - \boldsymbol{\omega} \times \boldsymbol{I}\Delta t + O(\Delta t^2)\right)\left(-d\boldsymbol{I} + \boldsymbol{v}\boldsymbol{n}^\top \Delta t + O(\Delta t^2)\right) \\ &= -d\boldsymbol{I} + \left(\boldsymbol{v}\boldsymbol{n}^\top + d\boldsymbol{\omega} \times \boldsymbol{I}\right)\Delta t + O(\Delta t^2). \end{aligned} \tag{12.172}$$

Comparing this with eq. (12.167) and noting the normalization (12.170), we see that the flow matrix $\boldsymbol{W}$ has the following form:

$$\boldsymbol{W} = -\frac{1}{d}\left(\boldsymbol{v}\boldsymbol{n}^\top - \frac{1}{3}(\boldsymbol{v}, \boldsymbol{n})\boldsymbol{I}\right) - \boldsymbol{\omega} \times \boldsymbol{I}. \tag{12.173}$$

Alternatively, eq. (12.169) can be obtained from the general optical flow equation (12.28) by noting that if space point $\boldsymbol{r} = Z(\boldsymbol{x})\boldsymbol{x}$ is on space plane $(\boldsymbol{n}, \boldsymbol{r}) = d$, the depth is given by

$$Z(\boldsymbol{x}) = \frac{d}{(\boldsymbol{n}, \boldsymbol{x})}. \tag{12.174}$$

Substituting this into eq. (12.28) and noting that $\boldsymbol{Q}_{\boldsymbol{x}}\boldsymbol{x} = \boldsymbol{0}$, we obtain

$$\begin{aligned} \dot{\boldsymbol{x}} &= -\boldsymbol{Q}_{\boldsymbol{x}}\left(\frac{(\boldsymbol{n}, \boldsymbol{x})\boldsymbol{v}}{d} + \boldsymbol{\omega} \times \boldsymbol{x}\right) = -\boldsymbol{Q}_{\boldsymbol{x}}\left(\frac{\boldsymbol{v}\boldsymbol{n}^\top}{d} + \boldsymbol{\omega} \times \boldsymbol{I}\right)\boldsymbol{x} \\ &= \boldsymbol{Q}_{\boldsymbol{x}}\boldsymbol{W}\boldsymbol{x} = \boldsymbol{W}\boldsymbol{x} - (\boldsymbol{k}, \boldsymbol{W}\boldsymbol{x})\boldsymbol{x}. \end{aligned} \tag{12.175}$$

12.8.2 Estimation of the flow matrix

In the presence of noise, the observed flow $\dot{\boldsymbol{x}}$ does not satisfy eq. (12.169) exactly. As in the case of finite motion analysis, the surface and motion parameters $\{\boldsymbol{n}, d\}$ and $\{\boldsymbol{v}, \boldsymbol{\omega}\}$ can be estimated by solving the following two subproblems:

Problem 12.5 *Estimate a matrix* $\boldsymbol{W}$ *of trace* 0 *such that*

$$\bar{\dot{\boldsymbol{x}}} = \boldsymbol{W}\boldsymbol{x} - (\boldsymbol{k}, \boldsymbol{W}\boldsymbol{x})\boldsymbol{x} \tag{12.176}$$

from the observed flow $\dot{\boldsymbol{x}}$.

Problem 12.6 *Decompose the matrix* $\boldsymbol{W}$ *into surface and motion parameters* $\{\boldsymbol{n}, d\}$ *and* $\{\boldsymbol{v}, \boldsymbol{\omega}\}$ *in such a way that*

$$\boldsymbol{W} = -\frac{1}{d}\left(\boldsymbol{v}\boldsymbol{n}^\top - \frac{1}{3}(\boldsymbol{v}, \boldsymbol{n})\boldsymbol{I}\right) - \boldsymbol{\omega} \times \boldsymbol{I}. \tag{12.177}$$

Problem 12.5 can be optimally solved by minimizing the integral of the squared Mahalanobis distance, i.e.,

$$J[\boldsymbol{W}] = \int_S \|\dot{\boldsymbol{x}} - \boldsymbol{W}\boldsymbol{x} + (\boldsymbol{k}, \boldsymbol{W}\boldsymbol{x})\boldsymbol{x}\|^2_{V_0[\dot{\boldsymbol{x}}]} dxdy \to \min, \tag{12.178}$$

under the constraint $\mathrm{tr}\boldsymbol{W} = 0$. The first variation of $J[\boldsymbol{W}]$ for an infinitesimal variation $\boldsymbol{W} \to \boldsymbol{W} + \delta\boldsymbol{W}$ is

$$\begin{aligned}\delta J[\boldsymbol{W}] &= 2\int_S (-\delta\boldsymbol{W}\boldsymbol{x} + (\boldsymbol{k}, \delta\boldsymbol{W}\boldsymbol{x})\boldsymbol{x}, V_0[\dot{\boldsymbol{x}}]\,(\dot{\boldsymbol{x}} - \boldsymbol{W}\boldsymbol{x} + (\boldsymbol{k}, \boldsymbol{W}\boldsymbol{x})\boldsymbol{x}))dxdy \\ &= -2(\int_S (V_0[\dot{\boldsymbol{x}}]\,(\dot{\boldsymbol{x}} - \boldsymbol{W}\boldsymbol{x} + (\boldsymbol{k}, \boldsymbol{W}\boldsymbol{x})\boldsymbol{x})\,\boldsymbol{x}^\top \\ &\quad -(\boldsymbol{x}, V_0[\dot{\boldsymbol{x}}]\,(\dot{\boldsymbol{x}} - \boldsymbol{W}\boldsymbol{x} + (\boldsymbol{k}, \boldsymbol{W}\boldsymbol{x})\boldsymbol{x}))\,\boldsymbol{k}\boldsymbol{x}^\top)\,dxdy; \delta\boldsymbol{W}).\end{aligned} \tag{12.179}$$

This must vanish for an arbitrary variation $\delta\boldsymbol{W}$ such that $\mathrm{tr}\delta\boldsymbol{W} = 0$, so we obtain

$$\begin{aligned}&\int_S V_0[\dot{\boldsymbol{x}}]\left((\boldsymbol{W}\boldsymbol{x} - (\boldsymbol{k}, \boldsymbol{W}\boldsymbol{x})\boldsymbol{x})\,\boldsymbol{x}^\top - (\boldsymbol{x}, V_0[\dot{\boldsymbol{x}}]\,(\boldsymbol{W}\boldsymbol{x} - (\boldsymbol{k}, \boldsymbol{W}\boldsymbol{x})\boldsymbol{x}))\boldsymbol{k}\boldsymbol{x}^\top\right) dxdy \\ &= \int_S \left(V_0[\dot{\boldsymbol{x}}]\,(\dot{\boldsymbol{x}} - (\boldsymbol{x}, V_0[\dot{\boldsymbol{x}}]\dot{\boldsymbol{x}}))\boldsymbol{k}\boldsymbol{x}^\top\right) dxdy + \lambda\boldsymbol{I},\end{aligned} \tag{12.180}$$

where λ is the Lagrange multiplier for the constraint $\mathrm{tr}\boldsymbol{W} = 0$. If a 3333-tensor $\mathcal{A} = (A_{ijkl})$ and a 33-matrix $\boldsymbol{B}$ are defined by

$$\begin{aligned}A_{ijkl} = \int_S \Big(V_0[\dot{\boldsymbol{x}}]_{ik} - &\sum_{m=1}^{3}(k_i V_0[\dot{\boldsymbol{x}}]_{km} + k_k V_0[\dot{\boldsymbol{x}}]_{im})x_m \\ &+ (\boldsymbol{x}, V_0[\dot{\boldsymbol{x}}]\boldsymbol{x})k_i k_k\Big)\, x_j x_l dxdy,\end{aligned} \tag{12.181}$$

$$\boldsymbol{B} = \int_S \left(V_0[\dot{\boldsymbol{x}}]\dot{\boldsymbol{x}}\boldsymbol{x}^\top - (\boldsymbol{x}, V_0[\dot{\boldsymbol{x}}]\dot{\boldsymbol{x}})\boldsymbol{k}\boldsymbol{x}^\top\right) dxdy, \tag{12.182}$$

eq. (12.180) can be written in the following form:

$$\mathcal{A}\boldsymbol{W} = \boldsymbol{B} + \lambda\boldsymbol{I}. \tag{12.183}$$

It is easily seen from eq. (12.181) that $\mathcal{A}\boldsymbol{I} = \boldsymbol{O}$. This is a consequence of the fact that the flow matrix $\boldsymbol{W}$ can be determined only up to a constant multiple of $\boldsymbol{I}$. It follows that the rank of tensor $\mathcal{A}$ is 8; its null space is $\{\boldsymbol{I}\}_L$. Multiplying the generalized inverse $\mathcal{A}^-$ of tensor $\mathcal{A}$ on both sides of eq. (12.183) and noting that $\mathcal{A}^-\boldsymbol{I} = \boldsymbol{O}$, we obtain the solution $\boldsymbol{W}$ in the following form:

$$\boldsymbol{W} = \mathcal{A}^-\boldsymbol{B}. \tag{12.184}$$

The generalized inverse $\mathcal{A}^-$ is obtained by casting the 3333-tensor $\mathcal{A}$ into a 99-matrix, computing its generalized inverse, and casting it back into a 3333-tensor (see Section 2.4.2).

12.8.3 Planarity test

Let $\hat{\boldsymbol{W}}$ be the optimal estimate given by eq. (12.184). Since $V[\dot{\boldsymbol{x}}]$ has rank 2 and the flow matrix $\boldsymbol{W}$ has eight degrees of freedom, the residual

$$J[\hat{\boldsymbol{W}}] = \int_S \|\dot{\boldsymbol{x}} - \hat{\boldsymbol{W}}\boldsymbol{x} + (\boldsymbol{k}, \hat{\boldsymbol{W}}\boldsymbol{x})\boldsymbol{x}\|^2_{V_0[\dot{\boldsymbol{x}}]} dxdy \tag{12.185}$$

is a χ^2 variable with $2(S-4)$ degrees of freedom if the noise is Gaussian, where S is the number of the pixels at which the optical flow is defined. This implies that optical flow must be defined at least at four pixels, as expected from the finite motion analysis for a planar surface (see Section 11.6.1). If the covariance matrix $V[\dot{\boldsymbol{x}}]$ is decomposed into the noise level ϵ and the normalized covariance matrix $V_0[\dot{\boldsymbol{x}}]$ in the form of eq. (12.74), an unbiased estimator of ϵ^2 is obtained in the form

$$\hat{\epsilon}^2 = \frac{J_0[\hat{\boldsymbol{W}}]}{2(S-4)}, \tag{12.186}$$

where $J_0[\hat{\boldsymbol{W}}]$ is the normalized residual obtained by replacing $V[\dot{\boldsymbol{x}}]$ by $V_0[\dot{\boldsymbol{x}}]$ in the expression for $J[\hat{\boldsymbol{W}}]$. The expectation and variance of $\hat{\epsilon}^2$ are given by

$$E[\hat{\epsilon}^2] = \epsilon^2, \qquad V[\hat{\epsilon}^2] = \frac{\epsilon^4}{S-4}. \tag{12.187}$$

Since S is usually very large, $\hat{\epsilon}^2$ is a very good estimate of ϵ^2.

This analysis can be used to test if the observed flow is due to a planar surface motion. Namely, the hypothesis that the object surface is planar is rejected with significance level $a\%$ if[6]

$$J[\hat{\boldsymbol{W}}] > \chi^2_{2(S-4),a}. \tag{12.188}$$

[6]Since S is usually very large, we can use the approximation $\chi^2_{2(S-4),a} \approx (N_a + \sqrt{4S-17})^2/2 \approx (N_a + 2\sqrt{S})^2/2$; see Footnote 15 in Section 3.3.4.

In terms of the estimate $\hat{\epsilon}^2$ computed by eq. (12.186), this planarity test takes the following form:

$$\frac{\hat{\epsilon}^2}{\epsilon^2} > \frac{\chi^2_{2(S-4),a}}{2(S-4)}. \tag{12.189}$$

This has the same interpretation as the planarity test for finite motion (see eq. (11.152)). Namely, we compare the estimate $\hat{\epsilon}$ computed under the hypothesis that the object surface is planar with the noise level ϵ expected from the accuracy of the image processing operations for detecting optical flow.

As in the case of finite motion analysis, the planar surface algorithm can be applied to general optical flow as well. Hence, it is safer to apply the planar surface algorithm than the general optical flow algorithm; it should be switched to the general optical flow algorithm *only when assuming planarity is not compatible with the observed optical flow to a convincing degree.* The opposite approach has often been adopted in practice: the general optical flow algorithm is used first and abandoned, say, when the second smallest eigenvalue of the moment tensor $\mathcal{M}$ given by eq. (12.77) is sufficiently close to its smallest eigenvalue, and the decision criterion is set arbitrarily. In contrast, the use of the planarity test as described above has a solid statistical foundation. In Chapter 13, we present an alternative criterion which does not involve the significance level.

12.8.4 Computation of surface and motion parameters

We now solve Problem 12.6. Unlike the decomposition of the flow matrix $\boldsymbol{F}$ for general optical flow, *no decomposability condition needs to be imposed.* In other words, there always exist $\{\boldsymbol{n}, d\}$ and $\{\boldsymbol{v}, \boldsymbol{\omega}\}$ that satisfy eq. (12.176) exactly, although the decomposition may not be unique. This is easily understood by counting the degrees of freedom. The flow matrix $\boldsymbol{W}$ has eight degrees of freedom since its trace is 0. The surface parameters $\{\boldsymbol{n}, d\}$ have three degrees of freedom; the motion parameters $\{\boldsymbol{v}, \boldsymbol{\omega}\}$ have five degrees of freedom. Thus, the matrix $\boldsymbol{W}$ has the same degrees of freedom as the surface and motion parameters.

The surface and motion parameters $\{\boldsymbol{n}, d\}$ and $\{\boldsymbol{v}, \boldsymbol{\omega}\}$ are computed by the following procedure:

1. Compute
$$\boldsymbol{w} = -\text{type}_3[A[\boldsymbol{W}]]. \tag{12.190}$$

2. If $S[\boldsymbol{W}] = \boldsymbol{O}$, then $\boldsymbol{v} = \boldsymbol{0}$ and $\boldsymbol{\omega} = \boldsymbol{w}$. The surface parameters $\{\boldsymbol{n}, d\}$ are indeterminate.

3. Else, let $\sigma_1 \geq \sigma_2 \geq \sigma_3$ be the eigenvalues of $S[\boldsymbol{W}]$, and $\{\boldsymbol{u}_1, \boldsymbol{u}_2, \boldsymbol{u}_3\}$ the orthonormal system of the corresponding eigenvectors.

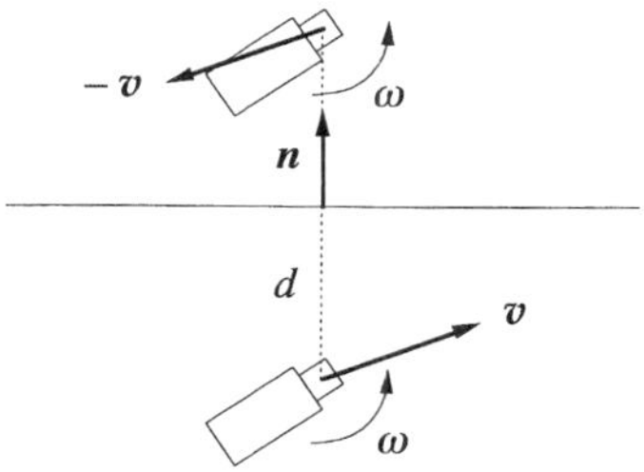

Fig. 12.17. Two solutions for the same image transformation.

4. The surface parameters are given by

$$n = \frac{\pm\sqrt{\sigma_1 - \sigma_2}u_1 + \sqrt{\sigma_2 - \sigma_3}u_3}{\sqrt{\sigma_1 - \sigma_3}}, \qquad d = \frac{1}{\sigma_1 - \sigma_3}. \tag{12.191}$$

5. The motion parameters are given by

$$v = -\frac{\pm\sqrt{\sigma_1 - \sigma_2}u_1 - \sqrt{\sigma_2 - \sigma_3}u_3}{\sqrt{\sigma_1 - \sigma_3}}, \qquad \omega = w - \frac{n \times v}{2d}, \tag{12.192}$$

   where the double sign $\pm$ corresponds to that in eqs. (12.191).

6. For each solution, another solution is obtained by changing the signs of n and v simultaneously.

Thus, *four* solutions are obtained. As in the case of finite motion, this ambiguity is partly due to the fact that the interpretation is solely based on the flow matrix W, not *the flow itself*, and the fact that the scene *behind* the camera can be seen (Fig. 11.13). Suppose one solution $\{n, d\}$ and $\{v, \omega\}$ is obtained. If we

1. move the camera to the other side of the planar surface, and

2. reverse the translation,

the new surface and motion parameters are $\{-n, d\}$ and $\{-v, \omega\}$, respectively (Fig. 12.17). We can see from eq. (12.177) that the flow matrix W does not change.

As in the case of finite motion, however, this ambiguity can be reduced if we also consider which part of the surface we are observing. In fact, if the condition that the visible part of the surface has positive depth is imposed, the number of solution reduces to at most two—in most cases one. This is because spurious interpretations are in most cases such that the surface is all behind the camera or some of part of it is behind the camera. In the latter

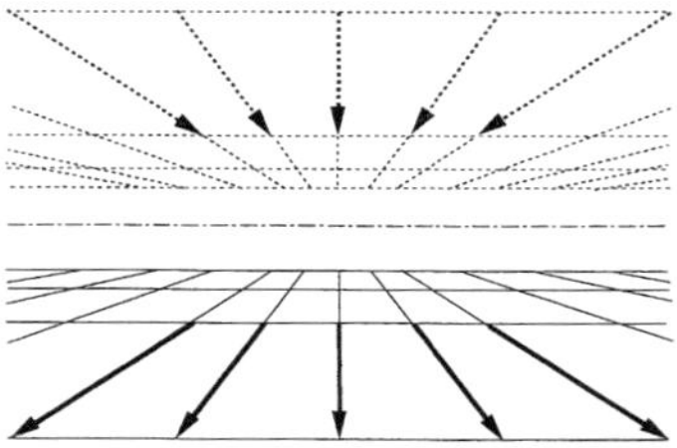

Fig. 12.18. The vanishing line cannot pass through the visible part of the planar surface; only one side of it is visible.

case, the *vanishing line* passes through the visible part, which is impossible because only one side of it is visible (Fig. 12.18; see Section 4.6.2).

From eq. (12.174), the condition to be imposed is

$$\frac{d}{(\boldsymbol{n}, \boldsymbol{x})} > 0. \tag{12.193}$$

The 3-D position $\boldsymbol{r}$ is given by

$$\boldsymbol{r} = \frac{d\boldsymbol{x}}{(\boldsymbol{n}, \boldsymbol{x})}. \tag{12.194}$$

Example 12.5 Fig. 12.19a shows a simulated optical flow image (512×512 pixels) of a planar grid in the scene viewed from a moving camera. The focal length is assumed to be $f = 600$ (pixels). We added Gaussian noise of mean 0 and standard deviation σ (pixels) to the x and y component of the flow at each point independently (the noise level $\epsilon = \sigma/f$ is not used in the reconstruction computation). We assume that the correct solution can be distinguished from spurious solutions. Fig. 12.19b shows the reconstructed grid for $\sigma = 1$, 2, 3. The true position is superimposed in dashed lines. We can see that the reconstruction error increases as the noise increases.

12.9 Camera Rotation and Information

12.9.1 Rotation estimation

So far, we have assumed that $\boldsymbol{v} \neq \boldsymbol{0}$. If $\boldsymbol{v} = \boldsymbol{0}$, no 3-D information can be obtained from the optical flow. Hence, we must test if $\boldsymbol{v} = \boldsymbol{0}$, i.e., if the camera motion is a pure rotation, before attempting 3-D reconstruction. In order to do this *rotation test*, we first hypothesize that $\boldsymbol{v} = \boldsymbol{0}$, i.e., the camera motion is a pure rotation around the center of the lens. If no image noise exists, eq. (12.28) implies that the flow should have the form $\dot{\boldsymbol{x}} = \boldsymbol{Q}_{\boldsymbol{x}}(\boldsymbol{x} \times \boldsymbol{\omega})$.

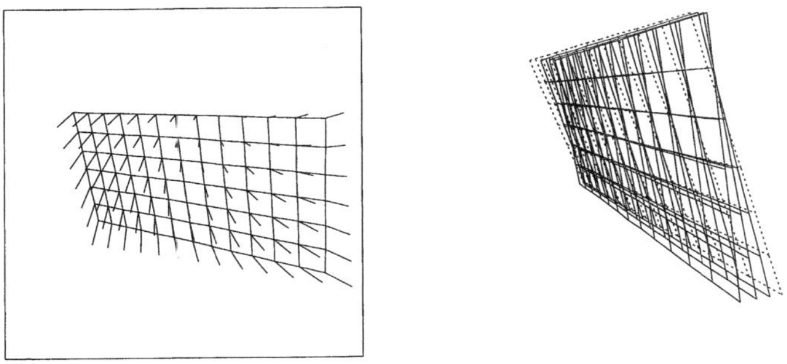

(a) (b)

Fig. 12.19. (a) Optical flow of a planar grid in the scene. (b) 3-D reconstruction for $\sigma = 1, 2, 3$.

Since the observed flow may be perturbed by image noise, we consider the following problem.

Problem 12.7 *Estimate a vector* $\boldsymbol{\omega}$ *such that*

$$\bar{\dot{\boldsymbol{x}}} = \boldsymbol{Q}_{\boldsymbol{x}}(\boldsymbol{x} \times \boldsymbol{\omega}) \tag{12.195}$$

from the observed flow $\dot{\boldsymbol{x}}$.

This problem can be solved in the same way as Problem 12.6: we minimizing the integral of the squared Mahalanobis distance, i.e.,

$$J[\boldsymbol{\omega}] = \int_S \|\dot{\boldsymbol{x}} - \boldsymbol{Q}_{\boldsymbol{x}}(\boldsymbol{x} \times \boldsymbol{\omega})\|^2_{V[\dot{\boldsymbol{x}}]} dxdy \to \min. \tag{12.196}$$

Noting the identities

$$\|\dot{\boldsymbol{x}} - \boldsymbol{Q}_{\boldsymbol{x}}(\boldsymbol{x} \times \boldsymbol{\omega})\|^2_{V[\dot{\boldsymbol{x}}]} = \|\dot{\boldsymbol{x}}\|^2_{V[\dot{\boldsymbol{x}}]} - 2(\dot{\boldsymbol{x}}, V[\dot{\boldsymbol{x}}]^- \boldsymbol{Q}_{\boldsymbol{x}}(\boldsymbol{x} \times \boldsymbol{\omega})) + \|\boldsymbol{Q}_{\boldsymbol{x}}(\boldsymbol{x} \times \boldsymbol{\omega})\|^2_{V[\dot{\boldsymbol{x}}]}, \tag{12.197}$$

$$(\dot{\boldsymbol{x}}, V[\dot{\boldsymbol{x}}]^- \boldsymbol{Q}_{\boldsymbol{x}}(\boldsymbol{x} \times \boldsymbol{\omega})) = (\boldsymbol{Q}_{\boldsymbol{x}}^\top V[\dot{\boldsymbol{x}}]^- \dot{\boldsymbol{x}}, \boldsymbol{x} \times \boldsymbol{\omega}) = -(\boldsymbol{x} \times \boldsymbol{Q}_{\boldsymbol{x}}^\top V[\dot{\boldsymbol{x}}]^- \dot{\boldsymbol{x}}, \boldsymbol{\omega}), \tag{12.198}$$

$$\|\boldsymbol{Q}_{\boldsymbol{x}}(\boldsymbol{x} \times \boldsymbol{\omega})\|^2_{V[\dot{\boldsymbol{x}}]} = (\boldsymbol{Q}_{\boldsymbol{x}}(\boldsymbol{x} \times \boldsymbol{\omega}), V[\dot{\boldsymbol{x}}]^- \boldsymbol{Q}_{\boldsymbol{x}}(\boldsymbol{x} \times \boldsymbol{\omega}))$$
$$= (\boldsymbol{x} \times \boldsymbol{\omega}, \boldsymbol{Q}_{\boldsymbol{x}}^\top V[\dot{\boldsymbol{x}}]^- \boldsymbol{Q}_{\boldsymbol{x}}(\boldsymbol{x} \times \boldsymbol{\omega})) = (\boldsymbol{\omega}, (\boldsymbol{x} \times \boldsymbol{Q}_{\boldsymbol{x}}^\top V[\dot{\boldsymbol{x}}]^- \boldsymbol{Q}_{\boldsymbol{x}} \times \boldsymbol{x})\boldsymbol{\omega}). \tag{12.199}$$

we can rewrite the function $J[\boldsymbol{\omega}]$ in the form

$$J[\boldsymbol{\omega}] = C - 2(\boldsymbol{b}, \boldsymbol{\omega}) + (\boldsymbol{\omega}, \boldsymbol{A}\boldsymbol{\omega}), \tag{12.200}$$

where

$$\boldsymbol{A} = \int_S \boldsymbol{x} \times \boldsymbol{Q}_{\boldsymbol{x}}^\top V[\dot{\boldsymbol{x}}]^- \boldsymbol{Q}_{\boldsymbol{x}} \times \boldsymbol{x} dxdy,$$

$$\boldsymbol{b} = -\int_S \boldsymbol{x} \times \boldsymbol{Q}_{\boldsymbol{x}}^\top V[\dot{\boldsymbol{x}}]^- \dot{\boldsymbol{x}} dxdy, \qquad C = \int_S \|\dot{\boldsymbol{x}}\|^2_{V[\dot{\boldsymbol{x}}]} dxdy. \tag{12.201}$$

Differentiating $J[\boldsymbol{\omega}]$ with respect to $\boldsymbol{\omega}$ and setting the result $\mathbf{0}$, we obtain

$$\boldsymbol{A}\boldsymbol{\omega} = \boldsymbol{b}. \tag{12.202}$$

The matrix $\boldsymbol{A}$ is nonsingular in general[7], so eq. (12.202) gives a unique solution.

12.9.2 Rotation test

Let $\hat{\boldsymbol{\omega}}$ be the resulting estimate. Since $V[\dot{\boldsymbol{x}}]$ generally has rank 2 and the vector $\boldsymbol{\omega}$ has three degrees of freedom, the residual

$$J[\hat{\boldsymbol{\omega}}] = \int_S \|\dot{\boldsymbol{x}} - \boldsymbol{Q}_{\boldsymbol{x}}(\boldsymbol{x} \times \hat{\boldsymbol{\omega}})\|^2_{V[\dot{\boldsymbol{x}}]} dxdy = \int_S \|\dot{\boldsymbol{x}}\|_{V[\dot{\boldsymbol{x}}]} dxdy - (\hat{\boldsymbol{\omega}}, \boldsymbol{A}\hat{\boldsymbol{\omega}}). \tag{12.203}$$

is a χ^2 variable with $2S$ degrees of freedom. If the covariance matrix $V[\dot{\boldsymbol{x}}]$ is decomposed into the noise level ϵ and the normalized covariance matrix $V_0[\dot{\boldsymbol{x}}]$ in the form

$$V[\dot{\boldsymbol{x}}] = \epsilon^2 V_0[\dot{\boldsymbol{x}}], \tag{12.204}$$

an unbiased estimator of ϵ^2 is obtained in the form

$$\hat{\epsilon}^2 = \frac{J_0[\hat{\boldsymbol{\omega}}]}{2S-3}, \tag{12.205}$$

where $J_0[\hat{\boldsymbol{\omega}}]$ is the normalized residual obtained by replacing $V[\dot{\boldsymbol{x}}]$ by $V_0[\dot{\boldsymbol{x}}]$ in the expression for $J[\hat{\boldsymbol{\omega}}]$. The expectation and variance of $\hat{\epsilon}^2$ are given by

$$E[\hat{\epsilon}^2] = \epsilon^2, \qquad V[\hat{\epsilon}^2] = \frac{2\epsilon^4}{2S-3}. \tag{12.206}$$

Since S is usually very large, $\hat{\epsilon}$ is a very good estimate of ϵ.

The above analysis is based on the hypothesis that $\boldsymbol{v} = \mathbf{0}$, i.e. the camera motion is a pure rotation. It follows that the rotation hypothesis can be tested by the standard χ^2 test: the hypothesis $\boldsymbol{v} = \mathbf{0}$ is rejected with significance level a% if

$$J[\hat{\boldsymbol{\omega}}] > \chi^2_{2S-3,a}. \tag{12.207}$$

In terms of the estimate $\hat{\epsilon}^2$ computed by eq. (12.205), this rotation test takes the following form:

$$\frac{\hat{\epsilon}^2}{\epsilon^2} > \frac{\chi^2_{2S-3}}{2S-3}. \tag{12.208}$$

This has the same interpretation as the planarity test (see eq. (12.189)). Namely, we compare the estimate $\hat{\epsilon}$ computed under the hypothesis $\boldsymbol{v} = \mathbf{0}$ with the noise level ϵ expected from the accuracy of the image processing operations for detecting optical flow.

[7] The matrix $\mathbf{A}$ is singular if the rank of $V[\dot{\mathbf{x}}]$ is less than 2 or the flow is defined only along a special type of curve on the image plane. We ignore such pathological cases

12.9.3 Information in optical flow

We can define the *information* of optical flow as in the case of finite motion (see Section 11.7.2). Let $\hat{\omega}$ be the rotation velocity estimated by the method described in Section 12.9.1. Define

$$\dot{\boldsymbol{x}}_u = \boldsymbol{Q}_{\boldsymbol{x}}(\boldsymbol{x} \times \hat{\boldsymbol{\omega}}). \tag{12.209}$$

This is the rotational flow that best "mimics" the observed flow $\dot{\boldsymbol{x}}$; it contains no information about the 3-D structure of the scene. Letting $\dot{\boldsymbol{x}}_i = \dot{\boldsymbol{x}} - \dot{\boldsymbol{x}}_u$, we can write

$$\dot{\boldsymbol{x}} = \dot{\boldsymbol{x}}_i + \dot{\boldsymbol{x}}_u. \tag{12.210}$$

This equation can be interpreted as decomposing the observed flow into the part that contains information about the 3-D scene and the part that contains no information. Let us call $\dot{\boldsymbol{x}}_i$ the *informative part* and $\dot{\boldsymbol{x}}_n$ the *uninformative part*. In general, the accuracy of 3-D interpretation decreases as the image noise increases. As in the case of finite motion, however, the negative effect of image noise cannot be measured simply by its magnitude but *it must be compared with the magnitude of the informative part of the optical flow.*

Let us call the normalized residual $J_0[\hat{\omega}]$ the *information* in the optical flow. We observe the following:

- The information is minimum if $\boldsymbol{v} = \boldsymbol{0}$; it generally increases as $\|\boldsymbol{v}\|$ increases.
- If $\boldsymbol{v}$ is the same, the information increases as the variation of the depth becomes larger.
- The information is unchanged if an arbitrary rotational velocity $\boldsymbol{\omega}$ is added to the camera motion.
- The information increases as the number pixels at which optical flow is defined increases.

Thus, 3-D interpretation can be disrupted by image noise of a very small magnitude if the information is very small, while 3-D interpretation is robust to image noise if the information is sufficiently large. If a rotational velocity is added to the camera motion, the flow magnitude increases, thereby decreasing the relative noise magnitude. However, such an apparent decrease of the relative noise magnitude does not increase the accuracy of 3-D interpretation, because camera rotation does not add any 3-D information to the optical flow. Thus, the relative noise magnitude is a very deceptive measure of the accuracy of 3-D interpretation. This viewpoint is further extended in a more general framework in Chapter 13.

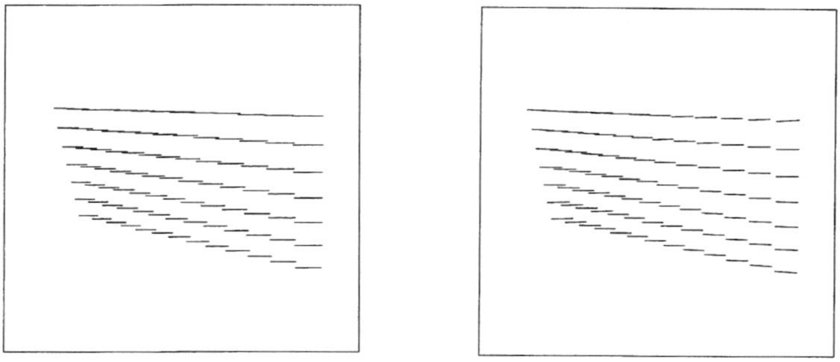

(a) (b)

Fig. 12.20. (a) Optical flow of a scene viewed from a translating camera. (b) Optical flow of the same scene viewed from a rotating camera.

Example 12.6 Fig. 12.20a shows an optical flow of a scene with a small depth variance viewed from a translating camera. It is widely known that 3-D interpretation of this type of optical flow is very difficult in the presence of noise, because a similar optical flow would be obtained by a camera rotating around the center of the lens (Fig. 12.20b), in which case 3-D interpretation is theoretically impossible. For flows such as shown in Fig. 12.20a, accurate 3-D interpretation is possible only if the image noise is very small. In other words, *the image noise must be small as compared with the information in the flow*.

Example 12.7 Fig. 12.21 shows the informative part and the uninformative part of the optical flow given in Fig. 12.19. The informative part of this flow has a sufficiently large magnitude, so we can expect a stable 3-D interpretation.

12.9.4 Midpoint flow approximation

In applying the theory described in this chapter to an optical flow detected from real images, we must note the discrepancy between the *theoretical optical flow* and the *real optical flow*. Theoretically, optical flow is defined as an *instantaneous velocity*: it is defined by *differentiation* of a continuous image motion. On the other hand, what can be detected by image processing operations is the *inter-frame displacement* of each pixel. These two definitions differ even if optical flow is exactly detected with subpixel accuracy.

This difference becomes apparent when the trajectory of an image point is not straight. In fact, the theoretical optical flow is *tangent* to the trajectory, while the interframe displacement is a *chord*, connecting two points on it (Fig. 12.22a). Suppose an image point $\boldsymbol{x}$ moves to $\boldsymbol{x}'$ in the next frame. A simple and effective way of avoiding the above discrepancy is to identify the

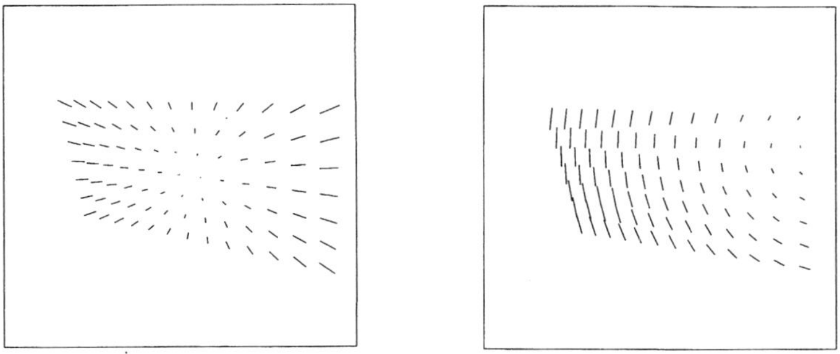

(a) (b)

Fig. 12.21. (a) Informative part of the flow shown in Fig. 12.19a. (b) Uninformative part.

displacement $\boldsymbol{x}' - \boldsymbol{x}$ with the flow at the *midpoint* $(\boldsymbol{x} + \boldsymbol{x}')/2$. Let us call this the *midpoint flow approximation* (Fig. 12.22a).

Example 12.8 Fig. 12.23 shows superimposition of simulated consecutive images of a planar grid in the scene viewed from a moving camera (no image noise is added). The image frame has 512×512 pixels, and the focal length is assumed to be $f = 600$ (pixels). Fig. 12.24a shows the 3-D shape reconstructed by identifying the displacement field with an optical flow. The true shape is superimposed in dashed lines. Fig. 12.24b shows the corresponding result obtained by the midpoint flow approximation. We can see that the midpoint flow approximation dramatically increases the accuracy of 3-D reconstruction.

(a) (b)

Fig. 12.22. (a) The difference between the theoretical optical flow and the real image optical flow. (b) Midpoint flow approximation.

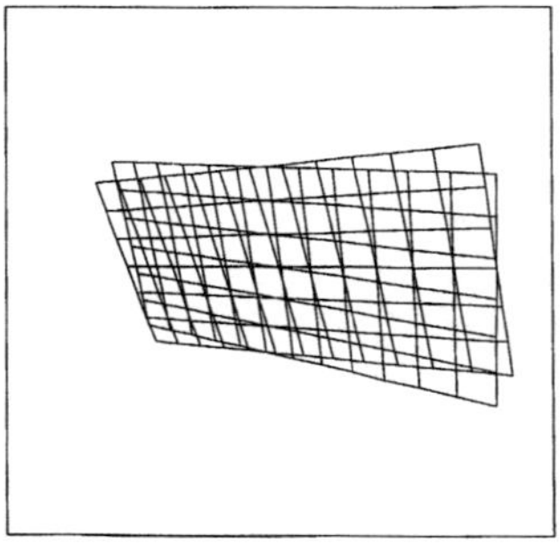

Fig. 12.23. Two consecutive image frames superimposed.

(a) (b)

Fig. 12.24. 3-D reconstruction from Fig. 12.23: (a) direct optical flow approximation; (b) midpoint flow approximation.

Chapter 13

Information Criterion for Model Selection

In order to apply geometric correction and parametric fitting presented in Chapters 5 and 7, we first need to know the *geometric model*, i.e., the constraints and hypotheses that should hold in the absence of noise. But how can we prefer one geometric model to another? In this chapter, the *AIC* introduced in Section 3.7 is modified so that it can be used for selecting a plausible geometric model, which is identified with a *manifold*. We show that the complexity of the model is evaluated by not only its *degrees of freedom* but also such invariant quantities as the *dimension* and *codimension* of the manifold. We also present a procedure for evaluating relative goodness of one model to another without using any arbitrarily set threshold such as the significance level of the χ^2 test. This comparison criterion is applied to point data in two and three dimensions, 3-D reconstruction by stereo vision, 3-D motion analysis, and 3-D interpretation of optical flow.

13.1 Model Selection Criterion

13.1.1 Model estimation

In the preceding chapters, we have formulated many types of statistical test, all of which have the form of comparing the *a posteriori* noise level $\hat{\epsilon}$ estimated under a hypothesis with the *a priori* value ϵ expected from the accuracy of the data acquisition process (image processing, range sensing, etc.). As has often been pointed out, however, it is very difficult to predict the noise level ϵ a priori in real situations. It can be estimated a posteriori *only if the hypothesis is true*. Then, is it not possible to test the hypothesis without using the a priori noise level ϵ? In this chapter, we show that although this is impossible in general, we can *compare* two hypotheses for the same data and determine which hypothesis is more plausible. In order to show this, we must first generalize the problem in abstract terms.

Let $\boldsymbol{a}_1$, ..., $\boldsymbol{a}_N$ be m-vector data sampled from an m'-dimensional manifold $\mathcal{A} \subset \mathcal{R}^m$, which we call the *data space*. We write

$$\boldsymbol{a}_\alpha = \bar{\boldsymbol{a}}_\alpha + \Delta\boldsymbol{a}_\alpha, \tag{13.1}$$

where $\bar{\boldsymbol{a}}_\alpha$ is the true position of datum $\boldsymbol{a}_\alpha$. The noise term $\Delta\boldsymbol{a}_\alpha$ is assumed to be an independent Gaussian random variable of mean $\boldsymbol{0}$ and covariance matrix $V[\boldsymbol{a}_\alpha]$. We assume that the covariance matrix $V[\boldsymbol{a}_\alpha]$ has range $T_{\bar{\boldsymbol{a}}_\alpha}(\mathcal{A})$ (=

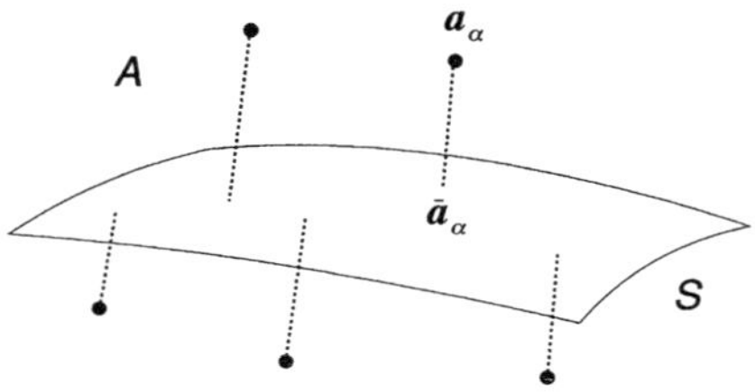

Fig. 13.1. Estimating a manifold $\mathcal{S} \subset \mathcal{A}$ and $\bar{\mathbf{a}}_\alpha \in \mathcal{S}$, $\alpha = 1, ..., N$, from the data $\{\mathbf{a}_\alpha\}$.

the tangent space to $\mathcal{A}$ at $\bar{\boldsymbol{a}}_\alpha$) and hence has rank m'. We want to find a manifold $\mathcal{S} \subset \mathcal{A}$ such that the true values $\bar{\boldsymbol{a}}_\alpha$ are all in it. This problem is stated as follows (Fig. 13.1):

Problem 13.1 *Estimate a manifold $\mathcal{S} \subset \mathcal{A}$ and m-vectors $\{\bar{\boldsymbol{a}}_\alpha\}$ such that $\bar{\boldsymbol{a}}_\alpha \in \mathcal{S}$, $\alpha = 1, ..., N$, from the data $\{\boldsymbol{a}_\alpha\}$.*

In order to solve this problem, we parameterize the manifold $\mathcal{S}$ by an n-vector $\boldsymbol{u}$ constrained to be in an n'-dimensional manifold $\mathcal{U} \subset \mathcal{R}^n$, which we call the *parameter space*. It follows that the manifold $\mathcal{S}$ has n' degrees of freedom, and Problem 13.1 reduces to *parametric fitting*: we want to compute an optimal value of the parameter $\boldsymbol{u} \in \mathcal{U}$. If $\mathcal{S}$ has 0 degrees of freedom, it is a fixed manifold, and Problem 13.1 can be identified with *geometric correction*: we want to correct each $\boldsymbol{a}_\alpha$ into $\bar{\boldsymbol{a}}_\alpha \in \mathcal{S}$ in an optimal manner. Thus, Problem 13.1 generalizes both geometric correction and parametric fitting.

We call a parameterized manifold a *(geometric) model.* If it is a d-dimensional manifold in an m'-dimensional data space and has n' free parameters, we say that the model has *dimension* d, *codimension* $r = m' - d$, and n' *degrees of freedom*. If the model is specified by L equations in the form

$$F^{(k)}(\boldsymbol{a}, \boldsymbol{u}) = 0, \qquad k = 1, ..., L, \tag{13.2}$$

the codimension r of the manifold was called the *rank* of the "hypothesis" (or "constraint" if the parameter $\boldsymbol{u}$ does not exist) in the preceding chapters (see Sections 5.1.1 and 7.1.1).

Example 13.1 Suppose N image points $\boldsymbol{x}_1, ..., \boldsymbol{x}_N$ are observed. The data space is the entire image plane $\mathcal{X} = \{\boldsymbol{x} \in \mathcal{R}^3 | (\boldsymbol{k}, \boldsymbol{x}) = 1\}$. Let $\bar{\boldsymbol{x}}_\alpha$ be the true position of $\boldsymbol{x}_\alpha$. We write $\boldsymbol{x}_\alpha = \bar{\boldsymbol{x}}_\alpha + \Delta\boldsymbol{x}_\alpha$ and regard the noise term $\Delta\boldsymbol{x}_\alpha$ as an independent Gaussian random variable of mean $\mathbf{0}$ and covariance matrix $V[\boldsymbol{x}_\alpha]$, which we assume has range $\mathcal{X}$ and rank 2. The following are typical models for the image point data $\{\boldsymbol{x}_\alpha\}$:

1. *Image point model*: the true positions $\bar{\boldsymbol{x}}_\alpha$, $\alpha = 1, ..., N$, all coincide (Fig. 13.2a). This model is an image point $p = \{\bar{\boldsymbol{x}}\}$; it has dimension

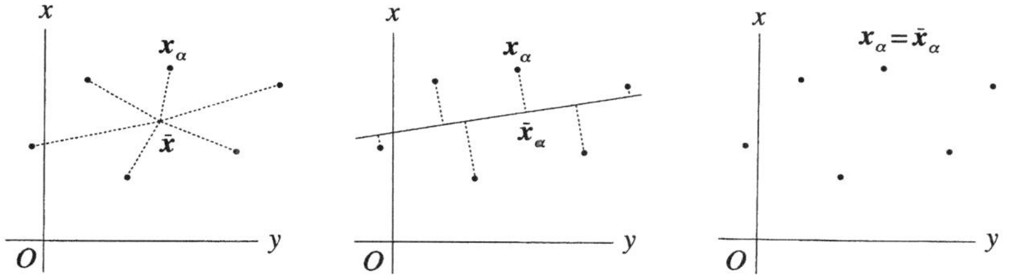

(a) (b) (c)

Fig. 13.2. (a) Image point model p. (b) Image line model l. (c) Image plane model $\mathcal{X}$.

0, codimension 2, and two degrees of freedom (an image point has two degrees of freedom; see Section 4.1.1). If p is constrained to be at the image origin, the constrained model $p' = \{\boldsymbol{k}\}$ has zero degrees of freedom.

2. *Image line model*: the true positions $\bar{\boldsymbol{x}}_\alpha$, $\alpha = 1, ..., N$, are all collinear (Fig. 13.2b). This model is an image line $l = \{(\boldsymbol{n}, \boldsymbol{x}) = 0\}$; it has dimension 1, codimension 1, and two degrees of freedom (an image line has two degrees of freedom; see Section 4.1.2). If l is constrained to pass through the image origin, the constrained model $l' = \{(\boldsymbol{n}, \boldsymbol{x}) = 0\text{—}\ (\boldsymbol{n}, \boldsymbol{k}) = 0\}$ has one degree of freedom.

3. *Image plane model*: no constraint is imposed on the true positions $\bar{\boldsymbol{x}}_\alpha$, $\alpha = 1, ..., N$ (Fig. 13.2c). This model is the entire image plane $\mathcal{X}$; it has dimension 2, codimension 0, and zero degrees of freedom.

Example 13.2 Suppose N space points $\boldsymbol{r}_1, ..., \boldsymbol{r}_N$ are observed. The data space is $\mathcal{R}^3$ itself. Let $\bar{\boldsymbol{r}}_\alpha$ be the true position of $\boldsymbol{r}_\alpha$. We write $\boldsymbol{r}_\alpha = \bar{\boldsymbol{r}}_\alpha + \Delta\boldsymbol{r}_\alpha$ and regard the noise term $\Delta\boldsymbol{r}_\alpha$ as an independent Gaussian random variable of mean $\boldsymbol{0}$ and covariance matrix $V[\boldsymbol{r}_\alpha]$, which we assume has range $\mathcal{R}^3$ and rank 3. The following are typical models for the space point data $\{\boldsymbol{r}_\alpha\}$:

1. *Space point model*: the true positions $\bar{\boldsymbol{r}}_\alpha$, $\alpha = 1, ..., N$, all coincide (Fig. 13.3a). This model is a space point $P = \{\bar{\boldsymbol{r}}\}$; it has dimension 0, codimension 3, and three degrees of freedom (a space point has three degrees of freedom; see Section 4.2.1). If P is constrained to be at the coordinate origin, the constrained model $P' = \{\boldsymbol{0}\}$ has zero degrees of freedom.

2. *Space line model*: the true positions $\bar{\boldsymbol{r}}_\alpha$, $\alpha = 1, ..., N$, are all collinear (Fig. 13.3b). This model is a space line $L = \{\boldsymbol{r} \times \boldsymbol{p} = \boldsymbol{n}\}$; it has dimension 1, codimension 2, and four degrees of freedom (a space line

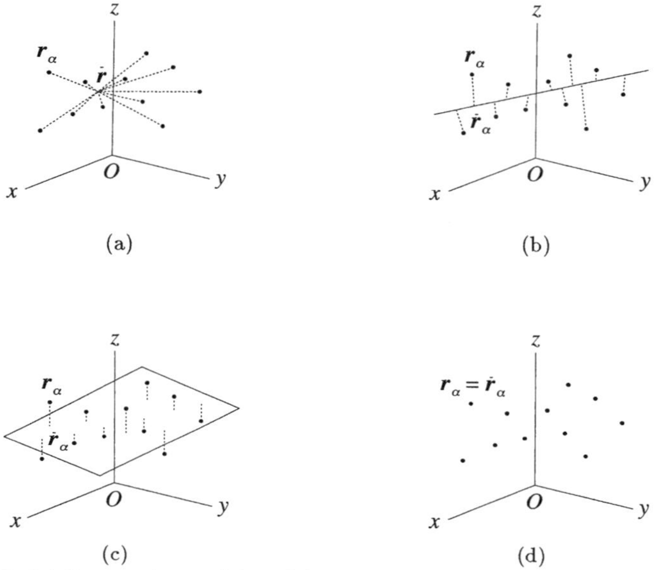

Fig. 13.3. (a) Space point model P. (b) Space line model L. (c) Space plane model Π. (c) Space model $\mathcal{R}^3$.

has four degrees of freedom; see Section 4.2.2). If L is constrained to pass through the coordinate origin, the constrained model $L' = \{\boldsymbol{r} \times \boldsymbol{p} = \boldsymbol{0}\}$ has two degrees of freedom.

3. *Space plane model*: the true positions $\bar{\boldsymbol{r}}_\alpha$, $\alpha = 1, ..., N$, are all coplanar (Fig. 13.3c). This model is a space plane $\Pi = \{(\boldsymbol{n}, \boldsymbol{r}) = d\}$; it has dimension 2, codimension 1, and three degrees of freedom (a space plane has three degrees of freedom; see Section 4.3.1). If Π is constrained to pass through the coordinate origin, the constrained model $\Pi' = \{(\boldsymbol{n}, \boldsymbol{r}) = 0\}$; has two degree of freedom.

4. *Space model*: no constraint is imposed on the true positions $\bar{\boldsymbol{r}}_\alpha$, $\alpha = 1, ..., N$ (Fig. 13.3d). This model is the entire space $\mathcal{R}^3$; it has dimension 3, codimension 0, and zero degrees of freedom.

13.1.2 Minimization of expected residual

Given a particular model $\mathcal{S}$, we can compute an optimal solution as described in Chapters 5 and 7. Namely, we can obtain the maximum likelihood solution of Problem 13.1 by minimizing the sum of the squared Mahalanobis distances

$$J = \sum_{\alpha=1}^{N} (\boldsymbol{a}_\alpha - \bar{\boldsymbol{a}}_\alpha, V[\boldsymbol{a}_\alpha]^- (\boldsymbol{a}_\alpha - \bar{\boldsymbol{a}}_\alpha)) \tag{13.3}$$

under the constraint that $\bar{\boldsymbol{a}}_\alpha \in \mathcal{S}$, $\alpha = 1, ..., N$ (see eq. (5.6), (7.5), and (7.12)). Let $\{\hat{\boldsymbol{a}}_\alpha\}$ and $\hat{\mathcal{S}}$ be the resulting maximum likelihood estimators of $\{\bar{\boldsymbol{a}}_\alpha\}$ and $\mathcal{S}$, respectively. Substituting them back into the function J, we write the *residual* in the following form:

$$J[\{\boldsymbol{a}_\alpha\}, \{\hat{\boldsymbol{a}}_\alpha\}, \hat{\mathcal{S}}] = \sum_{\alpha=1}^{N} \|\boldsymbol{a}_\alpha - \hat{\boldsymbol{a}}_\alpha\|^2_{V[\boldsymbol{a}_\alpha]}. \tag{13.4}$$

This quantity measures the minimum discrepancy between the assumed model $\mathcal{S}$ and the data $\{\boldsymbol{a}_\alpha\}$, so this appears to be a good measure of the goodness of the model. However, because $\{\hat{\boldsymbol{a}}_\alpha\}$ and $\hat{\mathcal{S}}$ are determined so as to minimize the residual *for the current data* $\{\boldsymbol{a}_\alpha\}$, the residual can be made arbitrarily small. In fact, if the manifold $\mathcal{S}$ has a sufficient number of parameters, we can make $\mathcal{S}$ pass through all the data $\{\boldsymbol{a}_\alpha\}$. Such an artificial model may explain the current data but may be unable to predict occurrence of the data to be observed in the future.

In order to measure the *predicting capacity* of the model, we adopt the criterion introduced in Section 3.7.1. Namely, we consider *future data* $\{\boldsymbol{a}^*_\alpha\}$ that have the same distribution as the current data $\{\boldsymbol{a}_\alpha\}$ and require that the residual for the future data

$$J[\{\boldsymbol{a}^*_\alpha\}, \{\hat{\boldsymbol{a}}_\alpha\}, \hat{\mathcal{S}}] = \sum_{\alpha=1}^{N} \|\boldsymbol{a}^*_\alpha - \hat{\boldsymbol{a}}_\alpha\|^2_{V[\boldsymbol{a}_\alpha]} \tag{13.5}$$

be small. Since this is a random variable, we take expectation to define a definitive value for the model:

$$I(\mathcal{S}) = E^*[E[J[\{\boldsymbol{a}^*_\alpha\}, \{\hat{\boldsymbol{a}}_\alpha\}, \hat{\mathcal{S}}]]]. \tag{13.6}$$

Here, $E^*[\,\cdot\,]$ and $E[\,\cdot\,]$ denote expectation with respect to the future data $\{\boldsymbol{a}^*_\alpha\}$ and the current data $\{\boldsymbol{a}_\alpha\}$, respectively. We call $I(\mathcal{S})$ the *expected residual* of model $\mathcal{S}$ and regard $\mathcal{S}$ as good if $I(\mathcal{S})$ is small.

Example 13.3 Consider the three models described in Example 13.1. For whatever data $\{\boldsymbol{x}_\alpha\}$, the residual of the image line model l is always no more than the image point model p, and the residual of the image plane model $\mathcal{X}$ is always 0 since the maximum likelihood estimator $\hat{\boldsymbol{x}}_\alpha$ of the true value $\bar{\boldsymbol{x}}_\alpha$ is the datum $\boldsymbol{x}_\alpha$ itself. The same can be said for Example 13.2: for whatever data $\{\boldsymbol{r}_\alpha\}$, the residual of the space line model L is always no more than

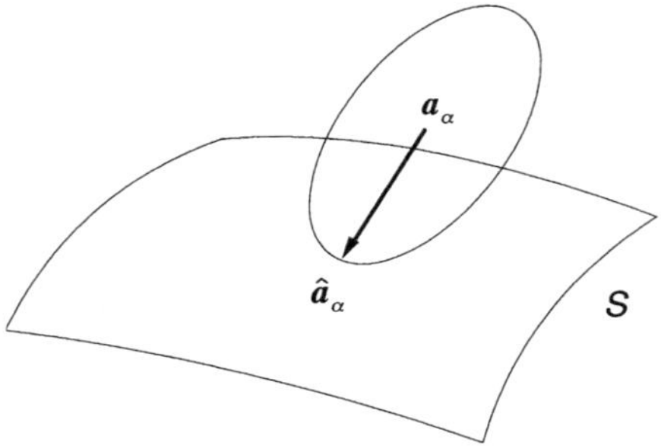

Fig. 13.4. The Mahalanobis projection of $\mathbf{a}_\alpha$ onto manifold $\mathcal{S}$ is the tangent point of the equilikelihood surface to $\mathcal{S}$.

the space point model P; the residual of the space plane model Π is always no more than the space line model L; the residual of the space model $\mathcal{R}^3$ is always 0. Thus, the residual is not a good measure of the goodness of the model.

13.2 Mahalanobis Geometry

13.2.1 Mahalanobis projection

First, we consider the case in which the model $\mathcal{S}$ has no free parameters (n' = 0). As discussed in Section 5.1.1, if the true value $\bar{\boldsymbol{a}}_\alpha$ of datum $\boldsymbol{a}_\alpha$ is in $\mathcal{S}$, the maximum likelihood estimator $\hat{\boldsymbol{a}}_\alpha$ of $\bar{\boldsymbol{a}}_\alpha$ is the point in $\mathcal{S}$ in the shortest distance from $\boldsymbol{a}_\alpha$ measured in the *Mahalanobis distance* with respect to the covariance matrix $V[\boldsymbol{a}_\alpha]$. Geometrically, $\hat{\boldsymbol{a}}_\alpha$ is the *tangent point* of the *equilikelihood surface* $(\boldsymbol{a} - \boldsymbol{a}_\alpha, V[\boldsymbol{a}_\alpha]^-(\boldsymbol{a} - \boldsymbol{a}_\alpha))$ = constant, $\boldsymbol{a} \in \mathcal{A}$, to the manifold $\mathcal{S}$ (Fig. 13.4). Let us call $\hat{\boldsymbol{a}}_\alpha$ the *Mahalanobis projection* of $\boldsymbol{a}_\alpha$ onto $\mathcal{S}$ with respect to $V[\boldsymbol{a}_\alpha]$. In the following, the proviso "with respect to $V[\boldsymbol{a}_\alpha]$" is omitted as understood.

We assume that the *noise is sufficiently small*, by which we mean that $V[\boldsymbol{a}_\alpha] = O(\epsilon^2)$, $\alpha = 1, ..., N$, for an appropriately defined noise level ϵ, as compared with which the data space $\mathcal{A}$ and the manifold $\mathcal{S}$ are both assumed to be sufficiently smooth[1]. The following proposition is easily obtained (Fig. 13.5):

Proposition 13.1 *The following equality holds to a first approximation:*

$$\|\boldsymbol{a}_\alpha - \bar{\boldsymbol{a}}_\alpha\|^2_{V[\boldsymbol{a}_\alpha]} = \|\boldsymbol{a}_\alpha - \hat{\boldsymbol{a}}_\alpha\|^2_{V[\boldsymbol{a}_\alpha]} + \|\hat{\boldsymbol{a}}_\alpha - \bar{\boldsymbol{a}}_\alpha\|^2_{V[\boldsymbol{a}_\alpha]}. \tag{13.7}$$

[1] A precise statement is that the *radius of curvature*, which is defined as the reciprocal of the *total curvature*, of the manifold is sufficiently small. This is the fundamental assumption for defining a (local) Gaussian distribution over a manifold (see Section 3.2.2). This assumption also played a fundamental role in Section 3.5.2.

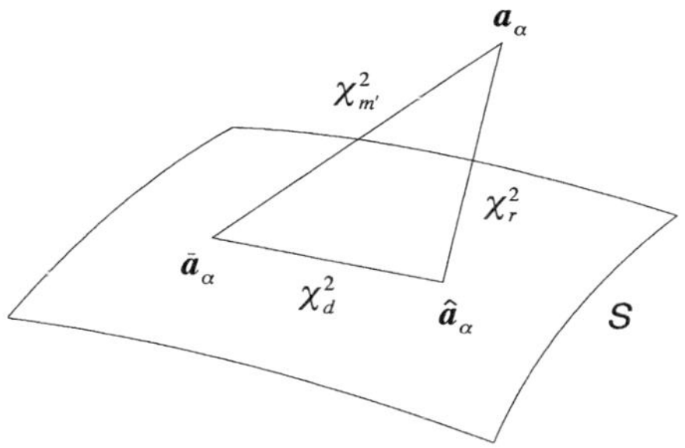

Fig. 13.5. The relation between datum $\mathbf{a}_\alpha$, its true value $\bar{\mathbf{a}}_\alpha \in \mathcal{S}$, and the Mahalanobis projection $\hat{\mathbf{a}}_\alpha$ of $\mathbf{a}_\alpha$ onto $\mathcal{S}$.

Proof. A point $\hat{\boldsymbol{a}}'_\alpha$ on the line that connects $\hat{\boldsymbol{a}}_\alpha$ and $\bar{\boldsymbol{a}}_\alpha$ can be expressed in the form

$$\hat{\boldsymbol{a}}'_\alpha = \hat{\boldsymbol{a}}_\alpha + t(\bar{\boldsymbol{a}}_\alpha - \hat{\boldsymbol{a}}_\alpha) \tag{13.8}$$

for a real number t. If t is small, point $\hat{\boldsymbol{a}}'_\alpha$ is on $\mathcal{S}$ to a first approximation. Hence,

$$\begin{aligned} \|\boldsymbol{a}_\alpha - \hat{\boldsymbol{a}}'_\alpha\|^2_{V[\boldsymbol{a}_\alpha]} &= \|(\boldsymbol{a}_\alpha - \hat{\boldsymbol{a}}_\alpha) - t(\bar{\boldsymbol{a}}_\alpha - \hat{\boldsymbol{a}}_\alpha)\|^2_{V[\boldsymbol{a}_\alpha]} \\ &= \|\boldsymbol{a}_\alpha - \hat{\boldsymbol{a}}_\alpha\|^2_{V[\boldsymbol{a}_\alpha]} - 2t(\boldsymbol{a}_\alpha - \hat{\boldsymbol{a}}_\alpha, V[\boldsymbol{a}_\alpha]^-(\bar{\boldsymbol{a}}_\alpha - \hat{\boldsymbol{a}}_\alpha)) \\ &\quad + t^2\|\bar{\boldsymbol{a}}_\alpha - \hat{\boldsymbol{a}}_\alpha\|^2_{V[\boldsymbol{a}_\alpha]}, \end{aligned} \tag{13.9}$$

which should take a minimum at $t = 0$ by the definition of the Mahalanobis projection $\hat{\boldsymbol{a}}_\alpha$. Hence,

$$(\boldsymbol{a}_\alpha - \hat{\boldsymbol{a}}_\alpha, V[\boldsymbol{a}_\alpha]^-(\bar{\boldsymbol{a}}_\alpha - \hat{\boldsymbol{a}}_\alpha)) = 0. \tag{13.10}$$

From this and

$$\begin{aligned} \|\boldsymbol{a}_\alpha - \bar{\boldsymbol{a}}_\alpha\|^2_{V[\boldsymbol{a}_\alpha]} &= \|(\boldsymbol{a}_\alpha - \hat{\boldsymbol{a}}_\alpha) - (\bar{\boldsymbol{a}}_\alpha - \hat{\boldsymbol{a}}_\alpha)\|^2_{V[\boldsymbol{a}_\alpha]} \\ &= \|\boldsymbol{a}_\alpha - \hat{\boldsymbol{a}}_\alpha\|^2_{V[\boldsymbol{a}_\alpha]} - 2(\boldsymbol{a}_\alpha - \hat{\boldsymbol{a}}_\alpha, V[\boldsymbol{a}_\alpha]^-(\bar{\boldsymbol{a}}_\alpha - \hat{\boldsymbol{a}}_\alpha)) \\ &\quad + \|\bar{\boldsymbol{a}}_\alpha - \hat{\boldsymbol{a}}_\alpha\|^2_{V[\boldsymbol{a}_\alpha]}, \end{aligned} \tag{13.11}$$

we obtain eq. (13.7). □

Proposition 13.1 can be interpreted as the *Pythagoras theorem*, stating that three points $\boldsymbol{a}_\alpha$, $\hat{\boldsymbol{a}}_\alpha$, and $\bar{\boldsymbol{a}}_\alpha$ define a "right-angled triangle," where the length is measured in the Mahalanobis distance and the orthogonality of vectors $\boldsymbol{u}$ and $\boldsymbol{v}$ is defined by $(\boldsymbol{u}, V[\boldsymbol{a}_\alpha]^-\boldsymbol{v}) = 0$. In this sense, $\hat{\boldsymbol{a}}_\alpha$ is the "foot" of the "perpendicular line" drawn from $\boldsymbol{a}_\alpha$ to $\mathcal{S}$.

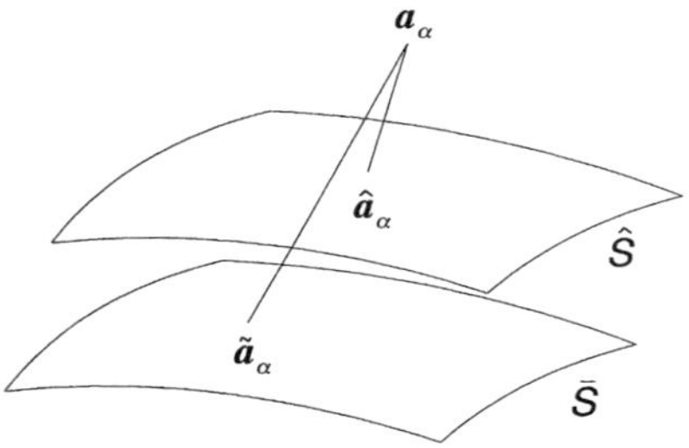

Fig. 13.6. The Mahalanobis projections $\tilde{\mathbf{a}}_\alpha$ and $\hat{\mathbf{a}}_\alpha$ of datum $\mathbf{a}_\alpha$ onto the true manifold $\bar{\mathcal{S}}$ and the the optimally fitted manifold $\hat{\mathcal{S}}$, respectively.

Proposition 13.2 *To a first approximation,* $\|\boldsymbol{a}_\alpha-\bar{\boldsymbol{a}}_\alpha\|^2_{V[\boldsymbol{a}_\alpha]}$, $\|\boldsymbol{a}_\alpha-\hat{\boldsymbol{a}}_\alpha\|^2_{V[\boldsymbol{a}_\alpha]}$, *and* $\|\hat{\boldsymbol{a}}_\alpha-\bar{\boldsymbol{a}}_\alpha\|^2_{V[\boldsymbol{a}_\alpha]}$ *are* χ^2 *variables with* m', r, *and* d *degrees of freedom, respectively.*

Proof. By definition, $\boldsymbol{a}_\alpha - \bar{\boldsymbol{a}}_\alpha$ is a Gaussian random variable of mean $\mathbf{0}$ and covariance matrix $V[\boldsymbol{a}_\alpha]$ of rank m' (see eq. (3.61)). Hence, $\|\boldsymbol{a}_\alpha-\bar{\boldsymbol{a}}_\alpha\|^2_{V[\boldsymbol{a}_\alpha]}$ is a χ^2 variable with m' degrees of freedom. The fact that $\|\boldsymbol{a}_\alpha-\hat{\boldsymbol{a}}_\alpha\|^2_{V[\boldsymbol{a}_\alpha]}$ is a χ^2 variable with r degrees with freedom was proved in Section 5.1.5 (see eqs. (5.6) and (5.35)). Since $\boldsymbol{a}_\alpha - \hat{\boldsymbol{a}}_\alpha$ is obtained from $\boldsymbol{a}_\alpha - \bar{\boldsymbol{a}}_\alpha$ by a linear mapping, $\hat{\boldsymbol{a}}_\alpha - \bar{\boldsymbol{a}}_\alpha = (\boldsymbol{a}_\alpha - \bar{\boldsymbol{a}}_\alpha) - (\boldsymbol{a}_\alpha - \hat{\boldsymbol{a}}_\alpha)$ is a χ^2 variable with $m' - r$ $(= d)$ degrees of freedom (see eq. (3.64)). □

Corollary 13.1 *To a first approximation,*

$$E[\|\boldsymbol{a}_\alpha-\bar{\boldsymbol{a}}_\alpha\|^2_{V[\boldsymbol{a}_\alpha]}] = m',$$

$$E[\|\boldsymbol{a}_\alpha-\hat{\boldsymbol{a}}_\alpha\|^2_{V[\boldsymbol{a}_\alpha]}] = r,$$

$$E[\|\hat{\boldsymbol{a}}_\alpha-\bar{\boldsymbol{a}}_\alpha\|^2_{V[\boldsymbol{a}_\alpha]}] = d. \qquad (13.12)$$

13.2.2 Residual of model fitting

Now, we consider the case in which the model $\mathcal{S}$ has n' free parameters. Let $\hat{\mathcal{S}}$ be its maximum likelihood estimators. We assume that this model contains the true manifold $\bar{\mathcal{S}}$ (Fig. 13.6). Let $\tilde{\boldsymbol{a}}_\alpha$ and $\hat{\boldsymbol{a}}_\alpha$ be the Mahalanobis projections of $\boldsymbol{a}_\alpha$ onto $\bar{\mathcal{S}}$ and $\hat{\mathcal{S}}$, respectively. Since each datum $\boldsymbol{a}_\alpha$ is independent, we obtain from Proposition 13.2 the following propositions:

Proposition 13.3 *The residual*

$$J[\{\boldsymbol{a}_\alpha\},\{\bar{\boldsymbol{a}}_\alpha\},\bar{\mathcal{S}}] = \sum_{\alpha=1}^{N} \|\boldsymbol{a}_\alpha-\bar{\boldsymbol{a}}_\alpha\|^2_{V[\boldsymbol{a}_\alpha]} \qquad (13.13)$$

is a χ^2 variable with $m'N$ degrees of freedom.

Proposition 13.4 *The residual*

$$J[\{\boldsymbol{a}_\alpha\}, \{\tilde{\boldsymbol{a}}_\alpha\}, \bar{\mathcal{S}}] = \sum_{\alpha=1}^{N} \|\boldsymbol{a}_\alpha - \tilde{\boldsymbol{a}}_\alpha\|^2_{V[\boldsymbol{a}_\alpha]} \tag{13.14}$$

is a χ^2 variable with rN degrees of freedom.

In other words, the degrees of freedom of the residual decrease from $m'N$ to rN if the true value $\bar{\boldsymbol{a}}_\alpha$ is replaced by the Mahalanobis projection $\tilde{\boldsymbol{a}}_\alpha$. However, if $\boldsymbol{a}_\alpha$ is projected onto the maximum likelihood estimator $\hat{\mathcal{S}}$, the degrees of freedom of the residual further decrease by n', as shown in Section 7.1.4 (Fig. 13.6). Namely,

Proposition 13.5 *The residual*

$$J[\{\boldsymbol{a}_\alpha\}, \{\hat{\boldsymbol{a}}_\alpha\}, \hat{\mathcal{S}}] = \sum_{\alpha=1}^{N} \|\boldsymbol{a}_\alpha - \hat{\boldsymbol{a}}_\alpha\|^2_{V[\boldsymbol{a}_\alpha]} \tag{13.15}$$

is a χ^2 variable with $rN - n'$ degrees of freedom.

This proposition implies that if the model $\mathcal{S}$ has rN or more independent parameters, the residual given by eq. (13.15) is generally 0, confirming the fact that the residual is not a good measure of the goodness of the model. From Propositions 13.3–13.5, we obtain the following corollary:

Corollary 13.2 *The following relationships hold:*

$$E[\sum_{\alpha=1}^{N} \|\boldsymbol{a}_\alpha - \bar{\boldsymbol{a}}_\alpha\|^2_{V[\boldsymbol{a}_\alpha]}] = m'N,$$

$$E[\sum_{\alpha=1}^{N} \|\boldsymbol{a}_\alpha - \tilde{\boldsymbol{a}}_\alpha\|^2_{V[\boldsymbol{a}_\alpha]}] = rN,$$

$$E[\sum_{\alpha=1}^{N} \|\boldsymbol{a}_\alpha - \hat{\boldsymbol{a}}_\alpha\|^2_{V[\boldsymbol{a}_\alpha]}] = rN - n'. \tag{13.16}$$

13.3 Expected Residual

13.3.1 Evaluation of the expected residual

We now evaluate the expected residual $I(\mathcal{S})$ defined by eq. (13.6). Let $\bar{\boldsymbol{a}}_\alpha \in \bar{\mathcal{S}}$ be the true position of datum $\boldsymbol{a}_\alpha$, and let $\tilde{\boldsymbol{a}}_\alpha$ and $\hat{\boldsymbol{a}}_\alpha$ be the Mahalanobis projections of $\boldsymbol{a}_\alpha$ onto $\bar{\mathcal{S}}$ and $\hat{\mathcal{S}}$, respectively. Let $\boldsymbol{a}^*_\alpha$ be the future datum corresponding to $\boldsymbol{a}_\alpha$ (Fig. 13.7).

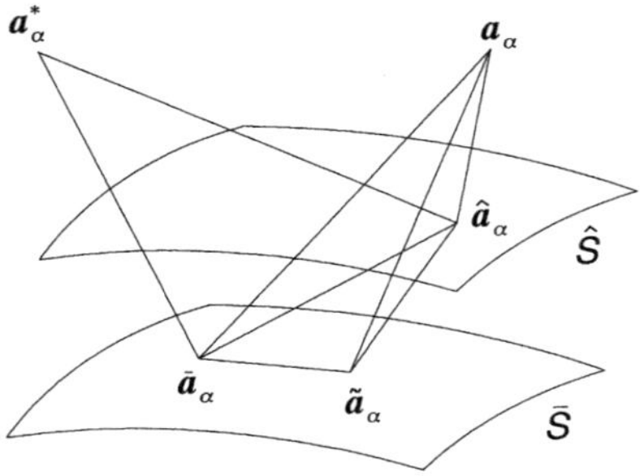

Fig. 13.7. Evaluation of the expected residual.

Lemma 13.1 *To a first approximation,*

$$I(\mathcal{S}) = E[J[\{\boldsymbol{a}_\alpha\},\{\hat{\boldsymbol{a}}_\alpha\},\hat{\mathcal{S}}]] + E[J[\{\hat{\boldsymbol{a}}_\alpha\},\{\tilde{\boldsymbol{a}}_\alpha\},\bar{\mathcal{S}}]] + 2dN + n'. \qquad (13.17)$$

Proof. Since the maximum likelihood estimators $\{\hat{\boldsymbol{a}}_\alpha\}$ are determined from the current data $\{\boldsymbol{a}_\alpha\}$, they are independent of the future data $\{\boldsymbol{a}^*_\alpha\}$. Hence, eq. (13.6) reduces to

$$\begin{aligned} I(\mathcal{S}) &= E^*[E[\sum_{\alpha=1}^N \|(\boldsymbol{a}^*_\alpha - \bar{\boldsymbol{a}}_\alpha) - (\hat{\boldsymbol{a}}_\alpha - \bar{\boldsymbol{a}}_\alpha)\|^2_{V[\boldsymbol{a}_\alpha]}]] \\ &= E^*[\sum_{\alpha=1}^N \|\boldsymbol{a}^*_\alpha - \bar{\boldsymbol{a}}_\alpha\|^2_{V[\boldsymbol{a}_\alpha]}] + E[\sum_{\alpha=1}^N \|\hat{\boldsymbol{a}}_\alpha - \bar{\boldsymbol{a}}_\alpha\|^2_{V[\boldsymbol{a}_\alpha]}]. \qquad (13.18) \end{aligned}$$

Since $\{\boldsymbol{a}^*_\alpha\}$ and $\{\boldsymbol{a}_\alpha\}$ have the same distribution, we have $E^*[\|\boldsymbol{a}^*_\alpha - \bar{\boldsymbol{a}}_\alpha\|^2_{V[\boldsymbol{a}_\alpha]}] = E[\|\boldsymbol{a}_\alpha - \bar{\boldsymbol{a}}_\alpha\|^2_{V[\boldsymbol{a}_\alpha]}]$. Hence, eq. (13.18) can be written as

$$I(\mathcal{S}) = E[\sum_{\alpha=1}^N \|\boldsymbol{a}_\alpha - \bar{\boldsymbol{a}}_\alpha\|^2_{V[\boldsymbol{a}_\alpha]}] + E[\sum_{\alpha=1}^N \|\hat{\boldsymbol{a}}_\alpha - \bar{\boldsymbol{a}}_\alpha\|^2_{V[\boldsymbol{a}_\alpha]}]. \qquad (13.19)$$

Consider the first term on the right-hand side. Applying Proposition 13.1, we can write

$$E[\sum_{\alpha=1}^N \|\boldsymbol{a}_\alpha - \bar{\boldsymbol{a}}_\alpha\|^2_{V[\boldsymbol{a}_\alpha]}] = E[\sum_{\alpha=1}^N \|\boldsymbol{a}_\alpha - \tilde{\boldsymbol{a}}_\alpha\|^2_{V[\boldsymbol{a}_\alpha]}] + E[\sum_{\alpha=1}^N \|\tilde{\boldsymbol{a}}_\alpha - \bar{\boldsymbol{a}}_\alpha\|^2_{V[\boldsymbol{a}_\alpha]}]. \qquad (13.20)$$

From Corollary 13.2, we have

$$E[\sum_{\alpha=1}^N \|\boldsymbol{a}_\alpha - \tilde{\boldsymbol{a}}_\alpha\|^2_{V[\boldsymbol{a}_\alpha]}] = E[\sum_{\alpha=1}^N \|\boldsymbol{a}_\alpha - \hat{\boldsymbol{a}}_\alpha\|^2_{V[\boldsymbol{a}_\alpha]}] + n'. \qquad (13.21)$$

Corollary 13.1 implies

$$E[\sum_{\alpha=1}^{N} \|\tilde{\boldsymbol{a}}_\alpha - \bar{\boldsymbol{a}}_\alpha\|^2_{V[\boldsymbol{a}_\alpha]}] = dN. \tag{13.22}$$

Hence, eq. (13.20) reduces to

$$E[\sum_{\alpha=1}^{N} \|\boldsymbol{a}_\alpha - \bar{\boldsymbol{a}}_\alpha\|^2_{V[\boldsymbol{a}_\alpha]}] = E[\sum_{\alpha=1}^{N} \|\boldsymbol{a}_\alpha - \hat{\boldsymbol{a}}_\alpha\|^2_{V[\boldsymbol{a}_\alpha]}] + dN + n'. \tag{13.23}$$

Next, consider the second term on the right-hand side of eq. (13.19). If the noise is small, the Mahalanobis projection of $\hat{\boldsymbol{a}}_\alpha \in \hat{\mathcal{S}}$ onto the true manifold $\bar{\mathcal{S}}$ coincides to a first approximation with the Mahalanobis projection $\tilde{\boldsymbol{a}}_\alpha$ of the datum $\boldsymbol{a}_\alpha$ onto $\bar{\mathcal{S}}$ (Fig. 13.7). Hence, we obtain from Proposition 13.1 and eq. (13.22)

$$\begin{aligned}\sum_{\alpha=1}^{N} \|\hat{\boldsymbol{a}}_\alpha - \bar{\boldsymbol{a}}_\alpha\|^2_{V[\boldsymbol{a}_\alpha]} &= \sum_{\alpha=1}^{N} \|\hat{\boldsymbol{a}}_\alpha - \tilde{\boldsymbol{a}}_\alpha\|^2_{V[\boldsymbol{a}_\alpha]} + \sum_{\alpha=1}^{N} \|\tilde{\boldsymbol{a}}_\alpha - \bar{\boldsymbol{a}}_\alpha\|^2_{V[\boldsymbol{a}_\alpha]} \\ &= \sum_{\alpha=1}^{N} \|\hat{\boldsymbol{a}}_\alpha - \tilde{\boldsymbol{a}}_\alpha\|^2_{V[\boldsymbol{a}_\alpha]} + dN. \end{aligned} \tag{13.24}$$

Substituting eqs. (13.23) and (13.24) into eq. (13.19), we obtain eq. (13.17).□

13.3.2 Accuracy of parametric fitting

In order to evaluate the second term on the right-hand side of eq. (13.17), we need an explicit representation of the model $\mathcal{S}$. Suppose $\mathcal{S}$ is given by L equations

$$F^{(k)}(\boldsymbol{a}, \boldsymbol{u}) = 0, \qquad k = 1, ..., L, \tag{13.25}$$

parameterized by an n-vector $\boldsymbol{u}$ constrained to be in an n'-dimensional parameter space $\mathcal{U} \subset \mathcal{R}^n$. As in Sections 5.1.1 and 7.1.1, we assume that the hypothesis (13.25) is *nonsingular*[2] and has *rank* r. This means that

1. only r of the L equations are algebraically independent,

2. each equation defines a manifold of codimension 1 in the m'-dimensional data space $\mathcal{A}$, and

3. the L manifolds intersect with each other *transversally* (see Section 3.2.1).

[2]The detailed mathematical argument will be given in the next chapter.

It follows that the intersection of the L manifolds is a manifold of codimension r.

Let $\bar{\boldsymbol{u}}$ be the true value of $\boldsymbol{u}$, i.e., the value that realizes the true manifold $\bar{\mathcal{S}}$. The *moment matrix* is defined as follows (see eq. (7.23)):

$$\bar{\boldsymbol{M}} = \sum_{\alpha=1}^{N} \sum_{k,l=1}^{L} \bar{W}_\alpha^{(kl)} \left(\boldsymbol{P}_{\bar{\boldsymbol{u}}}^{\mathcal{U}} \nabla_{\boldsymbol{a}} \bar{F}_\alpha^{(k)} \right) \left(\boldsymbol{P}_{\bar{\boldsymbol{u}}}^{\mathcal{U}} \nabla_{\boldsymbol{a}} \bar{F}_\alpha^{(l)} \right)^\top, \tag{13.26}$$

$$(\bar{W}_\alpha^{(kl)}) = \left((\nabla_{\boldsymbol{a}} \bar{F}_\alpha^{(k)}, V[\boldsymbol{a}_\alpha] \nabla_{\boldsymbol{a}} \bar{F}_\alpha^{(l)}) \right)^-. \tag{13.27}$$

Here, $\boldsymbol{P}_{\bar{\boldsymbol{u}}}^{\mathcal{U}}$ is the n-dimensional projection matrix onto the tangent space $T_{\bar{\boldsymbol{u}}}(\mathcal{U})$ to the manifold $\mathcal{U}$ at $\bar{\boldsymbol{u}}$. The symbol $\nabla_{\boldsymbol{a}}$ denotes differentiation with respect to the argument $\boldsymbol{a}$, and $\nabla_{\boldsymbol{a}} \bar{F}_\alpha^{(k)}$ is the abbreviation of $\nabla_{\boldsymbol{a}} F^{(k)}(\bar{\boldsymbol{a}}_\alpha, \bar{\boldsymbol{u}})$. We also use the abbreviated notation for generalized inverse introduced in Chapters 5 and 7: eq. (13.27) has the same meaning as eq. (7.8). Let $\hat{\boldsymbol{u}}$ be the maximum likelihood estimator of $\boldsymbol{u}$, i.e., the value that realizes the maximum likelihood estimator $\hat{\mathcal{S}}$ of $\mathcal{S}$.

Lemma 13.2 *To a first approximation,*

$$\sum_{\alpha=1}^{N} \| \hat{\boldsymbol{a}}_\alpha - \tilde{\boldsymbol{a}}_\alpha \|_{V[\boldsymbol{a}_\alpha]}^2 = (\hat{\boldsymbol{u}} - \bar{\boldsymbol{u}}, \bar{\boldsymbol{M}}(\hat{\boldsymbol{u}} - \bar{\boldsymbol{u}})). \tag{13.28}$$

Proof. The Mahalanobis projection $\tilde{\boldsymbol{a}}_\alpha$ of $\hat{\boldsymbol{a}}_\alpha \in \hat{\mathcal{S}}$ onto $\bar{\mathcal{S}}$ is given as follows (see eq. (5.17)):

$$\tilde{\boldsymbol{a}}_\alpha = \hat{\boldsymbol{a}}_\alpha - V[\boldsymbol{a}_\alpha] \sum_{k,l=1}^{L} \tilde{W}_\alpha^{(kl)} F^{(k)}(\hat{\boldsymbol{a}}_\alpha, \bar{\boldsymbol{u}}) \nabla_{\boldsymbol{a}} \tilde{F}_\alpha^{(l)}, \tag{13.29}$$

$$(\tilde{W}_\alpha^{(kl)}) = \left((\nabla_{\boldsymbol{a}} \tilde{F}_\alpha^{(k)}, V[\boldsymbol{a}_\alpha] \nabla_{\boldsymbol{a}} \tilde{F}_\alpha^{(l)}) \right)^-. \tag{13.30}$$

Here, $\nabla_{\boldsymbol{a}} \tilde{F}_\alpha^{(k)}$ is the abbreviation of $\nabla_{\boldsymbol{a}} F^{(k)}(\tilde{\boldsymbol{a}}_\alpha, \bar{\boldsymbol{u}})$. Since $\hat{\boldsymbol{a}}_\alpha \in \hat{\mathcal{S}}$, we have

$$F^{(k)}(\hat{\boldsymbol{a}}_\alpha, \hat{\boldsymbol{u}}) = 0, \qquad k = 1, ..., L. \tag{13.31}$$

Letting $\Delta \hat{\boldsymbol{u}} = \hat{\boldsymbol{u}} - \bar{\boldsymbol{u}}$ and taking a linear approximation, we obtain

$$F^{(k)}(\hat{\boldsymbol{a}}_\alpha, \bar{\boldsymbol{u}}) = F^{(k)}(\hat{\boldsymbol{a}}_\alpha, \hat{\boldsymbol{u}} - \Delta \hat{\boldsymbol{u}}) = -(\nabla_{\boldsymbol{u}} \hat{F}_\alpha^{(k)}, \Delta \hat{\boldsymbol{u}}), \tag{13.32}$$

where $\nabla_{\boldsymbol{u}}$ denotes differentiation with respect to the parameter $\boldsymbol{u}$, and $\nabla_{\boldsymbol{u}} \hat{F}_\alpha^{(k)}$ is the abbreviation of $\nabla_{\boldsymbol{u}} F^{(k)}(\hat{\boldsymbol{a}}_\alpha, \hat{\boldsymbol{u}})$. Since eq. (13.32) is a linear approximation in $\Delta \hat{\boldsymbol{u}}$, we can replace $\nabla_{\boldsymbol{u}} \hat{F}_\alpha^{(k)}$ by $\nabla_{\boldsymbol{u}} \bar{F}_\alpha^{(k)}$ (= the abbreviation of

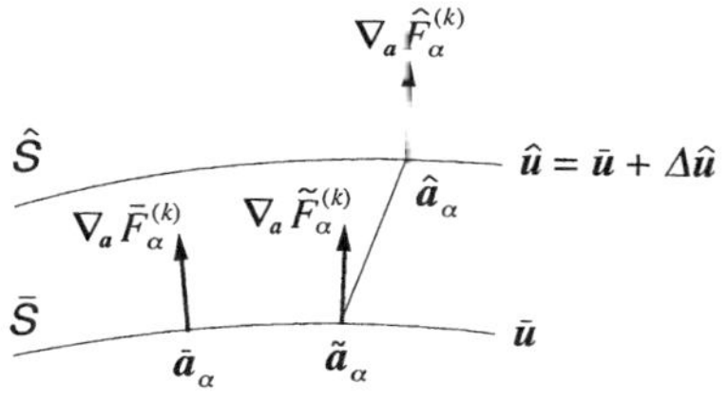

Fig. 13.8. Accuracy of the optimal fitting

$\nabla_{\boldsymbol{u}} F^{(k)}(\bar{\boldsymbol{a}}_\alpha, \bar{\boldsymbol{u}}))$ with errors of $O(\Delta\hat{\boldsymbol{u}})^2$ (see Fig. 13.8). Hence, eq. (13.29) is written to a first approximation as

$$\hat{\boldsymbol{a}}_\alpha - \tilde{\boldsymbol{a}}_\alpha = V[\boldsymbol{a}_\alpha] \sum_{k,l=1}^{L} \tilde{W}^{(kl)}_\alpha \nabla_{\boldsymbol{a}} \tilde{F}^{(k)}_\alpha \nabla_{\boldsymbol{u}} \bar{F}^{(l)\top}_\alpha \Delta\hat{\boldsymbol{u}}. \tag{13.33}$$

From this, we see that

$$\begin{aligned}
&\sum_{\alpha=1}^{N} \|\hat{\boldsymbol{a}}_\alpha - \tilde{\boldsymbol{a}}_\alpha\|^2_{V[\boldsymbol{a}_\alpha]} \\
&= \sum_{\alpha=1}^{N} (V[\boldsymbol{a}_\alpha] \sum_{k,l=1}^{L} \tilde{W}^{(kl)}_\alpha \nabla_{\boldsymbol{a}} \tilde{F}^{(k)}_\alpha \nabla_{\boldsymbol{u}} \bar{F}^{(l)\top}_\alpha \Delta\hat{\boldsymbol{u}}, \\
&\quad V[\boldsymbol{a}_\alpha]^- V[\boldsymbol{a}_\alpha] \sum_{m,n=1}^{L} \tilde{W}^{(mn)}_\alpha \nabla_{\boldsymbol{a}} \tilde{F}^{(m)}_\alpha \nabla_{\boldsymbol{u}} \bar{F}^{(n)\top}_\alpha \Delta\hat{\boldsymbol{u}}) \\
&= (\Delta\hat{\boldsymbol{u}}, \Big(\sum_{\alpha=1}^{N} \sum_{k,l,m,n=1}^{L} \tilde{W}^{(kl)}_\alpha \tilde{W}^{(mn)}_\alpha \nabla_{\boldsymbol{u}} \bar{F}^{(l)}_\alpha \nabla_{\boldsymbol{a}} \tilde{F}^{(k)\top}_\alpha \\
&\quad V[\boldsymbol{a}_\alpha] V[\boldsymbol{a}_\alpha]^- V[\boldsymbol{a}_\alpha] \nabla_{\boldsymbol{a}} \tilde{F}^{(m)}_\alpha \nabla_{\boldsymbol{u}} \bar{F}^{(n)\top}_\alpha \Big) \Delta\hat{\boldsymbol{u}}) \\
&= (\Delta\hat{\boldsymbol{u}}, \Big(\sum_{\alpha=1}^{N} \sum_{l,n=1}^{L} \left(\sum_{k,m=1}^{L} \tilde{W}^{(lk)}_\alpha (\nabla_{\boldsymbol{a}} \tilde{F}^{(k)}_\alpha, V[\boldsymbol{a}_\alpha] \nabla_{\boldsymbol{a}} \tilde{F}^{(m)}_\alpha) \tilde{W}^{(mn)}_\alpha \right) \\
&\quad \nabla_{\boldsymbol{u}} \bar{F}^{(l)}_\alpha \nabla_{\boldsymbol{u}} \bar{F}^{(n)\top}_\alpha \Big) \Delta\hat{\boldsymbol{u}}) \\
&= (\Delta\hat{\boldsymbol{u}}, \left(\sum_{\alpha=1}^{N} \sum_{l,n=1}^{L} \tilde{W}^{(ln)}_\alpha \nabla_{\boldsymbol{u}} \bar{F}^{(l)}_\alpha \nabla_{\boldsymbol{u}} \bar{F}^{(n)\top}_\alpha \right) \Delta\hat{\boldsymbol{u}}),
\end{aligned} \tag{13.34}$$

where we have used the defining equation (13.30) of the (LL)-matrix $\tilde{\boldsymbol{W}}_\alpha = (\tilde{W}^{(kl)}_\alpha)$ and the identities $V[\boldsymbol{a}_\alpha]V[\boldsymbol{a}_\alpha]^- V[\boldsymbol{a}_\alpha] = V[\boldsymbol{a}_\alpha]$ and $\tilde{\boldsymbol{W}}_\alpha \tilde{\boldsymbol{W}}^-_\alpha \tilde{\boldsymbol{W}}_\alpha =$

$\tilde{\boldsymbol{W}}_\alpha$. To a first approximation, $\tilde{W}_\alpha^{(kl)}$ can be replaced by $\bar{W}_\alpha^{(kl)}$ defined by eq. (13.27). Since $\Delta\hat{\boldsymbol{u}} \in T_{\bar{\boldsymbol{u}}}(\mathcal{U})$ to a first approximation and hence $\boldsymbol{P}_{\bar{\boldsymbol{u}}}^{\mathcal{U}}\Delta\hat{\boldsymbol{u}} = \Delta\hat{\boldsymbol{u}}$, we can write eq. (13.34) as $(\Delta\hat{\boldsymbol{u}}, \bar{\boldsymbol{M}}\Delta\hat{\boldsymbol{u}})$. □

Proposition 13.6

$$I(\mathcal{S}) = E[J[\{\boldsymbol{a}_\alpha\}, \{\hat{\boldsymbol{a}}_\alpha\}, \hat{\mathcal{S}}]] + 2(dN + n'). \tag{13.35}$$

Proof. As shown in Section 7.1.3, the a posteriori covariance matrix of the optimal estimate $\hat{\boldsymbol{u}}$ is given by the generalized inverse of $\bar{\boldsymbol{M}}$ (see eq. (7.29)):

$$\bar{V}[\hat{\boldsymbol{u}}] = \bar{\boldsymbol{M}}^-. \tag{13.36}$$

Since this matrix generally has rank n', the quadratic form

$$(\hat{\boldsymbol{u}} - \bar{\boldsymbol{u}}, \bar{\boldsymbol{M}}(\hat{\boldsymbol{u}} - \bar{\boldsymbol{u}})) = (\hat{\boldsymbol{u}} - \bar{\boldsymbol{u}}, \bar{V}[\hat{\boldsymbol{u}}]^-(\hat{\boldsymbol{u}} - \bar{\boldsymbol{u}})) \tag{13.37}$$

is a χ^2 variable with n' degrees of freedom (see eq. (3.61)). Hence, its expectation is n'. From Lemmas 13.1 and 13.2, we obtain eq. (13.35). □

13.4 Geometric Information Criterion

13.4.1 Model selection by AIC

Proposition 13.6 implies that

$$AIC(\mathcal{S}) = J[\{\boldsymbol{a}_\alpha\}, \{\hat{\boldsymbol{a}}_\alpha\}, \hat{\mathcal{S}}] + 2(dN + n') \tag{13.38}$$

is an unbiased estimator of the expected residual $I(\mathcal{S})$. We call this quantity the *geometric information criterion*, or *geometric AIC* for short, and use it as a measure of the goodness of the model: if $AIC(\mathcal{S}_1) < AIC(\mathcal{S}_2)$ for models $\mathcal{S}_1$ and $\mathcal{S}_2$, we prefer model $\mathcal{S}_1$ because $\mathcal{S}_1$ is expected to have more predicting capacity than model $\mathcal{S}_2$.

Eq. (13.38) formally coincides with eq. (3.172) up to additive constants. This is easily seen if we note the following:

1. The residual $J[\{\boldsymbol{a}_\alpha\}, \{\hat{\boldsymbol{a}}_\alpha\}, \hat{\mathcal{S}}]$ equals -2 times the log-likelihood of the Gaussian probability density plus some additive constants.

2. The manifold $\hat{\mathcal{S}}$ is specified by n' parameters. If a d-dimensional curvilinear coordinate system is defined in $\hat{\mathcal{S}}$, the maximum likelihood estimator $\hat{\boldsymbol{a}}_\alpha \in \hat{\mathcal{S}}$ is specified by d coordinates; in total dN parameters are necessary for $\{\hat{\boldsymbol{a}}_\alpha\}$. Thus, the number of parameters for describing the solution of Problem 13.1 is $dN + n'$.

However, there is a marked difference: eq. (3.172) is obtained in the *asymptotic limit* $N \to \infty$ by applying the *law of large numbers* and the *central limit theorem*, while eq. (13.38) is obtained in the limit of *small noise* for a *fixed* number N of the data[3]. This leads to the following distinctive features of the geometric AIC as compared with the usual AIC:

- The degree of freedom n' of the model has no significant effect for the geometric AIC if the number N of data is large, whereas it plays a dominant role in the usual AIC.
- The dimension d of the model manifold plays a dominant role in the geometric AIC, while no such geometric concepts are involved in the usual AIC.
- The number N of data explicitly appears in the expression for the geometric AIC, but it does not in the usual AIC.

Eq. (13.38) is not a convenient form for actual applications, because computing the residual $J[\{\boldsymbol{a}_\alpha\}, \{\hat{\boldsymbol{a}}_\alpha\}, \hat{\mathcal{S}}]$ requires knowledge of the covariance matrices $V[\boldsymbol{a}_\alpha]$. We decompose it into the *noise level* ϵ and the *normalized covariance matrix* $V_0[\boldsymbol{a}_\alpha]$ in the form

$$V[\boldsymbol{a}_\alpha] = \epsilon^2 V_0[\boldsymbol{a}_\alpha], \qquad (13.39)$$

and define the *normalized residual* by

$$J_0[\hat{\mathcal{S}}] = \sum_{\alpha=1}^{N} (\boldsymbol{a}_\alpha - \hat{\boldsymbol{a}}_\alpha, V_0[\boldsymbol{a}_\alpha]^-(\boldsymbol{a}_\alpha - \hat{\boldsymbol{a}}_\alpha)). \qquad (13.40)$$

Multiplying eq. (13.38) by ϵ^2, we define the *normalized geometric AIC* of model $\mathcal{S}$ by

$$AIC_0(\mathcal{S}) = J_0[\hat{\mathcal{S}}] + 2(dN + n')\epsilon^2. \qquad (13.41)$$

In the following, we call $AIC_0(\mathcal{S})$ and $J_0[\hat{\mathcal{S}}]$ simply the *AIC* and the *residual*, respectively.

Given a set of N data $\{\boldsymbol{a}_\alpha\}$ and two models $\mathcal{S}_1$ and $\mathcal{S}_2$, we regard model $\mathcal{S}_1$ better than $\mathcal{S}_2$ if $AIC_0(\mathcal{S}_1) < AIC_0(\mathcal{S}_2)$. If model $\mathcal{S}_1$ has dimension d_1, codimension r_1, and n'_1 degrees of freedom, and model $\mathcal{S}_2$ has dimension d_2, codimension r_2, and n'_2 degrees of freedom, this condition is written as

$$J_0[\hat{\mathcal{S}}_1] - J_0[\hat{\mathcal{S}}_2] < 2\left((d_2 - d_1)N + (n'_2 - n'_1)\right)\epsilon^2. \qquad (13.42)$$

Example 13.4 Consider Example 13.1. If each image coordinate is perturbed by independent Gaussian noise of mean 0 and standard deviation ϵ, the AICs of the models defined there are:

$$AIC_0(p) = J_0[\hat{p}] + 4\epsilon^2, \qquad AIC_0(p') = J_0[\hat{p}'], \qquad (13.43)$$

[3] This difference will be discussed in a more general framework in the next chapter.

$$AIC_0(l) = J_0[\hat{l}] + 2(N+2)\epsilon^2, \qquad AIC_0(l') = J_0[\hat{l}'] + 2(N+1)\epsilon^2, \tag{13.44}$$

$$AIC_0(\mathcal{X}) = 4N\epsilon^2. \tag{13.45}$$

- Comparing model l with model $\mathcal{X}$, we can infer that the true positions are collinear if

$$J_0[\hat{l}] < 2(N-2)\epsilon^2. \tag{13.46}$$

We can then infer that the image line on which the true positions lie passes through the image origin if

$$J_0[\hat{l}'] - J_0[\hat{l}] < 2\epsilon^2. \tag{13.47}$$

- Comparing model p with model l, we can infer that the true positions are identical if

$$J_0[\hat{p}] - J_0[\hat{l}] < 2N\epsilon^2. \tag{13.48}$$

We can then infer that the true position is at the image origin if

$$J_0[\hat{p}'] - J_0[\hat{p}] < 4\epsilon^2. \tag{13.49}$$

Let (x_α, y_α) be the image coordinates of the data, and $(\hat{x}, \hat{y})$ their sample average. The above condition can be rewritten as follows:

$$\sqrt{\hat{x}^2 + \hat{y}^2} < \frac{2\epsilon}{\sqrt{N}}. \tag{13.50}$$

In other words, the true position is inferred to be at the image origin if the sample average is within distance $2\epsilon/\sqrt{N}$ from the image origin.

Example 13.5 Consider Example 13.2. If each coordinate is perturbed by independent Gaussian noise of mean 0 and standard deviation ϵ, the AICs of the models defined there are:

$$AIC_0(P) = J_0[\hat{P}] + 6\epsilon^2, \qquad AIC_0(P') = J_0[\hat{P}'], \tag{13.51}$$

$$AIC_0(L) = J_0[\hat{L}] + 2(N+4)\epsilon^2, \quad AIC_0(L') = J_0[\hat{L}'] + 2(N+2)\epsilon^2, \tag{13.52}$$

$$AIC_0(\Pi) = J_0[\hat{\Pi}] + 2(2N+3)\epsilon^2, \quad AIC_0(\Pi') = J_0[\hat{L}'] + 4(N+1)\epsilon^2, \tag{13.53}$$

$$AIC_0(\mathcal{R}^3) = 6N\epsilon^2. \tag{13.54}$$

- Comparing model Π with model V, we can infer that the true positions are coplanar if

$$J_0[\hat{\Pi}] < 2(N-3)\epsilon^2. \tag{13.55}$$

We can then infer that the space plane on which the true positions lie passes through the coordinate origin if

$$J_0[\hat{\Pi}'] - J_0[\hat{\Pi}] < 2\epsilon^2. \tag{13.56}$$

- Comparing model L with model Π, we can infer that the true positions are collinear if
$$J_0[\hat{L}] - J_0[\hat{\Pi}] < 2(N-1)\epsilon^2. \tag{13.57}$$
We can then infer that the space line on which the true positions lie passes through the coordinate origin if
$$J_0[\hat{L}'] - J_0[\hat{L}] < 4\epsilon^2. \tag{13.58}$$

- Comparing model P with model L, we can infer that the true positions are identical if
$$J_0[\hat{P}] - J_0[\hat{L}] < 2(N+1)\epsilon^2. \tag{13.59}$$
We can then infer that the true position is at the coordinate origin if
$$J_0[\hat{P}'] - J_0[\hat{P}] < 6\epsilon^2. \tag{13.60}$$
Let $\hat{\boldsymbol{r}}$ be the sample average of the data $\{\boldsymbol{r}_\alpha\}$. The above condition can be rewritten as
$$\|\hat{\boldsymbol{r}}\| < \frac{\sqrt{6}\epsilon}{\sqrt{N}}. \tag{13.61}$$
In other words, the true position is inferred to be at the coordinate origin if the sample average is within distance $\sqrt{6}\epsilon/\sqrt{N}$ from the coordinate origin.

13.4.2 Model comparison by AIC

In order to apply the AIC criterion as described in the preceding subsection, we need to know the noise level ϵ, which is very difficult to predict. Note that *if model $\mathcal{S}$ is correct*, an unbiased estimator of ϵ^2 is obtained in the form

$$\hat{\epsilon}^2 = \frac{J_0[\hat{\mathcal{S}}]}{rN - n'}, \tag{13.62}$$

as shown in Section 7.1.4. However, we need the true noise level ϵ to judge if the model $\mathcal{S}$ is correct. This difficulty can be avoided if we focus on *comparing* two models such that one *implies* the other.

Let $\mathcal{S}_1$ be a model of dimension d_1 and codimension r_1 with n'_1 degrees of freedom, and $\mathcal{S}_2$ a model of dimension d_2 and codimension r_2 with n'_2 degrees of freedom. Suppose model $\mathcal{S}_2$ is obtained by adding an additional constraint to model $\mathcal{S}_1$. We say that model $\mathcal{S}_2$ is *stronger* than model $\mathcal{S}_1$, or model $\mathcal{S}_1$ is *weaker* than model $\mathcal{S}_2$, and write

$$\mathcal{S}_2 \succ \mathcal{S}_1. \tag{13.63}$$

Then, $J_0[\hat{\mathcal{S}}_2] \geq J_0[\hat{\mathcal{S}}_1]$ for whatever data $\{\boldsymbol{a}_\alpha\}$. This confirms our observation made earlier that the residual alone cannot measure the goodness of the model.

Suppose $\mathcal{S}_1$ is a general model which is assumed to be correct. Then, the squared noise level ϵ^2 is estimated by eq. (13.62) as long as $rN - n' \neq 0$. Substituting it to ϵ^2 in the expression for the geometric AIC, we obtain

$$AIC_0(\mathcal{S}_1) = J_0[\hat{\mathcal{S}}_1] + \frac{2(d_1N + n_1')}{r_1N - n_1'} J_0[\hat{\mathcal{S}}_1], \tag{13.64}$$

$$AIC_0(\mathcal{S}_2) = J_0[\hat{\mathcal{S}}_2] + \frac{2(d_2N + n_2')}{r_1N - n_1'} J_0[\hat{\mathcal{S}}_1]. \tag{13.65}$$

If $AIC_0(\mathcal{S}_2) < AIC_0(\mathcal{S}_1)$, the predicting capability is expected to increase by replacing the general model $\mathcal{S}$ by the strong model $\mathcal{S}_2$. The condition for this is

$$\frac{J_0[\hat{\mathcal{S}}_2]}{J_0[\hat{\mathcal{S}}_1]} < 1 + \frac{2(d_1 - d_2)N + 2(n_1' - n_2')}{r_1N - n_1'}. \tag{13.66}$$

In terms of the estimators $\hat{\epsilon}_1^2$ and $\hat{\epsilon}_2^2$ defined in the form of eq. (13.62) for models $\mathcal{S}_1$ and $\mathcal{S}_2$, respectively, the above condition can be written as follows:

$$\frac{\hat{\epsilon}_2^2}{\hat{\epsilon}_1^2} < \frac{(r_1 + 2(d_1 - d_2))N + n_1' - 2n_2'}{r_2N - n_2'}. \tag{13.67}$$

Example 13.6 Consider Example 13.1. We see that

$$p' \succ p, \qquad l' \succ l, \qquad \succ l \succ \mathcal{X}. \tag{13.68}$$

We assume that each image coordinate is perturbed by independent Gaussian noise of mean 0 and standard deviation ϵ.

- If the true positions are known to be collinear, we can infer that the image line on which the true positions lie passes through the image origin if

$$\frac{J_0[\hat{l}']}{J_0[\hat{l}]} < 1 + \frac{2}{N-2}, \tag{13.69}$$

  and infer that the true positions are identical if

$$\frac{J_0[\hat{p}]}{J_0[\hat{l}]} < 3 + \frac{2}{N-2}. \tag{13.70}$$

- If the true positions are known to be identical, we can infer that the true position is at the image origin if

$$\frac{J_0[\hat{p}']}{J_0[\hat{p}]} < 1 + \frac{2}{N-1}. \tag{13.71}$$

  In terms of the sample average $(\hat{x}, \hat{y})$, this condition can be written as follows:

$$\sqrt{\hat{x}^2 + \hat{y}^2} < \sqrt{\frac{2\sum_{\alpha=1}^N (x_\alpha^2 + y_\alpha^2)}{N(N+1)}}. \tag{13.72}$$

Note that application of the comparison criterion (13.67) is somewhat limited as compared with the direct use of the AIC criterion (13.42). For example, one cannot infer collinearity by comparing model l and model $\mathcal{X}$, since the residual for $\mathcal{X}$ is identically zero.

Example 13.7 Consider Example 13.2. We see that

$$P' \succ P, \quad L' \succ L, \quad \Pi' \succ \Pi, \quad P \succ L \succ \mathcal{R}^3. \tag{13.73}$$

We assume that each coordinate is perturbed by independent Gaussian noise of mean 0 and standard deviation ϵ.

- If the true positions are known to be coplanar, we can infer that the space plane on which the true positions lie passes through the coordinate origin when

$$\frac{J_0[\hat{\Pi}']}{J_0[\hat{\Pi}]} < 1 + \frac{2}{N-3}, \tag{13.74}$$

  and infer that the true positions are collinear when

$$\frac{J_0[\hat{L}]}{J_0[\hat{\Pi}]} < 3 + \frac{4}{N-3}. \tag{13.75}$$

- If the true positions are known to be collinear, we can infer that the space line on which the true positions lie passes through the coordinate origin when

$$\frac{J_0[\hat{L}']}{J_0[\hat{L}]} < 1 + \frac{2}{N-2}, \tag{13.76}$$

  and infer that the true positions are identical when

$$\frac{J_0[\hat{P}]}{J_0[\hat{P}]} < 2 + \frac{3}{N-2}. \tag{13.77}$$

- If the true positions are known to be identical, we can infer that the true position is at the coordinate origin when

$$\frac{J_0[\hat{P}']}{J_0[\hat{P}]} < 1 + \frac{2}{N-1}. \tag{13.78}$$

  In terms of the sample average $\hat{\boldsymbol{r}}$, this condition can be written as follows:

$$\|\hat{\boldsymbol{r}}\| < \sqrt{\frac{2\sum_{\alpha=1}^{N} \|\boldsymbol{r}_\alpha\|^2}{N(N+1)}}. \tag{13.79}$$

13.4.3 Model selection vs. testing of hypotheses

The model selection criterion given by eq. (13.66) has a positive implication in contrast to the negative meaning of the statistical *testing of hypotheses*, according to which the procedure is given as follows. Since $J_0[\hat{\mathcal{S}}_2]/\epsilon^2$ is a χ^2 variable with $r_2N - n_2'$ degrees of freedom, the *hypothesis* that "model $\mathcal{S}_2$ is correct" is rejected if

$$\frac{J_0[\hat{\mathcal{S}}_2]}{\epsilon^2} > \chi^2_{r_2N-n_2',a} \tag{13.80}$$

with *significance level* $a\%$, where $\chi^2_{p,a}$ is the *upper* $a\%$ *point* of the χ^2 distribution with p degrees of freedom. If the square noise level ϵ^2 is approximated by the estimator $\hat{\epsilon}^2$ given by eq. (13.62), we can rewrite (13.80) as[4]

$$\frac{J_0[\hat{\mathcal{S}}_2]}{J_0[\hat{\mathcal{S}}_1]} > \frac{\chi^2_{r_2N-n_2',a}}{r_1N - n_1'}. \tag{13.81}$$

The interpretation of this test is that if eq. (13.81) holds, the hypothesis that the model is $\mathcal{S}_2$ is very questionable with *confidence level* $(100 - a)\%$ because if the hypothesis is true, we are observing a very rare event that occurs only with a probability $a\%$. Hence, we decide that *there exists no reason to favor model* $\mathcal{S}_2$ *over* $\mathcal{S}_1$. In other words, a statistical test can only *reject* a hypothesis when the data do not support it within a specified allowance threshold. Its ultimate purpose is to *negate* a hypothesis (hence called the *null hypothesis*) in favor of a default hypothesis (called the *alternative hypothesis*).

After all, *any* hypothesis is rejected in the presence of noise if the significance level is lowered (or the confidence level is raised); the judgement is not definitive in this sense, and it does not address the issue of choosing one model in favor of another. In contrast, the criterion given by eq. (13.66) gives a positive and definitive assertion that model $\mathcal{S}_2$ is *preferable* to $\mathcal{S}_1$ with regard to the predicting capability; it requires no knowledge about the noise magnitude and no arbitrarily set thresholds.

13.5 3-D Reconstruction by Stereo Vision

13.5.1 General model

Consider a stereo system. Let $\{\boldsymbol{h}, \boldsymbol{R}\}$ be its motion parameters. Suppose we observe N pairs of corresponding image points $\boldsymbol{x}_\alpha$ and $\boldsymbol{x}'_\alpha$, $\alpha = 1, ..., N$. We can regard them as *six*-dimensional data $\boldsymbol{a}_\alpha = \boldsymbol{x}_\alpha \oplus \boldsymbol{x}'_\alpha \in \mathcal{R}^6$ sampled from

---

[4]The left-hand side equals the ratio of the logarithmic likelihoods of the two models, so this test belongs to a class called the *logarithmic likelihood ratio test*.

the *four*-dimensional data space

$$\mathcal{A} = \{ \begin{pmatrix} x \\ y \\ 1 \\ x' \\ y' \\ 1 \end{pmatrix} | x, y, x', y' \in \mathcal{R}\} \subset \mathcal{R}^6. \tag{13.82}$$

The condition that image points $\boldsymbol{x}$ and $\boldsymbol{x}'$ correspond to each other is given by the following *epipolar equation* (see eq. (6.8)):

$$(\boldsymbol{x}, \boldsymbol{G}\boldsymbol{x}') = 0. \tag{13.83}$$

The *essential matrix* $\boldsymbol{G}$ is defined as follows (see eq. (6.7)):

$$\boldsymbol{G} = \boldsymbol{h} \times \boldsymbol{R}. \tag{13.84}$$

The epipolar equation (13.83) describes the condition that the lines of sight of $\boldsymbol{x}$ and $\boldsymbol{x}'$ meet in the scene (see Fig. 6 4), or equivalently image point $\boldsymbol{x}$ is *on the epipolar of* $\boldsymbol{x}'$ in the first image and $\boldsymbol{x}'$ is *on the epipolar of* $\boldsymbol{x}$ in the second image (see Section 6.1.2).

The epipolar equation (13.83) defines a *three*-dimensional manifold $\mathcal{S}$ in the four-dimensional data space $\mathcal{A}$. It follows that 3-D reconstruction by stereo vision can be viewed as the following problem:

Problem 13.2 *Estimate the true positions* $\bar{\boldsymbol{a}}_\alpha = \bar{\boldsymbol{x}}_\alpha \oplus \bar{\boldsymbol{x}}'_\alpha$ *of the data* $\boldsymbol{a}_\alpha = \boldsymbol{x}_\alpha \oplus \boldsymbol{x}'_\alpha \in \mathcal{A}$ *in such a way that* $\bar{\boldsymbol{a}}_\alpha \in \mathcal{S}$.

Since no free parameters are involved, the manifold $\mathcal{S}$ is a model of dimension 3, codimension 1, and *zero* degrees of freedom; the maximum likelihood estimator $\hat{\mathcal{S}}$ of $\mathcal{S}$ is $\mathcal{S}$ itself. Let $V[\boldsymbol{x}_\alpha]$ and $V[\boldsymbol{x}'_\alpha]$ be the covariance matrices of $\boldsymbol{x}_\alpha$ and $\boldsymbol{x}'_\alpha$, respectively, and decompose them into the noise level ϵ and the normalized covariance matrices $V_0[\boldsymbol{x}_\alpha]$ and $V_0[\boldsymbol{x}'_\alpha]$ in the form

$$V[\boldsymbol{x}_\alpha] = \epsilon^2 V_0[\boldsymbol{x}_\alpha], \qquad V[\boldsymbol{x}'_\alpha] = \epsilon^2 V_0[\boldsymbol{x}'_\alpha]. \tag{13.85}$$

The noise is assumed to be Gaussian and independent for each datum. The AIC of this model is

$$AIC_0(\mathcal{S}) = J_0[\hat{\mathcal{S}}] + 6N\epsilon^2. \tag{13.86}$$

As shown in Section 6.2.2, the residual $J_0[\hat{\mathcal{S}}]$ is computed as follows[5] (see eq. (6.22)):

$$J_0[\hat{\mathcal{S}}] = \sum_{\alpha=1}^{N} \frac{(\boldsymbol{x}_\alpha, \boldsymbol{G}\boldsymbol{x}'_\alpha)^2}{(\boldsymbol{x}'_\alpha, \boldsymbol{G}^\top V_0[\boldsymbol{x}_\alpha]\boldsymbol{G}\boldsymbol{x}'_\alpha) + (\boldsymbol{x}_\alpha, \boldsymbol{G}V_0[\boldsymbol{x}'_\alpha]\boldsymbol{G}^\top \boldsymbol{x}_\alpha)}. \tag{13.87}$$

[5] Theoretically, $\mathbf{x}_\alpha$ and $\mathbf{x}'_\alpha$ in the denominator should be their true values $\bar{\mathbf{x}}_\alpha$ and $\bar{\mathbf{x}}'_\alpha$. They could be approximated by their corrected values $\hat{\mathbf{x}}_\alpha$ and $\hat{\mathbf{x}}'_\alpha$. Since $J_0[\hat{\mathcal{S}}] = O(\epsilon^2)$, the use of the data $\mathbf{x}_\alpha$ and $\mathbf{x}'_\alpha$ results in only a higher order difference.

If this model is correct, $J_0[\hat{\mathcal{S}}]/\epsilon^2$ is a χ^2 variable with N degrees of freedom. Hence, the squared noise level ϵ^2 can be estimated by

$$\hat{\epsilon}^2 = \frac{J_0[\hat{\mathcal{S}}]}{N}. \tag{13.88}$$

13.5.2 Planar surface model

If the object is a planar surface $(\boldsymbol{n}, \boldsymbol{r}) = d$, the constraint is given by

$$\boldsymbol{x}' \times \boldsymbol{A}\boldsymbol{x} = \boldsymbol{0}, \tag{13.89}$$

where the matrix $\boldsymbol{A}$ is defined as follows (see eqs. (6.67) and (6.71)):

$$\boldsymbol{A} = \boldsymbol{R}^\top(\boldsymbol{h}\boldsymbol{n} - d\boldsymbol{I}). \tag{13.90}$$

Eq. (13.89) describes the condition that the lines of sight of $\boldsymbol{x}$ and $\boldsymbol{x}'$ meet on the space plane $(\boldsymbol{n}, \boldsymbol{r}) = d$ (see Fig. 6.12). Since only two component equations of eq. (13.89) are independent, it defines a *two*-dimensional manifold $\mathcal{S}_\Pi$ in the *four*-dimensional data space $\mathcal{A}$. It follows that 3-D reconstruction of a planar surface by stereo vision can be identified with the following problem:

Problem 13.3 *Estimate the manifold $\mathcal{S}_\Pi$ and the true positions $\bar{\boldsymbol{a}}_\alpha = \bar{\boldsymbol{x}}_\alpha \oplus \bar{\boldsymbol{x}}'_\alpha$ of the data $\boldsymbol{a}_\alpha = \boldsymbol{x}_\alpha \oplus \boldsymbol{x}'_\alpha \in \mathcal{A}$ in such a way that $\bar{\boldsymbol{a}}_\alpha \in \mathcal{S}_\Pi$.*

The unknown surface parameters $\{\boldsymbol{n}, d\}$ have three degrees of freedom[6], so the manifold $\mathcal{S}_\Pi$ is a model of dimension 2, codimension 2, and *three* degrees of freedom. Let $\hat{\mathcal{S}}_\Pi$ be the maximum likelihood estimator of $\mathcal{S}_\Pi$. The AIC of this model is

$$AIC_0(\mathcal{S}_\Pi) = J_0[\hat{\mathcal{S}}_\Pi] + 2(2N+3)\epsilon^2. \tag{13.91}$$

Let $\{\hat{\boldsymbol{n}}, \hat{d}\}$ be the maximum likelihood estimators of the surface parameters $\{\boldsymbol{n}, d\}$; the computational scheme for this estimation by renormalization is described in Section 10.4. The residual $J_0[\hat{\mathcal{S}}_\Pi]$ is computed as follows[7] (see eqs. (6.77) and (6.80)):

$$J_0[\hat{\mathcal{S}}_\Pi] = \sum_{\alpha=1}^{N} (\boldsymbol{x}'_\alpha \times \hat{\boldsymbol{A}}\boldsymbol{x}_\alpha, \hat{\boldsymbol{W}}_\alpha(\boldsymbol{x}'_\alpha \times \hat{\boldsymbol{A}}\boldsymbol{x}_\alpha)), \tag{13.92}$$

$$\hat{\boldsymbol{W}}_\alpha = \left(\boldsymbol{x}'_\alpha \times \hat{\boldsymbol{A}}V_0[\boldsymbol{x}_\alpha]\hat{\boldsymbol{A}}^\top \times \boldsymbol{x}'_\alpha + (\hat{\boldsymbol{A}}\boldsymbol{x}_\alpha) \times V_0[\boldsymbol{x}'_\alpha] \times (\hat{\boldsymbol{A}}\boldsymbol{x}_\alpha)\right)^-_2. \tag{13.93}$$

Here, $\hat{\boldsymbol{A}}$ is the estimate of matrix $\boldsymbol{A}$ obtained by replacing $\{\boldsymbol{n}, d\}$ by their maximum likelihood estimators $\{\hat{\boldsymbol{n}}, \hat{d}\}$ in eq. (13.90). If this model is correct,

---

[6] Vector $\mathbf{n}$ is normalized into a unit vector; see Section 4.3.1.

[7] The approximation used in the general model is also used.

$J_0[\hat{\mathcal{S}}_{\rm II}]/\epsilon^2$ is a χ^2 variable with $2N-3$ degrees of freedom. Hence, the squared noise level ϵ^2 can be estimated by

$$\hat{\epsilon}_{\rm II}^2 = \frac{J_0[\hat{\mathcal{S}}_{\rm II}]}{2N-3}. \tag{13.94}$$

13.5.3 Infinity model

If the object we are viewing is infinitely far away, we have

$$\boldsymbol{x} \times \boldsymbol{R}\boldsymbol{x}' = \boldsymbol{0}, \tag{13.95}$$

which describes the condition that the lines of sight of $\boldsymbol{x}$ and $\boldsymbol{x}'$ are parallel (see eq. (6.100)). Like eq. (13.89), eq. (13.89) are independent, it defines a *two*-dimensional manifold $\mathcal{S}_\infty$ in the *four*-dimensional data space $\mathcal{A}$. Consider the following problem:

Problem 13.4 *Estimate the true positions $\bar{\boldsymbol{a}}_\alpha = \bar{\boldsymbol{x}}_\alpha \oplus \bar{\boldsymbol{x}}'_\alpha$ of the data $\boldsymbol{a}_\alpha = \boldsymbol{x}_\alpha \oplus \boldsymbol{x}'_\alpha \in \mathcal{A}$ in such a way that $\bar{\boldsymbol{a}}_\alpha \in \mathcal{S}_\infty$.*

Since no free parameters are involved, the manifold $\mathcal{S}_\infty$ is a model of dimension 2, codimension 2, and *zero* degrees of freedom; the maximum likelihood estimator $\hat{\mathcal{S}}_\infty$ of $\mathcal{S}_\infty$ is $\mathcal{S}_\infty$ itself. The AIC of this model is

$$AIC_0(\mathcal{S}_\infty) = J_0[\hat{\mathcal{S}}_\infty] + 4N\epsilon^2. \tag{13.96}$$

The residual $J_0[\hat{\mathcal{S}}_\infty]$ is computed as follows[8] (see eqs. (6.105) and (6.107)):

$$J_0[\hat{\mathcal{S}}_\infty] = \sum_{\alpha=1}^{N} (\boldsymbol{x}_\alpha \times \boldsymbol{R}\boldsymbol{x}'_\alpha, \hat{\boldsymbol{W}}_\alpha(\boldsymbol{x}_\alpha \times \boldsymbol{R}\boldsymbol{x}'_\alpha)), \tag{13.97}$$

$$\hat{\boldsymbol{W}}_\alpha = \left(\boldsymbol{x}_\alpha \times \boldsymbol{R}V_0[\boldsymbol{x}_\alpha]\boldsymbol{R}^\top \times \boldsymbol{x}_\alpha + (\boldsymbol{R}\boldsymbol{x}'_\alpha) \times V_0[\boldsymbol{x}'_\alpha] \times (\boldsymbol{R}\boldsymbol{x}'_\alpha)\right)_2^-. \tag{13.98}$$

If this model is correct, $J_0[\hat{\mathcal{S}}_\infty]/\epsilon^2$ is a χ^2 variable with $2N$ degrees of freedom. Hence, the squared noise level ϵ^2 can be estimated by

$$\hat{\epsilon}_\infty^2 = \frac{J_0[\hat{\mathcal{S}}_\infty]}{2N}. \tag{13.99}$$

13.5.4 Model comparison

Eq. (13.95) is obtained if $\boldsymbol{A}$ is replaced by $\boldsymbol{R}^\top$ in eq. (13.89). Alternatively, eq. (13.95) is obtained from eq. (13.89) by taking the limit $d \to \infty$.

[8]The approximation used in the general model is also used.

Eqs. (13.89) and (13.95) both imply the epipolar equation (13.83). From these observation, we have the following order in the strength of the three models:

$$\mathcal{S}_\infty \succ \mathcal{S}_\Pi \succ \mathcal{S}. \tag{13.100}$$

It follows that for whatever data $\boldsymbol{x}_\alpha$, $\boldsymbol{x}'_\alpha$, $\alpha = 1, ..., N$, we have $J_0[\hat{\mathcal{S}}_\infty] \geq J_0[\hat{\mathcal{S}}_\Pi] \geq J_0[\hat{\mathcal{S}}]$. If we apply the comparison criterion (13.66), we obtain the following test procedures:

1. *Planarity test*: Comparing the planar surface model $\mathcal{S}_\Pi$ with the general model $\mathcal{S}$, we infer that the object is a planar surface if

$$\frac{J_0[\hat{\mathcal{S}}_\Pi]}{J_0[\hat{\mathcal{S}}]} < 3 - \frac{6}{N}. \tag{13.101}$$

   In terms of the estimators $\hat{\epsilon}^2$ and $\hat{\epsilon}^2_\Pi$ defined by eqs. (13.88) and (13.94), the above condition can be written as follows:

$$\frac{\hat{\epsilon}^2_\Pi}{\hat{\epsilon}^2} < \frac{3N-6}{2N-3}. \tag{13.102}$$

2. *Infinity test*: Comparing the infinity model $\mathcal{S}_\infty$ with the general model $\mathcal{S}$, we infer that the object is infinitely far away if

$$\frac{J_0[\hat{\mathcal{S}}_\infty]}{J_0[\hat{\mathcal{S}}]} < 3. \tag{13.103}$$

   In terms of the estimators $\hat{\epsilon}^2$ and $\hat{\epsilon}^2_\infty$ defined by eqs. (13.88) and (13.99), the above condition can be written as follows:

$$\frac{\hat{\epsilon}^2_\infty}{\hat{\epsilon}^2} < \frac{3}{2}. \tag{13.104}$$

13.6 3-D Motion Analysis

13.6.1 General model

As discussed in Chapter 11, 3-D motion analysis from two views is essentially 3-D reconstruction by stereo vision; the only difference is that the motion parameters $\{\boldsymbol{h}, \boldsymbol{R}\}$ are unknown. Hence, the mathematical framework is the same except for the degrees of freedom of the model. As in the case of stereo vision, the corresponding image points $\boldsymbol{x}_\alpha$ and $\boldsymbol{x}'_\alpha$, $\alpha = 1, ..., N$, can be regarded as *six*-dimensional data $\boldsymbol{a}_\alpha = \boldsymbol{x}_\alpha \oplus \boldsymbol{x}'_\alpha \in \mathcal{R}^6$ sampled from the *four*-dimensional data space

$$\mathcal{A} = \{ \begin{pmatrix} x \\ y \\ 1 \\ x' \\ y' \\ 1 \end{pmatrix} | x, y, x', y' \in \mathcal{R}\} \subset \mathcal{R}^6. \tag{13.105}$$

The epipolar equation

$$|\boldsymbol{x}, \boldsymbol{h}, \boldsymbol{R}| = 0 \tag{13.106}$$

defines a three-dimensional manifold $\mathcal{S} \subset \mathcal{A}$. The problem is stated as follows:

Problem 13.5 *Estimate the manifold $\mathcal{S}$ and the true positions $\bar{\boldsymbol{a}}_\alpha = \bar{\boldsymbol{x}}_\alpha \oplus \bar{\boldsymbol{x}}'_\alpha$ of the data $\boldsymbol{a}_\alpha = \boldsymbol{x}_\alpha \oplus \boldsymbol{x}'_\alpha \in \mathcal{A}$ in such a way that $\bar{\boldsymbol{a}}_\alpha \in \mathcal{S}$.*

The unknown motion parameters $\{\boldsymbol{h}, \boldsymbol{R}\}$ have five degrees of freedom[9], so the manifold $\mathcal{S}$ is a model of dimension 3, codimension 1, and *five* degrees of freedom. As in the case of stereo vision, let $V[\boldsymbol{x}_\alpha]$ and $V[\boldsymbol{x}'_\alpha]$ be the covariance matrices of $\boldsymbol{x}_\alpha$ and $\boldsymbol{x}'_\alpha$, respectively, and decompose them into the noise level ϵ and the normalized covariance matrices $V_0[\boldsymbol{x}_\alpha]$ and $V_0[\boldsymbol{x}'_\alpha]$ in the form of eqs. (13.85). The noise is assumed to be Gaussian and independent for each datum. Let $\hat{\mathcal{S}}$ be the maximum likelihood estimator of $\mathcal{S}$. The AIC of this model is

$$AIC_0(\mathcal{S}) = J_0[\hat{\mathcal{S}}] + 2(3N+5)\epsilon^2. \tag{13.107}$$

Let $\{\hat{\boldsymbol{h}}, \hat{\boldsymbol{R}}\}$ be the maximum likelihood estimators of the motion parameters $\{\boldsymbol{h}, \boldsymbol{R}\}$; the computational scheme for this estimation by linearization and renormalization is described in Sections 11.2 and 11.3. The residual $J_0[\hat{\mathcal{S}}]$ is computed as follows[10] (see eqs. (11.11) and (11.12)):

$$J_0[\hat{\mathcal{S}}] = \sum_{\alpha=1}^{N} \frac{|\boldsymbol{x}_\alpha, \hat{\boldsymbol{h}}, \hat{\boldsymbol{R}}\boldsymbol{x}'_\alpha|^2}{(\hat{\boldsymbol{h}} \times \hat{\boldsymbol{R}}\boldsymbol{x}'_\alpha, V_0[\boldsymbol{x}_\alpha](\hat{\boldsymbol{h}} \times \hat{\boldsymbol{R}}\boldsymbol{x}'_\alpha)) + (\hat{\boldsymbol{h}} \times \boldsymbol{x}, \hat{\boldsymbol{R}}V_0[\boldsymbol{x}'_\alpha]\hat{\boldsymbol{R}}^\top(\hat{\boldsymbol{h}} \times \boldsymbol{x}))}. \tag{13.108}$$

If this model is correct, $J_0[\hat{\mathcal{S}}]/\epsilon^2$ is a χ^2 variable with $N-5$ degrees of freedom. Hence, the squared noise level ϵ^2 can be estimated by

$$\hat{\epsilon}^2 = \frac{J_0[\hat{\mathcal{S}}]}{N-5}. \tag{13.109}$$

---

[9]The translation $\mathbf{h}$ is normalized into a unit vector; see Section 11.1.1.

[10]As in the case of stereo vision, the denominator should be expressed in terms of the true values $\bar{\mathbf{x}}_\alpha$ and $\bar{\mathbf{x}}'_\alpha$ but they are approximated by the data $\mathbf{x}_\alpha$ and $\mathbf{x}'_\alpha$. Also, the second order term $\epsilon^2(V_0[\mathbf{x}_\alpha](\mathbf{h} \times \mathbf{R}); (\mathbf{h} \times \mathbf{R})V_0[\mathbf{x}'_\alpha])$ is omitted. Since $J_0[\hat{\mathcal{S}}] = O(\epsilon^2)$, these approximations result in only a higher order difference.

In Sections 11.3.2 and 11.4.1, we showed that the solution $\{\hat{\boldsymbol{h}},\ \hat{\boldsymbol{R}}\}$ of the motion parameters was not unique; we chose the one for which all the feature points have positive depths before and after the camera motion. The existence of multiple solutions is due to the fact that different motion parameters can define the same manifold $\mathcal{S}$. Hence, although the maximum likelihood estimators $\{\hat{\boldsymbol{h}},\ \hat{\boldsymbol{R}}\}$ of the motion parameters $\{\boldsymbol{h},\ \boldsymbol{R}\}$ may not be unique, the maximum likelihood estimator $\hat{\mathcal{S}}$ of the manifold $\mathcal{S}$ is generally unique.

13.6.2 Planar surface model

If the object is a planar surface $(\boldsymbol{n}, \boldsymbol{r}) = d$, the constraint is given by

$$\boldsymbol{x}' \times \boldsymbol{A}\boldsymbol{x} = \boldsymbol{0}, \tag{13.110}$$

$$\boldsymbol{A} = \boldsymbol{R}^\top(\boldsymbol{h}\boldsymbol{n} - d\boldsymbol{I}). \tag{13.111}$$

As in the case of stereo vision, eq. (13.110) defines a *two*-dimensional manifold $\mathcal{S}_\Pi$ in the *four*-dimensional data space $\mathcal{A}$. Hence, 3-D reconstruction of a planar surface can be identified with the following problem:

Problem 13.6 *Estimate the manifold $\mathcal{S}_\Pi$ and the true positions $\bar{\boldsymbol{a}}_\alpha = \bar{\boldsymbol{x}}_\alpha \oplus \bar{\boldsymbol{x}}'_\alpha$ of the data $\boldsymbol{a}_\alpha = \boldsymbol{x}_\alpha \oplus \boldsymbol{x}'_\alpha \in \mathcal{A}$ in such a way that $\bar{\boldsymbol{a}}_\alpha \in \mathcal{S}_\Pi$.*

The unknown surface parameters $\{\boldsymbol{n},\ d\}$ have three degrees of freedom, and the unknown motion parameters $\{\boldsymbol{h},\ \boldsymbol{R}\}$ have five degrees of freedom. Hence, the manifold $\mathcal{S}_\Pi$ is a model of dimension 2, codimension 2, and *eight* degrees of freedom. Let $\hat{\mathcal{S}}_\Pi$ be the maximum likelihood estimator of $\mathcal{S}_\Pi$. The AIC of this model is

$$AIC_0(\mathcal{S}_\Pi) = J_0[\hat{\mathcal{S}}_\Pi] + 2(2N+8)\epsilon^2. \tag{13.112}$$

Let $\{\hat{\boldsymbol{n}},\ \hat{d}\}$ and $\{\hat{\boldsymbol{h}},\ \hat{\boldsymbol{R}}\}$ be the maximum likelihood estimators of the surface and motion parameters $\{\boldsymbol{n},\ d\}$ and $\{\boldsymbol{h},\ \boldsymbol{R}\}$, respectively; the computational scheme for this estimation by renormalization is described in Section 11.6. The residual $J_0[\hat{\mathcal{S}}_\Pi]$ is computed as follows[11] (see eqs. (11.131) and (11.132)):

$$J_0[\hat{\mathcal{S}}_\Pi] = \sum_{\alpha=1}^{N} (\boldsymbol{x}'_\alpha \times \hat{\boldsymbol{A}}\boldsymbol{x}_\alpha, \hat{\boldsymbol{W}}_\alpha(\boldsymbol{x}'_\alpha \times \hat{\boldsymbol{A}}\boldsymbol{x}_\alpha)), \tag{13.113}$$

$$\hat{\boldsymbol{W}}_\alpha = \left(\boldsymbol{x}'_\alpha \times \hat{\boldsymbol{A}} V_0[\boldsymbol{x}_\alpha] \hat{\boldsymbol{A}}^\top \times \boldsymbol{x}'_\alpha + (\hat{\boldsymbol{A}}\boldsymbol{x}_\alpha) \times V_0[\boldsymbol{x}'_\alpha] \times (\hat{\boldsymbol{A}}\boldsymbol{x}_\alpha)\right)^-_2. \tag{13.114}$$

Here, $\hat{\boldsymbol{A}}$ is the estimate of matrix $\boldsymbol{A}$ obtained by replacing $\{\boldsymbol{n},\ d\}$ and $\{\boldsymbol{h},\ \boldsymbol{R}\}$ by their maximum likelihood estimators $\{\hat{\boldsymbol{n}},\ \hat{d}\}$ and $\{\hat{\boldsymbol{h}},\ \hat{\boldsymbol{R}}\}$, respectively,

[11] The approximations used in the general model are also used.

in eq. (13.90). If this model is correct, $J_0[\hat{\mathcal{S}}_\Pi]/\epsilon^2$ is a χ^2 variable with $2N-8$ degrees of freedom. Hence, the squared noise level ϵ^2 can be estimated by

$$\hat{\epsilon}_\Pi^2 = \frac{J_0[\hat{\mathcal{S}}_\Pi]}{2N-8}. \tag{13.115}$$

In Section 11.6.3, we showed that the solution $\{\hat{\boldsymbol{n}}, \hat{d}\}$ and $\{\hat{\boldsymbol{h}}, \hat{\boldsymbol{R}}\}$ of the surface and motion parameters was not unique and reduced the ambiguity by imposing the condition that the depths of the feature points should be all positive before and after the camera motion. As in the case of the general model $\mathcal{S}$, this ambiguity is due to the fact that different surface and motion parameters can define the same manifold. Hence, although the maximum likelihood estimators $\{\hat{\boldsymbol{n}}, \hat{d}\}$ and $\{\hat{\boldsymbol{h}}, \hat{\boldsymbol{R}}\}$ of the surface and motion parameters $\{\boldsymbol{n}, d\}$ and $\{\boldsymbol{h}, \boldsymbol{R}\}$ may not be unique, *the maximum likelihood estimator* $\hat{\mathcal{S}}_\Pi$ *of the manifold* $\mathcal{S}_\Pi$ *is generally unique.*

13.6.3 Rotation model

If the camera motion is a pure rotation (i.e., $\boldsymbol{h} = \boldsymbol{0}$), no 3-D information can be obtained (see Section 11.7.1). The camera is a pure rotation if and only if all corresponding image points $\boldsymbol{x}$ and $\boldsymbol{x}'$ satisfy the following equation (see eq. (11.173)):

$$\boldsymbol{x} \times \boldsymbol{R}\boldsymbol{x}' = \boldsymbol{0}. \tag{13.116}$$

Like eq. (13.110), eq. (11.173) defines a *two*-dimensional manifold $\mathcal{S}_R$ in the *four*-dimensional data space $\mathcal{A}$. Hence, estimation of the camera rotation $\boldsymbol{R}$ can be identified with the following problem:

Problem 13.7 *Estimate the manifold* $\mathcal{S}_R$ *and the true positions* $\bar{\boldsymbol{a}}_\alpha = \bar{\boldsymbol{x}}_\alpha \oplus \bar{\boldsymbol{x}}'_\alpha$ *of the data* $\boldsymbol{a}_\alpha = \boldsymbol{x}_\alpha \oplus \boldsymbol{x}'_\alpha \in \mathcal{A}$ *in such a way that* $\bar{\boldsymbol{a}}_\alpha \in \mathcal{S}_R$.

The unknown rotation $\boldsymbol{R}$ has three degrees of freedom, so the manifold $\mathcal{S}_R$ is a model of dimension 2, codimension 2, and *three* degrees of freedom. Let $\hat{\mathcal{S}}_R$ be the maximum likelihood estimator of $\mathcal{S}_R$. The AIC of this model is

$$AIC_0(\mathcal{S}_R) = J_0[\hat{\mathcal{S}}_R] + 2(2N+3)\epsilon^2. \tag{13.117}$$

Let $\hat{\boldsymbol{R}}$ be the maximum likelihood estimator of the rotation $\boldsymbol{R}$; the computational scheme for this estimation is described in Section 11.7.1. The residual $J_0[\hat{\mathcal{S}}_R]$ is computed as follows[12] (see eqs. (11.174) and (11.175)):

$$J_0[\hat{\mathcal{S}}_R] = \sum_{\alpha=1}^{N} (\boldsymbol{x}_\alpha \times \hat{\boldsymbol{R}}\boldsymbol{x}'_\alpha, \hat{\boldsymbol{W}}_\alpha(\boldsymbol{x}_\alpha \times \hat{\boldsymbol{R}}\boldsymbol{x}'_\alpha)), \tag{13.118}$$

[12]The approximations used in the general model are also used.

$$\hat{\boldsymbol{W}}_\alpha = \left(\boldsymbol{x}_\alpha \times \hat{\boldsymbol{R}} V_0[\boldsymbol{x}'_\alpha] \hat{\boldsymbol{R}}^\top \times \boldsymbol{x}_\alpha + (\hat{\boldsymbol{R}}\boldsymbol{x}'_\alpha) \times V_0[\boldsymbol{x}_\alpha] \times (\hat{\boldsymbol{R}}\boldsymbol{x}'_\alpha)\right)^-_2. \quad (13.119)$$

If this model is correct, $J_0[\hat{\mathcal{S}}_R]/\epsilon^2$ is a χ^2 variable with $2N-3$ degrees of freedom. Hence, the squared noise level ϵ^2 can be estimated by

$$\hat{\epsilon}_R^2 = \frac{J_0[\hat{\mathcal{S}}_R]}{2N-3}. \quad (13.120)$$

13.6.4 Model comparison

As was pointed out in Chapter 11, the general algorithm fails if all the feature points are coplanar in the scene because a planar surface is a degenerate critical surface (see Section 11.5). Also, the general and the planar surface algorithms both assume that the translation of the camera is not zero. If follows that the 3-D motion analysis must follow the following steps:

1. *Rotation test*: We test if the translation is 0. If so, output a warning message and stop.

2. *Planarity test*: We test if the object is a planar surface. If so, apply the planar surface algorithm.

3. Else, apply the general algorithm.

In Chapter 11, the rotation and planarity tests were formulated as χ^2 tests (see Sections 11.6.1 and 11.7.1). However, they have the following shortcomings:

- We need to estimate the noise level ϵ, but this is different from image to image.

- Even if ϵ is predicted, we need to set the *significance level*. The judgment differs if the significance level is set differently.

These difficulties can be avoided by the use of the geometric AIC. As in the case of stereo vision, eqs. (13.110) and (13.116) both imply the epipolar equation (13.83). Also, eq. (13.110) reduces to eq. (13.116) if the matrix $\boldsymbol{A}$ in eq. (13.89) is replaced by $\boldsymbol{R}^\top$. These observations imply the following order of the strength of the three models:

$$\mathcal{S}_R \succ \mathcal{S}_\Pi \succ \mathcal{S}. \quad (13.121)$$

It follows that for whatever data $\boldsymbol{x}_\alpha$, $\boldsymbol{x}'_\alpha$, $\alpha = 1, ..., N$, we have $J_0[\hat{\mathcal{S}}_R] \geq J_0[\hat{\mathcal{S}}_\Pi] \geq J_0[\hat{\mathcal{S}}]$. If we apply the comparison criterion (13.66), we obtain the following test procedures:

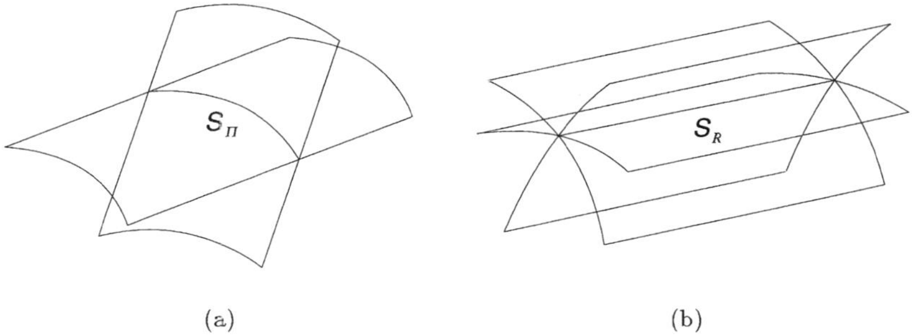

Fig. 13.9. (a) Multiple instances of the general model $\mathcal{S}$ can pass through the planar surface model $\mathcal{S}_\Pi$. (b) Infinitely many instances of the general model $\mathcal{S}$ can pass through the rotation model $\mathcal{S}_R$.

1. *Planarity test*: Comparing the planar surface model $\mathcal{S}_\Pi$ with the general model $\mathcal{S}$, we infer that the object surface is planar if

$$\frac{J_0[\hat{\mathcal{S}}_\Pi]}{J_0[\hat{\mathcal{S}}]} < 3 + \frac{4}{N-5}. \tag{13.122}$$

In terms of the estimators $\hat{\epsilon}^2$ and $\hat{\epsilon}^2_\Pi$ defined by eqs. (13.109) and (13.115), the above condition can be written as follows:

$$\frac{\hat{\epsilon}^2_\Pi}{\hat{\epsilon}^2} < \frac{3N-11}{2N-8}. \tag{13.123}$$

2. *Rotation test*: Comparing the rotation model $\mathcal{S}_R$ with the general model $\mathcal{S}$, we infer that the camera motion is a pure rotation if

$$\frac{J_0[\hat{\mathcal{S}}_R]}{J_0[\hat{\mathcal{S}}]} < 3 + \frac{14}{N-5}. \tag{13.124}$$

In terms of the estimators $\hat{\epsilon}^2$ and $\hat{\epsilon}^2_R$ defined by eqs. (13.109) and (13.120), the above condition can be written as follows:

$$\frac{\hat{\epsilon}^2_R}{\hat{\epsilon}^2} < \frac{3N-1}{2N-3}. \tag{13.125}$$

The fact that the 3-D reconstruction algorithm for the general model fails if the object shape is a planar surface or the camera motion is a pure rotation can be geometrically understood as follows. Note that in the *four*-dimensional data space $\mathcal{A}$ the intersection of two *three*-dimensional manifolds is in general a *two*-dimensional manifold.

- The general 3-D reconstruction algorithm fails for a planar surface because *multiple* instances of the three-dimensional general model $\mathcal{S}$ can pass through the planar surface model $\mathcal{S}_\Pi$ (Fig. 13.9).

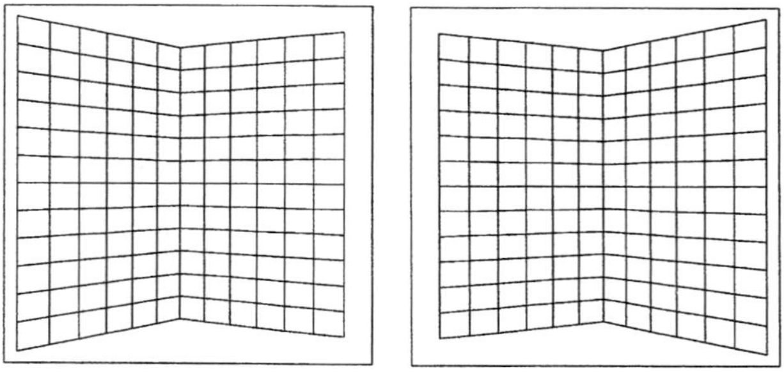

Fig. 13.10. Images of two planar grids hinged together in the scene ($\theta = 50°$).

- No 3-D information can be obtained if the camera motion is a pure rotation because *infinitely many* instances of the general model $\mathcal{S}$ can pass through the rotation model $\mathcal{S}_R$ (Fig. 13.9). In fact, we can see that

$$\{\boldsymbol{x} \oplus \boldsymbol{x}' | \boldsymbol{x} \times \boldsymbol{R}\boldsymbol{x}' = 0\} = \bigcap_{\boldsymbol{h} \in \mathcal{R}^3} \{\boldsymbol{x} \oplus \boldsymbol{x}' | |\boldsymbol{x}, \boldsymbol{h}, \boldsymbol{R}\boldsymbol{x}'| = 0\}. \tag{13.126}$$

Example 13.8 Two planar grids hinged together with angle $\pi - \theta$ were defined in the scene, and images viewed from different camera positions were generated. The image size and the focal length were assumed to be 512×512 (pixels) and 600 (pixels), respectively. Fig. 13.10a shows the images for $\theta = 50°$. We added random Gaussian noise of mean 0 and standard deviation σ (pixels) to the x- and y-coordinates of each grid point independently. Using the grid points as feature points, we conducted the planarity test 100 times, each time using different noise. Fig. 13.11 shows the percentage of the instances for which the object is judged as planar.

If $\sigma = 1.0$, the percentage is approximately 50% for $\theta = 22°$. Fig. 13.12 shows one instance for which the object is judged as planar; Fig. 13.13 shows one instance for which the object is judged as non-planar. Fig. 13.14a and Fig. 13.14b show the corresponding 3-D shapes reconstructed by the general and planar surface algorithms. The true shape also is superimposed. We can see that although the images look almost the same, the shape reconstructed by the general algorithm has little sense if the object is judged as planar, while the non-planar shape can be reconstructed fairly well if the object is judged as non-planar.

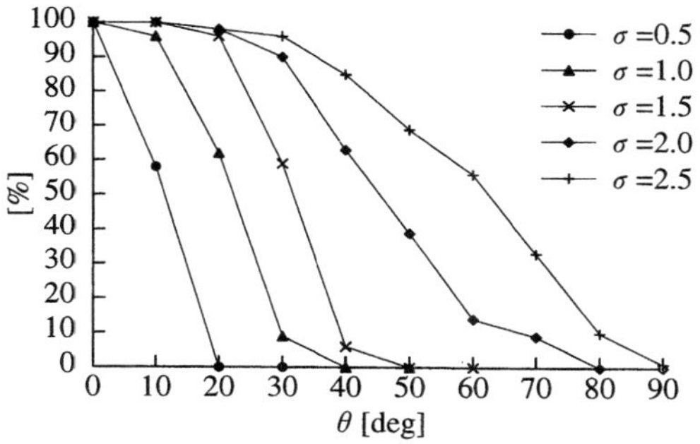

Fig. 13.11. The percentage of the instances judged as planar.

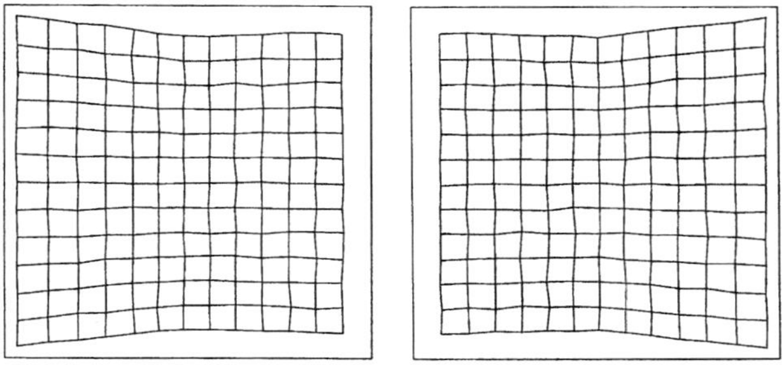

Fig. 13.12. An instance for which the object is judged as planar.

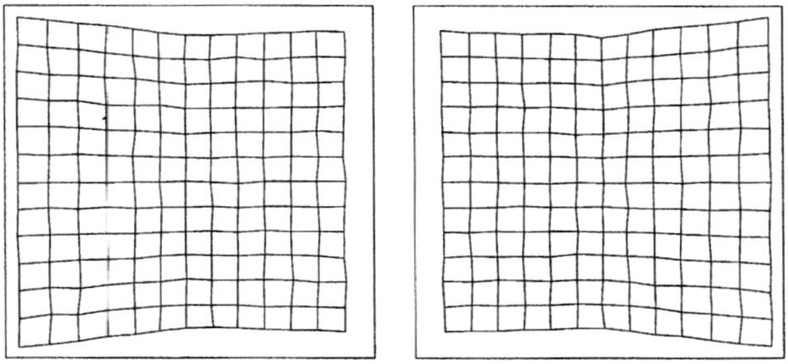

Fig. 13.13. An instance for which the object is judged as nonplanar.

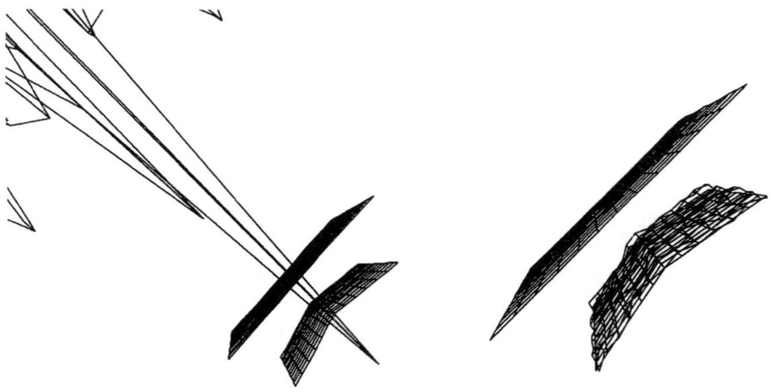

(a) (b)

Fig. 13.14. (a) 3-D shapes and the true shape reconstructed from Fig. 13.12. (a) 3-D shapes and the true shape reconstructed from Fig. 13.13.

13.7 3-D Interpretation of Optical Flow

13.7.1 General model

As argued in Chapter 12, the formalism of 3-D interpretation of optical flow can be obtained from finite motion analysis in the limit of infinitesimal camera motion. Hence, the mathematical structure is almost identical with that for finite motion analysis. However, a subtle difference exists. Given optical flow $\dot{\boldsymbol{x}}$ at image point $\boldsymbol{x}$, we can view the set of *6-vectors* $\boldsymbol{a} = \dot{\boldsymbol{x}} \oplus \boldsymbol{x} \in \mathcal{R}^6$ as data sampled from the *four*-dimensional data space

$$\mathcal{A} = \{ \begin{pmatrix} \dot{x} \\ \dot{y} \\ 0 \\ x \\ y \\ 1 \end{pmatrix} | \dot{x}, \dot{y}, x, y \in \mathcal{R}\} \subset \mathcal{R}^6. \tag{13.127}$$

However, *noise occurs only in* $\dot{x}$ *and* $\dot{y}$. Hence, the covariance matrix $V[\boldsymbol{a}]$ of $\boldsymbol{a}$ has rank 2, *not 4*, in contrast to the case of stereo vision and finite motion analysis. Because of this singularity, we must make appropriate modifications to the general argument given in Sections 13.1–13.4.

Let $\{\boldsymbol{v}, \boldsymbol{\omega}\}$ be the motion parameters. In the absence of noise, the optical flow $\dot{\boldsymbol{x}}$ must satisfy the following epipolar equation (see eqs. (12.32) and (12.33)):

$$|\boldsymbol{x}, \dot{\boldsymbol{x}}, \boldsymbol{v}| + (\boldsymbol{v} \times \boldsymbol{x}, \boldsymbol{\omega} \times \boldsymbol{x}) = 0. \tag{13.128}$$

This equation defines a *three*-dimensional manifold $\mathcal{S}$ in the four-dimensional data space $\mathcal{A}$. Hence, the problem of 3-D interpretation of optical flow is stated as follows:

Problem 13.8 *Estimate the manifold $\mathcal{S}$ and the true positions $\bar{\boldsymbol{a}} = \bar{\dot{\boldsymbol{x}}} \oplus \boldsymbol{x}$ of the data $\boldsymbol{a} = \dot{\boldsymbol{x}} \oplus \boldsymbol{x} \in \mathcal{A}$ in such a way that $\bar{\boldsymbol{a}} \in \mathcal{S}$.*

The the unknown motion parameters $\{\boldsymbol{v}, \boldsymbol{\omega}\}$ have five degrees of freedom[13], so the model $\mathcal{S}$ has dimension 3, codimension 1, and *five* degrees of freedom. Let $V[\dot{\boldsymbol{x}}]$ be the covariance matrix of optical flow $\dot{\boldsymbol{x}}$, and decompose it into the noise level ϵ and the normalized covariance matrix $V_0[\dot{\boldsymbol{x}}]$ in the form of

$$V[\dot{\boldsymbol{x}}] = \epsilon^2 V_0[\dot{\boldsymbol{x}}]. \tag{13.129}$$

The noise is assumed to be Gaussian and independent at each point. Let $\hat{\mathcal{S}}$ be the maximum likelihood estimator of $\mathcal{S}$. The AIC of this model is

$$AIC_0(\mathcal{S}) = J_0[\hat{\mathcal{S}}] + 2(S+5)\epsilon^2, \tag{13.130}$$

where S is the number of pixels where the optical flow is observed. Note that although the model $\mathcal{S}$ has dimension 3, the last term on the right-hand side of eq. (13.130) is not $2(3S+5)\epsilon^2$. This is because $\boldsymbol{x}$ *is not a random variable*: for a fixed position $\boldsymbol{x}$, the epipolar equation (13.129) defines a *one*-dimensional submanifold of $\mathcal{S}$ (see Section 12.5.2 and Fig. 12.8). Hence, if $\hat{\boldsymbol{a}}$ and $\hat{\mathcal{S}}$ are estimators of $\bar{\boldsymbol{a}}$ and $\mathcal{S}$, respectively, the estimator $\hat{\boldsymbol{a}} \in \hat{\mathcal{S}}$ has only *one degree of freedom* within $\hat{\mathcal{S}}$.

Let $\{\hat{\boldsymbol{v}}, \hat{\boldsymbol{\omega}}\}$ be the maximum likelihood estimators of the motion parameters $\{\boldsymbol{v}, \boldsymbol{\omega}\}$; the computational scheme for this estimation by linearization and renormalization is described in Sections 12.4 and 12.5. The residual $J_0[\hat{\mathcal{S}}]$ is computed as follows (see eqs. (12.49) and (12.50)):

$$J_0[\hat{\mathcal{S}}] = \int_S \frac{\left(|\boldsymbol{x}, \dot{\boldsymbol{x}}, \hat{\boldsymbol{v}}| + (\hat{\boldsymbol{v}} \times \boldsymbol{x}, \hat{\boldsymbol{\omega}} \times \boldsymbol{x})\right)^2 dxdy}{(\hat{\boldsymbol{v}}, (\boldsymbol{x} \times V_0[\dot{\boldsymbol{x}}] \times \boldsymbol{x})\hat{\boldsymbol{v}})}. \tag{13.131}$$

As in Chapter 12, integration $\int_S dxdy$ is a symbolic notation for summation over all pixels at which the optical flow is defined. If this model is correct, $J_0[\hat{\mathcal{S}}]/\epsilon^2$ is a χ^2 variable with $S-5$ degrees of freedom. Hence, the squared noise level ϵ^2 can be estimated by

$$\hat{\epsilon}^2 = \frac{J_0[\hat{\mathcal{S}}]}{S-5}. \tag{13.132}$$

In Section 12.5.3, we showed that the solution $\{\hat{\boldsymbol{v}}, \hat{\boldsymbol{\omega}}\}$ of the motion parameters was not unique and chose the one for which the depth is positive

[13]The translation $\mathbf{v}$ is normalized into a unit vector; see Section 12.2.3.

everywhere. As in the case of finite motion analysis, this ambiguity is due to the fact that the same manifold $\mathcal{S}$ admits different parameterizations by the motion parameters $\{\boldsymbol{v}, \boldsymbol{\omega}\}$ (see Section 12.8.4). Hence, although the maximum likelihood estimators $\{\hat{\boldsymbol{v}}, \hat{\boldsymbol{\omega}}\}$ of the motion parameters $\{\boldsymbol{v}, \boldsymbol{\omega}\}$ may not be unique, the maximum likelihood estimator $\hat{\mathcal{S}}$ of the manifold $\mathcal{S}$ is generally unique.

13.7.2 Planar surface model

If the object is a planar surface $(\boldsymbol{n}, \boldsymbol{r}) = d$, the optical flow has the form

$$\dot{\boldsymbol{x}} = \boldsymbol{W}\boldsymbol{x} - (\boldsymbol{k}, \boldsymbol{W}\boldsymbol{x})\boldsymbol{x}, \tag{13.133}$$

where $\boldsymbol{k} = (0,0,1)^\top$. The *flow matrix* $\boldsymbol{W}$ is defined as follows (see eqs. (12.176) and (12.177)):

$$\boldsymbol{W} = -\frac{1}{d}\left(\boldsymbol{v}\boldsymbol{n}^\top - \frac{1}{3}(\boldsymbol{v}, \boldsymbol{n})\boldsymbol{I}\right) - \boldsymbol{\omega} \times \boldsymbol{I}. \tag{13.134}$$

The third components of both sides of eq. (13.133) are identically 0. Hence, eq. (13.133) defines a *two*-dimensional manifold $\mathcal{S}_\Pi$ in the four-dimensional data space $\mathcal{A}$. Thus, 3-D reconstruction of a planar surface can be identified with the following problem:

Problem 13.9 *Estimate the manifold $\mathcal{S}_\Pi$ and the true positions $\bar{\boldsymbol{a}} = \dot{\bar{\boldsymbol{x}}} \oplus \boldsymbol{x}$ of the data $\boldsymbol{a} = \dot{\boldsymbol{x}} \oplus \boldsymbol{x} \in \mathcal{A}$ in such a way that $\bar{\boldsymbol{a}} \in \mathcal{S}_\Pi$.*

The unknown surface parameters $\{\boldsymbol{n}, d\}$ have three degrees of freedom, and the unknown motion parameters $\{\boldsymbol{v}, \boldsymbol{\omega}\}$ have five degrees of freedom. Hence, the manifold $\mathcal{S}_\Pi$ is a model of dimension 2, codimension 2, and *eight* degrees of freedom. Let $\hat{\mathcal{S}}_\Pi$ be the maximum likelihood estimator of $\mathcal{S}_\Pi$. The AIC of this model is

$$AIC_0(\mathcal{S}_\Pi) = J_0[\hat{\mathcal{S}}_\Pi] + 16\epsilon^2. \tag{13.135}$$

Note that although $\mathcal{S}_\Pi$ has dimension 2, the last term on the right-hand side of eq. (13.135) is not $2(2S+8)\epsilon^2$. This is because for a fixed position $\boldsymbol{x}$ and a fixed estimator $\hat{\mathcal{S}}_\Pi$ of $\mathcal{S}$ the planar surface optical flow equation (13.133) determines the optical flow $\dot{\boldsymbol{x}}$ *uniquely*[14].

Let $\{\hat{\boldsymbol{n}}, \hat{d}\}$ and $\{\hat{\boldsymbol{v}}, \hat{\boldsymbol{\omega}}\}$ be the maximum likelihood estimators of the surface and motion parameters $\{\boldsymbol{n}, d\}$ and $\{\boldsymbol{v}, \boldsymbol{\omega}\}$, respectively; the computational scheme for this estimation is described in Section 12.8. The residual $J_0[\hat{\mathcal{S}}_\Pi]$ is computed as follows (see eq. (12.178)):

$$J_0[\hat{\mathcal{S}}_\Pi] = \int_S \|\dot{\boldsymbol{x}} - \hat{\boldsymbol{W}}\boldsymbol{x} + (\boldsymbol{k}, \hat{\boldsymbol{W}}\boldsymbol{x})\boldsymbol{x}\|^2_{V_0[\dot{\boldsymbol{x}}]} dxdy, \tag{13.136}$$

[14] A problem of this special type corresponds to what is called *regression* in statistics.

Here, $\hat{\boldsymbol{W}}$ is the estimate of the flow matrix $\boldsymbol{W}$ obtained by replacing $\{\boldsymbol{n}, d\}$ and $\{\boldsymbol{v}, \boldsymbol{\omega}\}$ by their maximum likelihood estimators $\{\hat{\boldsymbol{n}}, \hat{d}\}$ and $\{\hat{\boldsymbol{v}}, \hat{\boldsymbol{\omega}}\}$, respectively, in eq. (13.134). If this model is correct, $J_0[\hat{\mathcal{S}}_\Pi]/\epsilon^2$ is a χ^2 variable with $2S-8$ degrees of freedom. Hence, the squared noise level ϵ^2 can be estimated by

$$\hat{\epsilon}^2_\Pi = \frac{J_0[\hat{\mathcal{S}}_\Pi]}{2S-8}. \tag{13.137}$$

In Section 12.8.4, we showed that the solution $\{\hat{\boldsymbol{n}}, \hat{d}\}$ and $\{\hat{\boldsymbol{v}}, \hat{\boldsymbol{\omega}}\}$ of the surface and motion parameters was not unique and reduced the ambiguity by imposing the condition that the depth should be positive everywhere. As in the case of finite motion analysis, this ambiguity is due to the fact that the same manifold $\mathcal{S}$ admits different parameterizations by the surface and motion parameters $\{\boldsymbol{n}, d\}$ and $\{\boldsymbol{v}, \boldsymbol{\omega}\}$. Hence, although the maximum likelihood estimators $\{\hat{\boldsymbol{n}}, \hat{d}\}$ and $\{\hat{\boldsymbol{v}}, \hat{\boldsymbol{\omega}}\}$ of the surface and motion parameters $\{\boldsymbol{n}, d\}$ and $\{\boldsymbol{v}, \boldsymbol{\omega}\}$ may not be unique, the maximum likelihood estimator $\hat{\mathcal{S}}_\Pi$ of the manifold $\mathcal{S}_\Pi$ is generally unique.

13.7.3 Rotation model

If the camera motion is a pure rotation (i.e., $\boldsymbol{v} = \boldsymbol{0}$), no 3-D information can be obtained (see Section 12.9). The camera motion is a pure rotation if and only if the optical flow $\dot{\boldsymbol{x}}$ has the following form (see eq. (12.195)):

$$\dot{\boldsymbol{x}} = \boldsymbol{Q}_{\boldsymbol{x}}(\boldsymbol{x} \times \boldsymbol{\omega}). \tag{13.138}$$

Here, $\boldsymbol{Q}_{\boldsymbol{x}} = \boldsymbol{I} - \boldsymbol{x}\boldsymbol{k}^\top$. The third components of both sides of eq. (13.138) are identically 0. Hence, eq. (13.138) defines a *two*-dimensional manifold $\mathcal{S}_R$ in the four-dimensional data space $\mathcal{A}$. Thus, estimation of the rotation velocity $\boldsymbol{\omega}$ can be identified with the following problem:

Problem 13.10 *Estimate the manifold $\mathcal{S}_R$ and the true positions $\bar{\boldsymbol{a}} = \bar{\dot{\boldsymbol{x}}} \oplus \boldsymbol{x}$ of the data $\boldsymbol{a} = \dot{\boldsymbol{x}} \oplus \boldsymbol{x} \in \mathcal{A}$ in such a way that $\bar{\boldsymbol{a}} \in \mathcal{S}_R$.*

The unknown rotation $\boldsymbol{\omega}$ has three degrees of freedom, so the manifold $\mathcal{S}_R$ is a model of dimension 2, codimension 2, and *three* degrees of freedom. Let $\hat{\mathcal{S}}_R$ be the maximum likelihood estimator of $\mathcal{S}_R$. The AIC of this model is

$$AIC_0(\mathcal{S}_R) = J_0[\hat{\mathcal{S}}_R] + 6\epsilon^2. \tag{13.139}$$

The last term on the right-hand side is not $2(2S+3)\epsilon^2$ by the same reason as in the case of the planar surface model.

Let $\hat{\boldsymbol{\omega}}$ be the maximum likelihood estimator of the rotation velocity $\boldsymbol{\omega}$; the computational scheme for this estimation is described in Section 12.9.1. The residual $J_0[\hat{\mathcal{S}}_R]$ is computed as follows (see eq. (12.196)):

$$J_0[\hat{\mathcal{S}}_R] = \int_S \|\dot{\boldsymbol{x}} - \boldsymbol{Q}_{\boldsymbol{x}}(\boldsymbol{x} \times \hat{\boldsymbol{\omega}})\|^2_{V_0[\dot{\boldsymbol{x}}]} dxdy. \tag{13.140}$$

If this model is correct, $J_0[\hat{\mathcal{S}}_R]/\epsilon^2$ is a χ^2 variable with $2S-3$ degrees of freedom. Hence, the squared noise level ϵ^2 can be estimated by

$$\hat{\epsilon}_R^2 = \frac{J_0[\hat{\mathcal{S}}_R]}{2S-3}. \tag{13.141}$$

13.7.4 Model comparison

Eq. (13.116) is obtained if $\boldsymbol{W}$ in eq. (13.89) is an antisymmetric matrix. Alternatively, eq. (13.116) is obtained from eq. (13.89) by taking the limit $d \to \infty$. Eqs. (13.89) and (13.116) both imply the epipolar equation (13.128). From these observations, we have the following order in the strength of the three models:

$$\mathcal{S}_R \succ \mathcal{S}_\Pi \succ \mathcal{S}. \tag{13.142}$$

It follows that for whatever optical flow $\dot{\boldsymbol{x}}$ we have $J_0[\hat{\mathcal{S}}_R] \geq J_0[\hat{\mathcal{S}}_\Pi] \geq J_0[\hat{\mathcal{S}}]$. If we apply the comparison criterion (13.66), we obtain the following test procedures:

1. *Planarity test*: Comparing the planar surface model $\mathcal{S}_\Pi$ with the general model $\mathcal{S}$, we infer that the object is a planar surface if

$$\frac{J_0[\hat{\mathcal{S}}_\Pi]}{J_0[\hat{\mathcal{S}}]} < 3 + \frac{4}{S-5}. \tag{13.143}$$

In terms of the estimators $\hat{\epsilon}^2$ and $\hat{\epsilon}_\Pi^2$ defined by eqs. (13.132) and (13.137), the above condition can be written as follows:

$$\frac{\hat{\epsilon}_\Pi^2}{\hat{\epsilon}^2} < \frac{3S-11}{2S-8}. \tag{13.144}$$

2. *Rotation test*: Comparing the infinity model $\mathcal{S}_\infty$ with the general model $\mathcal{S}$, we infer that the object is infinitely far away if

$$\frac{J_0[\hat{\mathcal{S}}_R]}{J_0[\hat{\mathcal{S}}]} < 3 + \frac{14}{S-5}. \tag{13.145}$$

In terms of the estimators $\hat{\epsilon}^2$ and $\hat{\epsilon}_R^2$ defined by eqs. (13.132) and (13.141), the above condition can be written as follows:

$$\frac{\hat{\epsilon}_R^2}{\hat{\epsilon}^2} < \frac{3S-1}{2S-3}. \tag{13.146}$$

The above results are formally identical with the case of finite motion. The singularity of 3-D interpretation also has the same geometric interpretation (see Fig. 13.9).

Chapter 14

General Theory of Geometric Estimation

This chapter gives a mathematical extension to the theory of geometric correction in Chapter 5 and the theory of parametric fitting in Chapter 7. Here, the error distribution is no longer assumed to be Gaussian: the role of the covariance matrix for a Gaussian distribution is played by the *Fisher information matrix*. We derive a theoretical lower bound, called the *Cramer-Rao lower bound*, on the covariance matrix of an unbiased estimator of the parameter to be determined and show that the maximum likelihood estimator attains this bound in the first order if the problem belongs to the *exponential family*. We then express the computation of maximum likelihood estimation in the form suitable for numerical computation in practical applications.

14.1 Statistical Estimation in Engineering

The subject of this chapter is essentially *statistical estimation*. However, the following treatment is very different from traditional statistics, whose main objective is to infer the structure of a random phenomenon by observing multiple data with a view to evaluating and comparing effects and procedures in domains that involve a large degree of uncertainty, such as medicine, biology, agriculture, manufacturing, sociology, economics, and politics. In such a domain, the problem is usually translated into the mathematical language as estimating *parameters involved in the probability distribution* from multiple independent samples from it (see Section 3.5 for the classical results). Although this framework covers almost all types of relevant applications in the above mentioned domains, geometric estimation problems in computer vision and robotics have many non-traditional elements.

In traditional statistics, errors are regarded as *uncontrollable*; the accuracy of estimation is improved only by repeated measurements. However, repeating measurements is costly. Hence, if the accuracy is the same, those methods which require a smaller number of data are more desirable. In other words, methods whose accuracy improves rapidly as the number of data increases are more desirable than those with slow increase of accuracy. Thus, the study of *asymptotic* properties of estimation in the limit of a large number of data has been one of the central subjects in traditional statistics (see Section 3.6.2 for the classical results).

In such engineering domains as computer vision and robotics, where electronic sensing devices are used, errors are usually small and called *noise*.

Moreover, they are *controllable*: the accuracy of sensing can be improved by using high-resolution devices and controlling the environment (lighting, dust, temperature, humidity, vibration, etc.). However, such control is costly. Hence, if the accuracy is the same, those methods which tolerate a higher level of noise are more desirable. In other words, methods whose accuracy improves rapidly as the noise level decreases are more desirable than those with slow increase of accuracy. Thus, the study of the accuracy of estimation *in the limit of small noise* is very important. In this sense, our approach of assuming local distributions and applying linear analysis, which has been the basic principle in the preceding chapters, can be justified.

In many engineering domains, repeating measurements under the same conditions (which is easy) often produces the same results because the sources of inaccuracy in the device and the environment are fixed (but unknown). Also, the number of independent data is usually fixed. Hence, the basic premise of traditional statistics that *independent samples from the same distribution can be observed as many times as desired* does not hold. This observation also underlies the statistical treatment in the preceding chapters.

14.2 General Geometric Correction

14.2.1 Definition of the problem

Let $F^{(k)}(\boldsymbol{u})$, $k = 1, ..., L$, be continuously differentiable scalar functions of argument $\boldsymbol{u} \in \mathcal{R}^n$. The domain of $\boldsymbol{u}$ is assumed to be an n'-dimensional manifold $\mathcal{U} \subset \mathcal{R}^n$, which we call the *data space*. *Geometric correction* is the problem of correcting a given datum $\boldsymbol{u}$ to such a value $\hat{\boldsymbol{u}}$ that satisfies the L equations

$$F^{(k)}(\hat{\boldsymbol{u}}) = 0, \qquad k = 1, ..., L. \tag{14.1}$$

In Chapter 5.1, we considered N variables $\boldsymbol{u}_1, ..., \boldsymbol{n}_N$, but the problem becomes equivalent if we regard their direct sum $\boldsymbol{u} = \boldsymbol{u}_1 \oplus \cdots \oplus \boldsymbol{u}_N$, $n = n_1 + \cdots + n_N$, as a new variable. As in Chapter 5.1, we call the L equations (14.1) simply the *constraint*, as opposed to which we call the constraint $\boldsymbol{u} \in \mathcal{U}$ the *inherent constraint* on $\boldsymbol{u}$.

Let $\bar{\boldsymbol{u}} \in \mathcal{U}$ be the true value of $\boldsymbol{u}$, which we assume is unknown. We regard the datum $\boldsymbol{u}$ as a random variable with probability density $p(\boldsymbol{u}; \bar{\boldsymbol{u}})$ parameterized by $\bar{\boldsymbol{u}}$. We assume that the distribution of $\boldsymbol{u}$ in the data space $\mathcal{U}$ is *local* (see Section 3.2.2) and define the "true value" $\bar{\boldsymbol{u}}$ by the relationship given by eq. (3.29). However, such an explicit characterization of $\bar{\boldsymbol{u}}$ will not be used in the subsequent analysis, so it can be thought of simply as *some value* in $\mathcal{U}$. The geometric correction problem is formally stated as follows (see Section 5.1.1):

Problem 14.1 *From a given datum $\boldsymbol{u} \in \mathcal{U}$, estimate the true value $\bar{\boldsymbol{u}} \in \mathcal{U}$ that satisfies the constraint*

$$F^{(k)}(\bar{\boldsymbol{u}}) = 0, \qquad k = 1, ..., L. \tag{14.2}$$

As in Sections 3.5 and 3.6, the probability density $p(\boldsymbol{u}; \bar{\boldsymbol{u}})$ is assumed to satisfy the following regularity condition:

Assumption 14.1 *The probability density $p(\boldsymbol{u}; \bar{\boldsymbol{u}})$ is continuously differentiable with respect to $\boldsymbol{u}$ and $\bar{\boldsymbol{u}}$ an arbitrary number of times, and*

$$p(\boldsymbol{u}; \bar{\boldsymbol{u}}) > 0 \tag{14.3}$$

for all $\boldsymbol{u} \in \mathcal{U}$. Furthermore, the integration operation $\int_{\mathcal{U}} d\boldsymbol{u}$ for any expression of $p(\boldsymbol{u}; \bar{\boldsymbol{u}})$ with respect to $\boldsymbol{u}$ is, if the integration exists, interchangeable with the differentiation operation $\nabla_{\bar{\boldsymbol{u}}}$ with respect to $\bar{\boldsymbol{u}}$.

The probability density $p(\boldsymbol{u}; \bar{\boldsymbol{u}})$ is not defined for $\bar{\boldsymbol{u}} \notin \mathcal{U}$. For the convenience of analysis, however, we extend it to $\bar{\boldsymbol{u}} \notin \mathcal{U}$ in such a way that

$$p(\boldsymbol{u}; \bar{\boldsymbol{u}} + \Delta\bar{\boldsymbol{u}}) = p(\boldsymbol{u}; \bar{\boldsymbol{u}}) + O(\Delta\bar{\boldsymbol{u}})^2 \tag{14.4}$$

for all $\bar{\boldsymbol{u}} \in \mathcal{U}$ and $\Delta\bar{\boldsymbol{u}} \in T_{\bar{\boldsymbol{u}}}(\mathcal{U})^{\perp}$, where $T_{\bar{\boldsymbol{u}}}(\mathcal{U}) \subset \mathcal{R}^n$ is the tangent space to the manifold $\mathcal{U}$ at $\bar{\boldsymbol{u}}$ (see Section 3.5.1). In precise terms, this assumption is stated as follows (see eq. (3.112)):

Assumption 14.2

$$\nabla_{\bar{\boldsymbol{u}}} p \in T_{\bar{\boldsymbol{u}}}(\mathcal{U}). \tag{14.5}$$

14.2.2 The rank of the constraint

The L equations (14.1) may be algebraically dependent. If only r of them are independent, they give only r effective constraints and define an $(n' - r)$-dimensional manifold $\mathcal{S}$, or a manifold $\mathcal{S}$ of *codimension* r, in the data space $\mathcal{U}$ except at possible singular points. We assume that $\bar{\boldsymbol{u}} \in \mathcal{S}$ is not a singular point and call r the *rank* of the constraint (14.1). A precise statement of this assumption, which also gives a precise definition of the rank r, is as follows (see Section 7.1.1):

Assumption 14.3 *The L equations of the constraint define a manifold $\mathcal{S}$ of codimension r in the data space $\mathcal{U}$ in the neighborhood of $\bar{\boldsymbol{u}}$.*

As in the preceding chapter, let us call the manifold $\mathcal{S}$ the (*geometric*) *model* of Problem 14.1. The solution is given by the *projection of the datum $\boldsymbol{u} \in \mathcal{U}$ onto the model $\mathcal{S} \subset \mathcal{U}$* (Fig. 14.1).

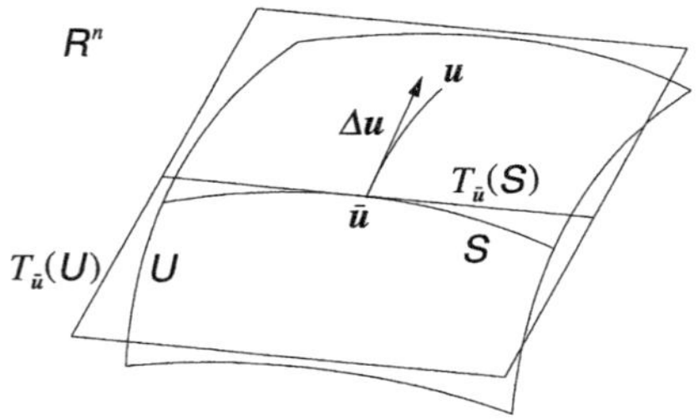

Fig. 14.1. Projecting $\mathbf{u} \in \mathcal{U}$ onto the model $\mathcal{S} \subset \mathcal{U}$.

If we write $\boldsymbol{u} = \bar{\boldsymbol{u}} + \Delta\boldsymbol{u}$, we have

$$\begin{pmatrix} F^{(1)}(\bar{\boldsymbol{u}} + \Delta\boldsymbol{u}) \\ \vdots \\ F^{(L)}(\bar{\boldsymbol{u}} + \Delta\boldsymbol{u}) \end{pmatrix} = \left(\nabla\bar{F}^{(1)}, ..., \nabla\bar{F}^{(L)}\right)^\top \Delta\boldsymbol{u} \tag{14.6}$$

to a first approximation in $\Delta\boldsymbol{u}$, where and hereafter $\nabla F^{(k)}(\bar{\boldsymbol{u}})$ is abbreviated as $\nabla\bar{F}^{(k)}$. The inherent constraint $\boldsymbol{u} \in \mathcal{U}$ requires that $\Delta\boldsymbol{u} \in T_{\bar{\mathbf{u}}}(\mathcal{U})$ to a first approximation. If $\Delta\boldsymbol{u}$ ranges over the entire tangent space $T_{\bar{\mathbf{u}}}(\mathcal{U})$, eq. (14.6) defines a *linear mapping* from $T_{\bar{\mathbf{u}}}(\mathcal{U})$ to a linear subspace

$$\mathcal{L}_{\bar{\mathbf{u}}} = \{\begin{pmatrix} (\nabla\bar{F}^{(1)}, \Delta\boldsymbol{u}) \\ \vdots \\ (\nabla\bar{F}^{(L)}, \Delta\boldsymbol{u}) \end{pmatrix} |\Delta\boldsymbol{u} \in T_{\bar{\mathbf{u}}}(\mathcal{U})\} \subset \mathcal{R}^L. \tag{14.7}$$

The dimension of this subspace equals the dimension of the subspace

$$\{\boldsymbol{P}^{\mathcal{U}}_{\bar{\mathbf{u}}}\nabla\bar{F}^{(1)}, ..., \boldsymbol{P}^{\mathcal{U}}_{\bar{\mathbf{u}}}\nabla\bar{F}^{(L)}\}_L \subset \mathcal{R}^n, \tag{14.8}$$

where $\boldsymbol{P}^{\mathcal{U}}_{\bar{\mathbf{u}}}$ be the n-dimensional projection matrix onto the tangent space $T_{\bar{\mathbf{u}}}(\mathcal{U})$. Consequently, the dimension of $\mathcal{L}_{\bar{\mathbf{u}}}$ is at most the rank r of the constraint (14.1), but we further assume the following[1]:

Assumption 14.4 *The dimension of the linear subspace $\mathcal{L}_{\bar{\mathbf{u}}}$ is equal to the rank r of the constraint.*

In mathematical terms, Assumption 14.4 states that each of the L equations (14.1) defines a manifold of *codimension 1* in the data space $\mathcal{U}$ such that they intersect each other *transversally* at $\bar{\boldsymbol{u}}$ (see Section 3.2.1). We say that the

[1] As a result, the L equations (14.1) cannot be replaced by a single equation, say $\sum_{k=1}^{L} F^{(k)}(\mathbf{u})^2 = 0$, which still has rank r but $\mathcal{L}_{\bar{\mathbf{u}}} = \{\mathbf{0}\}$.

intersection of the L manifolds is a *singular value* of the constraint (14.1) if the dimension of the subspace $\mathcal{L}_{\bar{\boldsymbol{u}}}$ is less than r, and an *ordinary value* otherwise. Assumption 14.4 requires that the true value $\bar{\boldsymbol{u}}$ should not be a singular value. We say that the constraint (14.1) is *nonsingular* if Assumption 14.4 is satisfied, and *singular* otherwise (see Section 5.1.1).

The linear subspace $\mathcal{L}_{\bar{\boldsymbol{u}}}$ defined by eq. (14.7) can alternatively be defined as the tangent space to the manifold

$$\mathcal{F} = \{\begin{pmatrix} F^{(1)}(\boldsymbol{u}) \\ \vdots \\ F^{(L)}(\boldsymbol{u}) \end{pmatrix} | \boldsymbol{u} \in \mathcal{U}\} \in \mathcal{R}^L \tag{14.9}$$

at $\bar{\boldsymbol{u}} \in \mathcal{U}$. However, $\bar{\boldsymbol{u}}$ can be a singular point of the mapping $(F^{(1)}(\boldsymbol{u}), ..., F^{(L)}(\boldsymbol{u}))^\top$: $\mathcal{U} \to \mathcal{R}^L$. In other words, the linear subspace $\mathcal{L}_{\boldsymbol{u}}$ defined by replacing $\bar{\boldsymbol{u}}$ by $\boldsymbol{u}$ ($\neq \bar{\boldsymbol{u}}$) in eq. (14.7) may have a higher dimension than r. If that is the case, we say that the constraint is *degenerate* (see Section 5.1.1).

14.2.3 Cramer-Rao lower bound for geometric correction

Let $\bar{\boldsymbol{l}}$ be the *score* of $\boldsymbol{u}$ with respect to $\bar{\boldsymbol{u}}$ (see eq. (3.113)):

$$\bar{\boldsymbol{l}} = \nabla_{\bar{\boldsymbol{u}}} \log p. \tag{14.10}$$

The *Fisher information matrix* $\boldsymbol{J}$ with respect to $\bar{\boldsymbol{u}}$ is defined as follows (see eq. (3.118)):

$$\boldsymbol{J} = E[\bar{\boldsymbol{l}}\bar{\boldsymbol{l}}^\top]. \tag{14.11}$$

The symbol $E[\cdot]$ denotes the expectation with respect to the probability density $p(\boldsymbol{u}; \bar{\boldsymbol{u}})$. Assumption 14.2 implies that $\bar{\boldsymbol{l}} \in T_{\bar{\boldsymbol{u}}}(\mathcal{U})$. We assume that the distribution $p(\boldsymbol{u}; \bar{\boldsymbol{u}})$ is *regular* with respect to $\bar{\boldsymbol{u}}$ in the sense that $\bar{\boldsymbol{l}}$ takes all orientations in $T_{\bar{\boldsymbol{u}}}(\mathcal{U})$ as $\boldsymbol{u}$ ranges over the entire data space $\mathcal{U}$ (see Section 3.5.1). In precise terms, this assumption is stated as follows:

Assumption 14.5 *The Fisher information matrix $\boldsymbol{J}$ is positive semi-definite and has range $T_{\bar{\boldsymbol{u}}}(\mathcal{U})$.*

Define an (LL)-matrix $\bar{\boldsymbol{V}} = (\bar{V}^{(kl)})$ by

$$(\bar{V}^{(kl)}) = \left((\nabla \bar{F}^{(k)}, \boldsymbol{J}^- \nabla \bar{F}^{(l)})\right), \tag{14.12}$$

and an (LL)-matrix $\bar{\boldsymbol{W}} = (\bar{W}^{(kl)})$ by $\bar{\boldsymbol{W}} = \bar{\boldsymbol{V}}^-$, or in the notation used in Chapter 5

$$(\bar{W}^{(kl)}) = \left((\nabla \bar{F}^{(k)}, \boldsymbol{J}^- \nabla \bar{F}^{(l)})\right)^-. \tag{14.13}$$

Define (nn)-matrices $\boldsymbol{N}$ and $\boldsymbol{S}$ by

$$\boldsymbol{N} = \sum_{k,l=1}^{L} \bar{W}^{(kl)} \left(\boldsymbol{J}^{-}\nabla\bar{F}^{(k)}\right)\left(\boldsymbol{J}^{-}\nabla\bar{F}^{(l)}\right)^{\top}, \tag{14.14}$$

$$\boldsymbol{S} = \boldsymbol{J}^{-} - \boldsymbol{N}. \tag{14.15}$$

Let $\hat{\boldsymbol{u}} = \hat{\boldsymbol{u}}(\boldsymbol{u})$ be an *unbiased estimator* of the true value $\bar{\boldsymbol{u}}$ determined as a function of the datum $\boldsymbol{u}$. Since $\bar{\boldsymbol{u}}$ is constrained to be in the model $\mathcal{S}$, it must satisfy $\hat{\boldsymbol{u}}(\boldsymbol{u}) \in \mathcal{S}$ for any $\boldsymbol{u} \in \mathcal{U}$. Its unbiasedness is defined by

$$\boldsymbol{P}^{\mathcal{S}}_{\bar{\boldsymbol{u}}} E[\hat{\boldsymbol{u}} - \bar{\boldsymbol{u}}] = \boldsymbol{0}, \tag{14.16}$$

where $\boldsymbol{P}^{\mathcal{S}}_{\bar{\boldsymbol{u}}}$ is the n-dimensional projection matrix onto the tangent space $T_{\bar{\boldsymbol{u}}}(\mathcal{S})$ to the model $\mathcal{S}$ at $\bar{\boldsymbol{u}}$ (see eq. (3.126)). Assuming that the distribution of $\hat{\boldsymbol{u}}$ is sufficiently localized around $\bar{\boldsymbol{u}}$ in the model $\mathcal{S}$, we define the covariance matrix of $\hat{\boldsymbol{u}}$ in the following form (see eq. (3.115)):

$$V[\hat{\boldsymbol{u}}] = \boldsymbol{P}^{\mathcal{S}}_{\bar{\boldsymbol{u}}} E[(\hat{\boldsymbol{u}} - \bar{\boldsymbol{u}})(\hat{\boldsymbol{u}} - \bar{\boldsymbol{u}})^{\top}]\boldsymbol{P}^{\mathcal{S}}_{\bar{\boldsymbol{u}}}. \tag{14.17}$$

We now prove the following theorem[2]:

Theorem 14.1

$$V[\hat{\boldsymbol{u}}] \succeq \boldsymbol{S}. \tag{14.18}$$

Since this is an analogue of the *Cramer-Rao inequality* (3.134) in Section 3.5.2, we call eq. (14.18) and the bound it gives the *Cramer-Rao inequality* and the *Cramer-Rao lower bound*, respectively, for geometric correction.

14.2.4 Proof of the main theorem

Lemma 14.1 *The (LL)-matrix $\bar{\boldsymbol{V}}$ is positive semi-definite and has range $\mathcal{L}_{\bar{\boldsymbol{u}}}$.*

Proof. Let $\boldsymbol{\eta} = (\eta_k)$ be an arbitrary L-vector. We have

$$(\boldsymbol{\eta}, \bar{\boldsymbol{V}}\boldsymbol{\eta}) = (\sum_{k=1}^{L} \eta_k \nabla\bar{F}^{(k)}, \boldsymbol{J}^{-}\sum_{l=1}^{L} \eta_l \nabla\bar{F}^{(l)}). \tag{14.19}$$

Since $\boldsymbol{J}$ is positive semi-definite, its generalized inverse $\boldsymbol{J}^{-}$ is also positive semi-definite. Hence, the right-hand side of eq. (14.19) is nonnegative for an arbitrary $\boldsymbol{\eta}$, meaning that $\bar{\boldsymbol{V}}$ is positive semi-definite. Suppose $\boldsymbol{\eta}$ belongs to the null space of $\bar{\boldsymbol{V}}$. Then, the left-hand side of eq. (14.19) is 0. It follows that $\sum_{k=1}^{L} \eta_k \nabla\bar{F}^{(k)}$ belongs to the null space of $\boldsymbol{J}^{-}$, which is identical with the

[2] Recall that the relation $\mathbf{A} \succeq \mathbf{B}$ means that matrix $\mathbf{A} - \mathbf{B}$ is positive semi-definite (see Section 3.5.2).

null space of $\boldsymbol{J}$. Assumption 14.5 implies that the null space of $\boldsymbol{J}$ coincides with $T_{\bar{\boldsymbol{u}}}(\mathcal{U})^{\perp}$. Hence,

$$\sum_{k=1}^{L} \eta_k (\nabla \bar{F}^{(k)}, \Delta \boldsymbol{u}) = 0 \tag{14.20}$$

for an arbitrary $\Delta \boldsymbol{u} \in T_{\bar{\boldsymbol{u}}}(\mathcal{U})$. From the definition of the subspace $\mathcal{L}_{\bar{\boldsymbol{u}}}$, this means that $\boldsymbol{\eta} \in \mathcal{L}_{\bar{\boldsymbol{u}}}^{\perp}$. Conversely, if $\boldsymbol{\eta} \in \mathcal{L}_{\bar{\boldsymbol{u}}}^{\perp}$, we obtain $(\boldsymbol{\eta}, \bar{\boldsymbol{V}}\boldsymbol{\eta}) = 0$ by following eqs. (14.20) and (14.19) backward. Hence, the null space of the (LL)-matrix $\bar{\boldsymbol{V}}$ coincides with the $(L-r)$-dimensional subspace $\mathcal{L}_{\bar{\boldsymbol{u}}}^{\perp} \subset \mathcal{R}^L$. Consequently, the range of $\bar{\boldsymbol{V}}$ coincides with $\mathcal{L}_{\bar{\boldsymbol{u}}}$. □

Consider the (LL)-matrix $\bar{\boldsymbol{W}}$ defined by eq. (14.13). Let $\boldsymbol{P}_{\bar{\boldsymbol{u}}}^{\mathcal{L}}$ be the L-dimensional projection matrix onto $\mathcal{L}_{\bar{\boldsymbol{u}}}$. Lemma 14.1 implies the following:

Corollary 14.1 *The (LL)-matrix $\bar{\boldsymbol{W}}$ is positive semi-definite and has range $\mathcal{L}_{\boldsymbol{u}}$, and the following relationship holds:*

$$\bar{\boldsymbol{V}}\bar{\boldsymbol{W}} = \boldsymbol{P}_{\bar{\boldsymbol{u}}}^{\mathcal{L}}. \tag{14.21}$$

Lemma 14.2

$$\boldsymbol{N}\boldsymbol{J}\boldsymbol{N} = \boldsymbol{N}. \tag{14.22}$$

Proof. Since $\boldsymbol{J}^{-}\boldsymbol{J}\boldsymbol{J}^{-} = \boldsymbol{J}^{-}$, we see that

$$\begin{aligned}
&\boldsymbol{N}\boldsymbol{J}\boldsymbol{N} = \\
&= \left(\sum_{k,l=1}^{L} \bar{W}^{(kl)} \left(\boldsymbol{J}^{-}\nabla \bar{F}^{(k)}\right) \left(\boldsymbol{J}^{-}\nabla \bar{F}^{(l)}\right)^{\top} \right) \boldsymbol{J} \\
&\quad \left(\sum_{m,n=1}^{L} \bar{W}^{(mn)} \left(\boldsymbol{J}^{-}\nabla \bar{F}^{(m)}\right) \left(\boldsymbol{J}^{-}\nabla \bar{F}^{(n)}\right)^{\top} \right) \\
&= \sum_{k,l,m,n=1}^{L} \bar{W}^{(kl)}\bar{W}^{(mn)} \left(\boldsymbol{J}^{-}\nabla \bar{F}^{(k)}\right) (\nabla \bar{F}^{(l)}, \boldsymbol{J}^{-}\boldsymbol{J}\boldsymbol{J}^{-}\nabla \bar{F}^{(m)}) \left(\boldsymbol{J}^{-}\nabla \bar{F}^{(n)}\right)^{\top} \\
&= \sum_{k,n=1}^{L} \left(\sum_{l,m=1}^{L} \bar{W}^{(kl)}\bar{V}^{(lm)}\bar{W}^{(mn)} \right) \left(\boldsymbol{J}^{-}\nabla \bar{F}^{(k)}\right) \left(\boldsymbol{J}^{-}\nabla \bar{F}^{(n)}\right)^{\top} \\
&= \sum_{k,n=1}^{L} \bar{W}^{(kn)} \left(\boldsymbol{J}^{-}\nabla \bar{F}^{(k)}\right) \left(\boldsymbol{J}^{-}\nabla \bar{F}^{(n)}\right)^{\top} = \boldsymbol{N},
\end{aligned} \tag{14.23}$$

where the identity $\bar{\boldsymbol{W}}\bar{\boldsymbol{V}}\bar{\boldsymbol{W}} = \bar{\boldsymbol{W}}\bar{\boldsymbol{W}}^{-}\bar{\boldsymbol{W}} = \bar{\boldsymbol{W}}$ is used. □

Lemma 14.3

$$\boldsymbol{S}\boldsymbol{J}\boldsymbol{S} = \boldsymbol{S}. \tag{14.24}$$

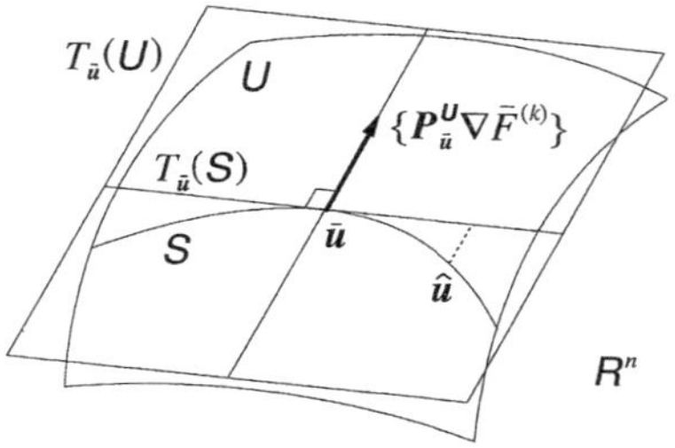

Fig. 14.2. The tangent space $T_{\bar{\mathbf{u}}}(\mathcal{S})$ to model $\mathcal{S}$ at $\bar{\mathbf{u}}$ and its orthogonal complement with respect to $T_{\bar{\mathbf{u}}}(\mathcal{U})$.

Proof. Since $\boldsymbol{J}^-\boldsymbol{J}\boldsymbol{J}^- = \boldsymbol{J}^-$, we see that

$$\begin{aligned}\boldsymbol{SJS} &= (\boldsymbol{J}^- - \boldsymbol{N})\boldsymbol{J}(\boldsymbol{J}^- - \boldsymbol{N}) = \boldsymbol{J}^-\boldsymbol{J}\boldsymbol{J}^- - \boldsymbol{J}^-\boldsymbol{J}\boldsymbol{N} - \boldsymbol{N}\boldsymbol{J}\boldsymbol{J}^- + \boldsymbol{N}\boldsymbol{J}\boldsymbol{N}\\ &= \boldsymbol{J}^- - \boldsymbol{P}^{\mathcal{U}}_{\bar{\boldsymbol{u}}}\boldsymbol{N} - \boldsymbol{N}\boldsymbol{P}^{\mathcal{U}}_{\bar{\boldsymbol{u}}} + \boldsymbol{N} = \boldsymbol{J}^- - \boldsymbol{N} = \boldsymbol{S},\end{aligned} \tag{14.25}$$

where the relationship $\boldsymbol{J}^-\boldsymbol{J} = \boldsymbol{J}\boldsymbol{J}^- = \boldsymbol{P}^{\mathcal{U}}_{\bar{\boldsymbol{u}}}$ implied by Assumption 14.5 and the relationship $\boldsymbol{P}^{\mathcal{U}}_{\bar{\boldsymbol{u}}}\boldsymbol{N} = \boldsymbol{N}\boldsymbol{P}^{\mathcal{U}}_{\bar{\boldsymbol{u}}} = \boldsymbol{N}$ implied by eq. (14.14) are used. □

The following lemma plays an essential role in proving Theorem 14.1.

Lemma 14.4 *The (nn)-matrix $\boldsymbol{S}$ is positive semi-definite and has range $T_{\bar{\boldsymbol{u}}}(\mathcal{S})$.*

Proof. By definition, $\boldsymbol{S}$ is a positive semi-definite matrix. From Lemma 14.3, we see that for an arbitrary n-vector $\boldsymbol{\xi}$ we see that

$$(\boldsymbol{\xi}, \boldsymbol{S}\boldsymbol{\xi}) = (\boldsymbol{\xi}, \boldsymbol{SJS}\boldsymbol{\xi}) = (\boldsymbol{S}\boldsymbol{\xi}, \boldsymbol{J}(\boldsymbol{S}\boldsymbol{\xi})). \tag{14.26}$$

Since $\boldsymbol{J}$ is positive semi-definite, the above expression is nonnegative for any $\boldsymbol{\xi} \in \mathcal{R}^n$. Hence, $\boldsymbol{S}$ is positive semi-definite. By definition, the tangent space $T_{\bar{\boldsymbol{u}}}(\mathcal{S})$ is the orthogonal complement of the linear subspace $\{\boldsymbol{P}^{\mathcal{U}}_{\bar{\boldsymbol{u}}}\nabla\bar{F}^{(k)}\}_L \subset T_{\bar{\boldsymbol{u}}}(\mathcal{U})$ (Fig. 14.2). From eqs. (14.14) and (14.15), we see that

$$\begin{aligned}&(\left(\boldsymbol{P}^{\mathcal{U}}_{\bar{\boldsymbol{u}}}\nabla\bar{F}^{(k)}\right), \boldsymbol{S}\left(\boldsymbol{P}^{\mathcal{U}}_{\bar{\boldsymbol{u}}}\nabla\bar{F}^{(k)}\right))\\ &\quad= (\boldsymbol{P}^{\mathcal{U}}_{\bar{\boldsymbol{u}}}\nabla\bar{F}^{(k)}, \boldsymbol{J}^-\boldsymbol{P}^{\mathcal{U}}_{\bar{\boldsymbol{u}}}\nabla\bar{F}^{(k)})\\ &\quad\quad - \sum_{l,m=1}^{L}\bar{W}^{(lm)}(\boldsymbol{P}^{\mathcal{U}}_{\bar{\boldsymbol{u}}}\nabla\bar{F}^{(k)}, \boldsymbol{J}^-\nabla\bar{F}^{(l)})(\boldsymbol{J}^-\nabla\bar{F}^{(m)}, \boldsymbol{P}^{\mathcal{U}}_{\bar{\boldsymbol{u}}}\nabla\bar{F}^{(k)})\\ &\quad= (\nabla\bar{F}^{(k)}, \boldsymbol{P}^{\mathcal{U}}_{\bar{\boldsymbol{u}}}\boldsymbol{J}^-\boldsymbol{P}^{\mathcal{U}}_{\bar{\boldsymbol{u}}}\nabla\bar{F}^{(k)})\\ &\quad\quad - \sum_{l,m=1}^{L}\bar{W}^{(lm)}(\nabla\bar{F}^{(k)}, \boldsymbol{P}^{\mathcal{U}}_{\bar{\boldsymbol{u}}}\boldsymbol{J}^-\nabla\bar{F}^{(l)})(\nabla\bar{F}^{(m)}, \boldsymbol{J}^-\boldsymbol{P}^{\mathcal{U}}_{\bar{\boldsymbol{u}}}\nabla\bar{F}^{(k)})\end{aligned}$$

$$= (\nabla\bar{F}^{(k)}, \boldsymbol{J}^{-}\nabla\bar{F}^{(k)}) - \sum_{l,m=1}^{L} (\nabla\bar{F}^{(k)}, \boldsymbol{J}^{-}\nabla\bar{F}^{(l)})\bar{W}^{(lm)}(\nabla\bar{F}^{(m)}, \boldsymbol{J}^{-}\nabla\bar{F}^{(k)})$$

$$= \bar{V}^{(kk)} - \sum_{l,m=1}^{L} \bar{V}^{(kl)}\bar{W}^{(lm)}\bar{V}^{(mk)} = 0, \tag{14.27}$$

where the identity $\bar{\boldsymbol{V}}\bar{\boldsymbol{W}}\bar{\boldsymbol{V}} = \bar{\boldsymbol{V}}\bar{\boldsymbol{V}}^{-}\bar{\boldsymbol{V}} = \bar{\boldsymbol{V}}$ is used. Eq. (14.27) implies that the linear subspace $\{\boldsymbol{P}_{\bar{\boldsymbol{u}}}^{\mathcal{U}}\nabla\bar{F}^{(k)}\}_L$ is included in the null space of $\boldsymbol{S}$. This means that the range of $\boldsymbol{S}$ is included in $T_{\bar{\boldsymbol{u}}}(\mathcal{S})$. Hence, the rank of $\boldsymbol{S}$ is at most $n' - r$. We now prove that the range of $\boldsymbol{S}$ exactly coincides with $T_{\bar{\boldsymbol{u}}}(\mathcal{S})$ by showing that $\boldsymbol{S}$ has at least rank $n' - r$.

Let $\mathcal{J}$ be the linear subspace of $\mathcal{R}^n$ obtained by mapping the r-dimensional subspace $\{\boldsymbol{P}_{\bar{\boldsymbol{u}}}^{\mathcal{U}}\nabla\bar{F}^{(k)}\}_L \subset T_{\bar{\boldsymbol{u}}}(\mathcal{U})$ by $\boldsymbol{J}^{-}$. Assumption 14.5 implies that $\boldsymbol{J}^{-}$ has range $T_{\bar{\boldsymbol{u}}}(\mathcal{U})$, so $\mathcal{J}$ is an r-dimensional subspace of $T_{\bar{\boldsymbol{u}}}(\mathcal{U})$. Let $\{\boldsymbol{v}_1, ..., \boldsymbol{v}_{n'-r}\}$ be a basis of the orthogonal complement of $\mathcal{J}$ with respect to $T_{\bar{\boldsymbol{u}}}(\mathcal{U})$. Since the vectors $\boldsymbol{v}_i$, $i = 1, ..., n' - r$, are orthogonal to vectors $\boldsymbol{J}^{-}\boldsymbol{P}_{\bar{\boldsymbol{u}}}^{\mathcal{U}}\nabla\bar{F}^{(k)}$, $k = 1, ..., L$, we obtain

$$\boldsymbol{S}\boldsymbol{v}_i = \boldsymbol{J}^{-}\boldsymbol{v}_i - \sum_{k,l=1}^{L} \bar{W}^{(kl)}(\boldsymbol{J}^{-}\nabla\bar{F}^{(l)}, \boldsymbol{v}_i)\boldsymbol{J}^{-}\nabla\bar{F}^{(k)} = \boldsymbol{J}^{-}\boldsymbol{v}_i. \tag{14.28}$$

By Assumption 14.5, matrix $\boldsymbol{J}^{-}$ is positive semi-definite and has range $T_{\bar{\boldsymbol{u}}}(\mathcal{U})$. Since vectors $\{\boldsymbol{v}_i\} \in T_{\bar{\boldsymbol{u}}}(\mathcal{U})$, $i = 1, ..., n' - r$, are linearly independent, vectors $\{\boldsymbol{J}^{-}\boldsymbol{v}_i\} = \{\boldsymbol{S}\boldsymbol{v}_i\}$, $i = 1, ..., n' - r$, are also linearly independent. Thus, $\boldsymbol{S}$ has at least rank $n' - r$. □

Corollary 14.2 *The variation* $\Delta\bar{\boldsymbol{u}}$ *of* $\bar{\boldsymbol{u}}$ *defined by*

$$\Delta\bar{\boldsymbol{u}} = \boldsymbol{S}\Delta\boldsymbol{\xi} \tag{14.29}$$

satisfies $\Delta\bar{\boldsymbol{u}} \in T_{\bar{\boldsymbol{u}}}(\mathcal{S})$ *for any* $\boldsymbol{\xi} \in \mathcal{R}^n$.

Lemma 14.5

$$E[\boldsymbol{P}_{\bar{\boldsymbol{u}}}^{\mathcal{S}}(\hat{\boldsymbol{u}} - \bar{\boldsymbol{u}})(\boldsymbol{S}\bar{\boldsymbol{l}})^{\top}] = \boldsymbol{S}. \tag{14.30}$$

Proof. Since the estimator $\hat{\boldsymbol{u}}$ is unbiased, eq. (14.16) holds identically for any $\bar{\boldsymbol{u}} \in \mathcal{S}$. It follows that the first variation of the left-hand side of eq. (14.16) should be $\boldsymbol{0}$ for any $\Delta\bar{\boldsymbol{u}} \in T_{\bar{\boldsymbol{u}}}(\mathcal{S})$. Hence, it should be $\boldsymbol{0}$ for $\Delta\bar{\boldsymbol{u}} = \boldsymbol{S}\Delta\boldsymbol{\xi}$ for any $\Delta\boldsymbol{\xi} \in \mathcal{R}^n$. Taking the first variation of the left-hand side of eq. (14.16) with respect to $\bar{\boldsymbol{u}}$, substituting eq. (14.29) into it, noting Assumption 14.1, and using the logarithmic differentiation formula (3.115), we obtain

$$\begin{aligned}\Delta(\boldsymbol{P}_{\bar{\boldsymbol{u}}}^{\mathcal{S}}E[\hat{\boldsymbol{u}} - \bar{\boldsymbol{u}}]) &= -\boldsymbol{P}_{\bar{\boldsymbol{u}}}^{\mathcal{S}}E[\Delta\bar{\boldsymbol{u}})] + \boldsymbol{P}_{\bar{\boldsymbol{u}}}^{\mathcal{S}}\int_{\mathcal{U}}(\hat{\boldsymbol{u}} - \bar{\boldsymbol{u}})(\nabla_{\bar{\boldsymbol{u}}}\log p, \Delta\bar{\boldsymbol{u}})pd\boldsymbol{u} \\ &= \boldsymbol{P}_{\bar{\boldsymbol{u}}}^{\mathcal{S}}\boldsymbol{S}\Delta\xi - \boldsymbol{P}_{\bar{\boldsymbol{u}}}^{\mathcal{S}}E[(\hat{\boldsymbol{u}} - \bar{\boldsymbol{u}})\bar{\boldsymbol{l}}^{\top}]\Delta\xi \\ &= -\left(\boldsymbol{S} - E[\boldsymbol{P}_{\bar{\boldsymbol{u}}}^{\mathcal{S}}(\hat{\boldsymbol{u}} - \bar{\boldsymbol{u}})(\boldsymbol{S}\bar{\boldsymbol{l}})^{\top}]\right)\Delta\xi. \end{aligned} \tag{14.31}$$

Here, we have used the assumption that $P^S_{\bar{u}+\Delta u}E[\hat{u}-\bar{u}] \approx P^S_{\bar{u}}E[\hat{u}-\bar{u}]$ (see Section 3.5.2) and the relationship $P^S_{\bar{u}}S = S$. Since the above expression should be $\mathbf{0}$ for any $\Delta\xi \in \mathcal{R}^n$, we obtain eq. (14.30). □

The proof of Theorem 14.1 is given as follows. From eq. (14.11) and Lemma 14.3, we see that

$$E[(S\bar{l})(S\bar{l})^\top] = SE[\bar{l}\bar{l}^\top]S = SJS = S. \tag{14.32}$$

From this and Lemmas 14.5, we obtain

$$E[\begin{pmatrix} P^S_{\bar{u}}(\hat{u}-\bar{u}) \\ S\bar{l} \end{pmatrix}\begin{pmatrix} P^S_{\bar{u}}(\hat{u}-\bar{u}) \\ S\bar{l} \end{pmatrix}^\top] = \begin{pmatrix} V[\hat{u}] & S \\ S & S \end{pmatrix}. \tag{14.33}$$

This matrix is symmetric and positive semi-definite. If matrix A is positive semi-definite and symmetric, so is $B^\top AB$ for any matrix B as long as the multiplication can be defined (see Section 2.2.3). Hence, the following matrix is positive semi-definite and symmetric:

$$\begin{pmatrix} P^S_{\bar{u}} & -P^S_{\bar{u}} \\ & P^S_{\bar{u}} \end{pmatrix}\begin{pmatrix} V[\hat{u}] & S \\ S & S \end{pmatrix}\begin{pmatrix} P^S_{\bar{u}} & \\ -P^S_{\bar{u}} & P^S_{\bar{u}} \end{pmatrix} = \begin{pmatrix} V[\hat{u}]-S & \\ & S \end{pmatrix}. \tag{14.34}$$

Here, we have used the fact that matrices $V[\hat{u}]$ and S share the same range $T_{\bar{u}}(\mathcal{S})$ and hence $P^S_{\bar{u}}S = SP^S_{\bar{u}} = S$ and $P^S_{\bar{u}}V[\hat{u}]P^S_{\bar{u}} = V[\hat{u}]$. Since S is positive semi-definite, matrix $V[\hat{u}] - S$ is also positive semi-definite. □

14.3 Maximum Likelihood Correction

14.3.1 Maximum likelihood estimator

Maximum likelihood estimation for this problem is to choose $\bar{u}$ that maximize the *likelihood*, i.e., the probability density $p(u;\bar{u})$ viewed as a function of $\bar{u}$ for a given datum u. The resulting estimate $\hat{u}$ is called the *maximum likelihood estimator* of $\bar{u}$ (see Section 3.6.1). In order to distinguish "variables" from their specific "values", we regard the symbol $\bar{u}$ as the true value and use x when it is viewed as a variable. Since maximizing the likelihood is equivalent to minimizing its logarithm with opposite sign, the problem reduces to the minimization

$$J = -2\log p(u;x) \to \min \tag{14.35}$$

with respect to $x \in \mathcal{S}$, i.e., under the constraint

$$F^{(k)}(x) = 0, \qquad k = 1, ..., L, \qquad x \in \mathcal{U}. \tag{14.36}$$

Define an (nn)-matrix $\bar{L}$ by

$$\bar{L} = -\nabla^2_x \log p|_{x=\bar{u}}. \tag{14.37}$$

With the expectation that the maximum likelihood estimator $\hat{\boldsymbol{u}}$ is located near $\bar{\boldsymbol{u}}$, we write

$$\boldsymbol{x} = \bar{\boldsymbol{u}} + \Delta\boldsymbol{x}. \tag{14.38}$$

Substituting this into the function J in eq. (14.35) and expanding it in the neighborhood of $\bar{\boldsymbol{u}}$, we obtain

$$J = \bar{c} - 2(\bar{\boldsymbol{l}}, \Delta\boldsymbol{x}) + (\Delta\boldsymbol{x}, \bar{\boldsymbol{L}}\Delta\boldsymbol{x}) + O(\Delta\boldsymbol{x})^3, \tag{14.39}$$

where $\bar{c}$ is the value of J evaluated at $\bar{\boldsymbol{u}}$. Assumption 14.2 implies that the rank of $\bar{\boldsymbol{L}}$ is no more than the dimension n' of the tangent space $T_{\boldsymbol{x}}(\mathcal{U})$. In order to guarantee unique existence of the value $\boldsymbol{x} \in \mathcal{U}$ that minimizes J, we assume the following:

Assumption 14.6 *For any $\boldsymbol{u} \in \mathcal{U}$, the (nn)-matrix $\bar{\boldsymbol{L}}$ is positive semi-definite and has range $T_{\bar{\boldsymbol{u}}}(\mathcal{U})$.*

Substituting eq. (14.38) into eq. (14.36) and taking a linear approximation, we obtain

$$(\nabla\bar{F}^{(k)}, \Delta\boldsymbol{x}) = 0. \tag{14.40}$$

The inherent constraint $\boldsymbol{x} \in \mathcal{U}$ requires that $\Delta\boldsymbol{x} \in T_{\bar{\boldsymbol{u}}}(\mathcal{U})$ to a first approximation. Introducing Lagrange multipliers $\bar{\lambda}^{(k)}$ to eq. (14.40), differentiating J with respect to $\Delta\boldsymbol{x}$, and ignoring higher order terms, we obtain the following linear equation in $\Delta\boldsymbol{x}$:

$$\bar{\boldsymbol{L}}\Delta\boldsymbol{x} = \bar{\boldsymbol{l}} + \sum_{k=1}^{L} \bar{\lambda}^{(k)} \boldsymbol{P}^{\mathcal{U}}_{\bar{\boldsymbol{u}}} \nabla\bar{F}^{(k)}. \tag{14.41}$$

By Assumptions 14.2 and 14.6, eq. (14.41) has the following unique solution (see Section 2.3.3):

$$\Delta\boldsymbol{x} = \bar{\boldsymbol{L}}^{-}\bar{\boldsymbol{l}} + \sum_{k=1}^{L} \bar{\lambda}^{(k)} \bar{\boldsymbol{L}}^{-}\nabla\bar{F}^{(k)}. \tag{14.42}$$

Substituting eq. (14.42) into eq. (14.40), we obtain

$$\bar{\boldsymbol{V}}'\bar{\boldsymbol{\lambda}} = -\Big((\nabla\bar{F}^{(k)}, \Delta\boldsymbol{u})\Big), \tag{14.43}$$

where we put $\bar{\boldsymbol{\lambda}} = (\bar{\lambda}^{(k)})$ and define the (LL)-matrix $\bar{\boldsymbol{V}}' = (\bar{V}'^{(kl)})$ by

$$(\bar{V}'^{(kl)}) = \Big((\nabla\bar{F}^{(k)}, \bar{\boldsymbol{L}}^{-}\nabla\bar{F}^{(l)})\Big). \tag{14.44}$$

Since matrix $\bar{\boldsymbol{L}}$ has range $T_{\boldsymbol{x}}(\mathcal{U})$, matrix $\bar{\boldsymbol{V}}'$ can be shown to be positive semi-definite and have range $\mathcal{L}_{\bar{\boldsymbol{u}}}$ by exactly the same argument as given in the proof of Lemma 14.6. Consequently, if $\boldsymbol{u}$ is an ordinary value (see Section 14.2.2), the solution of eq. (14.43) is uniquely given by

$$\bar{\lambda}^{(k)} = -\sum_{l=1}^{L} \bar{W}'^{(kl)}(\nabla\bar{F}^{(l)}, \Delta\boldsymbol{u}), \tag{14.45}$$

where $\bar{W}'^{(kl)}$ is the (kl) element of the (LL)-matrix $\bar{\boldsymbol{W}}' = \bar{\boldsymbol{V}}'^{-}$, i.e.,

$$(\bar{W}'^{(kl)}) = \left((\nabla\bar{F}^{(k)}, \bar{\boldsymbol{L}}^{-}\nabla\bar{F}^{(l)})\right)^{-}. \tag{14.46}$$

Substituting eq. (14.45) into eq. (14.42), we can express the maximum likelihood estimator $\hat{\boldsymbol{u}}$ in the form

$$\hat{\boldsymbol{u}} = \bar{\boldsymbol{u}} + \boldsymbol{S}'\bar{\boldsymbol{l}}, \tag{14.47}$$

where we define the (nn)-matrix $\boldsymbol{S}'$ by

$$\boldsymbol{S}' = \bar{\boldsymbol{L}}^{-} - \sum_{k,l=1}^{L} \bar{W}'^{(kl)}(\bar{\boldsymbol{L}}^{-}\nabla\bar{F}^{(k)})(\bar{\boldsymbol{L}}^{-}\nabla\bar{F}^{(l)})^{\top}. \tag{14.48}$$

By the argument similar to that in the proofs of Lemmas 14.2 and 14.3, we can show that $\boldsymbol{S}'\bar{\boldsymbol{L}}\boldsymbol{S}' = \boldsymbol{S}'$. Hence, the (nn)-matrix $\boldsymbol{S}'$ can be shown to be positive semi-definite and have range $T_{\bar{\boldsymbol{u}}}(\mathcal{S})$ by the same argument as given in the proof of Lemma 14.4. Consequently, the covariance matrix of the maximum likelihood estimator $\hat{\boldsymbol{u}}$ has the following form:

$$V[\hat{\boldsymbol{u}}] = E[\boldsymbol{S}'\bar{\boldsymbol{l}}\bar{\boldsymbol{l}}^{\top}\boldsymbol{S}']. \tag{14.49}$$

14.3.2 Geometric correction of the exponential family

If the distribution of $\boldsymbol{u}$ belongs to the *exponential family* (see Section 3.6.1), the probability density $p(\boldsymbol{u};\bar{\boldsymbol{u}})$ has the following expression (see eq. (3.146)):

$$p(\boldsymbol{u};\bar{\boldsymbol{u}}) = C(\bar{\boldsymbol{u}})\exp[(\boldsymbol{f}(\boldsymbol{u}),\bar{\boldsymbol{u}}) + g(\boldsymbol{u})]. \tag{14.50}$$

Here, $\boldsymbol{f}(\boldsymbol{u})$ is an n-dimensional function of $\boldsymbol{u} \in \mathcal{R}^n$, and $C(\boldsymbol{u})$ and $g(\boldsymbol{u})$ are scalar functions of $\boldsymbol{u} \in \mathcal{R}^n$. We call Problem 14.1 *geometric correction of the exponential family* if the probability density $p(\boldsymbol{u};\bar{\boldsymbol{u}})$ has the above expression.

For geometric correction of the exponential family, we see from eqs. (14.37) and (14.50) that the matrix $\bar{\boldsymbol{L}}$ does not depend on $\boldsymbol{u}$. Hence, the following equality holds (see eq. (3.122)):

$$\boldsymbol{J} = E[\bar{\boldsymbol{L}}] = \bar{\boldsymbol{L}}. \tag{14.51}$$

From this, we obtain the following proposition:

Proposition 14.1 *The maximum likelihood estimator $\hat{\boldsymbol{u}}$ for geometric correction of the exponential family is unbiased in the first order.*

Proof. If eq. (14.51) holds, the matrix $\boldsymbol{S}'$ defined by eq. (14.48) is a constant matrix. Since $E[\bar{\boldsymbol{l}}] = \boldsymbol{0}$ (see eq. (3.117)), we see from eq. (14.47) that

$$\boldsymbol{P}^S_{\bar{\boldsymbol{u}}} E[\hat{\boldsymbol{u}} - \bar{\boldsymbol{u}}] = \boldsymbol{S}E[\Delta\boldsymbol{u}] = \boldsymbol{0}. \tag{14.52}$$

□

An estimator is said to be *efficient* if its covariance matrix attains the Cramer-Rao lower bound (see Section 3.5.2). Extending this concept, we say that an estimator for geometric correction is *efficient* if its covariance matrix attains the Cramer-Rao lower bound given by Theorem 14.1.

Proposition 14.2 *The maximum likelihood estimator $\hat{\boldsymbol{u}}$ for geometric correction of the exponential family is efficient in the first order.*

Proof. Since $\bar{\boldsymbol{L}} = \boldsymbol{J}$, we see that $\bar{\boldsymbol{V}}' = \bar{\boldsymbol{V}}$ (see eqs. (14.12) and (14.44)) and hence $\bar{\boldsymbol{W}}' = \bar{\boldsymbol{W}}$ (see eqs. (14.13) and (14.46)). It follows that $\boldsymbol{S}' = \bar{\boldsymbol{S}}$ (see eqs. (14.15) and (14.48)). From eq. (14.49), we see that

$$V[\hat{\boldsymbol{u}}] = \boldsymbol{S}E[\bar{\boldsymbol{l}}\bar{\boldsymbol{l}}^\top]\boldsymbol{S} = \boldsymbol{SJS} = \boldsymbol{S}, \tag{14.53}$$

where we have used Lemma 14.3. □

The proofs of Propositions 14.1 an 14.2 are based on eq. (14.47), which has been obtained by approximating the function J in eq. (14.35) by a quadratic polynomial in the neighborhood of their true values and applying a linear approximation to the constraint (14.36). In other words, eq. (14.47) is an *expression in the limit of small noise*. Accordingly, Propositions 14.1 and 14.2 hold in this sense, which is what the phrase "in the first order" means. This type of first order analysis is compatible with our approach of studying statistical behavior for small noise (see Section 14.1).

14.3.3 Computation of maximum likelihood correction

We now derive an analytical expression for the maximum likelihood estimator $\hat{\boldsymbol{u}}$. First, we introduce a new assumption about the probability density $p(\boldsymbol{u}; \boldsymbol{x})$. We assume that viewed as a function of $\boldsymbol{x}$ it takes its maximum at $\boldsymbol{u}$. In other words, given a single datum $\boldsymbol{u}$, the datum $\boldsymbol{u}$ itself is the most likely value of the true value $\bar{\boldsymbol{u}}$ if no other information is given. To be specific, what we need is the following:

Assumption 14.7

$$\nabla_{\boldsymbol{x}} p(\boldsymbol{u}; \boldsymbol{x})|_{\boldsymbol{x}=\boldsymbol{u}} = \boldsymbol{0}. \tag{14.54}$$

For example, this assumption is satisfied if the probability density is expressed by a smooth scalar function $f(\cdot)$ that takes its maximum at the origin $\mathbf{0}$ in the form $p(\boldsymbol{u}; \bar{\boldsymbol{u}}) = f(\boldsymbol{u} - \bar{\boldsymbol{u}})$. The Gaussian distribution is a typical example.

Assuming that the solution $\boldsymbol{x}$ of the optimization (14.35) is close to the datum $\boldsymbol{u}$, we write $\boldsymbol{x} = \boldsymbol{u} + \Delta\boldsymbol{x}$. Substituting this into the function J, expanding it in the neighborhood of the datum $\boldsymbol{u}$, and using Assumption 14.7, we obtain

$$J = c - (\Delta\boldsymbol{x}, \boldsymbol{L}\Delta\boldsymbol{x}) + O(\Delta\boldsymbol{x})^3. \tag{14.55}$$

The constant c is the value of J evaluated at $\boldsymbol{u}$, and $\boldsymbol{L}$ is an (nn)-matrix defined by

$$\boldsymbol{L} = -\nabla^2_{\boldsymbol{x}} \log p|_{\boldsymbol{x}=\boldsymbol{u}}. \tag{14.56}$$

In order to guarantee unique existence of the value $\boldsymbol{x} \in \mathcal{U}$ that minimizes J, we assume that Assumption 14.6 also applies to $\boldsymbol{L}$:

Assumption 14.8 *For any $\boldsymbol{u} \in \mathcal{U}$, the (nn)-matrix $\boldsymbol{L}$ is positive semi-definite and has range $T_{\boldsymbol{u}}(\mathcal{U})$.*

The linear approximation of the constraint (14.36) is

$$(\nabla F^{(k)}, \Delta\boldsymbol{x}) = -F^{(k)}, \tag{14.57}$$

where $F^{(k)}$ and $\nabla F^{(k)}$ are the abbreviations of $F^{(k)}(\boldsymbol{u})$ and $\nabla F^{(k)}(\boldsymbol{u})$, respectively. The inherent constraint $\boldsymbol{x} \in \mathcal{U}$ requires that $\Delta\boldsymbol{x} \in T_{\boldsymbol{u}}(\mathcal{U})$ to a first approximation. Introducing Lagrange multipliers $\lambda^{(k)}$ to eq. (14.57), differentiating J with respect to $\Delta\boldsymbol{x}$, and ignoring higher order terms, we obtain the solution $\Delta\boldsymbol{x}$ uniquely in the following form:

$$\Delta\boldsymbol{x} = \sum_{k=1}^{L} \lambda^{(k)} \boldsymbol{L}^- \nabla F^{(k)}. \tag{14.58}$$

Substituting this into eq. (14.57) and putting $\boldsymbol{\lambda} = (\lambda^{(k)})$, we obtain

$$\boldsymbol{V}\boldsymbol{\lambda} = -(F^{(k)}), \tag{14.59}$$

where the (LL)-matrix $\boldsymbol{V} = (V^{(kl)})$ is defined by

$$(V^{(kl)}) = \left((\nabla F^{(k)}, \boldsymbol{L}^- \nabla F^{(l)})\right). \tag{14.60}$$

Here, a *computational problem* arises. If the L equations (14.36) are not algebraically independent but give only r constraints, the matrix $\bar{\boldsymbol{V}}'$ computed from the true value $\bar{\boldsymbol{u}}$ by eq. (14.44) has rank r. However, the rank of the matrix $\boldsymbol{V}$ computed from the datum $\boldsymbol{u}$ by eq. (14.60) may be larger than r. In other words, among the L linear equations in $\Delta\boldsymbol{x}$ given by eq. (14.57), more than r equations may be linearly independent if evaluated at $\boldsymbol{u} \neq \bar{\boldsymbol{u}}$. If this

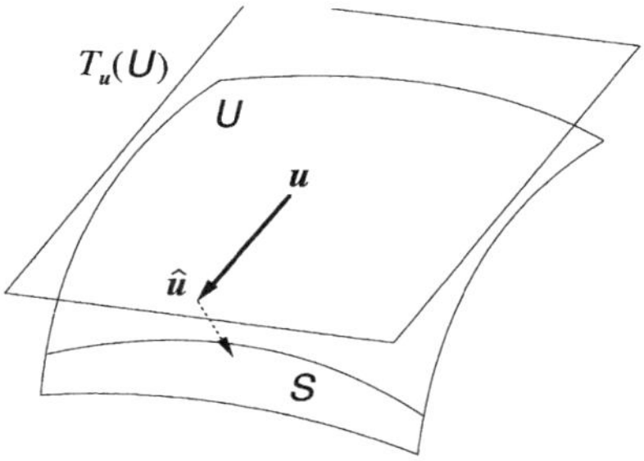

Fig. 14.3. The corrected value $\hat{\mathbf{u}}$ is in $T_{\mathbf{u}}(\mathcal{U})$ but may not be in $\mathcal{U}$. Hence, a higher order correction is necessary.

occurs, the constraint (14.36) is said to be *degenerate* (see Section 14.2.2). If the constraint (14.36) is degenerate, some of the positive eigenvalues of matrix $\boldsymbol{V}$ converge to 0 in the limit $\boldsymbol{u} \to \bar{\boldsymbol{u}}$. Hence, eq. (14.59) viewed as a linear equation in $\boldsymbol{\lambda}$ is *ill-conditioned* when the datum $\boldsymbol{u}$ is close to the true value $\bar{\boldsymbol{u}}$ (see Section 2.3.2).

In order to avoid this ill-posedness, we *project* both sides of eq. (14.59) onto the eigenspace defined by the largest r eigenvalues of $\boldsymbol{V}$. In other words, we effectively use only r constraints from among the L linear equations (14.57). The resulting solution may be slightly different from the exact solution of eq. (14.59) but does not effectively affect the subsequent analysis because eq. (14.59) is a first order approximation. The solution of eq. (14.59) given by this projection is

$$\lambda^{(k)} = -\sum_{l=1}^{L} W^{(kl)} F^{(l)}, \tag{14.61}$$

where $W^{(kl)}$ is the (kl) element of the rank-constrained generalized inverse $(\boldsymbol{V})_r^-$ (see eq. (2.132)), which we write

$$(W^{(kl)}) = \left((\nabla F^{(k)}, \boldsymbol{L}^- \nabla F^{(l)})\right)_r^- . \tag{14.62}$$

Substituting eq. (14.61) into eq. (14.58), we obtain the following conclusion:

Proposition 14.3 *The maximum likelihood estimator $\hat{\boldsymbol{u}}$ is computed to a first approximation in the form*

$$\hat{\boldsymbol{u}} = \boldsymbol{u} - \boldsymbol{L}^- \sum_{k,l=1}^{L} W^{(kl)} F^{(k)} \nabla F^{(l)}. \tag{14.63}$$

This gives only a first order solution under the linearized inherent constraint $\hat{\boldsymbol{u}} - \boldsymbol{u} \in T_{\boldsymbol{u}}(\mathcal{U})$ (Fig. 14.3). In actual computation, we need to *projected*

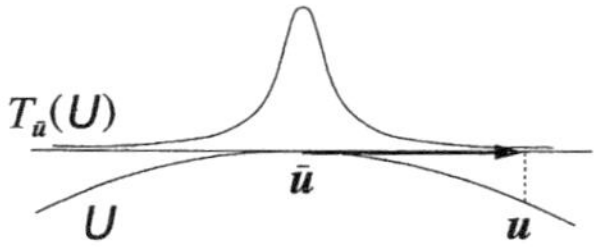

Fig. 14.4. Locally Gaussian distribution.

thus computed value $\hat{\boldsymbol{u}}$ onto the data space $\mathcal{U}$ by adding a higher order correction, as discussed in Section 5.1.3. The resulting solution still may not be exactly in the model $\mathcal{S} \subset \mathcal{U}$, so the correction is repeated until the constraint is sufficiently satisfied.

14.3.4 Locally Gaussian model

Suppose the distribution of $\boldsymbol{u}$ is locally Gaussian (Fig. 14.4; see Section 3.3.1): the probability density has the form

$$p(\boldsymbol{u}; \bar{\boldsymbol{u}}) = Ce^{-(\boldsymbol{u}-\bar{\boldsymbol{u}}, V[\boldsymbol{u}]^{-}(\boldsymbol{u}-\bar{\boldsymbol{u}}))/2}, \tag{14.64}$$

where C is a normalization constant. The covariance matrix $V[\boldsymbol{u}]$ is assumed to have range $T_{\bar{\boldsymbol{u}}}(\mathcal{U})$ and have the expression

$$V[\boldsymbol{u}] = \boldsymbol{P}^{\mathcal{U}}_{\bar{\boldsymbol{u}}} E[(\boldsymbol{u}-\bar{\boldsymbol{u}})(\boldsymbol{u}-\bar{\boldsymbol{u}})^{\top}]\boldsymbol{P}^{\mathcal{U}}_{\bar{\boldsymbol{u}}}. \tag{14.65}$$

This distribution belongs to the exponential family, and eq. (14.64) implies that $\boldsymbol{L} = \bar{\boldsymbol{L}} = \boldsymbol{J} = V[\hat{\boldsymbol{u}}]^{-}$. Hence, the minimization (14.35) is equivalent to

$$(\boldsymbol{x}-\boldsymbol{u}, V[\boldsymbol{u}]^{-}(\boldsymbol{x}-\boldsymbol{u})) \to \min. \tag{14.66}$$

In other words, the solution $\hat{\boldsymbol{u}}$ is the *Mahalanobis projection* of $\boldsymbol{u}$ onto the model $\mathcal{S}$, i.e., the nearest point to $\boldsymbol{u}$ in $\mathcal{S}$ measured in the Mahalanobis distance with respect to $V[\boldsymbol{u}]$ (see Section 13.2.1).

From Proposition 14.3, the solution is obtained in the form

$$\hat{\boldsymbol{u}} = \boldsymbol{u} - V[\boldsymbol{u}] \sum_{k,l=1}^{L} W^{(kl)} F^{(k)} \nabla F^{(l)}, \tag{14.67}$$

$$(W^{(kl)}) = \left((\nabla F^{(k)}, V[\hat{\boldsymbol{u}}] \nabla F^{(l)}) \right)_r^{-}, \tag{14.68}$$

which coincides with the expression derived in Section 5.1.2 (see eq. (5.17)).

As shown in Section 5.1.4, the a posteriori covariance matrix of this estimator has the following expression (see eqs. (5.31) and (5.32)):

$$V[\hat{\boldsymbol{u}}] = \hat{V}[\boldsymbol{u}] - \sum_{k,l=1}^{L} \hat{W}^{(kl)} \left(\hat{V}[\boldsymbol{u}]\nabla F^{(k)}\right)\left(\hat{V}[\boldsymbol{u}]\nabla F^{(l)}\right)^{\top}. \tag{14.69}$$

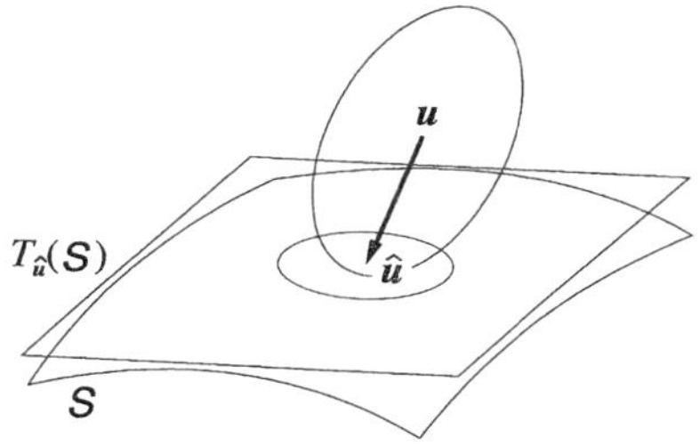

Fig. 14.5. The datum $\mathbf{u}$ is projected onto the tangent point $\hat{\mathbf{u}}$ of the equilikelihood surface of $\mathbf{u}$ to $\mathcal{U}$, and the standard confidence region is projected onto $T_{\hat{\mathbf{u}}}(\mathcal{U})$.

Here, $\hat{V}[\boldsymbol{u}]$ is the matrix obtained by projecting $V[\boldsymbol{u}]$ onto $T_{\hat{\boldsymbol{u}}}(\mathcal{U})$ (see eq. (5.26)), and $\hat{W}^{(kl)}$ is the (kl) element of the (LL)-matrix $\hat{\boldsymbol{W}} = (W^{(kl)})$ defined by

$$(\hat{W}^{(kl)}) = \left((\nabla\hat{F}^{(k)}, \hat{V}[\boldsymbol{u}]\nabla\hat{F}^{(l)})\right)_r^-, \tag{14.70}$$

where $\nabla\hat{F}^{(k)}$ is an abbreviation of $\nabla F^{(k)}(\hat{\boldsymbol{u}})$ (see eq. (5.33)). As discussed in Section 5.1.4, eq. (14.69) can be interpreted as the projection of the standard confidence region of $\boldsymbol{u}$ onto the tangent space $T_{\hat{\boldsymbol{u}}}(\mathcal{S})$ to the model $\mathcal{S}$ at $\hat{\boldsymbol{u}}$ (Fig. 14.5); it gives a first order approximation to the Cramer-Rao lower bound.

14.4 General Parametric Fitting

14.4.1 Definition of the problem

Let $F^{(k)}(\boldsymbol{a}, \boldsymbol{u})$, $k = 1, ..., L$, be continuously differentiable scalar functions of arguments $\boldsymbol{a} \in \mathcal{R}^m$ and $\boldsymbol{u} \in \mathcal{R}^n$. The domains of $\boldsymbol{a}$ and $\boldsymbol{u}$ are assumed to be an m'-dimensional manifold $\mathcal{A} \subset \mathcal{R}^m$ and an n'-dimensional manifold $\mathcal{U} \subset \mathcal{R}^n$, respectively; we call $\mathcal{A}$ the *data space* and $\mathcal{U}$ the *parameter space*. *Parametric fitting* is the problem of computing the parameter $\boldsymbol{u}$ in such a way that the L equations

$$F^{(k)}(\boldsymbol{a}, \boldsymbol{u}) = 0, \qquad k = 1, ..., L, \tag{14.71}$$

"fit" multiple instances $\{\boldsymbol{a}_\alpha\}$, $\alpha = 1, ..., N$, of the variable $\boldsymbol{a}$. We call the vectors $\{\boldsymbol{a}_\alpha\}$ and $\boldsymbol{u}$ the *data* and the *fitting parameter*, respectively. As in Chapter 7.1, let us call the above L equations the *hypothesis*.

Let $\bar{\boldsymbol{a}}_\alpha \in \mathcal{A}$ be the true value of $\boldsymbol{a}_\alpha$, which we assume is unknown. We regard each datum $\boldsymbol{a}_\alpha$ as a random variable with probability density $p_\alpha(\boldsymbol{a}_\alpha; \bar{\boldsymbol{a}}_\alpha)$ parameterized by $\bar{\boldsymbol{a}}_\alpha$. As in the case of geometric correction, we assume that the distribution of $\boldsymbol{a}_\alpha$ in the data space $\mathcal{A}$ is local and define the "true value"

$\bar{\boldsymbol{a}}_\alpha$ by the relationship given by eq. (3.29). However, such an explicit characterization of $\bar{\boldsymbol{a}}_\alpha$ will not be used in the subsequent analysis, so it can be thought of simply as *some value* in $\mathcal{A}$. The parametric fitting problem is formally stated as follows (see Section 7.1.1):

Problem 14.2 *Estimate the true value $\bar{\boldsymbol{u}}$ of the fitting parameter $\boldsymbol{u} \in \mathcal{U}$ from the data $\{\boldsymbol{a}_\alpha\}$, $\alpha = 1, ..., N$, in such a way that their (unknown) true values $\{\bar{\boldsymbol{a}}_\alpha\}$, $\alpha = 1, ..., N$, satisfy the hypothesis*

$$F^{(k)}(\bar{\boldsymbol{a}}_\alpha, \bar{\boldsymbol{u}}) = 0, \qquad k = 1, ..., L. \tag{14.72}$$

As in the case of geometric correction, the probability density $p_\alpha(\boldsymbol{a}_\alpha; \bar{\boldsymbol{a}}_\alpha)$ is assumed to satisfy the same regularity conditions (see Section 14.2.1):

Assumption 14.9 *The probability density $p_\alpha(\boldsymbol{a}_\alpha; \bar{\boldsymbol{a}}_\alpha)$ is continuously differentiable with respect to $\boldsymbol{a}_\alpha$ and $\bar{\boldsymbol{a}}_\alpha$ an arbitrary number of times, and*

$$p_\alpha(\boldsymbol{a}_\alpha; \bar{\boldsymbol{a}}_\alpha) > 0 \tag{14.73}$$

for all $\boldsymbol{a}_\alpha \in \mathcal{A}$. Furthermore, the integration operation $\int_{\mathcal{A}} d\boldsymbol{a}_\alpha$ with respect to $\boldsymbol{a}_\alpha$ for any expression of $p_\alpha(\boldsymbol{a}_\alpha; \bar{\boldsymbol{a}}_\alpha)$ is, if the integration exists, interchangeable with the differentiation operation $\nabla_{\bar{\boldsymbol{a}}_\alpha}$ with respect to $\bar{\boldsymbol{a}}_\alpha$.

The probability density $p_\alpha(\boldsymbol{a}_\alpha; \bar{\boldsymbol{a}}_\alpha)$ is not defined for $\bar{\boldsymbol{a}}_\alpha \notin \mathcal{A}$. For the convenience of analysis, however, we extend it to $\bar{\boldsymbol{a}}_\alpha \notin \mathcal{A}$ in the same way as in the case of geometric correction. Namely, we assume that

$$p_\alpha(\boldsymbol{a}_\alpha; \bar{\boldsymbol{a}}_\alpha + \Delta\bar{\boldsymbol{a}}_\alpha) = p_\alpha(\boldsymbol{a}_\alpha; \bar{\boldsymbol{a}}_\alpha) + O(\Delta\bar{\boldsymbol{a}}_\alpha)^2 \tag{14.74}$$

for all $\bar{\boldsymbol{a}}_\alpha \in \mathcal{A}$ and $\Delta\bar{\boldsymbol{a}}_\alpha \in T_{\bar{\boldsymbol{a}}_\alpha}(\mathcal{A})^\perp$, where $T_{\bar{\boldsymbol{a}}_\alpha}(\mathcal{A}) \subset \mathcal{R}^m$ is the tangent space to the manifold $\mathcal{A}$ at $\bar{\boldsymbol{a}}_\alpha$. In precise terms, we assume the following:

Assumption 14.10

$$\nabla_{\bar{\boldsymbol{a}}_\alpha} p_\alpha \in T_{\bar{\boldsymbol{a}}_\alpha}(\mathcal{A}). \tag{14.75}$$

14.4.2 The rank of the hypothesis

The L equations (14.71) may be algebraically dependent. If only r of them are independent, they generally give only r effective constraints and define an $(m' - r)$-dimensional manifold $\mathcal{S}$, or a manifold $\mathcal{S}$ of codimension r, in the data space $\mathcal{A}$ except at singular points. We assume that all $\{\bar{\boldsymbol{a}}_\alpha\}$ are not singular points and call r the *rank* of the hypothesis (14.71). A precise statement of this assumption, which also gives a precise definition of the rank r, is as follows:

Assumption 14.11 *The L equations of the hypothesis define a manifold $\mathcal{S}$ of codimension r in the data space $\mathcal{A}$ in the neighborhood of $\bar{\boldsymbol{a}}_\alpha$, $\alpha = 1, ..., N$, for $\boldsymbol{u} = \bar{\boldsymbol{u}}$.*

As in the preceding chapter, let us call the manifold $\mathcal{S}$ the (*geometric*) *model* of Problem 14.2.

If we write $\boldsymbol{a}_\alpha = \bar{\boldsymbol{a}}_\alpha + \Delta\boldsymbol{a}_\alpha$, we have

$$\begin{pmatrix} F^{(1)}(\bar{\boldsymbol{a}}_\alpha + \Delta\boldsymbol{a}_\alpha, \bar{\boldsymbol{u}}) \\ \vdots \\ F^{(L)}(\bar{\boldsymbol{a}}_\alpha + \Delta\boldsymbol{a}_\alpha, \bar{\boldsymbol{u}}) \end{pmatrix} = \left(\nabla_{\boldsymbol{a}}\bar{F}_\alpha^{(1)}, ..., \nabla_{\boldsymbol{a}}\bar{F}_\alpha^{(L)}\right)^\top \Delta\boldsymbol{a}_\alpha \tag{14.76}$$

to a first approximation in $\Delta\boldsymbol{a}_\alpha$, where and hereafter $\nabla_{\boldsymbol{a}}F^{(k)}(\bar{\boldsymbol{a}}_\alpha, \bar{\boldsymbol{u}})$ is abbreviated as $\nabla_{\boldsymbol{a}}\bar{F}_\alpha^{(k)}$. The inherent constraint $\boldsymbol{a}_\alpha \in \mathcal{A}$ requires that $\Delta\boldsymbol{a}_\alpha \in T_{\bar{\boldsymbol{a}}_\alpha}(\mathcal{A})$ to a first approximation. If $\Delta\boldsymbol{a}_\alpha$ ranges over the entire tangent space $T_{\bar{\boldsymbol{a}}_\alpha}(\mathcal{A})$, eq. (14.76) defines a *linear mapping* from $T_{\bar{\boldsymbol{a}}_\alpha}(\mathcal{A})$ to a linear subspace

$$\mathcal{L}_{\bar{\boldsymbol{a}}_\alpha} = \{\begin{pmatrix} (\nabla_{\boldsymbol{a}}\bar{F}_\alpha^{(1)}, \Delta\boldsymbol{a}_\alpha) \\ \vdots \\ (\nabla_{\boldsymbol{a}}\bar{F}_\alpha^{(L)}, \Delta\boldsymbol{a}_\alpha) \end{pmatrix} |\Delta\boldsymbol{a}_\alpha \in T_{\bar{\boldsymbol{a}}_\alpha}(\mathcal{A})\} \subset \mathcal{R}^L. \tag{14.77}$$

The dimension of this linear subspace equals the dimension of the subspace

$$\{\boldsymbol{P}_{\bar{\boldsymbol{a}}_\alpha}^{\mathcal{A}}\nabla_{\boldsymbol{a}}\bar{F}_\alpha^{(1)}, ..., \boldsymbol{P}_{\bar{\boldsymbol{a}}_\alpha}^{\mathcal{A}}\nabla_{\boldsymbol{a}}\bar{F}_\alpha^{(L)}\}_L \subset \mathcal{R}^m, \tag{14.78}$$

where $\boldsymbol{P}_{\bar{\boldsymbol{a}}_\alpha}^{\mathcal{A}}$ is the m-dimensional projection matrix onto $T_{\bar{\boldsymbol{a}}_\alpha}(\mathcal{A})$. Consequently, the dimension of $\mathcal{L}_{\bar{\boldsymbol{a}}_\alpha}$ is at most the rank r of the hypothesis (14.71), but we further assume the following[3]:

Assumption 14.12 *The dimension of the linear subspace $\mathcal{L}_{\bar{\boldsymbol{a}}_\alpha}$ is equal to the rank r of the hypothesis for all α.*

In mathematical terms, Assumption 14.12 states that each of the L equations of the hypothesis defines a manifold of *codimension 1* in the data space $\mathcal{A}$ such that they intersect each other *transversally* at $\bar{\boldsymbol{a}}_\alpha$, $\alpha = 1, ..., N$. We say that $\boldsymbol{a}_\alpha$ is a *singular datum* if the dimension $\mathcal{L}_{\bar{\boldsymbol{a}}_\alpha}$ is less than r, and an *ordinary datum* otherwise. Assumption 14.12 requires that singular data should be excluded from the fitting process. We say that the hypothesis (14.71) is *nonsingular* if Assumption 14.12 is satisfied, and *singular* otherwise (see Section 7.1.1).

The linear subspace $\mathcal{L}_{\bar{\boldsymbol{a}}_\alpha}$ defined by eq. (14.77) can alternatively be defined as the tangent space to the manifold

$$\mathcal{F} = \{\begin{pmatrix} F^{(1)}(\boldsymbol{a}, \bar{\boldsymbol{u}}) \\ \vdots \\ F^{(L)}(\boldsymbol{a}, \bar{\boldsymbol{u}}) \end{pmatrix} \in \mathcal{R}^L \,|\, \boldsymbol{a} \in \mathcal{A}\} \in \mathcal{R}^L \tag{14.79}$$

[3] As a result, the L equations (14.71) cannot be replaced by a single equation, say $\sum_{k=1}^L F^{(k)}(\mathbf{a}, \mathbf{u})^2 = 0$, which still has rank r but $\mathcal{L}_{\bar{\mathbf{a}}_\alpha} = \{\mathbf{0}\}$ for all α.

at $\bar{\boldsymbol{a}}_\alpha \in \mathcal{A}$. However, $\bar{\boldsymbol{a}}_\alpha$ can be a singular point of the mapping $(F^{(1)}(\boldsymbol{a}, \bar{\boldsymbol{u}}), ..., F^{(L)}(\boldsymbol{a}, \bar{\boldsymbol{u}}))^\top$: $\mathcal{A} \to \mathcal{R}^L$. In other words, the linear subspace $\mathcal{L}_{\boldsymbol{a}_\alpha}$ defined by replacing $\bar{\boldsymbol{a}}_\alpha$ by $\boldsymbol{a}_\alpha$ ($\neq \bar{\boldsymbol{a}}_\alpha$) in eq. (14.77) has a higher dimension that r. If this is the case, we say that the hypothesis is *degenerate* (see Section 7.1.1).

We further assume that no effective constraints exist on $\bar{\boldsymbol{u}}$ other than the inherent constraint $\bar{\boldsymbol{u}} \in \mathcal{U}$. Putting it differently, we assume that for an arbitrary variation $\Delta\boldsymbol{u}$ of $\bar{\boldsymbol{u}}$ there exists a variation $\Delta\bar{\boldsymbol{a}}_\alpha$ of $\bar{\boldsymbol{a}}_\alpha$ such that

$$F^{(k)}(\bar{\boldsymbol{a}}_\alpha + \Delta\bar{\boldsymbol{a}}_\alpha, \bar{\boldsymbol{u}} + \Delta\boldsymbol{u}) = 0, \qquad k = 1, ..., L, \tag{14.80}$$

for $\alpha = 1$, ..., N. The above equation is equivalent to

$$(\nabla_{\boldsymbol{u}} \bar{F}_\alpha^{(k)}, \Delta\boldsymbol{u}) = -(\nabla_{\boldsymbol{a}} \bar{F}_\alpha^{(k)}, \Delta\boldsymbol{a}_\alpha), \qquad k = 1, ..., L, \tag{14.81}$$

to a first approximation, where and hereafter $\nabla_{\boldsymbol{u}} F^{(k)}(\bar{\boldsymbol{a}}_\alpha, \bar{\boldsymbol{u}})$ is abbreviated as $\nabla_{\boldsymbol{u}} \bar{F}_\alpha^{(k)}$. Let $T_{\bar{\boldsymbol{u}}}(\mathcal{U})$ be the tangent space to the manifold $\mathcal{U}$ at $\bar{\boldsymbol{u}}$. The inherent constraint $\boldsymbol{u} \in \mathcal{U}$ requires that $\Delta\boldsymbol{u} \in T_{\bar{\boldsymbol{u}}}(\mathcal{U})$ to a first approximation. Similarly, the inherent constraint $\bar{\boldsymbol{a}}_\alpha \in \mathcal{A}$ requires that $\Delta\bar{\boldsymbol{a}}_\alpha \in T_{\bar{\boldsymbol{a}}_\alpha}(\mathcal{A})$ to a first approximation. From the definition of the linear subspace $\mathcal{L}_{\bar{\boldsymbol{a}}_\alpha}$, the above assumption is formally stated as follows:

Assumption 14.13 *For an arbitrary* $\Delta\boldsymbol{u} \in T_{\bar{\boldsymbol{u}}}(\mathcal{U})$,

$$\begin{pmatrix} (\nabla_{\boldsymbol{u}} \bar{F}_\alpha^{(1)}, \Delta\boldsymbol{u}) \\ \vdots \\ (\nabla_{\boldsymbol{u}} \bar{F}_\alpha^{(L)}, \Delta\boldsymbol{u}) \end{pmatrix} \in \mathcal{L}_{\bar{\boldsymbol{a}}_\alpha}. \tag{14.82}$$

14.4.3 Cramer-Rao lower bound for parametric fitting

Let $\bar{\boldsymbol{l}}_\alpha$ be the *score* of $\boldsymbol{a}_\alpha$ with respect to $\bar{\boldsymbol{a}}_\alpha$:

$$\bar{\boldsymbol{l}}_\alpha = \nabla_{\bar{\boldsymbol{a}}_\alpha} \log p_\alpha. \tag{14.83}$$

The *Fisher information matrix* $\boldsymbol{J}_\alpha$ with respect to $\bar{\boldsymbol{a}}_\alpha$ is defined by

$$\boldsymbol{J}_\alpha = E[\bar{\boldsymbol{l}}_\alpha \bar{\boldsymbol{l}}_\alpha^\top]. \tag{14.84}$$

The symbol $E[\,\cdot\,]$ denotes the expectation with respect to the joint probability density[4] $\prod_{\alpha=1}^N p_\alpha(\boldsymbol{a}_\alpha; \bar{\boldsymbol{a}}_\alpha)$. Assumption 14.10 implies that $\bar{\boldsymbol{l}}_\alpha \in T_{\bar{\boldsymbol{a}}_\alpha}(\mathcal{A})$. We assume that the distribution $p_\alpha(\boldsymbol{a}_\alpha; \bar{\boldsymbol{a}}_\alpha)$ is *regular* with respect to $\bar{\boldsymbol{a}}_\alpha$, i.e., $\bar{\boldsymbol{l}}_\alpha$ takes all orientations in $T_{\bar{\boldsymbol{a}}_\alpha}(\mathcal{A})$ as $\boldsymbol{a}_\alpha$ ranges over the entire data space $\mathcal{A}$. In precise terms, this assumption is stated as follows:

[4] Since each $\mathbf{a}_\alpha$ is assumed to be independent, the expectation $E[\,\cdot\,]$ of a quantity that depends on $\mathbf{a}_\alpha$ alone equals the expectation with respect to $p_\alpha(\mathbf{a}_\alpha; \bar{\mathbf{a}}_\alpha)$ alone.

Assumption 14.14 *The Fisher information matrix* $\boldsymbol{J}_\alpha$ *is positive semi-definite and has range* $T_{\bar{\boldsymbol{a}}_\alpha}(\mathcal{A})$.

Define an (LL)-matrix $\bar{\boldsymbol{V}}_\alpha = (\bar{V}_\alpha^{(kl)})$ by

$$(\bar{V}_\alpha^{(kl)}) = \Big((\nabla_{\boldsymbol{a}}\bar{F}_\alpha^{(k)}, \boldsymbol{J}_\alpha^- \nabla_{\boldsymbol{a}}\bar{F}_\alpha^{(l)})\Big), \tag{14.85}$$

and an (LL)-matrix $\bar{\boldsymbol{W}}_\alpha = (\bar{W}_\alpha^{(kl)})$ by $\bar{\boldsymbol{W}}_\alpha = \bar{\boldsymbol{V}}_\alpha^-$, or

$$(\bar{W}_\alpha^{(kl)}) = \Big((\nabla_{\boldsymbol{a}}\bar{F}_\alpha^{(k)}, \boldsymbol{J}_\alpha^- \nabla_{\boldsymbol{a}}\bar{F}_\alpha^{(l)})\Big)^-. \tag{14.86}$$

Define an (nn)-matrix $\bar{\boldsymbol{M}}$ by

$$\bar{\boldsymbol{M}} = \sum_{\alpha=1}^{N}\sum_{k,l=1}^{L} \bar{W}_\alpha^{(kl)} \Big(\boldsymbol{P}_{\bar{\boldsymbol{u}}}^{\mathcal{U}} \nabla_{\boldsymbol{u}}\bar{F}_\alpha^{(k)}\Big)\Big(\boldsymbol{P}_{\bar{\boldsymbol{u}}}^{\mathcal{U}} \nabla_{\boldsymbol{u}}\bar{F}_\alpha^{(l)}\Big)^\top, \tag{14.87}$$

where $\boldsymbol{P}_{\bar{\boldsymbol{u}}}^{\mathcal{U}}$ is the n-dimensional projection matrix onto the tangent space $T_{\bar{\boldsymbol{u}}}(\mathcal{U})$ to the manifold $\mathcal{U}$ at $\bar{\boldsymbol{u}}$. We call $\bar{\boldsymbol{M}}$ the *moment matrix* (see eq. (7.23)). We assume that the number of the data $\{\boldsymbol{a}_\alpha\}$ is sufficiently large and their true values $\{\bar{\boldsymbol{a}}_\alpha\}$ distribute "generally" in the model $\mathcal{S}$. In other words, we exclude the possibility that $\{\bar{\boldsymbol{a}}_\alpha\}$ have a special configuration (being identical, collinear, coplanar, etc.) such that the fitting parameter $\boldsymbol{u}$ is indeterminate. In precise terms, this assumption is stated as follows:

Assumption 14.15 *The moment matrix* $\bar{\boldsymbol{M}}$ *is positive semi-definite and has range* $T_{\bar{\boldsymbol{u}}}(\mathcal{U})$.

Let $\hat{\boldsymbol{u}} = \hat{\boldsymbol{u}}(\boldsymbol{a}_1, ..., \boldsymbol{a}_N)$ be an *unbiased estimator* of $\boldsymbol{u}$ determined from the data $\{\boldsymbol{a}_\alpha\}$. Since $\boldsymbol{u}$ is constrained to be in the parameter space $\mathcal{U}$, it must satisfy $\hat{\boldsymbol{u}}(\boldsymbol{a}_1, ..., \boldsymbol{a}_N) \in \mathcal{U}$ for any $\boldsymbol{a}_\alpha \in \mathcal{A}$, $\alpha = 1, ..., N$. Its unbiasedness is defined by

$$\boldsymbol{P}_{\bar{\boldsymbol{u}}}^{\mathcal{U}} E[\hat{\boldsymbol{u}} - \bar{\boldsymbol{u}}] = \boldsymbol{0}. \tag{14.88}$$

As in the case of geometric correction, we assume that the distribution of $\hat{\boldsymbol{u}}$ is sufficiently localized around $\bar{\boldsymbol{u}}$ in the parameter space $\mathcal{U}$ and define the covariance matrix of $\hat{\boldsymbol{u}}$ in the following form:

$$V[\hat{\boldsymbol{u}}] = \boldsymbol{P}_{\bar{\boldsymbol{u}}}^{\mathcal{U}} E[(\hat{\boldsymbol{u}} - \bar{\boldsymbol{u}})(\hat{\boldsymbol{u}} - \bar{\boldsymbol{u}})^\top] \boldsymbol{P}_{\bar{\boldsymbol{u}}}^{\mathcal{U}}. \tag{14.89}$$

We now prove the following theorem:

Theorem 14.2

$$V[\hat{\boldsymbol{u}}] \succeq \bar{\boldsymbol{M}}^-. \tag{14.90}$$

Since Theorem 14.2 is an analogue of the *Cramer-Rao inequality* (3.134) in Section 3.5.2, we call eq. (14.90) and the bound it gives the *Cramer-Rao inequality* and the *Cramer-Rao lower bound*, respectively, for parametric fitting.

14.4.4 Proof of the main theorem

Lemma 14.6 *The (LL)-matrix $\bar{\boldsymbol{V}}_\alpha$ is positive semi-definite and has range $\mathcal{L}_{\bar{\boldsymbol{a}}_\alpha}$.*

Proof. Let $\boldsymbol{\eta} = (\eta^{(k)})$ be an arbitrary L-vector. We have

$$(\boldsymbol{\eta}, \bar{\boldsymbol{V}}_\alpha \boldsymbol{\eta}) = (\sum_{k=1}^{L} \eta^{(k)} \nabla_{\boldsymbol{a}} \bar{F}_\alpha^{(k)}, \boldsymbol{J}_\alpha^{-} \sum_{l=1}^{L} \eta^{(l)} \nabla_{\boldsymbol{a}} \bar{F}_\alpha^{(l)}). \tag{14.91}$$

Since $\boldsymbol{J}_\alpha$ is positive semi-definite, its generalized inverse $\boldsymbol{J}_\alpha^-$ is also positive semi-definite. It follows that the right-hand side of eq. (14.91) is nonnegative for an arbitrary $\boldsymbol{\eta}$, meaning that $\bar{\boldsymbol{V}}_\alpha$ is positive semi-definite. Suppose $\boldsymbol{\eta}$ belongs to the null space of $\bar{\boldsymbol{V}}_\alpha$. Then, the left-hand side of eq. (14.91) is 0. Hence, $\sum_{k=1}^{L} \eta^{(k)} \nabla_{\boldsymbol{a}} \bar{F}_\alpha^{(k)}$ belongs to the null space of $\boldsymbol{J}_\alpha^-$, which is identical with the null space of $\boldsymbol{J}_\alpha$. Assumption 14.14 implies that the null space of $\boldsymbol{J}_\alpha$ coincides with $T_{\bar{\boldsymbol{a}}_\alpha}(\mathcal{A})^\perp$. Hence,

$$\sum_{k=1}^{L} \eta^{(k)} (\nabla_{\boldsymbol{a}} \bar{F}_\alpha^{(k)}, \Delta \boldsymbol{a}_\alpha) = 0 \tag{14.92}$$

for an arbitrary $\Delta \boldsymbol{a}_\alpha \in T_{\bar{\boldsymbol{a}}_\alpha}(\mathcal{A})$. From the definition of the subspace $\mathcal{L}_{\bar{\boldsymbol{a}}_\alpha}$, this means that $\boldsymbol{\eta} \in \mathcal{L}_{\bar{\boldsymbol{a}}_\alpha}^\perp$. Conversely, if $\boldsymbol{\eta} \in \mathcal{L}_{\bar{\boldsymbol{a}}_\alpha}^\perp$, we obtain $(\boldsymbol{\eta}, \bar{\boldsymbol{V}}_\alpha \boldsymbol{\eta}) = 0$ by following eqs. (14.92) and (14.91) backward. Hence, the null space of the (LL)-matrix $\bar{\boldsymbol{V}}_\alpha$ coincides with the $(L-r)$-dimensional subspace $\mathcal{L}_{\bar{\boldsymbol{a}}_\alpha}^\perp \subset \mathcal{R}^L$. Consequently, the range of $\bar{\boldsymbol{V}}_\alpha$ coincides with $\mathcal{L}_{\bar{\boldsymbol{a}}_\alpha}$. □

Consider the (LL)-matrix $\bar{\boldsymbol{W}}_\alpha$ defined by eq. (14.86). Let $\boldsymbol{P}_{\bar{\boldsymbol{a}}_\alpha}^{\mathcal{L}}$ be the L-dimensional projection matrix onto $\mathcal{L}_{\bar{\boldsymbol{a}}_\alpha}$. Lemma 14.6 implies the following:

Corollary 14.3 *The (LL)-matrix $\bar{\boldsymbol{W}}_\alpha$ is positive semi-definite and has range $\mathcal{L}_{\bar{\boldsymbol{a}}_\alpha}$, and the following relationship holds:*

$$\bar{\boldsymbol{V}}_\alpha \bar{\boldsymbol{W}}_\alpha = \boldsymbol{P}_{\bar{\boldsymbol{a}}_\alpha}^{\mathcal{L}}. \tag{14.93}$$

The following lemma plays an essential role in proving Theorem 14.2.

Lemma 14.7 *The variation $\Delta \bar{\boldsymbol{a}}_\alpha$ of $\bar{\boldsymbol{a}}_\alpha$ defined by*

$$\Delta \bar{\boldsymbol{a}}_\alpha = -\boldsymbol{J}_\alpha^- \sum_{k,l=1}^{L} \bar{W}_\alpha^{(kl)} (\nabla_{\boldsymbol{u}} \bar{F}_\alpha^{(k)}, \Delta \boldsymbol{u}) \nabla_{\boldsymbol{a}} \bar{F}_\alpha^{(l)} \tag{14.94}$$

satisfies $\Delta \bar{\boldsymbol{a}} \in T_{\bar{\boldsymbol{a}}_\alpha}(\mathcal{A})$ for any $\Delta \boldsymbol{u} \in T_{\bar{\boldsymbol{u}}}(\mathcal{U})$, and to a first approximation

$$F^{(k)}(\bar{\boldsymbol{a}}_\alpha + \Delta \bar{\boldsymbol{a}}_\alpha, \bar{\boldsymbol{u}} + \Delta \boldsymbol{u}) = 0, \qquad k = 1, ..., L. \tag{14.95}$$

Proof. Assumption 14.10 implies $\bar{\boldsymbol{l}}_\alpha \in T_{\bar{\boldsymbol{a}}_\alpha}(\mathcal{A})$. Hence, we have $\boldsymbol{P}^{\mathcal{U}}_{\bar{\boldsymbol{a}}_\alpha}\boldsymbol{J}_\alpha = \boldsymbol{J}_\alpha$ from the definition (14.84) of the Fisher information matrix $\boldsymbol{J}_\alpha$. Consequently, $\boldsymbol{P}^{\mathcal{U}}_{\bar{\boldsymbol{a}}_\alpha}\boldsymbol{J}^-_\alpha = \boldsymbol{J}^-_\alpha$, and eq. (14.94) implies $\boldsymbol{P}^{\mathcal{U}}_{\bar{\boldsymbol{a}}_\alpha}\Delta\bar{\boldsymbol{a}}_\alpha = \Delta\bar{\boldsymbol{a}}_\alpha$, i.e., $\Delta\bar{\boldsymbol{a}}_\alpha \in T_{\bar{\boldsymbol{a}}_\alpha}(\mathcal{A})$. Eq. (14.95) is proved by showing

$$(\nabla_{\boldsymbol{u}}\bar{F}^{(k)}_\alpha, \Delta\bar{\boldsymbol{a}}_\alpha) = -(\nabla_{\boldsymbol{a}}\bar{F}^{(k)}_\alpha, \Delta\boldsymbol{u}). \tag{14.96}$$

From eqs. (14.93) and (14.94), we see that

$$(\nabla_{\boldsymbol{a}}\bar{F}^{(k)}_\alpha, \Delta\bar{\boldsymbol{a}}_\alpha) = -\sum_{l=1}^{L}\left(\sum_{m=1}^{L}(\nabla_{\boldsymbol{a}}\bar{F}^{(k)}_\alpha, \boldsymbol{J}^-_\alpha\nabla_{\boldsymbol{a}}\bar{F}^{(m)}_\alpha)\bar{W}^{(ml)}_\alpha\right)(\nabla_{\boldsymbol{u}}\bar{F}^{(l)}_\alpha, \Delta\boldsymbol{u})$$
$$= -\sum_{l=1}^{L}(\boldsymbol{P}^{\mathcal{L}}_{\bar{\boldsymbol{a}}_\alpha})_{kl}(\nabla_{\boldsymbol{u}}\bar{F}^{(l)}_\alpha, \Delta\boldsymbol{u}), \tag{14.97}$$

where $(\boldsymbol{P}^{\mathcal{L}}_{\bar{\boldsymbol{a}}_\alpha})_{kl}$ is the (kl) element of the projection matrix $\boldsymbol{P}^{\mathcal{L}}_{\bar{\boldsymbol{a}}_\alpha}$. Assumption 14.13 implies that the L-vector $\left((\nabla_{\boldsymbol{u}}\bar{F}^{(k)}_\alpha, \Delta\boldsymbol{u})\right)$ belongs to the subspace $\mathcal{L}_{\bar{\boldsymbol{a}}_\alpha}$. Hence, the last term of eq. (14.97) is equal to $-(\nabla_{\boldsymbol{a}}\bar{F}^{(k)}_\alpha, \Delta\boldsymbol{u})$. □

Define an n-vector random variable $\boldsymbol{m}_\alpha$ by

$$\boldsymbol{m}_\alpha = \sum_{k,l=1}^{L}\bar{W}^{(kl)}_\alpha(\bar{\boldsymbol{l}}_\alpha, \boldsymbol{J}^-_\alpha\nabla_{\boldsymbol{a}}\bar{F}^{(k)}_\alpha)\boldsymbol{P}^{\mathcal{U}}_{\bar{\boldsymbol{u}}}\nabla_{\boldsymbol{u}}\bar{F}^{(l)}_\alpha. \tag{14.98}$$

Lemma 14.8

$$E[\boldsymbol{P}^{\mathcal{U}}_{\bar{\boldsymbol{u}}}(\hat{\boldsymbol{u}} - \bar{\boldsymbol{u}})\left(\sum_{\alpha=1}^{N}\boldsymbol{m}_\alpha\right)^\top] = -\boldsymbol{P}^{\mathcal{U}}_{\bar{\boldsymbol{u}}}. \tag{14.99}$$

Proof. Since the estimator $\hat{\boldsymbol{u}}$ is unbiased, eq. (14.88) holds identically for any $\bar{\boldsymbol{a}}_\alpha \in \mathcal{S}$, $\alpha = 1, \ldots, N$, and $\bar{\boldsymbol{u}} \in \mathcal{U}$ that satisfy eq. (14.72). This means that the first variation of the left-hand side of eq. (14.88) should be $\boldsymbol{0}$ for any $\Delta\boldsymbol{u} \in T_{\bar{\boldsymbol{u}}}(\mathcal{U})$ if the variation $\Delta\bar{\boldsymbol{a}}_\alpha$ defined by eq. (14.94) is added to each $\bar{\boldsymbol{a}}_\alpha$. Taking the first variation of the left-hand side of eq. (14.88), substituting eq. (14.94) into it, noting Assumption 14.9, and using the logarithmic differentiation formula (3.115), we obtain

$$\Delta\boldsymbol{P}^{\mathcal{U}}_{\bar{\boldsymbol{u}}}E[\hat{\boldsymbol{u}} - \bar{\boldsymbol{u}}] = E[\Delta\left(\boldsymbol{P}^{\mathcal{U}}_{\bar{\boldsymbol{u}}}(\hat{\boldsymbol{u}} - \bar{\boldsymbol{u}})\right)]$$
$$+ \int_{\mathcal{A}^N}\boldsymbol{P}^{\mathcal{U}}_{\bar{\boldsymbol{u}}}(\hat{\boldsymbol{u}} - \bar{\boldsymbol{u}})\prod_{\beta=1}^{N}p_\beta\sum_{\alpha=1}^{N}(\nabla_{\bar{\boldsymbol{a}}_\alpha}\log p_\alpha, \Delta\bar{\boldsymbol{a}}_\alpha)d\boldsymbol{a}_1\cdots d\boldsymbol{a}_N$$
$$= -\Delta\boldsymbol{u} - E[\boldsymbol{P}^{\mathcal{U}}_{\bar{\boldsymbol{u}}}(\hat{\boldsymbol{u}} - \bar{\boldsymbol{u}})$$
$$\left(\sum_{\alpha=1}^{N}\sum_{k,l=1}^{L}\bar{W}^{(kl)}_\alpha(\bar{\boldsymbol{l}}_\alpha, \boldsymbol{J}^-_\alpha\nabla_{\boldsymbol{a}}\bar{F}^{(l)}_\alpha)\nabla_{\boldsymbol{u}}\bar{F}^{(k)}_\alpha\right)^\top]\Delta\boldsymbol{u}. \tag{14.100}$$

Here, we have used the assumption that $\boldsymbol{P}^{\mathcal{U}}_{\bar{\boldsymbol{u}}+\Delta\boldsymbol{u}}E[\hat{\boldsymbol{u}}-\bar{\boldsymbol{u}}] \approx \boldsymbol{P}^{\mathcal{U}}_{\bar{\boldsymbol{u}}}E[\hat{\boldsymbol{u}}-\bar{\boldsymbol{u}}]$ (see Section 3.5.2). The above expression should be $\boldsymbol{0}$ for any $\Delta\boldsymbol{u} \in T_{\bar{\boldsymbol{u}}}(\mathcal{U})$. If we let $\Delta\boldsymbol{u} = \boldsymbol{P}^{\mathcal{U}}_{\bar{\boldsymbol{u}}}\Delta\boldsymbol{\xi}$, we have $\Delta\boldsymbol{u} \in T_{\bar{\boldsymbol{u}}}(\mathcal{U})$ for any $\Delta\boldsymbol{\xi} \in \mathcal{R}^n$. Hence,

$$E[\boldsymbol{P}^{\mathcal{U}}_{\bar{\boldsymbol{u}}}(\hat{\boldsymbol{u}}-\bar{\boldsymbol{u}})\left(\sum_{\alpha=1}^{N}\sum_{k,l=1}^{L}\bar{W}_{\alpha}^{(kl)}(\bar{\boldsymbol{l}}_{\alpha},\boldsymbol{J}_{\alpha}^{-}\nabla_{\boldsymbol{a}}\bar{F}_{\alpha}^{(l)})\nabla_{\boldsymbol{u}}\bar{F}_{\alpha}^{(k)}\right)^{\top}]\boldsymbol{P}^{\mathcal{U}}_{\bar{\boldsymbol{u}}}\Delta\boldsymbol{\xi} = -\boldsymbol{P}^{\mathcal{U}}_{\bar{\boldsymbol{u}}}\Delta\boldsymbol{\xi}. \tag{14.101}$$

Since this is an identity for $\Delta\boldsymbol{\xi} \in \mathcal{R}^n$, we obtain eq. (14.99). □

Lemma 14.9

$$\bar{\boldsymbol{M}} = E[\left(\sum_{\alpha=1}^{N}\boldsymbol{m}_{\alpha}\right)\left(\sum_{\beta=1}^{N}\boldsymbol{m}_{\beta}\right)^{\top}]. \tag{14.102}$$

Proof. Since the expectation of the score $\bar{\boldsymbol{l}}_{\alpha}$ is $\boldsymbol{0}$ (see eq. (3.117)), eq. (14.98) implies that $E[\boldsymbol{m}_{\alpha}] = \boldsymbol{0}$. Noting that each $\boldsymbol{a}_{\alpha}$ is independent, we see that

$$E[\left(\sum_{\alpha=1}^{N}\boldsymbol{m}_{\alpha}\right)\left(\sum_{\beta=1}^{N}\boldsymbol{m}_{\beta}\right)^{\top}] = \sum_{\alpha,\beta=1}^{N}E[\boldsymbol{m}_{\alpha}\boldsymbol{m}_{\beta}^{\top}] = \sum_{\alpha=1}^{N}E[\boldsymbol{m}_{\alpha}\boldsymbol{m}_{\alpha}^{\top}]. \tag{14.103}$$

From eq. (14.98), we obtain

$$\begin{aligned}
E[\boldsymbol{m}_{\alpha}\boldsymbol{m}_{\alpha}^{\top}] &= \sum_{k,l,m,n=1}^{L}\bar{W}_{\alpha}^{(kl)}\bar{W}_{\alpha}^{(mn)}(\nabla_{\boldsymbol{a}}\bar{F}_{\alpha}^{(k)},\boldsymbol{J}_{\alpha}^{-}E[\bar{\boldsymbol{l}}_{\alpha}\bar{\boldsymbol{l}}_{\alpha}^{\top}]\boldsymbol{J}_{\alpha}^{-}\nabla_{\boldsymbol{a}}\bar{F}_{\alpha}^{(m)}) \\
&\qquad \left(\boldsymbol{P}^{\mathcal{U}}_{\bar{\boldsymbol{u}}}\nabla_{\boldsymbol{u}}\bar{F}_{\alpha}^{(l)}\right)\left(\boldsymbol{P}^{\mathcal{U}}_{\bar{\boldsymbol{u}}}\nabla_{\boldsymbol{u}}\bar{F}_{\alpha}^{(n)}\right)^{\top} \\
&= \sum_{l,n=1}^{L}\left(\sum_{k,m=1}^{L}\bar{W}_{\alpha}^{(lk)}(\nabla_{\boldsymbol{a}}\bar{F}_{\alpha}^{(k)},\boldsymbol{J}_{\alpha}^{-}\nabla_{\boldsymbol{a}}\bar{F}_{\alpha}^{(m)})\bar{W}_{\alpha}^{(mn)}\right) \\
&\qquad \left(\boldsymbol{P}^{\mathcal{U}}_{\bar{\boldsymbol{u}}}\nabla_{\boldsymbol{u}}\bar{F}_{\alpha}^{(l)}\right)\left(\boldsymbol{P}^{\mathcal{U}}_{\bar{\boldsymbol{u}}}\nabla_{\boldsymbol{u}}\bar{F}_{\alpha}^{(n)}\right)^{\top} \\
&= \sum_{l,n=1}^{L}\bar{W}_{\alpha}^{(ln)}\left(\boldsymbol{P}^{\mathcal{U}}_{\bar{\boldsymbol{u}}}\nabla_{\boldsymbol{u}}\bar{F}_{\alpha}^{(l)}\right)\left(\boldsymbol{P}^{\mathcal{U}}_{\bar{\boldsymbol{u}}}\nabla_{\boldsymbol{u}}\bar{F}_{\alpha}^{(n)}\right)^{\top} = \bar{\boldsymbol{M}},
\end{aligned} \tag{14.104}$$

where we have used the defining equation (14.86) of the matrix $\bar{\boldsymbol{W}}_{\alpha}$ and the identity $\bar{\boldsymbol{W}}_{\alpha}\bar{\boldsymbol{V}}_{\alpha}\bar{\boldsymbol{W}}_{\alpha} = \bar{\boldsymbol{W}}_{\alpha}\bar{\boldsymbol{W}}_{\alpha}^{-}\bar{\boldsymbol{W}}_{\alpha} = \bar{\boldsymbol{W}}_{\alpha}$. □

The proof of Theorem 14.2 is as follows. From eq. (14.89) and Lemmas 14.8 and 14.9, we see that

$$E[\begin{pmatrix}\boldsymbol{P}^{\mathcal{U}}_{\bar{\boldsymbol{u}}}(\hat{\boldsymbol{u}}-\bar{\boldsymbol{u}}) \\ \sum_{\alpha=1}^{N}\boldsymbol{m}_{\alpha}\end{pmatrix}\begin{pmatrix}\boldsymbol{P}^{\mathcal{U}}_{\bar{\boldsymbol{u}}}(\hat{\boldsymbol{u}}-\bar{\boldsymbol{u}}) \\ \sum_{\beta=1}^{N}\boldsymbol{m}_{\beta}\end{pmatrix}^{\top}] = \begin{pmatrix}V[\hat{\boldsymbol{u}}] & -\boldsymbol{P}^{\mathcal{U}}_{\bar{\boldsymbol{u}}} \\ -\boldsymbol{P}^{\mathcal{U}}_{\bar{\boldsymbol{u}}} & \bar{\boldsymbol{M}}\end{pmatrix}. \tag{14.105}$$

This matrix is symmetric and positive semi-definite. If matrix $\boldsymbol{A}$ is positive semi-definite and symmetric, so is $\boldsymbol{B}^\top \boldsymbol{A}\boldsymbol{B}$ for any matrix $\boldsymbol{B}$ as long as the multiplication can be defined (see Section 2.2.3). Hence, the following matrix is positive semi-definite and symmetric:

$$\begin{pmatrix} \boldsymbol{P}^{\mathcal{U}}_{\bar{\boldsymbol{u}}} & \bar{\boldsymbol{M}}^- \\ & \bar{\boldsymbol{M}}^- \end{pmatrix} \begin{pmatrix} V[\hat{\boldsymbol{u}}] & -\boldsymbol{P}^{\mathcal{U}}_{\bar{\boldsymbol{u}}} \\ -\boldsymbol{P}^{\mathcal{U}}_{\bar{\boldsymbol{u}}} & \bar{\boldsymbol{M}} \end{pmatrix} \begin{pmatrix} \boldsymbol{P}^{\mathcal{U}}_{\bar{\boldsymbol{u}}} & \\ \bar{\boldsymbol{M}}^- & \bar{\boldsymbol{M}}^- \end{pmatrix}$$

$$= \begin{pmatrix} V[\hat{\boldsymbol{u}}] - \bar{\boldsymbol{M}}^- & \\ & \bar{\boldsymbol{M}}^- \end{pmatrix}. \tag{14.106}$$

Here, we have noted that the covariance matrix $V[\hat{\boldsymbol{u}}]$ and the moment matrix $\bar{\boldsymbol{M}}$ share the same range $T_{\bar{\boldsymbol{u}}}(\mathcal{U})$ and hence $\bar{\boldsymbol{M}}\bar{\boldsymbol{M}}^- = \bar{\boldsymbol{M}}^-\bar{\boldsymbol{M}} = \boldsymbol{P}^{\mathcal{U}}_{\bar{\boldsymbol{u}}}$. Since $\bar{\boldsymbol{M}}^-$ is positive semi-definite, matrix $V[\hat{\boldsymbol{u}}] - \bar{\boldsymbol{M}}^-$ is also positive semi-definite. □

Suppose the probability densities $p_\alpha(\boldsymbol{a}_\alpha; \bar{\boldsymbol{a}})$ of the data $\{\boldsymbol{a}_\alpha\}$ are expressed in terms of a common function $p(\,\cdot\,;\,\cdot\,)$ in the form $p_\alpha(\boldsymbol{a}_\alpha; \bar{\boldsymbol{a}}_\alpha) = p(\boldsymbol{a}_\alpha; \bar{\boldsymbol{a}}_\alpha)$. Then, the weight $\bar{W}^{(kl)}_\alpha$ defined by eq. (14.86) is also expressed in terms of a common function $\bar{W}^{(kl)}(\,\cdot\,)$ in the form $\bar{W}^{(kl)}_\alpha = \bar{W}^{(kl)}(\bar{\boldsymbol{a}}_\alpha)$. Suppose the true values $\{\bar{\boldsymbol{a}}_\alpha\}$ are chosen from a bounded subset $\mathcal{S}'$ of the model $\mathcal{S}$ with density $P(\boldsymbol{a})$ (the number of data is given by $N = \int_{\mathcal{S}'} P(\boldsymbol{a})d\boldsymbol{a}$). Define the following matrix (cf. eq. (14.87)):

$$\bar{\boldsymbol{M}}_\infty = \int_{\mathcal{S}'} \sum_{k,l=1}^{L} \bar{W}^{(kl)}(\boldsymbol{a}) \left(\boldsymbol{P}^{\mathcal{U}}_{\bar{\boldsymbol{u}}} \nabla_{\boldsymbol{u}} F^{(k)}(\boldsymbol{a}, \bar{\boldsymbol{u}})\right) \left(\boldsymbol{P}^{\mathcal{U}}_{\bar{\boldsymbol{u}}} \nabla_{\boldsymbol{u}} F^{(l)}(\boldsymbol{a}, \bar{\boldsymbol{u}})\right)^\top P(\boldsymbol{a})d\boldsymbol{a}. \tag{14.107}$$

If N is sufficiently large, the Cramer-Rao lower bound is approximated by $\bar{\boldsymbol{M}}^-_\infty$, which we call the *asymptotic approximation* of the Cramer-Rao lower bound.

14.5 Maximum Likelihood Fit

14.5.1 Maximum likelihood estimator

Maximum likelihood estimation for this problem is to choose $\{\bar{\boldsymbol{a}}_\alpha\}$ and the fitting parameter $\boldsymbol{u}$ that maximize the *likelihood*, i.e., the joint probability density $\prod_{\alpha=1}^{N} p_\alpha(\boldsymbol{a}_\alpha; \bar{\boldsymbol{a}}_\alpha)$ viewed as a function of $\{\bar{\boldsymbol{a}}_\alpha\}$ for given data $\{\boldsymbol{a}_\alpha\}$. The resulting estimates $\{\hat{\boldsymbol{a}}_\alpha\}$ and $\hat{\boldsymbol{u}}$ are called the *maximum likelihood estimators* of $\{\bar{\boldsymbol{a}}_\alpha\}$ and $\boldsymbol{u}$, respectively. In order to distinguish "variables" from their specific "values", we regard the symbol $\bar{\boldsymbol{a}}_\alpha$ as the true value and use $\boldsymbol{x}_\alpha$ when it is viewed as a variable. Since maximizing the likelihood is equivalent to minimizing its logarithm with opposite sign, the problem reduces to the minimization

$$J = -2\sum_{\alpha=1}^{N} \log p_\alpha(\boldsymbol{a}_\alpha; \boldsymbol{x}_\alpha) \to \min \tag{14.108}$$

with respect to $\boldsymbol{x}_\alpha \in \mathcal{S}$ and $\boldsymbol{u} \in \mathcal{U}$, i.e., under the hypothesis

$$F^{(k)}(\boldsymbol{x}_\alpha, \boldsymbol{u}) = 0, \quad k = 1, ..., L, \qquad \boldsymbol{x}_\alpha \in \mathcal{A}, \quad \boldsymbol{u} \in \mathcal{U}. \tag{14.109}$$

Define an (mm)-matrix $\bar{\boldsymbol{L}}_\alpha$ by

$$\bar{\boldsymbol{L}}_\alpha = -\nabla^2_{\boldsymbol{x}_\alpha} \log p_\alpha|_{\boldsymbol{x}_\alpha = \bar{\boldsymbol{a}}_\alpha}. \tag{14.110}$$

With the expectation that the maximum likelihood estimators $\{\hat{\boldsymbol{a}}_\alpha\}$ and $\hat{\boldsymbol{u}}$ are located near $\{\bar{\boldsymbol{a}}_\alpha\}$ and $\bar{\boldsymbol{u}}$, respectively, we write

$$\boldsymbol{x}_\alpha = \bar{\boldsymbol{a}}_\alpha + \Delta\boldsymbol{x}_\alpha, \qquad \boldsymbol{u} = \bar{\boldsymbol{u}} + \Delta\boldsymbol{u}. \tag{14.111}$$

Substituting the first equation into the function J in eq. (14.108) and expanding it in the neighborhood of $\bar{\boldsymbol{a}}_\alpha$, we obtain

$$J = \bar{c} - 2\sum_{\alpha=1}^{N} (\bar{\boldsymbol{l}}_\alpha, \Delta\boldsymbol{x}_\alpha) + \sum_{\alpha=1}^{N} (\Delta\boldsymbol{x}_\alpha, \bar{\boldsymbol{L}}_\alpha \Delta\boldsymbol{x}_\alpha) + O(\Delta\boldsymbol{x}_\alpha)^3, \tag{14.112}$$

where $\bar{c}$ is the value of J evaluated at $\bar{\boldsymbol{a}}_\alpha$, $\alpha = 1, ..., N$. Assumption 14.10 implies that the rank of $\bar{\boldsymbol{L}}_\alpha$ is no more than the dimension m' of the tangent space $T_{\boldsymbol{x}_\alpha}(\mathcal{A})$. In order to guarantee unique existence of the values $\boldsymbol{x}_\alpha \in \mathcal{A}$, $\alpha = 1, ..., N$, that minimize J, we assume the following:

Assumption 14.16 *For any $\boldsymbol{a}_\alpha \in \mathcal{A}$, matrix $\bar{\boldsymbol{L}}_\alpha$ is positive semi-definite and has range $T_{\bar{\boldsymbol{a}}_\alpha}(\mathcal{A})$.*

Substituting eqs. (14.111) into the hypothesis (14.109) and taking a linear approximation, we obtain

$$(\nabla_{\boldsymbol{a}} \bar{F}^{(k)}_\alpha, \Delta\boldsymbol{x}_\alpha) = -(\nabla_{\boldsymbol{u}} \bar{F}^{(k)}_\alpha, \Delta\boldsymbol{u}). \tag{14.113}$$

The inherent constraint $\boldsymbol{x}_\alpha \in \mathcal{A}$ requires that $\Delta\boldsymbol{x}_\alpha \in T_{\bar{\boldsymbol{a}}_\alpha}(\mathcal{A})$ to a first approximation. We seek the value $\Delta\boldsymbol{x}_\alpha$ that minimizes J for a fixed value of $\Delta\boldsymbol{u}$ under the linearized hypothesis (14.113). Introducing Lagrange multipliers $\bar{\lambda}^{(k)}_\alpha$ to eq. (14.113), differentiating J with respect to $\Delta\boldsymbol{x}_\alpha$, and ignoring higher order terms, we obtain the following linear equation in $\Delta\boldsymbol{x}_\alpha$:

$$\bar{\boldsymbol{L}}_\alpha \Delta\boldsymbol{x}_\alpha = \bar{\boldsymbol{l}}_\alpha + \sum_{k=1}^{L} \bar{\lambda}^{(k)}_\alpha \boldsymbol{P}^{\mathcal{A}}_{\bar{\boldsymbol{a}}_\alpha} \nabla_{\boldsymbol{a}} \bar{F}^{(k)}_\alpha. \tag{14.114}$$

By Assumptions 14.10 and 14.16, eq. (14.114) has the following unique solution:

$$\Delta\boldsymbol{x}_\alpha = \bar{\boldsymbol{L}}_\alpha^- \bar{\boldsymbol{l}}_\alpha + \sum_{k=1}^{L} \bar{\lambda}_\alpha^{(k)} \bar{\boldsymbol{L}}_\alpha^- \nabla_{\boldsymbol{a}} \bar{F}_\alpha^{(k)}. \tag{14.115}$$

Substituting eq. (14.115) into eq. (14.113) and putting $\bar{\boldsymbol{\lambda}}_\alpha = (\bar{\lambda}_\alpha^{(k)})$, we obtain

$$\bar{\boldsymbol{V}}'_\alpha \bar{\boldsymbol{\lambda}}_\alpha = -\Big((\nabla_{\boldsymbol{u}} \bar{F}_\alpha^{(k)}, \Delta\boldsymbol{u})\Big) - \Big((\nabla_{\boldsymbol{a}} \bar{F}_\alpha^{(k)}, \bar{\boldsymbol{L}}_\alpha^- \bar{\boldsymbol{l}}_\alpha)\Big), \tag{14.116}$$

where the (LL)-matrix $\bar{\boldsymbol{V}}'_\alpha = (\bar{V}'^{(kl)}_\alpha)$ is defined by

$$(\bar{V}'^{(kl)}_\alpha) = \Big((\nabla_{\boldsymbol{a}} \bar{F}_\alpha^{(k)}, \bar{\boldsymbol{L}}_\alpha^- \nabla_{\boldsymbol{a}} \bar{F}_\alpha^{(l)})\Big). \tag{14.117}$$

By Assumption 14.16, matrix $\bar{\boldsymbol{V}}'_\alpha$ can be shown to be positive semi-definite and have range $\mathcal{L}_{\bar{\boldsymbol{a}}_\alpha}$ by exactly the same argument as given in the proof of Lemma 14.6. Consequently, if $\boldsymbol{a}_\alpha$ is an ordinary datum (see Section 14.4.2), the solution of eq. (14.116) is uniquely given by

$$\bar{\lambda}_\alpha^{(k)} = -\sum_{l=1}^{L} \bar{W}'^{(kl)}_\alpha \Big((\nabla_{\boldsymbol{u}} \bar{F}_\alpha^{(l)}, \Delta\boldsymbol{u}) + (\nabla_{\boldsymbol{a}} \bar{F}_\alpha^{(l)}, \bar{\boldsymbol{L}}_\alpha^- \bar{\boldsymbol{l}}_\alpha)\Big), \tag{14.118}$$

where $\bar{W}'^{(kl)}_\alpha$ is the (kl) element of the (LL)-matrix $\bar{\boldsymbol{W}}'_\alpha$ defined by $\bar{\boldsymbol{W}}'_\alpha = \bar{\boldsymbol{V}}'^{-}_\alpha$, i.e.,

$$(\bar{W}'^{(kl)}_\alpha) = \Big((\nabla_{\boldsymbol{a}} \bar{F}_\alpha^{(k)}, \bar{\boldsymbol{L}}_\alpha^- \nabla_{\boldsymbol{a}} \bar{F}_\alpha^{(l)})\Big)^-. \tag{14.119}$$

Substituting eq. (14.118) into eq. (14.115), substituting the resulting expression for $\Delta\boldsymbol{x}_\alpha$ into eq. (14.112), and ignoring higher order terms, we obtain the following form:

$$J = \bar{c} + \sum_{\alpha=1}^{N} \Bigg(\sum_{k,l=1}^{L} \bar{W}'^{(kl)}_\alpha \Big((\nabla_{\boldsymbol{u}} \bar{F}_\alpha^{(k)}, \Delta\boldsymbol{u}) + (\nabla_{\boldsymbol{a}} \bar{F}_\alpha^{(k)}, \bar{\boldsymbol{L}}_\alpha^- \bar{\boldsymbol{l}}_\alpha)\Big)$$
$$\Big((\nabla_{\boldsymbol{u}} \bar{F}_\alpha^{(l)}, \Delta\boldsymbol{u}) + (\nabla_{\boldsymbol{a}} \bar{F}_\alpha^{(l)}, \bar{\boldsymbol{L}}_\alpha^- \bar{\boldsymbol{l}}_\alpha)\Big) - (\bar{\boldsymbol{l}}_\alpha, \bar{\boldsymbol{L}}_\alpha^- \bar{\boldsymbol{l}}_\alpha)\Bigg). \tag{14.120}$$

We now seek the value $\Delta\boldsymbol{u}$ that minimizes eq. (14.120). The inherent constraint $\boldsymbol{u} \in \mathcal{U}$ requires that $\Delta\boldsymbol{u} \in T_{\bar{\boldsymbol{u}}}(\mathcal{U})$ to a first approximation. Differentiating eq. (14.120) with respect to $\Delta\boldsymbol{u}$, we obtain the following linear equation in $\Delta\boldsymbol{u}$:

$$\boldsymbol{M}'\Delta\boldsymbol{u} = -\sum_{\alpha=1}^{N} \sum_{k,l=1}^{L} \bar{W}'^{(kl)}_\alpha (\nabla_{\boldsymbol{a}} \bar{F}_\alpha^{(k)}, \bar{\boldsymbol{L}}_\alpha^- \bar{\boldsymbol{l}}_\alpha) \boldsymbol{P}^{\mathcal{U}}_{\bar{\boldsymbol{u}}} \nabla_{\boldsymbol{u}} \bar{F}_\alpha^{(l)}. \tag{14.121}$$

Here, the matrix $\boldsymbol{M}'$ is defined as follows (see eq. (14.87)):

$$\boldsymbol{M}' = \sum_{\alpha=1}^{N} \sum_{k,l=1}^{L} \bar{W}_{\alpha}^{\prime\,(kl)} \left(\boldsymbol{P}_{\bar{\boldsymbol{u}}}^{\mathcal{U}} \nabla_{\boldsymbol{u}} \bar{F}_{\alpha}^{(k)}\right) \left(\boldsymbol{P}_{\bar{\boldsymbol{u}}}^{\mathcal{U}} \nabla_{\boldsymbol{u}} \bar{F}_{\alpha}^{(l)}\right)^{\top}. \tag{14.122}$$

We assume that Assumption 14.15 also applies to $\boldsymbol{M}'$:

Assumption 14.17 *The matrix $\boldsymbol{M}'$ is positive semi-definite and has range $T_{\bar{\boldsymbol{u}}}(\mathcal{U})$.*

Then, the solution of eq. (14.121) is uniquely given by

$$\Delta \boldsymbol{u} = -\boldsymbol{M}'^{-} \sum_{\alpha=1}^{N} \sum_{k,l=1}^{L} \bar{W}_{\alpha}^{\prime\,(kl)} (\nabla_{\boldsymbol{a}} \bar{F}_{\alpha}^{(k)}, \bar{\boldsymbol{L}}_{\alpha}^{-} \bar{\boldsymbol{l}}_{\alpha}) \boldsymbol{P}_{\bar{\boldsymbol{u}}}^{\mathcal{U}} \nabla_{\boldsymbol{u}} \bar{F}_{\alpha}^{(l)}. \tag{14.123}$$

Since the maximum likelihood estimator of $\boldsymbol{u}$ is given by $\hat{\boldsymbol{u}} = \bar{\boldsymbol{u}} + \Delta \boldsymbol{u}$, the covariance matrix $V[\hat{\boldsymbol{u}}]$ (see eq. (14.89)) is evaluated as follows:

$$V[\hat{\boldsymbol{u}}] = E[\boldsymbol{M}'^{-} \boldsymbol{M}'' \boldsymbol{M}'^{-}]. \tag{14.124}$$

Here, the matrix $\boldsymbol{M}''$ is defined by

$$\boldsymbol{M}'' = \sum_{\alpha,\beta=1}^{N} \sum_{k,l,m,n=1}^{L} \bar{W}_{\alpha}^{\prime\,(kl)} \bar{W}_{\beta}^{\prime\,(mn)} (\nabla_{\boldsymbol{a}} \bar{F}_{\alpha}^{(k)}, \bar{\boldsymbol{L}}_{\alpha}^{-} \bar{\boldsymbol{l}}_{\alpha}) (\nabla_{\boldsymbol{a}} \bar{F}_{\beta}^{(m)}, \bar{\boldsymbol{L}}_{\beta}^{-} \bar{\boldsymbol{l}}_{\beta}) \left(\boldsymbol{P}_{\bar{\boldsymbol{u}}}^{\mathcal{U}} \nabla_{\boldsymbol{u}} \bar{F}_{\alpha}^{(l)}\right) \left(\boldsymbol{P}_{\bar{\boldsymbol{u}}}^{\mathcal{U}} \nabla_{\boldsymbol{u}} \bar{F}_{\beta}^{(n)}\right)^{\top}. \tag{14.125}$$

14.5.2 Fitting of the exponential family

If the distribution of each $\boldsymbol{a}_\alpha$ belongs to the *exponential family*, the probability density $p_\alpha(\boldsymbol{a}_\alpha; \bar{\boldsymbol{a}}_\alpha)$ has the following expression (see eq. (3.146)):

$$p_\alpha(\boldsymbol{a}_\alpha; \bar{\boldsymbol{a}}_\alpha) = C_\alpha(\bar{\boldsymbol{a}}_\alpha) \exp[(\boldsymbol{f}_\alpha(\boldsymbol{a}_\alpha), \bar{\boldsymbol{a}}_\alpha) + g_\alpha(\boldsymbol{a}_\alpha)]. \tag{14.126}$$

Here, $\boldsymbol{f}_\alpha(\boldsymbol{a})$ is an m-dimensional function of $\boldsymbol{a} \in \mathcal{R}^m$, and $C_\alpha(\boldsymbol{a})$ and $g_\alpha(\boldsymbol{a})$ are scalar functions of $\boldsymbol{a} \in \mathcal{R}^m$. We call Problem 14.2 *parametric fitting of the exponential family* if the probability densities $p_\alpha(\boldsymbol{a}_\alpha; \bar{\boldsymbol{a}}_\alpha)$ have the above expression.

For parametric fitting of the exponential family, we see from eqs. (14.110) and (14.126) that the matrix $\bar{\boldsymbol{L}}_\alpha$ does not depend on $\boldsymbol{a}_\alpha$. Consequently, we have the following relationship:

$$\boldsymbol{J}_\alpha = E[\bar{\boldsymbol{L}}_\alpha] = \bar{\boldsymbol{L}}_\alpha. \tag{14.127}$$

From this, we obtain the following proposition:

Proposition 14.4 *The maximum likelihood estimator* $\hat{\boldsymbol{u}}$ *for parametric fitting of the exponential family is unbiased in the first order.*

Proof. If eq. (14.127) holds, the matrix $\boldsymbol{M}'$ defined by eq. (14.107) is a constant matrix. Since $E[\bar{\boldsymbol{l}}] = \boldsymbol{0}$, we see from eq. (14.123) that

$$\boldsymbol{P}^{\mathcal{U}}_{\bar{\boldsymbol{u}}} E[\hat{\boldsymbol{u}} - \bar{\boldsymbol{u}}] = E[\Delta \boldsymbol{u}] = \boldsymbol{0}. \tag{14.128}$$

□

As in the case of geometric correction, we say that an estimator for parametric fitting is *efficient* if its covariance matrix attains the Cramer-Rao lower bound given by Theorem 14.2.

Proposition 14.5 *The maximum likelihood estimator* $\hat{\boldsymbol{u}}$ *for parametric fitting of the exponential family is efficient in the first order.*

Proof. Since $\bar{\boldsymbol{L}}_\alpha = \boldsymbol{J}_\alpha$, we see that $\bar{\boldsymbol{V}}'_\alpha = \bar{\boldsymbol{V}}_\alpha$ (see eqs. (14.85) and (14.117)) and hence $\bar{\boldsymbol{W}}'_\alpha = \bar{\boldsymbol{W}}_\alpha$ (see eqs. (14.86) and (14.119)). It follows that $\boldsymbol{M}' = \bar{\boldsymbol{M}}$ (see eqs. (14.87) and (14.122)). Taking the expectation of eq. (14.125) and noting the independence of each $\boldsymbol{a}_\alpha$, we see that

$$\begin{aligned}
E[\boldsymbol{M}''] &= \sum_{\alpha,\beta=1}^{N} \sum_{k,l,m,n=1}^{L} \bar{W}^{(kl)}_\alpha \bar{W}^{(mn)}_\alpha (\nabla_{\boldsymbol{a}} \bar{F}^{(k)}_\alpha, \boldsymbol{J}^-_\alpha E[\bar{\boldsymbol{l}}_\alpha \bar{\boldsymbol{l}}^\top_\beta] \boldsymbol{J}^-_\beta \nabla_{\boldsymbol{a}} \bar{F}^{(m)}_\beta) \\
&\qquad \left(\boldsymbol{P}^{\mathcal{U}}_{\bar{\boldsymbol{u}}} \nabla_{\boldsymbol{u}} \bar{F}^{(l)}_\alpha\right) \left(\boldsymbol{P}^{\mathcal{U}}_{\bar{\boldsymbol{u}}} \nabla_{\boldsymbol{u}} \bar{F}^{(n)}_\beta\right)^\top \\
&= \sum_{\alpha=1}^{N} \sum_{k,l,m,n=1}^{L} \bar{W}^{(kl)}_\alpha \bar{W}^{(mn)}_\alpha (\nabla_{\boldsymbol{a}} \bar{F}^{(k)}_\alpha, (\boldsymbol{J}^-_\alpha \boldsymbol{J}_\alpha \boldsymbol{J}^-_\alpha) \nabla_{\boldsymbol{a}} \bar{F}^{(m)}_\alpha) \\
&\qquad \left(\boldsymbol{P}^{\mathcal{U}}_{\bar{\boldsymbol{u}}} \nabla_{\boldsymbol{u}} \bar{F}^{(l)}_\alpha\right) \left(\boldsymbol{P}^{\mathcal{U}}_{\bar{\boldsymbol{u}}} \nabla_{\boldsymbol{u}} \bar{F}^{(n)}_\alpha\right)^\top \\
&= \sum_{\alpha=1}^{N} \sum_{l,n=1}^{L} \left(\sum_{k,m=1}^{L} \bar{W}^{(lk)}_\alpha (\nabla_{\boldsymbol{a}} \bar{F}^{(k)}_\alpha, \boldsymbol{J}^-_\alpha \nabla_{\boldsymbol{a}} \bar{F}^{(m)}_\alpha) \bar{W}^{(mn)}_\alpha \right) \\
&\qquad \left(\boldsymbol{P}^{\mathcal{U}}_{\bar{\boldsymbol{u}}} \nabla_{\boldsymbol{u}} \bar{F}^{(l)}_\alpha\right) \left(\boldsymbol{P}^{\mathcal{U}}_{\bar{\boldsymbol{u}}} \nabla_{\boldsymbol{u}} \bar{F}^{(n)}_\alpha\right)^\top \\
&= \sum_{\alpha=1}^{N} \sum_{l,n=1}^{L} \bar{W}^{(ln)}_\alpha \left(\boldsymbol{P}^{\mathcal{U}}_{\bar{\boldsymbol{u}}} \nabla_{\boldsymbol{u}} \bar{F}^{(l)}_\alpha\right) \left(\boldsymbol{P}^{\mathcal{U}}_{\bar{\boldsymbol{u}}} \nabla_{\boldsymbol{u}} \bar{F}^{(n)}_\alpha\right)^\top = \bar{\boldsymbol{M}},
\end{aligned} \tag{14.129}$$

where we have used the identities $\boldsymbol{J}^-_\alpha \boldsymbol{J}_\alpha \boldsymbol{J}^-_\alpha = \boldsymbol{J}^-_\alpha$ and $\bar{\boldsymbol{W}}_\alpha \bar{\boldsymbol{V}}_\alpha \bar{\boldsymbol{W}}_\alpha = \bar{\boldsymbol{W}}_\alpha \bar{\boldsymbol{W}}^-_\alpha \bar{\boldsymbol{W}}_\alpha = \bar{\boldsymbol{W}}_\alpha$. Thus, eq. (14.124) is written as $V[\hat{\boldsymbol{u}}] = \bar{\boldsymbol{M}}^- \bar{\boldsymbol{M}} \bar{\boldsymbol{M}}^- = \bar{\boldsymbol{M}}^-$. □

Recall that an estimator is *consistent* if it converges to the true value as the number of data increases (see Section 3.6.2).

Proposition 14.6 *The maximum likelihood estimator for parametric fitting of the exponential family is consistent in the first order if all* $\bar{\boldsymbol{a}}_\alpha$, $\alpha = 1, ..., N$, *are in a bounded subset of the model* $\mathcal{S}$.

Proof. If all $\bar{\boldsymbol{a}}_\alpha$ are bounded, we have $\bar{\boldsymbol{M}} = O(N)$ from the definition of the moment matrix $\bar{\boldsymbol{M}}$ (see eq. (14.87)). Hence, $V[\hat{\boldsymbol{u}}] = \bar{\boldsymbol{M}}^- = O(1/N)$. □

The proofs of Propositions 14.4, 14.5, and 14.6 are based on eq. (14.123), which is obtained by approximating the function J in eq. (14.108) by a quadratic polynomial in the neighborhood of their true values and applying a linear approximation to the hypothesis (14.109). In other words, eq. (14.123) is an *expression in the limit of small noise*. Accordingly, Propositions 14.4, 14.5, and 14.6 hold in this sense, which is what the phrase "in the first order" means. As in the case of geometric correction, this type of first order analysis is compatible with our approach of studying statistical behavior for small noise (see Section 14.1).

14.5.3 Computation of maximum likelihood fit

We now derive a computational scheme for the maximum likelihood estimator $\hat{\boldsymbol{u}}$ by eliminating the estimators $\{\hat{\boldsymbol{a}}_\alpha\}$, $\alpha = 1, ..., N$. To this end, we first compute the solution $\boldsymbol{x}_\alpha$ of the optimization (14.108) under the hypothesis (14.109) for a *fixed* $\boldsymbol{u}$ and then substitute the resulting expression back into eq. (14.108), reducing it to a function of $\boldsymbol{u}$ alone.

As in the case of geometric correction, we first introduce a new assumption about the probability density $p_\alpha(\boldsymbol{a}_\alpha; \boldsymbol{x}_\alpha)$. We assume that viewed as a function of $\boldsymbol{x}_\alpha$ it takes its maximum at $\boldsymbol{a}_\alpha$. In other words, given a single datum $\boldsymbol{a}_\alpha$, the datum $\boldsymbol{a}_\alpha$ itself is the most likely value of the true value $\bar{\boldsymbol{a}}_\alpha$ if no other information is given. To be specific, what we need is the following:

Assumption 14.18

$$\nabla_{\boldsymbol{x}_\alpha} p_\alpha(\boldsymbol{a}_\alpha; \boldsymbol{x}_\alpha)|_{\boldsymbol{x}_\alpha = \boldsymbol{a}_\alpha} = \boldsymbol{0}. \tag{14.130}$$

For example, this assumption is satisfied if the probability density is expressed by a smooth scalar function $f(\,\cdot\,)$ that takes its maximum at the origin $\boldsymbol{0}$ in the form $p_\alpha(\boldsymbol{a}_\alpha; \bar{\boldsymbol{a}}_\alpha) = f(\boldsymbol{a}_\alpha - \bar{\boldsymbol{a}}_\alpha)$. The Gaussian distribution is a typical example.

Assuming that the solution $\boldsymbol{x}_\alpha$ of the optimization (14.108) is close to the data value $\boldsymbol{a}_\alpha$, we write $\boldsymbol{x}_\alpha = \boldsymbol{a}_\alpha + \Delta\boldsymbol{x}_\alpha$. Substituting this into the function J, expanding it in the neighborhood of the datum $\boldsymbol{a}_\alpha$, and using Assumption

14.18, we obtain

$$J = c + \sum_{\alpha=1}^{N}(\Delta \boldsymbol{x}_\alpha, \boldsymbol{L}_\alpha \Delta \boldsymbol{x}_\alpha) + \sum_{\alpha=1}^{N} O(\Delta \boldsymbol{x}_\alpha)^3. \tag{14.131}$$

The constant c is the value of J evaluated at $\boldsymbol{a}_\alpha$, $\alpha = 1, ..., N$, and $\boldsymbol{L}_\alpha$ is an (mm)-matrix defined by

$$\boldsymbol{L}_\alpha = -\nabla^2_{\boldsymbol{x}_\alpha} \log p_\alpha|_{\boldsymbol{x}_\alpha = \boldsymbol{a}_\alpha}. \tag{14.132}$$

In order to guarantee unique existence of the values $\boldsymbol{x}_\alpha \in \mathcal{A}$, $\alpha = 1, ..., N$, that minimize J, we assume that Assumption 14.16 also applies to $\boldsymbol{L}_\alpha$:

Assumption 14.19 *For any $\boldsymbol{a}_\alpha \in \mathcal{A}$, matrix $\boldsymbol{L}_\alpha$ is positive semi-definite and has range $T_{\bar{\boldsymbol{a}}_\alpha}(\mathcal{A})$.*

The linear approximation of the hypothesis (14.109) is

$$(\nabla_{\boldsymbol{a}} F^{(k)}_\alpha, \Delta \boldsymbol{x}_\alpha) = -F^{(k)}_\alpha, \tag{14.133}$$

where $F^{(k)}_\alpha$ and $\nabla_{\boldsymbol{a}} F^{(k)}_\alpha$ are the abbreviations of $F^{(k)}(\boldsymbol{a}_\alpha, \boldsymbol{u})$ and $\nabla_{\boldsymbol{a}} F^{(k)}(\boldsymbol{a}_\alpha, \boldsymbol{u})$, respectively. The inherent constraint $\boldsymbol{x}_\alpha \in \mathcal{A}$ requires that $\Delta \boldsymbol{x}_\alpha \in T_{\boldsymbol{a}_\alpha}(\mathcal{A})$ to a first approximation. Introducing Lagrange multipliers $\lambda^{(k)}_\alpha$ to the linearized hypothesis (14.133), differentiating J with respect to $\Delta \boldsymbol{x}_\alpha$, and ignoring higher order terms, we obtain the solution $\Delta \boldsymbol{x}_\alpha$ uniquely in the following form:

$$\Delta \boldsymbol{x}_\alpha = \sum_{k=1}^{L} \lambda^{(k)}_\alpha \boldsymbol{L}^-_\alpha \nabla_{\boldsymbol{a}} F^{(k)}_\alpha. \tag{14.134}$$

Substituting this into eq. (14.133) and putting $\boldsymbol{\lambda} = (\lambda^{(k)}_\alpha)$, we obtain

$$\boldsymbol{V}_\alpha \boldsymbol{\lambda}_\alpha = -(F^{(k)}_\alpha), \tag{14.135}$$

where the (LL)-matrix $\boldsymbol{V}_\alpha = (V^{(kl)}_\alpha)$ is defined by

$$(V^{(kl)}_\alpha) = \Big((\nabla_{\boldsymbol{a}} F^{(k)}_\alpha, \boldsymbol{L}^-_\alpha \nabla_{\boldsymbol{a}} F^{(l)}_\alpha)\Big) \tag{14.136}$$

As in the case of geometric correction, a *computational problem* arises. Namely, if the L equations (14.109) are not algebraically independent but give only r constraints, the matrix $\bar{\boldsymbol{V}}'_\alpha$ computed from the true value $\bar{\boldsymbol{a}}_\alpha$ by eq. (14.117) has rank r. However, the rank of the matrix $\boldsymbol{V}_\alpha$ computed from the datum $\boldsymbol{a}_\alpha$ by eq. (14.136) may be larger than r. In other words, among the L linear equations in $\Delta \boldsymbol{x}_\alpha$ given by eq. (14.133), more than r equations may be linearly independent if evaluated at the datum $\boldsymbol{a}_\alpha \neq \bar{\boldsymbol{a}}_\alpha$. If this occurs, the hypothesis (14.109) is said to be *degenerate* (see Section 14.4.2). If the hypothesis (14.109) is degenerate, some of the positive eigenvalues of matrix $\boldsymbol{V}_\alpha$ converge to 0 in the limit $\boldsymbol{a}_\alpha \to \bar{\boldsymbol{a}}_\alpha$. Hence, eq. (14.135) viewed as a linear equation in $\boldsymbol{\lambda}_\alpha$ is *ill-conditioned* when the data $\{\boldsymbol{a}_\alpha\}$ are close to the true values $\{\bar{\boldsymbol{a}}_\alpha\}$ (see Section 2.3.2).

This ill-posedness can be avoided in the same way as in the case of geometric correction: we *project* both sides of eq. (14.135) onto the eigenspace defined by the largest r eigenvalues of $\boldsymbol{V}_\alpha$. In other words, we effectively use only r constraints from among the L linear equations (14.133). The resulting solution may be slightly different from the exact solution of eq. (14.135) but does not effectively affect the subsequent analysis because eq. (14.135) is a first order approximation. The solution of eq. (14.135) given by this projection is

$$\lambda_\alpha^{(k)} = -\sum_{l=1}^{L} W_\alpha^{(kl)} F_\alpha^{(l)}, \tag{14.137}$$

where $W_\alpha^{(kl)}$ is the (kl) element of the rank-constrained generalized inverse $(\boldsymbol{V}_\alpha)_r^-$, which we write

$$(W_\alpha^{(kl)}) = \Big((\nabla_{\boldsymbol{a}} F_\alpha^{(k)}, \boldsymbol{L}_\alpha^- \nabla_{\boldsymbol{a}} F_\alpha^{(l)})\Big). \tag{14.138}$$

Substituting eq. (14.137) into eq. (14.134), substituting the resulting expression into eq. (14.131), and ignoring higher order terms, we obtain the following expression:

$$J = c + \sum_{\alpha=1}^{N} \sum_{k,l=1}^{L} W_\alpha^{(kl)} F_\alpha^{(k)} F_\alpha^{(l)}. \tag{14.139}$$

Since the constant c is irrelevant to optimization, we obtain the following conclusion:

Proposition 14.7 *The maximum likelihood estimator* $\hat{\boldsymbol{u}}$ *of* $\boldsymbol{u}$ *is given as the solution of the nonlinear optimization*

$$J[\boldsymbol{u}] = \sum_{\alpha=1}^{N} \sum_{k,l=1}^{L} W_\alpha^{(kl)} F_\alpha^{(k)} F_\alpha^{(l)} \to \min, \qquad \boldsymbol{u} \in \mathcal{U}. \tag{14.140}$$

This proposition gives a rigorous mathematical foundation to the formulation of parametric fitting as presented Chapter 7, upon which the subsequent applications are all based. Let $\hat{\boldsymbol{u}}$ be the computed maximum likelihood estimator of $\boldsymbol{u}$. If the model $\mathcal{S}$ has codimension r in the m'-dimensional data space $\mathcal{A}$ parameterized by $\boldsymbol{u} \in \mathcal{U}$, it is a model of dimension $d = m' - r$, codimension r, and n' degrees of freedom. The argument given in Sections

13.2 and 13.3 can be generalized to data with non-Gaussian distributions, and the *geometric information criterion* can be defined as follows:

$$AIC(\mathcal{S}) = J[\hat{\boldsymbol{u}}] + 2(dN + n'). \tag{14.141}$$

Bibliography

[1] G. Adiv, Determining three-dimensional motion and structure from optical flow generated by several moving objects, *IEEE Transactions on Pattern Analysis and Machine Intelligence*, **7**-4 (1985), 384–401.

[2] H. Akaike, A new look at the statistical model identification, *IEEE Transactions on Automatic Control*, **19**-6 (1974), 716–723.

[3] J. Aisbett, An iterated estimation of the motion parameters of a rigid body from noisy displacement vectors, *IEEE Transactions on Pattern Analysis and Machine Intelligence*, **12**-11 (1990), 1092–1098.

[4] A. Albano, Representation of digitized contours in terms of conic arcs and straight-line segments, *Computer Graphics and Image Processing*, **3**-1 (1974), 23–33.

[5] Y. Aloimonos (ed.), *Active Perception*, Lawrence Erlbaum Associates, Hillsdale, NJ, U.S.A., 1993.

[6] J. (Y.) Aloimonos and D. Shulman, *Integration of Visual Modules: An Extension of the Marr Paradigm*, Academic Press, San Diego, CA, U.S.A., 1989.

[7] S. Amari, *Differential-Geometrical Methods in Statistics*, Springer, Berlin, F.R.G., 1985.

[8] S. Amari and M. Kumon, Estimation in the presence of infinitely many nuisance parameters—Geometry for estimation functions, *Annals of Statistics*, **16**-3 (1988), 1044–1068.

[9] E. B. Andersen, Asymptotic properties of conditional maximum likelihood estimators, *Journal of the Royal Society of London*, B**32**-2 (1970), 283–301.

[10] B. D. O. Anderson and J. B. Moore, *Optimal Filtering*, Prentice-Hall, Englewood Cliffs, NJ, U.S.A., 1979.

[11] K. S. Arun, T. S. Huang and S. D. Blostein, Least-squares fitting of two 3-D point sets, *IEEE Transactions on Pattern Analysis and Machine Intelligence*, **9**-5 (1987), 698–700.

[12] H. Ballard and C. M. Brown, *Computer Vision*, Prentice-Hall, Englewood Cliffs, NJ, U.S.A., 1982.

[13] A. Bani-Hashemi, A Fourier approach to camera orientation, *IEEE Transactions on Pattern Analysis and Machine Intelligence*, **15**-11 (1993), 1197–1202.

[14] O. E. Barndorff-Nielsen, *Information and Exponential Families in Statistical Theory*, John Wiley and Sons, New York, NY, U.S.A., 1978.

[15] O. E. Barndorff-Nielsen, *Parametric Statistical Models and Likelihood*, Lecture Notes in Statistics 50, Springer, Berlin, F.R.G., 1988.

[16] J. L. Barron, D. J. Fleet and S. S. Beauchemin, Performance of optical flow techniques, *International Journal of Computer Vision*, **12**-1 (1994), 43–77.

[17] J. Bhanja and J. K. Ghosh, Efficient estimation with many nuisance parameters, Parts 1, 2, 3, *Sankhyā*, A**54**-1,2,3 (1992), 1–39, 135–156, 297–308.

[18] P. J. Bickel, C. A. J. Klaassen, Y. Ritov and J. A. Wellner, *Efficient and Adaptive Estimation for Semiparametric Models*, Johns Hopkins University Press, Baltimore, MD, U.S.A., 1993.

[19] S. D. Blostein and T. S. Huang, Error analysis in stereo determination of 3-D point positions, *IEEE Transactions on Pattern Analysis and Machine Intelligence*, **9**-6 (1987), 752–765 (corrected in **10**-5 (1988), 765).

[20] R. M. Bolle and B. C. Verumi, On three-dimensional surface reconstruction methods, *IEEE Transactions on Pattern Analysis and Machine Intelligence*, **9**-1 (1991), 1–13.

[21] F. L. Bookstein, Fitting conic sections to scattered data, *Computer Graphics and Image Processing*, **9**-1 (1979), 56–71.

[22] K. L. Boyer, J. Mirza and G. Ganguly, The robust sequential estimator: A general approach and its application to surface organization in range data, *IEEE Transactions on Pattern Analysis and Machine Intelligence*, **16**-10 (1994), 987–1001.

[23] B. Brillault-O'Mahony, New method for vanishing point detection, *CVGIP: Image Understanding*, **54**-2 (1991), 289–300.

[24] B. Brillault-O'Mahony, High level 3D structure from a single view, *Image and Vision Computing*, **10**-7 (1992), 508–520.

[25] T. J. Broida, S. Chandrashekhar and R. Chellappa, Recursive estimation of 3-D motion from a monocular image sequence, *IEEE Transactions on Aerospace and Electronic Systems*, **26**-4 (1990), 639–656.

[26] T. J. Broida and R. Chellappa, Estimation of object motion parameters from noisy images, *IEEE Transactions on Pattern Analysis and Machine Intelligence*, **8**-1 (1986), 90–99.

[27] T. J. Broida and R. Chellappa, Performance bounds for estimating three-dimensional motion parameters from a sequence of noisy images, *Journal of the Optical Society of America*, A**6**-6 (1989), 879-898.

[28] T. J. Broida and R. Chellappa, Estimating the kinematics and structure of a rigid object from a sequence of monocular images, *IEEE Transactions on Pattern Analysis and Machine Intelligence*, **13**-6 (1991), 497–513.

[29] R. C. Brown, *Introduction to Random Signal Analysis and Kalman Filtering*, John Wiley and Sons, New York, NY, U.S.A., 1983.

[30] A. R. Bruss and B. K. P. Horn, Passive navigation, *Computer Vision, Graphics, and Image Processing*, **21**-1 (1983), 3–20.

[31] M. Campani and A. Verri, Motion analysis from first-order properties of optical flow, *CVGIP: Image Understanding*, **56**-1 (1992), 90–107.

[32] B. Caprile and V. Torre, Using vanishing points for camera calibration, *International Journal of Computer Vision*, **4**-2 (1990), 127–140.

[33] C. K. Chui and G. Chen, *Kalman Filtering with Real-Time Applications*, Springer, Berlin, F.R.G., 1991.

[34] D. R. Cooper and N. Yalabik, On the computational cost of approximating and recognizing noise-perturbed straight lines and quadratic arcs in the plane, *IEEE Transactions on Computers*, **25**-10 (1976), 1020–1032.

[35] I. J. Cox, A review of statistical data association techniques for motion correspondence, *International Journal of Computer Vision*, **10**-1 (1993), 53–66.

[36] I. J. Cox, J. M. Rehg and S. Hingorani, A Bayesian multiple-hypothesis approach to edge grouping and contour segmentation, *International Journal of Computer Vision*, **11**-1 (1993), 5–24.

[37] D. R. Cox and D. V. Hinkley, *Theoretical Statistics*, Chapman and Hall, London, U.K., 1974.

[38] H. Cramér, *The Elements of Probability Theory and Some of Its Applications*, Almqvist and Wiksell/Gebers, Stockholm, Sweden, 1955.

[39] J. L. Crowley, P. Bobet and C. Schmid, Auto-calibration by direct observation of objects, *Image and Vision Computing*, **11**-2 (1993), 67–81.

[40] K. Daniilidis and H.-H. Nagel, Analytical results on error sensitivity of motion estimation from two views, *Image and Vision Computing*, **8**-4 (1990), 287–303.

[41] E. R. Davies, Finding ellipses using the generalised Hough transform, *Pattern Recognition Letters*, **9**-1 (1989), 87–96.

[42] A. J. Dobson, *An Introduction to Generalized Linear Models*, Chapman and Hall, London, U.K., 1990.

[43] L. S. Dreschler and H.-H. Nagel, Volumetric model and 3-D trajectory of a moving car derived from monocular TV-frame sequences of a street scene, *Computer Graphics and Image Processing*, **20**-3 (1982), 199–228.

[44] T. Echigo, A camera calibration technique using three sets of parallel lines, *Machine Vision and Applications*, **3**-3 (1990), 159–167.

[45] T. Ellis, A. Abboot and B. Brillault, Ellipse detection and matching with uncertainty, *Image and Vision Computing*, **10**-5 (1992), 271–276.

[46] T. Endoh, T. Toriu and N. Tagawa, A superior estimator to the maximum likelihood estimator on 3-D motion estimation from noisy optical flow, *IEICE Transactions on Information and Systems*, **E77-D**-11 (1994), 1240–1246.

[47] O. Faugeras, *Three-Dimensional Computer Vision: A Geometric Viewpoint*, MIT Press, Cambridge, MA, U.S.A., 1993.

[48] O. D. Faugeras and S. Maybank, Motion from point matches: Multiplicity of solutions, *International Journal of Computer Vision*, **4**-3 (1990), 225–246.

[49] W. Feller, *Probability Theory and Its Applications*, Vols. 1, 2, John Wiley and Sons, New York, NY, U.S.A., 1950, 1966.

[50] D. Forsyth, J. L. Mundy and A. Zisserman, Transformation invariance—A primer, *Image and Vision Computing*, **10**-1 (1992), 39–45.

[51] D. Forsyth, J. L. Mundy, A. Zisserman, C. Coelho, A. Heller and C. Rothwell, Invariant descriptors for 3-D object recognition and pose, *IEEE Transactions on Pattern Analysis and Machine Intelligence*, **13**-10 (1991), 971–991.

[52] J. E. Freund, *Mathematical Statistics*, Prentice-Hall, Englewood Cliffs, NJ, U.S.A., 1962.

[53] J. E. Freund, *Modern Elementary Statistics*, 3rd Ed., Prentice-Hall, Englewood Cliffs, NJ, U.S.A., 1967.

[54] A. Gelb (ed.), *Applied Optimal Estimation*, MIT Press, Cambridge, MA, U.S.A., 1974.

[55] R. Gnanadesikan, *Methods for Statistical Data Analysis of Multivariate Observations*, John Wiley and Sons, New York, NY, U.S.A., 1977.

[56] W. I. Grosky and L. A. Tamburino, A unified approach to the linear camera calibration problem, *IEEE Transactions on Pattern Analysis and Machine Intelligence*, **12**-7 (1990), 663–671.

[57] R. M. Haralick and L. G. Shapiro, *Computer and Robot Vision*, Vols. 1, 2, Addison-Wesley, Reading, MA, U.S.A., 1992.

[58] J. C. Hay, Optical motions and space perception—An extension of Gibson's analysis, *Psychological Review*, **73**-6 (1966), 550–565.

[59] G. Healey and R. Kondepudy, Radiometric CCD camera calibration and noise estimation, *IEEE Transactions on Pattern Analysis and Machine Intelligence*, **16**-3 (1994), 267–275.

[60] D. J. Heeger and A. D. Jepson, Subspace methods for recovering rigid motion I: Algorithm and implementation, *International Journal of Computer Vision*, **7**-2 (1992), 95–117.

[61] Y. Hel-Or and M. Werman, Pose estimation by fusing noisy data of different dimensions, *IEEE Transactions on Pattern Analysis and Machine Intelligence*, **17**-2 (1995), 195–201.

[62] P. G. Hoel, *Elementary Statistics*, 2nd Ed., John Wiley and Sons, New York, NY, U.S.A., 1966.

[63] P. G. Hoel, *Introduction to Mathematical Statistics*, 4th Ed., John Wiley and Sons, New York, NY, U.S.A., 1971.

[64] R. Horaud, R. Mohr and B. Lorecki, On single-scanline camera calibration, *IEEE Transactions on Robotics and Automation*, **9**-1 (1993), 71–74.

[65] B. K. P. Horn, *Robot Vision*, MIT Press, Cambridge, MA, U.S.A., 1986.

[66] B. K. P. Horn, Motion fields are hardly ever ambiguous, *International Journal of Computer Vision*, **1**-3 (1987), 259–274.

[67] B. K. P. Horn, Closed-form solution of absolute orientation using unit quaternions, *Journal of the Optical Society of America*, A**4**-4 (1987), 629–642.

[68] B. K. P. Horn, Relative orientation, *International Journal of Computer Vision*, **4**-1 (1990), 59–78.

[69] B. K. P. Horn, H. M. Hilden and S. Negahdaripour, Closed-form solution of absolute orientation using orthonormal matrices, *Journal of the Optical Society of America*, A**5**-5 (1988), 1127–1135.

[70] B. K. P. Horn and B. Schunck, Determining optical flow, *Artificial Intelligence*, **17**-1/2/3 (1981), 185–203.

[71] X. Hu and N. Ahuja, Motion and structure estimation using long sequence motion models, *Image and Vision Computing*, **11**-9 (1993), 549–569.

[72] X. Hu and N. Ahuja, Necessary and sufficient conditions for a unique solution of planar motion and structure, *IEEE Transactions on Robotics and Automation*, **11**-2 (1995), 304–308.

[73] T. S. Huang and O. D. Faugeras, Some properties of the E matrix in two-view motion estimation, *IEEE Transactions on Pattern Analysis and Machine Intelligence*, **11**-12 (1989), 1310–1312.

[74] A. H. Jazwinski, *Stochastic Processes and Filtering Theory*, Academic Press, New York, NY, U.S.A., 1970.

[75] C. Jerian and R. Jain, Polynomial methods for structure from motion, *IEEE Transactions on Pattern Analysis and Machine Intelligence*, **12**-12 (1990), 1150–1165.

[76] C. P. Jerian and R. Jain, Structure from Motion—A critical analysis of methods, *IEEE Transactions on Systems, Man, and Cybernetics*, **21**-3 (1991), 572–588.

[77] J.-M. Jolion, P. Meer and S. Bataouche, Robust clustering with applications in computer vision, *IEEE Transactions on Pattern Analysis and Machine Intelligence*, **13**-8 (1991), 791–802.

[78] R. E. Kalman, A new approach to linear filtering and prediction problems, *Transactions of ASME, Journal of Basic Engineering*, **82D**-1 (1960), 35–45.

[79] R. E. Kalman and R. S. Bucy, New results in linear filtering and prediction theory, *Transactions of ASME, Journal of Basic Engineering*, **83D**-1 (1961), 95–108.

[80] B. Kamgar-Parsi and R. D. Eastman, Calibration of a stereo system with small relative angles, *Computer Vision, Graphics, and Image Processing*, **51**-1 (1990), 1–19.

[81] B. Kamgar-Parsi, B. Kamgar-Parsi and N. Netanyahu, A nonparametric method for fitting a straight line to a noisy image, *IEEE Transactions on Pattern Analysis and Machine Intelligence*, **11**-9 (1989), 998–1001.

[82] K. Kanatani, Distribution of directional data and fabric tensors, *International Journal of Engineering Science*, **22**-2 (1984), 149–164.

[83] K. Kanatani, Structure and motion from optical flow under orthographic projection, *Computer Vision, Graphics, and Image Processing*, **35**-2 (1986), 181–199.

[84] K. Kanatani, Structure and motion from optical flow under perspective projection, *Computer Vision, Graphics, and Image Processing*, **38**-2 (1987), 122–146.

[85] K. Kanatani, *Group-Theoretical Methods in Image Understanding*, Springer, Berlin, F.R.G., 1990.

[86] K. Kanatani, Computational projective geometry, *CVGIP: Image Understanding*, **54**-3 (1991), 333–348.

[87] K. Kanatani, Hypothesizing and testing geometric properties of image data, *CVGIP: Image Understanding*, **54**-3 (1991), 349–357.

[88] K. Kanatani, Statistical analysis of focal length calibration using vanishing points, *IEEE Transactions on Robotics and Automation*, **8**-6, (1992), 767–775.

[89] K. Kanatani, Unbiased estimation and statistical analysis of 3-D rigid motion from two views, *IEEE Transactions on Pattern Analysis and Machine Intelligence*, **15**-1 (1993), 37–50.

[90] K. Kanatani, *Geometric Computation for Machine Vision*, Oxford University Press, Oxford, U.K., 1993.

[91] K. Kanatani, Statistical analysis of geometric computation, *CVGIP: Image Understanding*, **59**-3 (1994), 286–306.

[92] K. Kanatani, Statistical bias of conic fitting and renormalization, *IEEE Transactions on Pattern Analysis and Machine Intelligence*, **16**-3 (1994), 320–326.

[93] K. Kanatani, Analysis of 3-D rotation fitting, *IEEE Transactions on Pattern Analysis and Machine Intelligence*, **16**-5 (1994), 543–549.

[94] K. Kanatani, Statistical foundation for hypothesis testing of image data, *CVGIP: Image Understanding*, **60**-2 (1994), 382–391.

[95] K. Kanatani, Computational cross ratio for computer vision, *CVGIP: Image Understanding*, **60**-2 (1994), 371–381.

[96] K. Kanatani, Renormalization for motion analysis: Statistically optimal algorithm, *IEICE Transactions on Information and Systems*, **E77-D**-11 (1994), 1233–1239.

[97] K. Kanatani and W. Liu, 3-D interpretation of conics and orthogonality, *CVGIP: Image Understanding*, **58**-3 (1993), 286–301.

[98] K. Kanatani and Y. Onodera, Anatomy of camera calibration using vanishing points, *IEICE Transactions on Information and Systems*, **E74**-10 (1991), 3369–3378.

[99] K. Kanatani and S. Takeda, 3-D motion analysis of a planar surface by renormalization, *IEICE Transactions on Information and Systems*, **E78-D**-8 (1995), 1074–1079.

[100] Y. Kanazawa and K. Kanatani, Direct reconstruction of planar surfaces by stereo vision, *IEICE Transactions on Information and Systems*, **E78-D**-7 (1995), 917–922.

[101] Y. Kanazawa and K. Kanatani, Reliability of 3-D reconstruction by stereo vision, *IEICE Transactions on Information and Systems*, **E78-D**-10 (1995), 1301–1306.

[102] Y. Kanazawa and K. Kanatani, Reliability of fitting a plane to range data, *IEICE Transactions on Information and Systems*, **E78-D**-12 (1995), 1630–1635.

[103] A. Kara, D. M. Wilkes and K. Kawamura, 3D structure reconstruction from point correspondences between two perspective projections, *CVGIP: Image Understanding*, **60**-3 (1994), 392–387.

[104] M. G. Kendall and A. Stuart, *The Advanced Theory of Statistics*, Vols. 1 (3rd Ed.), 2 (3rd Ed.), 3 (2nd Ed.), Griffin, London, U.K., 1969, 1973, 1968.

[105] Y. C. Kim and J. K. Aggarwal, Determining object motion in a sequence of stereo images, *IEEE Journal of Robotics and Automation*, **3**-6 (1987), 599–614.

[106] N. Kiryati and A. M. Bruckstein, What's in a set of points?, *IEEE Transactions on Pattern Analysis and Machine Intelligence*, **14**-4 (1992), 496–500.

[107] E. Kreyszig, *Introductory Mathematical Statistics: Principle and Methods*, John Wiley and Sons, New York, NY. U.S.A., 1970.

[108] R. Kumar, Robust methods for estimating pose and a sensitivity analysis, *CVGIP: Image Understanding*, **60**-3 (1994), 313–432.

[109] M. Kumon and S. Amari, Estimation of a structural parameter in the presence of a large number of nuisance parameters, *Biometrika*, **71**-3 (1984), 445–459.

[110] J. Z. C. Lai, On the sensitivity of camera calibration, *Image and Vision Computing*, **11**-10 (1993), 656–664.

[111] S. M. LaValle and S. A. Hutchinson, A Bayesian segmentation methodology for parametric image models, *IEEE Transactions on Pattern Analysis and Machine Intelligence*, **17**-2 (1995), 211–217.

[112] C.-H. Lee, Time-varying images: The effect of finite resolution on uniqueness, *CVGIP: Image Understanding*, **54**-3 (1991), 325–332.

[113] C.-H. Lee, Computing three-dimensional motion parameters: A hypothesis testing approach, *Image and Vision Computing*, **11**-3 (1993), 145–154.

[114] S. Lee and Y. Kay, A Kalman filter approach for accurate 3-D motion estimation from sequence of stereo images, *CVGIP: Image Understanding*, **54**-2 (1991), 244–258.

[115] E. L. Lehman, *Testing Statistical Hypotheses*, 2nd Ed., John Wiley and Sons, New York, NY. U.S.A., 1986.

[116] R. K. Lenz and R. Y. Tsai, Techniques for calibration of the scale factor and image center for high-accuracy 3-D machine vision metrology, *IEEE Transactions on Pattern Analysis and Machine Intelligence*, **10**-5 (1988), 713–720.

[117] R. K. Lenz and R. Y. Tsai, Calibrating a Cartesian robot with eye-on-hand configuration independent of eye-to-hand relationship, *IEEE Transactions on Pattern Analysis and Machine Intelligence*, **11**-9 (1989), 916–928.

[118] L. Li and J. H. Duncan, 3-D translational motion and structure from binocular image flow, *IEEE Transactions on Pattern Analysis and Machine Intelligence*, **15**-7 (1993), 657-667.

[119] B. G. Lindsay, Conditional score functions: Some optimality results, *Biometrika*, **69**-3 (1982), 503–512.

[120] B. G. Lindsay, Using empirical partially Bayes inference for increased efficiency, *Annals of Statistics*, **13**-3 (1985), 914–931.

[121] W. Liu and K. Kanatani, Interpretation of conic motion and its applications, *International Journal of Computer Vision*, **10**-1 (1993), 67–84.

[122] H. C. Longuet-Higgins, A computer algorithm for reconstructing a scene from two projections, *Nature*, **293**-10 (1981), 133–135.

[123] H. C. Longuet-Higgins, The reconstruction of a scene from two projections—Configurations that defeat the 8-point algorithm, *Proceedings of the First IEEE Conference on Artificial Intelligence Applications*, Denver, CO, U.S.A., December 1984, pp. 395–397.

[124] H. C. Longuet-Higgins, The visual ambiguity of a moving plane, *Proceedings of the Royal Society of London*, B**223** (1984), 165–175.

[125] H. C. Longuet-Higgins, The reconstruction of a plane surface from two perspective projections, *Proceedings of the Royal Society of London*, B**227** (1986), 399–410.

[126] H. C. Longuet-Higgins, Multiple interpretations of a pair of images of a surface, *Proceedings of the Royal Society of London*, A**418** (1988), 1–15.

[127] H. C. Longuet-Higgins and K. Prazdny, The interpretation of a moving retinal image, *Proceedings of the Royal Society of London*, B**208** (1980), 385–397.

[128] S. D. Ma, Conics-based stereo, motion estimation, and pose determination, *International Journal of Computer Vision*, **10**-1 (1993), 7–25.

[129] K. V. Mardia, *Statistics of Directional Data*, Academic Press, London, U.K., 1972.

[130] D. Marr, *Vision: A Computational Investigation into the Human Representation and Processing of Visual Information*, W. H. Freeman, San Francisco, CA, U.S.A., 1982.

[131] L. Matthies, T. Kanade and R. Szeliski, Kalman filter-based algorithms for estimating depth from image sequences, *International Journal of Computer Vision*, **3**-3 (1989), 209–236.

[132] S. Maybank, The angular velocity associated with the optical flowfield arising from motion through rigid environment, *Proceedings of the Royal Society of London*, A**401** (1985), 317–326.

[133] S. J. Maybank, The projective geometry of ambiguous surface, *Philosophical Transactions of the Royal Society of London*, A**332** (1990), 1–47.

[134] S. J. Maybank, Ambiguity in reconstruction from image correspondences, *Image and Vision Computing*, **9**-2 (1991), 93–99.

[135] S. Maybank, *Theory of Reconstruction from Image Motion*, Springer, Berlin, F.R.G., 1993.

[136] S. J. Maybank and O. D. Faugeras, A theory of self-calibration of a moving camera, *International Journal of Computer Vision*, **8**-2 (1992), 123–151.

[137] P. McCullagh and J. A. Nelder, *Generalized Linear Models*, 2nd edition, Chapman and Hall, London, U.K., 1989.

[138] J. S. Meditch, *Stochastic Optimal Linear Estimation and Control*, McGraw-Hill, New York, NY, U.S.A., 1969.

[139] P. Meer, D. Mintz and A. Rosenfeld, Robust regression methods for computer vision: A review, *International Journal of Computer Vision*, **6**-1 (1991), 59–70.

[140] J. Mendel, *Lessons in Digital Estimation Theory*, Prentice-Hall, Englewood Cliffs, NJ, U.S.A., 1987.

[141] E. De Micheli, V. Torre and S. Uras, The accuracy of the computation of optical flow and the recovery of motion parameters, *IEEE Transactions on Pattern Analysis and Machine Intelligence*, **15**-5 (1993), 434–447.

[142] K. M. Mutch, Determining object translation information using stereoscopic motion, *IEEE Transactions on Pattern Analysis and Machine Intelligence*, **8**-6 (1986), 750–755.

[143] M. K. Murray and J. W. Rice, *Differential Geometry and Statistics*, Chapman and Hall, London, U.K., 1993.

[144] H.-H. Nagel, Representation of moving rigid objects based on visual observations, *Computer*, **14**-8 (1981), 29–39.

[145] Y. Nakagawa and A. Rosenfeld, A note on polygonal and elliptical approximation of mechanical parts, *Pattern Recognition*, **11**-2 (1979), 133–142.

[146] S. Negahdaripour, Critical surface pairs and triplets, *International Journal of Computer Vision*, **3**-4 (1989), 293–312.

[147] S. Negahdaripour, Multiple interpretations of the shape and motion of objects from two perspective images, *IEEE Transactions on Pattern Analysis and Machine Intelligence*, **12**-11 (1990), 1025–1030.

[148] S. Negahdaripour, Closed-form relationship between the two interpretations of a moving plane, *Journal of the Optical Society of America*, A**7**-2 (1990), 279–285.

[149] S. Negahdaripour and S. Lee, Motion recovery form image sequences using only first order optical flow information, *International Journal of Computer Vision*, **9**-3 (1992), 163–184.

[150] A. N. Netravali, T. S. Huang, A. S. Krishnakumar and R. J. Holt, Algebraic methods in 3-D motion estimation from two-view point correspondences, *International Journal of Imaging Systems and Technology*, **1**-1 (1989), 78–99.

[151] R. Nevatia, *Machine Perception*, Prentice-Hall, Englewood Cliffs, NJ, U.S.A., 1982.

[152] J. Neyman and E. L. Scott, Consistent estimates based on partially consistent observations, *Econometrica*, **16**-1 (1948), 1–32.

[153] Y. Nomura, M. Sagara, H. Naruse and A. Ide, Simple calibration algorithm for high-distortion-lens camera, *IEEE Transactions on Pattern Analysis and Machine Intelligence*, **14**-12 (1992), 1095–1099.

[154] N. Ohta and K. Kanatani, Optimal structure from motion algorithm for optical flow, *IEICE Transactions on Information and Systems*, **E78-D**-12 (1995), 1559–1566.

[155] K. A. Paton, Conic sections in chromosome analysis, *Pattern Recognition*, **2**-1 (1970), 39–51.

[156] M. A. Penna, Camera calibration: A quick and easy way to determine the scale factor, *IEEE Transactions on Pattern Analysis and Machine Intelligence*, **13**-12 (1991), 1240–1245.

[157] J. Philip, Estimation of three-dimensional motion of rigid objects from noisy observations, *IEEE Transactions on Pattern Analysis and Machine Intelligence*, **13**-1 (1991), 61–66.

[158] T. Poggio and C. Koch, Ill-posed problems in early vision: From computational theory to analogue networks, *Proceedings of the Royal Society of London*, B**226** (1985), 303–323.

[159] T. Poggio, V. Torre and C. Koch, Computational vision and regularization theory, *Nature*, **317** (1985), 314–319.

[160] R. Poli, G. Coppini and G. Valli, Recovery of 3D closed surfaces from sparse data, *CVGIP: Image Understanding*, **60**-1 (1994), 1–25.

[161] S. B. Pollard, J. Porrill and J. E. Mayhew, Experiments in vehicle control using predictive feed-forward stereo, *Image and Vision Computing*, **8**-1 (1990), 63–70.

[162] S. B. Pollard, T. P. Pridmore, J. Porrill, J. E. Mayhew and J. P. Frisby, Geometric modeling from multiple stereo views, *International Journal of Robotics Research*, **8**-4 (1989), 3–32.

[163] J. Porrill, Optimal combination and constraints for geometrical sensor data, *International Journal of Robotics Research*, **7**-6 (1988), 66–77.

[164] J. Porrill, Fitting ellipses and predicting confidence envelopes using a bias corrected Kalman filter, *Image and Vision Computing*, **8**-1 (1990), 37–41.

[165] J. Porrill, S. B. Pollard and J. E. W. Mayhew, Optimal combination of multiple sensors including stereo vision, *Image and Vision Computing*, **5**-2 (1987), 174–180.

[166] T. Poston and I. Stewart, *Catastrophe Theory and Its Applications*, Pitman, London, U.K., 1978.

[167] V. Pratt, Direct least-squares fitting of algebraic surfaces, *Computer Graphics*, **21**-4 (1987), 145–152.

[168] K. Prazdny, Determining the instantaneous direction of motion from optical flow generated by a curvilinearly moving observer, *Computer Graphics and Image Processing*, **17**-3 (1981), 238–248.

[169] L. Quan, P. Gros and R. Mohr, Invariants of a pair of conics revisited, *Image and Vision Computing*, **10**-5 (1992), 319–323.

[170] C. R. Rao, *Linear Statistical Inference and Its Applications*, 2nd Ed., John Wiley and Sons, New York, NY, U.S.A., 1973.

[171] C. R. Rao and S. K. Mitra, *Generalized Inverse of Matrices and Its Applications*, John Wiley and Sons, New York, NY, U.S.A., 1971.

[172] J. H. Rieger and D. T. Lawton, Processing differential image motion, *Journal of the Optical Society of America*, A**2**-2 (1985), 354–360.

[173] J. Rissanen, Modeling by shortest data description, *Automatica*, **14**-5 (1978), 465–471.

[174] J. Rissanen, A universal prior for integers and estimation by minimum description length, *Annals of Statistics*, **11**-3 (1983), 416–431.

[175] J. Rissanen, Universal coding, information, prediction, and estimation, *IEEE Transactions on Information Theory*, **30**-4 (1984), 629–636.

[176] J. Rissanen, *Stochastic Complexity in Statistical Inquiry*, World Scientific, Singapore, 1987.

[177] J. W. Roach and J. K. Aggarwal, Determining the movement of objects from a sequence of images, *IEEE Transactions on Pattern Analysis and Machine Intelligence*, **2**-6 (1980), 544–562.

[178] C. A. Rothwell, A. Zisserman, C. I. Marinos, D. A. Forsyth and J. L. Mundy, Relative motion and pose from arbitrary plane curves, *Image and Vision Computing*, **10**-4 (1992), 250–262.

[179] R. Safaee-Rad, I. Tchoukanov, B. Benhabib and K. C. Smith, Accurate parameter estimation of quadratic curves from grey-level images, *CVGIP: Image Understanding*, **54**-2 (1991), 259–274.

[180] R. Safaee-Rad, I. Tchoukanov, K. C. Smith and B. Benhabib, Constraints on quadratic-curved features under perspective projection, *Image and Vision Computing*, **19**-8 (1992), 532–548.

[181] R. Safaee-Rad, I. Tchoukanov, K. C. Smith and B. Benhabib, Three-dimensional location estimation of circular features for machine vision, *IEEE Transactions on Robotics and Automation*, **8**-5 (1992), 624–640.

[182] P. D. Sampson, Fitting conic sections to "very scattered" data: An iterative refinement of the Bookstein algorithm, *Computer Graphics and Image Processing*, **18**-1 (1982), 97–108.

[183] J. G. Semple and G. T. Kneebone, *Algebraic Projective Geometry*, Clarendon Press, Oxford, U.K., 1952 (reprinted 1979).

[184] J. G. Semple and L. Roth, *Introduction to Algebraic Geometry*, Clarendon Press, Oxford, U.K., 1949 (reprinted 1987).

[185] Y. Shirai, *Three-Dimensional Computer Vision*, Springer, Berlin, F.R.G., 1987.

[186] S. Smith, Note on small angle approximations for stereo disparity, *Image and Vision Computing*, **11**-6 (1993), 395–398.

[187] M. A. Snyder, The precision of 3-D parameters in correspondence based techniques: The case of uniform translation motion in a rigid environment, *IEEE Transactions on Pattern Analysis and Machine Intelligence*, **11**-5 (1989), 523–528.

[188] M. E. Spetsakis, A linear algorithm for point and line-based structure from motion, *CVGIP: Image Understanding*, **56**-2 (1992), 230–241.

[189] M. Spetsakis and J. (Y.) Aloimonos, A multi-frame approach to visual motion perception, *International Journal of Computer Vision*, **6**-3 (1991), 245–255.

[190] M. E. Spetsakis and Y. Aloimonos, Optimal visual motion estimation: A note, *IEEE Transactions on Pattern Analysis and Machine Intelligence*, **14**-9 (1992), 959–964.

[191] M. Spetsakis, Models of statistical visual motion estimation, *CVGIP: Image Understanding*, **60**-1 (1994), 300–312.

[192] M. Subbarao and A. M. Waxman, Closed form solution to image flow equations for a planar surface in motion. *Computer Vision, Graphics, and Image Processing*, **36**-2/3 (1986), 208–228.

[193] M. Subbarao, Interpretation of image flow: Rigid curved surfaces in motion, *International Journal of Computer Vision*, **2**-1 (1988), 77–96.

[194] M. Subbarao, Bounds on time-to-collision and rotational component from first-order derivatives of image flow, *Computer Vision, Graphics, and Image Processing*, **50**-3 (1990), 329–341.

[195] N. Tagawa, T. Toriu and T. Endoh, Un-biased linear algorithm for recovering three-dimensional motion from optical flow, *IEICE Transactions on Information and Systems.*, **E76**-10 (1993), 1263–1275.

[196] N. Tagawa, T. Toriu and T. Endoh, Estimation of 3-D motion from optical flow with unbiased objective function, *IEICE Transactions on Information and Systems.*, **E77-D**-10 (1994), 1148–1161.

[197] H. Takeda, C. Facchinetti and J.-C. Latombe, Planning the motions of a mobile robot in a sensory uncertainty field, *IEEE Transactions on Pattern Analysis and Machine Intelligence*, **16**-19 (1994), 1002–1017.

[198] T. N. Tan, K. D. Baker and G. D. Sullivan, 3D structure and motion estimation from 2D image sequences, *Image and Vision Computing*, **11**-4 (1993), 203–210.

[199] K. Tarabanis, R. Y. Tsai and D. S. Goodman, Calibration of a computer controlled robotic vision sensor with a zoom lens, *CVGIP: Image Understanding*, **59**-2 (1994), 226–241.

[200] G. Taubin, F. Cukierman, S. Sullivan, J. Ponce and D. J. Kriegman, Parameterized families for bounded algebraic curve and surface fitting, *IEEE Transactions on Pattern Analysis and Machine Intelligence*, **16**-3 (1994), 287–303.

[201] R. Thom, *Stabilité Structurelle et Morphogénèse*, Benjamin, New York, NY, 1972.

[202] J. I. Thomas, Refining 3D reconstructions: A theoretical and experimental study of the effect of cross-correlations, *CVGIP: Image Understanding*, **60**-3 (1994), 359–370.

[203] C. Tomasi and T. Kanade, Shape and motion from image streams under orthography—A factorization method, *International Journal of Computer Vision*, **9**-2 (1992), 137–154.

[204] P. H. S. Torr and D. W. Murray, Statistical detection of independent movement from a moving camera, *Image and Vision Computing*, **11**-4 (1993), 180–187.

[205] H. P. Trivedi, Estimation of stereo and motion parameters using a variational principle, *Image and Vision Computing*, **5**-2 (1987), 181–183.

[206] R. Y. Tsai and T. S. Huang, Estimating three-dimensional motion parameters of a rigid planar patch, *IEEE Transactions on Acoustics, Speech, and Signal Processing*, **29**-6 (1981), 1147–1152.

[207] R. Y. Tsai and T. S. Huang, Uniqueness and estimation of three-dimensional motion parameters of rigid objects with curved surfaces, *IEEE Transactions on Pattern Analysis and Machine Intelligence*, **6**-1 (1984), 13–27.

[208] T. Y. Tsai and T. S. Huang, Estimating 3-D motion parameters of a rigid planar patch III. Finite point correspondences and the three-view problem, *IEEE Transactions on Acoustics, Speech, and Signal Processing*, **32**-2 (1984), 213–220.

[209] R. Y. Tsai, T. S. Huang and W.-L. Zhu, Estimating three-dimensional motion parameters of a rigid planar patch, II: Singular value decomposition, *IEEE Transactions on Acoustics, Speech, and Signal Processing*, **30**-3 (1982), 525–534.

[210] R. Y. Tsai and R. K. Lenz, A new technique for fully autonomous and efficient 3D robotics hand/eye calibration, *IEEE Transactions on Robotics and Automation*, **5**-3 (1989), 345–358.

[211] S. Ullman, *The Interpretation of Visual Motion*, MIT Press, Cambridge, MA, U.S.A., 1979.

[212] S. Umeyama, Least-squares estimation of transformation parameters between two point patterns, *IEEE Transactions on Pattern Analysis and Machine Intelligence*, **13**-4 (1991), 376–380.

[213] T. Viéville, Auto-calibration of visual sensor parameters on a robotic head, *Image and Vision Computing*, **12**-4 (1994), 227–237.

[214] C.-C. Wang, Extrinsic calibration of a vision sensor mounted on a robot, *IEEE Transactions on Robotics and Automation*, **8**-2 (1992), 161–175.

[215] L.-L. Wang and W.-H. Tsai, Camera calibration by vanishing lines for 3-D computer vision, *IEEE Transactions on Pattern Analysis and Machine Intelligence*, **13**-4 (1991), 370–376.

[216] A. M. Waxman and J. H. Duncan, Binocular image flows: Steps toward stereo-motion fusion, *IEEE Transactions on Pattern Analysis and Machine Intelligence*, **8**-6 (1986), 715–729.

[217] A. M. Waxman, B. Kamgar-Parsi and M. Subbarao, Closed-form solutions to image flow equations for 3D structure and motion, *International Journal of Computer Vision*, **1**-3 (1987), 239–258.

[218] G.-Q. Wei and S. D. Ma, Implicit and explicit camera calibration: Theory and experiments, *IEEE Transactions on Pattern Analysis and Machine Intelligence*, **16**-5 (1994), 469–480.

[219] I. Weiss, Line fitting in a noisy image, *IEEE Transactions on Pattern Analysis and Machine Intelligence*, **11**-3 (1989), 325–329.

[220] R. Weiss, H. Nakatani and E. M. Riseman, An error analysis for surface orientation from vanishing points, *IEEE Transactions on Pattern Analysis and Machine Intelligence*, **12**-12 (1990), 1179–1185.

[221] J. Weng, N. Ahuja and T. S. Huang, Motion and structure from point correspondences with error estimation: Planar surfaces, *IEEE Transactions on Signal Processing*, **39**-12 (1991), 2691–2717.

[222] J. Weng, N. Ahuja and T. S. Huang, Optimal motion and structure estimation, *IEEE Transactions on Pattern Analysis and Machine Intelligence*, **15**-9 (1993), 864–884.

[223] J. Weng, P. Cohen and M. Herniou, Camera calibration with distortion models and accuracy evaluation, *IEEE Transactions on Pattern Analysis and Machine Intelligence*, **14**-10 (1992), 965–980.

[224] J. Weng, P. Cohen and N. Rebibo, Motion and structure estimation from stereo image sequences, *IEEE Transactions on Robotics and Automation*, **8**-3 (1992), 362–382.

[225] J. Weng, T. S. Huang and N. Ahuja, 3-D motion estimation, understanding, and prediction from noisy image sequences, *IEEE Transactions on Pattern Analysis and Machine Intelligence*, **9**-3 (1987), 370–389.

[226] J. Weng, T. S. Huang and N. Ahuja, Motion and structure from two perspective views: Algorithms, error analysis, and error estimation, *IEEE Transactions on Pattern Analysis and Machine Intelligence*, **11**-5 (1989), 451–467.

[227] J. Weng, T. S. Huang and N. Ahuja, *Motion and Structure from Image Sequences*, Springer, Berlin, F.R.G., 1993.

[228] M. Werman and Z. Greyzel, Fitting a second degree curve in the presence of error, *IEEE Transactions on Pattern Analysis and Machine Intelligence*, **17**-2 (1995), 207–211.

[229] P. H. Winston (ed), *Psychology of Computer Vision*, McGraw-Hill, New York, NY, U.S.A., 1975.

[230] T. H. Wonnacott and R. J. Wonnacott, *Introductory Statistics*, 2nd Ed., John Wiley and Sons, New York, NY, U.S.A., 1972.

[231] J. J. Wu, R. E. Rink, T. M. Caelli and V. G. Gourishankar, Recovery of the 3-D location and motion of a rigid object through camera image (an extended Kalman filter approach), *International Journal of Computer Vision*, **2**-4 (1989), 373–394.

[232] Y. Yasumoto and G. Medioni, Robust estimation of three-dimensional motion parameters from a sequence of image frames using regularization, *IEEE Transactions on Pattern Analysis and Machine Intelligence*, **8**-4 (1986), 464–471.

[233] G.-S. J. Young and R. Chellappa, Statistical analysis of inherent ambiguities in recovering 3-D motion from a noisy flow field, *IEEE Transactions on Pattern Analysis and Machine Intelligence*, **14**-10 (1992), 995–1013.

[234] Z. Zhang and O. D. Faugeras, Three-dimensional motion computation and object segmentation in a long sequence of stereo frames, *International Journal of Computer Vision*, **7**-3 (1992), 211–241.

[235] Y. T. Zhou, V. Venkateswar and R. Chellappa, Edge detection and linear feature extraction using a 2-D random field model, *IEEE Transactions on Pattern Analysis and Machine Intelligence*, **11**-1 (1989), 84–95.

[236] X. Zhuang, A simplification to linear two-view motion algorithm, *Computer Vision, Graphics, and Image Processing*, **46**-2 (1989), 175–178.

[237] X. Zhuang, T. S. Huang, N. Ahuja and R. M. Haralick, A simplified linear optical flow-motion algorithm, *Computer Vision, Graphics, and Image Processing*, **42**-3 (1988), 334–344.

[238] X. Zhuang, T. S. Huang and R. M. Haralick, Two-view motion analysis: A unified algorithm, *Journal of the Optical Society of America*, A**3**-9 (1986), 1492–1500.

Index

A CATALOG OF SELECTED

DOVER BOOKS

IN SCIENCE AND MATHEMATICS

Astronomy

BURNHAM'S CELESTIAL HANDBOOK, Robert Burnham, Jr. Thorough guide to the stars beyond our solar system. Exhaustive treatment. Alphabetical by constellation: Andromeda to Cetus in Vol. 1; Chamaeleon to Orion in Vol. 2; and Pavo to Vulpecula in Vol. 3. Hundreds of illustrations. Index in Vol. 3. 2,000pp. $6\frac{1}{8}$ x $9\frac{1}{4}$.
Vol. I: 23567-X
Vol. II: 23568-8
Vol. III: 23673-0

EXPLORING THE MOON THROUGH BINOCULARS AND SMALL TELESCOPES, Ernest H. Cherrington, Jr. Informative, profusely illustrated guide to locating and identifying craters, rills, seas, mountains, other lunar features. Newly revised and updated with special section of new photos. Over 100 photos and diagrams. 240pp. $8\frac{1}{4}$ x 11. 24491-1

THE EXTRATERRESTRIAL LIFE DEBATE, 1750–1900, Michael J. Crowe. First detailed, scholarly study in English of the many ideas that developed from 1750 to 1900 regarding the existence of intelligent extraterrestrial life. Examines ideas of Kant, Herschel, Voltaire, Percival Lowell, many other scientists and thinkers. 16 illustrations. 704pp. $5\frac{3}{8}$ x $8\frac{1}{2}$. 40675-X

THEORIES OF THE WORLD FROM ANTIQUITY TO THE COPERNICAN REVOLUTION, Michael J. Crowe. Newly revised edition of an accessible, enlightening book recreates the change from an earth-centered to a sun-centered conception of the solar system. 242pp. $5\frac{3}{8}$ x $8\frac{1}{2}$. 41444-2

A HISTORY OF ASTRONOMY, A. Pannekoek. Well-balanced, carefully reasoned study covers such topics as Ptolemaic theory, work of Copernicus, Kepler, Newton, Eddington's work on stars, much more. Illustrated. References. 521pp. $5\frac{3}{8}$ x $8\frac{1}{2}$.
65994-1

A COMPLETE MANUAL OF AMATEUR ASTRONOMY: Tools and Techniques for Astronomical Observations, P. Clay Sherrod with Thomas L. Koed. Concise, highly readable book discusses: selecting, setting up and maintaining a telescope; amateur studies of the sun; lunar topography and occultations; observations of Mars, Jupiter, Saturn, the minor planets and the stars; an introduction to photoelectric photometry; more. 1981 ed. 124 figures. 26 halftones. 37 tables. 335pp. $6\frac{1}{2}$ x $9\frac{1}{4}$.
42820-6

AMATEUR ASTRONOMER'S HANDBOOK, J. B. Sidgwick. Timeless, comprehensive coverage of telescopes, mirrors, lenses, mountings, telescope drives, micrometers, spectroscopes, more. 189 illustrations. 576pp. $5\frac{5}{8}$ x $8\frac{1}{4}$. (Available in U.S. only.)
24034-7

STARS AND RELATIVITY, Ya. B. Zel'dovich and I. D. Novikov. Vol. 1 of *Relativistic Astrophysics* by famed Russian scientists. General relativity, properties of matter under astrophysical conditions, stars, and stellar systems. Deep physical insights, clear presentation. 1971 edition. References. 544pp. $5\frac{5}{8}$ x $8\frac{1}{4}$. 69424-0

Chemistry

THE SCEPTICAL CHYMIST: The Classic 1661 Text, Robert Boyle. Boyle defines the term "element," asserting that all natural phenomena can be explained by the motion and organization of primary particles. 1911 ed. viii+232pp. 5⅜ x 8½. 42825-7

RADIOACTIVE SUBSTANCES, Marie Curie. Here is the celebrated scientist's doctoral thesis, the prelude to her receipt of the 1903 Nobel Prize. Curie discusses establishing atomic character of radioactivity found in compounds of uranium and thorium; extraction from pitchblende of polonium and radium; isolation of pure radium chloride; determination of atomic weight of radium; plus electric, photographic, luminous, heat, color effects of radioactivity. ii+94pp. 5⅜ x 8½. 42550-9

CHEMICAL MAGIC, Leonard A. Ford. Second Edition, Revised by E. Winston Grundmeier. Over 100 unusual stunts demonstrating cold fire, dust explosions, much more. Text explains scientific principles and stresses safety precautions. 128pp. 5⅜ x 8½. 67628-5

THE DEVELOPMENT OF MODERN CHEMISTRY, Aaron J. Ihde. Authoritative history of chemistry from ancient Greek theory to 20th-century innovation. Covers major chemists and their discoveries. 209 illustrations. 14 tables. Bibliographies. Indices. Appendices. 851pp. 5⅜ x 8½. 64235-6

CATALYSIS IN CHEMISTRY AND ENZYMOLOGY, William P. Jencks. Exceptionally clear coverage of mechanisms for catalysis, forces in aqueous solution, carbonyl- and acyl-group reactions, practical kinetics, more. 864pp. 5⅜ x 8½. 65460-5

ELEMENTS OF CHEMISTRY, Antoine Lavoisier. Monumental classic by founder of modern chemistry in remarkable reprint of rare 1790 Kerr translation. A must for every student of chemistry or the history of science. 539pp. 5⅜ x 8½. 64624-6

THE HISTORICAL BACKGROUND OF CHEMISTRY, Henry M. Leicester. Evolution of ideas, not individual biography. Concentrates on formulation of a coherent set of chemical laws. 260pp. 5⅜ x 8½. 61053-5

A SHORT HISTORY OF CHEMISTRY, J. R. Partington. Classic exposition explores origins of chemistry, alchemy, early medical chemistry, nature of atmosphere, theory of valency, laws and structure of atomic theory, much more. 428pp. 5⅜ x 8½. (Available in U.S. only.) 65977-1

GENERAL CHEMISTRY, Linus Pauling. Revised 3rd edition of classic first-year text by Nobel laureate. Atomic and molecular structure, quantum mechanics, statistical mechanics, thermodynamics correlated with descriptive chemistry. Problems. 992pp. 5⅜ x 8½. 65622-5

FROM ALCHEMY TO CHEMISTRY, John Read. Broad, humanistic treatment focuses on great figures of chemistry and ideas that revolutionized the science. 50 illustrations. 240pp. 5⅜ x 8½. 28690-8

Engineering

DE RE METALLICA, Georgius Agricola. The famous Hoover translation of greatest treatise on technological chemistry, engineering, geology, mining of early modern times (1556). All 289 original woodcuts. 638pp. 6¾ x 11. 60006-8

FUNDAMENTALS OF ASTRODYNAMICS, Roger Bate et al. Modern approach developed by U.S. Air Force Academy. Designed as a first course. Problems, exercises. Numerous illustrations. 455pp. 5⅜ x 8½. 60061-0

DYNAMICS OF FLUIDS IN POROUS MEDIA, Jacob Bear. For advanced students of ground water hydrology, soil mechanics and physics, drainage and irrigation engineering, and more. 335 illustrations. Exercises, with answers. 784pp. 6⅛ x 9¼. 65675-6

THEORY OF VISCOELASTICITY (Second Edition), Richard M. Christensen. Complete, consistent description of the linear theory of the viscoelastic behavior of materials. Problem-solving techniques discussed. 1982 edition. 29 figures. xiv+364pp. 6⅛ x 9¼. 42880-X

MECHANICS, J. P. Den Hartog. A classic introductory text or refresher. Hundreds of applications and design problems illuminate fundamentals of trusses, loaded beams and cables, etc. 334 answered problems. 462pp. 5⅜ x 8½. 60754-2

MECHANICAL VIBRATIONS, J. P. Den Hartog. Classic textbook offers lucid explanations and illustrative models, applying theories of vibrations to a variety of practical industrial engineering problems. Numerous figures. 233 problems, solutions. Appendix. Index. Preface. 436pp. 5⅜ x 8½. 64785-4

STRENGTH OF MATERIALS, J. P. Den Hartog. Full, clear treatment of basic material (tension, torsion, bending, etc.) plus advanced material on engineering methods, applications. 350 answered problems. 323pp. 5⅜ x 8½. 60755-0

A HISTORY OF MECHANICS, René Dugas. Monumental study of mechanical principles from antiquity to quantum mechanics. Contributions of ancient Greeks, Galileo, Leonardo, Kepler, Lagrange, many others. 671pp. 5⅜ x 8½. 65632-2

STABILITY THEORY AND ITS APPLICATIONS TO STRUCTURAL MECHANICS, Clive L. Dym. Self-contained text focuses on Koiter postbuckling analyses, with mathematical notions of stability of motion. Basing minimum energy principles for static stability upon dynamic concepts of stability of motion, it develops asymptotic buckling and postbuckling analyses from potential energy considerations, with applications to columns, plates, and arches. 1974 ed. 208pp. 5⅜ x 8½. 42541-X

METAL FATIGUE, N. E. Frost, K. J. Marsh, and L. P. Pook. Definitive, clearly written, and well-illustrated volume addresses all aspects of the subject, from the historical development of understanding metal fatigue to vital concepts of the cyclic stress that causes a crack to grow. Includes 7 appendixes. 544pp. 5⅜ x 8½. 40927-9

Mathematics

FUNCTIONAL ANALYSIS (Second Corrected Edition), George Bachman and Lawrence Narici. Excellent treatment of subject geared toward students with background in linear algebra, advanced calculus, physics, and engineering. Text covers introduction to inner-product spaces, normed, metric spaces, and topological spaces; complete orthonormal sets, the Hahn-Banach Theorem and its consequences, and many other related subjects. 1966 ed. 544pp. 6⅛ x 9¼. 40251-7

ASYMPTOTIC EXPANSIONS OF INTEGRALS, Norman Bleistein & Richard A. Handelsman. Best introduction to important field with applications in a variety of scientific disciplines. New preface. Problems. Diagrams. Tables. Bibliography. Index. 448pp. 5⅜ x 8½. 65082-0

VECTOR AND TENSOR ANALYSIS WITH APPLICATIONS, A. I. Borisenko and I. E. Tarapov. Concise introduction. Worked-out problems, solutions, exercises. 257pp. 5⅜ x 8¼. 63833-2

THE ABSOLUTE DIFFERENTIAL CALCULUS (CALCULUS OF TENSORS), Tullio Levi-Civita. Great 20th-century mathematician's classic work on material necessary for mathematical grasp of theory of relativity. 452pp. 5⅜ x 8¼. 63401-9

AN INTRODUCTION TO ORDINARY DIFFERENTIAL EQUATIONS, Earl A. Coddington. A thorough and systematic first course in elementary differential equations for undergraduates in mathematics and science, with many exercises and problems (with answers). Index. 304pp. 5⅜ x 8½. 65942-9

FOURIER SERIES AND ORTHOGONAL FUNCTIONS, Harry F. Davis. An incisive text combining theory and practical example to introduce Fourier series, orthogonal functions and applications of the Fourier method to boundary-value problems. 570 exercises. Answers and notes. 416pp. 5⅜ x 8½. 65973-9

COMPUTABILITY AND UNSOLVABILITY, Martin Davis. Classic graduate-level introduction to theory of computability, usually referred to as theory of recurrent functions. New preface and appendix. 288pp. 5⅜ x 8½. 61471-9

ASYMPTOTIC METHODS IN ANALYSIS, N. G. de Bruijn. An inexpensive, comprehensive guide to asymptotic methods–the pioneering work that teaches by explaining worked examples in detail. Index. 224pp. 5⅜ x 8½ 64221-6

APPLIED COMPLEX VARIABLES, John W. Dettman. Step-by-step coverage of fundamentals of analytic function theory–plus lucid exposition of five important applications: Potential Theory; Ordinary Differential Equations; Fourier Transforms; Laplace Transforms; Asymptotic Expansions. 66 figures. Exercises at chapter ends. 512pp. 5⅜ x 8½. 64670-X

INTRODUCTION TO LINEAR ALGEBRA AND DIFFERENTIAL EQUATIONS, John W. Dettman. Excellent text covers complex numbers, determinants, orthonormal bases, Laplace transforms, much more. Exercises with solutions. Undergraduate level. 416pp. 5⅜ x 8½. 65191-6

TENSOR CALCULUS, J.L. Synge and A. Schild. Widely used introductory text covers spaces and tensors, basic operations in Riemannian space, non-Riemannian spaces, etc. 324pp. 5⅜ x 8¼. 63612-7

ORDINARY DIFFERENTIAL EQUATIONS, Morris Tenenbaum and Harry Pollard. Exhaustive survey of ordinary differential equations for undergraduates in mathematics, engineering, science. Thorough analysis of theorems. Diagrams. Bibliography. Index. 818pp. 5⅜ x 8½. 64940-7

INTEGRAL EQUATIONS, F. G. Tricomi. Authoritative, well-written treatment of extremely useful mathematical tool with wide applications. Volterra Equations, Fredholm Equations, much more. Advanced undergraduate to graduate level. Exercises. Bibliography. 238pp. 5⅜ x 8½. 64828-1

FOURIER SERIES, Georgi P. Tolstov. Translated by Richard A. Silverman. A valuable addition to the literature on the subject, moving clearly from subject to subject and theorem to theorem. 107 problems, answers. 336pp. 5⅜ x 8½. 63317-9

INTRODUCTION TO MATHEMATICAL THINKING, Friedrich Waismann. Examinations of arithmetic, geometry, and theory of integers; rational and natural numbers; complete induction; limit and point of accumulation; remarkable curves; complex and hypercomplex numbers, more. 1959 ed. 27 figures. xii+260pp. 5⅜ x 8½. 42804-4

POPULAR LECTURES ON MATHEMATICAL LOGIC, Hao Wang. Noted logician's lucid treatment of historical developments, set theory, model theory, recursion theory and constructivism, proof theory, more. 3 appendixes. Bibliography. 1981 ed. ix+283pp. 5⅜ x 8½. 67632-3

CALCULUS OF VARIATIONS, Robert Weinstock. Basic introduction covering isoperimetric problems, theory of elasticity, quantum mechanics, electrostatics, etc. Exercises throughout. 326pp. 5⅜ x 8½. 63069-2

THE CONTINUUM: A Critical Examination of the Foundation of Analysis, Hermann Weyl. Classic of 20th-century foundational research deals with the conceptual problem posed by the continuum. 156pp. 5⅜ x 8½. 67982-9

CHALLENGING MATHEMATICAL PROBLEMS WITH ELEMENTARY SOLUTIONS, A. M. Yaglom and I. M. Yaglom. Over 170 challenging problems on probability theory, combinatorial analysis, points and lines, topology, convex polygons, many other topics. Solutions. Total of 445pp. 5⅜ x 8½. Two-vol. set.
Vol. I: 65536-9 Vol. II: 65537-7

INTRODUCTION TO PARTIAL DIFFERENTIAL EQUATIONS WITH APPLICATIONS, E. C. Zachmanoglou and Dale W. Thoe. Essentials of partial differential equations applied to common problems in engineering and the physical sciences. Problems and answers. 416pp. 5⅜ x 8½. 65251-3

THE THEORY OF GROUPS, Hans J. Zassenhaus. Well-written graduate-level text acquaints reader with group-theoretic methods and demonstrates their usefulness in mathematics. Axioms, the calculus of complexes, homomorphic mapping, *p*-group theory, more. 276pp. 5⅜ x 8½. 40922-8

Math–Decision Theory, Statistics, Probability

ELEMENTARY DECISION THEORY, Herman Chernoff and Lincoln E. Moses. Clear introduction to statistics and statistical theory covers data processing, probability and random variables, testing hypotheses, much more. Exercises. 364pp. 5⅜ x 8½. 65218-1

STATISTICS MANUAL, Edwin L. Crow et al. Comprehensive, practical collection of classical and modern methods prepared by U.S. Naval Ordnance Test Station. Stress on use. Basics of statistics assumed. 288pp. 5⅜ x 8½. 60599-X

SOME THEORY OF SAMPLING, William Edwards Deming. Analysis of the problems, theory, and design of sampling techniques for social scientists, industrial managers, and others who find statistics important at work. 61 tables. 90 figures. xvii +602pp. 5⅜ x 8½. 64684-X

LINEAR PROGRAMMING AND ECONOMIC ANALYSIS, Robert Dorfman, Paul A. Samuelson and Robert M. Solow. First comprehensive treatment of linear programming in standard economic analysis. Game theory, modern welfare economics, Leontief input-output, more. 525pp. 5⅜ x 8½. 65491-5

PROBABILITY: An Introduction, Samuel Goldberg. Excellent basic text covers set theory, probability theory for finite sample spaces, binomial theorem, much more. 360 problems. Bibliographies. 322pp. 5⅜ x 8½. 65252-1

GAMES AND DECISIONS: Introduction and Critical Survey, R. Duncan Luce and Howard Raiffa. Superb nontechnical introduction to game theory, primarily applied to social sciences. Utility theory, zero-sum games, n-person games, decision-making, much more. Bibliography. 509pp. 5⅜ x 8½. 65943-7

INTRODUCTION TO THE THEORY OF GAMES, J. C. C. McKinsey. This comprehensive overview of the mathematical theory of games illustrates applications to situations involving conflicts of interest, including economic, social, political, and military contexts. Appropriate for advanced undergraduate and graduate courses; advanced calculus a prerequisite. 1952 ed. x+372pp. 5⅜ x 8½. 42811-7

FIFTY CHALLENGING PROBLEMS IN PROBABILITY WITH SOLUTIONS, Frederick Mosteller. Remarkable puzzlers, graded in difficulty, illustrate elementary and advanced aspects of probability. Detailed solutions. 88pp. 5⅜ x 8½. 65355-2

PROBABILITY THEORY: A Concise Course, Y. A. Rozanov. Highly readable, self-contained introduction covers combination of events, dependent events, Bernoulli trials, etc. 148pp. 5⅜ x 8¼. 63544-9

STATISTICAL METHOD FROM THE VIEWPOINT OF QUALITY CONTROL, Walter A. Shewhart. Important text explains regulation of variables, uses of statistical control to achieve quality control in industry, agriculture, other areas. 192pp. 5⅜ x 8½. 65232-7

Math–Geometry and Topology

ELEMENTARY CONCEPTS OF TOPOLOGY, Paul Alexandroff. Elegant, intuitive approach to topology from set-theoretic topology to Betti groups; how concepts of topology are useful in math and physics. 25 figures. 57pp. 5⅜ x 8½. 60747-X

COMBINATORIAL TOPOLOGY, P. S. Alexandrov. Clearly written, well-organized, three-part text begins by dealing with certain classic problems without using the formal techniques of homology theory and advances to the central concept, the Betti groups. Numerous detailed examples. 654pp. 5⅜ x 8½. 40179-0

EXPERIMENTS IN TOPOLOGY, Stephen Barr. Classic, lively explanation of one of the byways of mathematics. Klein bottles, Moebius strips, projective planes, map coloring, problem of the Koenigsberg bridges, much more, described with clarity and wit. 43 figures. 210pp. 5⅜ x 8½. 25933-1

CONFORMAL MAPPING ON RIEMANN SURFACES, Harvey Cohn. Lucid, insightful book presents ideal coverage of subject. 334 exercises make book perfect for self-study. 55 figures. 352pp. 5⅝ x 8¼. 64025-6

THE GEOMETRY OF RENÉ DESCARTES, René Descartes. The great work founded analytical geometry. Original French text, Descartes's own diagrams, together with definitive Smith-Latham translation. 244pp. 5⅜ x 8½. 60068-8

PRACTICAL CONIC SECTIONS: The Geometric Properties of Ellipses, Parabolas and Hyperbolas, J. W. Downs. This text shows how to create ellipses, parabolas, and hyperbolas. It also presents historical background on their ancient origins and describes the reflective properties and roles of curves in design applications. 1993 ed. 98 figures. xii+100pp. 6½ x 9¼. 42876-1

THE THIRTEEN BOOKS OF EUCLID'S ELEMENTS, translated with introduction and commentary by Thomas L. Heath. Definitive edition. Textual and linguistic notes, mathematical analysis. 2,500 years of critical commentary. Unabridged. 1,414pp. 5⅜ x 8½. Three-vol. set. Vol. I: 60088-2 Vol. II: 60089-0 Vol. III: 60090-4

GEOMETRY OF COMPLEX NUMBERS, Hans Schwerdtfeger. Illuminating, widely praised book on analytic geometry of circles, the Moebius transformation, and two-dimensional non-Euclidean geometries. 200pp. 5⅝ x 8¼. 63830-8

DIFFERENTIAL GEOMETRY, Heinrich W. Guggenheimer. Local differential geometry as an application of advanced calculus and linear algebra. Curvature, transformation groups, surfaces, more. Exercises. 62 figures. 378pp. 5⅜ x 8½. 63433-7

CURVATURE AND HOMOLOGY: Enlarged Edition, Samuel I. Goldberg. Revised edition examines topology of differentiable manifolds; curvature, homology of Riemannian manifolds; compact Lie groups; complex manifolds; curvature, homology of Kaehler manifolds. New Preface. Four new appendixes. 416pp. 5⅜ x 8½. 40207-X

History of Math

THE WORKS OF ARCHIMEDES, Archimedes (T. L. Heath, ed.). Topics include the famous problems of the ratio of the areas of a cylinder and an inscribed sphere; the measurement of a circle; the properties of conoids, spheroids, and spirals; and the quadrature of the parabola. Informative introduction. clxxxvi+326pp; supplement, 52pp. 5⅜ x 8½. 42084-1

A SHORT ACCOUNT OF THE HISTORY OF MATHEMATICS, W. W. Rouse Ball. One of clearest, most authoritative surveys from the Egyptians and Phoenicians through 19th-century figures such as Grassman, Galois, Riemann. Fourth edition. 522pp. 5⅜ x 8½. 20630-0

THE HISTORY OF THE CALCULUS AND ITS CONCEPTUAL DEVELOPMENT, Carl B. Boyer. Origins in antiquity, medieval contributions, work of Newton, Leibniz, rigorous formulation. Treatment is verbal. 346pp. 5⅜ x 8½. 60509-4

THE HISTORICAL ROOTS OF ELEMENTARY MATHEMATICS, Lucas N. H. Bunt, Phillip S. Jones, and Jack D. Bedient. Fundamental underpinnings of modern arithmetic, algebra, geometry, and number systems derived from ancient civilizations. 320pp. 5⅜ x 8½. 25563-8

A HISTORY OF MATHEMATICAL NOTATIONS, Florian Cajori. This classic study notes the first appearance of a mathematical symbol and its origin, the competition it encountered, its spread among writers in different countries, its rise to popularity, its eventual decline or ultimate survival. Original 1929 two-volume edition presented here in one volume. xxviii+820pp. 5⅜ x 8½. 67766-4

GAMES, GODS & GAMBLING: A History of Probability and Statistical Ideas, F. N. David. Episodes from the lives of Galileo, Fermat, Pascal, and others illustrate this fascinating account of the roots of mathematics. Features thought-provoking references to classics, archaeology, biography, poetry. 1962 edition. 304pp. 5⅜ x 8½. (Available in U.S. only.) 40023-9

OF MEN AND NUMBERS: The Story of the Great Mathematicians, Jane Muir. Fascinating accounts of the lives and accomplishments of history's greatest mathematical minds–Pythagoras, Descartes, Euler, Pascal, Cantor, many more. Anecdotal, illuminating. 30 diagrams. Bibliography. 256pp. 5⅜ x 8½. 28973-7

HISTORY OF MATHEMATICS, David E. Smith. Nontechnical survey from ancient Greece and Orient to late 19th century; evolution of arithmetic, geometry, trigonometry, calculating devices, algebra, the calculus. 362 illustrations. 1,355pp. 5⅜ x 8½. Two-vol. set. Vol. I: 20429-4 Vol. II: 20430-8

A CONCISE HISTORY OF MATHEMATICS, Dirk J. Struik. The best brief history of mathematics. Stresses origins and covers every major figure from ancient Near East to 19th century. 41 illustrations. 195pp. 5⅜ x 8½. 60255-9

Physics

OPTICAL RESONANCE AND TWO-LEVEL ATOMS, L. Allen and J. H. Eberly. Clear, comprehensive introduction to basic principles behind all quantum optical resonance phenomena. 53 illustrations. Preface. Index. 256pp. 5⅜ x 8½. 65533-4

QUANTUM THEORY, David Bohm. This advanced undergraduate-level text presents the quantum theory in terms of qualitative and imaginative concepts, followed by specific applications worked out in mathematical detail. Preface. Index. 655pp. 5⅜ x 8½. 65969-0

ATOMIC PHYSICS: 8th edition, Max Born. Nobel laureate's lucid treatment of kinetic theory of gases, elementary particles, nuclear atom, wave-corpuscles, atomic structure and spectral lines, much more. Over 40 appendices, bibliography. 495pp. 5⅜ x 8½. 65984-4

A SOPHISTICATE'S PRIMER OF RELATIVITY, P. W. Bridgman. Geared toward readers already acquainted with special relativity, this book transcends the view of theory as a working tool to answer natural questions: What is a frame of reference? What is a "law of nature"? What is the role of the "observer"? Extensive treatment, written in terms accessible to those without a scientific background. 1983 ed. xlviii+172pp. 5⅜ x 8½. 42549-5

AN INTRODUCTION TO HAMILTONIAN OPTICS, H. A. Buchdahl. Detailed account of the Hamiltonian treatment of aberration theory in geometrical optics. Many classes of optical systems defined in terms of the symmetries they possess. Problems with detailed solutions. 1970 edition. xv+360pp. 5⅜ x 8½. 67597-1

PRIMER OF QUANTUM MECHANICS, Marvin Chester. Introductory text examines the classical quantum bead on a track: its state and representations; operator eigenvalues; harmonic oscillator and bound bead in a symmetric force field; and bead in a spherical shell. Other topics include spin, matrices, and the structure of quantum mechanics; the simplest atom; indistinguishable particles; and stationary-state perturbation theory. 1992 ed. xiv+314pp. 6⅛ x 9¼. 42878-8

LECTURES ON QUANTUM MECHANICS, Paul A. M. Dirac. Four concise, brilliant lectures on mathematical methods in quantum mechanics from Nobel Prize–winning quantum pioneer build on idea of visualizing quantum theory through the use of classical mechanics. 96pp. 5⅜ x 8½. 41713-1

THIRTY YEARS THAT SHOOK PHYSICS: The Story of Quantum Theory, George Gamow. Lucid, accessible introduction to influential theory of energy and matter. Careful explanations of Dirac's anti-particles, Bohr's model of the atom, much more. 12 plates. Numerous drawings. 240pp. 5⅜ x 8½. 24895-X

ELECTRONIC STRUCTURE AND THE PROPERTIES OF SOLIDS: The Physics of the Chemical Bond, Walter A. Harrison. Innovative text offers basic understanding of the electronic structure of covalent and ionic solids, simple metals, transition metals and their compounds. Problems. 1980 edition. 582pp. 6⅛ x 9¼. 66021-4

QUANTUM MECHANICS: Principles and Formalism, Roy McWeeny. Graduate student–oriented volume develops subject as fundamental discipline, opening with review of origins of Schrödinger's equations and vector spaces. Focusing on main principles of quantum mechanics and their immediate consequences, it concludes with final generalizations covering alternative "languages" or representations. 1972 ed. 15 figures. xi+155pp. 5⅜ x 8½. 42829-X

INTRODUCTION TO QUANTUM MECHANICS WITH APPLICATIONS TO CHEMISTRY, Linus Pauling & E. Bright Wilson, Jr. Classic undergraduate text by Nobel Prize winner applies quantum mechanics to chemical and physical problems. Numerous tables and figures enhance the text. Chapter bibliographies. Appendices. Index. 468pp. 5⅜ x 8½. 64871-0

METHODS OF THERMODYNAMICS, Howard Reiss. Outstanding text focuses on physical technique of thermodynamics, typical problem areas of understanding, and significance and use of thermodynamic potential. 1965 edition. 238pp. 5⅜ x 8½. 69445-3

TENSOR ANALYSIS FOR PHYSICISTS, J. A. Schouten. Concise exposition of the mathematical basis of tensor analysis, integrated with well-chosen physical examples of the theory. Exercises. Index. Bibliography. 289pp. 5⅜ x 8½. 65582-2

THE ELECTROMAGNETIC FIELD, Albert Shadowitz. Comprehensive undergraduate text covers basics of electric and magnetic fields, builds up to electromagnetic theory. Also related topics, including relativity. Over 900 problems. 768pp. 5⅝ x 8¼. 65660-8

GREAT EXPERIMENTS IN PHYSICS: Firsthand Accounts from Galileo to Einstein, Morris H. Shamos (ed.). 25 crucial discoveries: Newton's laws of motion, Chadwick's study of the neutron, Hertz on electromagnetic waves, more. Original accounts clearly annotated. 370pp. 5⅜ x 8½. 25346-5

RELATIVITY, THERMODYNAMICS AND COSMOLOGY, Richard C. Tolman. Landmark study extends thermodynamics to special, general relativity; also applications of relativistic mechanics, thermodynamics to cosmological models. 501pp. 5⅜ x 8½. 65383-8

STATISTICAL PHYSICS, Gregory H. Wannier. Classic text combines thermodynamics, statistical mechanics, and kinetic theory in one unified presentation of thermal physics. Problems with solutions. Bibliography. 532pp. 5⅜ x 8½. 65401-X

Paperbound unless otherwise indicated. Available at your book dealer, online at **www.doverpublications.com**, or by writing to Dept. GI, Dover Publications, Inc., 31 East 2nd Street, Mineola, NY 11501. For current price information or for free catalogs (please indicate field of interest), write to Dover Publications or log on to **www.doverpublications.com** and see every Dover book in print. Dover publishes more than 500 books each year on science, elementary and advanced mathematics, biology, music, art, literary history, social sciences, and other areas.